P9-AGS-174

THE MEDITERRANEAN
AND THE MEDITERRANEAN WORLD
IN THE AGE OF PHILIP II

FERNAND BRAUDEL

The Mediterranean

and the Mediterranean World in the Age of Philip II

VOLUME II

Translated from the French
by Siân Reynolds

HARPER & ROW, PUBLISHERS
NEW YORK, EVANSTON, SAN FRANCISCO, LONDON

First published in France under the title
La Méditerranée et le Monde Méditerranéen à l'Époque de Philippe II, 1949, second revised edition 1966

First U.S. Edition

Standard Book Number: 06-010456-2

Library of Congress Catalog Card Number: 72-138708

Contents

Part Two

COLLECTIVE DESTINIES AND GENERAL TRENDS
(continued)

Part Three

EVENTS, POLITICS AND PEOPLE

List of Illustrations

List of Figures

Part Two (continued)

CHAPTER IV

Empires

We must go far back in time, to the beginning of a long process of political evolution, before we can achieve a valid perspective on the sixteenth century.

At the end of the fourteenth century, the Mediterranean belonged to its towns, to the city-states scattered around its shores. There were of course already, here and there, a few territorial states, fairly homogeneous in character and comparatively large, bordering the sea itself: the Kingdom of Naples – 'il Reame' – the outstanding example; the Byzantine Empire; or the possessions united under the Crown of Aragon. But in many cases, these states were merely the extensions of powerful cities: Aragon in the broad sense was a by-product of the dynamic rise of Barcelona; the Byzantine Empire consisted almost entirely of the extended suburbs of two cities, Constantinople and Salonica.

By the fifteenth century, the city-state was already losing ground; first signs of the crisis could be detected in Italy during the early years of the century. In fifty years, the map of the Peninsula was entirely redrawn, to the advantage of some cities and the detriment of others. It was only a partial eclipse. The upheaval failed to achieve what may have been at issue – though I doubt it – the unification of the Italian Peninsula. Naples, Venice and Milan in turn proved unequal to the task. The attempt would in any case have been premature: too many particular interests were at stake, too many cities eager for an individual existence stood in the way of this difficult birth. So it is only partly true that there was a decline in the power of the city-state. The Peace of Lodi, in 1454, confirmed both a balance of power and a deadlock: the political map of Italy, although simplified, was still a patchwork.

Meanwhile, a similar crisis was becoming apparent throughout the rest of the Mediterranean. Everywhere the city-state, precarious and narrow-based, stood revealed inadequate to perform the political and financial tasks now facing it. It represented a fragile form of government, doomed to extinction, as was strikingly demonstrated by the capture of Constantinople in 1453, the fall of Barcelona in 1472 and the collapse of Granada in 1492.[1]

It was becoming clear that only the rival of the city-state, the territorial state,[2] rich in land and manpower, would in future be able to meet the

[1] See above, Vol. I, p. 339.

[2] I have deliberately avoided the term nation-state.

expense of modern warfare; it could maintain paid armies and afford costly artillery; it was soon to indulge in the added extravagance of full-scale naval wars. And its advance was long to be irreversible. Examples of the new pattern emerging at the end of the fifteenth century are Aragon under John II; Louis XI's expansion beyond the Pyrenees; Turkey under Muḥammad II, the conqueror of Constantinople; later France under Charles VIII with his Italian ambitions and Spain in the age of the Catholic Kings. Without exception, these states all had their beginnings far inland, many miles from the Mediterranean coast,[3] usually in poor regions where there were fewer cities to pose obstacles. In Italy by contrast, the wealth and very density of the cities maintained weaknesses and divisions as modern structures emerged only with difficulty from the grip of the past, particularly when that past had been a glorious one and much of its brilliance remained. Past glory could mean present weakness, as was revealed by the first Turco-Venetian war, from 1463 to 1479, in the course of which the Signoria, inadequately protected by her small territory, was eventually obliged, despite her technical superiority, to abandon the struggle;[4] it was demonstrated once more during the tragic occupation of Otranto by the Turks in 1480[5] and appeared even more strikingly in the beginnings of the storm unleashed by Charles VIII's invasion of Italy in 1494. Was there ever a more extraordinary military display than that swift march on Naples, when, according to Machiavelli, the invader had merely to send his billeting officers ahead to mark with chalk the houses selected for his troops' lodgings? Once the alarm was over, it was easy to make light of it, even to taunt the French ambassador Philippe de Commynes, as Filippo Tron, a Venetian patrician, did at the end of July, 1495. He added that he was not deceived by the intentions attributed to the king of France, 'desiring to go to the Holy Land when he really wanted to become no less than *signore di tutta l'Italia*'.[6]

Such bravado was all very well, but the event marked the beginning of a train of disasters for the Peninsula, the logical penalty for its wealth, its position at the epicentre of European politics and, undoubtedly the key factor, the fragility of its sophisticated political structures, of the intricate mechanisms which went to make up the 'Italian equilibrium'. It was no accident if from now on Italian thinkers, schooled by disaster and the daily lesson of events, were to meditate above all upon politics and the destiny of the state, from Machiavelli and Guicciardini in the early part of the century to Paruta, Giovanni Botero or Ammirato at the end.

Italy: that extraordinary laboratory for statesmen. The entire nation was preoccupied with politics, every man to his own passion, from the porter in the market-place to the barber in his shop or the artisans in the

[3] A. Siegfried, *op. cit.*, p. 184.

[4] H. Kretschmayr, *op. cit.*, II, p. 382.

[5] See studies by Enrico Perito, E. Carusi and Pietro Egidi (nos. 2625, 2630 and 2626 in Sánchez Alonso's bibliography).

[6] A.d.S., Modena, Venezia VIII, Aldobrandino Guidoni to the Duke, Venice, 31st July, 1495.

taverns;[7] for *raggione di stato, raison d'état,*[8] an Italian rediscovery, was the result not of isolated reflection but of collective experience. Similarly, the frequent cruelty in political affairs, the betrayals and renewed flames of personal vendettas are so many symptoms of an age when the old governmental structures were breaking up and a series of new ones appearing in rapid succession, according to circumstances beyond man's control. These were days when justice was frequently an absent figure and governments were too new and too insecure to dispense with force and emergency measures. Terror was a means of government. *The Prince* taught the art of day-to-day survival.[9]

But even in the fifteenth century and certainly by the sixteenth, a formidable newcomer confronted the mere territorial or nation-state. Larger, monster states were now appearing, through accumulation, inheritance, federation or coalition of existing states: what by a convenient though anachronistic term one could call empires in the modern sense – for how else is one to describe these giants? In 1494, the threat to Italy from beyond the Alps came not merely from the kingdom of France but from a French Empire, as yet hypothetical it is true. Its first objective was to capture Naples; then, without becoming immobilized at the centre of the Mediterranean, to speed to the East, there to defend the Christian cause in reply to the repeated appeals of the Knights of Rhodes, and to deliver the Holy Land. Such was the complex policy of Charles VIII, whatever Filippo Tron may have thought: it was a crusading policy, designed to span the Mediterranean in one grand sweep. For no empire could exist without some mystique and in western Europe, this mystique was provided by the crusade, part spiritual, part temporal, as the example of Charles V was soon to prove.

And indeed Spain under Ferdinand and Isabella, was no 'mere nation-state': it was already an association of kingdoms, states and peoples united in the persons of the sovereigns. The sultans too ruled over a combination of conquered peoples and loyal subjects, populations which had either been subjugated or associated with their fortunes. Meanwhile, maritime exploration was creating, for the greater benefit of Portugal and Castile, the first modern colonial empires, the importance of which was not fully grasped at first by even the most perspicacious observers. Machiavelli himself stood too close to the troubled politics of Italy to see beyond them – a major weakness in a commentator otherwise so lucid.[10]

[7] M. Seidlmayer, *op. cit.*, p. 342.

[8] Its first use is usually attributed to Cardinal Giovanni Della Casa, *Orazione di Messer Giovanni della Casa, scritta a Carlo Quinto intorno alla restitutione della città di Piacenza*, published in the *Galateo* by the same author, Florence, 1561, p. 61. For a full treatment of the topic of *raison d'état*, see F. Meinecke, *Die Idee der Staatsräson in der neueren Geschichte*, Munich, 1925 (published in English as *Machiavellism. The doctrine of raison d'état and its place in modern history*, transl. by Douglas Scott, London, 1957).

[9] Pierre Mesnard, *L'essor de la philosophie politique au XVIe siècle.*

[10] A. Renaudet, *Machiavel*, p. 236.

The story of the Mediterranean in the sixteenth century is in the first place a story of dramatic political growth, with the leviathans taking up their positions. France's imperial career, as we know, misfired almost immediately, for several reasons: external circumstances in part, a still backward economy and perhaps also temperamental factors, prudence, a characteristic preference for safe investments and a distaste for the grandiose. What had failed to occur was by no means an impossibility. It is not entirely fanciful to imagine a French Empire supported by Florence in the same way that the Spanish Empire (though not at first it is true) was supported by Genoa. And the imperial career of Portugal, a Mediterranean country only by courtesy in any case, developed (apart from a few Moroccan possessions) outside the Mediterranean region.

So the rise of empires in the Mediterranean means essentially that of the Ottoman Empire in the East and that of the Habsburg Empire in the West. As Leopold von Ranke long ago remarked, the emergence of these twin powers constitutes a single chapter in history and before going any further let us stress that accident and circumstance did not preside alone at the birth of these simultaneous additions to the great powers of history. I cannot accept that Sulaimān the Magnificent and Charles V were merely 'accidents' (as even Henri Pirenne has argued) – their persons, by all means, but not their empires. Nor do I believe in the preponderant influence of Wolsey,[11] the inventor of the English policy of the Balance of Power who, by supporting Charles V in 1521 (against his own principles) when the latter was already ruler of the Netherlands and Germany, that is by supporting the stronger power instead of François I, the weaker, is said to have been responsible for Charles's rapid victory at Pavia and the subsequent surrender of Italy to Spanish domination for two hundred years.

For without wishing to belittle the role played by individuals and circumstances, I am convinced that the period of economic growth during the fifteenth and sixteenth centuries created a situation consistently favourable to the large and very large state, to the 'super-states' which today are once again seen as the pattern of the future as they seemed to be briefly at the beginning of the eighteenth century, when Russia was expanding under Peter the Great and when a dynastic union at least was projected between Louis XIV's France and Spain under Philip V.[12] *Mutatis mutandis*, the same pattern was repeated in the East. In 1516, the sultan of Egypt laid siege to the free city of Aden and captured it, in accordance with the laws of logical expansion. Whereupon in obedience to the very same laws, the Turkish sultan in 1517 seized the whole of Egypt.[13] Small states could always expect to be snapped up by a larger predator.

The course of history is by turns favourable or unfavourable to vast

[11] G. M. Trevelyan, *op. cit.*, p. 293.

[12] Baudrillart (Mgr.), *Philippe V et la Cour de France*, 1889–1901, 4 vols., Introduction, p. 1.

[13] See below, p. 667 ff.

political hegemonies. It prepares their birth and prosperity and ultimately their decline and fall. It is wrong to suppose that their political evolution is fixed once for all, that some states are irremediably doomed to extinction and others destined to achieve greatness come what may, as if marked by fate 'to devour territory and prey upon their neighbours'.[14]

Two empires in the sixteenth century gave evidence of their formidable might. But between 1550 and 1600, advance signs can already be glimpsed of what was in the seventeenth century to be their equally inexorable decline.

I. THE ORIGIN OF EMPIRES

A word of warning: when discussing the rise and fall of empires, it is as well to mark closely their rate of growth, avoiding the temptation to telescope time and discover too early signs of greatness in a state which we know will one day be great, or to predict too early the collapse of an empire which we know will one day cease to be. The life-span of empires cannot be plotted by events, only by careful diagnosis and auscultation – and as in medicine there is always room for error.

Turkish ascendancy:[15] *from Asia Minor to the Balkans.* Behind the rise of Turkey to greatness lay three centuries of repeated effort, of prolonged conflict and of miracles. It was on the 'miraculous' aspect of the Ottoman Empire that western historians of the sixteenth, seventeenth and eighteenth centuries tended to dwell. It is after all an extraordinary story, the emergence of the Ottoman dynasty from the fortunes of war on the troubled frontiers of Asia Minor, a rendezvous for adventurers and fanatics.[16] For Asia Minor was a region of unparalleled mystical enthusiasm: here war and religion marched hand in hand, militant confraternities abounded and the janissaries were of course attached to such powerful sects as the Ahīs and later the Bektāshīs. These beginnings gave the Ottoman state its style, its foundations among the people and its original exaltation. The miracle is that such a tiny state should have survived the accidents and disturbances inherent in its geographical position.

But survive it did, and put to advantage the slow transformation of the Anatolian countryside. The Ottoman success was intimately connected with the waves of invasion, often silent invasion, which drove the peoples of Turkestan westwards. It was brought about by the internal transformation of Asia Minor[17] from Greek and Orthodox in the thirteenth century to Turkish and Moslem, following successive waves of infiltration and indeed of total social disruption; and also by the extraordinary propaganda

[14] Gaston Roupnel, *Histoire et destin*, p. 330.

[15] On the greatness of Turkey, see R. de Lusinge, *De la naissance, durée et chute des États*, 1588, 206 p. Ars. 8° H 17337, quoted by G. Atkinson, *op. cit.*, p. 184–185, and an unpublished diplomatic report on Turkey (1576), Simancas E° 1147.

[16] Fernand Grenard, *Décadence de l'Asie*, p. 48.

[17] See above, Vol. I, p. 178.

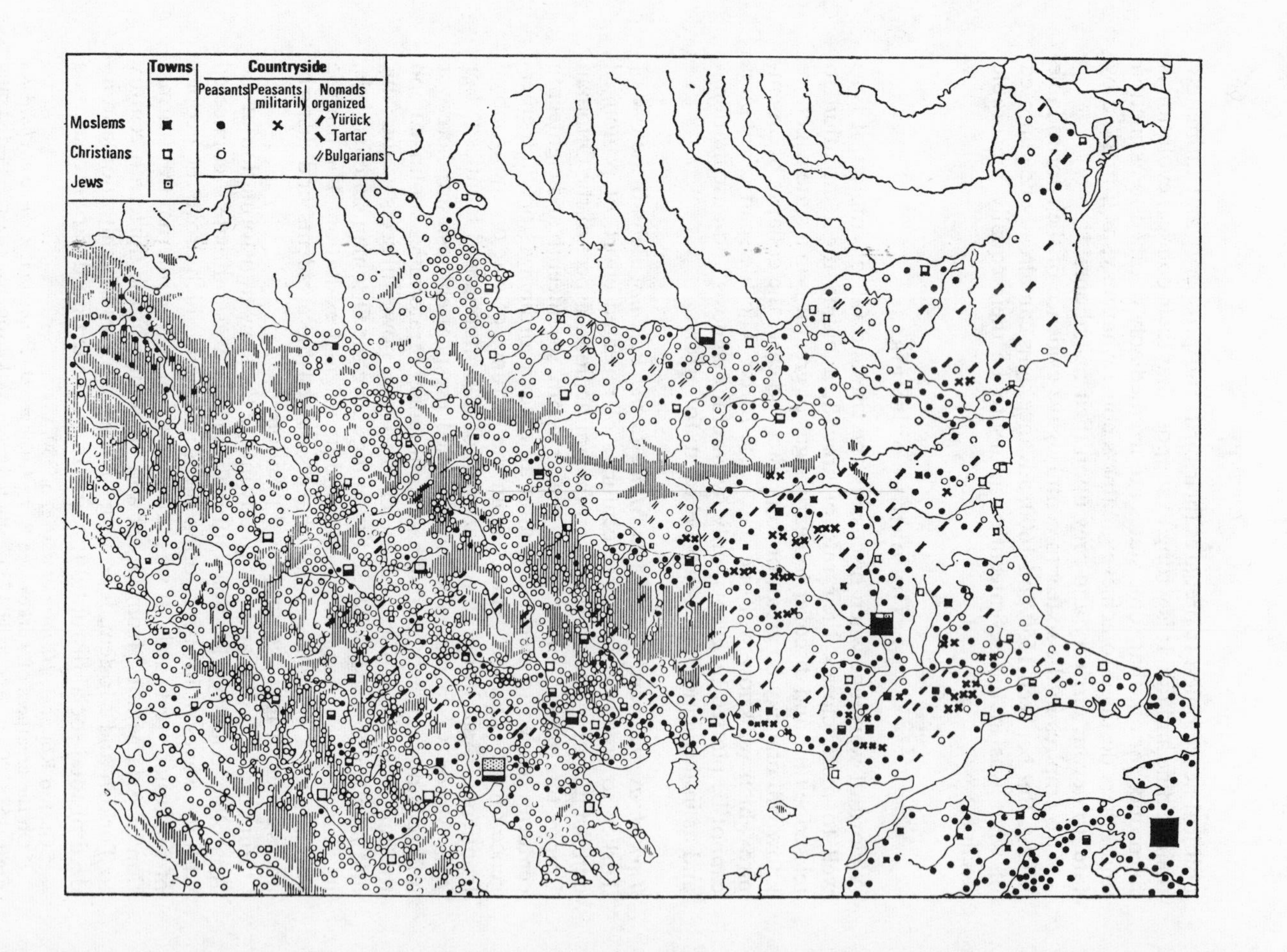

Towns
Countryside
Peasants
Peasants militarily organized
Nomads
Moslems
Christians
Jews
Yürück
Tartar
Bulgarians

of the Moslem orders, some of which were revolutionary, 'communist, like the Bābāīs, Ahīs and Abdālān; others more mystical and pacifist, for example the Mawlawīs of Konya. Following G. Huart, Koprülüzade has recently drawn attention to their proselytism'.[18] Their poetry – their propaganda – marked the dawn of western Turkish literature.

Beyond the straits, the Turkish conquest was largely favoured by circumstances. The Balkan Peninsula was far from poor, indeed in the fourteenth and fifteenth centuries it was comparatively wealthy. But it was divided: Byzantines, Serbs, Bulgars, Albanians, Venetians, and Genoese fought amongst themselves. There was religious conflict between the Orthodox and the Roman Churches; and socially the Balkan world was extremely fragile – a mere house of cards. So it should not be forgotten that the Turkish conquest of the Balkans was assisted by an extraordinary social revolution. A seignorial society, exploiting the peasants, was surprised by the impact and collapsed of its own accord. The conquest, which meant the end of the great landowners, absolute rulers on their own estates, was in its way 'a liberation of the oppressed'.[19] Asia Minor had been conquered patiently and slowly after centuries of effort; the Balkan Peninsula *seems* not to have offered any resistance to the invader. In Bulgaria, where the Turks made such rapid progress, the countryside had already been unsettled, well before their arrival, by violent rural disturbances.[20] Even in Greece there had been a social revolution. In Serbia, the native aristocracy was wiped out and some of the Serbian villages were incorporated into the *wakf* (possessions of the mosques) or distributed to the *sipāhis*.[21] And the *sipāhis*, soldiers whose titles were held only for life,

(Opposite) Fig. 55: *The population of the Balkan Peninsula at the beginning of the sixteenth century*

Missing from this map, compiled by Ömer Lutfi Barkan from Ottoman population counts, are the figures for Istanbul, which do not appear to have survived. The Turks controlled their acquisitions from frontier posts and above all from the key towns. Note the large implantation of Yürük nomads in the plains, but also in the highlands, in the Rhodope for example and in the mountains to the east of the Struma and Vardar. A line running roughly from the island of Thásos through Sofia divides a predominantly Christian zone, only partially colonized by the Turks, from a zone of strong Moslem implantation in Thrace and through to Bulgaria. Subsequent research by Ömer Lutfi Barkan and his pupils has now analysed virtually all the sixteenth century censuses, revealing a large population increase and confirming what was already thought: that Moslems predominated in the population of Anatolia. Every symbol on the map represents 250 families, that is over 1000 people. Note the density of the Moslem population in Bosnia and the large Jewish colony in Salonica.

---

[18] *Annuaire du monde musulman*, 1923, p. 323.

[19] An expression used by Mr. B. Truhelka, archivist at Dubrovnik, during our many discussions on this subject.

[20] Cf. in particular Christo Peyeff, *Agrarverfassung und Agrarpolitik*, Berlin, 1927, p. 69; I. Sakazov, *op. cit.*, p. 19; R. Busch-Zantner, *op. cit.*, p. 64 ff. However, according to an article by D. Anguelov, (Bulgarian) *Historical Review*, IX, 4, p. 374–398, Bulgarian resistance to the Turks was stronger than I have suggested.

[21] Jos. Zontar, 'Hauptprobleme der jugoslavischen Sozial- und Wirtschaftsgeschichte' in *Vierteljahrschr. für Sozial- und Wirtschaftsgeschichte*, 1934, p. 368.

at first asked for rents in money, not in kind. It was some time before the condition of the peasants once again became intolerable. And in Bosnia, around Sarajevo, there were mass conversions, due in part to the flourishing Bogomilian heresy.[22] The situation was even more complicated in Albania.[23] Here the landowners were able to take refuge in the Venetian *presidios*: Durazzo for example, which remained in Venetian possession until 1501. When these fortresses fell, the Albanian nobility fled to Italy, where some of their descendants remain to the present day. The Musachi family did not survive: its last member died in Naples in 1600. It left behind however the *Historia della Casa Musachi*, published in 1510 by Giovanni Musachi, a valuable record of the family fortunes which tells us much about the country and its ruling caste. The name of this ancient family is preserved in the Muzekie region of Albania[24] where it once had immense holdings.[25] The story of these exiles and their wanderings is an astonishing one. The same path was not trodden by all nobles and landowners in the Balkans. But whatever their fate, even when they succeeded in maintaining themselves for a while, by abjuring or otherwise, the general pattern was the same: before the Turkish advance an entire society fell into ruins, partly of its own accord, seeming to confirm yet again Albert Grenier's opinion that 'to be conquered, a people must have acquiesced in its own defeat.'

Social conditions in the Balkans help to explain the invader's success and the ravages it brought. The Turkish cavalry, ranging rapidly far and wide, blocking roads, ruining crops and disrupting economic life, went ahead of the main army and prepared the ground for an easy victory. Only the mountainous regions were for a while protected from the relentless invasion. Bowing to the geography of the Balkans, the Turks took control first of the principal highways, along the river valleys leading down to the Danube; the Maritsa, the Vardar, the Drin and the Morava. In 1371, they triumphed at Chernomen on the Maritsa; in 1389 they won the battle of Kossovo Polje, 'the Field of Blackbirds', from which flow the Vardar, Maritsa and Morava. In 1459, this time north of the Iron Gates, the Turk was victorious at Smederevo 'at the very point where the Morava meets the Danube and which as much as Belgrade commands the approaches to the Hungarian plains'.[26]

Conquest was rapid too in the wide spaces of the eastern plains.[27] In 1365, the Turk settled his capital at Adrianople, by 1386 all Bulgaria had been subdued, to be followed by Thessaly.[28] Victory came more slowly in the mountainous west and was often more apparent than real. In Greece, Athens was occupied in 1456, the Morea in 1460, Bosnia in 1462–1466,[29]

[22] J. W. Zinkeisen, *op. cit.*, II, p. 143; R. Busch-Zantner, *op. cit.*, p. 50.
[23] R. Busch-Zantner, *op. cit.*, p. 65.
[24] *Ibid.*, p. 55.
[25] *Ibid.*, p. 65 and references to studies by K. Jireček and Sufflay.
[26] *Ibid.*, p. 23.
[27] W. Heyd, *op. cit.*, II, p. 258.
[28] *Ibid.*, II, p. 270.
[29] *Ann. du monde musulman*, 1923, p. 228.

Herzegovina in 1481,[30] despite the resistance of some 'mountain kings'. Venice herself was unable for long to prevent the Turk from reaching the Adriatic: Scutari was captured in 1479 and Durazzo in 1501. Military victory was followed by another, more leisurely conquest: the construction of roads and fortified posts, the organization of camel trains, the setting in motion of all the supply and transport convoys (often handled by Bulgarian carriers) and finally, most important of all, that conquest which operated through those towns which the Turks had subdued, fortified or built. These now became major centres of diffusion of Turkish civilization: they calmed, domesticated and tamed at least the conquered regions, where it must not be imagined that an atmosphere of constant violence reigned.

In the early days, the Turkish conquest took a heavy toll of the subjugated peoples: after the battle of Kossovo, thousands upon thousands of Serbs were sold as slaves as far away as the markets of Christendom[31] or recruited as mercenaries; but the conqueror was not deficient in political wisdom, as can be seen from Muhammad II's concessions to the Greeks summoned to Constantinople after 1453. Eventually Turkey created, throughout the Balkans, structures within which the peoples of the Peninsula gradually found a place, collaborating with the conqueror and here and there curiously re-creating the patterns of the Byzantine Empire. This conquest brought a new order, a *pax turcica*. Let us take the word of the anonymous Frenchman who wrote in 1528: 'the country is safe and there are no reports of brigands or highwaymen. . . . The Emperor does not tolerate highwaymen or robbers.'[32] Could as much have been said of Catalonia or Calabria at the same period? There must have been some truth in this optimistic picture, since for many years the Turkish Empire remained to Christian eyes an extraordinary, incomprehensible and disconcerting example of orderliness; its army astonished westerners by its discipline and silence as much as by its courage, abundant munitions and the high quality and sobriety of its soldiers. Not that their astonishment prevented the Christians from hating these Infidels, 'far worse than dogs in all their works' as one writer put it in 1526.[33]

Gradually however, westerners came to take a more balanced view of the Turks. They were of course a scourge sent by the Lord: Pierre Viret, the Protestant reformer of French Switzerland, wrote of them in 1560: 'we must not be amazed if God is now punishing the Christians through the Turks as he once punished the Jews when they forsook their faith . . . for the Turks are today the Assyrians and Babylonians of the Christians and the rod and scourge and fury of God'.[34] But from mid-century on, others

[30] H. Hochholzer, *art. cit.*, p. 57.

[31] J. Zontar, in *Vierteljahrschr. für Sozial- und Wirtschaftsgeschichte*, 1934, p. 369.

[32] Quoted by G. Atkinson, *op. cit.*, p. 179.

[33] *Ibid.*, p. 211.

[34] *Ibid.*, p. 397. The same idea is expressed in 1544 by Jérôme Maurand, *Itinéraire de . . . d'Antibes à Constantinople* (1544), published by Léon Durez, 1901, p. 69, the victories of the Turks are a punishment for the sins of the Christians.

like Pierre Belon were to recognize the virtues of the Turks; and in later years this strange and contrary land was to exercise great fascination over Europeans, for whom it was a convenient place to escape to in imagination from the restrictions of western society.

It was at least an advance to recognize that Turkish actions could be explained by the faults and weaknesses of Europe.[35] A Ragusan citizen said as much to Maximilian I:[36] while the western nations are divided, 'all supreme authority in the Turkish Empire rests in a single man, all obey the sultan, he alone rules; he receives all revenues, in a word he is master and all other men are his slaves'. In substance, this was what was explained to Ferdinand's ambassadors in 1533 by Aloysius Gritti, a singular character, the son of a Venetian and a slave-girl, and for many years the favourite of the Grand Vizier, Ibrāhīm Pasha. Charles V should not risk his strength against that of Sulaimān: *Verum esse Carolum Cesarem potentem sed cui non omnes obediant, exemplo esse Germaniam et lutheranorum pervicaciam.*[37]

It is certainly true that Turkish strength was drawn, as if by a mechanical process, into the complex of European weaknesses. The bitter internal dissensions of Europe permitted and even encouraged the Turkish invasion of Hungary. 'It was the capture of Belgrade [29th August, 1521],' Busbecq quite rightly says,[38] 'which loosed that multitude of evils under whose weight we continue to groan. This threw open the flood-gates through which the barbarians entered to devastate Hungary, it brought about the death of King Louis, the loss of Buda and the enslavement of Transylvania. If the Turks had not captured Belgrade, they would never have entered Hungary, once one of the most flourishing kingdoms of Europe and now a desolate and ravaged land.'

In fact 1521, the year of Belgrade, also saw the beginning of the long struggle between François I and Charles V. One consequence was the battle of Mohács in 1526; another the siege of Vienna in 1529. Bandello, who wrote his *Novelle* not long after this event,[39] paints a picture of Christendom preparing for the worst, 'reduced to a canton of Europe, as a result of the discords which appear every day more profound between the Christian Princes'. Unless, that is, Europe[40] was less concerned with barring the way to the Turk than with other, brighter prospects in the Atlantic and elsewhere in the world, as some historians have suggested.[41] The time has surely come to turn on its head that hoary and misleading explanation,

[35] F. Babinger, *Suleiman der Prächtige* (*Meister der Politik*), 1923, p. 446–447.
[36] J. W. Zinkeisen, *op. cit.*, III, p. 19.
[37] Quoted by J. W. Zinkeisen, *op. cit.*, III, p. 20, n. 1, following Anton von Gevay, *Urkunden und Actenstücke zur Geschichte der Verhältnisse zwischen Österreich, Ungarn und der Pforte im XVI. und XVII. Jahrhundert*, 1840–1842, p. 31.
[38] Busbecq, *Lettres . . .*, p. 42; cf. *The Turkish Letters*, p. 14.
[39] *Op. cit.*, VIII, p. 305.
[40] F. Grenard, *op. cit.*, p. 86.
[41] Émile Bourgeois, *Manuel historique de politique étrangère*, Vol. I, 1892, Introduction, p. 2 ff.

still sometimes encountered, that it was the Turkish conquest which stimulated the great discoveries, whereas the reverse in fact occurred, for the great discoveries robbed the Levant of much of its appeal, enabling the Turks to extend their influence and settle there without too much difficulty. After all,the Turkish occupation of Egypt in January, 1517 did not occur until twenty years after Vasco da Gama had sailed round the Cape of Good Hope.

The Turks in Syria and Egypt. And surely the major event in the rise of the Ottoman Empire, more significant even than the capture of Constantinople (a mere 'episode' as Richard Busch-Zanter rather deflatingly calls it[42]) was indeed the conquest first of Syria in 1516, then of Egypt in 1517, both achieved in a single thrust. This was the first glimpse of the future greatness of the Ottoman state.[43] In itself, the conquest was not particularly remarkable and posed few difficulties. The disputed frontiers of northern Syria and in particular an attempt by the Mamlūk sultan to act as mediator between Turks and Persians served as a pretext when the right moment came. The Mamlūks, who considered artillery a dishonourable weapon, could not withstand the fire of Selīm's cannon on 24th August, 1516, outside Aleppo. Syria fell overnight into the hands of the conqueror who entered Damascus on 26th September. When the new Mamlūk ruler refused to recognize Turkish sovereignty, Selīm's army advanced into Egypt. The Mamlūk forces were again shattered by Turkish cannon[44] in January, 1517 outside Cairo. Artillery had once more created a major political power, as it had already done in France, in Muscovy[45] and at Granada in 1492.[46]

Egypt succumbed with hardly a struggle, and with a minimal disturbance of the established order. The Mamlūks, who retained their vast estates, very quickly regained effective power: Bonaparte, arriving in Egypt three centuries later, found them there still. The Baron de Tott was no doubt correct when he wrote: 'In examining the canons or code of Sultan Selim, one would imagine that this Prince capitulated with the Mamlūks, rather than conquered Egypt. It is evident, in fact, that by suffering the twenty-four Beys to subsist, who governed that kingdom, he only aimed at ballancing [*sic*] their authority, by that of a Pasha, whom he constituted Governor General and President of the Council . . .'.[47] We should be warned by such comments not to over-dramatize the conquest of 1517.

All the same, it was a landmark in Ottoman history. From the Egyptians, Selīm gained much of value. In the first place, the annual tribute, originally modest,[48] grew steadily. Through Egypt, the Ottoman Empire was able to

[42] '. . . eine Episode, kein Ereignis', p. 22.
[43] V. Hassel, *op. cit.*, p. 22–23.
[44] F. Grenard, *op. cit.*, p. 79.
[45] See above, Vol. I, p. 180.
[46] J. Dieulafoy, *Isabelle la Catholique, Reine de Castille*, 1920; Fernand Braudel, 'Les Espagnols . . .' in *Revue Africaine*, 1928, p. 216, n. 2.
[47] Tott, *Memoirs . . . on the Turks and the Tartars*, London, 1785, Vol. II, p. 29.
[48] Brockelmann, *History of the Islamic Peoples* (tr. Carmichael and Perlmann), London, 1959, p. 289.

participate in the traffic in African gold which passed through Ethiopia and the Sudan and in the spice trade with Christian countries. Mention has already been made of the gold traffic and of the revived importance of the Red Sea route in Levantine trade. By the time the Turks occupied Egypt and Syria, long after Vasco da Gama's voyage of discovery, these two countries were no longer exclusive gateways to the Far East but nevertheless remained important. The Turkish barrier between Christendom and the Indian Ocean[49] was now completed and consolidated, while at the same time a link was established between the hungry metropolis of Constantinople and an extensive wheat, bean and rice-producing region. On many subsequent occasions, Egypt was to be a crucial factor in the fortunes of the Turkish Empire and one might even say a source of corruption. It has been claimed with some plausibility that it was from Egypt that there spread to the far corners of the Ottoman Empire that venality of office[50] which has so frequently undermined a political order.

But Selīm derived from his victory something else quite as precious as gold. Even before becoming ruler of the Nile, he had had prayers said in his name and fulfilled the role of Caliph,[51] Commander of the Faithful. Now Egypt provided consecration for this role. Legend had it – that it was no more than a legend seemed not to matter – that the last of the Abbāsids, having taken refuge in Egypt with the Mamlūks, ceded to Selīm the caliphate over all true Moslem believers. Legend or not, the sultan returned from Egypt radiating an aura of immense prestige. In August, 1517, he received from the son of the Shaikh of Mecca the key to the Ka'ab itself.[52] It was from this date that the élite corps of horseguards was granted the privilege of carrying the green banner of the Prophet.[53] There can be no doubt that throughout Islam, the elevation of Selīm to the dignity of Commander of the Faithful in 1517 was as resounding an event as the famous election, two years later, of Charles of Spain as Emperor was in Christendom. This date at the dawn of the sixteenth century marked the arrival of the Ottomans as a world power and perhaps inevitably, of a wave of religious intolerance.[54]

Selīm died shortly after his victories, in 1520, on the road to Adrianople. His son Sulaimān succeeded him unchallenged. To Sulaimān was to fall the honour of consolidating the might of the Ottoman Empire, despite the pessimistic forecasts voiced concerning his person. In the event the man proved equal to the task. He arrived at an opportune moment, it is true.

[49] J. Mazzei, *op. cit.*, p. 41.

[50] *Annuaire du monde musulman*, p. 21.

[51] The sultan did not officially assume this title until the eighteenth century, Stanford J. Shaw, 'The Ottoman view of the Balkans' in *The Balkans in transition*, ed. C. and B. Jelavich, 1963, p. 63.

[52] J. W. Zinkeisen, *op. cit.*, III, p. 15.

[53] Brockelmann, *op. cit.*, p. 302.

[54] Stanford J. Shaw, *art. cit.*, p. 66, remarks upon the role played by the fanatica *ulema* class in the newly conquered Arab provinces and the Ottoman reaction to increased Franciscan missionary activity in the Balkans (supported by the Habsburgs and Venetians).

In 1521, he seized Belgrade, the gateway to Hungary; in July, 1522, he laid siege to Rhodes, which fell in December of the same year: once the formidable and influential fortress of the Knights of St. John had fallen, the entire eastern Mediterranean lay open to his youthful ambition. There was now no reason why the master of so many Mediterranean shores should not build a fleet. His subjects and the Greeks, including those who inhabited Venetian islands,[55] were to provide him with the indispensable manpower. Would the reign of Sulaimān, ushered in by these brilliant victories, have been so illustrious had it not been for his father's conquest of Egypt and Syria?

The Turkish Empire seen from within. As western historians, we have seen only the outer face of the Turkish Empire and it is as outsiders, only partially aware of its true workings, that we have tried to explain it. This narrow view is gradually being changed by utilizing the extremely rich archives of Istanbul and the rest of Turkey. In order to understand the strengths and also, for they soon became apparent, the weaknesses[56] and irregularities of this immense machine, it must be viewed from the inside. It will mean reconsidering a style of government which was also a way of life, a composite and complex heritage, a religious order as well as a social order and several different economic periods. The imperial career of the Ottomans covers centuries of history and therefore a series of different, sometimes contradictory experiences. It was a 'feudal' regime which expanded from Asia Minor into the Balkans (1360), a few years after the battle of Poitiers, during the early stages of what we know as the Hundred Years' War; and it was a feudal system (based on benefices and fiefs) which the Ottomans introduced to the conquered regions of Europe, creating a landed aristocracy controlled only with varying success by the sultans, and against which they were later to wage an untiring and eventually effective campaign. But that dominant class in Ottoman society, the slaves of the sultan, was constantly being recruited from new sources. Its struggle for power punctuated the internal rhythm of the imperial story, as we shall see.

Spanish unity: the Catholic Kings. In the East the Ottomans; in the West the Habsburgs. Before the rise of the latter, the Catholic Kings, the original authors of Spanish unity, played as vital a part in imperial history as the sultans of Bursa and Adrianople had in the genesis of the Ottoman Empire – if not more. Their achievement was furthered and assisted by the general temper of the fifteenth century after the Hundred Years' War. We should not take literally everything written about Ferdinand and Isabella by the historiographers. The achievement of the Catholic Kings, which I have no intention of belittling, had the times and the

[55] See above, Vol. I, p. 115 and note 43.

[56] Stanford J. Shaw, 'The Ottoman view of the Balkans' in *The Balkans in transition*, ed. Jelavich, p. 56–80.

desires of men in its favour. It was a development desired, demanded even, by the urban bourgeoisie, weary of civil war and anxious for domestic stability, for the peaceful renewal of trade and for security. The original *Hermandad* was an urban phenomenon: its alarum bells rang out from city to city, proclaiming a new age. The cities, with their astonishing reserves of democratic tradition, ensured the triumph of Ferdinand and Isabella.

So let us not exaggerate the part, admittedly an important one, played by the principal actors in this drama. Some historians have even suggested that the union between Castile and Aragon, which became a powerful reality through the marriage of 1469, could well have been replaced by a union between Castile and Portugal.[57] Isabella had the choice between a Portuguese husband and an Aragonese, between the Atlantic and the Mediterranean. In fact the unification of the Iberian Peninsula was already in the air, a logical development of the times. It was a question of choosing a Portuguese or an Aragonese formula, neither being necessarily superior to the other and both within easy reach. The decision finally reached in 1469 signalled the re-orientation of Castile towards the Mediterranean, an undertaking full of challenge and not without risk, in view of the traditional policies and interests of the kingdom, but which was nevertheless accomplished in the space of a generation. Ferdinand and Isabella were married in 1469; Isabella succeeded to the Crown of Castile in 1474 and Ferdinand to that of Aragon in 1479; the Portuguese threat was finally eliminated in 1483; the conquest of Granada was accomplished in 1492; the acquisition of Spanish Navarre in 1512. It is not possible even for a moment to compare this rapid unification with the slow and painful creation of France from its cradle in the region between the Loire and the Seine. The difference was not one of country but of century.

It would be surprising if this rapid unification of Spain had *not* created the necessity for a mystique of empire. Ximénez' Spain, at the height of the religious revival at the end of the fifteenth century, was still living in the age of crusades; hence the unquestionable importance of the conquest of Granada and the first steps, taken a few years later, towards expansion in North Africa. Not only did the occupation of southern Spain complete the reconquest of Iberian soil; not only did it present the Catholic Kings with a rich agricultural region, a region of rich farming land and industrious and populous towns: it also liberated for foreign adventures the energies of Castile, so long engaged in an endless combat with the remnants of Spanish Islam which refused to die – and these were youthful energies.[58]

Almost immediately however, Spain was distracted from African conquest. In 1492, Christopher Columbus discovered America. Three years later, Ferdinand was engrossed in the complicated affairs of Italy. Ferdinand, the over-cunning Aragonese, has been bitterly criticized, by Carlos Pereyra,[59] for being thus diverted towards the Mediterranean and so ne-

[57] Angel Ganivet, *Idearium español*, ed. Espasa, 1948, p. 62 ff.
[58] Pierre Vilar, *La Catalogne . . .*, I, p. 509 ff.
[59] *Imperio español*, p. 43.

glecting the true future of Spain which lay outside Europe, in the barren deserts of Africa and in America too, an unknown world abandoned by Spain's rulers to the worst kind of adventurer. And yet it was this handing over of the *Ultramar* to private enterprise which made possible the astonishing feats of the *Conquistadores*. I earlier criticized Machiavelli for failing to recognize the potential importance of the maritime discoveries; even as late as the seventeenth century, the Count Duke Olivares, the not always unsuccessful rival of Richelieu and very nearly a great man, had still not grasped the significance of the Indies.[60]

In the circumstances, nothing could have been more natural than Aragonese policy, with the weight of tradition behind it. Aragon was drawn towards the Mediterranean by her past and by her experience, intimately acquainted with its waters through her seaboard, her shipping and her possessions (the Balearics, Sardinia and Sicily) and not unnaturally attracted, like the rest of Europe and the Mediterranean, by the rich lands of Italy. When his commander Gonzalvo de Córdoba captured Naples in 1503, Ferdinand the Catholic became master of a vital position and a wealthy kingdom, his victory marking a triumph for the Aragonese fleet and, under the Great Captain, the creation, no less, of the Spanish *tercio*, an event which can rank in world history on a level with the creation of the Macedonian phalanx or the Roman legion.[61] To understand the attraction of the Mediterranean for Spain we must not let our image of Naples at the beginning of the sixteenth century be coloured by what it had become by the end – a country struggling to survive, hopelessly in debt. By then the possession of Naples had become a burden. But in 1503, in 1530 even,[62] the *Reame* of Naples afforded both a valuable strategic position and a substantial source of revenue.

The final point to note about the Aragonese policy to which Spain became committed, is its opposition to the advance of Islam: the Spaniards preceded the Turks in North Africa; in Sicily and Naples, Spain stood on one of the foremost ramparts of Christendom. Louis XII might boast: 'I am the Moor against whom the Catholic King is taking up arms',[63] but that did not prevent the Catholic King, by the mere location of the territories he possessed, from coming more and more to fulfil the role of Crusader and defender of the faith with all the duties as well as the privileges that implied. Under Ferdinand, the crusading ardour of Spain moved out of the Peninsula, not to plunge into the barren continent of

[60] R. Konetzke, *op. cit.*, p. 245; Erich Hassinger, 'Die weltgeschichtliche Stellung des XVI. Jahrhunderts', in *Geschichte in Wissenschaft und Unterricht*, 1951, refers to the book by Jacques Signot, *La division du monde* . . ., 1st ed. 1539 (other editions followed: 5th ed. 1599) which makes no mention of America.

[61] Well described by Angel Ganivet in *Idearium español*, 1948, p. 44–45.

[62] Naples had run into deficit by at least 1532, E. Albèri, *op. cit.*, I, 1, p. 37. From the time of Charles V, ordinary expenditure in his states, not counting the cost of war, exceeded his revenue by two million ducats. Guillaume du Vair, *Actions oratoires et traités*, 1606, p. 80–88.

[63] Ch. Monchicourt, 'La Tunisie et l'Europe. Quelques documents relatifs aux XVIe, XVIIe et XVIIIe siècles' in *Revue Tunisienne*, 1905, off-print, p. 18.

Africa on the opposite shore, nor to lose itself in the New World, but to take up a position in the sight of the whole world, at the very heart of what was then Christendom, the threatened citadel of Italy: a traditional policy, but a glorious one.

Charles V. Charles V succeeded Ferdinand in Spain. As Charles of Ghent, he became Charles I in 1516. With his coming, western politics took on new and more complicated dimensions, a development comparable to what was happening at the other end of the sea under Sulaimān the Magnificent. Spain now found herself little more than a background for the spectacular reign of the Emperor. Charles of Ghent became Charles V in 1519; he hardly had time to be Charles of Spain. Or, rather curiously, not until much later, at the end of his life, for reasons of sentiment and health. Spain was not prominent in the career of Charles V, though she contributed handsomely to his greatness.

It would certainly be unjust to overlook the contribution made by Spain to the imperial career. The Catholic Kings had after all carefully prepared the inheritance of their grandson. Had they not been active on every possible front which might prove useful, in England, Portugal, Austria and the Netherlands, staked throw after throw on the lottery of royal marriages? The notion of surrounding France, of neutralizing this dangerous neighbour, prefigured the later anomalous shape of the Habsburg Empire, with a gaping hole at its centre. The possibility of Charles V's reign was a calculated gamble on the part of Spain. An accident might of course have changed the course of history. Spain might have refused to recognize Charles as long as his mother, Joanna the Mad, was still alive (and she did not die, at Tordesillas, until 1555); or the preference might have gone to his brother Ferdinand, who had been brought up in the Peninsula. For that matter, Charles might not have won the imperial election in 1519. But for all that, Europe might not have escaped the great imperial experience. France, which had begun moving in that direction in 1494, might very well have begun again and succeeded. And it should also be remembered that behind the fortunes of Charles V for a long time lay the economic power of the Netherlands, already associated with the new Atlantic world, the crossroads of Europe, an industrial and commercial centre requiring markets and outlets for its trade and the political security which a disorganized German Empire would have threatened.

Since Europe was moving of its own accord towards the construction of a vast state, the imperial drama would have been played out sooner or later: only the *dramatis personae* would have been different had Charles' fortunes taken another turn. The electors of Frankfurt could hardly, in 1519, have pronounced in favour of a national candidate. As German historians have pointed out, Germany could not have borne the weight of such a candidacy: it would have meant taking on single-handed the two other candidates, François I and Charles. By electing Charles,

Germany chose the lesser of two evils and not merely, although this is sometimes suggested, the man who held Vienna and therefore protected her threatened eastern frontier. In 1519, it should be remembered, Belgrade was still a Christian possession and between Belgrade and Vienna lay the protective barrier of the kingdom of Hungary. Not until 1526 was the Hungarian frontier violated. Then – and only then – the situation was totally altered. The threads of the Habsburg and Ottoman destinies are tangled enough in reality without confusing them any further. The following popular rhyme about the Emperor would not have been heard in 1519:

Das hat er als getane
Allein für Vaterland
Auf das die römische Krone
Nit komm in Turkenhand.

In fact Charles never used Germany as a base of operations. By 1521, Luther was launched on his career. And soon after his coronation at Aix-la-Chapelle, in September, 1520, the Emperor renounced, in favour of his brother Ferdinand, his own marriage to the Hungarian princess Anna; in Brussels on 7th February, 1522, he secretly ceded the *Erbland* to his brother.[64] By so doing he was abdicating from personal intervention on any large scale in Germany.

Nor was it possible in the nature of things for Charles to make Spain his headquarters. Too far from the centre of Europe, Spain did not yet offer the compensating advantage of riches from the New World: not until 1535 was this to be a major consideration. In his struggle against France, the chief occupation of his life after 1521, Charles V was obliged to rely upon Italy and the Netherlands. Along this central axis of Europe, the Emperor concentrated his effort. Gattinara, the Grand Chancellor, advised Charles whatever else he did to hold on to Italy. Meanwhile in the Netherlands, in peace time at least, Charles V benefited from substantial revenues, the possibility of loans, as in 1529, and a budget surplus. It was frequently said during his reign that the Netherlands paid all his bills, an opinion even more widely voiced after 1552. For now the same fate befell the Netherlands which had already befallen Sicily, Naples and even Milan: although wealthy states, all of them had seen their budget surpluses practically wiped out. This development may have been hastened by the fact that both Charles and Philip II concentrated their military efforts on the Netherlands, thus damaging its trade. Large sums of money were of course imported from Spain, especially during Philip's reign. But complaints were still being voiced in 1560. The Netherlands claimed to have suffered more than Spain, 'for the latter remained free from all material harm and had continued to trade with France under cover of safe-conducts'.[65] So she could not complain over much of her sufferings during this war which, she admitted, had only been waged so that the king of

[64] Gustav Turba, *Geschichte des Thronfolgerechtes in allen habsburgischen Ländern* . . ., 1903, p. 153 ff.

[65] Granvelle to Philip II, Brussels, 6th October, 1560, *Papiers d'état*, VI, p. 179.

Spain could 'have a foothold in Italy'.[66] It was a sterile debate, but one which turned to the disadvantage of Flanders. Philip II took up residence in Spain and in 1567 one of the aims of the Duke of Alva was to make the rebel provinces disgorge their wealth. A reliable history of the finances of the Netherlands would therefore be very useful.[67] The Venetians, in 1559, described it as a rich and populous country but one where the cost of living was staggeringly high: 'what costs two in Italy and three in Germany costs four or five in Flanders'.[68] Was it the price rise, following the influx of American bullion and then the war, which finally dislocated the fiscal machinery of the Netherlands? Soriano in his *Relazione* of 1559 writes: 'these lands are the treasuries of the King of Spain, his mines and his Indies, they have financed the enterprises of the Emperor for so many years in the wars of France, Italy and Germany . . . '.[69] But even as he wrote, this had ceased to be true.

Italy and the Netherlands then, provided the vital double formula for Charles V's policy, with from time to time contributions from Spain and Germany. To the historian of the reign of Philip II, Charles's Empire appears cosmopolitan in character, open to Italians, Flemings and Burgundians who might, naturally, find themselves rubbing shoulders with Spaniards in the Emperor's entourage. The age of Charles V, coming as it does between the Spain of the Catholic Kings and the Spain of Philip II, carries something of a universal temper. The idea of crusade itself was modified,[70] losing its purely Iberian character and moving away from the ideals of the *Reconquista*. After his election in 1519, Charles V's policy took flight and was carried away in dreams of Universal Monarchy. 'Sire', wrote Gattinara to the Emperor shortly after his election, 'now that God in His prodigious grace has elevated Your Majesty above all Kings and Princes of Christendom, to a pinnacle of power occupied before by none except your mighty predecessor Charlemagne, you are on the road towards Universal Monarchy and on the point of uniting Christendom under a single shepherd.'[71] This notion of Universal Monarchy was the continuous inspiration of Charles's policy, which also had affinities with the great humanist movement of the age. A German writer, Georg Sauermann, who happened to be in Spain in 1520, dedicated to the Emperor's secretary, Pedro Ruiz de la Mota, his *Hispaniae Consolatio*, in which he tried to convert Spain herself to the idea of a pacific Universal Monarchy, uniting all Christendom against the Turk. Marcel Bataillon

[66] Granvelle to Philip II, Brussels, 6th October, 1560, *Papiers d'état*, VI, p. 179.

[67] See F. Braudel, 'Les emprunts de Charles-Quint sur la place d'Anvers' in *Charles-Quint et son temps*, Paris, 1959, graph, p. 196.

[68] E. Albèri, II, *op. cit.*, III, p. 357 (1559).

[69] *Ibid.*

[70] For a valuable discussion of this point see R. Menéndez Pidal, *Idea imperial de Carlos V*, Madrid, 1940; for a comprehensive review of the questions involved, Ricardo Delargo Ygaray, *La idea de imperio en la política y la literatura españolas*, Madrid, 1944.

[71] Quoted by E. Hering, *op. cit.*, p. 156.

has shown how dear this notion of Christian unity was to Erasmus and his friends and disciples.[72] In 1527, after the sack of Rome, Vives wrote to Erasmus in extremely revealing terms: 'Christ has granted an extraordinary opportunity to the men of our age to realize this ideal, thanks to the great victory of the Emperor and the captivity of the Pope,'[73] – a sentence which illuminates the true colours of the ideological mist, the vision in which the policy of the emperor was surrounded and the frequent source of motives for his actions. This is by no means the least fascinating aspect of what was the major political drama of the century.

Philip II's Empire. The work of Charles V was continued during the second half of the sixteenth century by that of Philip II, ruler over an empire too, but a very different one. Emerging from the heritage of the great emperor during the crucial years 1558–1559, this later empire was even more vast, coherent and solid than that of Charles V, less committed to Europe, more exclusively concentrated on Spain and thus drawn towards the Atlantic. It had the substance, the extent, the disparate resources and the wealth of an empire, but its sovereign ruler lacked the coveted title of Emperor which would have united and, as it were, crowned the other countless titles he held. Charles V's son had been excluded, after much hesitation, from the imperial succession which had in theory, but in theory only, been reserved for him at Augsburg in 1551.[74] And he sorely missed the prestige this title would have conferred on him, if only in the minor but irritating war of precedence with the French ambassadors at Rome, that vital stage on which all eyes were fixed. So in 1562, the Prudent King thought of seeking an Imperial crown. In January, 1563 it was rumoured that he would be declared Emperor of the Indies.[75] A similar rumour circulated in April, 1563[76] to the effect that Philip would be proclaimed 'King of the Indies and of the New World'. The rumours persisted the following year, in January, 1564, when once again the title of Emperor of the Indies was mentioned.[77] About twenty years later, in 1583, it was whispered at Venice that Philip II once more aspired to the highest title. 'Sire', wrote the French ambassador to Henri III, 'I have learnt from these Lords that Cardinal Granvelle is coming to Rome in September this year to have the title of emperor conferred upon his master.'[78]

Perhaps this was no more than Venetian gossip. Even so it is interesting. The same causes were to produce the same effects when Philip III in his turn was a candidate for the empire. It was not merely a question of vanity.

[72] *Op. cit.*, see all of Chapter VIII, p. 395 ff.

[73] According to R. Konetzke, *op. cit.*, p. 152.

[74] See below, p. 913.

[75] G. Micheli to the Doge, 30th January, 1563, G. Turba, *op. cit.*, I, 3, p. 217.

[76] *Ibid.*, p. 217, n. 3.

[77] 13th January, 1564, Saint-Sulpice, E. Cabié, *op. cit.*, p. 217, that is if Cabié is right about the date.

[78] H. de Maisse to the King, Venice, 6th June, 1583, A.E. Venice 81, f° 28 v°. Philip II was apparently thinking of applying for the Imperial Vicariate in Italy, 12th February, 1584, Longlée, *Dépêches diplomatiques* . . ., p. 19.

In a century preoccupied with prestige, governed by appearances, a merciless war of precedence was waged between the ambassadors of the Most Christian King of France and His Catholic Majesty of Spain. In 1560, in order to put an end to this irritating and unresolved conflict, Philip II proposed to the Emperor that they should appoint a joint ambassador to the Council of Trent. By not being emperor, Philip II was deprived, at the honorary level of appearances, of the front rank which should have been his among Christian powers and which no one had been able to dispute Charles V or his representatives during his entire lifetime.

The fundamental characteristic of Philip II's empire was its Spanishness – or rather Castilianism – a fact which did not escape the contemporaries of the Prudent King, whether friend or foe: they saw him as a spider sitting motionless at the centre of his web. But if Philip, after returning from Flanders in September, 1559, never again left the Peninsula, was it simply from inclination, from a pronounced personal preference for things Spanish? Or might it not also have been largely dictated by necessity? We have seen how the states of Charles V, one after another, silently refused to support the expense of his campaigns. Their deficits made Sicily, Naples, Milan and later the Netherlands, burdens on the empire, dependent places where it was no longer possible for the Emperor to reside. Philip II had had personal experience of this in the Netherlands where, during his stay between 1555 and 1559, he had relied exclusively on money imported from Spain or on the hope of its arrival. And it was now becoming difficult for the ruler to obtain such assistance without being in person close to its original source. Philip II's withdrawal to Spain was a tactical withdrawal towards American silver. His mistake, if anything, was not to go as far as possible to meet the flow of silver, to the Atlantic coast, to Seville or even later to Lisbon.[79] Was it the attraction of Europe, the need to be better and more quickly informed of what was happening in that buzzing hive, which kept the king at the geometrical centre of the Peninsula, in his Castilian retreat, to which personal inclination in any case drew him?

That the centre of the web lay in Spain bred many consequences; in the first place the growing, blind affection of the mass of Spanish people for the king who had chosen to live among them. Philip II was as much beloved of the people of Castile as his father had been by the good folk of the Low Countries. A further consequence was the logical predominance of Peninsular appointments, interests and prejudices during Philip's reign; of those harsh haughty men, the intransigent nobles of Castile whom Philip employed on foreign missions, while for the conduct of everyday affairs and bureaucratic routine he showed a marked preference for commoners. Charles V was forced to be a homeless traveller in his scattered empire: all his life he had had to negotiate the obstacle of an unfriendly France, in order to bring to his dominions one by one the warmth of his personal presence. Philip II's refusal to move encouraged the growth of a sedentary

[79] As suggested by Jules Gounon Loubens, see above, Vol. I, p. 351, note 402.

administration whose bags need no longer be kept light for travelling. The weight of paper became greater than ever. The other parts of the empire slipped imperceptibly into the role of satellites and Castile into that of the metropolitan power: a process clear to see in the Italian provinces. Hatred of the Spaniard began to smoulder everywhere, a sign of the times and a warning of storms ahead.

That Philip II was not fully aware of these changes, that he considered himself to be continuing the policy initiated by Charles V, his father's disciple as well as successor, is certainly true; the disciple was if anything a little too mindful of the lessons he had learnt, over-conscious of precedent in his dealings and encouraged in this by his immediate advisers, the Duke of Alva or Cardinal Granvelle, the walking legend and living dossier respectively of the defunct imperial policy. It is true that Philip not infrequently found himself in situations analogous at least in appearance to those experienced by the Emperor his father. As ruler of the Netherlands why should he not, like Charles V, seek peace with England, whose goodwill was vital to the security of the commercial centre of the North? Or since, like his father, he had several states on his hands, why should he not pursue the same prudent and delaying tactics, his object being to control from a distance the not always harmonious concert of his far-off possessions?

But circumstances were to dictate radical changes. Only the trappings of empire survived. The grandiose ambitions of Charles V were doomed by the beginning of Philip's reign, even before the treaty of 1559, and brutally liquidated by the financial disaster of 1557. The machinery of empire had to be overhauled and repaired before it could be started again. Charles V had never in his headlong career been forced to brake so sharply: the drastic return to peace in the early years of Philip's reign was the sign of latter-day weakness. Grand designs were not revived until later and then less as the result of the personal desires of the sovereign than through force of circumstance. Little by little, the powerful movement towards Catholic reform, misleadingly known as the Counter-Reformation, was gathering strength and becoming established. Born of a lengthy series of efforts and preparations, already by 1560 a force strong enough to sway the policy of the Prudent King, it exploded violently in opposition to the Protestant North in the 1580s. It was this movement which pushed Spain into the great struggles of the end of Philip's reign and turned the Spanish king into the champion of Catholicism, the defender of the faith. Religious passions ran higher in this struggle than in the crusade against the Turks, a war entered upon almost unwillingly and in which Lepanto seems to have been an episode without sequel.

And there was another compelling factor: after the 1580s, shipments of bullion from the New World reached an unprecedented volume. The time was now ripe for Granvelle to return to the Spanish court. But it would be wrong to think that the imperialism which appeared at the end of the reign was solely the result of his presence. The great war which began in the 1580s was fundamentally a struggle for control of the Atlantic Ocean, the new centre of gravity of the world. Its outcome would decide whether the Atlantic was in future to be ruled by Catholics or Protestants, northerners or Iberians, for the Atlantic was now the prize coveted by all. The mighty Spanish Empire with its silver, its armaments, ships, cargoes and political conceptions, now turned towards that immense battlefield. At the same moment in time, the Ottomans turned their backs firmly on the Mediterranean to plunge into conflict on the Asian border. This should remind us, if a reminder is needed, that the two great Mediterranean Empires beat with the same rhythm and that at least during the last twenty years of the century, the Mediterranean itself was no longer the focus of their ambitions and desires. Did the decline of empire sound earlier in the Mediterranean than elsewhere?

Accident and political explanation. That a historian should today combine politics and economics in his arguments will not seem remarkable. So much of what we have to discuss – though not of course everything – was dictated by the population increase, by the pronounced acceleration of trade and later by economic recession. It is my contention that a correlation can be established between the reversal of the secular trend and the series of difficulties which confronted the giant political combinations built up by the Ottomans and the Habsburgs. To make this connection clearer, I have deliberately excluded explanations advanced by historians who have concentrated on the outstanding personalities and events of the time, explanations which can sometimes be a distorting prism through which to view reality. I have also neglected what is rather more interesting to us, the long-term political explanation: politics and institutions can themselves contribute to the understanding of politics and institutions.

The controversy is rather curiously taken up again in a brief paragraph in the last book by the great economist Josef A. Schumpeter[80] whose views are in part the opposite of those expressed above. Schumpeter recognizes only one unbroken line of development, one constant: the rise of capitalism. Everything else in politics or economics is merely a matter of accident, circumstance, chance or detail. It was an accident 'that the conquest of South America produced a torrent of precious metals', without which the triumph of the Habsburgs would have been inconceivable. It was an accident that the 'price revolution' occurred to make social and political tensions more explosive; and yet another accident that the expanding states (and empires too of course) found the way clear before them in the sixteenth century. How an accident? Because the great political powers of the past collapsed of their own accord, the German Holy Roman Empire with the death of Frederick II in 1250; the Papacy about the same time, for its triumph was but a Pyrrhic victory. And well before 1453, the Byzantine Empire had fallen into decline.

Such a view of history (although the passage in Schumpeter is very

[80] *History of Economic Analysis*, London, 1954, p. 144 ff.

short) should in fairness be discussed point by point rather than dismissed abruptly by the historian. To be brief however, I would only say this, that the natural collapse of the Papacy and the Empire in the thirteenth century was no accident, nor was it the result of a blind pursuit of self-destruction. Economic growth in the thirteenth century made possible certain political developments just as it did in the sixteenth and prepared the way for large-scale political change. It was followed by a period of recession, the effects of which were universally felt. The series of collapses during the following century can be attributed to an economic depression of long duration: the 'waning of the Middle Ages' during which all rotten trees were marked out for destruction, from the Byzantine Empire to the Kingdom of Granada including the Holy Roman Empire itself. From start to finish this was a slow and natural process.

With the recovery which became apparent from roughly halfway through the fifteenth century, a further round of destruction, innovation and renewal could be expected. The Papacy was not seriously damaged until after the Lutheran revolt and the setback of the Diet of Augsburg (1530). It would have been possible for Rome to pursue a different policy, one more resolutely eirenical and conciliatory. Let it be remembered though that the Papacy nevertheless remained a great power, even in the political arena throughout the sixteenth century and indeed up to the Treaty of Westphalia (1648).

To return to the other points: the price revolution – as Schumpeter himself observes[81] – *preceded* the massive influx of precious metals from the New World. Similarly the expansion of the territorial states (of Louis XI, Henry VII of Lancaster, John of Aragon, Muḥammad II) preceded the discovery of America. And finally, if the mines of the New World did become a crucial factor, it was only because Europe possessed the means of exploiting them, for the operation of the mines was in itself costly. Castile, it has been said, won America in a lottery; only in a manner of speaking, for Castile had afterwards to turn her acquisition to account and this was frequently a mundane matter of balancing input and output. Furthermore, if the New World had not offered easy access to gold and silver mines, western Europe's need for expansion would have found other outlets and brought home other spoils. In a recently published study, Louis Dermigny[82] suggests that the West, by choosing the New World, where almost all facilities had to be created by the Old, may possibly have neglected another option – that of the Far East where so many facilities already existed, where wealth was more accessible, and perhaps other options too: the gold of Africa, the silver of central Europe, assets momentarily grasped but soon abandoned. The single decisive factor was the restless energy of the West.

In short, Schumpeter's thesis is merely a restatement of the kind of

[81] *Op. cit.*, p. 144.

[82] *La Chine et l'Occident. Le commerce à Canton au XVIIIe siècle* (*1719–1833*), 4 vols., 1964, Vol. I, p. 429 ff.

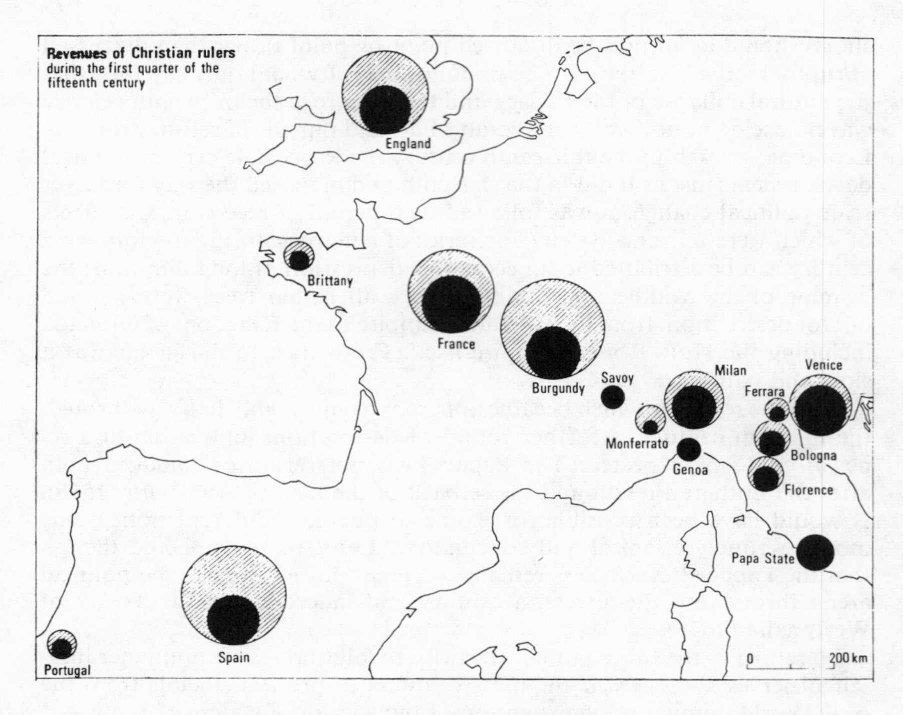

Fig. 56: *State finance and the general price situation*

These rather curious Venetian estimates (*Bilanci generali*, Vol. 1, Bk. 1, Venice, 1912, p. 98–99) are certainly not strictly accurate, but they give an idea of the universal decrease in the financial resources of the European states between 1410 and 1423 (1410 level represented by shaded circle, 1423 level by black circle). English revenues fell from 2 million ducats to 700,000; French from 2 million to 1 million; Spanish from 3 million to 800,000; Venetian from 1,100,000 to 800,000, etc. Even if these figures were accurate one would still have to calculate real income as one does real wages. In general, the state seems always to have lagged a little behind changes in the economic situation, both during upward and downward trends, that is to say its resources declined less quickly than others during a depression – and this was an advantage – and rose less quickly during periods of growth. Unfortunately such a theory cannot be verified either from the document in question here, or from others cited below. One thing is certain: the resources of the state fluctuated according to the prevailing economic conditions.

argument frequently heard in the days when accident was much in vogue as a theory among historians. It ignores or underestimates the importance of the state – and yet the state, quite as much as capitalism, was the product of a complex evolutionary process. The historical conjuncture, in the very widest sense of the term, carries within it the foundations of all political power; it breathes life or death into them. Yesterday's losers may be today's winners, for fortune rarely comes twice in succession.

2. THE STATE: RESOURCES AND WEAKNESSES

Of that rise of state and 'empire' witnessed by the sixteenth century, the effects are a great deal more visible than the causes. The modern state had a difficult birth. One of the more obvious of the new phenomena which accompanied it was the multiplication of the instruments and agents of state power: one problem among many.

The 'civil servant'.[83] The corridors of political history are suddenly thronged with the long procession of those men whom we may conveniently if anachronistically, call 'civil servants'. Their arrival marks a political revolution coupled with a social revolution.

Once summoned to power, the government official quickly appropriated for himself a share in public authority. In the sixteenth century at least, he was invariably of humble origin. In Turkey he often had the additional handicap of being born a Christian, a member of a conquered race, or no less often a Jew. According to H. Gelzer,[84] of the forty-eight grand viziers between 1453 and 1623, five were of 'Turkish' birth, including one Circassian; ten were of unknown origin; and thirty-three were renegades, including six Greeks, eleven Albanians or Yugoslavs, one Italian, one Armenian and one Georgian. The number of Christians who thus succeeded in reaching the peak of the Turkish hierarchy indicates the scale on which they had penetrated the ranks of the servants of the Ottoman Empire. And if ultimately the latter was to resemble the Byzantine Empire, rather than say a Mongol Empire,[85] it was because of this large-scale recruiting of civil servants.

In Spain, where we are better acquainted with the government employee, he would typically come from the urban lower classes, or even of peasant stock, which did not prevent him (far from it) from claiming descent from a *hidalgo* family (as who did not in Spain?). At any rate the social advancement of such men escaped no one's attention, least of all that of one of their declared enemies, the diplomat and soldier Diego Hurtado de Mendoza, representative of the high military aristocracy, who notes in his history of

[83] This title is obviously an anachronism – I use it purely for convenience as a term to cover the *officiers* of France, *letrados* of Spain and so on. Julio Caro Baroja, *op. cit.*, p. 148 ff uses 'bureaucracy', but that too is an anachronism.

[84] *Geistliches und Weltliches aus dem griechisch-türkischen Orient*, p. 179, quoted by Brockelmann, *op. cit.*, p. 316.

[85] F. Lot, *op. cit.*, II, p. 126.

the War of Granada:[86] 'The Catholic Kings put the management of justice and public affairs in the hands of the *letrados*, men of middling condition, neither high nor very low born, offending neither the one nor the other and whose profession was the study of the law', 'cuya profesión eran letras legales'. The *letrados* were brothers beneath the skin of the *dottori in legge* mentioned in Italian documents and the sixteenth century French lawyers, graduates of the University of Toulouse or elsewhere, whose notions of Roman law contributed so much to the absolutism of the Valois. With clear-sighted hatred, Hurtado de Mendoza enumerates the entire tribe, the *oídores* of civil cases, *alcaldes* of criminal cases, *presidentes*, members of the *Audiencias* (similar to the French *parlements*) and crowning all, the supreme court of the *Consejo Real*. According to these men their competence extends to all matters, being no more nor less than the 'ciencia de lo que es justo y injusto'. Jealous of other men's offices, they are always ready to encroach on the competence of the military authorities (in other words the great aristocratic families). And this scourge is not confined to Spain: 'this manner of governing has spread throughout Christendom and stands today at the pinnacle of its power and authority'.[87] In this respect, Hurtado de Mendoza was not mistaken. As well as those *letrados* who had already reached positions of authority, we must also imagine the army of those preparing to embark upon a government career who flocked in ever-growing numbers to the universities of Spain (and before long to those of the New World): 70,000 students at least, as is calculated at the beginning of the next century by an irritated Rodrigo Vivero, Marquis del Valle,[88] another nobleman and a Creole from New Spain; among these men are the sons of shoemakers and ploughmen! And who is responsible for this state of affairs if not the Church and state, which by offering offices and livings draws more students to the universities than ever did the thirst for knowledge. For the most part these *letrados* had graduated from Alcalà de Henares or Salamanca. Be that as it may and even if one remembers that the figure of 70,000, which seemed so enormous to Rodrigo Vivero, was modest in relation to the total population of Spain, there can be no doubt of the immense political significance of the rise of a new social category beginning in the age of construction under Ferdinand and Isabella. Even then there were appearing those 'royal clerks' of extremely modest origin, such as Palacios Rubios,[89] the lawyer who drafted the *Leyes de Indias* and was not even the son of a *hidalgo*! Or later, under Charles V, the secretary Gonzalo Pérez, a commoner whispered to have been of Jewish origin.[90] Or again, in the reign of Philip II, there was Cardinal Espinosa,

[86] *De la guerra de Granada comentarios por don Diego Hurtado de Mendoza*, edited by Manuel Gómez Moreno, Madrid, 1948, p. 12.

[87] *Ibid.*

[88] B.M. Add. 18 287.

[89] Eloy Bullon, *Un colaborador de los Reyes Católicos: el doctor Palacios Rubios y sus obras*, Madrid, 1927.

[90] R. Konetzke, *op. cit.*, p. 173, Gregorio Marañón, *Antonio Pérez*, 2 vols., 2nd ed., Madrid, 1948, I, p. 14 ff. Angel González Palencia, *Gonzalo Pérez secretario de Felipe II*, 2 vols., Madrid, 1946, does not mention this.

who when he died of apoplexy in 1572, combined in his person a wealth of titles, honours and functions, and left his house piled high with dossiers and papers which he had never had time to examine and which had lain there sometimes for years. Gonzalo Pérez was a cleric, like Espinosa and like Don Diego de Covarrubias de Leyva, details of whose career we know from the fairly long memoir written by a relative, Sebastián de Covarrubias de Leyva, in 1594:[91] we are told that Don Diego was born at Toledo of noble parents, originally from Viscaya, that he studied first at Salamanca, went on to become a professor at the College of Oviedo, was later magistrate at the *Audiencia* of Granada, next bishop of Ciudad Rodrigo, then archbishop of Santo Domingo 'en las Indias' and finally President of the Court of Castile and presented with the bishopric of Cuenca (in fact he died at Madrid on 27th September, 1577, at the age of sixty-seven, before he could take possession of it). If proof were needed, his story proves that it was possible to combine a career in the Church with a career in state service. And the Church, particularly in Spain, offered many openings to the sons of poor men.

In Turkey, the reign of Sulaimān was at the same time an age of victorious warfare, of widespread construction and of substantial legislative activity. Sulaimān bore the title of *Ḳānūnī*, or law-maker, indicative of a revival of law studies and the existence of a special class of jurists in the states under his rule and above all at Constantinople. His legal code so successfully regulated the judicial machinery that it was said that Henry VIII of England sent a legal mission to Constantinople to study its workings.[92] His *Ḳānūn-nāme* is to the East what the Justinian Code is to the West[93] and the *Recopilación de las Leyes* to Spain. All the legal machinery established by Sulaimān in Hungary was the work of the jurist Abu'l-Su'ūd; such a major achievement of legislation was it on the question of property that many of its detailed provisions remain in force to the present day. And the jurist Ibrāhīm Al-Halabī, author of a handbook on legal procedures, the *multaḳa*,[94] can be ranked alongside the most eminent western jurists of the sixteenth century.

The more one thinks about it, the more convinced one becomes of the striking similarities, transcending words, terminology and political appearances, between East and West, worlds very different it is true, but not always divergent. Experts in Roman law and learned interpreters of the Koran formed a single vast army, working in the East as in the West to enhance the prerogative of princes. It would be both rash and inaccurate to attribute the progress made by monarchy entirely to the zeal, calculations and devotions of these men. All monarchies remained charismatic. And there was always the economy. Nevertheless, this army of lawyers, whether eminent or modest, was fighting on the side of the large

[91] Cuenca, 13th May, 1594, Copy, B. Com. Palermo, Qq G 24, f° 250.
[92] P. Achard, *op. cit.*, p. 183 ff.
[93] Franz Babinger, *Suleiman der Prächtige* (*Meister der Politik*), 1923, p. 461.
[94] F. Babinger, *op. cit.*

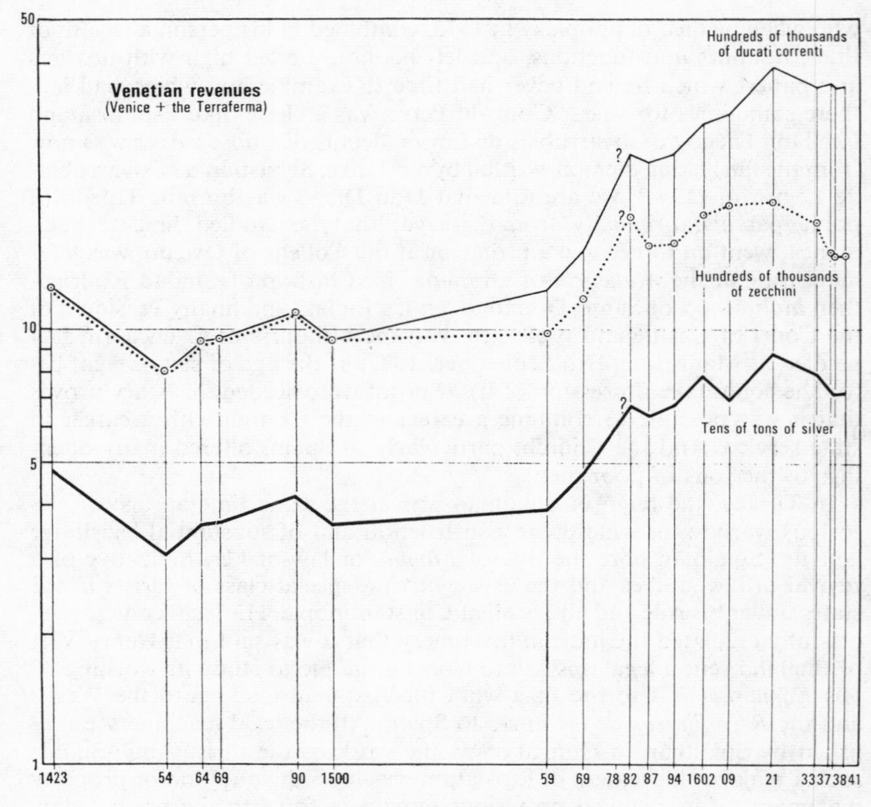

1. *The case of Venice*

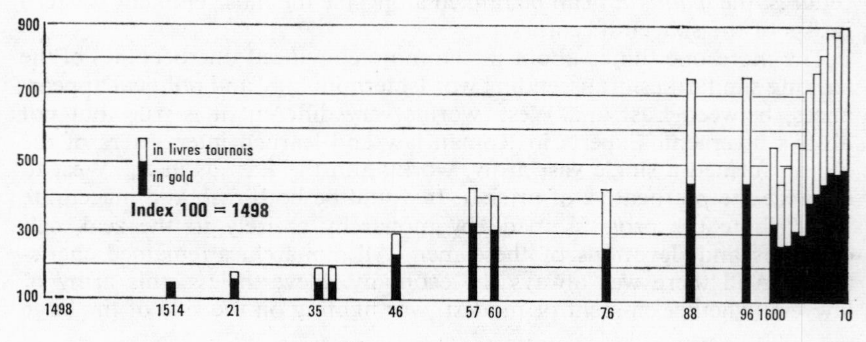

2. *The case of France*

state. It detested and strove to destroy all that stood in the way of state expansion. Even in America, where civil servants from Spain or Portugal often abused their office, the services rendered by these humble men devoted to the prince cannot be gainsaid. Turkey, now becoming partly against her own will a modern state, appointed to the conquered eastern provinces of Asia, increasing numbers of half-pay tax-farmers, who lived off the revenues they collected but transmitted the bulk of it to Istanbul; she also appointed increasing numbers of paid civil servants who, in exchange for specific service, preferably in the towns easiest to administer, would receive a salary from the imperial treasury. More and more of these officials tended to be renegade Christians, who thus gradually infiltrated the ruling class of the Ottoman Empire. They were recruited through the *dewshirme*, a sort of 'tribute which consisted of taking away from their homes in the Balkans a certain number of Christian children, usually under the age of five'.[95] And the word *dewshirme* designated both a political and a social category. These new agents of the Ottoman state were to reduce almost to ruin the *timarli* of the Balkans (the holders of *tīmārs* or fiefs) and for many years to sustain the renewed might of the empire.[96]

Without always explicitly seeking to do so, sixteenth century states moved their 'civil servants' about,[97] uprooting them to suit their convenience. The greater the state, the more likely was this to be the case. One rootless wanderer was Cardinal Granvelle, a son of the Franche-Comté who claimed to have no homeland. One might be inclined to dismiss him as an exception, but there is no shortage of similar cases in Spain – the *licenciado* Polomares for example, first employed at the *Audiencia* of the Grand Canary Island and ending his career in the *Audiencia* of Valladolid.[98] Even more than civilians, military officers in the king's service travelled abroad, with or without the army. From Nantes where, at the end of the century, he was an efficient servant of the

(Opposite) Fig. 57: *State budgets and the general price situation*

Venice's revenues came from three sources: the City, the mainland (*Terraferma*) and the empire. The empire has been omitted from this graph since the figures are often unsubstantiated. Graph constructed by Mlle Gemma Miani, chiefly from the *Bilanci generali*. The three curves correspond to the total receipts of Venice and the *Terraferma*: nominal figures (in *ducati correnti*); the figures in gold (expressed in sequins) and in silver (in tens of tons of silver). The figures for France (compiled by F. C. Spooner) are unfortunately far from complete (nominal figures in *livres tournois* and figures in gold). Despite their lacunae, these curves show that fluctuations in state revenues corresponded to fluctuations in the price sector.

---

[95] R. Mantran, *op. cit.*, p. 107, note 2.

[96] See the excellent exposition by Stanford J. Shaw, *art. cit.*, p. 67 ff, 'Decline of the Timar System and Triumph of the Devshirme Class'.

[97] Many examples can be found in the life-histories of patricians, engineers and soldiers in the service of the Venetian republic – or of Turkish agents who, we know, moved about in a similar manner.

[98] His dossier is in E° 137 at Simancas. This odd figure was the author of the long report to Philip II (Valladolid, October, 1559, E° 137) referred to below, p. 958.

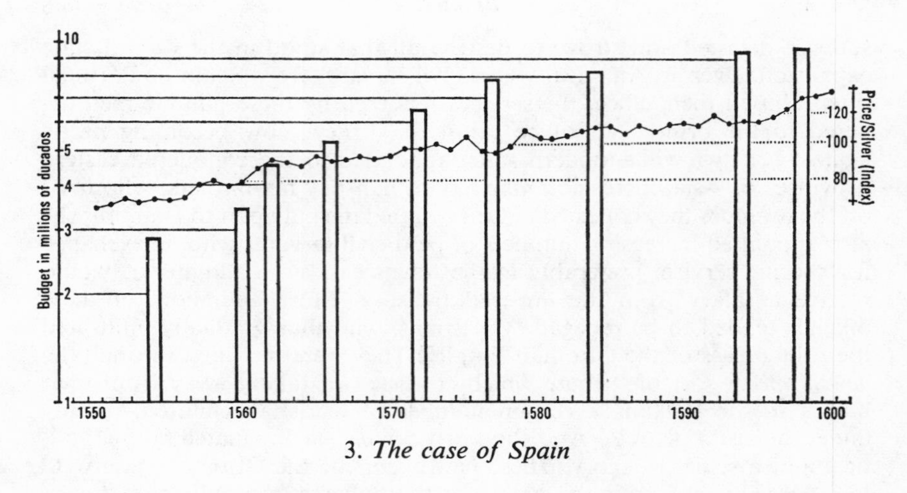

3. *The case of Spain*

Fig. 58: *State budgets and the general price situation*

The index of prices in silver is taken from Earl J. Hamilton. The budgets are expressed in millions of Castilian *ducados*, the money of account which did not alter during the period in question. Budget estimates from unpublished research by Alvaro Castillo Pintado. This time, despite the imperfect calculation of receipts, the coincidence between the price situation and the trend in fiscal revenues is much clearer than in the preceding examples. It will be noted that I use the term 'budget' although it is not strictly speaking applicable. Our knowledge of expenditure is always inadequate. To my knowledge, only the archives of Simancas and possibly archives in England contain sufficient material to construct a true budget. Tentative graphs similar to those given here could be calculated for Sicily and the Kingdom of Naples and even for the Ottoman Empire, a project already begun by Ömer Lutfi Barkan and his research group.

Crown, the Spanish representative Don Diego Mendo de Ledesma sent to Philip II,[99] in support of his request for 'assistance' in his financial difficulties, a long record of his loyal service to the king. Of undoubtedly noble birth, he was admitted as a child, along with his brother, to the ranks of pages to Queen Elizabeth, the 'Queen of Peace' (Catherine de Medici's daughter and Philip II's third wife). When still a boy he had served during the War of Granada, then had followed Don John of Austria to Italy. With his two brothers, on the conquest of Portugal in 1580, he had persuaded the town of Zamora to rally to His Catholic Majesty, adding his own vassals to the militia. Later when the same town was reluctant to accept the increase in its fixed payment for the *alcabala*, thus setting a bad example to other cities, the government dispatched Don Diego to bring the citizens to see reason. 'As soon as I entered the *Ayuntamiento*,' he writes, 'I smoothed the way and released all minds from perplexity. . . .'

[99] Mendo de Ledesma to Philip II, Nantes, 21st December, 1595, A.N., K 1597, B 83.

There could hardly be a better way to gain favour. Soon afterwards he was *corregidor* at Málaga. It was through the *corregidores*, figures of authority in the cities and powerful individuals, that the state then controlled its subjects. They were the equivalent of the *intendants* in France. In his new capacity, Don Diego supervised, amongst other things, the construction of the harbour mole. When called upon, he went to the immediate assistance of Tangiers and Ceuta which were under threat from Drake – and, he takes pains to point out, this help cost the king not a single *real*! Don Diego himself however was ruined financially by the incident. For in his official capacity, he had been obliged, during the relief of the *presidios*, to keep table for over sixty horsemen and other persons of quality. Next, he became governor of Ceuta and as such was required to investigate the administration of his predecessor. He flatters himself that he exercised such fair judgement on this occasion that the former incumbent was reinstated. Although satisfied that he had done the right thing, Don Diego was once more without a post and back home near Zamora, where he was greeted by the justified clamour of his wife and children, who were living in poverty. At this point he had agreed to leave for six months in Brittany. But the six months have now lasted five long years, in the course of which both his elder brother and his elder brother's wife have died without his being able to claim the least part of their estate for himself. And since judgement always goes against an absentee, he has already lost two cases at law. It is true that since his posting to Brittany, the king has allowed him a commandery of 1500 ducats income, with four years' back pay, but, he says, what is that set against the enormous expense he shoulders, his personal poverty and the poverty of his family?

The Spanish archives contain thousands of similar complaints and records. The historian is not obliged to take literally all the grievances thus voiced, but there can be no doubt at all that the civil servants of modern Spain were poorly and irregularly paid and constantly dispatched to all corners of the Spanish Empire, uprooted and cut off from home ties. That they often underwent hardship there can be little doubt. At Madrid there lived a large unemployed population, in search of posts, pensions, and back pay, and an army of wounded military men waiting to be granted an audience. Meanwhile, wives and daughters prostituted themselves to pay the rent. These were the casualties of history, the unemployed servants of the state, whiling away their hours of waiting on the Calle Mayor, the street of wealthy merchants, seeking the summer shade or the winter sunshine on the Prado San Hierónymo or mingling aimlessly with the throng of evening strollers.[100]

Reversion and sale of office. All these servants were attached to their offices by ties of loyalty, honour or self-interest. They gradually conceived the desire to hold them in perpetuity for their descendants. As the years passed this tendency was to become more marked. Sale of office was

[100] Pedro de Medina, *op. cit.*, p. 204 to 205 v°.

a widespread evil. France, notorious for the practice, was by no means an exception. If sixteenth and seventeenth century states everywhere permitted the abuse to thrive, is the true explanation the dwindling of real national income? In Spain at any rate, the *Recopilación de las Leyes*[101] enables us to follow the progressive dispossession of the state for private gain and the consequent rise of a new privileged caste. For full details of how this actually took place, one has to search among the enormous body of documents at Simancas relating to the *renuncias*[102] or 'reversions'. *Renunciar*, to resign, meant to dispose of one's office to another person – one example among hundreds is the *alguazil* of the Barcelona Inquisition, who requested permission in June, 1558, to pass his office on to his son.[103] Another example in the same year is the government's compliance with the demands made by the *regidores* or magistrates, who henceforth had the right of reversion in favour of anyone they pleased, even if the beneficiary were under eighteen years of age, with permission to exercise this right of nomination in their lifetime, on their deathbed or in their will. The reversion was to be valid even if they died before the statutory interval of twenty days had elapsed.[104]

These details, so reminiscent of France during the same period, give a notion of the problem without in any way resolving it. I have no doubt that a systematic study of the phenomenon in Spain will one day reveal a situation not unlike that revealed by French historians about their own country during the same period. The most curious aspect of this development in the Iberian Peninsula, in my view, is its extremely precocious appearance. The first signs can be glimpsed even before Ferdinand and Isabella, during the troubled reigns of John II and Henry IV[105] and probably as early as the beginning of the fifteenth century,[106] at least in the sphere of municipal offices, many of which were already *renunciables*. True the Crown could and often did reassert its rights, either by force, or by adhering strictly to the statutory time-limit on reversion of office, which was binding both upon the holder (who had to remain living for the stated period, at least twenty days[107]) and upon the new incumbent, who had to present himself and claim his rights in person during the thirty days following the act of reversion.[108] In 1563, the Cortes petitioned Philip II, unsuccessfully as it happened, to extend this interval from thirty to sixty days,[109] proof if it were needed that the old procedure remained

[101] *Recopilación de las leyes destos reynos hecha por mandado del Rey*, Alcalá de Henares, 1581, 3 vols. fol.: B.N. Paris, Fr. 4153 to 4155.

[102] Cámara de Castilla, series VIII, Renuncias de oficios.

[103] 9th June, 1558, A.H.N. Barcelona Inquisition, Libro I, f° 337.

[104] Manuel Danvila, *El poder civil en España*, Madrid, 1885, V, p. 348–351.

[105] *Recopilación*, I, f° 77.

[106] *Ibid.*, f° 73 and 73 v°.

[107] *Ibid.*, f° 79 v° (Law of Toledo, 1480).

[108] *Ibid.*, thirty days starting from the day of the reversion (laws of Burgos, 1515; Corunna, 1518; Valladolid, 1542). 60 days (Pragmatic of Granada, 14th September, 1501) to present the deeds 'en regimientos', *ibid.* But it is not clear whether this was the same thing.

[109] *Actas*, I, p. 339.

in force as a constant menace and a potential source of family discord (for purchasers of office often used for payment the precious sums set aside for dowries[110]). Gradually a substantial number of offices became *renunciables*. The numerous prohibitions on reversions unless from father to son[111] or on the sale of judicial or other office[112] testify in their way to the spread of the practice.[113] The king himself was guilty in so far as he created or sold offices.[114] Antonio Pérez[115] is often accused of encouraging this massive venality; but in such a century, the individual secretary is not wholly to blame. Even the offices of municipal *alcaldes* and the *escrivanias* of the Chancelleries and the Royal Council became *renunciables*.[116] As in France, this growing venality developed in an almost feudal atmosphere, or rather, as Georg Friederici[117] has pointed out, bureaucracy and paternalism now went hand in hand. Clearly the monarchy was the loser as offices continued to be sold and corruption increased. Such abuses created fresh obstacles to the authority of the Crown, which was very far, in the time of Philip II, from the absolute power of a Louis XIV. It is true that in Castile, venality was for the most part confined to minor appointments and was widespread only among holders of municipal office. It was precisely at this level that there flourished a sturdy urban patriciate, supported by the Cortes, attentive to its local interests and not amenable to discipline by the *corregidores*. After all the towns were hardly insignificant. Any worthwhile study of fiscal history must concern itself with the urban situation.[118]

Venality, detrimental to the state, was likewise embedded in Turkish institutions. I have already quoted the opinion that the farming of revenues throughout Turkey may have originated in Egypt.[119] The need to *corteggiare* one's superiors, to offer them substantial gifts, obliged every state servant to reimburse himself regularly, at the expense of his inferiors and of the localities he administered, and so on down the scale. The organized misappropriation of public funds operated throughout the hierarchy. The Ottoman Empire was the victim of these insatiable office-holders, obliged by the very tyranny of usage to be insatiable. The individual who drew most benefit from this daylight robbery was the grand

[110] *Ibid.*, p. 345–346. [111] *Recopilación*, I, f° 79 (Guadalajara, 1436).

[112] *Ibid.*, 73 v°, Valladolid, 1523.

[113] At what point did office become a saleable commodity? This question preoccupied Georges Pagès and we still do not know the precise answer. Mention is made however in the Pragmatic of Madrid, 1494 (I, f°s 72 and 72 verso) of those who have made reversion of their (municipal) offices for money, '. . . los que renuncian por dineros'.

[114] There was a curious two-way traffic in offices sold by the state and by individuals, as for example in the case of the office of *alcalde* at Málaga, D. Sancho de Cordova to Philip II, 18th January, 1559, Sim. E° 137, f° 70. At Segovia in 1591 (Cock, *Jornada de Tarrazona*, p. 11) municipal offices were sold or given away by the king 'cuando no se resignan en tiempo para ello limitado'.

[115] R. B. Merriman, *op. cit.*, IV, p. 325.

[116] *Actas*, I, p. 345–346 (1563).

[117] *Op. cit.*, I, p. 453–454. People of the middle classes . . .

[118] Jacob van Klaveren, *op. cit.*, p. 47, 49 ff.

[119] See above, p. 668, note 50.

vizier, as the Venetians never tired of pointing out and as Gerlach noted in his *Tagebuch* with reference to Muhammad Sokolli: an obscure youth living near Ragusa when he was carried off at the age of eighteen by the sultan's recruiting officers and who, many years later, in June, 1565, became grand vizier, a post he held until his assassination in 1579. He derived a huge income from gifts offered to him by candidates for public office. 'In the average year', writes the Venetian Garzoni, 'it amounts to a million gold ducats, as has been sworn to me by persons worthy of trust.'[120] Gerlach himself notes: 'Muhammad Pasha has an incredible treasure store of gold and precious stones. . . . Any man wishing to obtain a post must first bring him a gift of several hundred or several thousand ducats, or else bring him horses or children. . . .' The memory of Muḥammad Sokolli cannot be defended against these charges: great man though he undoubtedly was, with respect to the money of others, whether his own inferiors or foreign powers, he willingly bowed to the custom of the time.

In the Turkish Empire however, the enormous fortune of a vizier was always at the disposal of the sultan, who could seize it upon the death, whether natural or otherwise, of his minister. In this way, the Turkish state participated in the habitual peculation of its officers. Not everything could be recovered by such straightforward means: religious foundations, of which much architectural evidence survives, provided ministers with a safe keeping-place for their wealth. It was one way of preserving some of the misappropriated gold for the future or the security of a family.[121] Let us recognize that the western system was on the whole less thoroughgoing than eastern custom, but in both East and West, venality of office was one area where the power of the state was curiously undermined. As for dating this revealing disorganization, what we have in the sixteenth century is no more than a series of warning signs of things to come.

In any case, in the Turkish Empire as in other European states,[122] the sixteenth century saw a considerable rise in the number of public officials. In 1534, in the European part of Turkey, at the top of the hierarchy stood a *beglerbeg* [governor-general] and below him thirty *sandjak-begi* [governors]. In Asia there were six *beglerbegi* and sixty-three *sandjak-begi*. In 1533, a separate *beglerbeg* was appointed, the Kapudan Pasha or Lord High Admiral, known in the Spanish documents as the 'General of the Sea'. This 'Admiralty', besides commanding the fleets, administered the ports of Gallipoli, Kavalla and Alexandria. So with the *beglerbeg* of Cairo, created in 1534, there were nine supreme *beglerbegi*. By 1574, forty years later, there were twenty 'governments' [*eyālet-s*]: three in Europe – Sofia, Temesvar [Timisoara] and Buda; thirteen in Asia; three and later

[120] J. W. Zinkeisen, *op. cit.*, III, p. 100, note 1.

[121] Jean Sauvaget, *Alep. Essai sur le développement d'une grande ville syrienne des origines au milieu du XIXe siècle*, 1941, p. 212–214.

[122] At Venice, just after Agnadello, the Great Council made the decision to sell offices (10th March, 1510). The remarkable text of the resolution can be found in *Bilanci generali*, 2nd series, Vol. I, tome I, p. CCIV. Subsequent wars also led to increased sales of offices.

four in Africa – Cairo, Tripoli, Algiers and before long Tunis; plus the General of the Sea. As the heavy concentration of government officials there shows, Asia was the focus of Turkish preoccupations and military effort. This trend was to continue. Under Murād III, the total figure rose from twenty-one to forty governments, twenty-eight of them in Asia alone, where the war with Persia resulted in the conquest and administration of vast frontier zones. The increase was therefore justified by necessity. But neither can one ignore the new and growing desire in Turkey for titles, the ever-increasing thirst for public office. The *su-bashi* dreamt of becoming *sandjak-beg*, the *sandjak-beg* of rising to *beglerbeg*. And every one of them regularly lived above his station.

Developments similar to those in Spain had begun to work upon Turkey even before they were felt in the far-off Iberian Peninsula. For it was not until after the ascetic reign of Philip II that Spain could unrestrainedly indulge in a display of luxury and good living. In the East, the change occurred immediately after the death of Sulaimān in 1566. Silken garments, woven with cloth of silver and gold, forbidden under the reign of the old emperor who always wore cotton, made a sudden reappearance. As the century drew to a close, feast succeeded sumptuous feast in Constantinople, their flickering lights casting a lurid glow through even the arid pages of Hammer's history. The luxury of the Seray was unbelievable: the couches were upholstered in cloth of gold; in summer the habit was adopted of sleeping between sheets of the finest silk. Exaggerating only a little, contemporary observers declared that the slipper of a Turkish woman cost more than the entire costume of a Christian princess. In winter, silken sheets were replaced by precious furs. The luxury of even the tables of Italy was surpassed.[123] We must believe the evidence of the touchingly naive comment made by the first Dutch envoy to Constantinople, Cornelius Haga, who declared after his reception in May, 1612, 'it seemed that this was a day of public celebration'.[124] What is one to say of the great feasts in the time of Murād IV, while the country itself was bled white by the twin perils of war and famine? How strange it is that Turkey at almost exactly the same time as Spain, should have abandoned herself to the delights and revels of a 'Golden Age' at the very moment when this lavish display was in flagrant contradiction to all rules of good management and the relentless realities of the nation's budget.

Local autonomy: some examples. The spectacle of these gigantic political machines can be misleading. Alongside the typical fifteenth century state, the dimensions of sixteenth century powers are easily exaggerated. But it is merely a matter of degree: by comparison with the present day

[123] L. von Ranke, *Die Osmanen und die spanische Monarchie* . . ., Leipzig, 1877, p. 74, following Businello, *Relations historiques touchant la monarchie ottomane*, ch. XI.

[124] 'Es schien ein Tag des Triumphes zu sein', 1st May, 1612, quoted by H. Wätjen, *op. cit.*, p. 61.

and the size of modern bureaucracies, the number of office-holders in the sixteenth century is insignificant. And indeed, for want of sufficient personnel, the mighty state with its 'absolute' power had a good deal less than total control. In everyday matters the state's authority was imperfect and often ineffective. It was confronted with innumerable bastions of local autonomy against which it was powerless. In the great Spanish Empire, many cities retained their freedom of movement. By contracting for payment of fixed sums to the state, they were able to control indirect taxation. Seville and Burgos, something of whose institutions we know, had considerable liberties, as a Venetian ambassador noted in 1557:[125] in Spain, he writes 'si governa poi ciascuna signoria e communità di Spagna da se stessa, secondo le particolari leggi'. Or again, outside the Peninsula but still within the Spanish Empire, Messina remained until 1675 a republic, a thorn in the flesh of all the viceroys who, like Marcantonio Colonna in 1577, administered Sicily. 'Your Majesty knows', writes Colonna in June of that year,[126] 'how great are the privileges of Messina, and how many outlaws and *matadores* are harboured in the city, partly for the ease with which they can pass into Calabria. It is therefore of the utmost importance that the *stratico* (the governing magistrate) fulfil his functions honourably. Matters have already come to such a pass that the said office brings more profit in two years than the viceroyalty of the whole island in ten; I am informed amongst other things that there is not a man imprisoned for some capital offence who does not secure his liberty if he can offer good guarantees, and then if bail is broken, the *stratico* seizes the property. The city is today so surrounded by thieves, that even inside its walls, people are kidnapped and then held to ransom.'

Both in the Peninsula and outside it then, whole districts, towns, sometimes regions with their own *fueros*, or privileges, escaped the jurisdiction of the Spanish state. This was true of all distant and peripheral zones: the Kingdom of Granada for example until 1570; after 1580 and for many years until the final rupture of 1640 it was true of Portugal, in every sense a 'dominion' with its own liberties and rights which the invader dared not touch. It was always to be true of the tiny Basque provinces and of the possessions of the Crown of Aragon, with whose privileges, even after the rising and troubles of 1591, Philip II dared not interfere. Merely on crossing the Aragonese border, the most unobservant traveller could not fail to remark that as he left Castile he was entering a completely different society. Here he found semi-independent lords, exercising many rights over their long-suffering subjects and entrenched in castles fortified with artillery, only a stone's throw from their neighbour Castile, now humbled and disarmed. Socially privileged, politically privileged and fiscally privileged, the Aragonese bloc governed itself and paid only a fraction of its royal

[125] E. Albèri, I, III, p. 254.

[126] Palermo, 10th June, 1577, Simancas E° 1147. *Matadores* = assassins. On the city itself, read the book by Massimo Petrocchi, *La rivoluzione cittadina messinese del 1674*, Florence, 1954, although it chiefly concerns a later period.

taxes. But the real reason for such impunity was the proximity of the French border and the knowledge that the first signs of unrest would be exploited by her hostile neighbour to force open the imperfectly closed gates of Spain.[127]

It was for no other reason that, in the Turkish Empire this time, the sultan's authority was visibly weaker in Europe, on the northern perimeter of his states, in Moldavia, Wallachia, Transylvania, in the kingdom of the Crimean Tartars. And we have already seen how geography created multiple autonomous zones in the mountains of the Balkans, in Albania and the Morea.

Resistance to the state could take the most diverse forms. In the Kingdom of Naples for instance, as well as still unsubdued Calabria, both the pastoral associations and the city of Naples itself were prominent in this respect. Through the shepherd associations, the peasant could escape both his overlord and his king. Or if he settled in Naples, the 'city air' made him a free man. Further south, in Sicily, the power of the secular authorities could be escaped by allegiance to the Sicilian Inquisition, whose influence was thus singularly extended. Possibly the monstrous growth of the capital in Turkey was a similar type of response. In the provinces, nothing stood between the individual and the rapaciousness of the local *beglerbegi*, *sandjak-begi* and *su-bashis* and, most hated and feared of all, their agents the *voivodes*. At least in Constantinople one was assured of a minimum of justice and of relative tranquillity.

There can be no doubt that corruption among state officials was widespread in the sixteenth century, in Islam as in Christendom, in southern as in northern Europe. 'There is no case before a court, whether civil or criminal', wrote the Duke of Alva from Flanders in 1573,[128] 'which is not sold like meat in the butcher's shop . . . most of the councillors sell themselves daily to anyone who will pay the money.' This ubiquitous corruption was a curb on the will of rulers and by no means an easy one to remove. Corruption had become an insidious hydra-headed presence, a power unto itself;[129] one of the powers behind which the individual could shelter from the law. It was the old combination of force and cunning. 'The laws of Spain', wrote the ageing Rodrigo Vivero in about 1632,[130] 'are like spiders' webs, catching only flies and gnats.' The rich and powerful escaped their toils and the only ones who became entangled were the unprivileged and the poor, 'los desfavorecidos y los pobres', a state of affairs not confined to the sixteenth and seventeenth centuries.

Finance and credit in the service of the state. A further indication of weakness was that the large state was not directly in contact with the mass of

[127] B.N. Paris, Dupuy, 22, f° 122 ff.
[128] Quoted by Jakob van Klaveren, *op. cit.*, p. 49, note 5.
[129] Cf. the series of articles by the same writer in *Vierteljahrschr. für Sozial- und Wirtschaftsgeschichte*, 1957, 1958, 1960, 1961.
[130] B.M. Add, 18 287, f° 23.

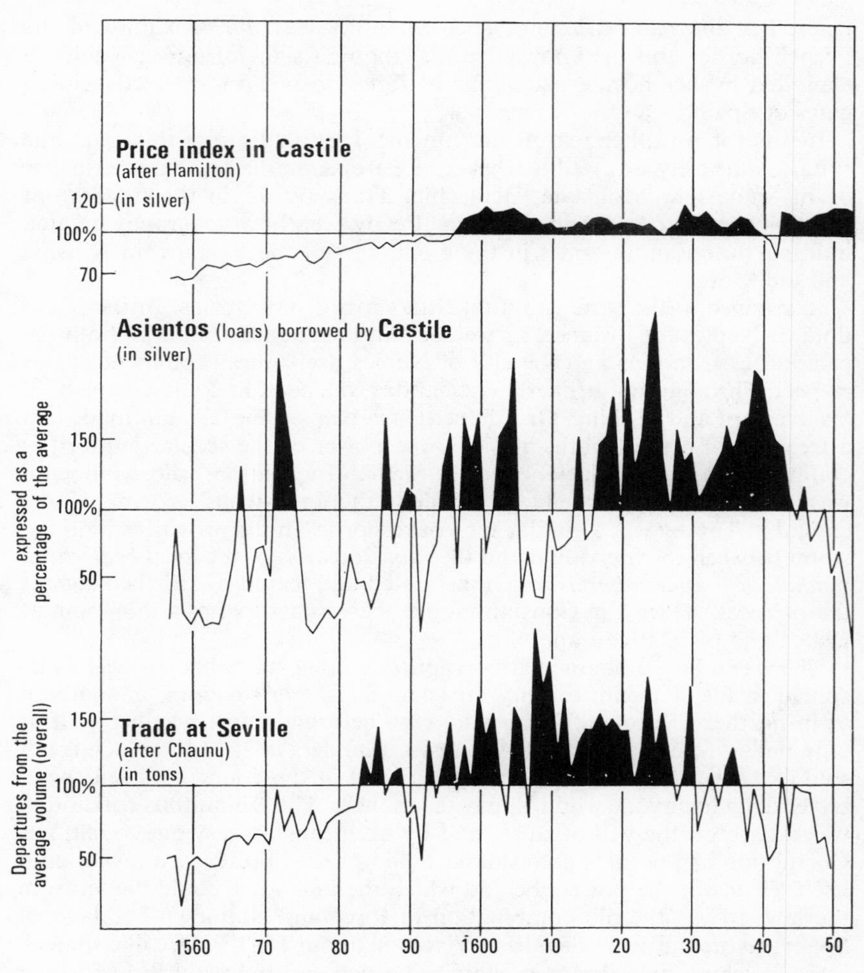

Fig. 59: *The 'asientos' and economic life in Castile*

Compared on one hand with the variations in the price index, according to Earl J. Hamilton, which as can be seen were quite modest, and on the other hand with the pattern of Sevilian trade, an enormous boom followed by a substantial recession, the graph representing the *asientos*, that is the short term national debt, has the frantic appearance of a seismograph. It does however present certain analogies with the price curve, particularly at Seville, naturally enough, since it was upon the import of silver from America that the whole system of the advance and repayment of *asientos* depended. By and large the curve rises above the 100 per cent mark in time of war and falls below it during periods of peace or withdrawal (except for the conquest of Portugal). Note the large scale borrowing during the so-called Thirty Years' War. Graph showing *asientos* constructed by Alvaro Castillo Pintado.

taxpayers and was therefore unable to exploit them as much as it would have wished: it could thus be placed under a fiscal and eventually a financial handicap. With the exception of the Italian examples mentioned earlier, the Mediterranean states at the very end of the sixteenth century possessed neither a treasury nor a state bank. In 1583 there was some talk in Philip II's entourage, of setting up a state bank[131] but the project was never realized. From the metropolitan centre of the great Spanish Empire, appeals went out to the moneylenders whom we call by the rather too modern name of bankers. The king could not do without them. After his return to Spain in September, 1559, Philip II's major preoccupation for the next ten years was to be the restoring of order to the finances of his realm. Advice poured in from every side, always in the last analysis recommending that he apply to the Affaitati or the Fuggers, or the Genoese, or even, during Eraso's outbursts of nationalism, to homegrown Spanish bankers like the Malvenda of Burgos.

The scattered nature of Philip's possessions and those of his father before him, meant inevitably that revenues had to be collected and payments made over a wide area and thus encouraged the use of international merchant firms. Their assistance was required merely to transfer money from place to place. But they could do more than this: they could advance money, thus mobilizing the resources of the state before they were due. In this capacity, they were frequently concerned in the direct collection of state taxes as reimbursement, which brought them into contact with the taxpayers. The Spanish fiscal system then was organized by the moneylenders to suit themselves. In 1564, Philip II assigned to the Genoese the monopoly of sales of playing cards. On a subsequent occasion, he handed over certain of the Andalusian salt-pans. Another time he might follow his father's example and delegate to the Fugger family the exploitation of the mines at Almadén or the administration of the estates belonging to the Military Orders – a step which in this case meant placing extensive arable and pasture land, tolls and peasant dues in foreign hands. The Fuggers' German factors and agents were everywhere in Spain, conscientious, methodical and zealous. Even when tax-gathering did not devolve upon a foreign firm it was just as likely to be carried out by the intermediary powers, the towns or the Cortes. Financially then, state control was, to say the least, incomplete.

In France, although the transfer of coin was not the vital necessity it had become for Spain, bankers and moneylenders nevertheless had a part to play, as they did in Turkey, where businessmen made free with state funds. Gerlach notes in his *Tagebuch*:[132] 'there are in Constantinople, many Greeks who have grown rich through trade or other means of making money, but who go about clad in mean clothes so that the Turks shall not know of their riches and steal them. . . .' The richest among them was one

[131] E. J. Hamilton, 'The Foundation of the Bank of Spain' in *Journal of Political Economy*, 1945, p. 97.

[132] P. 61, quoted by Zinkeisen, *op. cit.*, III, p. 368.

Michael Cantacuzenus: the devil's son according to the Turks, this pseudo-Greek was rumoured, unconvincingly, to be of English origin. Whatever his beginnings, his fortune was immense and was curiously connected with the services he rendered the Turkish State. For Cantacuzenus had the monopoly of all the saltworks in the empire, farmed innumerable customs duties, trafficked in offices and, like a vizier, could depose Greek patriarchs or metropolitans at his pleasure. In addition to this, he controlled the revenues of entire provinces, Moldavia for instance or Wallachia, and held enough feudal villages to be able to provide crews for twenty or thirty galleys. His palace at Anchioli rivalled the Seray in splendour. This successful adventurer is not to be confused with his humbler Greek compatriots in Galata or elsewhere; he flaunted his wealth before them and, lacking their prudence, was arrested in July, 1576; obliged to forfeit his possessions, he escaped with his life through the intercession of Muhammad Sokolli. On his release, he began once again, this time dealing not in salt but in furs, and as in the past had a personal arrangement in Moldavia and Wallachia. Finally the inevitable happened: on 13th March, 1578, on the orders of the sultan and without any form of trial, he was hanged from the gates of his own palace at Anchioli and his wealth was confiscated.[133]

A comparable, but even more extraordinary story is that of a figure mysterious in many ways, the Portuguese Jew Joseph (or Juan) Nasi, known variously as Migues or Micas and who at the end of his life bore the lofty title of Duke of Naxos. After spending much of his early life wandering aimlessly about Europe, to the Netherlands, Besançon,[134] and for a while Venice, he arrived in Constantinople in about 1550. Already a rich man, he celebrated his marriage with a great display of wealth and reverted to Judaism. The friend and confidant of Sultan Selīm before his accession – discreetly provided him with choice wines – he farmed the customs duty on the wines of the islands. It was he who urged the sultan to attack Cyprus in 1570. Most astonishing of all perhaps is that fact that he died of natural causes in 1579, still in possession of a large fortune. There has been an injudicious effort to rehabilitate the memory of this striking individual, but even after hearing the evidence we are little wiser about this Fugger of the East.[135] Spanish documents describe him as favourably disposed towards Spain, even slightly in league with His Catholic Majesty, but he was not the kind of man who could be classed once and for all as either pro-Spanish or anti-French. To do so would be to ignore the constantly shifting political loyalties in Constantinople. It would be interesting above all to know, as in the case of Cantacuzenus, exactly how he fitted into Turkish state finances. And our knowledge of these finances themselves is very scanty and likely to remain so for some time.

[133] According to Gerlach, quoted by Zinkeisen, *op. cit.*, III, p. 366–368.

[134] His stay at Besançon is mentioned in a note by Lucien Febvre.

[135] There is a not very readable labour of rehabilitation by J. Reznik, *Le duc Joseph de Naxos*, 1936; a more recent work by Cecil Roth, *The Duke of Naxos*, 1948, and above all the remarkable article by I. S. Revah, 'Un historien des "Sefardim" ', in *Bull. Hisp.*, 1939, on the work of Abraham Galante.

Conspicuously absent from Turkish finances as compared with those of Christian powers was the recourse to public credit, whether long or short term – the government loan, a polite and fairly painless way of appropriating the money of private citizens, of investors both large and small. It was a method universally practised in the West, where every state created some formula to attract its citizens' money. In France it was the famous *rentes sur l'Hôtel de Ville*.[136] In Spain, as we have seen, it was the *juros*, which by the end of Philip II's reign represented no less than 80 million ducats.[137] Such paper depreciated quickly and occasioned frantic speculation. At prices then current, the state might end up paying as much as seventy per cent interest. One character in Cervantes' short story *La Gitanilla*[138] talks revealingly of saving money and holding on to it, 'como quien tiene un juro sobre las yerbas de Extremadura', like a man who holds a bond on the pasture lands of Extremadura (presumably a good investment, for they could vary). In Italy, the appeal to the public was usually made through the *Monti di Pietà*. As Guicciardini had already noted, 'either Florence will be the undoing of the *Monte di Pietà*, or the *Monte di Pietà* will be the undoing of Florence',[139] a forecast which was to be even more true of the seventeenth century than of the sixteenth.[140] In his economic history of Italy, Alfred Doren maintains that these massive investments in state bonds were both symptom and cause of the decline of Italy at the beginning of the sixteenth century. Investors were reluctant to take risks.

Perhaps nowhere was the appeal for credit more frequently reiterated than in Rome, capital of that singular dominion, both dwarf and giant, the Papal States. In the fifteenth century, after the Council of Constance, the Papacy, finding itself the victim of the growing particularism of other states and reduced to the immediate resources of the Papal States, set about extending and recovering the latter. It was no accident that during the last years of the fifteenth century and the first years of the sixteenth, the Supreme Pontiffs resembled temporal rulers rather than spiritual leaders: papal finances obliged them to it. Towards the middle of the sixteenth century, the situation remained much the same: almost eighty per cent of papal revenues came from the Patrimony or Papal States. Hence the relentless struggle against financial immunity. Victory in this struggle led to the absorption by the Papal States of the urban finances of, for instance, Viterbo, Perugia, Orvieto or the medium-sized towns of Umbria. Only Bologna was able to preserve her autonomy. But these victories left unaltered the old and often archaic systems of tax collection: the sources of revenue were unblocked as it were, 'but only exceptionally did the Papal States come into contact with its taxpayers'.[141]

No less important than this fiscal war was the appeal to public credit.

[136] Bernard Schnapper, *Les rentes au XVIe siècle. Histoire d'un instrument de crédit*, 1957.

[137] See above, Vol. I, p. 533.

[138] 'La Gitanilla', *Novelas Ejemplares*, ed. Marin, 1941, p. 44.

[139] Gustav Fremerey, *Guicciardinis finanzpolitische Anschauungen*, Stuttgart, 1931.

[140] R. Galluzzi, *op. cit.*, III, p. 506 ff.

[141] Clemens Bauer, *art. cit.*, p. 482.

Clemens Bauer rightly says that from now on the history of the papal finances is a 'history of debt',[142] the short-term debt which took the familiar form of borrowing from bankers and the long-term debt, repayment of which was effected through the *Camera Apostolica.* Its origin is the more noteworthy in that it testifies to the venality of offices reserved for laymen. There was from the start confusion between office-holders and creditors of the Apostolic Chamber. The officer-shareholders formed colleges: the purchase price of office was invested capital but as interest they received fixed salaries. In the College of the *Presidentes annonae* for example, founded in 1509 and comprising 141 offices, sold for the total sum of 91,000 ducats, interest-salaries of 10,000 ducats were payable from the revenues of the *Salara di Roma.* Later on, through the creation of the *Societates officiorum*, the papacy succeeded in sharing out these bonds among small investors and the title of officer conferred on shareholders became purely honorary. This was already the case with the series of Colleges of Cavalieri begun in 1520 with the foundation of the *Cavalieri di San Pietro* in 1520; later came the *Cavalieri di San Paolo* and the *Cavalieri di San Giorgio.* Finally a conventional system of government bonds was instituted, with *Monti* probably on the model of Florence, by Clement VII, a Medici Pope. The principle was identical with that of the *rentes sur l'Hôtel de Ville*, the granting of a fixed and guaranteed income in return for the payment of a capital sum. Shares in the loan were called *luoghi di monti*, negotiable bonds often in fact negotiated both in Rome and elsewhere, usually above par. Thus were created, through the accident of circumstance or necessity, the *Monte Allumiere*, guaranteed by the alum mines at Tolfa, the *Monte S. Buonaventura*, the *Monte della Carne*, the *Monte della fede* and others. Over thirty are recorded.

Usually these were repayable loans; the *Monte novennale* for instance, set up in 1555, was in theory repayable after nine years. But there were also permanent loans, which could be passed on by inheritance. In fact one way for the papal finances to make a short-term profit, was to convert life-holdings into perpetual holdings, '*vacabili*' into '*non vacabali*', since it meant the rate of interest was lowered. Through these and other details it can be seen that the Roman *Monti* were by no means behind the times; they could stand comparison with those of Venice or Florence, with the Casa di San Giorgio and *a fortiori* with the *juros* of Castile. Any calculation in this sector is nevertheless difficult: between 1526 and 1601, the papacy seems to have borrowed for its own purposes (and sometimes on behalf of representatives of the Roman nobility) 13 million crowns. This figure may not impress a twentieth century reader; let me add however that out of these sums so persistently extracted from private citizens, Sixtus V was able to put by 26 tons of silver and three tons of gold to be locked up in his treasury in the Castel Sant'Angelo, thus gratifying a peasant's urge to hoard by extremely modern means. Since the *Monti* were meant for an international clientele, it is not surprising that the public debt at Genoa

[142] Clemens Bauer, *art. cit.*, p. 476.

stopped growing 'just as it was reaching impressive proportions at Rome'.[143] Was Leopold von Ranke right to call Rome 'perhaps the principal money-market in Europe',[144] at least of investors' money? Although possible, this is by no means certain. The crucial point however is not the scale of government borrowing in Rome, but the enormous expansion of the credit market on which every state drew, prudent and reckless alike, and on which innumerable stockholders did very satisfying business. This infatuation cannot be explained simply by the economic situation at least as we understand it. Is it to be explained in terms of collective psychology, of a mass urge for security? In Genoa for example, where between 1570 and 1620 inflation 'reached such a point', writes Carlo M. Cipolla,[145] 'that historians have described it as a price revolution, by some paradox there was a visible decrease in interest rates', which had varied between 4 and 6 per cent since 1522 but now dropped to 2 and even to 1·2 per cent, at least during their lowest ebb, between 1575 and 1588. This drop corresponded with the influx into Genoa of bullion, both gold and silver, which it was at this time difficult to invest. 'This was the first time in the history of Europe since the fall of the Roman Empire that capital was made available at such low rates and this was indeed an extraordinary revolution.' It would be interesting, if possible, to analyse the situation of other markets, to discover whether interest rates, as is quite probable, made business brisk in some places and slack in others as they do on modern stock exchanges. In any case the boom in government bonds, the sudden popularity of shares in the national debt smoothed the way for sixteenth century governments and made their task easier.

Given the common concerns of all states, there is good reason to suppose that the brutal tribute taken by the Turkish State both from fiefs and office-holders was at any rate partly the result of the impossibility in Turkey of appealing to investors, whether large or small, to aid the state, on the western model. Credit certainly existed in Ottoman countries: we have already mentioned the recognition of debts by merchants before the *cadis*[146] and bills of exchange were passed between merchants who were subjects of the Grand Turk. Recent publications have shown, if any doubt remained on this score, that Jewish merchants[147] used bills of exchange among themselves and sometimes with their co-religionists in the West. The rumour even circulated in the middle of the century and is mentioned by Jean Bodin,[148] that Turkish pashas participated in the

[143] All these aspects of the Roman *Monti* are admirably explained by J. Delumeau, *op. cit.*, II, p. 783 ff. I have merely summarized his account. [144] *Ibid.*, p. 821.

[145] 'Note sulla storia del saggio d'interesse, corso e sconto dei dividendi del banco di San Giorgio nel secolo XVI', in *Economia Internazionale*, 1952, p. 13–14.

[146] From information provided by Mr. Halil Sahillioglu.

[147] Aser Habanel and Eli Eškenasi, *Fontes hebraici ad res œconomicas socialesque terrarum balkanicarum sæculo XVI pertinentes*, I, Sofia, 1958 (remarkable).

[148] '. . . in their Factors' names for about five hundred thousand crowns', *The Six Books*, tr. Knolles, ed. Macrae, Harvard U.P., 1962, p. 674. This passage is cited in Atkinson, *op. cit.*, p. 342.

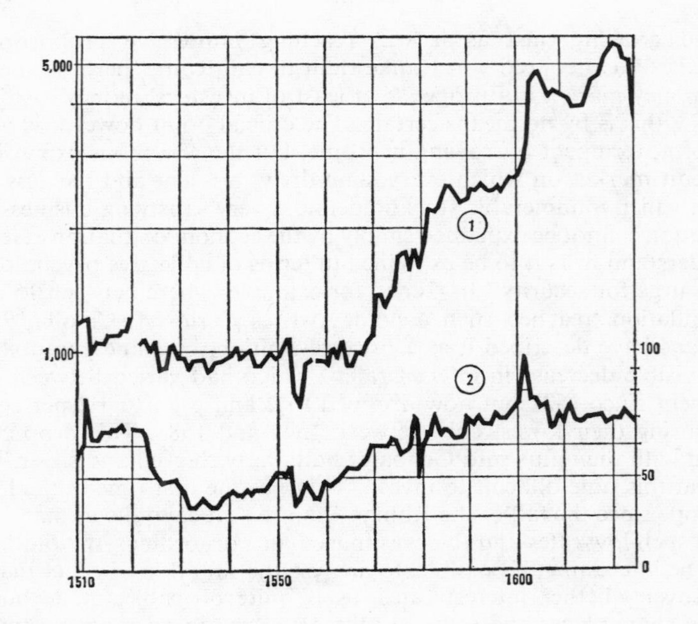
5,000
1,000
100
50
0
1
2
1510
1550
1600

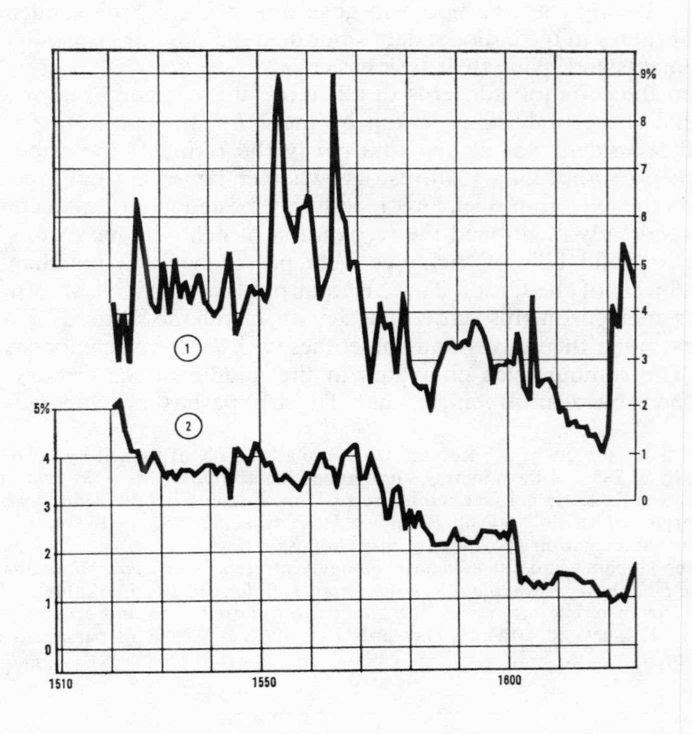
9%
8
7
6
5
4
3
2
1
0
5%
4
3
2
1
0
1
2
1510
1550
1600

speculation at Lyons on the *Grand Party*. While all this is quite possible, public credit did not exist in Turkey.

1600–1610: the comeback of the smaller state? Did the great states enter into a period of sickness, or at any rate fatigue, during the last years of the sixteenth century and the first of the seventeenth? This is the impression given by contemporary would-be physicians, who hastened to the bedside of the mighty invalids, each with his own diagnosis and remedy to offer. In Spain there was never any lack of *arbitristas*, givers of advice, both Spaniards and foreigners;[149] they form a distinct social category in themselves. As the first years of the seventeenth century go by, their number increases and their voice grows louder as they press forward to the benevolent tribunal of history. In Portugal, they were equally active.

In the face of all their testimony, how can one fail to believe in the decline of the Hispanic monarchy? Everything points to it: events and witnesses, the gloomy picture painted by Tomé Cano in 1612;[150] or the fascinating collection under the title *Historia tragico-maritima*,[151] a meticulous record of the catastrophes of Portuguese shipping on the routes to Brazil and the Indies. It is a catalogue of unmitigated disaster, decadence and exhaustion, of enemy victories, of the perils of the sea, of shipwreck on the reefs of Mozambique or vessels lost on the route round the

(Opposite) Fig. 60: *The 'Luoghi' of the Casa di San Giorgio, 1509–1625*

These four graphs summarize the findings of the important article by Carlo M. Cipolla (see note 145). The *luoghi* were shares in the public debt of the Republic of Genoa, issued at 100 *lire* (2000 *soldi*) nominal value. These were perpetual annuities with varying interest rates (in Venice by contrast, interest rates were fixed): they depended on the profits made by the *Casa di San Giorgio* which held as security the taxes it collected on behalf of the Signoria. The number of *luoghi* increased a great deal between 1509 (193,185) and 1544 (477,112) hence the drop in prices; the number later settled down (437,708 in 1597 and 476,706 in 1681). The first curve represents the price of negotiable *luoghi* on the market (left-hand axis = 1000 to 5000 *soldi*). The second curve represents the interest, *reddito*, of the *luoghi* (right-hand axis = 40 to 100 *soldi*). There is a clear rise during the second half of the sixteenth century and a falling off in the following century. Now interest on the *luoghi* was never paid immediately; half was paid four years later and the other half after a further year's interval. If a shareholder wanted to be paid straight away he had to have his dividend discounted, so the *sentio* varied on the market (top curve, second graph). It is therefore possible, taking into account both the delay and the *sentio*, to calculate the real interest on the *luoghi*: this is shown in the last curve which reveals a clear decline after 1570 becoming more marked still after 1600. 'So for one reason or another', concludes the author, 'capital was invested at Genoa at the beginning of the seventeenth century, at 1·2 per cent.' What we do not know is whether this abnormal situation is an indication of the healthy condition or otherwise of the financial centre.

---

[149] The obvious example is Anthony Sherley, cf. Xavier A. Flores, '*El peso politico de todo el mundo*' *d'Anthony Sherley, ou un aventurier anglais au service de l'Espagne*, Paris, 1963.

[150] *Op. cit.*, for reference see above, Vol. 1, p. 104, note 6.

[151] See reference above, Vol. I, p. 59, note 182.

Cape of Good Hope. As brigands multiplied on the roads of the Peninsula and plague decimated the population it seemed clear that the mighty frame of the Spanish Empire was rotting away. From outside of course, Spain still seemed a great power: although threatened, she was still a threat to others. Meanwhile, in Madrid at least, the most dazzling society of seventeenth century Europe held its revels.

But at the same time, the luxury of the sultan's court at Istanbul was unbelievable.

And yet here too, cracks were beginning to appear, the many unmistakable signs of fatigue. The Ottoman Empire was beginning to fall apart like a badly assembled boat as a series of open or concealed revolts convulsed it from Algiers to the Persian border, from the Tartar lands to southern Egypt. To western observers, ever wishful thinkers, it appeared that the Ottoman machinery was broken beyond repair. With remarkable zeal, Jesuits and Capuchins rallied to storm this tottering civilization. The time was surely ripe to chase the Infidel from Europe and share out his territory. Iñigo de Mendoza, Spain's ambassador at Venice, untiringly advanced this cause. It is true that this exalted individual was preparing to leave the diplomatic service to enter the Society of Jesus. And he was not the only visionary who swelled the ranks of the first battalion on the long road of history, ready to carve up the Ottoman Empire. Others were to follow: Father Carlo Lucio in 1600; a Frenchman Jean Aimé Chavigny in 1606; another Frenchman, Jacques Espinchard in 1609; Giovanni Motti in October, 1609; an anonymous Italian in December of the same year; a Capuchin, Francesco Antonio Bertucci in 1611. Not to mention Sully's Grand Design or the equally grandiose project of Charles Gonzaga, Duke of Nevers, and of Père Joseph (1613–1618). For any one name mentioned here, scholarship could probably with a little application find ten; in fact the number should be multiplied a hundredfold: spurred on by religious fervour, Europe was counting on the 'sick man's' inheritance from the beginning of the seventeenth century. But the eager heirs were mistaken: the sick man was not yet ready to die. He was to linger for many years, though never recovering his former vigour. Turkey had achieved an empty victory over Persia in 1590; in 1606 after an exhausting war the sultan had to be content with a profitless peace with Germany, in other words with the West.

So it was true: the wheel had turned. The early years of the century had favoured the large state, which represented what an economist would call the political enterprise of optimum dimensions. With the passing of the century, for reasons not fully clear to us, the leviathans were betrayed by circumstances. Was this a superficial or a structural crisis? Mere passing weakness or permanent decline? Whatever it was, by the beginning of the seventeenth century the only vigorous states were those of moderate size: France under Henri IV, a reign of sudden glory; or Elizabethan England, combining adventure with literary brilliance; Holland centred on Amsterdam; or Germany which experienced an age of material prosperity from

1555 until the first shots of the Thirty Years' War when it fell into the abyss. In the Mediterranean, there was Morocco, now once more rich in gold, or the Regency of Algiers, a city on the way to becoming a territorial state. And then there was Venice, radiant and resplendent with luxury, beauty and intelligence; or Tuscany under the Grand Duke Ferdinand. It is as if everything conspired in the new century to help these smaller states, of a size to keep their own houses in order. Many minor Colberts[152] rose to eminence in these modest states, men skilled in diagnosing their country's economy, capable of raising tariffs or encouraging the enterprises of their subjects and at the same time keeping a close guard over them. It is this series of small fortunes rather than the complex and sometimes obscure story of the empires which tells us that the wheel of history had indeed turned.

In other words, empires suffered more than smaller states from the long depression of 1595–1621; and afterwards these massive hulls were not refloated as quickly as their lighter rivals by the rising tide, which, it is true, did not rise very high nor last very long; for after the middle of the seventeenth century Europe plunged into a prolonged political crisis. Certainly the powers which were to emerge from it in the eighteenth century and to take advantage of the great economic revival were not the great empires of the sixteenth century, neither the Turkish nor the Spanish. The decline of the Mediterranean, some will say: with reason. But it was more than that. For Spain had every opportunity to turn whole-heartedly towards the Atlantic. Why did she choose not to?

[152] Ammintore Fanfani, *Storia del Lavoro . . .*, p. 32.

CHAPTER V

Societies

The evolution of Mediterranean societies in the sixteenth century followed an apparently simple course; provided that is, one takes a comprehensive view, disregarding for the moment the local exceptions and anomalies, the numerous missed opportunities and the disturbances which were more spectacular than significant, rising like bubbles to the surface and disappearing.

Such upheavals certainly cannot be ignored. But the development of society, based in the sixteenth century on land tenure, was slow, lagging behind political and economic advance. Moreover social change is like all change, tending first in one direction then in another. In the long run, change may be neutralized and development may be difficult to discern at all. It would probably be true to say that in France for example, the pendulum swung first one way then the other. The early sixteenth century was an age of horizontal social mobility, when poor men could move from one city or one region to another, without being destroyed in the process,[1] and it was also a time of vertical mobility, at every point on the social ladder, as the rich lost their wealth and were replaced by the new rich. This movement was checked about the years 1550–1560, beginning again, only to come to another standstill, perhaps[2] as early as 1587 in Burgundy, or in about 1595[3] as, all over the world, the secular trend was being reversed. Social change then passed through successive phases of acceleration, slowdown, revival and stagnation, culminating in the visible but short-lived triumph everywhere of the aristocracy and the partial freezing of society at the very end of the century. But this conclusion was no more than an episode in a continuing pattern of change, one which might be wiped out or neutralized by the next swing of the pendulum.

In short, the sixteenth century, despite its hesitations, or perhaps because of them, did not challenge the foundations of society. On the whole

[1] This horizontal mobility can also be a revealing index of the openness of societies. Gaston Roupnel, *La ville et la campagne au XVIIe siècle. Étude sur les populations du pays dijonnais*, 1955, 2nd ed., p. 99. 'In the sixteenth century, a beggar would be fed and cared for before he was expelled [from the towns]. At the beginning of the seventeenth century, he would have his head shaven. Later he was more likely to be whipped and by the end of the century the full weight of a repressive society set him to forced labour.'

[2] Henri Drouot, *Mayenne et la Bourgogne*, 1937, I, especially p. 48. 'These men of law, who a century previously had overthrown an old social order, had already, by 1587, become a conservative body. They sought to maintain the regime which had enabled them to rise in society and the bread which would guarantee their future. They were also tending to become isolated as a class on the summit they had reached.'

[3] See below, p. 897 ff.

it accepted them as it found them passed down from previous generations; as the seventeenth century too was to accept them in its turn. Antonio Domínguez Ortiz' excellent book, *La sociedad española en el siglo XVII*,[4] describes a social situation already suspected by historians: a nobility harassed by constant financial difficulties, but surviving nevertheless; a modern state failing to accomplish its mission or to assert itself as a social revolution (remaining content with compromise and hopeful of co-existence); a bourgeoisie subject to constant defection by its members (but whether it was aware of itself as a class at all is doubtful); and finally the mass of people, discontented, disturbed and restless, but lacking any true revolutionary consciousness.

I. NOBLE REACTION

In Christendom as in Islam, the nobility occupied the first rank and intended to keep it. It was the most conspicuous social group in France, in Spain and elsewhere. Everywhere the aristocracy monopolized symbols of prestige: precedence, lavish costumes, silks woven with thread of gold or silver, satins and velvets, Flemish tapestries, thoroughbred horses, luxurious residences, huge households and, by the end of the century, carriages. These were, it is true, so many roads to ruin. It was said that in the time of Henri II, the French nobility was importing four million *livres*' worth of clothes from Italy a year.[5] But appearances are not always deceptive: they may indicate a solid foundation of power and wealth. Over wide areas, these aristocracies still thrived and drew sustenance from their strong feudal roots. An ancient order had placed these privileged families at the summit of society and maintained them there still. The only exceptions were in and around large cities, the corrupters of traditional hierarchies, in commercial centres (though less than one would expect) in regions which had grown rich early, like the Netherlands and above all Italy – though not all Italy, as one might guess.

These exceptions are mere pinpoints and narrow strips on the map. On a Mediterranean or European scale theirs is clearly the history of a minority. Of the region as a whole, we can say what Lucien Romier said of France under Catherine de Medici: everything becomes clear 'when restored to its natural setting, a vast, semi-feudal kingdom'.[6] Everywhere the creation of the state, a social revolution (though barely begun) as well as a political one, had to contend with these 'owners of fiefs, lords over village, field and road, custodians of the immense rural population'.[7] To contend, or in other words to come to terms with them, dividing them and yet preserving them, for it is impossible to control a society without the complicity of a ruling class. The modern state took this tool as it came to hand, for destroying the nobility would have meant rebuilding society from

[4] Volume I only, Madrid, 1963.
[5] Lucien Romier, *Le Royaume de Catherine de Médicis*, 1925, I, 3rd ed., p. 177.
[6] *Ibid.*, p. 207–208. [7] *Ibid.*, p. 207.

scratch. And to re-establish a social order is no small undertaking; certainly it was never seriously contemplated during the sixteenth century.

Nobilities and feudal powers then had all the weight of custom behind them, the strength of positions long ago acquired, not to speak of the comparative weakness of the states or the lack of revolutionary imagination of the century.

Landlords and peasants. If we are to believe many much-quoted accounts, the sixteenth century reduced the aristocracy to poverty. These accounts are often true. But by no means all nobles were in such a plight, nor were they without exception victims either of the king or of war, of the lack of employment in peace time[8] or of overweening luxury. To say, as some historians have, that 'the feudal regime was collapsing as a result of currency depreciation following the discovery of precious metals in the Americas',[9] is as much an over-simplification, as it is to say that 'the acids of capitalism',[10] unaided, dissolved or at least profoundly altered all the structures of society – or that feudalism ended in Castile the day that Ferdinand acquired the grand masterships of the great military orders; or that Corsican feudalism received its death blow in 1511,[11] on the defeat of Giovanni Paolo de Lecca and the death of Renuccio della Rocca (who fell into an ambush laid by his own relatives[12]). It is as illusory to expect a precise explanation or chronology as it is to expect the words themselves to have precise meanings: the word *feudalism* alone is full of pitfalls. Only time can accomplish such transformations which rarely take a straight or a simple course.

At any rate in the essential confrontation between landlord and peasant, while the latter may occasionally have got the upper hand, as in Languedoc between 1480 and 1500,[13] and perhaps in Catalonia in the fifteenth century (at least certain well-to-do peasants may have) this was very much the exception.[14] Usually the nobleman had the last word, either in the long or short run, and in some places (Aragon, Sicily) he always had it. The price revolution, too often invoked, was not a worker of democratic miracles. It lightened the burden of peasant dues payable in money and assessed well before the discovery of America. In fact feudal dues from peasant holdings were often light, sometimes next to nothing. But not always. And above all, the landlord also collected payment in kind which kept pace with current prices. A register of the revenues of the Cardinal

[8] Lucien Romier, *Le Royaume de Catherine de Médicis*, 1925, I, 3rd ed., p. 193–203; Henri Drouot, *op. cit.*, I, p. 40.

[9] François de Ramel, *Les Vallées des Papes d'Avignon*, 1954, p. 142.

[10] Josef A. Schumpeter, *op. cit.*, p. 144.

[11] Carl J. von Hefele, *Le cardinal Ximénès*, p. 364.

[12] R. Russo, *art. cit.*, p. 421.

[13] E. Le Roy Ladurie, *op. cit.* In fact I would be inclined to say that there were not two, but three competitors, the third, opposed both to the nobleman and the small peasant, being the rich peasant. According to Le Roy Ladurie, the rich peasant made progress in Languedoc between 1550 and 1600.

[14] P. Vilar, *op. cit.*, I, p. 575 ff.

Duke of Lerma, in March, 1622, mentions his poultry, wheat, wine, the wheat at fixed rates, the wine at four reals, 'el pan a la tasa, el vino a quatro reales'.[15] Moreover, in the Mediterranean as in Europe, land division was never fixed once and for all. Peasant cunning might be met by aristocratic cunning and, on occasion, brutality. The landlord dispensed justice, held supreme rights over peasant holdings and the lands separating or surrounding them. The end of the fifteenth century and the early years of the sixteenth are remarkable for the construction or reconstruction of whole villages, always according to the landlord's desires. This was true of the Gâtine in Poitou,[16] of the Jura[17] where *granges* were founded; in Haut-Poitou,[18] where a noble family in difficulties might restore its finances by letting off vast tracts of heath previously unoccupied and where peasants were settled. In Spain, villages were founded by charter, the *cartas pueblas*,[19] and often land occupied for many years by peasants would pass into the landlord's hands. In Provence, charters of freedom and habitation become frequent after 1450: these often meant the re-population, on their old sites, of villages previously destroyed or abandoned, and only rarely the creation of completely new villages (Vallauris 1501, Mouans-Sartoux 1504, Valbonne 1519). The initiative every time came from the local landlord, 'anxious to see once neglected land peopled and farmed again'[20] and recruiting 'from the neighbourhood or more often from places further away, Liguria, the Genoese Riviera, Piedmont, farmers who were willing to settle . . . on his land'.[21] He might grant them advantageous terms, but the bargain was in his own interest too.[22]

Such 'colonization' was the clear consequence of bursts of economic growth and endemic over-population. In the Kingdom of Naples for instance, where each seignorial 'state' (some of them very large, notably in the Abruzzi, the County of Albi and Tagliacozzo) contained a certain number of urban and village communities, all with their own individual privileges and never therefore totally open to exploitation, the nobles took to creating 'new' towns and seeking to attract settlers to them. The Spanish authorities in Naples sought in vain to limit this practice, decreeing first by law in 1559 and again, a century later, in 1653, that new towns founded without government permission would be incorporated, without further ado, into the royal domains. It appears to have been easy to escape the

[15] Antonio Domínguez Ortíz, *op. cit.*, I, p. 364.

[16] Dr. L. Merle, *La métairie et l'évolution agraire de la Gâtine Poitevine de la fin du Moyen Age à la Révolution*, 1959.

[17] Lucien Febvre, *Philippe II et la Franche-Comté*, 1912, p. 201 ff.

[18] Gabriel Debien, *En Haut-Poitou: défricheurs au travail* (*XVe–XVIIe siècles*), Cahiers des Annales, 1952.

[19] Manuel Torres-López, 'El origen del Señorio Solariego de Benameji y su *carta-puebla* de 1549', in *Boletín de la Universidad de Granada*, 1932, no. 21; reviewed by Marc Bloch in *Annales hist. écon. et sociale*, 1934, p. 615.

[20] Robert Livet, *op. cit.*, p. 147 and 148.

[21] *Ibid.*

[22] R. Aubenas, *Chartes de franchises et actes d'habitation*, Cannes, 1943.

rigours of these laws, or to obtain the necessary authorization, since the towns and villages of the Kingdom of Naples continued to increase in number: from 1563 in Charles V's day to 1619 in 1579; then to 1973 in 1586. And the towns and villages belonging to the Church or the nobility (the majority: 1556 in 1579; 1904 in 1586) increased at the same rate as the meagre royal possessions, which over the same period rose from 53 to 69 (figures from the nineteenth century account by Bianchini). In short, the half-hearted policy of Spain was powerless to restrain the advance of the aristocracy, whether in Naples or Sicily. Nor was it a consistent policy, since towns, villages and lands within or returning to the royal domain were constantly being offered to new purchasers.[23]

A landlord then, whether old or new to the job, was likely to take full advantage of his rights and dues, his mills, hunting grounds and all the many areas in which he opposed the peasant, who was himself aware of the commercial possibilities of grain, wool and livestock. Bernardino de Mendoza, Philip II's ambassador in Paris – and for that reason alone beset with financial problems – made sure, from his distant post, of selling the wheat of the preceding harvest on his Spanish estate.[24] For the ambassador was a grain-farmer and even stockpiler. To take another example, cattle-raising was often introduced by large landowners, in the Roman Campagna and elsewhere.[25] Julius Klein has described the role played by certain nobles (and grandees) in the sheep-rearing of the *Mesta*.[26] In Andalusia in the seventeenth century, the nobility and the Church were to appropriate vast areas of land and, by their extensive agriculture, to depopulate the lowlands.[27] Plentiful documents concerning the farming of these estates await the historian's attention and several valuable studies have already shown what wealth they conceal.[28] The remarkable series of the *Sommaria* in Naples alone reveals the activities and speculation carried on by these great landowners, producing and marketing their grain, wool, oil and wood.[29] To cultivate one's land and sell its produce did not signify derogation, quite the contrary.

In the life of the aristocratic classes, the old feudal revenues, although reduced, still retained a certain importance.[30] They might be increased; or attempts might be made to increase them. Hence a number of disputes, lawsuits and revolts; several such revolts are recorded, unfortunately without any clear indication of their causes. The fresh contracts by which they were ended, or averted, should be put under the magnifying glass. In 1599, the community of Villarfocchiardo in Piedmont, made an agree-

[23] L. Bianchini, *op. cit.*, I, p. 260 ff.
[24] 8th October, 1585, A.N., K 1563.
[25] See above, Vol. I, p. 81. [26] *Op. cit.*, p. 354.
[27] G. Niemeyer, *op. cit.*, p. 51.
[28] Aldo de Maddalena, 'I bilanci dal 1600 al 1647 di una azienda fondiaria lombarda', in *Rivista internazionale di Scienze economiche e commerciali*, 1955.
[29] For some of the hundred possible examples, see A.d.S., Naples, Sommaria Partium 249, f° 181, 219 v°, 220, 247 (1544 and 1545).
[30] A. de Maddalena, *art. cit.*, p. 29, their 'drastic' drop after 1634.

ment with its overlords concerning feudal dues.[31] It would be interesting to discover how the agreement operated in the long run and whom it benefited most, as it would in a hundred other cases, for there is no doubt that adjustments were numerous. Of disputes and lawsuits, countless records remain. Most commonly, vassals would apply for integration to the royal estates, in Sicily, Naples, Castile and Aragon, probably because the crown was less alert to its interests than the aristocracy, and less swift to revise ancient contracts, on the pretext, whether valid or not, of economic change.

The price rise warns us in advance what was to be the general tenor of disputes between lord and peasant. During the summer of 1558, the subjects of the marquisate of Finale, near Genoa, rebelled against the excessive demands made by their overlord, Alphonso de Carreto. What were his demands? Was it not, as Carreto himself says, the fact that he had instigated a reassessment of the possessions of his vassals and proposed raising their rents? Since the Finale question was soon taken out of the hands of the marquis (Genoa and Spain were both far too interested in this vital territory not to turn the occasion to profit)[32] the practical origin of the affair is usually forgotten.

Many in the end were the noblemen who succeeded in maintaining their direct contact with the land and income from land, who thus survived, although not always unscathed,[33] the tempest of the price revolution. Nor was this protection their only resource.

In Castile: Grandes *and* Títulos *versus the king*. It has been said quite rightly that the modern state was the enemy of the nobility and of feudal powers. But that is not the whole story: the state was both enemy and protector, even associate. To bring the aristocracy to heel was its first objective, and one never completely attained: the next was to use it as an instrument of government, over it and through it to control the *peuple vulgaire* as they said in Burgundy.[34] It relied on the nobles for the maintenance of peace and public order, for the defence of the regions in which they owned estates or castles, for levying and commanding the *ban* and the *arrière-ban*, still something of importance in Spain: in 1542 for the siege of Perpignan; in 1569 for the war of Granada; in 1580 for the invasion of Portugal. More often, the king merely alerted his vassals when danger threatened; as in 1562[35] or 1567.[36] In 1580, the lords on the Portuguese border raised small armies at their own expense, a total of 30,000 men[37]

[31] Fr. Saverio Provana di Collegno, 'Notizie e documenti d'alcune certose del Piemonte', in *Miscellanea di storia italiana*, 1901, Vol. 37, 3rd series, Vol. 2, p. 393–395.

[32] His subjects rebelled again in 1566 (Simancas E° 1395, 7th February, 1566) and the rebellion was still unsubdued in 1568 (*ibid.*, 11th January, 1568).

[33] Carmelo Viñas Y Mey, *El problema de la tierra en los siglos XVI–XVII*, Madrid, 1941, p. 30, suggests that noble revenues rose less rapidly than general price levels.

[34] Henri Drouot, *op. cit.*, II, p. 477.

[35] *Correspondance de Saint-Sulpice*, published by E. Cabié, p. 37.

[36] *Dépêches de Fourquevaux*, I, p. 365.

[37] R. B. Merriman, *op. cit.*, IV, p. 365.

who were hardly used at all. These were almost always levies for frontier duty, but drew on many sources and were undoubtedly a heavy burden.

In addition, the king would keep prominent nobles constantly informed of his intentions, orders and important news; he would ask their advice and oblige them to lend him large sums of money. But the advantages granted in return by the monarchy were by no means negligible. When we think of the Spanish state, we must think above all of the *Grandes* and *Títulos*,[38] the king's principal interlocutors, that tiny minority of privileged persons through whom the monarchy from time to time indirectly governed, to avoid losing control of potential outbreaks of regional dissidence, for behind each of these great noblemen lay a huge clientele, as there did behind the Guises or the Montmorency family in France. When a royal judge (in 1664 it is true) was about to arrest the *corregidor* of Jeréz, the Duke of Arcos,[39] intervening, did not even trouble to see the judge himself, but had a word with his secretary: 'Tell him that the *corregidor* of Jeréz is of my house, that will suffice.' The nobility paled like the stars of the morning before the rising sun of the monarchy, as the imagery of the day described it. But the stars continued to shine in the firmament.

Castile provides the outstanding example. Here the unconcealed conflict took on a multitude of forms, not the least effective of which was the permanent hostility of the officers of royal justice to the pretensions of baronial justice and to the persons of the nobles themselves. Nothing was easier for example than to divide the nobles amongst themselves on the occasion of inheritance and property disputes. It was an opportunity to score off them. In 1572, Ruy Gómez was overjoyed at the news that the Duke of Medina Sidonia had won his suit against the Count of Alba, a nephew of the Prior Don Antonio, over the county of Niebla, a property, according to the Tuscan ambassadors, worth 60,000 ducats' income.[40] After winning his case – was it purely coincidence? – the Duke of Medina Sidonia married Ruy Gómez' daughter. And, infrequently it is true, but occasionally, the king's justice would uphold the claims of vassals against their lords. In July, 1568, the Duke of Infantado arrived at court. For many years the wealthiest of Castilian nobles (until as late as 1560[41]) he had yielded this distinction to the Duke of Medina Sidonia, perhaps because his pre-eminence had singled him out for attack and prompted him prudently to eclipse himself. At any rate, in 1568 he came to court on the occasion of a suit brought against him by the subjects of the marquisate of Santillana who claimed to owe allegiance directly to the crown. Fourquevaux, our source of information on this case,[42] adds: 'There are other grandees in the Kingdom who are similarly drawn into litigation,

[38] L. Pfandl, *Philippe II*, p. 315; S. Minguijón, *Historia del derecho español*, Barcelona, 1933, p. 370.

[39] A. Domínguez Ortíz, *op. cit.*, p. 222.

[40] Nobili and del Caccia to the Prince, Madrid, 12th March, 1572, A.d.S., Florence, Mediceo 4903.

[41] *C.S.P., Venetian*, VII, p. 178.

[42] *Lettres de Fourquevaux*, I, p. 295.

some of them having already lost valuable estates and others likely soon to do so.'

As for the justice meted out by the feudal nobles, it was strictly supervised from above, never allowed to escape observation. Their sentences, notes a Venetian in 1558, are referred to the *Chancillerías*.[43] Paolo Tiepolo confirms this in 1563: 'The nobles of Castile own huge estates and fairly good land, but their jurisdiction and power is much limited; in fact they do not dispense justice; they cannot levy any tribute from their people and have neither fortresses nor soldiers nor many weapons . . . unlike the lords of Aragon who, although of less consequence, yet usurp greater authority.'[44]

These minor victories of the monarchy – or even such a major success as the king's recovery in 1559 of the *diezmos de la mar*, the customs posts along the Cantabrian coast, on the death of the Almirante of Castile who had held them as a hereditary concession[45] – should not create any illusions. The power of the nobility had lost little of its vigour. In 1538, all the might of Charles V[46] had been unable, in the face of resistance from the representatives of the nobility, to obtain from the Cortes the establishment of a general tax on consumption. 'When Charles V tried to destroy their privileges', Michele Suriano later wrote,[47] 'he was opposed by all the grandees and most of all by the Grand Constable of Castile, although he was greatly devoted to His Majesty.' In 1548, in Charles' absence and again in 1555, in the absence of Philip II, the Spanish grandees attempted concerted action to get their own back: in 1558–1559,[48] Princess Joanna proceeded in Philip's name to the alienation of certain *lugares*, villages belonging to the cities. The cities defended themselves, some successfully, others not. But all the purchasers whose names are known are prominent nobles, singled out precisely because they were the very people, the most powerful in the land, whom the crown was pushing aside, or trying to. The crown did not wish for instance to see the Almirante of Castile purchase Tordesillas: nor the Marquis de las Navas acquire a substantial portion of the domain of Segovia;[49] nor the Duke of Alcalá gain possession of 1500 vassals of Seville,[50] sold for 150,000 ducats (that is 100 ducats for each vassal and his family). But for every one thwarted, ten nobles gained their ends and if there were no vassals of the cities to be had, they might buy vassals of the Church also auctioned off by the crown. The high Spanish nobility, as its archives testify, relentlessly bought up land, revenues and fiefs, even town houses.

However as years went by, royal authority gained in effectiveness and

[43] E. Albèri, *op. cit.*, I, III, p. 263.
[44] *Ibid.*, I, V, p. 19–20.
[45] See above, Vol. I, p. 293, note 86.
[46] Richard Konetzke, *op. cit.*, p. 146.
[47] E. Albèri, *op. cit.*, I, III, p. 338–339.
[48] According to the report by the *licenciado* Polomares, cf. above, p. 685, note 98.
[49] Simancas E° 137, f° 213, 9th June, 1559.
[50] *Ibid.*, 13th July, 1559.

certainly in severity, revealed in many signs: the arrest by the king's agents of the son of Hernán Cortés, Marquis del Valle,[51] on a charge of seeking to make himself independent ruler in New Spain; or that of the Grand Master of Montesa,[52] in Valencia through the intermediary of the Inquisition in 1572, on a charge of either heresy or sodomy – public opinion being a little uncertain on that point; the banishment of the Duke of Alva himself to his estates in 1579; or the disgrace and punishment in 1580[53] of the influential widow of Ruy Gómez, the Princess of Eboli, though only it is true after much royal heart-searching; then there was the arrest of the Almirante of Castile, Count of Modica, in his father's house in April, 1582;[54] the accused was, it is true, guilty of murdering his rival in love ('and this execution', noted a Venetian correspondent, 'has much saddened all the nobility and in particular the grandees of Spain, for they see that they are now no more respected than common mortals'); or again, the Prudent King in September, 1586 brought to heel the extravagant *jeunesse dorée* of Madrid without any form of trial.[55] In the time of Philip III and later of Philip IV, such acts of authority were frequently repeated. In December, 1608, the Duke of Maqueda and his brother Don Jaime were condemned to death for striking a notary and alcalde of the Royal Council. The furore eventually died down, but at the time feelings ran high,[56] as they did in April, 1621 at the abrupt disgrace of the Dukes of Osuna, Lerma and Uceda, which astonished even the French ambassador.

The nobility then was brought to heel, sometimes with its own consent. Great families had in fact already begun to live at court during the reign of Philip II; they settled at Madrid not without hesitation and a little repugnance at first, camping out in houses 'che sono infelici rispetto a quelle d'Italia', as Cardinal Borghese noted in 1597.[57] Sumptuous tapestries and silver did not prevent them from living 'porcamente senza una minima pollitia, che entrare nelle case loro par proprio d'entrare in tante stalle'. There is little point in defending them against these scornful Italian judgements: they really did live like the peasants they were, often violent and uncouth, although there were brilliant exceptions. And then of course, these houses in Madrid were only temporary residences, the pied-à-terre in the capital. It was on their country estates that all important festivities and ceremonies were held.[58] The outstandingly rich Dukes of Infantado had a magnificent palace at Guadalajara, the finest in Spain declared Navagero in 1525,[59] and it was here that the marriage of Philip II

[51] A.d.S., Florence, Mediceo 4903, 29th September, 1571.
[52] *Ibid.*, 19th June, 1572.
[53] A.d.S., Florence, Mediceo 4911, 15th February, 1580.
[54] A.d.S., Venice, Senato Dispacci Spagna, Matteo Zane to the Doge, Madrid, 21st April, 1582.
[55] A.d.S., Genoa, Spagna 15, Madrid, 27th December, 1608.
[56] Naples, Library of the Storia Patria, XXVIII, B 11, f° 114 v°, 30th April, 1621.
[57] A. Morel Fatio, *L'Espagne au XVI*e *et au XVII*e *siècle*, Heilbronn, 1878, p. 177.
[58] A.d.S., Florence, Mediceo 4903, 22 January, 1571.
[59] A. Navagero, *op. cit.*, p. 6.

to Elizabeth of France was celebrated. Most of the noble palaces were set in the heart of the countryside. It was at Lagartera, a village in the Sierra de Gredos, not far from Oropesa,[60] a village where, not so very long ago, 'the peasant women . . . still wore their traditional costumes, stockings in the form of gaiters and heavy embroidered skirts',[61] that the Dukes of Frías had their country seat, with its Renaissance windows cut in the thick walls, its spacious courtyard, broad staircases, moulded ceilings, exposed beams and huge fireplaces.

Gradually the lords yielded to the lure of the city. The Duke of Infantado was already at Guadalajara. In Seville, urban palaces sprang up in the sixteenth century. In Burgos, some of these houses are still standing today, with their Renaissance doors and porticoes and large armorial bearings supported by sculpted figures.[62] In 1545, Pedro de Medina was admiring the number and the distinction of the noble mansions in Valladolid.[63]

When Philip II's reign ended, the aristocracy flocked first to Madrid then to Valladolid, once more for a short period capital of Spain, towards the life of display of the *Corte*, the *fiestas* and the bullfights on the *Plaza Mayor*. The nobility built up round the monarch that ever-thickening screen which was to divide him from his people.[64] Taking advantage of the weakness of Philip III, this class filled all the chief posts of government with its own men, bringing its own factions and passions to the capital. This was the age of favourites, of *válidos*, the heyday of the courtier. They began to enjoy the luxury and relaxed atmosphere of Madrid, the long walks through the streets and the nocturnal life of the city: theatres, merry widows, courtesans who began to dress in silks for the fine company, scandalizing the virtuous. Intoxicated by this new style of life, the nobility took a certain perverse pleasure in descending to low society, mingling with the sophisticated crowd of the big city. The Duke of Medina Sidonia, the unfortunate hero of the Spanish Armada, is traditionally supposed to have founded the Cabaret of the Seven Devils.[65] At any rate, Madrid was not only the home of the king, the theatre and the underworld, it was also the city of the nobles, the scene of their extravagance, their eventful lives and their quarrels, whether settled in person or by their henchmen at street corners: murders occurred, according to observers, at the rate of more than one a day.[66]

[60] Baltasar Porreño, *Dichos y hechos del señor rey don Philipe segundo, el prudente . . .*, Cuenca, 1621, p. 6.

[61] R. Recouly, *Ombre et soleil d'Espagne*, 1934, p. 97.

[62] Théophile Gautier, *Voyage en Espagne*, 1879, p. 39.

[63] *Op. cit.*, in particular the house of the Count of Benavente, near Pisuerga, p. 229 vº.

[64] L. Pfandl, *op. cit.*, p. 132.

[65] Victor Hugo, *William Shakespeare*, 1882, p. 25, mentions the cabaret *El Puño en rostro*.

[66] A.d.S., Naples, Farnesiane 48, Canobio to the Duke, Madrid, 7th September, 1607. 'De quatro mesi in qua passa cosa qua contra il solito et mai più e intervenuto che siano state amazate in Madrid più di trecenti huomini et non si sa come ne perche delli più.'

But the nobility also thronged to Madrid to watch over the royal government and to take advantage of it. Having been kept so long at a distance during the interminable reign of The Prudent King, it took its revenge by imposing its own rule on his successor. Commoners continued to occupy pen-pushing jobs in the councils and to advance slowly along the road to honours. Grandees and *Títulos* went in search of royal favours, substantial gifts, profitable appointments, *ayudas de costa* [royal expense accounts], concessions of the *encomiendas* [commanderies] of the different Military Orders. They solicited on their own behalf or on that of their families. An appointment to a viceroyalty in Italy or America meant an assured fortune. Nominally, the income of the higher nobility did not cease to grow as inheritances and property regularly tended to be concentrated within a few hands. It kept pace unfailingly with the cost of living. Between them all, *Grandes* and *Títulos* totalled an income of 1,100,000 ducats, in 1525, according to the Venetians,[67] the Duke of Medina Sidonia alone being in receipt annually of 50,000 ducats. In 1558,[68] the duke's income reached 80,000; in 1581, twenty-two dukes, forty-seven counts and thirty-six marquises between them disposed of 3 million ducats and the Duke of Medina Sidonia alone of 150,000.[69]

Or so it appeared and was popularly believed. Actually these brilliant fortunes were mortgaged to the hilt; even in Philip II's day, families were regularly running up catastrophic debts and noble revenues were often assigned, like those of the crown, to pay off past debts. We know for instance that the Martelli[70] of Florence founded a firm in 1552 under the name of Francesco Lotti and Carlo Martelli, specializing until 1590 in loans at interest to great nobles (who were always behindhand with their payments) as well as to humbler persons (who were not). The list of bad debtors is a magnificent roll-call: 'Alonso Osorio, son of the Marquis of Astorga, Don Miguel de Velasco, Don Juan de Saavedra, Don Gabriel de Zapata, Don Diego Hurtado de Mendoza, Don Luis de la Cerda, Don Francisco de Velasco, Don Juan de Acuña, Don Luis de Toledo, son of the viceroy of Naples, Don Bernardino de Mendoza, Don Ruy Gómez da Silva, Don Bernardino Manrique de Lara, Don Garcilaso de la Vega, father of the Count of Palma, the Marquis de las Navas, the Count of Niebla . . .': a very exclusive clientele. Many of these debts having been contracted on the king's business, by his ambassadors for example, it is understandable that the monarch sometimes intervened and obliged the creditors to come to terms.[71] In the following century, when we have more information about these princely yet bankrupt treasuries, the same difficulties continued. A royal favour, a timely inheritance, a substantial dowry or a loan authorized by the king on an entailed estate might restore

[67] E. Albèri, *op. cit.*, I, I, 35–36, 16th November, 1525.
[68] *Ibid.*, I, III, p. 263.
[69] *Ibid.*, I, V, p. 288.
[70] Felipe Ruiz Martín, Introduction to *Lettres marchandes de Florence*, *op. cit.*
[71] B. Bennassar, *op. cit.*

a precarious budget.[72] But the balancing act soon began again. In the end this made the monarch's task easier. Cut off from active economic life, the nobility was doomed to resort to moneylenders: it did so to excess.

And the king had yet another means of pressure at his command. It was in about 1520 that there first emerged the distinct category of a select higher nobility, that of the *Grandes* and *Títulos*, 20 grandees and 35 *Títulos*. They numbered about 60 in 1525; 99 at the end of Philip II's reign (18 dukes, 38 marquises, 43 counts); Philip III created 67 new marquises and 25 counts.[73] So there was a promotion ladder. In 1533 and 1539, for instance, the Navas and Olivares families, both recently ennobled, were moved up in rank. Later came the division of the high nobility into three classes. By such means the king could govern and remain master.

Hidalgos *and* regidores *in Castile*. The high nobility of Castile consisted, at the end of Philip II's reign, of 100 persons, with their wives and children 400, or 500 at most. Any estimate of the total number of nobles in Castile must be very tentative, possibly 130,000,[74] that is a quarter of a million people out of a total population of six or seven million, a category which by its very size must have included a large proportion of poor and poverty-stricken aristocrats. In their thousands of sometimes tumbledown houses, often bearing 'gigantic scutcheons carved in stone',[75] dwelt a race that sought to live 'nobly', not soiling its hands with dishonourable work, serving king and Church, sacrificing everything, even its life, to this ideal. If there was noble folly, it was at its worst in Castile, despite the misery it could bring and the popular ridicule which expressed itself in so many sayings:[76] 'to catch what a *hidalgo* owes you, send out your greyhound'; 'on the *hidalgo*'s table is much linen but little food'; 'God save you from the poor *hidalgo* and the rich peasant.'

Such ridicule was to be expected: there was a contradiction in wishing but being unable to live like a noble, for lack of that money which justified almost anything. Certain towns even went so far as to refuse entry to the *hidalgos* who did not pay their share of the common fiscal burden. Written in letters of gold, it is said, in the *Ayuntamiento* of Gascueña, a village in the province of Cuenca, were the words: 'No consienten nuestras leyes hidalgos, frailes, ni bueyes.'[77] The oxen (dullards?) must be there for the rhyme: 'our laws do not allow for *hidalgos*, monks or oxen.' A great many other towns and villages refused to allow a distinction to be made between *hidalgos*, minor nobles, and *pecheros* or ordinary taxpayers. Yet most of the time, the two 'peoples' shared equally between them

[72] See the excellent third chapter, *La posición económica de la nobleza*, in A. Domínguez Ortíz, *op. cit.*, p. 223 ff.

[73] See L. Pfandl, *op. cit.*, p. 313 and A. Domínguez Ortíz, *op. cit.*, p. 215 ff.

[74] A. Domínguez Ortíz, *op. cit.*, p. 168.

[75] Théophile Gautier, *op. cit.*, p. 27.

[76] These examples are taken from A. Domínguez Ortíz, *op. cit.*, p. 224.

[77] A. Domínguez Ortíz, *ibid.*, p. 255 ff.

municipal offices and duties,[78] which put the smaller group at an advantage. And in many large cities, the vital port of Seville for instance,[79] the nobility had taken over all the posts of command. We have already noted the prevalence of venality which brought profit to families already holding positions, controlling the offices of *regidores* sold by the crown and resold by the incumbents. It was never, we may be sure, a question of mere vanity, but one of vital and often sordid necessity. Unable to plunder the whole of Castile as the grandees did, local nobles plundered urban and village revenues within their reach and depended on them for a living. Disputes, tension, class struggles were rarely absent from these troubled microcosms; incidents within them might be tragic or ridiculous but they were never meaningless.

Such meagre pickings were usually scorned by the high nobility. Similarly at court, minor appointments and humble offices were open to *hidalgos*, even before the social reversal which was to follow the death of Philip II. The latter did not, as is sometimes suggested, prefer plebeians or bourgeois, whose number even in the *Consejo de Hacienda* was limited (if one excepts clerics); but he preferred the lesser to the greater nobility. This preference, borne out by recent research, must affect previous general explanations.[80] The aristocratic reaction had thus begun with the sixteenth century, although of course not all noble families found a comfortable position straight away. Many impecunious aristocrats were content to serve in the ranks of the attendants, *criados*, of great seigneurs, without forgetting to wear on occasions the red crosses of the Orders of Santiago or Calatrava.[81]

Reaction to this widespread and penetrating movement is hard to find, almost nonexistent. Such as can be located therefore seems the more significant. Medina del Campo for instance, the old market city, refused to yield to the *hidalgos* half her municipal *oficios*, in spite of a judgement passed against the city in 1598, obtained a reprieve and finally had her way in 1635 on payment of the sum of 25,000 ducats.[82] Medina de Rio Seco was to resist and triumph in similar circumstances in 1632, this time in return for an enormous sum of money.[83] From these details it can at least be noted that the mercantile community was still resisting the nobility.

Other nobilities. Mutatis mutandis, the Castilian pattern is repeated elsewhere, in France,[84] even in Catalonia[85] and Valencia. In these two distinctive provinces of Spain, the king's authority was weak, and the nobles took such advantage of the fact that foreign observers credited them with intentions even more subversive than they actually had. In August, 1575,

[78] A. Domínguez Ortíz, *op. cit.*, p. 255 ff. [79] *Ibid.*
[80] *Ibid.*, p. 270.
[81] *Ibid.*, p. 277. [82] *Ibid.*, p. 263. [83] *Ibid.*, p. 262–263.
[84] Lucien Romier, *Le Royaume de Catherine de Médicis*, 3rd ed., 1925, I, p. 160–239.
[85] Pierre Vilar, *La Catalogne dans l'Espagne moderne*, 1962, I, p. 573, short notes; A. Domínguez Ortíz, *op. cit.*, p. 303 ff. The Catalan nobility was represented by only a few families.

when there was talk of sending to Flanders with Escovedo either the Duke of Gandía (who was in fact ill) or the Count of Aytona, either one of whom, said the Genoese Sauli, could claim to be 'citizen of a republic', since one is from Valencia, the other from Barcelona'.[86] 'Huomo di Repubblica' – a fine advertisement. Even more significant, at Valencia in April, 1616, the viceroy, the Duke of Feria, to punish a nobleman for some practical joke, had him paraded on a mule through the streets of the city. All the nobles immediately closed their houses, went into mourning and some even travelled to Madrid to protest to the king.[87]

In Naples, the violent invasions of Charles VIII and Louis XII had brought a whole series of tragedies to noble families. Great magnates like the Princes of Salerno and Taranto, or the Duke of Bari disappeared. Their 'states' were divided up. But in the process, lesser nobles enlarged their properties and fairly large states survived, the counties of Albi and Tagliacozzo, Matera and Cellano for instance. In 1558,[88] a Venetian report lists in the Kingdom of Naples 24 dukes, 25 marquises, 90 counts, about 800 barons and of these, 13 noblemen had incomes of between 16,000 and 45,000 crowns. These figures later rose. In 1580 there were 11 princes, 25 dukes, 37 marquises;[89] in 1597, 213 'titolati', consisting of 25 princes, 41 dukes, 75 marquises, 72 counts and 600 or more barons.[90] Later one loses count altogether of this small fry. In 1594, some noblemen had incomes of between 50,000 and 100,000 ducats.[91] How could the state, which sold patents of nobility through the intermediary of the *Sommaria*, lead the fight against its own clients?

Oppose them it did however, but never wholeheartedly. In 1538 and again later, Charles V made it known that he would not permit his feudatories in Naples to exercise the *mero* and *misto impero*[92] unless this right was duly specified in their privileges or established by a legitimate prescription; any feudatory infringing this ruling would be charged with usurping the right of jurisdiction. The emperor also tried to rescue land owned by the community and the freedom of vassals from the goodwill and pleasure of the lords: he attempted to restrict the number of 'services' to those fixed by custom. His efforts were in vain. To the baronage everything was fair game: forests, common grazing lands, the labour of their subjects (over whom they claimed total rights: Bianchini even talks of chairs covered with human skin[93]), the rights of the sovereign and sometimes even the money due in royal taxes. The crown, it is true, often abdicated its rights and fiscal revenues, which were sold in advance and

[86] A.d.S., Genoa, Spagna 6, 2415, Madrid, 4th August, 1575.

[87] A.d.S., Venice, Senato Dispacci Spagna, P° Vico to the Doge, Madrid, 27th April, 1616.

[88] E. Albèri, *op. cit.*, I, V, p. 276.

[89] *Ibid.*, II, V, p. 464.

[90] *Ibid.*, p. 316.

[91] *Arch. storico italiano*, IX, p. 247.

[92] L. Bianchini, *op. cit.*, II, p. 249, 252–253, 260, 299.

[93] *Op. cit.*, p. 249.

resold if necessary. In consequence most feudatories had almost sovereign authority in matters of justice and dues; they had their own liegemen, no government could subdue them. Short perhaps of coining money, they were free to do as they pleased. Only their extravagance, their preference, in Naples, for living near the viceroy and in the atmosphere of the big city, their vanity and their dependence on Spain in the struggle against the Turk and the 'popular', prevented them from giving too much trouble. Add to this perhaps the fact that among them were several foreigners, Spanish or Genoese, who had acquired fiefs by paying for them.

As a class, the baronage continued to grow. The last, disastrous years of the century undoubtedly ruined more than one nobleman, particularly in the city. Pressing debts led to the sale and sequestration of property by the *Sommaria*. But these were everyday occurrences in the life of the aristocracy; in Naples and elsewhere, certain risks were attendant upon gentle birth. The nobles survived. The individual might lose everything but the nobility as a class grew and prospered. Looking right ahead to the dramatic years of the mid-seventeenth century, we can see, behind the picturesque images and personalities of the revolution of Naples in Masaniello's time (1647), the completion of an unequivocal social revolution, from which the reactionary class of seigneurs emerged triumphant.[94]

The nobility had won for many years to come, and not only in Naples. It had won the battle in Milan[95] and in Tuscany,[96] in Genoa,[97] Venice[98] and Rome.[99] There is an abundance of dossiers to choose from and each tells the same story.

The successive aristocracies of Turkey. By far the most astonishing is the dossier on the Turkish Empire. While we have little or no direct knowledge of the countries of Islam, we are reasonably well informed about the social situation in Anatolia and are even better acquainted with the Balkans. And the picture here, contrary to what has been repeated so often,[100] was not at all the reverse of the West. The resemblances and analogies strike one immediately. Similar causes, one might say, produce similar effects, since a social order does not present unlimited structural possibilities and after all in both cases we are considering societies un-

[94] Cf. two excellent articles by Rosario Villari, 'Baronaggio e finanze a Napoli alla vigilia della rivoluzione del 1647–1648' in *Studi Storici*, 1962; 'Note sulla rifeudalizzazione del Regno di Napoli alla vigilia della rivoluzione di Masaniello' in *Studi Storici*, 1963.

[95] *Storia di Milano*, X, *L'età dei Borromei*, 1957, social problems indirectly touched on, p. 353 ff.

[96] See below, p. 728.

[97] Vito Vitale, *Breviario della storia di Genova*, 1955, I, p. 235 ff.

[98] James C. Davis, *The Decline of the Venetian Nobility as a Ruling Class*, Baltimore, 1962.

[99] J. Delumeau, *op. cit.*, II, p. 433 ff.

[100] P. Milioukov, Charles Seignobos and Louis Eisenmann, *Histoire de Russie*, 1932, I, p. XIII; Henri Pirenne, *Les villes du Moyen Age . . .*, p. 52; Henri Sée, *Esquisse d'une histoire du régime agraire aux XVIIIe et XIXe siècles*, 1921, p. 180.

equivocally based on land tenure, states still in their infancy, however spectacular and, in their immaturity at least, comparable to one another.

The studies of the last fifteen years have revealed, if not everything we would like to know, at least enough to make discernible certain patterns and valid 'models', the essential condition being to distinguish carefully between different periods. Too many western historians are inclined, when talking of Turkey, to confuse landscapes which emerged only over a period of centuries; while societies rarely progress with giant's strides, major transformations can be accomplished with the passing of time. There were three if not four successive ruling classes in Turkey, and the last, which boldly seized power towards the end of the sixteenth century, was, so to speak, the least legitimate; it overran the totalitarian Ottoman State, weakening its fabric if it did not altogether cause its ultimate collapse. For if the same causes and the same effects were everywhere visible, general economic conditions must take a major share of responsibility and they were everywhere instrumental in change.

The earliest Turkish nobility can be found by exploring the darkness of the fourteenth century; it settled in Anatolia either just before or just after the first Ottoman victories (from the capture of Bursa, 1326, to that of Adrianople, 1360). As described by historians,[101] this ruling caste was oppressive, presenting a menacing united front, at the same time slave-owning, feudal and seignorial and yet free, too free in fact, vis-à-vis the sultan who was no more than *primus inter pares*. Land was constantly being bought and sold without any pretence of control by the state. And it was a society recognizing private property, in the *mulk* or allodial lands, and *wakf* or family-owned estates, more precisely pious foundations, of which the founder and his descendants retained both control and profit, so that in some ways they provided stable repositories of wealth, like entailed estates in the West.

The second Turkish nobility was not exclusive to the Ottoman possessions in Europe during the fifteenth century, but it is here that we have the clearest picture of it as it takes root and flourishes.

Far in advance of the rapid Turkish conquest of the Balkans, a whole society had already been brought to the point of collapse by a series of peasant revolts. Before the final assault, from the capture of Belgrade in 1521 to the invasion of Hungary (Mohács, 1526), the Hungarian peasants rose up; the Christian ruling class suppressed them but the effort proved fatal.[102] A series of ancient feudal regimes crumbled away almost of its own accord, all long-standing regimes composed of mixed elements (Greeks, Slavs, even westerners). On the eve of conquest, the Balkans were socially comparable to the West by reason of their wealth and even in such

[101] Ömer Lütfi Barkan, *Aperçu sur l'histoire agraire des pays balkaniques*, offprint p. 141 ff.

[102] Nicoara Beldiceanu, 'La région de Timok-Morava dans les documents de Mehmed II et de Selim Ier' in *Revue des Études Roumaines*, 1957, p. 116–119 and reference to article by V. Papacostea.

details as the movement of nobles into towns neighbouring their estates (an *inurbamento* as in Italy). The Musachi, a great Albanian family, settled in Durazzo, in fortified palaces similar to those of Bologna and Florence. And more than one inland city had its street of rich noblemen's dwellings, Tirnovo its *bojarska mahala*, Vidin its *bojarska ulica*,[103] all this luxury founded on the *latifundia* and a harshly exploited peasantry. It was this system which tumbled like cardboard scenery before the Turkish advance.

In the wake of conquest came devastation, the retreat of the native populations to the forbidding mountains, but also to some extent the liberation of the peasants. They remained grouped in their communities, masters of their land, not entirely free certainly, for they were subject to taxation from which no one was exempt and reorganized into new seignorial estates modelled on the old, fiefs or rather benefices, the *tīmārs* into which the conquered territory and population were divided. The peasant still paid rent in money and rent in kind, the latter far less substantial than the former, but was now liberated from the traditional unpaid labour. In these early years the Turks were the more anxious to be conciliatory since the conquest had yet to be completed and the advantage of peasant disturbances in still unconquered areas had to be maintained; moreover the sultans were already suspicious of the old Anatolian aristocracy, now enriched by the distribution of fiefs in Europe and especially of the *Jandarli*,[104] the great families who were trying to take over control of affairs. The suspicions of the central government towards its feudal barons were never allayed, hence the abundance of prudent safeguard measures. From the start, the favours granted to the Christian nobility of the Balkans, who were generously provided with *tīmārs*, had basically no other motive.[105]

These *tīmārs* were no ordinary fiefs, despite their resemblances to western feudal estates; like the latter, they consisted of villages, land both farmed and uncultivated, waterways, tolls and sometimes rights on the market of a neighbouring town, as at Kostur, a small Bulgarian town.[106] But these domains served to maintain soldiers, horsemen, the *sipāhis* (indeed the *tīmārs* were often known as *sipāhiliks*). In fact they were conditional fiefs, a kind of salary, in return for which the holder was obliged to report when requisitioned, with a band of cavalry proportionate to the size of the *tīmār*, to the *sandjak-beg* of the province. Failure to obey the summons

[103] R. Busch-Zantner, *op. cit.*, p. 60–61 and references.

[104] Stanford J. Shaw, in *The Balkans in Transition*, 1963, p. 64.

[105] *Ibid.*, p. 64–65.

[106] This detail and much of the following material is taken from the article by Bistra A. Cvetkova, 'L'évolution du régime féodal turc de la fin du XVIe siècle jusqu'au milieu du XVIIIe siècle' in *Études historiques* (of the Bulgarian Academy of Sciences) *à l'occasion du XIe Congrès International des Sciences Historiques*, Stockholm, August, 1960, reference to Kostur on p. 176. For a bibliography of this historian see *Journal of Economic and Social History of the Orient*, 1963, p. 320–321; see in particular her important article, 'Nouveaux documents sur la propriété foncière des Sipahis à la fin du XVIe siècle', in *USSR Academy of Sciences, Institutum Populorum Asiae, Fontes Orientales*, 1964, summary in French, p. 220–221.

would mean forfeiting the *tīmār*. These revocable domains, granted only for life tenure, were benefices in the Carolingian sense rather than true fiefs. But from a very early time, the *tīmārs* began to pass from father to son, a move towards the hereditary fief rather than the benefice. From 1375, a legal provision recognized the right of succession of sons of timariots.[107]

The normal income from a *tīmār* was modest, never more than 20,000 aspers, the upper limit of the top category which was rarely reached. Records for the region of Vidin and Berkovitsa, between 1454 and 1479, list 21 *tīmārs* having revenues ranging from 1416 to 10,587 aspers (from 20 to 180 ducats), the bulk of the group falling between 2500 and 8000 aspers. This rather modest yield nevertheless marked the golden age of the system. The timariot was unable, under the watchful eyes of the local authorities, to supplement his income from the meagre earnings of the peasants and the only way to become rich was through the spoils of war, the rewards of the profitable age of conquest which lasted until mid-sixteenth century.[108]

That the income of the timariots was not high could be surmised merely from their numbers: possibly 200,000[109] towards the end of the century, that is one million persons out of a total population of 16 or 20 million. As an aristocracy it was too large to be rich. But within its ranks there were privileged people: a high nobility was very early established. There were in fact three categories of *tīmārs*,[110] the ordinary *tīmār*, yielding up to 20,000 aspers income; the middle category, or *zi'āmets* up to 100,000 and the *khāṣṣ* above this figure. In 1530, the Grand Vizier Ibrāhīm Pasha possessed in Rumelia a *khāṣṣ* yielding an income of 116,732 aspers; Ayās Pasha one worth 407,309 and Ḳāsim Pasha one worth 432,990. These were very large estates and there were also the *wakf* and *mulk* properties (the latter known as *khāṣṣa* or *khāṣṣa čiftliks* as opposed to peasant property, *ra'iyyet čiftliks*). Can we call these estates aristocratic preserves? Some of these preserves in the second half of the fifteenth century in Greece, comprised olive groves, vineyards, orchards and mills.[111] To a greater or lesser extent, private property was introduced very early then, almost always to the benefit of the great feudatories, and threatened to undermine the mighty institution of a landed aristocracy brought up in the tradition of public service and in accordance with the basic doctrine of the Turkish state that *all* national wealth was the exclusive property of the sultan.

That the latter should have felt the threat from the great nobility, too wealthy and now too free, explains the reaction, precocious on the part of Muḥammad II, belated on the part of Sulaimān the Magnificent, in favour of a firm centralization of a system menaced by possible separatism and local autonomy. The aim of the *Fātih*, the Conqueror of Constantinople,

[107] J. W. Zinkeisen, *op. cit.*, III, p. 146–147.

[108] Bistra A. Cvetkova, *art. cit.*, p. 173.

[109] *La Méditerranée* . . ., 1st ed., p. 639: 'The feudal army of unsalaried *sipāhis* amounted to 230,000 horsemen.'

[110] Bistra A. Cvetkova, *art. cit.*, p. 172.

[111] *Ibid.*, p. 173–175.

was to dissolve all *mulk* and *wakf* properties and incorporate them into the system of *sipāhiliks*.[112] The grand ruling of Sulaimān in 1530[113] was a general reorganization, characteristic of the age of the 'Lawmaker'. From now on, military fiefs would be distributed only in the capital (provincial *beglerbegi* preserving only the right to nominate to minor holdings). The compensation granted to sons of *sipāhis* was now fixed: it would vary depending on whether the father died on the field of battle or in his bed, and whether or not the heir already possessed a fief. But these measures like all such authoritarian attempts to legislate in social matters had doubtful results, the chief one being that, since it was centred on Istanbul, the system was now to depend on palace intrigues rather than on its own virtues. At any rate large estates had become an established pattern which remained unaffected. They were given further encouragement by the colonization of the Balkan interior, by the population increase, by the export boom in raw materials to the West. Between 1560 and 1570, many large landowners grew rich in the wheat trade: the Grand Vizier Rustem Pasha trafficked in wheat.[114]

The third age of the Turkish nobility, after 1550–1570 approximately, was not as novel as is sometimes claimed. It was characterized by the development of the large estate, but this dates from before the middle of the sixteenth century. The new element however was the halting of the profitable age of Turkish conquest even before the end of the too-glorious reign of Sulaimān the Magnificent (1566), and the consequent obligation felt by all nobles, of whatever standing or origin, to turn back to the peasantry and relentlessly and shamelessly to grind the faces of the poor, as money rents became worthless with the repeated devaluations of the asper.[115] The Ottoman State was itself placed in some difficulty: 'the receipts of the public treasury no longer cover its expenses', noted the Ottoman chronicler Muṣṭafā Selānīkī at the end of the sixteenth century.[116] Fiscal measures and alienation of revenues were the logical result. Inflation dealt the final blow to the old order. Contemporaries blamed the low morals of the government, the favours granted to court nobles and their servants and hangers-on. It is true that the Seray had become the source of distribution of *tīmārs* and that it reserved them for the courtiers and servants surrounding the sultan and his ministers, *cha'ushs*, scribes, controllers of tax-farms (*muteferriḳas*) the valets of dignitaries, palace pages,[117] not to mention viziers or the *Walide Sultān* (mother of the reigning sultan). This handout of fiefs went far beyond anything of the kind in the West. Letters of nobility distributed in France were nothing compared to the *fermāns* given out often twice over and shamelessly granting

[112] Bistra A. Cvetkova, 'Sur certaines réformes du régime foncier au temps de Mehmed II', in *Journal of Economic and Social History of the Orient*, 1963.
[113] J. W. Zinkeisen, *op. cit.*, III, p. 154–158.
[114] See above, Vol. I, p. 591.
[115] See above, Vol. I, p. 539 ff.
[116] Bistra A. Cvetkova, 'L'évolution du régime féodal . . .', p. 177.
[117] *Ibid.*, p. 184.

noble privileges even to *ecnebi*, 'outsiders'[118] (that is those who did not belong to the ruling Ottoman class). 'Vagabonds, brigands, gypsies, Jews, Lazis, Russians and townsfolk' was how an Ottoman chronicler[119] described what in the West would have been known as the new nobility. The days of 'ignominy' had come,[120] and they were to be prolonged as traditional values were trampled underfoot. The money economy propelling all before it, vast estates grew up, creeping plants against which there was no remedy. Under false names, a timariot could amass twenty or thirty domains.[121] Lesser estates were swallowed up by larger ones and nobles demoted or threatened with demotion from their rank soon become prominent figures in the peasant uprisings of the end of the century and the next century.

As well as the heyday of parvenu nobles, this third age ushered in the era of usurers, of 'financiers' exploiting simultaneously state, nobility and peasants. After 1550, the Ottoman State was in fact to resort to the old practice of selling fiscal revenues, the system of *muḳāṭa'a* and *iltizam* (tax-farming) operated earlier by the Seldjuk Turks and by Byzantium.[122] It was a system already extensively practised in the West, alike in Naples, Venice, Paris or Spain. In Naples, fiscal revenues were sold; in Venice, a certain tax or customs duty might be farmed out for two or three years. The Turkish government now followed suit, asking tax-farmers to advance immediately a sum corresponding to the taxes to be gathered. The farmer would then, in spite of the vigilance of the controllers, reimburse himself generously. At the tolls for instance, one asper was due for every two head of sheep; up to eight aspers per head was sometimes levied. In addition, before he accepted the contract, the tax-farmer could set his own terms which might be very high. Very often too, large landowners also farmed out their estates and so there sprang up unopposed a network of Jewish or Greek moneylenders,[123] who before long held Turkey fast in a noose. The development of the money economy and inflation carried them to commanding positions. Under such conditions, the old military regime of the *sipāhis* ceased to function, not surprisingly. Military service was evaded and the checks ordered by the central government became a mockery.

On this count, the evidence of 'Ain-i 'Alī, financial Intendant of the Sultan Aḥmad, is explicit:[124] 'Most of the feudal lords today', he wrote, 'release themselves from their military obligations, so that on campaign, when military service is required, from ten *tīmārs* not a man will turn up.'

[118] *Ibid.* [119] *Ibid.*, p. 184 ff. [120] *Ibid.* [121] *Ibid.*

[122] Bistra A. Cvetkova, 'The System of Tax-farming (iltizam) in the Ottoman Empire during the 16th–18th Centuries, with reference to the Bulgarian Lands', in Bulgarian, summary in English in *Izvestia na institouta za pravni naouki*, Sofia, XI–2.

[123] Bistra A. Cvetkova, 'L'évolution du régime féodal . . .', p. 184. All these observations and conclusions are confirmed by Ömer Lutfi Barkan's *leçons*, (typescript, *École des Hautes Études, VIe Section*, Paris).

[124] J. W. Zinkeisen, *op. cit.*, III, p. 153–154. Cf. J. von Hammer, *op. cit.*, I, p. 372.

The warrior mentality from which the institution had drawn its solidity was no more. Koči Beg, a native of Korytsa in southern Albania, says as much in his writings published in 1630.[125] But even before this, an obscure Bosnian, Mallah Hassan Elkjadi, had sounded a similar warning in 1596.[126] The decadence was visible even to foreigners.[127] In the seventeenth century, the *sipāhis* were to leave their country residences and move into the towns. It was at this time that the Toptani family for instance left the fortified castle of Kruya for the open city of Tirana, surrounded by its orchards and gardens.[128] The migration to the towns was one sign among many of the formation of an aristocracy with strong local roots, confident of its future.

The Čiftliks. With the coming of the seventeenth century there can be detected a further and no less considerable change, that is if Richard Busch-Zantner's studies are correct.[129] His stimulating book met with a cool and perhaps undeserved response from the world of scholarship.[130] Busch-Zantner was, it is true, attracted by the literature and the example of agrarian reform following the first World War, by the Yugoslav publications of Frangês and Iusić, but was that necessarily a fault? He has also been criticized for imprecise terminology, but this is something which seems to me unavoidable. For any western historian approaching the world of the East, all the words at his disposal are ambiguous and past definitions (for example the categorical opposition established by C. Beckers between the Turkish fief and the western fief) or general explanations (such as those of J. Cvijić) are not the best guide. The increasing use of original Turkish sources will in any case lead to the re-statement of these problems and inevitably to the total revision of our ideas of Turkish history.

It seems probable then that the *čiftlik* represents a new and important phenomenon. The word appears originally to have designated the acreage of land that could be ploughed in a day[131] (a counterpart of the Byzantine *zeugarion*, the German *Morgen* or *Joch* and the *jour* or *journal* of some regions of France). It later seems to have come to mean privately owned property, whether that of the peasants (*ra'iyyet čiftlik*) or of great noblemen (*khāṣṣa čiftlik*) and finally the large estate of modern times, a sort of colonial plantation or *Gutsherrschaft*. We cannot adequately explain this

[125] Franz Babinger, in *Encyclopedia of Islam*, 1913, Vol. II, p. 1055.

[126] Ludwig von Thalloczy, 'Eine unbekannte Staatschrift eines bosnischen Mohammedaners', quoted by R. Busch-Zantner, *op. cit.*, p. 15.

[127] To the Venetian L. Bernardo for example in 1592, B. A. Cvetkova, *art. cit.*, p. 193. Cf. J. W. Zinkeisen, *op. cit.*, III, p. 167, note 1.

[128] R. Busch-Zantner, *op. cit.*, p. 60.

[129] *Aus dem Grundherr wurde der Gutsherr*, *op. cit.*, p. 84.

[130] Carl Brinckmann, in *Vierteljahrschrift für Sozial- und Wirtschaftsgeschichte*, 1939, p. 173–174; Marc Bloch in *Mélanges d'histoire sociale*, I, p. 120.

[131] Traian Stoyanovitch, 'Land tenure and related sectors of the Balkan economy' in *Journal of Economic History*, 1953, p. 398 and 401.

evolution, but the word is certainly used in this sense after about 1609–1610.[132]

The existence of these tyrannically organized but productive modern estates would indicate that when considering the evolution of the Turkish nobility, other factors besides the social and political must be taken into account. It was not a story of unrelieved destruction and deterioration either, as the pessimistic accounts of the chroniclers would have us believe. These estates are reminiscent of the highly productive colonial plantation, or the impressive estates of the *Ostelbien*[133] or Poland. In the centre stood the lord's manor, stone-built as in the south Albanian plain of Korytsa, and, with its tower-like appearance, typical of the *kula* or fortress-dwelling, always several storeys high,[134] dwarfing the pitiful clay huts of the peasants. As a rule, the *čiftlik* converted the low-lying lands of the plains, the marshes between Lárisa and Volos for example, along the muddy banks of Lake Jezero[135] or the wet river-valleys. It was a conqueror's type of farming. The *čiftliks* produced cereals, first and foremost. And cereal-growing, in Turkey as in the Danube provinces or in Poland, when linked to a huge export trade, created from the first the conditions leading to the 'new serfdom'[136] observable in Turkey. These large estates everywhere debased the peasantry and took advantage of its debasement. At the same time, they were economically efficient, producing first wheat, then rice, before long maize, later cotton, and were from the start characterized by the use of sophisticated irrigation techniques and the spread of yoked buffalo teams.[137] The conversion of the Balkan plains closely resembles what was taking place in the West, in the Venetias for example. These were unquestionably massive and far-reaching land improvement schemes. As in the West, large proprietors put to use the depopulated land whose possibilities had never been fully explored by previous generations of nobles and peasants. The price of progress, here as elsewhere, was clearly social oppression. Only the poor gained nothing, could hope for nothing from this progress.

2. THE DEFECTION OF THE BOURGEOISIE

The bourgeoisie in the sixteenth century, committed to trade and the service of the crown, was always on the verge of disappearing. Financial ruin was not the only cause. If it grew too rich and tired of the risks inherent in commercial life, the bourgeoisie would buy offices, government bonds, titles or fiefs, succumb to the temptations, the prestige and carefree indolence of the aristocratic life. The king's service was a short cut to nobility; this course, which did not exclude others, was one way of

[132] *Ibid.*, p. 401. [133] R. Busch-Zantner, *op. cit.*, p. 86.
[134] A. Boué, *op. cit.*, II, p. 273.
[135] R. Busch-Zantner, *op. cit.*, p. 80–90.
[136] G. I. Bratianu, *op. cit.*, p. 244.
[137] T. Stoyanovitch, 'Land tenure . . .', p. 403.

thinning the ranks of the bourgeoisie. The bourgeois turned class traitor the more readily since the money which distinguished rich from poor was in the sixteenth century already appearing to be an attribute of nobility.[138] Moreover, at the turn of the century trade slumped, sending wise men back to the land as a safe investment. And land ownership was almost by definition a characteristic of the aristocracy.

'Among the principal Florentine merchants dispersed in the market cities of Europe', writes the historian Galluzzi,[139] 'many were those who [at the end of the sixteenth century] repatriated their fortunes to Tuscany to invest them in agriculture, such as the Corsini and the Gerini who returned from London, the Torrigiani who left Nuremberg and the Ximenez, Portuguese merchants who became Florentines' – a revealing picture of the move back to the land by wealthy merchants barely a century after Lorenzo the Magnificent. Turn the pages and in 1637, on the occasion of a new reign there emerges a different Tuscany, one of stiff, unbending court nobles,[140] Stendhal's Italy, foreseeable for a long time, yet still somehow shocking in the city where once beat the brave heart of the Renaissance. An old order had fallen away.

It is hardly an exaggeration, particularly looking ahead to the later seventeenth century, to talk of the bankruptcy of the bourgeoisie. As a class, its fortunes were linked with those of the towns; and the towns began the sixteenth century with a series of political crises: the revolt of the *Comuneros* in 1521, the capture of Florence in 1530. In the process the freedom of the city-states suffered severely. Then came economic crisis, temporary at first, but becoming persistent with the seventeenth century, which seriously crippled their prosperity. Change was inevitable and change was on the way.

Bourgeoisies of the Mediterranean. In Spain what was vanishing had hardly existed in the first place. Gustav Schnürer[141] claims that Spain, or at any rate Castile, was without a bourgeoisie from the time of the *Comuneros*: a rather sweeping statement but not a fundamentally wrong one. Insufficiently urbanized, the Peninsula was obliged to entrust essential commercial functions to intermediary groups unable to identify with the true interests of the country yet playing a vital part in the economy, as has been the case in certain South American countries for instance in 1939. In the Middle Ages these functions had been carried out by the Jewish communities which provided merchants, moneylenders and tax-collectors. After their expulsion (1492), the gaps were filled somehow. In the towns and villages in the sixteenth century, retail trade was often handled by Moriscos, new Christians, who were frequently accused of conspiring against public safety, of trafficking in arms and of hoarding.

[138] Antonio Domínguez Ortíz, *op. cit.*, p. 173 and 174.
[139] *Op. cit.*, III, p. 280–281.
[140] *Ibid.*, p. 497.
[141] *Op. cit.*, p. 168.

The wholesale trade, in Burgos in particular, was often represented by converted Jews.[142]

Such complaints, prejudices and suspicions, for lack of better evidence, would suggest that there did exist here and there a native Spanish bourgeoisie, in Seville, Burgos, Barcelona (awakened at the end of the century from its long sleep). We know of certain wealthy Spanish merchants such as the Malvenda of Burgos or Simón Ruiz of Medina del Campo.

On the other hand one cannot properly term 'bourgeois' the host of civil servants, the *letrados*[143] in the king's service, habitually prefixing their names with a *Don*, lesser nobles or aspirant nobles far more than true bourgeois. It is remarkable in that most remarkable of countries, Spain, to find even the illegitimate sons of the clergy acquiring the title of *hidalgo.* Not so remarkable after all perhaps, if one remembers the dishonour attached in Spain to manual work and trade, and the innumerable violations of the fragile borderline between the very minor nobility and commoners: of the seven hundred *hidalgos* in a modest town near the Portuguese frontier, perhaps three hundred are genuine, says a remonstrance of 1651,[144] not to mention the *hidalgos de gotera* [gutter aristocracy] and the fathers of twelve children, who were granted fiscal exemption without being noble and were vulgarly known as *hidalgos de bragueta* [codpiece aristocracy].[145] In Spain, the bourgeoisie was hemmed in on all sides by a fast-multiplying and fast-encroaching nobility.

In Turkey the urban bourgeoisie – essentially a merchant class – was foreign to Islam: Ragusan, Armenian, Jewish, Greek and western. In Galata and on the islands there still survived pockets of Latin culture. Symptomatic of change here is the rapid decline of those who had once been the great merchants of the empire – the Venetians, Genoese and Ragusans. Two foreign businessmen were prominent in the sultan's entourage: one, Michael Cantacuzenus, was a Greek,[146] the other, Micas, a Jew.[147] Iberian Jews, both Spanish and Portuguese, immigrants at the end of the fifteenth century, gradually came to occupy leading positions in trade (especially the Portuguese), in Cairo, Alexandria, Aleppo, Tripoli in Syria, Salonica, Constantinople. They were prominent among the tax-farmers (and even the bureaucrats) of the empire. How many times does one find the Venetians complaining of the bad faith of the Jews who retailed Venetian merchandise! Soon, no longer content with the role of retailers, they were to enter into direct competition with the Ragusans and Venetians. Already in the sixteenth century, they were handling a

[142] Julio Caro Baroja, *La sociedad criptojudia en la Corte de Felipe IV* (Lecture given on admission to the Academia de la Historia), 1963, p. 33 ff.
[143] On the lack of esteem for the *nobleza de letras*, see A. Domínguez Ortíz, *op. cit.*, p. 194.
[144] A. Domínguez Ortíz, *ibid.*, p. 266, note 38.
[145] *Ibid.*, p. 195.
[146] For references concerning Cantacuzenus, see Traian Stoyanovitch, 'Conquering Balkan Orthodox Merchant', in *Journal of Economic History*, 1960, p. 240–241.
[147] See below, p. 1148.

large seaborne trade with Messina, Ragusa, Ancona, and Venice. One of the most profitable ventures of Christian pirates in the Levant became the search of Venetian, Ragusan or Marseillais vessels, for Jewish merchandise, the *ropa de judíos* as the Spanish called it, likening it to contraband, a convenient pretext for the arbitrary confiscation of goods.[148] And the Jews themselves soon met competition from the Armenians who, in the seventeenth century, began to freight ships for the West, travelled there themselves and became the agents of the commercial expansion of the Shah 'Abbās.[149] These men were the successors, in the Levant, of that rich Italian merchant bourgeoisie which once controlled the entire Mediterranean.

In Italy itself, the situation was complex. Once more the heart of the problem lies here, where once had flourished the vital bourgeoisies and their cities. The splendour of Florence in the days of Lorenzo the Magnificent coincided with the heyday of an opulent and cultivated higher bourgeoisie, lending confirmation to Hermann Hefele's theory of the Renaissance,[150] as a coincidence of an intellectual and artistic explosion and a powerful movement towards social change from which Florence emerged an altered and more open society. The Florentine Renaissance represents the achievement of a bourgeois order: that of the *Arti Maggiori*[151] which both controlled the avenues of power, disdaining none of the necessary tasks of commerce, industry and banking, and also paid homage to the refinements of luxury, intelligence and art. As a class it lives again before our eyes, through the work of the painters it befriended, in that series of portraits in Florence which is sufficient witness in itself to a bourgeoisie at the height of its power.[152] But a few steps in the Uffizi take the visitor to Bronzino's portrait of Cosimo de' Medici in his long red robe: another age with its princes and court nobles. However a Spanish merchant living in Florence could still write in March, 1572: 'in this city, by an ancient tradition, merchants are held in high regard.'[153] It is true that he was speaking of the highest category of merchants, the *hombres de negocios*, many of whom were in fact nobles. To make the transformation complete, these men had only to abandon their commercial activities and live, as was perfectly possible, off their land and revenues.

Elsewhere too the scene was changing. In 1528, Genoa received the aristocratic constitution which was to last until the troubles of 1575–1576. In Venice, the merchant aristocracy at the end of the century resolutely turned its back on trade. In central and southern Italy the trend was very

[148] See below, p. 822.

[149] See above, Vol. I, p. 50.

[150] Hermann Hefele, *Geschichte und Gestalt, Sechs Essays*, 1940: the chapter 'Zum Begriff der Renaissance', p. 294 ff, was published in article form in *Hist. Jahrbuch*, Vol. 49, 1929.

[151] Alfred von Martin, *Sociología del Renacimiento*, 1946, p. 23.

[152] Marcel Brion, *Laurent le Magnifique*, 1937, p. 29 ff.

[153] Antonio de Montalvo to Simón Ruiz, Florence, 23rd September, 1572. Ruiz Archives, Valladolid, 17, f° 239, quoted by F. Ruiz Martín, *Introduction . . ., op. cit.*

similar. In Rome the bourgeoisie received its chastening in 1527. In Naples the only refuge left for it was the practice of law: chicanery was now to be its bread and butter.[154] Everywhere its role was being curtailed. In Lentini in Sicily, the town magistrates were recruited exclusively from the nobility;[155] Francesco Grimaldi and Antonio Scammacca for instance, syndics of the town who, in 1517, successfully applied for its reintegration to the royal domains, or Sebastian Falcone who, in 1537, in his capacity as *giurato e sindaco* saw to it, in return for a payment of 20,000 gold crowns to Charles V, that the town was not allocated to the feudal barons and that it was granted the privilege, already consecrated by long usage, of having reserved to the nobles of the town the functions of *capitano* of Lentini. So we need not imagine that there was a fight to the death between the lords and the domanial towns in Sicily. Even when the towns were still controlled by their bourgeoisie, and this was rare, the latter were only too ready to compromise with the nobles and their clientele. The time was past when *consoli* and *sindici* of the guilds fought against them for control of the city. In some towns the process had gone even further: in Aquila, in the north of the Kingdom of Naples, the *sindaco dell'Arte della lana* itself became after 1550 virtually a noble prerogative.[156] If we take these happenings chronologically, one by one, it becomes quite evident that the process had begun very early.

The defection of the bourgeoisie. If the social order seems to have been modified, the change was sometimes more apparent than real. The bourgeoisie was not always pushed out, brutally liquidated. It turned class traitor.

This was not a conscious betrayal, for there was no bourgeois class truly aware of itself as such; possibly because it was numerically so small. Even in Venice, the *cittadini* constituted at most only 5 or 6 per cent of the city's population at the end of the century.[157] And everywhere, rich bourgeois of every origin were irresistibly drawn towards the aristocracy as if towards the sun. One can see from their correspondence the curious attitudes of Simón Ruiz and Baltasar Suárez towards those who lived like gentlemen and sponged on occasions off prudent and thrifty merchants.[158] The chief ambition of these pseudo-bourgeois was to reach the ranks of the aristocracy, to be absorbed into it, or at the very least to marry their richly-dowered daughters to a nobleman.

From the beginning of the century at Milan, *mésalliances*, although always giving rise to scandal, were no less frequent for that. And our informant, Bandello, for all his liberal opinions, expresses his indignation. He writes for instance of a noblewoman, who having married a merchant

[154] Benedetto Croce, *Storia del regno di Napoli*, 3rd ed., Bari, 1944, p. 129–130.
[155] Matteo Gaudioso, 'Per la storia . . . di Lentini', *art. cit.*, p. 54.
[156] Cf. Vol. I, p. 340, note 335.
[157] D. Beltrami, *op. cit.*, p. 72: 5·1 per cent in 1586; 7·4 per cent in 1624.
[158] F. Ruiz Martín, *Introduction . . .*, *op. cit.*

of undistinguished ancestry, on her husband's death makes her son withdraw from his father's business and tries, by removing him from the world of commerce, to have him reinstated to noble rank.[159] Such conduct did not invite ridicule; at the time it was considered perfectly reasonable. On the other hand much spiteful fun was poked at a whole series of unequal matches, shameful blots on illustrious escutcheons which were nevertheless restored to their former financial glory by them. A relative of Azzo Vesconte marries the daughter of a butcher, with a dowry of 12,000 ducats. The narrator does not wish to attend the marriage: 'I have seen the bride's father', he writes, 'in his white smock, as is the habit of our butchers, bleeding a calf, his arms red with blood up to the elbow. . . . For myself if I were to take such a woman to wife, it would seem to me ever afterwards that I stank of the butcher's shop. I do not think I would ever be able to hold up my head again.'[160] And alas, it is not an isolated case: a Marescotto has taken to wife the daughter of a gardener (at least he has the excuse of being in love with her); and Count Lodovico, one of the Borromeo counts, a great feudal family of the empire, has married the daughter of a baker; the Marquis of Saluzzo has wed a simple peasant girl. Love perhaps, money certainly multiplied such unequal matches. 'I have heard it several times said', continues the narrator, 'by Count Andrea Mandello di Caorsi, that if a woman had but a dowry of over 4000 ducats, one might marry her without hesitation, even if she were one of the women who sell their bodies behind the cathedral in Milan. Believe me, the man who has money and plenty of it is noble; the man who is poor is not.'[161]

Even in Milan, which in the early years of the century had a liberal reputation, *mésalliances* might be the subject of jokes, but there was a serious undercurrent which could lead to sudden tragedy, as it did in Ancona in 1566. A doctor,[162] the son of a humble tailor, was called in to treat one of the daughters of a noble young widow (she had seven children and 5000 crowns). When she announced that she wished to marry the doctor, Mastro Hercule, the reaction was dramatic: the doctor was arrested, only just escaping with his life after paying a fine of 200 ducats, and then only thanks to the intervention of a protector who came to the rescue from Ravenna with a band of horsemen. The family meanwhile was strengthened in its opposition to the widow's marrying 'un consorte di bassa conditione e figliolo di persone infime'. And since it was feared that the doctor, once at liberty, would elope with his beloved, one of her sons assassinated him in broad daylight.

In Spain, dramatic situations could always result from tragic preoccupations with honour and dishonour. On the other hand, read the *Tizón de la Nobleza española*[163] (which gave Maurice Barrès to reflect on

[159] Vol. II, novella XX, p. 47 ff.
[160] *Ibid.*, VIII, novella LX, p. 278–279.
[161] *Ibid.*, p. 280.
[162] Marciana, Ital., 6085, f° 42 ff, 1556.
[163] Attributed to Francisco Mendoza y Bobadilla, 1880 edition, *El Tizón de la Nobleza española*.

Toledan Spain). The pamphlet is incorrectly attributed to Cardinal de Mendoza. Without taking literally every allegation in this, or any of the other *libros verdes*,[164] we should not reject or refuse to believe the dramatic revelations they make, the crimes against *limpieza de la sangre* [purity of the blood][165] committed in the very highest society. Marriages to the daughters of rich *marranos* [converted Jews], the everyday drama of a match across class barriers, took on a tragic aspect in rank-conscious Spain. But they existed none the less.

Nobility for sale. For anyone determined to enter the aristocracy, there were rapid ways of doing so and they became more numerous as the century wore on. Patents of nobility and fiefs were easily acquired in Swabia for instance, where they brought in very little in the way of income; or in Naples, where they were usually a burden and often, when the purchaser was incapable of administering them, the occasion of spectacular bankruptcies. But vanity was always gratified by such acquisitions and that without delay. At Boisseron, near Lunel, which Thomas Platter[166] passed through on 3rd August, 1598, stood a chateau and a village, both belonging to 'M. Carsan, a simple citizen of Uzès; he had just given it to his son, who by this act became Baron of Boisseron for the title goes with the land.' Thousands of similar cases are recorded. In Provence, from the fifteenth century, the purchase of land constituted, for a bourgeoisie enriched 'by trade, shipping, the exercise of the law and diverse offices' at the same time 'a profitable and safe investment, the creation of a family patrimony, proof of success and finally the pretext for an elevation to the nobility which was often swiftly obtained'. In about 1560, the Guadagni, Italian merchants established at Lyons, possessed 'about twenty domains in Burgundy, the Lyonnais, Forez, Dauphiné and Languedoc'.[167] In October the same year, the lawyer François Grimaudet was declaring to the assembly of the Third Estate at Angers:[168] 'infinite is the number of false nobles, the fathers and forebears of whom have borne arms and proved their knightly valour in the shops of cornmerchants, vintners and drapers, at the mill and on the farms of noblemen.' 'Many men have thrust themselves among the nobility,' writes another contemporary, 'merchants aping and panting after the ancient marks of gentlemen.'[169]

Whose fault was it? Hardly a state in the sixteenth century, hardly a prince, did not trade patents of nobility for cash. In Sicily after 1600, marquisates, counties and principalities were put up for sale, at low prices

[164] This was the name given to clandestine manuscripts listing the *mésalliances* of great families, A. Domínguez Ortíz, *op. cit.*, p. 163, note 11.

[165] Albert A. Sicroff, *Les controverses des statuts de 'pureté de sang' en Espagne du XVe au XVIIe siècle*, 1960.

[166] *Journal of a Younger Brother*, the life of Thomas Platter as a medical student in Montpellier, transl. by Seàn Jennett, London, 1963, p. 173.

[167] Lucien Romier, *op. cit.*, I, p. 184.

[168] *Ibid.*, p. 185–186.

[169] *Ibid.*, p. 186, quoting Noël du Fail.

and to all comers, whereas until then only rarely had titles been conceded.[170] The age of false money was also the age of false titles. In Naples, according to a long Spanish report written in about 1600,[171] the number of title-holders, *titolati*, had increased out of all measure. And like all goods in plentiful supply, the titles had depreciated, if not those of counts at least those of marquises. There had even been 'created several dukes and princes who could have been done without'. So nobility was on the market everywhere, in Rome, Milan, in the empire, in the Franche-Comté,[172] in France, in Poland,[173] even in Transylvania, where there were any number of 'parchment gentlemen'.[174] In Portugal,[175] concessions had begun in the fifteenth century in imitation of the English. The first dukes appeared in 1415, the first marquis in 1451, the first baron in 1475. Even in Spain, the Crown, which soon began increasing the number of grandees, was quite undiscriminating at lower levels. As its need for money grew, it sold *hidalguías* and *hábitos* or gowns of the orders to anyone who could afford them, *indianos* or *peruleros* grown rich in the Indies trade, or even worse, upstart usurers.[176] Why not get used to it? If the Crown is short of money, the Conde de Orgaz advised the secretary Mateo Vázquez, in a letter addressed to him from Seville on 16th April, 1586, it will have to resign itself to selling *hidalguías*, even if it means breaking promises that no more will be put up for sale.[177] The Cortes complained in Castile of course,[178] but the Crown could not afford to listen. Sales continued and reached the point in 1573 where Philip II's government was obliged to promulgate ordinances concerning the *feudos nuevos*.[179]

It has been said that the fashion – soon to become a mania – for titles originated in Spain, that she exported it along with tight-fitting clothes for men, *bigotes*, perfumed gloves and the themes of her comedies. But the new fashion was more than mere vanity. The bourgeoisie knew what it was doing and no small measure of calculation entered into its purchases. It turned to land as a prime safe investment, thus reinforcing a social order based on aristocratic privilege. In short, as with states so with men, quarrels of precedence often masked extremely precise and down-to-earth pretensions. But at first sight, vanity alone struck the observer. In 1560, Nicot, the French ambassador at Lisbon,[180] noted of the Portuguese lords: 'the extravagance of the people here is so great in the

[170] L. Bianchini, *op. cit.*, I, p. 151.

[171] B.N., Paris, Esp. 127.

[172] Lucien Febvre, *Philippe II et la Franche-Comté*, 1911, p. 275.

[173] From the fifteenth century on, A. Tymienecki, 'Les nobles bourgeois en Grande Pologne au XVe siècle' 1400–1475' in *Miesiecznik Heraldyczny*, 1937.

[174] *Revue d'histoire comparée*, 1946, p. 245.

[175] F. de Almeida, *op. cit.*, III, p. 168 ff.

[176] G. Schnürer, *op. cit.*, p. 148.

[177] El conde de Orgaz to Mateo Vázquez, Seville, 16th April, 1586, B.M. Add. 28 368, f° 305.

[178] *Actas*, III, p. 368–369, petition XVI, 1571.

[179] Simancas E° 156.

[180] *Correspondance de Jean Nicot*, p. 117.

numbers of superfluous *criades* [attendants] that the squire wishes to live in the style of a duke and the duke in that of the king; which is the reason why they are always bankrupt.' The Bishop of Limoges makes a similar comment on Spain in 1561.[181] There is talk of ennobling five hundred 'rich seasoned warriors' on condition that they take up arms and serve for three months a year on the Spanish frontiers. And the bishop goes on to wonder at 'the vanity among the men of this land, who are puffed up with conceit so long as they are taken for nobles and can wear the habits and appearance of nobility'.

But in 1615 the same spectacle could be found in France. 'It is at present impossible,' writes Montchrestien[182] of his native land, 'to distinguish men by their appearance. The shopkeeper dresses like a gentleman. Moreover who can fail to see that this conformity of apparel introduces corruption to our old discipline? . . . Insolence will increase in the cities and tyranny in the fields. Men will grow effeminate with too many luxuries, and women, in their desire to parade themselves, will neglect both their chastity and their households,' an outburst worthy of a preacher, but it testifies to an age, in France at least, of discontent with the existing social order.

Hostility to the new nobles. As several quotations have already indicated, no one applauded the fortune of the new nobles. Any pretext would do to pick a quarrel with them; no opportunity to humiliate them was passed by. In 1559, at the Estates of Languedoc, the barons were ordered to send as representatives only 'des gentilshommes de race et de robe courte'[183] [i.e. gentlemen of high birth, entitled to wear a sword]. On the least occasion their enemies would vent their spleen: in France for instance throughout the Ancien Régime and even later. It was the same everywhere in the seventeenth century, for the class barrier was always being crossed and social censure was ever on the watch. An incident at Naples illustrates it well:[184] a financier of the city, extremely rich but of humble birth, Bartolomeo d'Aquino, sought in 1640, with the approval of the viceroy himself, to marry Anna Acquaviva, sister of the Duke of Conversano. The bride-to-be was carried off by a band of armed horsemen belonging to the nobility, determined to prevent by force that 'a mano di vile uomo la gentil giovina pervenisse'. She was taken to a convent at Benevento where she was doubly safe since Benevento was a papal possession. Similar incidents abound in chroniclers' accounts but they had no effect on a growing trend. With the exception of the Venetian aristocracy, which barricaded itself behind locked doors, all the aristocracies of Europe could be penetrated by outsiders, all received an infusion of new

[181] The Bishop of Limoges to the Queen, Madrid, 28th November, 1561, B.N., Paris, fr. 16103, f° 104, copy.

[182] *Traite d'économie politique*, 1615, ed. Th. Funck-Brentano, 1889, p. 60, quoted by François Simiand, *Les fluctuations économiques à longue période et la crise mondiale*, 1932, p. 7.

[183] Lucien Romier, *op. cit.*, I, p. 187.

[184] Rosario Vilari, *art. cit.*, in *Studi Storici*, 1963, p. 644 ff.

blood. In Rome, seat of the Church (and undoubtedly the most liberal of all western societies), the Roman nobility advanced along this path even more quickly than any other, through the regular promotion to the nobility and indeed to the high nobility, of the relations of each new Pope, who was not necessarily himself of illustrious extraction.[185] All aristocracies were moving with the times, shedding a certain amount of dead wood and accepting the new rich, who brought solid wealth to prop up the social edifice. It put the nobility at a great advantage. Instead of having to struggle against the Third Estate, it found the latter only too eager to join it, and to part with its wealth in order to do so.

This continuing trend could of course be hastened. In Rome, the papacy contributed to the rejuvenation of the nobility. In England, after the Rising of the Northern Earls failed in 1569, the old ruling families were to some extent replaced by a more recent aristocracy which was to govern England down to the present day, the Russells, Cavendishes and Cecils.[186] In France, two series of wars, the first terminated by the treaty of Cateau-Cambrésis (1st–3rd April, 1559) and the second by the peace of Vervins (2nd May, 1598) hastened the collapse of the old nobility and cleared the way towards social power for the new rich.[187] This was how one of Philip II's advisers viewed the situation of the French aristocracy in 1598: 'the greater number of these seigneurs being deprived of their rents and revenues (which they have signed away) have no money to maintain their estate and find themselves deeply in debt; almost all the nobility is in the same pass, so much so that on the one hand they cannot be employed without being given great reward and payment, which is quite impossible, and on the other it is to be feared that if they do not have some relief from the evils and ruin of war, they will be forced into some form of protest.'[188]

3. POVERTY AND BANDITRY

Of the life of the poor, history gives us few glimpses, but the little that is recorded had its own way of compelling the attention of the powers of the day, and through them reaches us. Uprisings, riots and disturbances, the alarming spread of 'vagabonds and vagrants', increasing attacks by bandits, the sounds of violence, although frequently muffled, all tell us something about the extraordinary rise of poverty towards the end of the sixteenth century, which was to become even more marked in the next.

The lowest point in this collective distress can probably be located in about 1650. So at any rate it appears from the unpublished journal of G. Baldinucci[189] to which I have already referred more than once. Poverty

[185] Jean Delumeau, *op. cit.*, I, p. 458 ff.

[186] Lytton Strachey, *Elizabeth and Essex*, 2nd ed., 1941, p. 9.

[187] Pierre Goubert, *Beauvais et le Beauvaisis de 1600 à 1730*, 1960, *passim* and p. 214 ff.

[188] Oration of M. Aldigala (actually by Guarnix), Public Record Office, 30/25, no. 168, f° 133 ff.

[189] Marciana, G. Baldinucci, *Giornale di Ricordi*, 10th April, 1650.

was such in Florence in April, 1650 that it was impossible to hear mass *in pace*, so much was one importuned during the service by wretched people, 'naked and covered with sores', 'ignudi et pieni di scabbia'. Prices in the city were terrifyingly high and 'the looms stand idle'; on the Monday of Carnival week, to crown the misfortunes of the townsfolk, a storm destroyed the olive trees as well as mulberry and other fruit-trees.

Unfinished revolutions. Pauperization and oppression by the rich and powerful went hand in hand. The result was not far to seek. And the underlying cause is also clear to see: the correlation between over-population and economic depression, an unrelieved double burden which dictated social conditions. In an article written in 1935, Americo Castro[190] suggested that Spain had never experienced a revolution, a rather rash statement perhaps in general terms, but not untrue of Spain at least in the sixteenth century, which witnessed far more aspirations to social revolution than successful revolutions. Only the brief explosion of the *Comuneros* can be counted an exception; even that has been[191] and will no doubt continue to be disputed.[192]

For by contrast with the north of Europe, where so-called wars of religion masked a series of social revolutions, the Mediterranean in the sixteenth century, although hot-blooded enough, for some reason never managed to bring off a successful revolution. It was not for want of trying. But the Mediterranean seems to have been the victim of some spell. Was it because the cities were overthrown so early that the strong state was irresistibly led to assume the function of keeper of the peace? The result in any case is plain to see. It is possible to imagine a huge book recording the succession of riots, disturbances, assassinations, reprisals and revolts in the Mediterranean, the story of a perpetual and multiple social tension. But there is no final cataclysm. A history of revolution in the Mediterranean might run to many pages, but the chapters would be disconnected and the basic conception of the book is after all questionable[193] – the very title is misleading.

For these disturbances broke out regularly, annually, daily even, like mere traffic accidents which no one any longer thought worth attention, neither principals nor victims, witnesses nor chroniclers, not even the states themselves. It is as if everyone had resigned himself to these endemic troubles, whether the banditry occurred in Catalonia or in Calabria or the Abruzzi. And for every recorded incident, ten or a hundred are unknown to us and some of them will never be known. Even the most important are so minor, so obscure and so difficult to interpret. What exactly was the

[190] 'Intento de rebellión social durante el siglo XVI' in *La Nación*, August, 1935.

[191] Gregorio Marañón, *Antonio Pérez*, Madrid, 1957, 2nd ed., English translation, abridged, by C. D. Ley, London, 1954.

[192] Jose Antonio Maravall, 'Las communidades de Castilla, una primera revolucion moderna' in *Revista de Occidente*, 19th October, 1963.

[193] See the doubts expressed by Pierre Vilar about the nature of Catalan banditry, *op. cit.*, I, p. 579 ff.

revolt of Terranova in Sicily in 1516?[194] What was the real significance of the so-called Protestant rebellion in Naples in 1561–1562, the occasion of a punitive expedition sent out by the Spanish authorities against the Vaudois of the Calabrian mountains, when several hundred men were left with their throats cut like animals?[195] Or even of the war of Corsica (1564–1569) from start to finish, and the war of Granada towards the end, both of them disintegrating into indecisive episodes, wars of poverty rather than of foreign or religious intervention? And how much is known about the troubles in Palermo in 1560,[196] or the 'Protestant' conspiracies of Mantua in 1569?[197] In 1571, the subjects of the Duke of Urbino rose up against the oppressive demands of their overlord, Francesco Maria, but this little-known episode remains hard to explain: the Duchy of Urbino was a land of mercenary soldiers, so whose was the hidden hand?[198] The internal crisis of Genoa in 1575–1576 is hardly clearer. In Provence in 1579, the *jacquerie* of insurgent peasants – the *Razas* – their capture of the chateau of Villeneuve and massacre of the local landlord, Claude de Villeneuve,[199] is swallowed up in the confused history of the Wars of Religion, like so many other social disorders, like the *jacquerie* of 1580 in the Dauphiné, Protestant but democratic too, inspired by the example of the Swiss Cantons and directed against the nobility. It is comparable to the revolutionary and destructive outbursts of the Protestants in Gascony, a few years earlier in the time of Monluc, or the troubles, many years later far away in the Cotentin (1587).[200] So too is the revolt of the Aragonese peasants of the county of Ribagorza, as a result of which they were eventually incorporated into the crown domains. The following year, the subjects of the Duke of Piombino, on the Tuscan coast, in turn rebelled.[201] The Calabrian Insurrection, the occasion of Campanella's arrest in 1599, was no more than a trivial incident magnified.[202] Equally numerous were the revolts throughout the Turkish Empire during the years 1590–1600, not to mention the endemic unrest among the Arabs and nomads of North Africa and Egypt, the disturbing uprisings of the 'Black Scribe' and his followers in Asia Minor, on which Christendom was to pin such wild hopes; the rioting among the Serbian peasants in 1594 in the Banat district of what is today Hungary, in 1595 in Bosnia and Herzegovina and

[194] Pino Branca, *op. cit.*, p. 243.

[195] *Archivio storico italiano*, Vol. IX, p. 193–195.

[196] Palmerini, B. Comunale Palermo, Oq. D. 84.

[197] Luciano Serrano, *Correspondencia diplomática entre España y la Santa Sede*, Madrid, III, 1914, p. 94, 29th June, 1569.

[198] J. de Zuñiga to the Duke of Alcalá, 15th March, 1571, Simancas E° 1059, f° 73. They were still in revolt in February, 1573: Silva to Philip II, Venice, 7th February, 1573, Simancas E° 1332, six thousand rebels with artillery; the Duke declared himself master of the situation, his state was *quieto*, 10th April, 1573.

[199] Jean Héritier, *Catherine de Médicis*, 1940, p. 565.

[200] A.N., K 1566, 8th January, 1587.

[201] Simancas E° 109, the governor of Piombino to Philip II, 6th October, 1598, R. Galluzzi, *op. cit.*, III, p. 28 ff.

[202] Léon Blanchet, *Campanella*, 1920, p. 33 ff.

again in 1597 in Herzegovina.[203] If to this very incomplete list we add the fantastic wealth of incidents relating to brigandage, what we shall have is not so much a book as a huge collection of reports.

So it seems at first sight certainly: but are these incidents and accidents, trivial happenings in themselves, the surface signs of a valid social history which, lacking any other means of expression, is forced to communicate in this confusing, clumsy, sometimes misleading language? Is this evidence meaningful at some deeper level? That is the historian's problem. To answer yes, as I intend to do, means being willing to see correlations, regular patterns and general trends where at first sight there appears only incoherence, anarchy, a series of unrelated happenings. It means accepting for instance that Naples, 'where there are robberies and crossed swords [every day] as soon as darkness falls', was the theatre of a perpetual social war, something going far beyond the limits of ordinary crime. It means accepting the same thing of Paris, already the scene of political (but also social) fanaticism in spring of 1588. The Venetian ambassador explains that 'the Duke of Guise entered the city with only a few men, that it was gradually discovered that the prince was totally without money, had large debts and that since he was unable to bear the cost of war in the field with large forces [who would of course have to be paid] he judged it safer to seize the good opportunity offered to him by this city which is inflamed with emotions from top to bottom'.[204] It was a social war and was therefore waged both cruelly and cheaply, using pre-existent conflicts and passions.

Cruelty: all the incidents mentioned above also bear the marks of relentless cruelty on both sides. The rural crimes which began around Venice with the coming of the century, were as pitiless as was the repression which followed. Inevitably the chroniclers, or those who inscribed these events on public records, were opposed to the trouble-makers who are regularly painted in the blackest colours. In the Crema region, during the winter of 1506–1507, a band broke into the house of a certain Caterina de Revoglara and 'per vim ingressi, fractibus foribus, ipsam violaverunt et cum ea rem contra naturam habuere',[205] relates the scribe of the Senate. In all these reports, the imperfectly identified enemy is guilty before being tried. These men are *ladri*, of a 'malignity and an iniquity ever increasing', they are blackguards, especially the peasants who, one day in winter 1507 failed to kill the patrician Leonardo Mauroceno in his country residence but took their revenge on his orchards.[206] As the years go by, the tone of the documents hardly alters at all. The 'cursed of God' looted property near Portogruaro in spring 1562[207] chopping down trees and vines. Was all fear of God vanished? Or all pity? An *avviso* dated the end of September, 1585 impassively reports: 'This year in Rome, we have seen more

[203] J. Cvijić, *op. cit.*, p. 131.

[204] B.N. Paris, ital., 1737, Giovanni Mocenigo to the Doge of Venice, Paris, 11th May, 1588, copy.

[205] A.d.S., Venice, Senato Terra, 16, f° 92, 29th January, 1506.

[206] *Ibid.*, 15, f° 188, 16th December, 1507.

[207] *Ibid.*, 37, Portogruaro, 9th March, 1562.

heads [of bandits] on the Ponte Sant'Angelo than melons on the market place.'[208] This sets the tone of a kind of journalism still in its infancy. When a celebrated brigand chief, the Sienese Alfonso Piccolomini, was betrayed and captured by the agents of the Grand Duke of Tuscany on 5th January, 1591,[209] then on 16th March hanged 'al faro solito del palagio del Podesta',[210] every opportunity was taken to demean his miserable end, by insinuating that the bandit 'si lasció vilmente far prigione',[211] without offering resistance. Such venom in the written word, such cruelty, both in the acts committed and the reprisals they drew – place the seal of authenticity on these scattered happenings, make them meaningful episodes in the endless subterranean revolution which marked the whole of the sixteenth century and then the whole of the seventeenth.

Class struggle? Can we call it a class struggle? I imagine that Boris Porchnev,[212] the admirable historian of the popular disturbances in France in the seventeenth century, would not hesitate to use the term. After all, we historians are forever using terms we have invented ourselves, *feudalism*, *bourgeoisie*, *capitalism*, without always keeping an exact record of the disparate realities such terms may cover in different centuries. It is a question of terminology. If by *class struggle* we simply mean the fratricidal vengeance, the lies, and one-sided justice which appear from this record, why not use it – as an expression it is quite as good as that of 'social tension' suggested by sociologists. But if the term implies, as I think it must, some degree of consciousness, while a class struggle may be apparent to the historian, he must remember that he is looking at this bygone age with twentieth century eyes: it did not seem nearly so obvious to the men of the sixteenth century who were certainly far from lucid on this point.

The examples noted by a single historian in the course of his personal research necessarily constitute an inadequate sample: I have found a few glimmerings of class-consciousness only during the first half of the sixteenth century. There are for instance the astonishing words of Bayard (or of the Loyal Serviteur) before besieged Padua in 1509;[213] or the report of *nobeli* taking up arms *contra li villani*[214] in October, 1525 in Friuli,

[208] J. Delumeau, *op. cit.*, II, p. 551.

[209] *Diario fiorentino di Agostino Lapini dal 252 al 1596*, published by G. O. Corazzini, 1900, p. 310: he arrived in Florence on 11th January.

[210] *Ibid.*, p. 314.

[211] *Ibid.*, p. 315, note.

[212] *Les soulèvements populaires en France, de 1623 à 1648*, 1963.

[213] *Le Loyal Serviteur, op. cit.* (1872 ed.), p. 179. Bayard did not agree with the Emperor Maximilian's request that the French *gendarmerie* be dismounted and put alongside the *Landsknechte* to force the breach: 'Does the Emperor think it a reasonable thing to put so many nobles in peril and danger along with foot-soldiers, one of whom may be a shoemaker, another a smith, another a baker, and rough mechanicals who do not hold their honour as dear as does a gentleman?' This passage is singled out for particular mention in Giuliano Procacci, 'Lotta di classe in Francia sotto l'Ancien Régime (1484–1559)' in *Società*, September, 1951, p. 14–15.

[214] M. Sanudo, *op. cit.*, XL, column 59, 9th October, 1525.

which had been infected by the German peasants' revolt; or, in December, 1528, the peasants near Aquila in the Abruzzi, who, dying of starvation and anger, tried to rise up against the 'traitors' and 'tyrants' to the cry of *Viva la povertà!* without apparently being sure (according to the suspect account of the chronicler) who exactly were the tyrants to be punished;[215] or in Lucca in 1531–1532, the insurrection known as that of the *Straccioni* (the ragged men) which is described as a 'battaglia di popolo contro la nobiltà'.[216] After that, there is nothing, at least as far as I know. So if this extremely scanty survey proves anything, it is that between the first and the second half of the century there was a decline in lucidity or if we are prepared to use the term, in revolutionary consciousness, without which there can be no significant revolution with a serious chance of success.

And indeed the earlier part of the century, the spring whose flowering was brought to a close by the harsh years 1540–1560, seems to have been particularly restless: the *Comuneros* in 1521, the *Germanias* of Valencia in 1525–1526; the uprisings in Florence, the crisis in Genoa in 1528; the peasant rebellion in Guyenne in 1548. Much, much later, in the seventeenth century, were to come the internal revolts of the Ottoman Empire, the French disturbances studied by Porchnev, the secession of Catalonia and Portugal, the great rebellion of Naples in 1647, the Messina uprising of 1674.[217] Between these two series of conspicuous social disturbances, the long half-century from 1550 to 1600 (or even 1620 to 1630) comes off rather poorly for revolutions, which hardly ever seem to reach boiling point and have to be divined like underground streams. In fact, and this makes the historian's task more difficult, these revolts and revolutions were not directed exclusively against the privileged orders but also against the state, the protector of the rich and pitiless collector of taxes, the state which was itself both social reality and social edifice. Indeed the state headed the list of popular dislikes. It is possible then – and this takes us back to the general remarks of Hans Delbrück[218] and the opinions of political historians – that the solidity of the state in the time of Philip II explains this muted tone, this discretion on the part of the people. The keeper of the peace held his own, although often beaten, often derided and inefficient – and more often still an accomplice in crime.

Against vagrants and vagabonds. In the cities there now began to appear that silent and persistent expression of poverty, the 'vagrants and vagabonds' as they were called by the Consuls and Magistrates of Marseilles, who decided in a council meeting of 2nd January, 1566,[219] to proceed to all quarters of the city and expel these undesirables. It would not have

[215] Bernardino Cirillo, *Annali della città dell'Aquila*, Rome, 1570, p. 124 v°.

[216] *Orazioni politiche*, ed. and chosen by Pietro Dazzi, 1866, oration of Giovanni Guidiccioni to the Republic of Lucca, p. 73 ff. This oration was apparently never delivered.

[217] Massimo Petrocchi, *La rivoluzione cittadina messinense del 1674*, 1954.

[218] *Weltgeschichte*, III, p. 251.

[219] A. Communales, Marseille, BB 41, f° 45.

seemed an inhuman decision to the spirit of the times. Towns were obliged to act as their own watchmen and for their own good periodically to drive out the poor: beggars, madmen, genuine or simulated cripples, idlers who gathered in public squares, taverns and round the doors of convents where soup was distributed. Expelled, they would return, or others would arrive in their place. The expulsions, gestures of rage, are a measure of the impotence of the respectable towns in the face of this constant invasion.

In Spain, vagrants cluttered the roads, stopping at every town: students breaking bounds and forsaking their tutors to join the swelling ranks of *picardía*, adventurers of every hue, beggars and cutpurses. They had their favourite towns and within them their headquarters: San Lucar de Barrameda, near Seville; the Slaughterhouse in Seville itself; the Puerta del Sol in Madrid. The *mendigos* formed a brotherhood, a state with its own *ferias* and sometimes met together in huge gatherings.[220] Along the roads to Madrid moved a steady procession of poor travellers,[221] civil servants without posts, captains without companies, humble folk in search of work, trudging behind a donkey with empty saddle bags, all faint with hunger and hoping that someone, in the capital, would settle their fate. Into Seville streamed the hungry crowd of emigrants to America, impoverished gentlemen hoping to restore their family fortunes, soldiers seeking adventure, young men of no property hoping to make good,[222] and along with them the dregs of Spanish society, branded thieves, bandits, tramps all hoping to find some lucrative activity overseas, debtors fleeing pressing creditors and husbands fleeing nagging wives.[223] To all of them, the Indies represented the promised land, 'the refuge and protection of all the *desperados* of Spain, the church of rebels and sanctuary of murderers': so says Cervantes at the beginning of one of his most delightful tales, *El celoso extremeño*, the story of one of these returned travellers from the Indies, now rich, who invests his money, buys a house, settles down to a respectable living and, alas, takes a wife.[224]

Habitual wayfarers too were the soldiers, old timers and new recruits, picaresque characters whose life of wandering and chance encounters might lead them to ruin in the fleshpots, the *casas de carne*, perhaps accompanied on the road by some submissive girl. One fine day they might follow the recruiter's drum and find themselves embarking, at Málaga or some other port, in a crowd where raw youths rubbed shoulders with old soldiers, deserters, murderers, priests and prostitutes, setting sail on the orders of the administration, for the fair land of Italy or the living prison

[220] Federico Rahola, *Economistas españoles de los siglos XVI y XVII*, Barcelona, 1885, p. 28–29, B.N., Paris, Oo 1017, in-16.

[221] M. Aleman, *Guzmán de Alfarache*, *op. cit.*, I, II, p. 254, poor travellers arriving at Madrid 'tras un asnillo cargado de buena dicha'; Madrid, a city where one went to seek one's fortune, Pedro de Medina, *op. cit.*, p. 204 ff.

[222] Fernand Braudel, 'Vers l'Amérique', in *Annales E.S.C.*, 1959, p. 733.

[223] Stefan Zweig, *Sternstunden der Menschheit*, Frankfurt, 1959, p. 10.

[224] *Novelas Ejemplares*, ed. Francisco Rodríguez Marín, 1943, II, p. 87 ff.

of the African *presidios*. Among the crowds passing up the gangway were a few honest men, such as Diego Suárez, who as a young man had travelled from master to master across the whole of Spain, from Oviedo to Cartagena, where he embarked for Oran in 1575: he was to stay there a third of a century, proof that it was a great deal easier to get into the African barracks than it was to get out of them.[225]

A universal danger, vagrancy in Spain was a threat to both town and countryside. In the north of the Peninsula, in Viscaya, vagabonds were constantly drawn to the *Señorio*. The authorities tried to take action, in 1579,[226] against impostors mingling with the crowds of pilgrims: 'if they are not old, or ill, or legitimately handicapped, let them be sent to prison . . . and let physicians and surgeons examine them.' But as usual, such decisions had little effect; the evil increased with the years and sterner counter-measures were adopted in vain. In Valencia, on 21st March, 1586 – and the measure applied to all towns and villages in the kingdom as well as to the city – the viceroy took firm action against idlers.[227] They were given three days to find a master, and otherwise would be expelled,[228] especially the *brivons* and *vagamundos* who spent every working day gaming in the public squares and refused to do any work, offering the likely excuse that they could find none. The viceroy also informed *jornales* of no fixed abode that if they were caught playing any manner of game, they would be taken into custody[229] and the same went for the so-called beggars and foreigners, and all people who were unwilling to work for their living. Unexpectedly, the severity of Valencia had some effect. Around Saragossa, says a Venetian letter of 24th July, 1586, 'one has to travel in blazing heat and in great peril from cut-throats who are in the countryside in great numbers, all because at Valencia they have issued an order expelling from the kingdom all vagrants after a certain time limit, with the threat of greater penalties; some have gone to Aragon and some to Catalonia. Another reason for travelling by day and with a strong bodyguard!'[230]

Here is proof, though hardly needed, that vagrants and bandits were brothers in hardship and might change places. Proof too that to expel the poor from one place was merely to drive them to another. Unless, that is, one followed the example of Seville. In October, 1581, all vagabonds rounded up in a police raid were forcibly embarked on Sotomayor's ships, bound for the Strait of Magellan. It was intended that on arrival they should work as labourers, *guastatori*, but four vessels foundered and one thousand men were drowned.[231]

All these incidents raise the question of the urban underworld: every

[225] See p. 854ff.
[226] *Gobierno de Viscaya*, II, p. 64–65, 4th August, 1579.
[227] B.N. Paris, esp. 60, f° 55 (printed).
[228] *Ibid.*, art. 60. [229] *Ibid.*, art. 61.
[230] A.d.S., Venice, Senato Dispacci Spagna, V° Gradenigo to the Doge, Saragossa, 24th July, 1586.
[231] *Ibid.*, Zane to the Doge, Madrid, 30th October, 1581.

town had its *Cour des Miracles*. From *Rinconete y Cortadillo*,[232] that hardly 'exemplary' novel, we can, with the aid of scholarly commentaries, discover a great deal about the Sevilian underworld: the prostitutes, merry widows, double-crossing alguazils, genuine wanderers, the true *pícaros* of literature, the *peruleros*, comic dupes – they are all there. And it was just the same elsewhere, in Madrid or Paris. Italy was completely overrun with delinquents, vagabonds and beggars, all characters destined for literary fame.[233] Forever being expelled, they always came back. Only the statutory authorities ever believed in the efficacy of official measures – which always took the same unvarying course.

At Palermo, in February, 1590, energetic measures were taken against the 'vagabonds, drunkards and spies of this kingdom'.[234] Two incorruptible proctors, with a salary of 200 crowns, were to divide the town between them. Their task was to pursue this lazy, good-for-nothing fraternity which spent workdays gambling, wallowing in all the vices, 'destroying their goods and what is more their souls'. Gamble they certainly did – but then so did everyone else. Anything served as a pretext for a wager and not only card-games: at Palermo there were wagers on the price of wheat, the sex of unborn children and, as everywhere else, on the number of cardinals to be named by the Holy Father. In a packet of commercial correspondence at Venice, I once found a lottery ticket left behind by accident. In their battle against the alliance of wine, gaming and idleness, the Palermo authorities ordered police raids on inns, *fonduks*, taverns and furnished rooms and inquiries into the suspicious persons who frequented them. Their place of origin, nationality and source of income were all to be investigated.

This game of cops and robbers, of respectable township versus vagrant, had no beginning and no end: it was a continuous spectacle, a 'structure' of the times. After a raid, things would be quiet for a while, then thefts, attacks on passers-by and murders would begin to increase once more. In April, 1585, in Venice,[235] the Council of Ten threatened to intervene. In July, 1606, there was once more too much crime in Naples and the authorities carried out a series of nocturnal raids on inns and hostelries, resulting in 400 arrests; many of the prisoners were soldiers from Flanders, *avvantaggiati*, that is 'over-paid'.[236] In March, 1590, there were expelled from Rome in a week, 'li vagabondi, zingari, sgherri e bravazzi', the vagabonds, gypsies, cut-throats and bravos.[237]

It would be interesting to tabulate all these expulsion orders to see whether they are interconnected like the dates of the trade fairs; where did

[232] *Novelas Ejemplares*, ed. Marín, 1948, I, p. 133 ff.

[233] Cf. the success in Italy of the book by Giacinto Nobili (real name Rafaele Frianoro), *Il vagabundo*, Venice, 1627.

[234] Simancas E° 1157, Palermo, 24th February, 1590.

[235] Marciana, Memorie politiche dall'anno 1578 al 1586, 23rd April, 1585.

[236] *Archivio storico italiano*, Vol. IX, p. 264.

[237] A.d.S., Mantua, A. Gonzaga, series E 1522, Aurelio Pomponazzi to the Duke, Rome, 17th March, 1590.

they come from, the vagabonds whom the towns hurriedly set back on the road, and where did they go? They arrived in Venice from as far away as Piedmont. In March, 1545, over 6000 'di molte natione' beleaguered the city. Some returned to their native villages, others took boat, 'lo resto per esser furfanti, giotti, sari, piemontesi et de altre terre et loci alieni sono stati mandati fuora della città': the rest were expelled since they were wastrels and idlers from Piedmont and other foreign cities and places.[238] Five years earlier, in 1540, a year of great hunger, the city had been invaded by a throng of unfortunate family men, 'assaissimi poveri capi di caxa' who arrived by boat with their wives and children and took up residence under the bridges, along the banks of the canals.[239]

Soon the problem of the poor had progressed beyond the narrow confines of the unsympathetic towns, reaching nation-wide and European dimensions. By the beginning of the seventeenth century, men like Montchrestien were aghast at the swelling of the ranks of the needy; if he and others in France like him were 'colonialist', it was in order to find a way to be rid of this silent and terrifying army of proletarians.[240] Throughout Europe, too densely populated for its resources and no longer riding a wave of economic growth, even in Turkey, the trend was towards the pauperization of considerable masses of people in desperate need of daily bread. This was the humanity which was about to plunge into the horrors of the Thirty Years' War, pitilessly drawn by Callot and only too accurately chronicled by Grimmelshausen.[241]

Brigands everywhere. But police records of city life pale beside the bloodstained history of banditry in the Mediterranean, banditry on land that is, the counterpart of piracy on sea, with which it had many affinities. Like piracy and just as much as piracy, it was a long established pattern of behaviour in the Mediterranean. Its origins are lost in the mists of time. From the time when the sea first harboured coherent societies, banditry appeared, never to be eliminated. Even today it is very much alive.[242] Let us not be tempted then as historians sometimes are who rarely look outside 'their' specialist century, to say that banditry made its first appearance in Corsica say in the fifteenth century or in Naples in the fourteenth. Nor should we unsuspectingly hail as new a phenomenon which we see breaking out on all sides in the sixteenth century, with what is certainly new, or renewed vigour. The instructions given by Queen Joanna of Naples on 1st August, 1343, to the Captain of Aquila,[243] to 'procedere

[238] A.d.S., Venice, Senato Terra 1, 26th March, 1545.

[239] *Ibid.*, Brera, 51, f° 312 v° 1540.

[240] *Traité d'économie politique*, ed. Funck-Bretano, 1889, p. 26.

[241] Just as in England, the Poor Law drove the poor off the streets, G. M. Trevelyan, *op. cit.*, p. 285.

[242] *Mercure de France*, 15th July, 1939, 'La Sicile aux temps préfascistes connut des jacqueries dignes du Moyen Age'.

[243] G. Buzzi, 'Documenti angioni relativi al comune di Aquila dal 1343 al 1344', in *Bollettino della Reggia Deputazione abruzzese di storia patria*, 1912, p. 40.

rigorosamente contro i malandrini' could have been issued by the Duke of Alcala or Cardinal Granvelle in the sixteenth century. In different ages, brigandage might change its name or the form it took, but *malandrini*, *masnadieri*, *ladri*, *fuorusciti banditi* (*masnadieri* were originally mercenary soldiers, *fuorusciti* and *banditi* outlaws) – they were all brigands – or as we should call them, misfits, rebels against society.

No region of the Mediterranean was free from the scourge. Catalonia, Calabria and Albania, all notorious regions in this respect, by no means had a monopoly of brigandage. It cropped up everywhere in various guises, political, social, economic, terrorist; at the gates of Alexandria in Egypt or of Damascus and Aleppo; in the countryside round Naples, where watch towers were built to warn of brigands[244] and in the Roman Campagna, where brush fires were sometimes ordered to smoke out bands of robbers who found abundant cover there; even in a state so apparently well-policed as Venice.[245] When the sultan's army marched along the Stambul road to Adrianople, Niš, Belgrade and on into Hungary, it left behind along the roadside scores of hanged brigands whom it had disturbed in their lairs.[246] There were brigands and brigands of course. Their presence on the main highway of the Turkish Empire, famed for its security, is sober evidence of the quality of public safety in the sixteenth century.

At the other end of the Mediterranean, in Spain, the scene is the same. I have already several times mentioned the notorious condition of the roads in Aragon and Catalonia. It is quite unthinkable, writes a Florentine in 1567, to take the post between Barcelona and Saragossa. Beyond Saragossa yes, but not between these two cities. His solution was to join a caravan of armed noblemen.[247] In one of his novels, Cervantes imagines his small band of heroes set upon by *bandoleros* near Barcelona. It was in fact an everyday occurrence. And yet through Barcelona ran one of the major routes of Imperial Spain, her main line of communication with the Mediterranean and Europe. It frequently happened that official couriers were robbed or even failed to get through, as was the case in June, 1565,[248] the verv year that the route from Madrid to Burgos, Spain's other main artery towards Europe and the Ocean, was closed as a result of plague.[249] Here stands revealed one of the thousand weak points in the unwieldy Spanish Empire. But there were as many *bandouliers* in Languedoc as there were *bandoleros* in Catalonia. All the farms in the Lower Rhône valley[250] were fortified houses like the peasant fortresses of Catalonia we have already mentioned. In Portugal,[251] Valencia, even in

[244] E. Albèri, *op. cit.*, II, V, p. 409.

[245] L. von Pastor, *op. cit.*, X, p. 59.

[246] See below, p. 1037.

[247] A.d.S., Florence, Mediceo 4898, Scipione Alfonso d'Appiano to the Prince, Barcelona, 24th January, 1567.

[248] *Ibid.*, Mediceo 4897, 1st June, 1565, f° 110 v° and 119. On other disruptions of communications see *La Méditerranée . . .*, 1st ed., p. 650, note 3.

[249] *Ibid.*

[250] P. George, *op. cit.*, p. 576.

[251] D. Peres, *Historia de Portugal*, V, p. 263.

Venice, throughout Italy and in every corner of the Ottoman Empire, robber bands, states in miniature with the great advantage of mobility – could pass unobtrusively from the Catalan Pyrenees to Granada, or from Albania to the Black Sea: these tiny forces irritated established states and in the end wore them down. Like the guerrilla forces of modern popular wars, they invariably had the people on their side.

Between 1550 and 1600, then, the Mediterranean was consumed by agile, cruel, everyday war, a war hardly noticed by traditional historians, who have left what they consider a secondary topic to essayists and novelists. Stendhal, in his writings on Italy, makes some pertinent observations on the subject.

Banditry and the state. Banditry was in the first place a revenge upon established states, the defenders of a political and even social order. 'Naturally a populace harassed by the Baglioni, the Malatesta, the Bentivoglio, the Medici etc., loved and respected their enemies. The cruelties of the petty tyrants who succeeded the first usurpers, the cruelties for instance of Cosimo, the first grand duke [of Tuscany],[252] who had the Republicans who had fled to Venice and even to Paris, slain, furnished recruits to these brigands.'[253] 'These brigands were the opposition to the vile governments which in Italy took the place of the medieval Republics.'[254] Thus Stendhal. As it happened, his judgement was based on the evidence before his own eyes, for banditry was still flourishing in Italy in his lifetime. 'Even in our own day,' he writes, 'everyone dreads unquestionably an encounter with brigands; but when they are caught and punished, everyone is sorry for them. The fact is that this people, so shrewd, so cynical, which laughs at all the publications issued under the official censure of its masters, finds its favourite reading in the little poems which narrate with ardour the lives of the most renowned brigands. The heroic element that it finds in these stories thrills the artistic vein that still survives in the lower orders . . . in their heart of heart, the people were for them, and the village girls preferred to all the rest the boy who once in his life had been obliged *andare alla macchia*, that is to say, to flee to the woods.'[255] In Sicily, the exploits of brigands were sung by the *urvi*, blind wandering minstrels who accompanied their songs on a 'kind of small, dusty violin'[256] and who were always surrounded by an enthusiastic crowd under the trees of the avenues. Spain, particularly Andalusia, notes Théophile Gautier,[257] 'has remained Arab on this count and bandits are easily elevated into heroes'. Yugoslav and Rumanian folklore too is full

[252] Not, as the text says, Florence. On the struggle against the state, the self-defence of a peasant 'civilization', see Carlo Levi's admirable *Christ stopped at Eboli* (Eng. transl. by Frenaye, New York, 1947).

[253] Stendhal, *Abbesse de Castro*, Pléiade ed., 1952, p. 564 (English version by Scott Moncrieff, *The Abbess of Castro*, 1926, p. 11–15).

[254] *Ibid.*, p. 561.

[255] *Ibid.*, p. 562.

[256] Lanza del Vasto, *La baronne de Carins*, 'Le Génie d'Oc', 1946, p. 196.

[257] *Op. cit.*, p. 320.

of stories of *haiduks* and outlaws. A form of vengeance upon the ruling class and its lopsided justice, banditry has been at all times, more or less everywhere, a righter of wrongs. Within living memory, a Calabrian brigand 'defended himself before the Court of Assizes by claiming to be a Robin Hood, robbing the rich to pay the poor. He told his rosary every day and village priests used to give him their blessing. In order to accomplish his private social justice, he had by the age of thirty already killed about thirty people.'[258]

Enemies of power, bandits usually lurked in zones where state authority was weak: in the mountains where troops could not attack them and where the government had no rights; often in frontier zones, along the Dalmatian highlands between Venice and Turkey; in the vast frontier region of Hungary, one of the major haunts of bandits in the sixteenth century;[259] in Catalonia, in the Pyrenees near the French border; in Messina, a frontier too in the sense that Messina, being a free city, was a refuge; around Beneventum, a papal enclave in the Kingdom of Naples, for by moving from one zone of jurisdiction to another, one could foil one's pursuers; between the papal states and Tuscany; between Milan and Venice; between Venice and the hereditary possessions of the archdukes. All these border lands offered ideal conditions (in a later age and without such bloodthirsty intentions, Voltaire made the same use of Ferney). It was possible of course for states to make mutual pacts but such agreements were usually short-lived. In 1561, the king of France proposed to Philip II,[260] that they take joint action against the Pyrenean *bandouliers*; a brief moment of common sense, but it had no effect. The agreements between Rome and Naples on the subject of Beneventum were equally unproductive. In 1570, Venice made a formal agreement with Naples[261] and in 1572 signed a pact with Milan, renewed in 1580,[262] at a time when the ravages by brigands were creating a climate of widespread insecurity in the Venetian State.[263] Each of the two governments was authorized to pursue criminals up to six miles beyond the frontier. In 1578, when the Marquis of Mondéjar tried to subdue the Calabrian *fuorusciti*, he alerted all his neighbours, including Malta and the Lipari Islands.[264] In 1585, Sixtus V did the same, on the eve of his campaign against brigands in the papal states.[265]

But such negotiations, which involved national sovereignty, were slow, painful and often conducted in bad faith: what ruler in Italy did not secretly rejoice at his neighbour's difficulties? Extraditions were extremely rare, apart from exchange of prisoners. When Marcantonio Colonna,

[258] Armando Zanetti, *L'Ennemi*, Geneva, 1939, p. 84.

[259] Busbecq, *The Turkish Letters* . . . , p. 12.

[260] Report by the Bishop of Limoges, 21st July, 1561, B.N., Paris, fr. 16 110, f° 12 v° and 13.

[261] Simancas E° 1058, f° 107, Notas de los capitulos . . . (1570–1571).

[262] Simancas E° 1338.

[263] Salazar to Philip II, Venice, 29th May, 1580, Simancas E° 1337.

[264] Simancas E° 1077.

[265] L. von Pastor, *op. cit.*, X, p. 59 ff.

viceroy of Sicily, obtained from Cosimo the extradition of a prominent bandit, Rizzo di Saponara, who had been active in Naples and Sicily for twenty-five years, it was in exchange for a knight of the house of Martelli, charged with conspiracy against the grand duke. And then the bandit was poisoned when he arrived at Palermo under escort of two galleys.

As a rule, each state policed its own territory. And that was no small matter. In places where banditry had taken deep root, it was a never ending task. In 1578, the Duke of Mondéjar, viceroy of Naples, decided to launch a fresh offensive against the *fuorusciti* of Calabria. Ever since his arrival he had been informed of their crimes: of the lands they had ravaged, the travellers they had killed, the blocked roads, desecrated churches, arson, kidnapping for ransom, not to mention 'many other grave, infamous and atrocious misdeeds'. The measures ordered by Cardinal Granvelle had been ineffective and indeed, writes the viceroy, 'the number of *fuorusciti* has increased, their crimes have multiplied and their power and insolence grown so great that in a thousand parts of the Kingdom, it is impossible to journey without great risk and peril'. Where better to deal with the problem than in Calabria, in the provinces of 'Calabria citra et ultra'? (Ten years earlier, the Abruzzi would have been the worst hit area.)

In Calabria, if our sources are to be relied on,[266] outlaws flourished, aided by circumstances and the terrain. Their crimes were more frequent and more horrifying than elsewhere and their audacity knew no bounds, so that 'one day, in broad daylight, they entered the town of Reggio, brought in cannon, attacked a house, breaking in and killing the occupants, while the governor of the town stood by helpless, for the townsmen refused to obey and come to his aid'. But tackling Calabria was no easy matter, as Mondéjar was to find to his cost. After the Reggio incident, the exact date of which is not given, the search conducted by the town governor, reinforced for the occasion by a judge commissioner, accomplished nothing except to increase the energy and activity of the brigands. Nor did the efforts of Count Briatico, appointed to the provisional government of the Calabrian provinces, meet with any more success. Repression merely exasperated the bandits. They forced castles, entered cities in broad daylight, dared 'to kill their enemies even in church, captured prisoners and held them to ransom'. Their atrocities caused widespread panic; 'they laid waste the land, killed the flocks of any who resisted them or pursued them on the order and instruction of the governors, the latter being too frightened to do it themselves'. In short 'they had lost the respect, fear and obedience which justice must inspire'. In conclusion of his report, from which these quotations are extracts, the viceroy indicates that a military expedition has been organized against them under the command of his son, Don Pedro de Mendoza, at the moment commander of the infantry of the kingdom. He has deferred the operation for as long as possible, so that the provinces may be spared the damage invariably caused by troops however disciplined. But if he waits any longer,

[266] See esp. Dollinger, *op. cit.*, p. 75, Rome, 5th June, 1547.

there is a risk that next spring he will have to muster an army to deal with the brigands, whereas a small expeditionary force will suffice at the present time.[267]

The expeditionary force[268] was composed of nine companies of Spaniards (who were to be quartered in villages suspected of aiding the *fuorusciti*) and three companies of light horse: three frigates were to operate off the coast, so the suspect provinces were to be completely surrounded. As usual, a price was set on the heads of brigands, 30 ducats for a member of a band and 200 for a chief. Don Pedro left Naples on 8th January and on 9th April, the viceroy announced that his mission had been successfully completed.[269] In February, the heads of seventeen brigands had been sent to Naples and nailed over the city gates, for the greater satisfaction – so it was claimed – of the populace.[270] And some had been taken prisoner to be handed over to justice by Don Pedro on his return to Naples.

Was it as great a success as the official and paternal pronouncements of the viceroy claimed? In fact life in Calabria, over-populated, unhappy, producing brigands as well as silk, carried on as before, or very little changed. Such a small expedition, operating during three winter months, cannot have been very productive. In 1580, a Venetian agent[271] reported that the entire kingdom was infested with bandits, that the highwaymen ruled the land in Apulia and especially in Calabria. The trouble was that by seeking to avoid these perilous roads, one risked falling into the hands of pirates who infested coastal waters as far north as the Roman shores of the Adriatic.

About twenty years later[272] the situation was even worse. Brigands were making incursions as far as the port of Naples and the authorities were coming to prefer negotiation or indirect measures to open conflict. Thus it was that the huge band of Angelo Ferro, which had been terrorizing the Terra di Lavoro, was sent to Flanders to serve under the Spanish flag. Bands were also played off one against the other: at Sessa, one group of bandits wiped out a neighbouring band. *Fuorusciti* were accepted into the army on condition that they helped the government in its fight against similar outlaws. And there was always the possibility of punitive billeting. Since brigands were always in touch with certain villages where they had relatives and a food supply, it would first be suggested to the relatives that they 'find a remedy', that is deliver up 'their' brigand. If they refused, a company of Spanish soldiers would be quartered in the village, preferably in the houses of the relatives and of the richer villagers. It was then up to the latter to persuade the former to find a 'remedy'. Since they

[267] Viceroy of Naples to Philip II, A.N., K, 3rd January, 1578, Simancas E° 107.

[268] *Sumario de las provisiones que el Visorey de Napoles ha mandado hacer*; undated, *ibid.*

[269] Viceroy of Naples to Philip II, 9th April, 1578 (received 29th May), Simancas E° 1077.

[270] The same to the same, 17th February, 1578, *ibid.*

[271] E. Albèri, *op. cit.*, II, V, p. 469.

[272] B.N., Paris, esp., 127 f° 65 v° to 67.

were rich and had influence, either the wanted man was handed over without further ado, or his departure from the kingdom was arranged. An indemnity was then charged for the crimes committed by the exile and other expenses; the troops would withdraw and things went back to normal. At least that was how it worked according to the optimistic report which describes such methods as an example of the art of governing in Naples.

In fact there was nothing new in such methods: they were tried and traditional. A Venetian document shows that they were used on Crete, where in 1555,[273] a free pardon was granted to every brigand (there were, it was said, about two hundred on the island at the time) who would kill such of his life-companions as had committed more murders than he had himself. Sixtus V too had resorted to such inducements during his 1585 campaign against the brigands of the Roman countryside. It was a way of breaking up bands from within. Pardons and rewards 'fanno il loro frutto', noted an agent of the Gonzaga at Rome.[274] Genoa meanwhile granted pardons to all bandits in Corsica (apart from a few notorious criminals) who enlisted in the army. This solution rid the troubled island of some of its disturbing elements, the pardoned men gave guarantees for good behaviour and became, for the moment, Genoa's protectors instead of her enemies.[275] A similar procedure was adopted by the Turks in Anatolia.[276]

But we should beware of exaggerating the effect of such practices, which are a measure of weakness as much as of cunning. In the end neither a firm hand nor guile, neither money nor the passionate determination of say Sixtus V, who threw himself into the struggle with peasant stubbornness, could get the better of this elusive enemy who sometimes had powerful backing.

Bandits and nobles. Sea-pirates were aided and abetted by powerful towns and cities. Pirates on land, bandits, received regular backing from nobles. Robber bands were often led, or more or less closely directed, by some genuine nobleman: Count Ottavio Avogadro for instance, whom a French correspondent in Venice reported as operating with his band against the Venetians in June, 1583.[277] 'Count Ottavio, Sire, continues to trouble the lords of Sanguene, to which place, since I last wrote to Your Majesty, he has twice returned and burned several houses in the Veronese.' The Venetians pursued him and arranged that Ferrara and Mantua, where he usually took refuge, close their gates to him.[278] But they never caught him: two years later he reappeared at the Court of Ferdinand of

[273] 28th March, 1555, V. Lamansky, *op. cit.*, p. 558.

[274] 22nd June, 1585, L. von Pastor, *op. cit.*, X, p. 59.

[275] A. Marcelli, 'Intorno al cosidetto mal governo genovese', *art. cit.*, p. 147, September, 1578 and October, 1586.

[276] See above, Vol. I, p. 99.

[277] H. de Maisse to the king, Venice, 20th June, 1583, A.E., Venice, 31, f° 51 and 51 v°.

[278] *Ibid.*, f° 56 v°, 11th July, 1583.

Tyrol.[279] Among the bands roaming the papal states, the rendezvous of thieves and murderers from both northern and southern Italy, not to mention indigenous bandits who were legion, another noble brigand and one of the most desperate in the time of Gregory XIII, was the Duke of Montemarciano, Alfonso Piccolomini, whom we have already met.[280] His life was saved *in extremis* by the Grand Duke of Tuscany, who had for some time been the unseen hand manipulating this curious character. Alfonso, escaping with his life, fled to France – where he found regular war very different from guerrilla war and not at all to the taste of a leader of *masnadieri*, so he was soon lending an ear to invitations and promises and before long he was back in Italy, in Tuscany this time, operating, imprudently and ungratefully, against the grand duke. From his lair in the mountains of Pistoia, far from the fortresses and garrisons, he was in a position to 'sollevare i popoli' to carry out 'delle scorrerie' all the more so since where he was that year, 1590, a very poor year for the wheat harvest, 'la miseria potea più facilmente indurre gli uomini a tentare di variar condizione', extraordinarily clairvoyant words.[281] With the arrival in the heart of Tuscany of this trouble-maker, the worst might be feared, particularly since he was in touch with the Spanish *presidios* and all the enemies of the *Casa Medici*. If he moved into Siena and its Maremma, there could be real trouble. However his bands of men, who were unskilled at strategic warfare, failed to capture any key-positions, fell back before the guards of Tuscany or Rome and the prince had the last word. On 16th March, 1591, Piccolomini was executed at Florence.[282] Thus ended a curious domestic war, watched with attention by the outside world, for the threads of the plot led back to some very distant hands, some to the Escorial, others to Lesdiguières in the Dauphiné.[283]

These are famous cases where international politics intrude. Humbler examples would suit our purpose better, but they are the least easy to discover. There were undoubtedly however connections between the Catalan nobility and brigandage in the Pyrenees, between the Neapolitan or Sicilian nobility[284] and banditry in southern Italy, between the *signori* and *signorotti* of the papal states and the brigands round Rome. The nobility played its part everywhere, whether politically or socially. Money was the key: the aristocracy was often economically unsound. Impoverished gentlemen, some ruined, others the younger sons of families of small fortune, were very often the leaders of this disguised social war, a never-ending, 'hydra-headed'[285] conflict. They were driven to make a living from adventure and rapine, driven (as La Noue said of France where the spectacle was the same) 'à la désespérade', to desperate means.[286] The same social mechanism can frequently be seen at work, in much later times.

[279] G. Schnürer, *op. cit.*, p. 102.
[280] R. Galluzzi, *op. cit.*, II, passim, and Vol. III, p. 44 ff. See above, p. 738.
[281] *Ibid.*, III, p. 44. [282] *Ibid.*, III, p. 53. [283] *Ibid.*, II, p. 443.
[284] L. Bianchini, *op. cit.*, I, p. 60.
[285] Marciana, 5837, Notizie del mondo, Naples, 5th March, 1587.
[286] Quoted by E. Fagniez, *L'Économie sociale de la France sous Henri IV*, 1897, p. 7.

In the eighteenth century, Turkey had a problem with a nobility too numerous to be rich, the *Krjalis* of Bulgaria.[287] In Brazil at the beginning of the nineteenth century, bandits were the henchmen, the *cabras*, of wealthy landowners, all threatened to some extent by modern development and obliged to defend themselves.[288]

But we should beware of making the picture too simple. A many-sided and complex affair, brigandage, while serving the interests of certain nobles, might be directed against others: as is illustrated by the exploits in Lombardy, of Alexio Bertholoti, 'a notorious bandit and rebel against the Marquis of Castellon'. On 17th August, 1597, with over 200 men, he scaled the walls of the castle of Solferino, and seized the marquis's mother and his son, a boy of thirteen. He took the prisoners to Castellon where he tried to make the old marquessa open the doors, in the hope of capturing the marquis himself, but in vain. When she refused he savagely wounded her, killed the child then went on a rampage of looting and 'barbaric cruelty' according to the report made by the governor of Milan.[289]

And then, banditry had other origins besides the crisis in noble fortunes: it issued from peasantry and populace alike. This was a groundswell – 'a flood tide'[290] as an eighteenth-century historian called it, which stirred up a variety of waters. As a political and social (though not religious)[291] reaction, it had both aristocratic and popular components (the 'mountain kings' in the Roman Campagna and around Naples were more often than not peasants and humble folk). It was a latent form of the *jacquerie*, or peasant revolt, the product of poverty and overpopulation; a revival of old traditions, often of brigandage in its 'pure' state of savage conflict between man and man. But we should beware of reducing it merely to this last aspect, the one most stressed by the rich and powerful who trembled for their possessions, their positions and their lives.

However, even allowing for exaggeration, can so much brutality be entirely passed over? It is true that human life was not valued highly in the sixteenth century. The career of Alonso de Contreras, as narrated by himself, one of the finest picaresque novels ever written, because it is true, contains at least ten murders. Benvenuto Cellini's escapades would have taken him to prison or to the scaffold in modern times. From these models, we may imagine the scruples of men whose profession was to kill. And then there are the remarks attributed to Charles V on the occasion of the siege of Metz, by Ambroise Paré, physician to the beseiged. 'The

[287] R. Busch-Zantner, *op. cit.*, p. 32.

[288] Gilberto Freyre, *Sobrados e mucambos*, p. 80 ff.

[289] Simancas E° 1283, the Constable of Castile to Philip II, Milan, 25th August, 1597.

[290] R. Galluzzi, *op. cit.*, II, p. 441.

[291] *Diario fiorentino di Agostino Lapini* . . ., 1591, p. 317: tells of the Pope elected by the *banditti* of the Forli region, one Giacomo Galli: they obeyed him as if he were the Holy Father himself. He was later hanged wearing a golden hat. The anecdote is as much political as religious. There are no other recorded details of the kind. The partisans of order certainly accused bandits of violating both divine and human laws, but that was no more than a manner of speaking.

Emperor asked what manner of people were dying, whether they were gentlemen and men of note; he was told that they were all poor soldiers. Then he said it was no harm if they died, comparing them to caterpillars, insects and grubs which eat buds and other fruits of the earth and that if they were men of property they would not be in his camp for six *livres* a month.'[292]

The increase in banditry. Towards the end of the sixteenth century at any rate, banditry was on the increase. Italy, a mosaic of states, was a brigands' paradise: driven out of one place, they would take refuge somewhere else, reappear further away, aided by the mutual support of these underground networks if sometimes weakened by the bitter feuds between them. Mecatti, the reliable eighteenth-century historian, tells us how Italy was overrun, in the 1590s, by these brigands who often used the convenient disguise of Guelph and Ghibeline for their own internal conflicts.[293] The unvarying background to all these actions was hunger. The raids by brigands descending from the mountains were often comparable to the *rezzous* of Morocco, where not so long ago, the men of the unconquered mountains would swoop down to the lowlands rich in wheat and livestock. So a strange climate reigned in Italy at the end of the century. Whole regions were tormented by hunger,[294] brigandage spread everywhere, from Sicily to the Alps, from the Tyrrhenian to the Adriatic, an endless succession of thefts, arson, murder and atrocities comparable to those committed by pirates at sea. It was universally deplored. Antonio Serra, the Neapolitan economist, says that in 1613 more crimes, thefts and murders were committed in Naples than in any other region of Italy.[295] The situation was almost equally grave in Sicily and the papal states, where during periods of interregnum, brigands swarmed,[296] the borders of Naples and the Romagnas affording them ample scope for their activities.[297] Their ranks were swelled by a bizarre collection of adventurers, professional assassins, peasants, nobles, priests breaking their ban, monks no longer willing to submit to the Holy See. Their origins can be deduced from the processions of galley slaves whom the papal states delivered up, for instance, to Gian Andrea Doria, records of whom have sometimes survived. In Sardinia and Corsica, there were considerable numbers of bandits. Tuscany's difficulties during the reign of Francesco (1574–1587) were their doing.[298] In 1592–1593, Italy was con-

[292] Ambroise Paré, *Œuvres complètes*, 1598, p. 1208.

[293] G. Mecatti, *op. cit.*, II, p. 780. Party strife in the papal States in the time of Pius V, L. von Pastor, *op. cit.*, p. XV.

[294] *Ibid.*, p. 782.

[295] *Op. cit.*, p. 145.

[296] 28th March, 1592, Simancas E° 1093, f° 12; G. Mecatti, *op. cit.*, II, p. 781 (1590).

[297] G. Mecatti, *op. cit.*, II, p. 784 (1591); Amedeo pellegrini, *Relazioni inedite di ambasciatori lucchesi alla corte di Roma, sec. XVI–XVII*, Rome, 1901: in 1591 increased outbreaks of banditry near the frontier between Rome and Naples, ineffective repressive measures.

[298] H. Wätjen, *op. cit.*, p. 35.

sidering ridding the country of these unwelcome persons by a general amnesty on condition they joined the Venetian army and left for Dalmatia.[299]

But Italy was not the only country grappling with this problem. In North Africa, where the roads had never been safe, prudent voyagers (the merchants of Constantine for instance) travelled in groups; the cleverest among them, says Haëdo, go accompanied by Marabouts.[300] Turkey was infested by brigands and robbers. In the seventeenth century, according to Tavernier,[301] '. . . *Turkie* . . . is full of Thieves, that keep in Troops together and way-lay the Merchants upon the Roads.' Even in the sixteenth century, in Moldavia and Wallachia, travelling merchants formed themselves into caravans for protection, and camped in groups, visible from a distance by their great fires.[302] The merchant with his bales of goods ran as much risk on land as he did in his roundships on the sea.

No country better illustrates the rise of brigandage during the latter years of the sixteenth century and the early part of the seventeenth, than Spain, which after the death of the old king in the Escorial, was to experience that extraordinary flowering of art and intelligence, luxury and *fiestas* known as the Golden Age, in that new and rapidly growing city, the Madrid of Velásquez and Lope de Vega, the divided city, where the rich were extravagantly rich and the poor miserably poor: beggars sleeping at street corners, rolled up in their cloaks, bodies over whom the nobles had to step to reach their palaces, *serenos* guarding the doors of the rich, the disquieting underworld of ruffians, soldiers, hungry valets, gamblers fingering their greasy cards, cunning prostitutes, guitar-playing students forgetting to return to their universities, a city of many colours fed by the rest of Spain and invaded every morning by the peasants of the nearby countryside, coming to town to sell their bread. During the major part of the reign of the Prudent King, apart from the serious episode of Granada and English raids on the sea-ports, Spain had known peace at home, a domestic tranquillity which foreigners often envied. As for bandits, they only existed in sizeable numbers in the eastern Pyrenees, where they were connected with the minor Catalan nobility and nearby France. But during the last years of the reign, banditry increased all over the Peninsula. There were brigands on the road to Badajoz, which was concerned in the campaign against Portugal in 1580.[303] At Valencia, violent quarrels, sometimes culminating in murder, broke out among the great noble families. The danger was so evident in 1577, that it was the subject of a new *real pragmática*.[304]

Here as elsewhere, what good were official remedies? Ineffective, they were forever being re-applied. In 1599, 1603, 1605,[305] new pragmatics

[299] G. Mecatti, *op. cit.*, II, p. 786–787. [300] *Op. cit.*, p. 32.
[301] J. B. Tavernier, *The Six Voyages*, 1677, First Book, p. 1.
[302] Angelescu, *op. cit.*, I, p. 331.
[303] 11th October, 1580, *CODOIN*, XXXIII, p. 136.
[304] B.N. Paris, Esp. 60, f° 112 v° to 123 v° (undated), 1577.
[305] *Ibid.*, f° 350 to 359.

were issued against the 'bandolers de les viles [of the kingdom] que van divagant per le present regne amb armes prohibides pertubant la quietud de aquell'. The problem of the 'malefactors'[306] was under consideration on the very eve of the massive expulsion of the Moriscos, in the years 1609–1614, which was to offer them so many golden opportunities.[307] Corruption among lesser functionaries was another factor: they were often hand in glove with the offenders.[308]

Slaves. One final feature characterizes these Mediterranean societies. In spite of their pretensions to modernity, they all retained slavery, in the West as in the East. It was the sign of a curious attachment to the past and also perhaps of a certain degree of wealth, for slaves were expensive, entailed responsibilities and competed with the poor and destitute, even at Istanbul. It was shortage of labour, combined with the high yield of the mines and sugar-plantations which was to allow the slavery of antiquity to be revived in the New World, that huge and traumatic step backwards. At all events, slavery, which had been virtually wiped out in northern Europe and France, survived in the Mediterranean west,[309] in Italy and in Spain, in the fairly persistent form of domestic slavery. The ordinances of the Consulate of Burgos in 1572 regulated the conditions for the insurance of black slaves for transportation to the New World, but also to Portugal and 'a estos Reynos', that is to Spain.[310] Guzmán, the picaresque hero, while in the service of a lady whose husband is in the Indies, falls in love, to his great shame and dishonour, with a white slave belonging to the lady, 'una esclava blança que yo, mucho tiempo, crei ser libre'.[311] At Valladolid, which towards 1555 was still capital of Castile, slaves waited at table in great houses, 'were well fed on kitchen leftovers', and often given their liberty in their masters' wills.[312] In 1539, in Roussillon, a Turk, wandering without a master, and what was more caught stealing, was arrested and sold as a slave to a notary.[313] In Italy, a series of acts indicates the survival of domestic slavery, in the Mezzogiorno in particular

[306] Malhechores de Valencia, 1607–1609, Simancas E° 2025.

[307] A.d.S., Venice, Senato Dispacci Spagna, P° Priuli to the Doge, Madrid, 21st October, 1610.

[308] Jacob van Klaveren, *op. cit.*, p. 54, note 16.

[309] Georg Friederici, *op. cit.*, I, p. 307. What would Sancho Panza do with black vassals? why, sell them of course. On domestic slavery, R. Livi, *La schiavitù domestica nei tempi di mezzo e nei moderni*, Padua, 1928. Charles Verlinden's masterly book, *L'esclavage dans l'Europe médiévale*, I, *Péninsule ibérique, France*, 1955, takes the topic up to the fifteenth century. On black household slaves in Granada, Luis de Cabrera, *op. cit.*, I, p. 279; on Gibraltar, *Saco* . . ., p. 51, 77, 79. Slavery began to disappear from France after the thirteenth century, Pardessus, *op. cit.*, V, p. 260; Gaston Zeller, *Les institutions de la France*, 1948, p. 22; slaves sold as merchandise in Sicily, Pardessus, *op. cit.*, V, p. 437.

[310] E. Garcia de Quevedo, *Ordenanzas del Consulado de Burgos*, 1905, p. 206, note.

[311] *Op. cit.*, II, III, VII, p. 450.

[312] Villalón, *Viaje de Turquia*, 1555, p. 78.

[313] Archives Départementales, Pyrénées-Orientales, B 376, 'por esser latru e sens amo'.

but elsewhere as well. In Naples, notarial documents[314] announce slave sales (at 35 ducats 'apiece' as a rule during the first half of the sixteenth century); similar references are found in the minute books of Venetian lawyers[315] and also in the correspondence of the Gonzaga, who bought little Negro boys[316] presumably for the amusement of their court. At Leghorn, the *portate* from time to time mention the arrival aboard ship of a few black slaves.[317]

This uninterrupted traffic fully reveals itself only on exceptional occasions. The capture of Tripoli in 1510[318] for instance, put so many slaves on the Sicilian market that they had to be sold off cheaply, 3 to 25 ducats each, and the galleys of the western powers soon had a full complement of oarsmen. In 1549 (and he was not the only one) the Grand Duke of Tuscany sent an agent to Segna to buy Turkish slaves or *morlachi*.[319] Slavery was a structural feature of Mediterranean society, so stern towards the poor, in spite of the widespread movement towards piety and religious charity which gathered strength at the end of the century. It was by no means exclusive to the Atlantic and the New World.

Possible conclusions. A slow, powerful and deep-seated movement seems gradually to have twisted and transformed the societies of the Mediterranean between 1550 and 1600. It was a lengthy and painful metamorphosis. The general and growing malaise was not translated into open insurrection; but it nevertheless modified the entire social landscape. This was an upheaval of unquestionably social character. All doubts on this score must vanish after reading Jean Delumeau's precise account of Rome and the Roman Campagna in the sixteenth century, which has the great merit of drawing on the many *avvisi* of the 'journalists', the *fogliottanti* of the Eternal City. This evidence amply confirms what we have already discovered. There can be no doubt that society was tending to polarize into, on the one hand, a rich and vigorous nobility reconstituted into powerful dynasties owning vast properties and, on the other, the great and growing mass of the poor and disinherited, 'caterpillars and grubs', human insects, alas too many. A deep fissure split open traditional society, opening up gulfs which nothing would ever bridge, not even, as I have already remarked, the astonishing move towards charity in the Catholic world at the end of the sixteenth century. In England, France, Italy, Spain, and Islam, society was undermined by this dramatic upheaval, the full horror of which was to be revealed in the seventeenth century. The creeping evil reached states as well as societies, societies as

[314] A.d.S., Naples, Notai, Sezione Giustizia, 51, f° 5 (one black slave, male, 36 ducats, 1520) f° 244 (one black slave, female, 35 ducats, 1521).

[315] Alberto Tenenti, 'Gli schiavi di Venezia alla fine del Cinquecento' in *Rivista storica italiana*, 1955.

[316] A.d.S., Mantua, E. Venezia, 16th June, 1499.

[317] A.d.S., Florence, Mediceo, 2080.

[318] Sanudo, *op. cit.*, XI, col. 468, Palermo, 3rd September, 1510.

[319] A.d.S., Florence, Mediceo 2077, f° 34, 9th April, 1549.

well as civilizations. This crisis coloured the lives of men. If the rich stooped to debauchery, mingling with the crowd they despised, it was because society stood on two banks facing each other: on one side the houses of nobles, over-populated with servants; on the other *picardía*, the world of the black market, theft, debauchery, adventure, but above all poverty, just as the purest, the most exalted religious passion coexisted with the most incredible baseness and brutality. Here, some will say, are the astonishing and marvellous contradictions of the Baroque. Not so: these were the contradictions, not of the Baroque but of the society which produced it and which it only imperfectly conceals. And at the heart of that society lay bitter despair.

Is the explanation for all this once more that the Mediterranean was proving unequal to its task of distributing goods and services, wealth and even pleasure in living? That the ancient glory and prosperity was running out as the peoples living on the shores of the sea exhausted their ultimate reserves, as is quite possible? Or, if we return yet again to our sources, will our repeated inquiries show the true reason to be that the whole world, the Mediterranean included, was sooner or later to enter the extraordinary depression of the seventeenth century? Can it be be possible that François Simiand was right?[320]

[320] François Simiand, 1873–1935, philosopher, sociologist, historian, the *maître à penser* of French historians, who along with Marcel Mauss, was one of the major influences in the social sciences in his country. Principal works: *Cours d'économie politique*, 3 vols., 1928–1930; *Le salaire, l'évolution sociale et la monnaie*, 3 vols., 1932; *Recherches anciennes et nouvelles sur le mouvement général des prix du XVIe au XIXe siècle*, 1932; *Les fluctuations économiques à longue période et la crise mondiale*, 1932.

CHAPTER VI

Civilizations

Of all the complex and contradictory faces of the Mediterranean world, its civilizations are the most perplexing. No sooner does the historian think he has isolated the particular quality of a civilization than it gives proof of the exact opposite. Civilizations may be fraternal and liberal, yet at the same time exclusive and unwelcoming; they receive visits and return them; they can be pacific yet militant; in many ways astonishingly stable, they are nevertheless constantly shifting and straying, their surface disturbed by a thousand eddies and whirpools, the tiny particles of their daily life subject to random 'Brownian movements'. Civilizations, like sand dunes, are firmly anchored to the hidden contours of the earth; grains of sand may come and go, blown into drifts or carried far away by the wind, but the dunes, the unmoving sum of innumerable movements, remain standing.

The brief remarks of Marcel Mauss[1] on this subject had the indubitable merit of drawing attention to the dynamics, the capacity for change possessed by civilizations. Possibly he did not strongly enough stress their equally remarkable qualities of permanence. It would be agreed, I think, that the changing and shifting element in a civilization is not necessarily the sum of its existence, nor even the best part. Once again we shall encounter that dialogue between structure and conjuncture, the moment in time and the long or very long term. Neither by brute force, whether conscious or unconscious, nor by casual force, allowing itself to be directed by the accidents of history, nor even by propaganda, however widely disseminated and eagerly received, can any one civilization make appreciable inroads on the territory of another. A pattern is there too firmly from the start. If North Africa 'betrayed' the West, it was not in March, 1962,[2] but long ago, in the eighth century[3] if not even before the birth of Christ, with the building of Carthage, city of the East.

I. MOBILITY AND STABILITY OF CIVILIZATIONS

Movement and immobility complement and explain one another. We may choose without risk either approach to the civilizations of the Mediterranean, even if we take what is at first sight their least significant aspect,

[1] 'Civilisation, éléments et formes' in *Première Semaine Internationale de Synthèse*, Paris, 1929, p. 81–108.

[2] 'Betrayal' in the sense used by Lucien Febvre in a lecture given at the University of Buenos Aires in October, 1937.

[3] Charles-André Julien, *Histoire de l'Afrique du Nord*, 1931, p. 20.

that miscellany of trivia and daily happenings which rises like a cloud of dust from any living civilization.

The significance of anecdote. These apparently trivial details[4] tell us more than any formal description about the life of Mediterranean man – a wandering life, tossed in every direction by the winds of fortune. A Ragusan sea-captain, on board ship somewhere in the Mediterranean in 1598, is taken into the confidence of a Genoese traveller from Santa Margherita, the legal executor of another Ragusan who has died rich at Potosí and entrusted him with the task of finding his heirs in Mezzo, the small island off Ragusa which was the nursery of all her sailors and ocean-going captains. And the impossible happens: inquiries are made and the rightful heirs are found.[5] We know less about another Ragusan, Blas Francisco Conich, who had also settled in Peru and in whom Venice was interested because he owned half-shares, towards the end of the year 1611,[6] in a boat, the *Santa Maria del Rosario e quatr'occhi* which the Signoria had seized in reprisals. Another incident concerns the proceedings for certification of death, again in Ragusa. The man presumed dead, a ship's captain, had been lost with the armada sent against England by Philip II in 1596. The court heard a letter written by the missing captain to his wife before setting sail, dated 15th October at Lisbon; it reads like a last farewell. 'We leave today for Ireland. God knows who will return.' He was one who did not.[7] At Genoa, on 8th June, 1601, the captain *Pompeus Vassalus quodnam Jacobi*, as he is described in the Latin text, is giving evidence before the grand *Magistrato del Riscatto dei Schiavi* about the presumed death of Matteo Forte of Portofino. 'Being in Egypt last year', he says, 'from the month of May to 11th September, I inquired there of several persons, if Matteo Forte, formerly a slave in the galleys of the *bailo* of Alexandria, were still living, for the said Matteo owns a house near my own which I wished to buy.' But 'all those who knew him told me that he had died several months before, and there were there some slaves from Rapallo who had known him'.[8]

A commonplace story too was that of a Genoese from Bogliasco, Gieronimo Campodimeglio, a captive in Algiers. He was about fifty years old in 1598 and no record appears of the date of his capture or of the name of his former master in Algiers who on his death, left his shop to his slave. Meanwhile he has been seen in the street 'vestito de Turcho'; someone testifies that he has married a Moslem woman. 'I think he has

[4] They can be found in any series of documents; see in particular in Ragusa the *Diversa di Cancelleria* and the *Diversa de Foris*; in Genoa the *Magistrato del Riscatto dei Schiavi*; in Venice the *Quarantia Criminale*.

[5] Ragusa Archives, *Diversa de Foris*, VII, f° 62–66, October, 1598.

[6] A.d.S., Venice, Dispacci Senato Spagna, P. Priuli to the Doge, Madrid, 3rd December, 1611.

[7] Ragusa Archives, *Diversa de Foris*, V, f°s 152 v° and 153, Lisbon, 15th October, 1596.

[8] 8th June, 1601, A.d.S., Genoa, Atti 659.

abjured his faith and will not want to come back':[9] a more frequent conclusion than one might expect. In fact Christians went over to the Turks and Islam by the thousand, as even a contemporary writer[10] testifies. Great civilizations – or strong governments – might resist and struggle, buying back their lost sheep; individuals were usually more accommodating. As time went by, statutes were drawn up against them. In the sixteenth century, they did not even forfeit their civil status. A renegade living in Tunis was able to dispose of his estate to his brother in Syracuse.[11] In 1568, Fray Luis de Sandoval[12] even proposed a massive programme of redemption by the Christian princes of the Mediterranean: a pardon would be offered to these lost souls and an end thus be put to the measureless harm they were inflicting on Christianity. In the meantime it was possible for any renegade to return without danger, as did the Venetian Gabriel Zucato, captured and enslaved by the conquerors of Cyprus in 1572, who on his return to Venice and 'la sanctissima fede', thirty-five years later, in 1607, applied for a post as *sansaro* or broker, a petition favourably received by the *Cinque Savii* in view of his poverty and his knowledge of Greek, Arabic and Turkish, 'which he can even write'; and yet 'si feci turco', he had abjured[13] his faith.

In any case, the two great Mediterranean civilizations, warring neighbours, were frequently drawn, by circumstances and chance encounters, into fraternization. During the unsuccessful attack on Gibraltar by the Algerines in 1540, eighty Christians were in the hands of the corsairs. After the alarm was over, there were as usual negotiations. A sort of armistice was agreed and bargaining began. The Algerine ships entered the port, their sailors went ashore and met old acquaintances, their former captives or masters, before going off to eat in the *bodegones*. Meanwhile the civilian population helped to transport casks of fresh water for the supply of the enemy fleet.[14] There was an exchange of goodwill and a familiarity comparable perhaps to fraternization between soldiers in the trenches. Between the two enemy religions, it would be unrealistic to imagine a watertight barrier. Men passed to and fro, indifferent to frontiers, states and creeds. They were more aware of the necessities of shipping and trade, the hazards of war and piracy, the opportunities for complicity or betrayal provided by circumstances. Hence many an adventurous life story, like that of Melek Jasa, a Ragusan converted to Islam, whom we find in India at the beginning of the sixteenth century, and in charge of the defence of Diu against the Portuguese – a post he was to hold for many

[9] *Ibid.*, Atti 659.

[10] H. Porsius, *Brève Histoire*, Arsenal 8° H 17458, quoted by Atkinson, *op. cit.*, p. 244.

[11] 25th September, 1595, P. Grandchamp, *op. cit.*, p. 73. Cf. the fictional biography of the father of Guzmán de Alfarache, M. Alemán, *op. cit.*, I, 1, 1, p. 8–9.

[12] A.d.S., Florence, Mediceo 5037, f° 124, Fray Luis de Sandoval to the Grand Duke of Tuscany, Seville, 1st August, 1568.

[13] A.d.S., Venice, *Cinque Savii*, Riposte, 142, f°s 9 v° and 10, 25th May, 1607.

[14] *Saco*, *op. cit.*, p. 101.

years.[15] Then there were the three Spaniards who were picked up in 1581 at Derbent, on the Caspian Sea, by the little English ship freighted every two or three years by the Muscovy Company, on its way back from Astrakhan. The Spaniards were no doubt renegades, deserters from the Turkish army, who had been taken prisoner at La Goletta seven years earlier.[16] Their remarkable story has a no less remarkable parallel. In 1586, the English ship *Hercules* brought back to Turkey twenty Turks whom Drake had liberated in the West Indies, a detail mentioned briefly in the account of the voyage of this sailing ship to the Levant.[17]

Similar adventures occur at the beginning of the seventeenth century. In 1608 there was still imprisoned in the Castle of S. Julião da Barra at Lisbon, a certain Francisco Julião who had received baptism and who had been in command of the Turkish galleys off Malindi when he was captured.[18] Then in 1611, the Persians captured from the ranks of the Turkish army led by the Grand Vizier Murād Pasha, three Frenchmen and a German (how they got there is anyone's guess – through Constantinople in any case) and a Greek from Cyprus; their lives were spared by their captors and they were eventually taken in by the Capuchin friars of Ispahan.[19]

One last example: towards the end of the seventeenth century, we hear of the travels of a Greek adventurer, Constantine Phaulkon, originally from Cephalonia, who described himself as the son of a Venetian noble and became a favourite of the king of Siam: 'everything passed through his hands'.[20]

How cultural exports travelled. Men travelled; so did cultural possessions, both the everyday and the unexpected, following expatriates round the world. Arriving with a group of travellers one year they might be carried further by others a year or a century later, ferried from place to place, left behind or taken up again, often by ignorant hands. The first printing-presses in the Danube lands, which were to produce Orthodox devotional works, were brought in at the beginning of the sixteenth century by Montenegrin pedlars from Venice or from Venetian possessions.[21] The Jews expelled from Spain in 1492 organized in Salonica and Constantinople markets for everything they found lacking there: they opened ironmongery shops,[22] introduced the first printing-presses with Latin, Greek or Hebrew characters (not until the eighteenth century[23] did the first

[15] V. L. Mazuranic, *art. cit.*, summary in Zontar, *art. cit.*, p. 369. For another complicated renegade story, 10th November, 1571, see L. Serrano, *op. cit.*, IV, p. 514–515.

[16] R. Hakluyt, *op. cit.*, II, p. 282.

[17] *Ibid.*, II, p. 282–285.

[18] *Boletim de Filmoteca Ultramarina Portuguesa*, no. 16, p. 692, Madrid, 8th May, 1608.

[19] B. M. Royal, 14 A XXIII, f° 14 v° ff.

[20] Abbé Prévost, *Histoire générale des voyages*, IX, p. 135–136, from the voyage of Tachard (1685).

[21] N. Iorga, *Ospiti Romeni . . .*, p. 24.

[22] Pierre Belon, *op. cit.*, p. 182.

[23] *Annuaire statistique du monde musulman*, 1923, p. 21. Moslem priests who earned their living by copying manuscripts, Belon, *op. cit.*, p. 194.

presses with Arabic characters appear); they began producing woollen cloth[24] and brocade and, it is said, built the first wheeled gun-carriages,[25] thus providing the army of Sulaimān the Magnificent with its field artillery, one of the secrets of its success. And it seems that they took as their model the gun-carriages of Charles VIII's artillery in Italy in 1494.[26]

But most cultural transfers were the work of anonymous carriers. So many were they, some moving quickly, others so slowly, that it is almost impossible to find one's way through this immense baggage hall in perpetual confusion. For every piece of cultural baggage recognized, a thousand are untraceable: identification labels are missing and sometimes the contents or their wrappings have vanished too. In the case of works of art, the corner-stones of Bayeux cathedral,[27] for instance, a Catalan painting found at Sinai,[28] iron work of Barcelona manufacture identified in Egypt, or the curious paintings of Italian or German inspiration which were executed in the sixteenth century in the monasteries of Mount Athos, the origin may be discovered without too much trouble. It is still possible when we are dealing with tangible entities like words, whether geographical names or everyday vocabulary: we may be reasonably if not absolutely sure of their provenance. But when it comes to ideas, attitudes, techniques, the margin of error is very wide. Is it possible to say that Spanish mysticism in the sixteenth century can be traced back to Moslem Sufism through such intermediaries as the eclectic genius of Ramón Lull?[29] Is it true that the use of rhyme in the West owed its origin to the Moslem poets of Spain?[30] That the *chansons de geste* (as is quite probable) borrowed from Islam? We should be equally wary of those who are too positive in their identification of cultural phenomena (for example the borrowings from Arabic by French troubadours)[31] and of those who by reaction deny all borrowings between civilization and civilization, when in the Mediterranean to live is to exchange – men, ideas, ways of life, beliefs – or habits of courtship.

Lucien Febvre[32] in an entertaining article, imagined how astonished

[24] See above, Vol. I, p. 436.

[25] J. W. Zinkeisen, *op. cit.*, III, p. 266.

[26] *Ibid.*, note 2.

[27] Paper given to *Académie des Inscriptions et Belles Lettres* by Marcel Aubert, 1943.

[28] Conyat Barthoux, *Une Peinture catalane du XVe siècle trouvée au monastère du Sinaï.*

[29] Or by different paths. For a comparison between Ibn 'Abbād and St. John of the Cross, see Asin Palacios, 'Un précurseur hispano-musulman de San Juan de la Cruz' in *Al Andalous*, 1933; J. Baruzi, *Problèmes d'histoire des religions*, p. 111 ff. But the problem remains: was this an influence, a case of parallelism or merely a coincidence? J. Berque, 'Un mystique . . .', *art. cit.*, p. 759, n. 1.

[30] Abbé Massieu, *Histoire de la Poësie françoise avec une défense de la Poësie*, 1739, notice in *Journal de Trévoux*, February–March, 1740, p. 277–314, 442–476. Viardot, *op. cit.*, II, p. 191–193. A. González Palencia, 'Precedentes islámicos de la leyenda de Garin', in *Al Andalous*, I, 1933. Maxime Rodinson, 'Dante et l'Islam d'après les travaux récents' in *Revue de l'histoire des religions*, October–December, 1951.

[31] J. Sauvaget, *Introduction*, p. 186; for the opposite view, see R. Konetzke, *op. cit.*, p. 64.

[32] 'Pataie et pomme de terre', in *Ann. d'hist. soc.*, January, 1940, II, p. 29 ff; the article is reproduced in *Pour une histoire à part entière*, Paris, 1962, p. 643–645.

Herodotus would be if he were to repeat his itinerary today, at the flora which we think of as typically Mediterranean: orange, lemon and mandarin trees imported from the Far East by the Arabs; cactus from America; eucalyptus trees from Australia (they have invaded the whole region from Portugal to Syria and airline pilots say they can recognize Crete by its eucalyptus forests); cypresses from Persia; the tomato, an immigrant perhaps from Peru; peppers from Guyana; maize from Mexico; rice, 'the blessing brought by the Arabs'; the peach-tree, 'a Chinese mountain-dweller who came to Iran', the bean, the potato, the Barbary fig-tree, tobacco – the list is neither complete nor closed. A veritable saga could be written about the migrations of the cotton-plant, native to Egypt[33] from which it emerged to sail the seas. It would be interesting too to have a study of the arrival in the sixteenth century of maize, an American plant which Ignacio de Asso wrongly supposed, in the eighteenth century, to have come from two sources, the New World and the East Indies, brought from the latter by Arabs in the twelfth century.[34] The coffee-shrub was growing in Egypt by 1550: coffee had arrived in the East towards the middle of the fifteenth century; certain African tribes ate grilled coffee beans. As a beverage it was known in Egypt and Syria from that time on. In Arabia in 1556, it was forbidden at Mecca, as being a drink of dervishes. It reached Constantinople in about 1550. The Venetians imported it to Italy in 1580; it appeared in England between 1640 and 1660; in France it was first seen at Marseilles in 1646, then at court in 1670.[35] As for tobacco, it arrived in Spain from Santo Domingo and through Portugal 'the exquisite herb nicotiana' reached France[36] in 1559, or possibly even in 1556, with Thevet. In 1561, Nicot sent some powdered tobacco from Lisbon to Catherine de Medici as a remedy for migraine.[37] The precious plant had soon crossed the Mediterranean; by 1605 it had reached India;[38] it was quite often forbidden in Moslem countries but in 1664, Tavernier saw the Sophy himself smoking a pipe.[39]

It is tempting to prolong the list: the plane tree of Asia Minor appeared in Italy in the sixteenth century;[40] rice cultivation spread, again in the sixteenth century, around Nice and along the Provençal coast;[41] the kind of lettuce known as 'roman' or cos was brought to France by a traveller called Rabelais; and it was Busbecq whose Turkish letters I have fre-

[33] A. Philippson, *op. cit.*, p. 110. [34] *Ibid.*, p. 110.

[35] J. Kulischer, *op. cit.*, II, p. 26–27. The abundance of literature on coffee defies description. For all that the chronology is still highly conjectural. A. Franklin, *Le Café, le thé, le chocolat*, 1893; William H. Ukers, *All about Coffee*, New York, 1922; Jean Leclant, 'Le café et les cafés de Paris (1644–1693)' in *Annales E.S.C.*, 1951; Günther Schiedlausky, *Tee, Kaffee, Schokolade, ihr Eintritt in die europäische Gesellschaft*, 1961.

[36] Olivier de Serres, *Le Théâtre d'Agriculture*, Lyon, 1675, p. 557, 783, 839; Otto Maull; *Géographie der Kulturlandschaft*, Berlin, Leipzig, 1932, p. 23.

[37] According to a scholar of the Charente, Robert Gaudin.

[38] Otto Maull, see above, note 36. [39] *Op. cit.*, I, p. 451.

[40] Rabelais to Jean du Bellay, Lyon, 31st August, 1534, 'unicam platanum vidimus ad speculum Dianae Aricinae'. [41] Quiqueran de Beaujeu, *op. cit.*, p. 329.

quently quoted, who brought back to Vienna from Adrianople the first lilacs which, with the aid of the wind, soon covered the Viennese countryside. But further identification can add nothing to what is already plain: the extent and immensity of the intermingling of Mediterranean cultures, all the more rich in consequences since in this zone of exchanges cultural groups were so numerous from the start. In one region they might remain distinctive, exchanging and borrowing from other groups from time to time. Elsewhere they merged to produce the extraordinary charivari suggestive of eastern ports as described by romantic poets: a rendezvous for every race, every religion, every kind of man, for everything in the way of hairstyles, fashions, foods and manners to be found in the Mediterranean.

Théophile Gautier, in his *Voyage à Constantinople*, gives a minute description, at every port of call, of the spectacle of this overwhelming carnival. At first one shares his enthusiasm, then one finds oneself skipping the inevitable description – because it is always the same; everywhere he finds the same Greeks, the same Armenians, Albanians, Levantines, Jews, Turks and Italians. And when one considers the still lively (though now less picturesque) sights of the harbour quarters in Genoa, Algiers, Marseilles, Barcelona and Alexandria, one has an impression of great cultural instability. But the historian can very easily go astray in seeking to unsnare the tangled threads. He might have thought the saraband for instance was an ancient traditional Spanish dance; then he discovers that it had only just appeared in Cervantes' time.[42] He had thought of tunny fishing perhaps as a specific activity of Genoese seamen, of the fishermen of Naples, Marseilles or Cape Corse; in fact it was already practised by the Arabs who passed on the skill in the tenth century.[43] He might almost in the end be tempted to say with Gabriel Audisio[44] that the essential Mediterranean race is that which inhabits its extravagant cosmopolitan ports: Venice, Algiers, Leghorn, Marseilles, Salonica, Alexandria, Barcelona, Constantinople, to name only the largest – a single race embracing all others. But this is patently absurd. The very multiplicity of colour indicates a diversity of elements: the variety proves there has been no amalgamation, that distinct elements remain and can be isolated and recognized as one moves away from the big centres where they are hopelessly tangled.

Cultural diffusion and resistance. The mark of a living civilization is that it is capable of exporting itself, of spreading its culture to distant places. It is impossible to imagine a true civilization which does not export its people, its ways of thinking and living. There was once an Arab civilization: its importance and its decline are well known. There was once a Greek civilization, of which the substance at least has been saved. In the sixteenth century there was a Latin civilization (I have reservations about

[42] *El celoso extremeño, Novelas ejemplares*, II, p. 25.
[43] R. Lacoste, *La Colonisation maritime en Algérie*, Paris, 1931, p. 113.
[44] *Jeunesse de la Méditerranée*, 1935, p. 10, 15, 20 . . ., *Le Sel de mer*, *op. cit.*, p. 118.

calling it Christian) the sturdiest of all the civilizations to struggle for command of the sea: expanding, it advanced through the Mediterranean region and beyond, into the depths of Europe, towards the Atlantic and the Iberian *Ultramar*. The spread of Latin culture, dating back over several centuries, was felt as much in shipbuilding – which the Italians, masters of the craft, taught the men of Portugal and even the Baltic – as it was in silk manufacture, of which the Italians were first the pupils and then the demonstrators; or in accounting techniques, which the Venetians, Genoese and Florentines, merchants from the old days, had perfected well before the northerners. And of course Latin culture was spread by the mighty reverberations of the Renaissance, the child of Italy and the Mediterranean whose path can be followed across Europe.

A living civilization must be able not only to give but to receive and to borrow. Borrowing is more difficult than it seems: it is not every man who can borrow wisely, and put an adopted implement to as good use as its original master. One of the great borrowings of Mediterranean civilization was undoubtedly the printing press, which German master-printers introduced to Italy, Spain, Portugal and as far away as Goa.

But a great civilization can also be recognized by its refusal to borrow, by its resistance to certain alignments, by its resolute selection among the foreign influences offered to it and which would no doubt be forced upon it if they were not met by vigilance or, more simply, by incompatibility of temper and appetite. Only Utopians (and there were several remarkable Utopians – Guillaume Postel for example – in the sixteenth century) could dream of uniting all religions in one: religion, precisely one of the most personal and inalienable elements in that complex of possessions, forces and systems which goes to make up every civilization. It is possible to combine parts of religions, to transfer a certain idea from one to another, even in some cases an item of dogma or ritual: unity is quite another matter.

Refusals to borrow: of these the sixteenth century provides one of the most striking examples. After the Hundred Years' War, Catholicism was assaulted by a tidal wave of religious passions. Under the pressure of these waters it broke apart like a tree splitting its bark. In the North the Reformation spread through Germany, Poland, Hungary, the Scandinavian countries, England and Scotland. Throughout the South spread what is traditionally called the Counter-Reformation and the civilization known as the Baroque.

There had of course always been a gulf between the North and the Mediterranean, two worlds bound together but quite distinct, each with its own horizons, its own heart and, religiously speaking, its own soul. For in the Mediterranean religious sentiment is expressed in a way which even today still shocks the northerner as it once shocked Montaigne in Italy,[45] or the ambassador Saint-Gouard[46] in Spain, and as at first it

[45] *Voyage en Italie*, *op. cit.*, p. 127–128.

[46] *Sources inédites . . . du Maroc*, *France*, I, p. 322, Saint-Gouard to Charles IX, Madrid, 14th April, 1572.

shocked the whole of western Europe when it was introduced by the Jesuits and the Capuchins, the poor man's Jesuits. Even in a region as profoundly Catholic as the Franche-Comté, the processions of penitents, the new devotional practices, the sensual, dramatic and, to the French mind, excessive element in southern piety, scandalized many serious, reflective and reasonable men.[47]

Protestantism did succeed in pushing into the Austrian Alps,[48] the Massif Central, the French Alps and the Pyrenees of Béarn. But in the end it failed, universally, to cross the frontiers of the Mediterranean states. After a few moments of hesitation and enthusiasm which make its final refusal even more noteworthy, Latin civilization said no to the Reformation 'from over the mountains'. If certain Lutheran and, later, Calvinist notions gained some converts in Spain and Italy, they had currency only among a few isolated individuals and small groups. Almost always these were men who had lived abroad for some time, clerics, students, booksellers, artisans and merchants who would bring back forbidden books concealed in their bales of merchandise; or else, as Marcel Bataillon has shown in *Érasme et l'Espagne*, men who plunged the roots of their faith into a soil personal to themselves, borrowing from no-one, the soil tilled in Spain by the Erasmians and in Italy by the Valdesians.

Was the lack of success of the Reformation south of the Pyrenees and the Alps a question of government, as has been frequently suggested, the results of well-managed repression? No one will underestimate the efficacy of systematic persecution carried out over a long period. The example of the Netherlands, very largely won back to Catholicism by the ruthlessness of the Duke of Alva and his successors, is sufficient in itself to guard us from such an error. But neither should one over-estimate the impact of 'heresies' whether Spanish or Italian; they cannot seriously be compared to the powerful movements in the North. To mention only one difference, Protestantism in the Mediterranean hardly touched the masses at all. It was a movement confined to an élite and frequently, in Spain, this was a Reformation accomplished within the Church. Neither the disciples of Erasmus in Spain, nor the little group of Valdesians in Naples sought a rupture with the Church, any more than the circle around Marguerite de Navarre in France.

If the Italian Reformation, as Emmanuel Rodocanachi has said, 'was not a true religious revolt'; if it remained 'humble, meditative, in no way aggressive towards the Papacy'; if it was opposed to violence,[49] it is because much more than a 'Reformation', it was a Christian revival. The word Reformation is inappropriate. Danger, or the semblance of danger, loomed only in Piedmont, with the Vaudois[50] (but how Italian was

[47] Noted by Lucien Febvre.

[48] G. Turba, *op. cit.*, I, 3, 12th January, 1562.

[49] *La Réforme en Italie*, p. 3.

[50] In 1561, Emmanuel Philibert had signed a truce with his Vaudois. '. . . e como dire uno interim', writes Borromeo, J. Susta, *op. cit.*, I, p. 97. Since 1552, the Vaudois had been connected with the Reformed Church at Basle, and the French Protestants

Piedmont?); in Ferrara, at the court of Renée of France; in Lucca, where the opplent aristocracy of silk-manufacturers welcomed the Reformation in 1525;[51] in Cremona, where a few assemblies gathered at about the same period;[52] in Venice, which welcomed northerners and where in about 1529 Franciscan or Augustinian friars founded small groups where artisans were fairly well represented.[53] Elsewhere in Italy, the Reformation was a matter of individuals; its history one of scandals such as that of the 'senoys' Ochino, formerly a celebrated Catholic preacher in Italy and now, writes de Selve who saw him arrive in England in 1547,[54] converted 'to the new opinions of the Germans'. Frequently it was carried by travelling preachers[55] who passed through, sowing the seed as they went, but the harvest was disappointing. These were solitaries, thinkers, men marked for unusual destinies, whether an obscure figure like the Umbrian Bartolomeo Bartoccio,[56] settled in Geneva as a merchant, who was arrested on one of his trips to Genoa, handed over to the Roman inquisition and burned on 25th May, 1569, or an illustrious victim like Giordano Bruno,[57] who was burned on the Campo dei Fiori in 1600.[58]

---

of Dauphiné and Provence. F. Hayward, *Histoire de la Maison de Savoie*, 1941, II, p. 34–35. The duke made fresh concessions to the Vaudois in 1565, Nobili to the Duke, Avignon, 7th November, 1565, Mediceo 4897, f° 152. Towards 1600 there were fresh disturbances, as foreign heretics, chiefly French, attacked Catholics and convents. The Carthusians asked permission in 1600 to move down from Montebenedetto to Banda . . . Fra Saverio Provana di Collegno, 'Notizie e documenti d'alcune certose del Piemonte', in *Miscellanea di Storia Italiana*, 1901, t. 37, third series, Vol. 2, *art. cit.*, p. 233.

[51] Arturo Pascal, 'Da Lucca a Ginevra', a remarkable article in *Riv. st. ital.*, 1932–1935, 1932, p. 150–152.

[52] Federico Chabod, *Per la storia religiosa dello stato di Milano*, Bologna, 1938, see the many references in the index, p. 292.

[53] A. Renaudet, *Machiavel*, p. 194.

[54] 23rd November, 1547, p. 258.

[55] *Archivio storico italiano*, IX, p. 27–29, about 1535; Alonso de la Cueva to Philip III, Venice, 17th October, 1609, A.N., K 1679.

[56] M. Rosi, *La Riforma religiosa in Liguria e l'eretico umbro Bartolomeo Bartoccio, Atti della Soc. Ligure di storia patria*, 1892, reviewed in *Bol. della Soc. umbra di storia patria I*, fasc. II, 1895, p. 436–437.

[57] On Giordano Bruno, see: Virgilio Salvestrini, *Bibliografia di Giordano Bruno*, 1581–1950, 2nd ed. (posthumous) by Luigi Firpo, Florence, 1958; as far as one can tell from sampling, this bibliography is exhaustive for the period covered. To bring it up to date, the following books published since 1950 should be noted: Paul-Henri Michel, *Giordano Bruno, philosophe et poète*, 1952 (extract from *Le Collège philosophique: Ordre, désordre, lumière*); A. Corsano, *Il Pensiero di Giordano Bruno nel suo svolgimento storico*, Florence, 1955; Nicola Badaloni, *La Filosofia di Giordano Bruno*, Florence, 1955; Adam Raffy, *Wenn Giordano Bruno ein Tagebuch geführt hätte*, Budapest, 1956; John Nelson, *Renaissance Theory of Love, the Context of Giordano Bruno's 'Eroici furori'*, New York, 1958; Augusto Guzzo, *Scritti di storia della filosofia, II, Giordano Bruno*, Turin, 1960; Paul-Henri Michel, *La Cosmologie de Giordano Bruno*, Paris, 1962.

[58] These were frequently ordinary judiciary actions, as in the case of the heretic Alonso Biandrato, who took refuge in Saluzzi under French protection and whom the Pope wanted handed over, Cardinal de Rambouillet to Catherine de Medicis, Rome, 9th December, 1568, B.N., Fr. 17.989, f°s 29 v° to 30 v°, copy.

Finally, let us beware of judging the Protestant threat in Italy by Catholic, Papal or Spanish anxiety, ever ready to magnify it. Such anxiety was so acute that during the summer of 1568, it was even feared that the French Huguenots were preparing to invade Italy where, it was said, they would find the whole Peninsula dangerously corrupted from within.[59] It is as if one were to judge the threat from Protestantism in Spain and the virtues or crimes of the Inquisition from the works of Gonzalo de Illescas, Paramo, Llorente, Castro or Thomas M'Crie.[60]

The Reformation in Spain (if Reformation there was at all) was concentrated in two cities: Seville and Valladolid. After the repression of 1557–1558 there were only isolated instances. Sometimes they were merely madman like Hernández Díaz to whom the shepherds of the Sierra Morena spoke of the Protestants of Seville; he retained enough of what they said to be arrested by the Toledo Inquisition in 1563[61] – a contented madman as it happened, who cheerfully boasted that in prison he ate more meat than at home. A few genuine Spanish Protestants travelled through Europe, hunted from refuge to refuge, like the famous Michael Servetus, or the dozen or so exiles who in 1578 'were studying the sect' in Geneva and who were denounced to the ambassador, Juan de Vargas Mexia, because they were said to be preparing either to go and preach in Spain or to send works of propaganda to the Indies.[62]

The Spanish authorities were indeed deeply disturbed by these erring souls and pursued them relentlessly. The Inquisition received popular support in its persecution of them. The trial *in absentia* of Michael Servetus was followed with passionate attention, for the honour of the nation was concerned.[63] The same sentiments inspired Alonso Díaz when, at Neuberg on the Danube, in 1546, he had a servant execute his own brother, Juan, a disgrace not only to the family but to all Spain.[64] It is as unrealistic therefore to talk of the Reformation in Spain as it is to generalize about the Reformation in Ragusa on the strength of the heretic of the city of St Blaise, Francisco Zacco, who in 1540 refused to believe in either heaven or hell, or of the 'Protestant tendencies' which, according to the writer of the latter part of Razzi's history of Ragusa, apparently manifested themselves in 1570.[65] Such injections of Protestantism apparently served chiefly as vaccinations.

One Italian historian, Delio Cantimori,[66] has suggested that the history

[59] Philip II to the Prince of Florence, Aranjuez, 2nd June, 1568, Sim. E° 1447; the Grand Commander of Castile to Philip II, Cartagena, 10th June, 1568, Sim. E° 150, f°s 18, 19; Don John of Austria to Philip II, Cartagena, 10th June, 1568, *ibid.*, f° 17.

[60] E. Schäfer, *op. cit.*, I, p. 134–136.

[61] *Ibid.*, I, p. 34–36.

[62] Relacion de cartas de J. de Vargas Mexia para S. M., 29th December, 1578, 21st January, 1579, A.N., K 1552, B 48, no. 15.

[63] Marcel Bataillon, 'Honneur et Inquisition, Michel Servet poursuivi par l'Inquisition espagnole' in *Bulletin Hispanique*, 1925, p. 5–17.

[64] R. Konetzke, *op. cit.*, p. 146; Marcel Bataillon, *Érasme et l'Espagne*, p. 551.

[65] *Op. cit.*, p. 258.

[66] 'Recenti studi intorno alla Riforma in Italia ed i Riformatori italiani all'estero, 1924–1934', in *Rivista storica italiana*, 1936, p. 83–110.

of the Reformation in Italy, which has always been approached from a biographical angle, would become much clearer when related, as it has been in France and Germany, to the social environment from which it sprang. Edgar Quinet[67] expressed similar thoughts long ago, but the problem will be tackled even more effectively if it is treated as a cultural one. Was not the refusal of Italy to accept the Reformation, like that of Spain, a refusal to *borrow* in the anthropological sense, one of the fundamental criteria of civilization? I do not mean to suggest that Italy was 'pagan' as so many superficial observers have claimed to discover, but that the sap which rose in the old trees of Catholicism in Italy produced the fruit and flowers of Italy; not of Germany. What became known as the Counter-Reformation was, if you like, the Italian Reformation. It has been remarked that the southern countries were less drawn than those of the North to the reading of the Old Testament[68] and thus escaped being submerged by that incredible wave of witchcraft which unfurled from Germany to the Alps and northern Spain towards the end of the sixteenth century.[69] Possibly because of an ancient substratum of polytheism, Mediterranean Christendom remained, even in its superstitions, attached to the cult of the saints. Is it pure coincidence that devotion to the saints and the Virgin increased in intensity just as outside assaults became more vigorous? To view it as some manœuvre by Rome or the Jesuits is idle. In Spain it was the Carmelites who propagated the cult of Saint Joseph. Everywhere the popular associations of the Rosary fostered and exalted the passionate cult of the Mother of Jesus: witness the Neapolitan heretic, Giovanni Micro, who in 1564 declared that he had lost his faith in many articles of religion including the saints and holy relics, but clung to his belief in the Virgin;[70] this at the very moment when Spain was bringing to perfection her own resplendent and militant saints, Saint George and Saint James.[71] Others were to follow: Saint Emiliano, Saint Sebastian and the peasant saint Isidore whose fame spread even to Catalonia.[72]

The refusal then was deliberate and categorical. It has been said of the Reformation that it 'burst upon the Platonic and Aristotelian theology of the Middle Ages as the barbarian Germans burst upon Greco-Roman civilization'.[73] All one can say is that what remained of the Roman Empire on the shores of the Latin sea put up more resistance in the sixteenth century than it had in the fifth.

[67] Edgar Quinet, *Les Révolutions d'Italie*, Brussels, 1853, p. 235 ff.

[68] Herbert Schoffler, *Abendland und Altes Testament*, 2nd ed., Frankfurt-am-Main, 1943.

[69] See the enormous literature on this subject, esp. G. Schnürer, *op. cit.*, p. 266.

[70] E. Rodocanachi, *op. cit.*, I, p. 24.

[71] Gilberto Freyre, *Casa Grande*, *op. cit.*, p. 298.

[72] See above, Vol. I, p. 163–164.

[73] Julius Schmidhauser, *Der Kampf um das geistige Reich*, 1933, quoted by Jean-Édouard Spenlé, *La Pensée allemande de Luther à Nietzsche*, 1934, p. 13, note 1.

Greek civilization: did it survive? Greek civilization itself was not dead in the sixteenth century. How do we know? Because it too was capable of equally categorical and no less dramatic 'refusals'. Although dying, or rather threatened with death in the fifteenth century, it had refused to unite with the Roman Church. In the sixteenth century the problem arose once more and the refusal was equally firm. Unfortunately we are hardly better informed about the Orthodox countries at this period than we are about Turkey. And yet a series of remarkable documents (found at Venice and published by Lamansky in his extraordinarily rich collection) is still waiting, after all these years, for some historian to unravel their significance. This collection of texts sheds light on the astonishing situation of the Greeks in the sixteenth century in the face of Roman Catholicism.[74]

In 1570, a Greek, a gentleman of Crete or Morea, addressed several long reports to Venice. Offering his services, he explains that the time has come for the Greek countries to rise up against the Turk. Such a rebellion can only be undertaken with the support of the Christian countries, notably Venice. But first of all Christendom must try to understand the Greeks. It has never understood them. How many stupid vexations have the Greek bishops had to bear! The Catholic clergy in Venetian possessions has always adopted a scornful attitude towards them, seeking to rescue them from their 'error' often by force, forbidding or imposing certain rituals, trying to ban the Greek language from the churches. Rather than submit to the Catholic religion, the Greeks would prefer to go over to the Turks and indeed that is what they have done. They have almost always been allies of the Turks against the Venetians or western corsairs. Why? Because on the whole the Turks were tolerant, never seeking to proselytize and never interfering with the practice of the Orthodox religion. So the Orthodox clergy has regularly been one of the most determined adversaries of Venice and westerners in general. Its members have intervened every time that rebellion against the rule of Constantinople was planned, to cool tempers and to explain that the survival of the Greek people depended on their continuing to cause no trouble.

If today the banner of revolt is ready to be raised, continues our informant, it is because since about 1570 a wave of intolerance has been sweeping through the Turkish possessions. Churches have been looted, monasteries burned, priests molested.[75] The time has come for Venice to act, but there is only one way to succeed: to make peace with the metropolitans, to give them assurances that the Catholic clergy will be instructed never in the future to trouble the Greek clergy. The writer offers himself to act as mediator provided that Venice is really prepared to keep her word, in which case he feels victory is assured.

One has only to read, again in Lamansky's collection, the documents relating to the many incidents provoked in Crete and Cyprus by over-zealous Venetian priests or monks, to see the reality of the grievances

[74] In particular the long report by Gregorio Malaxa, V. Lamansky, *op. cit.*, p. 083 ff.
[75] *Ibid.*, p. 087.

expressed by the Greek Church. One can understand the collusion and 'betrayal' with which the West reproached Cretans and other Greeks in the Ægean. There are other explanations too of course: a Greek seaman going ashore from a Turkish ship to visit his family would learn from them full details of the Venetian fleet which had just sailed past or the western pirate-vessel which had anchored there the day before; even if the Turkish ship was itself a corsair and the port a Venetian possession (as was often the case). But the fundamental reason was the hostility between two civilizations, the Latin and the Orthodox.

Survivals and cultural frontiers. In fact, through all the changes which have altered or overturned civilizations, there are areas of astonishing permanence. Men as individuals may be false to them but civilizations live on in their own way, anchored to a few fixed and seemingly unalterable points.

Thinking of the obstacle imposed by mountain ranges, J. Cvijić says that they are a barrier 'less to ethnic penetration than to movements resulting from human activity and cultural currents'.[76] Correctly interpreted and possibly modified, this observation seems to me to be true. To man as an individual, no feat of exploration, no odyssey is impossible. Nothing can stop him or the belongings, both material and spiritual, which he carries, as long as he travels alone and on his own account. Mass removal, by a group or a society, is more difficult. A civilization cannot simply transplant itself, bag and baggage. By crossing a frontier, the individual becomes a foreigner. He 'betrays' his own civilization by leaving it behind.

For at bottom, a civilization is attached to a distinct geographical area and this is itself one of the indispensable elements of its composition. Before it can manifest that identity which is expressed in its art and which Nietzsche saw as its major truth (possibly because, with others of his generation, he made the word a synonym for quality) a civilization exists fundamentally in a geographical area which has been structured by men and history. That is why there are cultural frontiers and cultural zones of amazing permanence: all the cross-fertilization in the world will not alter them.

So the Mediterranean is criss-crossed with cultural frontiers, both major and minor, scars which are never completely healed and each with a role to play. J. Cvijić has identified three cultural zones in the Balkans.[77] And who can fail to be aware, in Spain, of the sharp contrast between the parts north and south of the parallel running through Toledo, the divided city which is the true heart of Spain? To the north lies the barren landscape peopled by small, semi-independent peasants and noblemen exiled in their small provincial towns; to the south the exploited colony, the

[76] *La Péninsule balkanique*, p. 27.

[77] *Ibid.*, the Mediterranean or Italian zones, the Greek or Byzantine zone and the patriarchal zone. Cf. R. Busch-Zantner's not entirely relevant criticism, *op. cit.*, p. 38–39.

Spain of history, where reconquering Christians found not only a sophisticated agricultural system and huge organized estates but a mass of hard-working *fellahin*, an extraordinary legacy which they did not destroy.

But there are even greater divisions both within the Mediterranean region and on its borders. One vital frontier for the Mediterranean world remains the ancient border of the Roman Empire, the line running along the Rhine and the Danube which was to be the advance line of the Catholic revival of the sixteenth century: the new *limes* along which there soon appeared the cupolas and accolades of the Jesuit churches and colleges. The split between Rome and Reformation occurred exactly along this ancient divide. It was this, rather than enmity between states, which conferred its 'solemn'[78] character upon the frontier of the Rhine. France in the sixteenth century, caught between this Roman front line and the line of the Pyrenees, the latter broken into at the very edge by the Protestant advance – France, torn in two directions, was once more to suffer the consequences of her geographical position.

But the most extraordinary division in the Mediterranean was one which, like the maritime frontiers we have already mentioned, separated East from West: that immutable barrier which runs between Zagreb and Belgrade, rebounds from the Adriatic at Lesh (Alessio) at the mouth of the Drin and the angle of the Dalmatian and Albanian coastline,[79] then passes through the ancient cities of Naissus, Remesiana and Ratiara to the Danube.[80] The entire Dinaric bloc came under Latin influence, from the Pannonian plains and the wide upland valleys embraced by the western part of the empire,[81] to the coastal fringe and the islands which look towards Italy. 'The last family to speak a Latin dialect on the island of Veglia' (another island, we may note in passing) died out in the first decade of the twentieth century.[82] In Croatia there survives to this day, though mingled with many other legacies, a way of life which owes much to Italy,[83] – that is, to the Italy of long-ago.

An example of a secondary cultural frontier: Ifrīqiya. A less spectacular example, that of a cultural subdivision, deserves some attention. It must be remembered that the three great Mediterranean civilizations, Latin, Islamic and Greek, are in fact groupings of sub-cultures, juxtapositions of separate civilizations linked by a common destiny. In North Africa no cultural bloc is more clearly delimited than the old urban region of ancient Africa, the Arab Ifrīqiya, present-day Tunisia.

Nature laid its foundations. To the north and east, the Tunisian plains are bordered by the sea. To the south they merge with the Sahara prolonging its landscapes of wormwood and esparto-grass, and harbour the

[78] Madame de Staël's expression.
[79] A. Philippson, 'Das byzantinische Reich', *art. cit.*, p. 445.
[80] Konstantin Jireček, *Die Romanen in den Städten Dalmatiens*, 1902, p. 9.
[81] A. Philippson, see note 79 above.
[82] J. Cvijić, *op. cit.*, p. 89.
[83] H. Hochholzer, 'Bosnien u. Herzegovina', *art. cit.*, p. 57.

nomadic tribes, pastoral and undisciplined, whom the towns do their best to tame. To the west, lies its distinctive physical feature: above the hot dry plains of Tunisia rises a series of hostile and forbidding reliefs[84] – hills, high plateaux, ridges and finally mountains, leading into what used to be Numidia, the cold Constantine region[85] of today, reminding the traveller perhaps of central Sicily, mountainous Andalusia or the Sardinian interior.

The mountain barrier between Tunisia and the central Maghreb roughly follows a line running from Cape Takouch through Oued el Kebir, Oued Cherif, Aïn Beida, Tafrent and Guentia. Charles Monchicourt has described the contrast between the regions lying to either side of this frontier: in the west, the storks, the ash and elm trees, the roofs of heavy brown tiles under a sky of mountain storms; in the east the terraced roofs, the white domes of the *qoubbas* declare the kinship between the cities of Tunisia and the cities of the East, Cairo or Beirut. 'Kairouan (Quayrawān) is a single vast white cube . . . the antithesis of Constantine', the latter still in many ways a large *chaouia* village with undistinguished rustic houses.[86] What history tells us is that Ifrīqiya, both in ancient and modern times, made this line its boundary, its western frontier, shaped by the obstacles which have sometimes prevented, sometimes grudgingly allowed but always hindered the imperialist ventures of the smiling and tempting plain.[87]

This dense and rustic region formed a screen to the west for the sophisticated civilization of Tunisia. The Constantine merchant[88] who travelled down into Tunisia in the sixteenth century, found, in this vision of white terraced houses and sunbaked cities, a rich land, looking to the East, trading regularly with Alexandria and Constantinople: a land of ordered civilization, where the Arabic tongue was predominant both in town and countryside.

During the same period, the central Maghreb, as far as Tlemcen (a city both of Morocco and the Sahara) was amazingly uncivilized. Algiers was to grow up in a country as yet without a leavening of culture, a virgin land, peopled by camel-drovers, shepherds and goatherds. The Levant by contrast had ancient traditions. The King of Tunis, Mawlāy Ḥasan, one of the last of the Hafsids, who, after being dethroned and blinded by his son, took refuge in Sicily and Naples in 1540, left everyone who met him with the impression of a prince of great distinction, a lover of beauty, a connoisseur of perfumes and philosophy: an 'Averroist' as Bandello,[89]

[84] A. E. Mitard, 'Considérations sur la subdivision morphologique de l'Algérie orientale' in *3e Congrès de la Fédération des Sociétés Savantes de l'Afrique du Nord*, p. 561–570. [85] On the Constantine region, R. Brunschvig, *op. cit.*, I, p. 290 ff.

[86] The contrast between rustic roofs and terraced houses also exists in southern Spain behind Almeria and the Alpujarras. What is the explanation? Julio Caro Baroja, *Los Moriscos del Reino de Granada*, Madrid, 1957.

[87] *Revue africaine*, 1938, p. 56–57.

[88] Leo Africanus, *op. cit.*, French ed., 1830, II, p. 11.

[89] M. Bandello, *op. cit.*, IX, p. 48.

his contemporary, describes him. A philosopher prince: one such would never have been found in the central Maghreb, even in Algiers, a city of upstarts and uncouth adventurers. Undoubtedly the horror displayed by Tunis when the Turks held the city for a while in 1534, once more in 1569 and permanently after 1574, was the revulsion felt by an ancient city, pious and highly organized, at a barbarian invasion.

What else can one conclude but that the fundamental reality of any civilization must be its geographical cradle. Geography dictates its vegetational growth and lays down often impassable frontiers. Civilizations are regions, zones not merely as anthropologists understand them when they talk about the zone of the two-headed axe or the feathered arrow; they are areas which both confine man and undergo constant change through his efforts. The example of Tunisia is after all fundamentally the opposition of a lowland complex to a highland complex of very different character.

The slow pace of change and transfer. The force of resistance of civilizations anchored to the soil explains the exceptional slowness of certain movements. Civilizations are transformed only over very long periods of time, by imperceptible processes, for all their apparent changeability. Light travels to them as it were from distant stars, relayed with sometimes unbelievable delays on the way: from China to the Mediterranean, from the Mediterranean to China, or from India and Persia to the inland sea.

Who knows how long it took for Indian numerals – the so-called Arabic numerals – to travel from their native land to the western Mediterranean through Syria and the outposts of the Arab world, North Africa or Spain?[90] Who knows how long it took, even after their arrival, for them to displace Roman numerals, which were considered more difficult to falsify? In 1299, the *Arte di Calimala* forbade their use in Florence; even in 1520, the 'new figures' were forbidden in Fribourg; they only came into use in Antwerp towards the end of the sixteenth century.[91] The apologue travelled far from its home in India or Persia, before it was first taken up by the Greek and Latin fabulists who were La Fontaine's sources, and then went on to enjoy perennial popularity in Atlantic Mauritania. Who can tell how many centuries passed before bells, originally from China, became the voice of Christian music and were placed on top of churches?[92] Some sources say it cannot have been before bell-towers were built in the West, in imitation of Asia Minor. The journey of paper took almost as long. Invented in China in 105 A.D., where it was made from vegetable matter,[93] the secret of its manufacture is said to have been disclosed at Samarkand in 751 by Chinese prisoners of war. The Arabs replaced the vegetable matter by rags, and cloth paper seems to have begun its career

[90] Lucien Febvre, *La Religion de Rabelais*, 1942, 2nd ed., 1947, p. 423.
[91] J. Kulischer, *op. cit.*, II, p. 297.
[92] Gen. Brémond, *op. cit.*, p. 339.
[93] Friedrich C. A. J. Hirth, *Chinesische Studien*, Munich, Vol. I, 1890, p. 266.

in Baghdad in 794.[94] From there it slowly travelled to the rest of the Moslem world. In the eleventh century its presence is recorded in Arabia[95] and Spain, but the first paperworks at Xativa (now San Felipe in Valencia) do not date from earlier than the twelfth century.[96] In the eleventh it was used in Greece[97] and in about 1350 it replaced parchment in the West.[98]

I have already remarked that according to G. I. Bratianu,[99] the sudden changes in men's costume in France towards 1340, the substitution for the flowing robes of the crusaders of the short, tight-fitting doublet, worn with clinging hose and pointed shoes, a new fashion introduced from Catalonia along with the goatee beard and moustache of the Trecento, in fact came from much further away; the Catalans had imported these styles from the eastern Mediterranean which in turn had received them from Bulgaria or even Siberia; while feminine costumes, notably the pointed headdress whose immediate source was the Lusignan court in Cyprus, actually dated far back in space and time to the China of the T'ang dynasty.

Time: unimaginable ages passed before these journeys were safely completed and before the innovations could take root in their new environment. The old tree-trunks of civilization by contrast remained astonishingly sturdy and resistant. Émile-Félix Gautier's argument, contested by other experts,[100] that in North Africa and Spain, Islam rediscovered the foundations of the ancient Punic civilization which had prepared the way for the Moslem invasion, is, to my mind, well within the boundaries of the possible. For there are ancient survivals, ancient cultural revivals everywhere in and around the Mediterranean region. The teachings of the religious centres of early Christianity, Alexandria and Antioch, survived in the sixteenth century in the Christianity of Abyssinia and the Nestorians. At Gafsa in North Africa, Latin was still spoken in the twelfth century according to al-Idrīsī. It was only in 1159, under the persecution of 'Abd al-Mu'min, that the last indigenous Christian communities disappeared from North Africa.[101] In 1159: that is four or five centuries after the Moslem conquest. But in the same North Africa, Ibn Khaldūn was still remarking upon 'idolaters' as late as the fourteenth century.[102] And the anthropological study conducted by Jean Servier in Kabylia in 1962, in the Soummam valley and elsewhere, also stresses, a thousand years later, the late arrival of Islam, which came 'not with the warrior horsemen of 'Okba but almost two hundred years later, in the ninth century, with the Shiite Fatimids who settled in Bougie, an Islam

[94] For different dates, see G. Marçais, *Histoire générale de Glotz*, *Moyen Age*, Vol. III, 1944, p. 365.

[95] *Chimie et industrie*, August, 1940.

[96] Berthold Bretholz, *Latein. Palaeographie*, Munich, 1912, 3rd ed., 1926, p. 16.

[97] See note 95. [98] *Ibid.*

[99] *Études byzantines*, 1938, p. 269 ff.

[100] Ch. André Julien, *Histoire de l'Afrique du Nord*, 1st ed., p. 320–327.

[101] Robert Brunschwig, *op. cit.*, I, p. 105.

[102] Gen. Brémond, *op. cit.*, p. 372, note 1.

given spiritual reinforcement by Iran, enriched by initiatory currents and which was inevitably brought into contact with the mystical symbolism of popular tradition'.[103] This book, moreover, which is very much concerned with the *twentieth* century, rooted in concrete reality, opens up a huge vista of popular tradition, of a fundamental religion surviving century after century and still alive today. There are no priests: every head of family, 'every mistress of the household' has 'the power to accomplish the rites . . . which strengthen on earth the human group in their charge'.[104] It is characterized by worship of the dead, of patron saints: 'When Saint Augustine said "Is not Africa strewn with the bodies of Holy Martyrs?" he was even then acknowledging the existence of these white tombs, the silent guardians of mountain passes who were later to become the recognized saints of Islam in the Maghreb.'[105]

So from the observatory of civilization we can and should look far out into the night of history and even beyond. As a historian of the sixteenth century, I consider relevant to my own studies for example the recently founded journal of protohistory, *Chthonia*,[106] which is concerned amongst other things with the ancient Mediterranean, Alpine and Nordic substrata and is informative on such topics as archaic revivals of the cult of the dead. What we call civilization is the distant and far distant past clinging to life, determined to impose itself and exerting as much influence over the habitat and agricultural practices of men as the all-important questions of relief, soil, water supply and climate – as a remarkable book on Provence, written by a geographer, confirms. According to Robert Livet whose particular interest is a form of 'geographical genetics' in which history is of fundamental importance, the distinctive hill habitation of Provence – so inadequately accounted for by conventional explanations, notably the theory of the defensive site – is unquestionably attached to what he has termed 'a civilization of the rocks', the foundations and traditions of which go back to 'the old Mediterranean civilizations which preceded Roman settlement'. It was eclipsed during Roman times, to be revived later and was very much alive at the beginning of the sixteenth century when so many upheavals affected the population of Provence.[107] It is a subject which leads us far away from the sixteenth century but not from its realities.

What will our conclusion be? A negative one, clearly: we shall not allow ourselves to repeat the often-voiced opinion that 'civilizations are mortal'. Mortal perhaps are their ephemeral blooms, the intricate and short-lived creations of an age, their economic triumphs and their social trials, in the short term. But their foundations remain. They are not indestructible, but they are many times more solid than one might imagine. They have

[103] Jean Servier, *op. cit.*, p. 17.
[104] *Op. cit.*, p. 21.
[105] *Op. cit.*, p. 20.
[106] First number July, 1963, Editorial Herder, Barcelona.
[107] *Op. cit.*, p. 221.

withstood a thousand supposed deaths, their massive bulk unmoved by the monotonous pounding of the centuries.

2. OVERLAPPING CIVILIZATIONS

If we now wish to turn from such vast perspectives to the history of civilization in the shorter term, a history of rapid but significant change, of more human and less universal proportions, we can do no better than consider the violent conflicts between neighbouring civilizations, one triumphant (or believing itself to be), the other subjugated (and dreaming of liberation). There is no lack of such conflicts in the Mediterranean in the sixteenth century. Islam, through its agents the Turks, captured the Christian strongholds of the Balkans. In the West, Spain under the Catholic Kings captured, with Granada, the last outpost of Islam on the Peninsula. What did either of the conquerors make of their victories?

In the East, the Turks were often to hold the Balkans with only a handful of men, much as the English used to hold India. In the West, the Spaniards mercilessly crushed their Moslem subjects. In so doing both powers were obeying the imperatives of their separate civilizations more than one might think: Christendom was over-populated, Islam short of men.

The Turks in the eastern Balkan plains. Turkish Islam took over, in the Balkans, the area directly or indirectly occupied by Byzantine civilization. To the north, it conquered the banks of the Danube, to the west, on one side it touched the borders of Latin civilization, at Ragusa, in Dalmatia, or around Zagreb in Croatia; on the other it stretched over the vast mountain cantons of 'patriarchal civilization' to borrow J. Cvijić's expression. Spread over a wide area, destined to survive for half a millennium, what richer and more thoroughly *colonial* experience can be imagined?

Unfortunately too little is still known about the Turkish past. Balkan historians and geographers have not always judged it with truly scientific detachment – even a Cvijić. And if the general histories of Zinkeisen and Hammer are out of date, Iorga is confused. Not only that, but the centuries of Turkish domination have been as it were wilfully neglected, in much the same way that the Moslem era was, until recently, neglected in Spain. This is an additional handicap when one seeks to investigate a world (for it is a whole world) unfamiliar even at the best of times.

And yet it is impossible to underestimate the impact of the Turkish experience, to ignore its contribution to the Balkan bloc upon which it left a profound cultural imprint.[108] The breath of Asia which is so manifest throughout the Balkans is a legacy of Turkish Islam. It disseminated the gifts it had itself received from the distant East. Through it, town and countryside alike were permeated with oriental culture. A significant detail is that at Ragusa, that island of Catholicism (and a particularly

[108] R. Busch-Zantner, *op. cit.*, *passim*, esp. p. 22; Otto Maull, *Südeuropa*, p. 391.

ardent form of Catholicism), women were still veiled and sequestered in the sixteenth century, and a husband did not see his wife before the marriage ceremony.[109] Western voyagers who came ashore on this narrow promontory were immediately aware that here began another world. But the Turks who set foot in the Balkans must have had exactly the same impression.

In fact, when studying the influence of the Turks in the Balkans it is essential to distinguish between two zones: the first embraces both the Slavonic west, bristling with mountain ranges, and the Greek south, again a mountainous region. Effective Turkish occupation of these areas was limited. It has been suggested, not entirely unreasonably, that in the Dinaric regions, even the Moslems were not of pure Turkish origin, but Slavs converted to Islam.[110] In short, the entire western half of the Balkans does not appear to have been profoundly reshaped by Islamic civilization: hardly surprising after all, since this was a mountain bloc, largely impervious to 'civilizing invasions' of whatever origin. As for its conversion to the Islamic faith, we have already noted the doubtful character of mountain 'conversions'.[111]

To the east by contrast, on the broad plains of Thrace, Rumelia and Bulgaria, the Turks settled many of their own people and heavily overlaid these regions with their own civilization. These lands, from the Danube to the Aegean, are as open to the north as they are to the south and from both directions, invaders poured in time after time. If the Turkish achievement can be judged – as success or failure – it is in these lands which it made as nearly as possible its own.

There it found what had become a homogeneous population, although composed of groups of diverse ethnic origin: the latest invaders, Bulgarians, Petchenegs and Kumans, from the North, had joined the Thracians, Slavs, Greeks, Aromani and Armenians, already long established there. But all these elements had virtually fused, the conversion to the Orthodox faith often being the decisive step towards assimilation of new arrivals. Again this is hardly astonishing, in a zone where Byzantium too had exercised strong influence. The whole of this region consists of large plains bound by the servitude of all large plains. Only in the Rhodope range and the chain of the Balkans, especially the Sredna Gora, have isolated pockets of independent mountain life survived, that of the *Balkandjis*, who are to this day a wandering nomadic race, one of the most original in Bulgaria.[112]

To these mountain refuges, in the Kyustendil and Kratovo regions, fled certain Bulgarian lords, during the Turkish conquest, in order to escape the enslavement suffered by those who remained in the plains (and in the end, on payment of tribute succeeded in retaining their former

[109] Davity, *op. cit.*, 1617, p. 637.
[110] J. Cvijić, *op. cit.*, p. 105; H. Hochholzer, *art. cit.*
[111] See above, Vol. I, p. 35 ff.
[112] J. Cvijić, *op. cit.*, p. 121.

privileges).[113] They formed a minute exception to the general rule: for the Turkish conquest placed the lowlands in serfdom, destroyed whatever might have preserved a Bulgarian community, killing or deporting to Asia the nobility, burning down the churches and almost immediately placing on the back of this peasant nation the heavy yoke of the *Sipāhilik* system, its warrior-nobility, soon to be transformed into a landed aristocracy. The latter lived comfortably off the labour of that patient and hard-working beast of burden, the Bulgarian peasant, the typical plainsman, slave of the rich, harshly disciplined, ground down by work, thinking only of his next meal, as his compatriots describe Baja Ganje, the peasant of Bulgarian folklore. Aleko Konstantinov paints him as coarse, 'brutal to the core'. 'The Bulgarians', he says, 'eat voraciously and are utterly preoccupied with the food they are absorbing. They would not interrupt their meal if three hundred dogs were killing each other all round them. Sweat stands out on their brows ready to fall into their plates.'[114] In 1917, a war correspondent penned a hardly more flattering portrait: 'They make excellent soldiers, disciplined, very brave without being foolhardy, obstinate without being enthusiastic. Theirs is the only army that has no marching songs. The men march forward, dogged, silent, uncomplaining, indifferent, cruel without violence and victorious without joy; they never sing. From their build and deportment, one has an immediate impression of obtuseness, of insensitivity and clumsiness. They look like unfinished human beings; as if they had not been created individually but as it were mass-produced in battalions. Slow of understanding, they are hardworking, persevering, eager for gain and very thrifty.'[115]

One could easily multiply these uncomplimentary descriptions, find even more unfair portraits, by seeking the comments of the people of the western mountains concerning the lowland peasant. The westerners are scathing about him, with the scorn of the free warrior for the clumsy peasant in his coarsely woven garments, feet firmly planted on the ground, accustomed from birth to teamwork, a man to whom individuality, fantasy and a taste for freedom had always been denied. To the north the Rumanian plains would have been enslaved in the same way had they not been protected from the Turks by distance and kept alert by the raids of nomadic Tartars; above all, the native dough was leavened by immigration from the vast Carpathian and Transylvanian mountains.

In the Bulgarian countryside, the Turkish conquest did not even have to force the peasants to the yoke. They were already bowed down and willing to obey – and to carry on working, for work they certainly did. Sixteenth and seventeenth century travellers describe rural Bulgaria as a rich land.[116] Paolo Giorgiu, in 1595, called it the granary of Turkey.[117]

[113] I. Sakazov, *op. cit.*, p. 192.
[114] *Baja Ganje*, p. 42, quoted by J. Cvijić, *op. cit.*, p. 481.
[115] Quoted by J. Cvijić, *op. cit.*, p. 481.
[116] I. Sakazov, *op. cit.*, p. 197.
[117] Antoine Juchereau de Saint-Denis, *Histoire de l'Empire ottoman, depuis 1792 jusqu'en 1844*, 4 vols., 1844, I, p. 30.

But the ravages of bandits, crueller here than anywhere else, the tolls exacted by the nobles and by the state, and not indeed the inactivity, but the poverty of the peasant with his rudimentary tools (he was still using the small wooden hand-plough, the *rolo*), left wide stretches uncultivated between the crops. Only on the largest estates were team-drawn ploughs in use. Such lands might be devoted either to extensive stockfarming or to the cultivation of soft and hard wheat. Rice, introduced by the Turks in the fifteenth century, was successfully grown in the regions of Philippopolis [Plovdiv] and Tatar Pazardzhik, and to some extent in the canton of Caribrod. Bulgarian production in the sixteenth century has been estimated at about 3000 tons. Sesame, introduced to the Maritsa valley and cotton in the regions of Adrianople, Kyustendil and in Macedonia, round Serres [Serrai], were Turkish imports of the sixteenth century.[118] This variety of crops was complemented by a little wine of indifferent quality, market gardening round the towns,[119] hemp, roses and the orchards near Üskub [Skopje]. Finally two other new crops, tobacco and maize, were about to make their first appearance, the precise date of which is unknown.

Most of these crops were grown on large plantations. Organized on Turkish lines (in *čiftliks*, the harshest in human terms of all patterns of farming in the Balkans) they resulted from the Turkish modification of the large estate. There followed, for the rural population, a series of fluctuations and removals towards the low-lying parts of the plains, removals which ceased when, in the nineteenth century, the large estate loosened its grip.[120] Above all, these changes led to the absolute rule of the Turks, based on an administration which the proximity of the capital made even sterner.

Alongside this entrenched and closely controlled rural society, several groups – the Vlachs and 'Arbanassi' amongst others – leading a pastoral and semi-nomadic agricultural life on the uncultivated lands, in temporary, makeshift villages, very different from the solid villages of the Slavs,[121] seem to have enjoyed a certain independence. But very often Asia caught up with them in the form of nomads who mingled and coexisted with them. The clearest example is that of the Yürüks who would periodically cross the straits to occupy the broad rich pasturelands of the Rhodope: they converted to Islam the strange Pomaks, poor Moslem Bulgarians carried along by the tide of Asiatic nomadism.

Asia seems to have left no corner of Bulgaria untouched, to have invaded the whole country with the heavy tramp of her soldiers and camels, submerging (with the aid of a few collaborators, especially the usurers, the *corbazi* of sinister repute who might inform upon their compatriots)

[118] F. de Beaujour, *Tableau du commerce de la Grèce*, 1800, I, p. 54 ff.

[119] According to Besolt, a sixteenth century traveller quoted by I. Sakazov, *op. cit.*, p. 202.

[120] J. Cvijić, *op. cit.*, p. 172.

[121] R. Busch-Zantner, *op. cit.*, p. 59; J. Burckhardt, 'Die thräkische Niederung und ihre anthropogeographische Stellung zwischen Orient und Okzident', in *Geogr. Anz.*, 1930, p. 241.

a people which by its blood, its origins and the nature of its territory was less well protected than most.

To this day, there are still traces in Bulgaria of its impregnation by an exotic, perfumed civilization of the East. To this day, its cities are steeped in memories of it: oriental cities, with long narrow streets closed in by blank walls, the inevitable bazaar and its close-packed shops behind wooden shutters; the shopkeeper puts down the shutter and crouches on it waiting for his customers, beside his *mangal*, the brazier of hot coals indispensable in these lands scoured by the snow-bearing winds from the north and the east. In the sixteenth century, a whole population of artisans worked in these shops for the caravan trade, blacksmiths, carpenters, pack and saddle-makers. Before their doors, in the shade of the poplar trees by the fountains, camels and horses would stop on market days amid the bustle of costumes, merchandise and people: Turks, lords of the *čiftliks*, returning briefly to their lands, Greeks from the Phanar travelling to the Danube provinces, spice-merchants or Aromani caravaneers, gypsy horse-dealers trusted by nobody.

For the Bulgarian people life was a succession of invasions. And yet the Bulgarian retained what was essential, for he remained himself. Whatever his borrowings during the long cohabitation, he did not allow himself to be swallowed up by the invading Turk, but safeguarded what was to preserve him from total assimilation: his religion and his language, guarantees of future resurrection. Firmly attached to the soil, he clung to it doggedly, always keeping the best regions of his dark earth. When the Turkish peasant from Asia Minor settled alongside the Bulgarian, he had to be content with the wooded slopes or marshy plots bordered with willows, down in the hollows, the only land left unoccupied by the *raia*.[122] When the Turks finally departed, the Bulgarian found himself a Bulgarian still, the same peasant who five centuries before had spoken his own language, prayed in his own churches and farmed the same land under the same Bulgarian sky.

Islam in Spain: the Moriscos. At the other end of the Mediterranean, the Spaniards too were at odds with an inassimilable population and the conflict ended in tragedy. Few other problems have so profoundly disturbed the Peninsula.

As the very name suggests, the problem of the Moriscos arose from a conflict between religions, in other words a conflict in the very strongest sense between civilizations, difficult to resolve and destined to last many years. The name Moriscos was given to the descendants of the Moslems of Spain who had been converted to Christianity, in 1501 in Castile and in 1526 in Aragon. By turns bullied, indoctrinated, courted, but always feared, they were in the end driven out of Spain during the great expulsions of the years 1609–1614.

[122] Herbert Wilhelmy, *Hochbulgarien*, Kiel, 1935; R. Busch-Zantner, *op. cit.*, p. 28; Wolfgang Stubenrauch, *Zur Kulturgeogr. des Deli Orman*, Berlin, 1933.

Investigating the problem amounts to studying the long survival, or rather the slow shipwreck, of Islam on the Iberian Peninsula after the fall of Granada in 1492. From the wreckage many things were still floating on the surface, even after the fateful date of 1609.[123]

Morisco problems. There was not one Morisco problem but several, as many as there were Morisco societies and civilizations in Spain on the way to destruction, no two of them at the same stage of decline and decay: a variety which is explained by the chronology of the *Reconquista* and conversion.

Moslem Spain, even at the height of its expansion, controlled only part of the Peninsula: the Mediterranean coast, Andalusia, the Tagus valley, the Ebro valley, the south and centre of Portugal. It left untouched the poorer regions of Old Castile and had little or no foothold in the Pyrenees or their continuation in the west, the Cantabrian ranges. For long ages, the *Reconquista* gathered strength in the semi-desert of Old Castile, where the Christians, in their endeavour to create a front line of vigilant cities, had to import all their requirements and build on virgin soil. It was not until the eleventh century that they began to win victories and encroach upon the living body of Spanish Islam: the capture of Toledo (1085) opened the path towards this coveted world. And even Toledo was, to Islam, no more than an outpost in the continental centre of the Peninsula.

Only gradually did the Christian kingdoms enter into possession of the populous valleys of Aragon, Valencia, Murcia, Andalusia. Saragossa fell to them in 1118, Córdoba in 1236, Valencia in 1238, Seville in 1248, Granada not until 1492. The stages of the Reconquest are hundreds of years apart.

Before 1085 then, the Reconquest settled its Christian population on unoccupied land, but after this date it began to incorporate lands peopled by *fellahin*, Moslem or Christian, and by townsmen more or less won over to Islam. This marked a transition from colonization by settlers to colonization by exploitation, posing the immediate problem, with a thousand variants, of the complex relationships between conqueror and conquered and, more fundamentally, between two opposing civilizations.

Since the conflict did not erupt simultaneously in the various regions of Moorish Spain which were won back by the Christians, the Morisco problems of the sixteenth century are far from identical. Spain presents a range of varying situations; different, yet inseparable from each other and mutually illuminating.

[123] Since the publication of the first edition of *La Méditerranée*, some important studies of the Morisco question have appeared: Tulio Halpérin Donghi, *Un conflicto nacional: Moriscos y Christianos viejos en Valencia*, Buenos Aires, 1955; 'Recouvrements de civilisations: les Morisques du Royaume de Valence au XVIe siècle' in *Annales E.S.C.*, 1956; Henri Lapeyre, *Géographie de l'Espagne Morisque*, 1959, resolves the difficult statistical problem of the expulsion of the Moriscos; the previously mentioned study by Julio Caro Baroja, *Los Moriscos del Reino de Granada*, is a masterpiece, one of the best books of history and cultural anthropology I have ever seen.

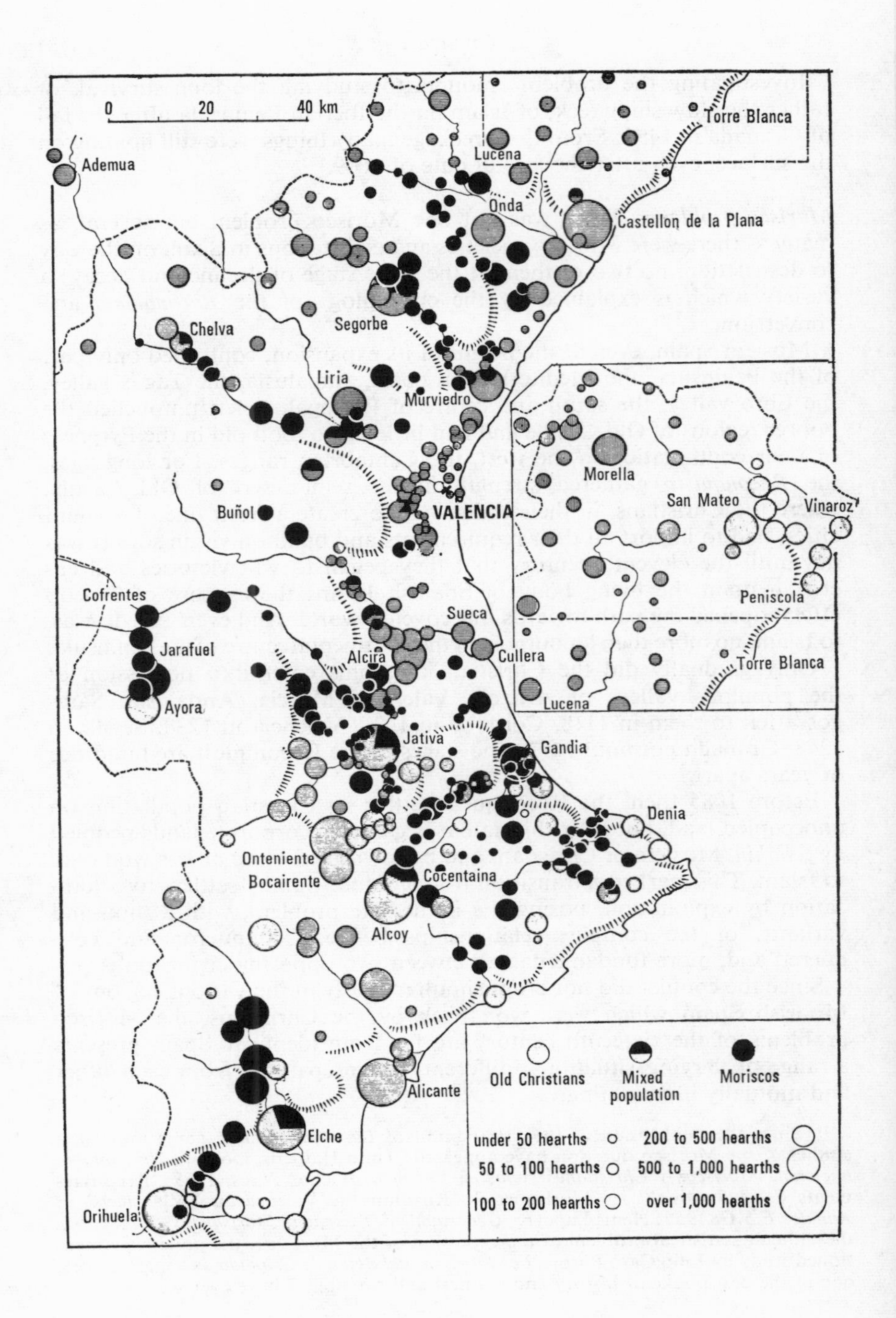

0
20
40 km
Ademua
Torre Blanca
Lucena
Onda
Castellon de la Plana
Chelva
Segorbe
Liria
Murviedro
Morella
San Mateo
Vinaroz
Buñol
VALENCIA
Peniscola
Cofrentes
Sueca
Jarafuel
Alcira
Cullera
Torre Blanca
Ayora
Lucena
Jativa
Gandia
Denia
Onteniente
Bocairente
Cocentaina
Alcoy
Alicante
Elche
Orihuela
Old Christians
Mixed population
Moriscos
under 50 hearths
50 to 100 hearths
100 to 200 hearths
200 to 500 hearths
500 to 1,000 hearths
over 1,000 hearths

The differences between them are themselves explanatory. The Moors of Granada for instance were converted in 1499 on government orders. Cardinal Cisneros decided on the step against the advice of the local authorities and in violation of the promise made by the Catholic Kings who in 1492, when the city surrendered, had given assurance of freedom of worship. This step, prepared with the complicity of several converted Moors, was preceded and accompanied by much ceremonial, including the auto-da-fé of many Korans and Arab manuscripts. The result was the uprising of the Albaicin, the Moslem quarter of Granada, and a rebellion which proved difficult to quell, in the Sierra Vermeja. In 1502, when this rebellion had at last been subdued, the Moors had to embrace Christianity or go into exile. There is no doubt, despite the official disavowals, that the Catholic Kings, who claimed to be surprised, were in fact acting in concert with the Archbishop of Toledo. His responsibility is their responsibility.[124]

This was the beginning of forced conversions in Spain. The measure enacted in Granada was applied throughout Castile. But its impact on the inhabitants of recently conquered Granada was very different from its effect among the few Moors in Castile, the *Mudejares* who had long lived among Christians and had until then been free to practise their own religion.

In the Aragon provinces (Aragon, Catalonia and Valencia) it meant something else again. Conversion here was belated and equally perfunctory, but was not ordered by the state. It was the Old Christians, among whom the Moors were dispersed, who, in the course of the crisis of the *Germanias* in 1525–1526, forcibly baptized their Moslem compatriots in the mass. Were these compulsory baptisms valid or not? The question was debated as far away as Rome, where, it is worth noting, compromise solutions found support far more often than they did in Spain.[125] Charles V, when he was asked for his advice in 1526, pronounced in favour of conversion, in order both to follow the precedent set at Granada and to give thanks to God for his victory at Pavia.[126] But his role in this drama was a minor one. It is quite clear that Valencia and Granada, two faces of Spain, one Aragonese the other Castilian, did not become

(Opposite) Fig. 61: I. *Moriscos and Christians in Valencia in 1609*

After T. Halperin Donghi, 'Les Morisques du royaume de Valence' in *Annales E.S.C.*, April–June, 1956.

Inset shows continuation of Valencia to the north. The fascinating situation revealed by this exceptional map is the extraordinary intermingling of the two populations, within a general context of demographic increase, as is shown by the next map on population changes between 1565 and 1609.

---

[124] H. Hefele and F. de Retana notwithstanding . . . A retrospective account, but one which categorically confirms my view, is Diego Hurtado de Mendoza, *De la guerra de Granada*, ed. Manuel Gómez-Moreno, Madrid, 1948, p. 8 ff; sensitively treated by J. C. Baroja, *op. cit.*, p. 5 ff.

[125] Even in 1609, Clement VIII opposed the expulsion of the Moriscos and the zeal of the holy prelate of Valencia, Juan de Ribera, G. Schnürer, *op. cit.*, p. 196.

[126] R. Konetzke, *op. cit.*, p. 57.

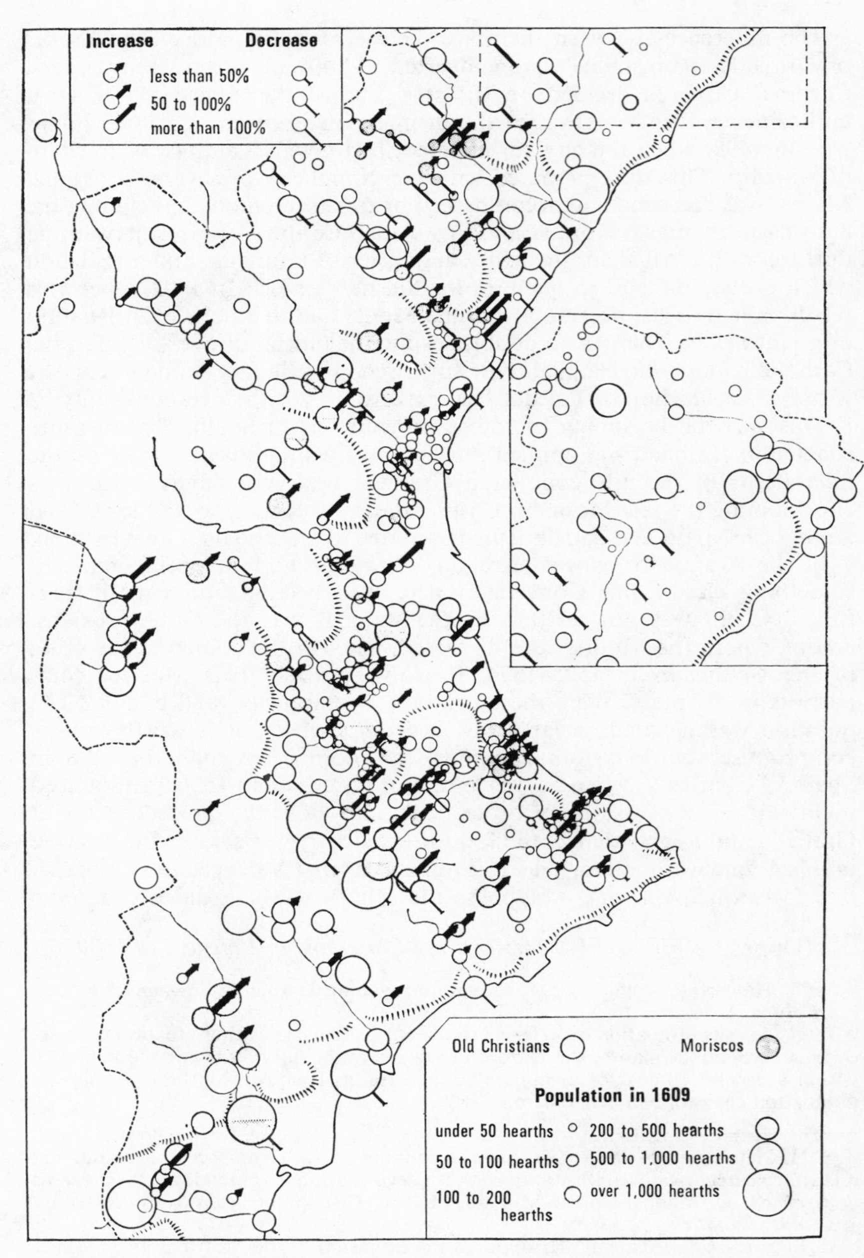

Fig. 62: II. *Population changes in Valencia between 1565 and 1609*

'Christian' (or as they were later called 'Morisco') provinces in the same circumstances. Here we have a distinction between at least two Morisco regions and problems.

A geography of Morisco Spain. Upon closer inspection, other zones and other distinctions become clear, depending on whether the Moriscos were present in greater or lesser numbers, whether they were more or less closely hemmed in or had been in contact for a longer or shorter period with the conquering civilization. In Viscaya, Navarre and the Asturias, the Morisco, while not entirely unknown – he might be an artisan or travelling merchant, perhaps selling powder for arquebuses[127] – was certainly not a frequent figure, except in the Ebro valley in Navarre, where there lived the descendants of *Moros*. There were more of them in Castile and their numbers seemed to grow the further one moved south. Every town had its Moriscos.[128] A traveller at the end of the fifteenth century, Doctor Hieronymus Münzer, notes that at Madrid, a town 'no bigger than Biberach', there are two *morerias*, Moorish ghettos.[129] The proportion was even larger in Toledo and south of Toledo in Andalusia, which was literally teeming with Moriscos, peasants and proletarians serving the big cities. In Aragon proper, they lived as artisans in urban surroundings (at Saragossa they were leather workers and manufactured arms and powder[130]) and, more numerous still, in the uplands[131] between the Ebro and the Pyrenees, they formed active agricultural and pastoral communities.[132] A few great nobles had in their dependence, in the 'lugares de moriscos', the major part of those who had remained on the land: the Count of Fuentes at Exca, for instance, one of the stormiest regions of Morisco Aragon, the Count of Aranda at Almonezil, or the Duke of Aranda at Torellas.[133]

In Catalonia by contrast, there were few or no Moriscos, hardly any trace at all of Spanish Islam. Old Catalonia had lived on the borders of Islam, which had made its influence felt only tc the south, near Tarragona and the Ebro. And in 1516, Catalonia expelled the Moriscos living in Tortosa.[134] It was exceptional for the Barcelona Inquisition to find itself interrogating a Morisco.[135]

[127] *Gobierno de Vizcaya*, II, p. 357. In 1582, the racial laws were invoked against them (*ibid.*, II, p. 223) and in 1585, in the name of the prohibitions laid down by the *fuero*, *ibid.*, p. 309; the prohibitions were enforced in 1572 too, in the neighbouring province of Navarre, Antonio Chavier, *Fueros de Navarra*, 1686, p. 142.

[128] Simancas Patronato Real, 15th August, 1543, for Arévalo and Medina del Campo.

[129] Quoted by Louis Pfandl, *Philippe II*, Madrid, p. 310–311, 'habet duas morerias cum Saracenis plenas'.

[130] I. de Asso, *op. cit.*, p. 219–220.

[131] Cabrera, quoted by R. Menéndez Pidal, *op. cit.*, I, p. 122.

[132] They formed a total of 20 per cent of the population of Aragon, H. Lapeyre, *op. cit.*, p. 96.

[133] Apuntamientos del Virrey de Aragon sobre prevenciones de aquel reyno contra los Moriscos, Simancas E° 335, undated (about March, 1575).

[134] *Geografía General de Catalunya*, *op. cit.*, p. 343.

[135] But see A.H.N. Barcelona Inquisition, Libro I, f° 21, 20th December, 1543.

Further south, rural Valencia was a typical colonial territory, taken over in the thirteenth century by the lords of Aragon and the Catalan merchants and since then subject to a number of movements and waves of immigration. Henri Lapeyre[136] sees a parallel to the situation in Valencia in Algeria before 1962. The proportions are not the same, but the two populations were similarly embedded one within the other, as is illustrated by Tulio Halperin Donghi's maps.[137] The general characteristics of geographical distribution are fairly clear: the towns were predominantly Christian, Moriscos tending to live, if anywhere, in the suburbs; the *regadio* regions, irrigated lands, were also predominantly Christian, except round Jativa and Gandía; *secano* regions on the other hand were peopled by Moriscos – poor land in hilly country. 'So it is hardly surprising that the two major rebellions should have broken out in the mountains, in 1526 in the Sierra de Espadán and in 1609 in the region of Mucla de Cortes on the right bank of the Júcar and in the Laguar valley, south of Gandía.'[138]

In 1609, the Moriscos represented almost one-third of the total population of Valencia, 31,715 hearths as against 65,016 'old' Christians,[139] but the latter held the commanding positions, completely controlling Valencia and its fertile *huerta*.

This relationship was of course the result of a long evolution dating back several centuries. The conquered society, still alive but reduced to a meagre existence, was as it were a threadbare fabric, in places quite worn through. There was no Moslem aristocracy, no élite of any kind to provide leadership for the conquered proletariat: consequently there are no examples of organized resistance in response to provocation. Everywhere, in town and countryside, the Morisco was oppressed by the victorious society. The strongest defenders of the *fellahin* were in fact the feudal landlords.[140] They were to protect the Moriscos much as in the United States southern plantation owners protected their slaves. But alongside them, there had grown up, the product of several centuries of Christian domination, a proletariat, both rural and urban, of Old Christians, fanatical and cruel, who can be compared, to pursue the analogy, to the poor whites of the southern states.

What Valencia must have been like in the thirteenth century can be seen in Granada in the sixteenth: Granada, where the Christian victory was still recent and had been won at the expense of a rich kingdom, betrayed by its lack of modern artillery more than by its obvious internal weaknesses.[141] If Moslem society was not totally unchanged – it would be

[136] H. Lapeyre, *op. cit.*, p. 27.

[137] See above, p. 782 and 784.

[138] H. Lapeyre, *op. cit.*, p. 26.

[139] *Ibid.*, p. 30.

[140] A fact of which there is much evidence, e.g. Castagna to Alessandrino, Madrid, 15th March, 1569, L. Serrano, *op. cit.*, III, p. 5, the Moriscos 'sono favoriti da tutti li signori di quel paese perché da loro cavano quasi tutta l'entrata che hanno . . .' with reference to Valencia and the Maestre of Montesa.

[141] J. C. Baroja, *op. cit.*, p. 2 ff and *passim*.

surprising if conquest had not brought some disruption – it was at least still recognizable in this land which man had disciplined and brought under the plough, had cultivated up to the highest terraces, rich in *vegas* of astonishing fertility, tropical oases in an African landscape. Christian nobles had settled in these rich lands, Juan Enríquez for example,[142] defender of the Moriscos in 1568, who owned property in the plain of Granada. Everywhere civil servants and clerics had moved in, categories whose integrity was sometimes in doubt and who were often downright dishonest, shamelessly exploiting their privileges. Everything that has been said or written about 'colonialism' at any time and in any country, is strangely true of the reconquered kingdom of Granada. On this subject, even official documents speak loud. The *licenciado* Hurtado[143] for instance, when making enquiries in the Alpujarras in spring 1561, recognized the merit of the *gente morisca*, who remained so uncomplaining although for twenty years there had been no justice worthy of the name in the province but instead a series of crimes, misdeeds, malpractice and countless thefts at the expense of this people. If the true culprits, who should be subjected to the full fury of the law, insist that the Moriscos are dangerous, the report continues, that they are hoarding rations, wheat, flour and arms with the intention of rising up one day against the Christians, it is merely in an attempt to excuse their inexcusable conduct.

Is it likely that he would have allowed himself to be misled? When Granada rose up, on Christmas night 1568, Philip II's ambassador in France, Francés de Alava, felt the need to unburden his conscience by making similar revelations. In October, 1569, he wrote at length to the secretary Zayas,[144] making it clear from the very first lines that he had been seven or eight times, in the course of the previous twelve years, to Granada, that he was acquainted with the competent civil, military and religious authorities. There is every reason to believe what he says. Why should an ambassador hasten to the defence of these poor wretches from whom he was hundreds of miles away at the time – unless it was to make the truth known?

The Moriscos are in revolt, he writes, but it is the Old Christians who have driven them to desperation, by their arrogance, their thefts and the insolence with which they take Moslem women. Even the priests are not blameless and he gives a precise example: an entire Morisco village petitioned the archbishop about its parish priest and the complaint was investigated. The villagers begged that the priest be removed – or else 'marry him to someone, for all our children are born with eyes as blue as his'. Not content with this story, which he reports as being absolutely true and not at all funny, the ambassador in fury and anguish had

[142] *Ibid.*, p. 154.

[143] Simancas E° 328, the *licenciado* Hurtado to His Majesty, las Alpujarras, 29th June, 1561.

[144] F. de Alava to Zayas, Tours, 29th October, 1569, A.N., K 1512, B 24, n° 138 b orig. dup. no. 138 a.

conducted his own private enquiry. He had discovered the habitual malpractice of petty officials, even those who were of Morisco origin but who were as relentless as the rest in exploiting their own people. He had entered churches on feast-days to see for himself how little trouble was taken to make divine worship respected and worthy of respect. At the moment of consecration, between the host and the cup, he had seen the priest turn round to make sure all his native congregation, men and women, were on their knees as they should be and then harangue his flock with insults – a thing 'so contrary to the worship of God', notes Don Francés, 'que me temblavan las carnes', 'that I shuddered from head to foot'.

Rapine, theft, injustice, murder, massive unlawful condemnations: it would not be difficult to draw up a list of charges against Christian Spain. But it is doubtful whether it was even aware of what was being perpetuated, often secretly, in its name, or supposedly in its name, in the rich South which drew all towards it in search of gain, benefit, land and employment and where Flemings and Frenchmen did not disdain to settle as artisans, as is proved by a document of the Granada Inquisition in 1572.[145] The physics of history, the inexorable law of a stronger force was at work. Alongside the Moslem city,[146] separate from it since 1498,[147] an official Christian city was growing up, near the Alhambra where the Captain General had his residence, in the University, founded in 1537, and in the *Chancillería* created in 1505, already all-powerful and aggressive in 1540.[148] It should not be forgotten if one is to understand – I do not say judge – the Spaniards, that they found themselves, like the French in Algeria, the Dutch in Batavia or the English in Calcutta in modern times, deeply immersed in a colonial undertaking, in a maelstrom of competing civilizations whose troubled waters refused to mingle.

Facing the sometimes heavy-handed colonialism of Spain stood a still coherent indigenous society with what Valencia lacked, or no longer possessed, a ruling class: the rich families of the Albaicin, a group of notables in silken robes, prudent, secretive, reigning over a people of market gardeners tending their plants and their silkworms, peasants whose skills were devoted to irrigating barren land with water channels and maintaining the supporting walls of their terraced plots; reigning too over a people of muleteers, small merchants and retailers, of artisans – weavers, dyers, shoemakers, masons and plumbers often competing with artisans from the North who brought their own methods and principles. All these poor and humble folk went dressed in cotton. That the nobles of the Albaicin did not possess unlimited courage was later to be proved in spectacular fashion: they were afraid of compromising themselves, of losing their *carmenes*, their country villas. Moreover, part of the Granada nobility, at

[145] A.H.N., Granada Inquisition, 2602, 20th March, 28th May, 17th July, 1572; 7th September, 1573.

[146] On the 'colonial' aspect of Granada, Pedro de Medina, *op. cit.*, p. 159 v°.

[147] J. C. Baroja, *op. cit.*, p. 13.

[148] *Ibid.*, p. 142.

least its most illustrious members, had left Spain shortly after the fall of Granada. Nevertheless this ruling class had preserved certain of its numbers, its traditions and also its unreasoning respect for lineage and the great dynasties, and the 1568 rebellion was to see inter-family conflicts very similar to those which had precipitated the fall of Granada.

This surviving aristocracy had seen growing up around it and above it, a Christian aristocracy of recent importation, richly favoured (if not quite as royally as in Valencia), shamelessly living off the labour of its Morisco peasants; the latter being a sober race and therefore all the easier to exploit. It was estimated that a Morisco consumed only half as much as a Christian, as local proverbs confirm: 'quien tiene Moro, tiene oro; a más Moros, más ganancias a más Moros más despojos', the Moors brought their masters gold, gain and profit.[149]

The Christian nobles were the protectors of Morisco peasants, indeed they had the right to offer sanctuary on their lands to refugees from neighbouring estates. Then the state, in its desire to restore order in Granada, suppressed this privilege and limited the right of sanctuary to only a few days, even in the churches. Moreover, noticeably since 1540 but even before then, the *letrados* of the *Audiencia* of Granada had been seeking to whittle away the rights of the nobility and its head, the Captain General of the Kingdom, in other words the powerful Mendoza family. There grew up what was in fact a civil government based on the Christian towns and the immigrant population in Granada, in opposition to the military and aristocratic government of the Mendoza. This political and social crisis was not responsible for the outbreak of war, but it aggravated tension and disorder. At the same time, Philip II's government, in search of fiscal revenues as usual, had begun to question titles of property, at least since 1559. Lastly, in Granada as in Valencia, the population was increasing; with economic hardship banditry reappeared. The brigands – the *monfíes* – now no longer able to take refuge on noble estates or in churches, took to the mountains and would swoop down on looting raids, with the connivance of the *gandules*, their urban accomplices, or of Barbary or Turkish corsairs.[150] In 1569, a few months after the outbreak of rebellion, and just after the punitive expedition sent by the Marquis of Mondéjar into the Alpujarras, a solution could still have been reached, once more through the good offices of the aristocracy, as Julio Caro Baroja specifically and quite rightly points out in his outstanding book on the Moriscos.[151] But would that have brought any true peace? Civilizations are less malleable than societies: cruel and unforgiving, their wrath lives on. It is the fearsome face of hate, cruelty and incomprehension that we must try to see here, putting aside for a moment the events of a war which we shall consider later.[152]

[149] *Ibid.*, p. 23.
[150] *Ibid.*, p. 166.
[151] *Ibid.*, p. 193 ff.
[152] See below, p. 1055 ff.

The drama of Granada. Any 'colonial' war means the inevitable clash of two civilizations and the intrusion of blind passions, violent and insidious. All pretence at reason vanishes, the more so in this case since Spanish policy had had everything its own way since 1502 in Granada, since 1526 in Valencia and all along in Aragon. Without the slightest trouble, it had succeeded in dividing its enemies, preventing sedition from spreading from one region to another. It never had to tackle more than one Morisco problem at a time: Granada in 1499–1502, Valencia in 1525–1526; then again briefly in 1563;[153] Aragon in 1575[154] (but the alarm was not serious); Castile in 1580;[155] Granada in 1584,[156] Valencia again in 1609;[157] Castile in 1610 and Aragon in 1614. The Spanish government also had the external frontiers closely watched and tried to close them to fugitive Moriscos on the Pyrenean as well as the Mediterranean side. Such vigilance did not prevent escapes but it made them more difficult, along the Valencia coast for instance after 1550.[158] These were only sensible moves, from Spain's point of view, made by an experienced and pragmatic administration. It was political wisdom too to listen to the advice of the landlords of Morisco peasants who were willingly given a hearing both in the Council of War and the Council of State.[159] Spain, in Morisco territory as everywhere else, was forced to use the high nobility as her intermediary.

But all the rules of good administration went by the board in an emergency. In 1568 and 1569, the advice of the Marquis of Mondéjar was disregarded in favour of the passionate feelings of Cardinal Espinosa and Don Pedro de Deza, the fanatical president of the *Audiencia* of Granada, both of whom represented the *letrados*, the *bonetes* who would soon, if they were not stopped, impose their rule on Spain. The cardinal was a stubborn man, 'resoluto en lo que no era de su profesión', as a chronicler describes him,[160] resolute in a matter outside his professional competence – a military operation. It almost seems as if everything was done to provoke the explosion, particularly in Madrid, in the belief that it was unlikely ever to happen. For over forty years, the Moriscos had been living peacefully, since the *Germanias* of 1526. The Pragmatic which was to set fire to the powder keg was passed on 17th November, 1566 and promulgated on 1st January, 1567; for over two years it was the subject of discussion, leaving the Moriscos and their protectors with the impression that a compromise was still possible and that a reprieve might conceivably be obtained in return for a substantial payment. What Philip II's advisers had decided in this paper was nothing less than the condemnation without

[153] Manuel Danvila y Collado, 'Desarme de los Moriscos en 1563' in *Boletín de la Real Academia de la Historia*, X, 1887, p. 275–306.

[154] Simancas E° 335.

[155] 6th July, 1580, A. E. Spain, f° 333, Moriscos of Castile were forbidden to enter Portugal.

[156] H. Lapeyre, *op. cit.*, p. 127.

[157] *Ibid.*, p. 162 ff.

[158] *Ibid.*, p. 29.

[159] J. C. Baroja, *op. cit.*, p. 154.

[160] *Ibid.*, p. 151, the writer was L. Cabrera de Córdoba.

appeal of an entire civilization, a whole way of life. Moorish costumes were forbidden both for men and women (women were to give up wearing the veil in the street), and so were closed houses in the Moorish fashion, the hiding place of clandestine Moslem ceremonies; Moorish baths were to be closed and the use of the Arabic tongue prohibited. Their intention was to eliminate any traces of Islam still suspected of persisting in Granada; or rather perhaps to threaten and intimidate the Morisco population. As discussions and bargaining dragged on, those inclined to violence had ample time and opportunity to conspire under cover of meetings of the Hospital and *Cofradía de la Resurrección* which the Moriscos maintained in Granada.[161]

Finally on Christmas night 1568, the *monfíes* penetrated into the Albaicin and tried to provoke an uprising. The Alhambra, just opposite, had less than fifty defenders, but the Alhambra was not attacked and the Moorish quarter did not rise up. War would break out in earnest only when popular passions and cruelty were aroused with the massacres of Christians and their priests in the Alpujarras, the raids on the plains and the manhunts soon to be instigated by both sides. There now began a vast orgy of killing, indiscriminate and disorganized, dispersed all over a huge, wild and trackless region. When the king finally granted the Old Christians the right to loot at will, giving them *campo franco*,[162] he gave fresh impetus to the war, driving it to even further extremes. The stealing of flocks, bales of silk, hidden treasure and jewels and the hunting of slaves became the everyday pattern of the war, not to mention constant marauding by soldiers and army quartermasters. At Saldas, near Almería, the Moriscos sold their Christian prisoners to the Barbary corsairs: 'un cristiano por una escopeta', 'a Christian for a carbine'.[163] At Granada meanwhile the market was flooded with Morisco slaves and the Christian population lived in hopes of descending upon the Moorish city and stripping it of all its treasures for good.[164] Passion, fear, panic and suspicion mingled in equal parts. Christian Spain, victorious but ill at ease, lived in terror of Turkish intervention, which was indeed discussed at Istanbul.[165] Spain consistently, both before and after 1568, over-estimated the threat from Islam.

In their attempt to re-create a Kingdom of Granada, the rebels were merely reviving a ghost. But their endeavours, the coronation ceremonies of the first king of the rebellion, the construction of a mosque in the Alpujarras, the desecration of churches, are all significant from the point of view of this chapter. This was indeed a civilization seeking to rise from its ashes before it finally fell back to the ground.

With the dearly-bought victories of Don John of Austria (who replaced the Marquis of Mondéjar as commander in chief on 13th April, 1569[166]) the extreme policy prevailed. Mass surrenders of the rebels had begun by April, 1570. To all practical intents the war was over; the

[161] *Ibid.*, p. 169. [162] *Ibid.*, p. 196. [163] *Ibid.*, p. 188.
[164] *Ibid.*, p. 199. [165] See below, p. 1066. [166] *Ibid.*, p. 199.

rebellion had rotted from within. Even in the preceding year, in June, 1569, expulsions had started: 3500 Moriscos from Granada (between 10 and 60 years of age) had been transported from the capital to the neighbouring province of La Mancha.[167] On 28th October, 1570,[168] the order was issued for the expulsion of all Moriscos; on 1st November, the unfortunate people were rounded up in long convoys and sent in chains into exile in Castile. From this point on the rebellion was doomed for lack of assistance from the local population, which had been outwardly peaceable but in fact in league with the rebels and providing them with supplies.[169] The highland revolt now consisted of little more than a few hundred *salteadores* waging a miniature war, as a Genoese correspondent describes it, 'a guisa di ladroni'.[170] It looked as if everything was over for good. Crowds of immigrants, *gallegos*, *asturianos* and Castilians, about 12,000 peasant families in all, moved into the deserted villages of Granada. Meanwhile the former possessions of the conquered people were sold off to nobles, monasteries and churches, to the immense profit, so it was said, of the royal exchequer. In fact nothing was solved by the war: peasant settlement was to end in a fiasco,[171] not all the Moriscos had left the unhappy kingdom, some returned and in 1584 new expulsion orders were issued,[172] as they were yet again in 1610.[173]

Aftermath of Granada. The Morisco problem had been eliminated from Granada merely to crop up again in Castile, and particularly in New Castile. Closing one file simply meant opening another. Refugees from Granada, grafted on to Castilian society, did not take long to multiply[174] and become rich – and therefore once more a threat. How could a hard-working people avoid becoming rich in a country flooded with precious metals and populated by an abundance of *hidalgos* for whom any work was dishonourable? So it came about that in the years 1580–1590, that is less than twenty years later, the Granada question had by the curious logic of fate, become the Castilian and Andalusian question: the peril had simply crept nearer to the heart of Spain. It was not so much for Granada now (where some Moriscos had of course remained) as for Seville, Toledo or Avila that fears began to be expressed and once again extreme measures were envisaged. During the summer of 1580, a major conspiracy with accomplices in Morocco was uncovered at Seville: it was indeed by the ambassadors of the Sharif, who was at that time anxious

[167] H. Lapeyre, *op. cit.*, p. 122. [168] *Ibid.*

[169] A.d.S., Florence, Mediceo 4903, Nobili to the prince, Madrid, 22nd January, 1571.

[170] A.d.S., Genoa, Spagna . . ., Sauli to the Republic of Genoa, Madrid, 11th January, 1571; there were over 2500 'bandolieri'.

[171] H. Lapeyre, *op. cit.*, p. 122 and n. 4.

[172] *Ibid.*, p. 127.

[173] *Ibid.*, p. 162 ff.

[174] In the Toledo region, they numbered 1500 in 1570 but 13,000 in 1608, according to the Cardinal of Toledo, J. C. Baroja, *op. cit.*, p. 214.

to preserve the backing of Spain, that the plot was revealed.[175] In spring, 1588, disturbances broke out, this time in Aragon.[176] They were the occasion for a debate in the Council of State in July,[177] during which the danger to Spain of a fast-multiplying domestic enemy was pointed out; His Majesty was urged not to repeat the mistake made in 1568 at Granada, but to take the initiative immediately. All this alarm was generated by the uprising of a few hundred Moriscos, following several clashes with Old Christians.[178] It soon subsided and the viceroy of Naples was so incredulous that he unhesitatingly declared in May that the uprising was merely a rumour put about by English propaganda.[179]

Perhaps the alarm besides being an indication of uneasiness also served as a pretext. For from November of the same year, the Spanish Church once more intervened. Its mouthpiece, the Cardinal of Toledo, had a seat on the Council of State and took as his text the reports of the commissioner for the Inquisition at Toledo, Juan de Carillo.[180] His evidence states that in the city, where the old colony of *mudejares* had been reinforced in 1570 by an influx of Moriscos from Granada, the latter, the deportees, speak Arabic amongst themselves, while the former, often public scribes, having mastered the Spanish tongue, seek to insinuate themselves into high office. Both groups however have become rich through trade. Diabolical unbelievers, they never go to mass, never accompany the Holy Sacrament through the streets, only go to confession for fear of sanctions against them. They marry amongst themselves, hide their children to avoid having them baptized, and when they do baptize them take the first passers-by on the church steps as god-parents. Extreme unction is never requested except for those who are practically dead and unable to receive it. And since the people responsible for supervising and educating these unbelievers do very little supervision or education, the latter do as they please. It is the immediate duty of the Council of State to give the matter urgent consideration.

And so it did, on Tuesday, 29th November, 1588, on the proposal of the Cardinal who presented his case.[181] Was the Council to remain indifferent to the disquieting proliferation of the Moriscos in Castile and in particular at Toledo, their 'alcazar and fortress', while the Old Christians, called into the 'militia', were diminishing in numbers and, since they were poorly armed, were in danger of being taken by surprise one day? At the very least – and the Council was unanimous on this point – the Inquisitors

[175] A.d.S., Florence, Mediceo 4911, Bernardo Canigiani, ambassador of the Grand Duke, Madrid, 27th June, 1580, thought at first that it was a fabrication, but letters from merchants of Seville confirmed the story.

[176] Longlée to the king, Madrid, 5th March, 1588, *Correspondance*, p. 352.

[177] Simancas E° 165, f° 347. Consulta del C° de Est°, 5th July, 1588.

[178] Longlée to the king, 5th June, 1588, p. 380.

[179] Simancas E° 1089, f° 268, Miranda to the king, Naples, 6th May, 1588.

[180] Sobre los moriscos, Council of State, 14th November, 1588, Simancas E° 165, f° 34.

[181] Los muchos nuevos christianos que ay por toda Castilla, Madrid, 30th November, 1588, Simancas E° 165, f° 348.

should be ordered to make all enquiries within their competence and to draw up a census of the Moriscos.

So fear, a bad counsellor, was introduced to the heart of Spain. During the raids by the English in the following year, 1589, it was feared that the Moriscos, who were numerous in Seville, would give assistance to the attackers.[182] In Valencia in 1596, anxiety was expressed about similar foreign contacts.[183] The presence of the enemy within preoccupied and deflected Spanish policy. Like a splinter the Morisco population was embedded right in the heart (or as Spanish says 'the kidney') of Spain.[184] In 1589 the Council of State was talking only in terms of a census. But events moved quickly. The following year the king was acquainted with several extreme proposals: that the Moriscos should be obliged to serve a certain term in the galleys, for pay: at least it would hold back their birth-rate; that children should be separated from their families and entrusted instead to noblemen, priests or artisans who would be responsible for their Christian education; that the most dangerous of them be executed; that the Granada Moriscos who had settled in Castile should be sent back to their original home and thus removed from the famous *riñón del país* and that they should also be driven from the towns to the countryside.[185] After 5th May, there was talk of expulsion pure and simple: after all, the Catholic Kings had expelled the Jews and won holy renown.[186] The undesirable Moriscos were already condemned in spirit by all the members of the Council, without exception. But the condemned people was to enjoy a fairly long period of grace.

The reason no doubt was that Spain, still involved in the Netherlands, at odds with both France and England, was preoccupied with other matters besides this domestic settling of accounts. It was not clemency but impotence, the curious consequence of Spain's imperialist policy, that saved the Moriscos, as a rope supports a hanging man. Around them anger and hatred continued to seethe. A report addressed to the king in February, 1596[187] criticizes the lack of severity of government policy towards these unbelievers and draws attention to their fabulous wealth: more than 20,000 Moriscos in Andalusia and the Kingdom of Toledo possess revenues of over 20,000 ducats. How can such a thing be tolerated? The report goes on to denounce one Francisco Toledano, a Morisco of Toledo now settled at Madrid, the richest iron merchant in the city, who is enabled by his business to trade in Viscaya and at Vitoria, taking advantage of this to deal in arms and arquebuses. His Majesty is besought to arrest this man without delay and discover the names of his customers and accomplices.

[182] A.d.S., Florence, Mediceo 4185, f°s 171 to 175, anonymous report.

[183] Marqués de Denia to Philip II, Valencia, 3rd August, 1596, Simancas E° 34, f° 42.

[184] Madrid, 22nd May, 1590, Simancas E° 165.

[185] 22nd May, 1590, see preceding note.

[186] 5th May, 1590, Simancas E° 165.

[187] Archives of former *Gouvernement-Général* of Algeria, Register 1686, f° 101.

In 1599, interminable discussions began once more in the Council of State. That the king should make his mind up, and quickly, was the conclusion of all the proposals. Among the signatories, there once again figured the Cardinal of Toledo, but alongside him Don Juan de Borja, Don Juan de Idiáquez, the Count of Chinchón, Pedro de Guevara.[188] In all the vast files of documents at Simancas concerning these deliberations, not a single plea in favour of the Moriscos is to be found.[189]

The epilogue to the debate was to be the expulsion of 1609–1614. A combination of circumstances finally made it possible: the return of peace (1598, 1604, 1609) and the silent mobilization of the entire Spanish war fleet, galleons and galleys,[190] to handle the embarkation and ensure the security of the operation. Julio Caro Baroja suggests that the victories of the sultan of Marrakesh in spring 1609 over the 'king' of Fez may have pushed Spain towards this radical course and this seems quite likely.[191]

Thus ended in failure the long experiment in the assimilation of Iberian Islam, a failure sharply felt at the time. 'Who will make our shoes now?' asked the Archbishop of Valencia at the time of the expulsion – of which he was nevertheless a strong supporter. Who will farm our land, the lords of the 'lugares de moriscos' were no doubt thinking. The expulsion, it was realized in advance, would leave serious wounds. The *Diputados del Reino* of Aragon had indeed voted against it. In 1613–1614, Juan Bautista Lobana, who was touring the kingdom for the purpose of mapping it, makes several references in his notes to the desolation of its deserted villages: at Longares, only 16 inhabitants are left out of 1000; at Miedas, 80 out of 700; at Alfamén, 3 out of 120; at Clanda, 100 out of 300.[192] Historians have frequently remarked that all these wounds healed in the long run, which is true.[193] Henri Lapeyre has shown that the expulsion affected at most 300,000 persons out of a total population of possibly 8 million.[194] But this was still a very considerable proportion for the time and for Spain, although well below the exaggerated figures which have been suggested in the past. At the same time, Lapeyre[195] thinks that its immediate effects were serious and the decline of the birth rate in the seventeenth century delayed recovery.

But the most difficult question to answer is not whether Spain paid a high price for the expulsion and the policy of violence it implied, or whether or not she was *right* to take such action. We are not here concerned with judging Spain in the light of present-day attitudes: all historians are of course on the side of the Moriscos. Whether Spain did well

[188] Consulta del C° de E°, 2nd February, 1599, Simancas E° 165, f° 356. See also C° de E° to the king, 10th August, 1600, A.N., K 1603.

[189] H. Lapeyre, *op. cit.*, p. 210, would modify this judgement slightly: 'This was true concerning the obstinate Morisco who rejected Christian civilization, but several pleas can be found in favour of the Morisco who might be called "right-thinking". '

[190] J. de Salazar, *op. cit.*, p. 16–17; Gen. Brémond, *op. cit.*, p. 304.

[191] J. C. Baroja, *op. cit.*, p. 231.

[192] I. de Asso, *op. cit.*, p. 338.

[193] E. J. Hamilton, *American treasure* . . ., p. 304–305.

[194] H. Lapeyre, *op. cit.*, p. 204.

[195] *Ibid.*, p. 71 and 212.

or badly to rid herself of the hardworking and prolific Morisco population is beside the point. Why did she do it?

Above all it was because the Morisco had remained inassimilable. Spain's actions were not inspired by racial hatred (which seems to have been almost totally absent from the conflict) but by religious and cultural enmity. And the explosion of this hatred, the expulsion, was a confession of impotence, proof that the Morisco after one, two or even three centuries, remained still the Moor of old, with his Moorish dress, tongue, cloistered houses and Moorish baths. He had retained them all. He had refused to accept western civilization and this was his fundamental crime. A few spectacular exceptions in the religious sphere, and the undeniable fact that the Moriscos who lived in the cities tended increasingly to adopt the dress of their conquerors,[196] could not alter this. The Morisco was still tied deep in his heart to that immense world, which as Spain was well aware,[197] stretched as far as distant Persia, a world of similar domestic patterns, similar customs and identical beliefs.

All the diatribes against the Moriscos are summarized in the declaration of the Cardinal of Toledo: they are 'true Mohammedans like those of Algiers'.[198] On this point the cardinal can be reproached for intolerance but not for untruthfulness, as the solutions suggested by the members of the Council confirm. It was not a matter of annihilating a hated race: but it seemed impossible to maintain an irreducible core of Islam right in the heart of Spain. What was suggested? The problem could be plucked out once and for all by removing the pre-requisite of any civilization: its human material; this was the solution finally adopted. Alternatively that assimilation which forced baptism had failed to achieve could be pursued at all costs. One suggestion was that all the children, minds as yet unmoulded, should be kept in Spain, while the departure of the adults to Barbary should be encouraged, provided it was done discreetly.[199] The Marquis of Denia proposed that the children be given a Christian education, that the men between the ages of fifteen and sixty be sent to the galleys for life and the women and old men be sent to Barbary.[200] Another opinion was that it would be sufficient merely to divide the Moriscos up among different villages at a ratio of one family to every fifty families of Old Christians, making it illegal for them to change their residence or to exercise any occupation other than farming – the disadvantage of industry, trade and transport being that they encouraged travel and contacts with the outside world.[201]

Of all the possible solutions, Spain chose the most radical: deportation, the uprooting of a civilization from its native soil.

But had Spain really seen the last of her Morisco population? Of course not. In the first place, it was not easy in some instances to distinguish between a Morisco and a non-Morisco. Mixed marriages were

[196] J. C. Baroja, *op. cit.*, p. 127.
[197] *Ibid.*, p. 107.
[198] Simancas, E° 165, 11th August, 1590.
[199] *Ibid.*
[200] 2nd February, 1599, for reference see note 188.
[201] *Ibid.*

frequent enough to be taken into account in the expulsion edict.[202] And there were certain interested parties who undoubtedly intervened on behalf of some who would otherwise have been deported. The Moriscos in the towns were expelled almost to a man; rather less were expelled of those who inhabited the *realengos* and there were even greater exceptions among the Morisco tenants of seignorial estates, mountain dwellers and isolated peasants.[203]

For the Morisco often stayed on, now only a face lost in the crowd, but leaving his own indelible mark upon the rest.[204] The Christian population of Spain and indeed the aristocracy, already bore the traces of its Moorish ancestry. Historians of America too have described in a variety of ways how the Morisco played a part in the settlement of the Americas.[205] One thing is certain: Moslem civilization, supported by the remnants of the Morisco population as well as by those elements of Islam which Spain had absorbed over the centuries, did not cease to contribute to the complex civilization of the Peninsula, even after the drastic amputation of 1609–1614.

That surge of hatred was unable to wipe out everything which had taken root in Iberian soil: the black eyes of the Andalusians, the hundreds of Arabic place-names, or the thousands of words embedded in the vocabulary of the former conquered race who had become the rulers of Spain. A dead heritage, some will say, unimpressed by the fact that culinary habits,[206] certain trades and structures of hierarchy still convey the voice of Islam in the everyday life of Spain and Portugal. And yet in the eighteenth century, when French influence was at its height, an artistic tradition survived, in the art of the *mudejares*, with its stuccoes, ceramics and the tender colours of its *azulejos*.[207]

[202] Gen. Brémond, *op. cit.*, p. 170.

[203] 'It is time to put aside the sentimental outpourings of a certain school of historians concerning what it calls "the odious and barbaric expulsion of the Moors from Spain". The amazing thing is that, despite the advice of the great Ximénez, the presence of a million or so Moriscos in a permanent state of internal and external conspiracy should have been tolerated for over a hundred years . . .', Henri Delmas de Grammont, *Relations entre la France et la Régence d'Alger au XVIIe siècle*, Algiers, 1879, Part I, note p. 2–3.

[204] As observant travellers did not fail to notice, Le Play, 1833, 'there is some Arab blood in all these people', p. 123; Théophile Gautier, *Voyage en Espagne*, p. 219–220; Edgar Quinet, *Vacances en Espagne*, p. 196, etc.

[205] See Carlos Pereyra for Spanish America. On Brazil, Nicolas D. Dabane, *L'Influence arabe dans la formation historique et la civilisation du peuple brésilien*, Cairo, 1911.

[206] On Portugal, see the *Arte de Cozinha* by Domingos Rodríguez, 1652, quoted by Gilberte Freyre in *Casa Grande e Senzala*, I, p. 394, a book on which I have also based my brief remarks concerning the eighteenth century. On the survival of 'Moorish' architecture and ornamentation in Toledo, until the sixteenth century and perhaps later, Royall Tyler, *Spain, a Study of her Life and Arts*, London, 1909, p. 505.

[207] There is still an immense amount of unpublished material on the Morisco problem, at Simancas, for example E° 2025 (Moriscos que pasaban a Francia, 1607–1609). Transport of Morisco refugees 'with their belongings' by a Marseilles boat, Bouches-du-Rhône Archives, Amirauté B IX, 14, 24th May, 1610. A remarkable document is

The supremacy of the West. But the Morisco question was only an episode in a wider conflict. In the Mediterranean the essential struggle was between East and West – for the latter there was always an 'eastern question' – fundamentally a cultural conflict, revived whenever fortune favoured one or other protagonist. With their alternating rise and fall, major cultural currents were established running from the richer to the poorer civilization, either from west to east or east to west.[208]

The first round was won by the West under Alexander the Great: Hellenism represents the first 'Europeanization' of the Middle East and Egypt, one which was to last until the age of Byzantium.[209] With the fall of the Roman Empire and the great invasions of the fifth century, the West and its ancient heritage collapsed: now it was the Byzantine and Moslem East which preserved or received the riches, transmitting them for centuries back to the barbaric West. Our Middle Ages were saturated, shot through with the light of the East, before, during and after the Crusades. 'Civilizations had mingled through their armies; a host of stories and narratives telling of these distant lands entered circulation: the Golden Legend teems with these tales; the story of St. Eustace, of St. Christopher, of Thaïs, of the Seven Sleepers of Ephesus, of Barlaam and Josaphat are eastern fables. The legend of the Holy Grail was grafted on to the memory of Joseph of Arimathea; the Romance of Huon of Bordeaux is a fantasmagoria sparkling with the spells of Oberon, spirit of dawn and morning; Saint Brendan's voyage is an Irish version of the adventures of Sindbad the Sailor.'[210] And these borrowings are only a fraction of the substantial volume of cultural exchange. 'A work composed in Morocco or Cairo', says Renan,[211] 'was known to readers in Paris or Cologne in less time than it takes, in our own day, for an important

---

hidden away in Eugenio Larruga, *Memorias políticas y económicas*, Vol. XVII, Madrid, 1792, p. 115–117. Some Moriscos, *desterrados*, had returned to Spain (1613), men alone, without women and children. Should they be employed in the mercury mines at Almaden? No, a search should be made in the galleys for any experienced miners and their place taken in the galleys by these vagrants, more guilty than the galley-slaves, 'pues han sido de apostasia y crimen loesae Majestatis'.

On the survival of Moslem civilization, see the energetic and often original arguments of Julio Caro Baroja, *op. cit.*, p. 758 ff. On the actual expulsion of the Moriscos and the enormous transport it called for, see Henri Lapeyre, *op. cit.*, *passim*. This admirable book deals only with one aspect of the problem (the statistical); it has yet to be situated in relation to the whole political, social, economic and international history of Spain and here the task is far from complete: '. . . the expulsion of the Moriscos does not seem to be the act of a state in decline', *ibid.*, p. 213 – possible, but not proved. Other factors may have been demographic pressure, *ibid.*, p. 29 ff, and the hatred for a class of artisans and merchants who were also prolific. Until there is more information, I will accept the traditional explanation (see above, p. 796): religion decided who should go.

[208] Alfred Hettner, *art. cit.*, p. 202, or the brilliant remarks of André Malraux, *La Lutte avec l'ange*, 1945.

[209] On this enormous topic, see the illuminating book by E.-F. Gautier, *Mœurs et coutumes des Musulmans* (reprinted, 1955).

[210] Louis Gillet, *Le Dante*, 1941, p. 80.

[211] Quoted by Louis Gillet, *ibid.*, p. 94.

German book to cross the Rhine. The history of the Middle Ages will not be complete until someone has compiled statistics on the Arabic works familiar to scholars of the thirteenth and fourteenth centuries.' Is it surprising then that Moslem literature should have provided some of the sources of the Divine Comedy; that to Dante the Arabs were shining examples to be imitated,[212] or that St. John of the Cross had notable Moslem precursors, one of whom, Ibn 'Abbād, the poet of Ronda, had developed well before him the theme of the 'Dark Night'?[213]

From the time of the Crusades onwards, a movement in the opposite direction was beginning. The Christian became ruler of the waves. His now was the superiority and wealth brought by control of routes and trade. Alfred Hettner noticed the pattern of alternance, but he is clearly wrong when he suggests that in the sixteenth, seventeenth and eighteenth centuries,[214] contacts between East and West diminished. On the contrary, 'from the middle of the seventeenth to the end of the eighteenth century, European travellers' tales appeared in great numbers and in every European language'. For now travel in the East 'was open to permanent embassies, consuls, merchant colonies, missions of economic enquiry, scientific missions, Catholic missions . . . to adventurers entering the service of the Grand Turk'.[215] There was suddenly an invasion of the East by the West: an invasion which contained elements of domination.

But to return to the West in the sixteenth century: at this period it was far ahead of the East. Of this there can be no question, Fernand Grenard's arguments notwithstanding. To say so does not imply any value judgement on either civilization: it is merely to record that in the sixteenth century, the pendulum had swung the other way, western civilization now being the more forceful and holding the Islamic world in its dependence.

Human migration alone is sufficient indication of the change. Men flocked from Christendom to Islam, which tempted them with visions of adventure and profit – and paid them to stay. The Grand Turk needed artisans, weavers, skilled shipbuilders, experienced seamen, cannon-founders, the 'iron-workers' (in fact handling all metals) who constituted the chief strength of any state; 'as the Turks and several other peoples very well know', writes Montchrestien,[216] 'and keep them whenever they can catch them'. The curious correspondence between a Jewish merchant in Constantinople and Murād Agha in Tripoli shows that the former was looking for Christian slaves able to weave velvet and damask.[217] For prisoners too supplied the East with labour.

Was it because it was over-populated and not yet fully committed to trans-Atlantic colonization, that Christendom did not attempt to reduce the flow of emigrants to the East? On coming into contact with Islamic countries, Christians were often seized with the urge to turn Moslem. In

[212] Fernard Grenard, *Grandeur et décadence de l'Asie*, p. 34.
[213] Louis Gillet, in *Revue des Deux Mondes*, 1942, p. 241.
[214] *Ibid.*, p. 202.
[215] J. Sauvaget, *Introduction*, p. 44–45.
[216] *Op. cit.*, p. 51.
[217] A.d.S., Florence, Mediceo 4279.

the *presidios* on the African coast, Spanish garrisons were decimated by epidemics of desertion. On Djerba in 1560, before the fort surrendered to the Turks, a number of Spaniards had already joined the enemy, 'abandoning their faith and their comrades'.[218] Not long afterwards at La Goletta, a plot to surrender the position to the Infidel was uncovered.[219] Small boats frequently left Sicily with cargoes of candidates for apostasy.[220] At Goa the same phenomenon was observable among the Portuguese.[221] The call was so strong that it even reached the clergy. The 'Turk' who accompanied one of His Christian Majesty's ambassadors back to France, and whom the Spanish authorities were advised to capture en route, was a former Hungarian priest.[222] It cannot have been such a very rare occurrence: in 1630, Père Joseph was asked to recall the Capuchins living in the Levant, 'lest they turn Turk'.[223] From Corsica, Sardinia, Sicily, Calabria, Genoa, Venice, Spain, from every point in the Mediterranean world, renegades converged on Islam. There was no comparable flow in the other direction.

Perhaps unconsciously, the Turks were opening doors just as Christendom was shutting them. Christian intolerance, the consequence of large numbers, did not welcome strangers, it repelled them. And all those expelled from its lands – the Jews of 1492, the Moriscos of the sixteenth century and 1609–1614 – joined the ranks of the voluntary exiles, all moving towards Islam where there was work and money to be had. The surest sign of this is the wave of Jewish emigration, particularly during the second half of the sixteenth century, from Italy and the Netherlands towards the Levant: a wave which was strong enough to alert the Spanish agents in Venice, for it was through that city that this curious migration was channelled.[224]

Through these immigrants, sixteenth-century Turkey completed its western education. 'The Turks', wrote Philippe de Canaye in 1573, 'have acquired, through the renegades, all the Christian superiorities.'[225] 'All' is an exaggeration. For hardly had the Turk acquired one kind than he became aware of another still outside his grasp.

It was a strange war, this cultural rivalry, fought with any means, large or small. One day the enemy needed a doctor, another time a bombardier from the experienced artillery schools of the West; another time perhaps a cartographer or a painter.[226] He might be in search of precious goods: gunpowder, yew wood, which it is true grew round the Black Sea

[218] Paolo Tiepolo, 19th January, 1563, E. Albèri, *op. cit.*, I, V, p. 18.
[219] *Ibid.*
[220] In 1596 for example, report on Africa, Palermo, 15th September, 1596, Simancas E° 1158.
[221] G. Atkinson, *op. cit.*, p. 244.
[222] 4th September, 1569, Simancas E° 1057, f° 75.
[223] E. de Vaumas, *op. cit.*, p. 121.
[224] Francisco de Vera to Philip II, Venice, 23rd November, 1590, A.N., K 1674.
[225] *Op. cit.*, p. 120.
[226] As early as the fifteenth century, Pisanello.

(since in the past the Venetians had fetched it from there to sell to the English[227]) but no longer in sufficient quantities for the long bows needed by the sixteenth-century Turkish army which now imported this wood from southern Germany.[228] In 1570, Ragusa was accused, by Venice ironically enough, of providing the Turks with gunpowder and oars, and what was more a Jewish surgeon[229] – Ragusa so often herself in search of Italian doctors.[230] Towards the end of the century some of the most important English exports to the East were lead, tin and copper.

Pieces of artillery cast in Nuremberg may have been delivered to the Turks. Constantinople also received supplies from her frontier zones, through Ragusa or the Saxon towns of Transylvania, whether of arms, men, or as a letter from a Vlach prince to the people of Kronstadt suggests, of doctors and medical supplies.[231] The Barbary states too performed a similar service for the capital, for poor and literally 'barbaric' as they were, they stood – among Moslem states that is – closest to new developments of western progress: through their immigrants, voluntary or otherwise, their position in the western half of the sea and before long their connections with the Dutch, they were the first to be informed of technical innovations. They had their workers: the abundant crop of prisoners harvested every year by the corsairs of Algiers, and the Andalusians, skilful craftsmen some of whom could manufacture and all of whom could handle the carbine.[232] Was it purely coincidence that the reconstruction of the Turkish war fleet after 1571, armed in the western style (with arquebuses instead of bows and arrows and considerably reinforced artillery aboard the galleys) was the feat of a Neapolitan, Euldj ['Ulūj] 'Alī, a renegade schooled by the corsairs of Algiers?

But cultural grafts did not always take. In 1548, the Turks had tried, during their campaign against Persia, to revolutionize the equipment of the *sipāhis*, providing them with pistols ('minores sclopetos quorum ex equis usus est', Busbecq specifies[233]); the attempt was a hopeless failure and the *sipāhis*, at Lepanto and even later, were still armed with bows and arrows.[234] This minor example shows how difficult it could be for the Turks to imitate their enemies. Had it not been for the divisions among the latter, their quarrels and betrayals, the Turks for all their discipline and fanaticism and the excellence of their cavalry, would never have been able to hold their own against the West.

[227] B.N., Paris, Fr 5599.

[228] Richard B. Hief, 'Die Ebenholz-Monopole des 16. Jahrhunderts' in *Vierteljahrschrift für Sozial- und Wirtschaftsgeschichte*, XVIII, 1925, p. 183 ff.

[229] L. Voinovitch, *Histoire de Dalmatie*, 1934, p. 30.

[230] The Rectors to Marino di Bona, Ragusan Consul at Naples, 8th March, 1593, Ragusa Archives, L.P. VII, f° 17. A 'Lombard' doctor at Galata, N. Iorga, *Ospiti romeni*, p. 39.

[231] N. Iorga, *Ospiti romeni*, p. 37, 39, 43.

[232] A fact frequently mentioned even by Bandello, *op. cit.*, IX, p. 50.

[233] *Epist.* III, p. 199.

[234] J. W. Zinkeisen, *op. cit.*, III, p. 173–174.

All these gifts from the other side[235] were still insufficient to keep the Turkish bark afloat: it was showing signs of sinking from the end of the sixteenth century. Until then, war had been a powerful means of obtaining the necessary goods, men, techniques or products of technical advance, of picking up crumbs from the affluent Christian table, on land, on sea or from the zone bounded by Russia, Poland and Hungary. Gassot, when visiting the Arsenal at Constantinople, saw artillery pieces lying in heaps, most of them acquired as spoils of war rather than by informed purchases or local manufacture.[236] War is perhaps, who knows, the great equalizer of civilizations. The war in question had ended in deadlock in the Mediterranean by 1574 and in 1606, on the battlefields of Hungary, in an inescapable stalemate. At this point there first appeared that inferiority which was to become plainer still in later years.

Many Christians were mistaken, it is true, about the Ottoman future, in the early years of the seventeenth century, years which saw the revival of the idea of crusade.[237] But was it not perhaps division in Europe and the beginning of the Thirty Years' War which fostered illusions of Ottoman strength and, for the time being, saved the mighty eastern empire?

3. ONE CIVILIZATION AGAINST THE REST: THE DESTINY OF THE JEWS

All the conflicts discussed so far have been confined to a dialogue between two civilizations. In the case of the Jews, every civilization was implicated and invariably found itself in a position of overwhelming superiority. Against such strength and numbers the Jews were but a tiny band of adversaries.

But these adversaries had unusual opportunities: one prince might persecute them, another protect them: one economy might ruin them, another make their fortunes; one civilization might reject them and another welcome them with open arms. Spain expelled them in 1492 and Turkey received them, glad perhaps of the opportunity to use them as

[235] And penetration by Europeans, Catholic or Protestant, G. Tongas, *op. cit.*, p. 69; H. Wätjen, *op. cit.*, p. 69; the role of Venice divided between Capuchins and Jesuits, E. de Vaumas, *op. cit.*, p. 135; the affair of the Holy Sites in 1625, *ibid.*, p. 199; the eventful story of the patriarch Cyril Lascaris, K. Bihlmeyer, *op. cit.*, III, p. 181, G. Tongas, *op. cit.*, p. 130. Even North Africa was touched by this bloodless crusade, R. Capot-Rey, 'La Politique française et le Maghreb méditerranéen 1648–1685' in *Revue Africaine*, 1934, p. 47–61.

[236] Jacques Gassot, *Le Discours du voyage de Venise à Constantinople*, 1550, 2nd ed. 1606, p. 11. In the foundry at Pera, 40 or 50 Germans '. . . are making pieces of artillery', 1544, *Itinéraire de J. Maurand d'Antibes à Constantinople*, ed. Léon Duriez, 1901, p. 204.

[237] See below, p. 844.

[238] For a more complete bibliography, see the basic works by Attilio Milano, *Storia degli ebrei in Italia*, Turin, 1963 and Julio Caro Baroja, *Los Judios en la España moderna y contemporanea*, Madrid, 3 vols., 1961. The chief problem for the historian in this domain is his own attitude: can he, like Julio Caro Baroja, remain outside the drama he is describing, a mere spectator? Michelet would not have taken such a course.

counters against the Greeks. It was also possible to exert pressure, to be the indirect source of action, as the Jews of Portugal amply demonstrated.[239] They were able to obtain the tolerance which money can buy and at Rome they had an ambassador generally sympathetic to their cause. So it was comparatively simple to see to it that the measures enacted against them by the Lisbon government remained a dead letter: they were unfailingly either repealed or rendered ineffective, as Luis Sarmiento[240] explained to Charles V in December, 1535. The *conversos*, converted Jews, had obtained from the Pope a bull pardoning them for their previous errors, which would impede government action against them, the more so since the *conversos* had lent money to the king of Portugal who was hopelessly in debt: 500,000 ducats, not counting the rest in Flanders 'and on the exchanges'. Meanwhile the populace continued to mutter against these merchants of *peixe seco* (the dried fish eaten by the poor) and muttered very bitterly, *fieramente*, as a late Venetian letter observes in October, 1604, over half a century after the establishment of the Portuguese Inquisition in 1536.[241]

And then there were the perennial resources of the weak: resignation, the subtle distinctions learned from the Talmud, cunning, obstinacy, courage and even heroism. To complicate their case even further to the historian, the Jews, wherever they might find themselves, have always seemed extremely capable of adapting to the prevailing environment. They have proved quick to acclimatize themselves, whether their encounter with a civilization is lasting or short-lived. Jewish artists and writers have been recognized as authentic artistic representatives of Castile, Aragon or wherever it may be. They have adapted equally quickly to the social situations offered to or imposed upon them, humble and brilliant alike. It would seem then that they might be perilously near cultural shipwreck, that loss of identity of which there is no lack of examples. But in most cases, they succeeded in preserving what sociologists and anthropologists would call their 'basic personality'. They remained enclosed within their beliefs, at the centre of a universe from which nothing could dislodge them. This obstinacy, these desperate refusals are the dominant features of their destiny. The Christians were not mistaken when they complained that the *marranos* (the pejorative name for converted Jews[242]) secretly persisted in practising Judaism. There

[239] Léon Poliakoff, *Histoire de l'antisémitisme*, II, *De Mahomet aux Marranes*, Paris, 1961, p. 235 ff.

[240] Simancas, Guerra Antigua, 7, f° 42, Luis Sarmiento to Charles V, Evora, 5th December, 1535.

[241] A.d.S., Venice, Senato Dispacci Spagna, Contarini to the Doge, Valladolid, 4th October, 1604, Simancas, E° Portugal 436 (1608–1614) 'Licenças a varios judeus e cristãos novos de Portugal para sairem do reino'. Proof that it was still possible at this period to come to terms with the Portuguese authorities. Permission for new Christians to leave was granted in 1601 and withdrawn in 1610, J. Lucio de Azevedo, *Historia dos christãos novos portugueses*, 1922, p. 498.

[242] On this word, see I. S. Revah, 'Les Marranes' in *Revue des Études Juives*, 3rd series, Vol. I, 1959–1960, p. 29–77; on the persistence of Judaic practices, see J.

was quite undoubtedly a Jewish civilization, so individual that it is not always recognized as an authentic civilization. And yet it exerted its influence, transmitted certain cultural values, resisted others, sometimes accepting, sometimes refusing: it possessed all the qualities by which we have defined civilization. True it was not or was only notionally rooted to any one place; it did not obey any stable and unvarying geographical imperatives. This was one of its most original features, but not the only one.

An unquestionable civilization. The matter of this civilization was dispersed, scattered, like tiny drops of oil, over the deep waters of other civilizations, never truly blending with them yet always dependent on them. So its movements were always the movements of others, and consequently exceptionally sensitive 'indicators'. Émile-Félix Gautier, trying to find an equivalent of the Jewish diaspora, proposed as a very humble example, the history of the Mozabites of North Africa, who were also dispersed in very small colonies.[243] Another possible parallel is the case of the Armenians, mountain peasants who at about the time of the Renaissance in western Europe, were becoming international merchants from the Philippines to Amsterdam; or there are the Parsees in India or the Nestorian Christians of Asia. It is essential then to accept that there are civilizations of the diaspora type, scattering their countless islands in foreign waters, and they are more numerous than one might imagine at first sight: for instance the Christian communities in North Africa, from the Moslem conquest of the eighth century until the Almohad persecutions, in the thirteenth, which all but put an end to their existence. In a sense, the same is true of European colonies in the Third World countries both before and after independence; not to mention the Moriscos, the repository of Moslem civilization, whom Spain brutally cast out, in a gesture of cold hatred, as we have already seen.

If these islands made contact with each other, the whole situation could be very different. In medieval Spain for instance, until the ferocious persecution of the fourteenth and fifteenth centuries, Jewish communities tended to form a more or less coherent network, a sort of confessional nation,[244] a *millet* as the Turks called it, a *mellah* as it was known in North Africa. Portugal owed its originality to the fact that in 1492 its Jewish population was massively reinforced by refugees from Spain. The Levant received a similar influx for the same reason. So too in the suddenly enlarged Poland of early modern times, from the fifteenth century on, there was an increased Jewish presence, the result of large numbers,

Caro Baroja, *op. cit.*, *passim* or a few pages on the minor case of Majorca in the nineteenth-century work, *Histoire des races maudites de la France et de l'Espagne*, Paris, 1847, Vol. II, p. 33 ff.

[243] *Mœurs et coutumes des Musulmans*, *op. cit.*, p. 212.

[244] Léon Poliakov, *Histoire de l'anti-sémitisme*, II, *De Mahomet aux Marranes*, p. 127 ff: the Jewish nation in Spain. I have drawn heavily on this honest and intelligent book.

indeed almost a Jewish nation within the nation, a state within the state, which was to be swept away by economic hardship and pitiless repression in the seventeenth century, with the Chmielnicki massacres of 1648.[245] Again, in the sparsely inhabited Brazil of early days, the Jews were less threatened than elsewhere until the end of the sixteenth century.[246] The relative density of the Jewish population was always a significant factor.

But even when numbers did not favour or exaggerate the Jewish presence, these small communities, primary cells, were linked by education, beliefs, the regular travels of merchants, rabbis and beggars (who were legion); by the uninterrupted flow of letters of business, friendship or family matters; and by printed books.[247] Printing served Jewish quarrels but it served the cause of Jewish unity even more. To burn or confiscate all these books, so vital and so easily reproduced, would have been an impossible task. The life stories of certain wanderers are illustrations of the vitality of these unifying links. Jacob Sasportas was born towards the beginning of the seventeenth century in Oran, which was then held by the Spanish: he became a rabbi first at Tlemcen, then at Marrakesh and Fez; imprisoned, he escaped and fled to Amsterdam where he was a professor at the Pintò Academy; he returned to Africa and in 1655 accompanied Manasseh ben Israel on his embassy to London; he once more officiated as rabbi, in particular in Hamburg from 1666–1673; then returned to Amsterdam, was called to Leghorn, and returned to Amsterdam where he died.[248] This network of connections both explains and reinforces the coherence of the Jewish destiny. Johann Gottfried von Herder, in his *Ideen zur Philosophie der Geschichte der Menschheit* (1785–1792), was even then saying that 'the Jews continue to be an Asiatic people in Europe, foreign to our part of the world, inextricably the prisoners of an ancient law given to them under a distant sky'.[249]

[245] *Ibid.*, *I, Du Christ aux Juifs de Cour*, 1955, p. 266 ff, esp. p. 277 ff. (English translation by Richard Howard, *The History of Anti-Semitism*, Vol. I (the only one so far complete), *From the Time of Christ to the Court Jews*, p. 256 ff).

[246] Plinio Barreto, 'Note sur les Juifs au Brésil' in *O. Estado de São Paolo*, 31st October, 1936; there is a rich and rewarding literature on this subject starting with the classic studies by Gilberto Freyre and Lucio de Azevedo; the fundamental documentary evidence remains the three volumes of the *Primeira Visitação do Santo Officio as Partes do Brasil* pelo Licenciado Heitor Furtado de Mendoça . . ., deputado do Sto Officio: I. *Confissôes da Bahia, 1591–1592.* Introducçâo de Capistrano de Abreu, São Paulo, 1922; *Denunciacôes de Bahia, 1591–1593*, São Paulo, 1925; *Denunciacôes de Pernambuco, 1593–1595. Introducçâo de Rodolpho Garcia*, São Paulo, 1929. On Portugal, Léon Poliakov, *op. cit.*, *De Mahomet aux Marranes*, p. 235 ff.

[247] Léon Poliakov, *Du Christ aux Juifs de Cour*, p. VI–VII; *De Mahomet aux Marranes*, p. 139; Joseph Ha Cohen, *Emek Habakha ou la Vallée des Pleurs; Chronique des souffrances d'Israël depuis sa dispersion*, 1575 and *Continuation de la Vallée des Pleurs*, ed. Julien Sée, Paris, 1881, p. 167. (This book will be referred to from now on under the name of Joseph Ha Cohen.)

[248] Hermann Kellenbenz, *Sephardim an der unteren Elbe. Ihre wirtschaftliche und politische Bedeutung vom Ende des 16. bis zum Beginn des 18. Jahr.*, 1958, p. 45.

[249] Quoted by J. Lucio de Azevedo, *op. cit.*, p. 52.

And yet the Jews are not a race:[250] all scientific studies prove the contrary. Their colonies are biologically dependent on the host-nations with which they have lived for centuries. German Jews or *Ashkenazim*, Spanish Jews or *Sephardim* are biologically at least half German or Spanish, for there was frequent intermarriage and Jewish communities often originated in local conversions to Judaism; they never cloistered themselves from the outside world to which, on the contrary, they were often wide open. It would in any case be surprising if the accumulation of sometimes very many centuries had not led to such intermingling of different populations. The Jews who left Sicily in 1492 had been there after all for over 1500 years.[251]

Moreover the Jews did not always live apart, nor wear any distinctive dress or sign, such as the yellow cap or the *rotella*, a yellow badge, the 'segno de tela zala in mezo del pecto' as a Venetian text described it in 1496.[252] They did not always inhabit a separate quarter of the city, the *ghetto* (from the name of the quarter which was assigned to them in Venice and whose name is supposed to be derived from its formerly being the foundry where iron was cast – *ghettare* for *getare* – into moulds to make cannon).[253] In August, 1540, the Jews of Naples for example, in their battle against that deep-rooted hostility towards them which finally triumphed a year later, were still protesting against orders obliging them to 'live together and wear a special badge', 'habitar junto y traer señal', which was contrary to their privileges.[254] And even when there was official segregation, it was very often infringed and disobeyed. In Venice, Jews passing through the city and others, says a senatorial debate of March, 1556, 'have recently been spreading throughout the city, staying in Christian houses, going wherever they please, by day and night'. It is essential that this scandal cease: they must be ordered to live in the ghetto 'and not to keep an inn in any other part of the city but that one'.[255] At about the same period, Jews from Turkey were arriving in Italy wearing white turbans (the privilege of the Turks) when they should have been wearing yellow. It is knavery on their part, writes Belon,[256] they are usurping the good faith of the Turks which is better established in the West than their own. In 1566, though this was not the first alarm, the Jews of Milan were obliged to wear a yellow cap.[257]

Frequently segregation was enforced at a late date and was only partially effective. At Verona in 1599 (although it had been mooted since at least 1593) the Jews who 'lived scattered, one here, another there', had to take

[250] Léon Poliakov, *op. cit.*, I, p. 307 ff. (Eng. trans. p. 283 ff.)

[251] A. Milano, *op. cit.*, p. 221.

[252] A.d.S., Venice, Senato Terra 12, f°s 135 and 135 v°, 26th March, 1496. Cf. M. Sanudo, *op. cit.*, I, col. 81, 26th March, 1496.

[253] Giuseppe Tassini, *Curiosità veneziane*, Venice, 1887, p. 319.

[254] Simancas, E° Napoles 1031, f° 155, Naples, 25th August, 1540. Many references to the Jews in this *legajo*.

[255] A.d.S., Venice, Senato Terra, 31, 29th March, 1556.

[256] *Op. cit.*, p. 181.

[257] September, 1566, Joseph Ha Cohen, *op. cit.*, p. 158.

up residence 'near the main square of the city',[258] 'where they sell wine', along the street running to the church of San Sebastiano, thereafter popularly known as the *via delli Hebrei*.[259] It was not until 1602 that a similar measure was enacted at Padua, where until then, the 'Israelites had for the most part lived scattered in all four corners of the town'.[260] In August, 1602, there were incidents at Mantua arising from the fact that Jews were walking about in black caps like anybody else.[261]

In Spain and Portugal, coexistence had been the rule for centuries. In Portugal, one of the most common popular demands concerned the obligation laid by the Pope on the Jews (who did not observe it) to wear distinctive marks upon their clothing, in order, the Cortes went so far as to say, to prevent the many attempts by Jews to seduce Christian women. Jewish tailors and shoemakers were frequently accused of seducing the wives and daughters of the peasants in whose houses they went to work.[262] In fact, in Portugal, the Jews had intermarried with the aristocracy even more than with the common people. In Turkey, the Jews had Christian slaves, both men and women, and 'use Christian slave-women with no more qualms over mixing with them than if they were Jewish women'.[263] Whichever side found itself proscribed, the isolation of the Jewish communities was the result not of racial incompatibility, as is often suggested, but of the hostility of others towards them and their own feelings of repugnance towards others. The root of it all was religion: isolation was the consequence of a whole complex of inherited habits, beliefs, even methods of preparing food. Of converted Jews, Bernaldez, the historiographer of the Catholic Kings,[264] says: 'they never lost the habit of eating in the Jewish manner, preparing their meat dishes with onions and garlic and frying them in oil, which they use instead of bacon fat' – to a modern reader a description of Spanish cooking today. But the use of pork fat in cooking was of course the way of the Old Christians and, as Salvador de Madariaga says, its eventual displacement by oil was a legacy of the Jews, a cultural transfer.[265] The converted Jew or *marrano* also gave himself away by carefully forgetting to light a fire in his house on Saturdays. An inquisitor once said to the governor of Seville: 'My lord, if you wish to see how the *conversos* keep the Sabbath, come up the tower with me.' When they reached the top, he pointed around, saying 'See how you can tell the houses of these *conversos*: however cold it is, you will never see smoke coming from their chimneys on a Saturday.'[266] This story, told by Ibn

[258] *Ibid.*, p. 207.

[259] According to Lodovico Moscardo, *op. cit.*, p. 441, the ghetto project dated back to 1593.

[260] Joseph Ha Cohen, *op. cit.*, p. 215–216.

[261] Museo Correr, Cicogna 1993, f° 261, 16th August, 1602.

[262] J. Lucio de Azevedo, *op. cit.*, p. 10.

[263] Pierre Belon, *op. cit.*, p. 180, 193 v°.

[264] Quoted by Léon Paliakov, *op. cit.*, II, p. 180.

[265] *Ibid.*, based on S. de Madariaga, *Spain and the Jews*, 1946.

[266] *Ibid.*, p. 191. Ibn Verga, *The Scourge of Juda*, quoted by L. Poliakov, *op. cit.*, Vol. II, p. 64, from the German translation by Wiener, Hanover, 1856.

Verga (about 1500) has a ring of truth; cold spells in Seville in winter were certainly only too real. There were other revealing details: in the Levant, the Jews 'will never eat of any meat prepared by a Turk, Greek or Frank and will accept no fat from either Christians or Turks, nor will they drink any wine sold by a Turk or Christian'.[267]

But all Jewish communities were obliged to engage in a dialogue, sometimes in dramatic circumstances when around them the entire nature of the dominant civilization changed. The Moslems replaced the Christians in Spain, then the Christians returned after the belated victories of the Reconquest. Jews who had spoken Arabic now had to learn Spanish. They were in the same unhappy position in Hungary when, with the imperial advance of 1593–1606, the Jews of Buda were caught between the twin perils of the Imperials and the Turks.[268] Changing circumstances made them the involuntary heirs of once-powerful civilizations, whose gifts they were to pass on in one direction or another. Unintentionally, they were until the thirteenth century and even later, the intermediaries through whom the West received Arab thought and science, as philosophers, mathematicians, doctors and cosmographers. In the fifteenth century they rapidly developed an enthusiasm for printing: the first book to be printed in Portugal was the Pentateuch (at Faro, in 1487, by Samuel Gacon). Not until about ten years later did German printers appear in Portugal.[269] When one remembers that printing was not introduced to Spain by the Germans until 1475, the haste with which the Jews set about printing sacred texts is the more striking. Expelled from Spain in 1492, the Jews took the art of printing with them to Turkey. By 1550, they had 'translated all manner of books into the Hebrew tongue'.[270] Founding a press was a work of devotion, undertaken for example by the widow of Joseph Micas in the countryside of Koregismi, near Constantinople.[271]

In 1573, Venice was preparing to drive out her Jews in accordance with the decision of 14th December, 1571.[272] But things had changed since Lepanto, and at this point Soranzo arrived from Constantinople, where he had held the office of *bailo*. According to a Jewish chronicler,[273] he addressed the Council of Ten in the following terms: 'What pernicious act is this, to expel the Jews? Do you not know what it may cost you in years to come? Who gave the Turk his strength and where else would he have found the skilled craftsmen to make the cannon, bows, shot, swords, shields and bucklers which enable him to measure himself against other powers, if not among the Jews who were expelled by the Kings of Spain?' An earlier French description of Constantinople[274] (about 1550) had al-

[267] Pierre Belon, *op. cit.*, p. 181. [268] *Ibid.*, p. 209–210.
[269] J. Lucio de Azevedo, *op. cit.*, p. 36. [270] Belon, *op. cit.*, p. 180 v°.
[271] J. Ha Cohen, *op. cit.*, p. 251, according to E. Carmoly, *Archives israélites de France*, 1857.
[272] A. Milano, *op. cit.*, p. 180 ff.
[273] The continuation of J. Ha Cohen, *op. cit.*, p. 181.
[274] B.N. Paris, Fr. 6121 (undated). See also L. Poliakov, *op. cit.*, II, p. 247, references to the voyage of G. d'Aramon and Nicolas de Nicolay.

ready noted as much: 'and these [the *marranos*] are the men who have made known to the Turks the manner of trading and handling those things we use mechanically.'

They had a further advantage: the Jews were, in the East, born interpreters of all speech and without their help much business would have been impossible or difficult. Belon[275] explains: 'those of them who left Spain, Germany, Hungary and Bohemia have taught the languages [of those countries] to their children; and their children have learnt the languages of the nations in which they have to live and speak, it might be Greek, Slavonic, Turkish, Arabic, Armenian or Italian' . . . The Jews who live in Turkey ordinarily speak four or five languages: and there are several who know ten or twelve.' He made a similar observation at Rosetta in Egypt, where the Jews, 'have so multiplied over all the lands ruled by the Turk, that there is no town or village where they do not live and increase. And so they speak every language, and have been of great service to us, not only in translating for us but in communicating to us how things are in that country.'[276]

Linguistically, it is curious that the Jews expelled from Germany in the fourteenth, fifteenth and even sixteenth centuries, who were to contribute to the fortunes of Polish Jewry, should have introduced their own language, Yiddish,[277] a form of German, just as the Spanish Jews who after 1492 formed the large colonies in Istanbul and above all Salonica, brought with them their own language, *ladino*, Renaissance Spanish, and preserved a genuine feeling for Spain, of which there is abundant evidence,[278] proof that the soil of a man's native land may cling to his shoes. Such memories survive in curious details: a student of Spanish literature of our own day has discovered among the Jews of Morocco knowledge of the words and melodies of medieval Spanish romances;[279] a historian has noted the reluctance and lack of facility with which the Sephardites of Hamburg adapted to the German tongue.[280] Loyalty to their origins persisted too in the names of the Jewish communities at Salonica – *Messina*, *Sicilia*, *Puglia*, *Calabria*.[281]

Such fidelity was not without its drawbacks: it created categories. Several Jewish nations could be distinguished and there was sometimes conflict between them. Venice for instance set up one after another, between 1516 and 1633, three ghettoes: the *vecchio*, the *nuovo* and the *nuovissimo*, linked islands where the houses stood sometimes as much as seven storeys high – for space was scarce and the density of the population here was the highest in the city. The *ghetto vecchio*, reserved for Jews from the Levant (*levantini*), had been under the control of the *Cinque Savii alla Mercanzia* since 1541; the *nuovo*, under the control of the *Cattaveri*,

[275] *Op. cit.*, p. 180 v°, p. 118. [276] *Ibid.*, p. 100 v°.

[277] L. Poliakov, *op. cit.*, I, p. 270–271 (Eng. trans. p. 250).

[278] *Ibid.*, p. 249 and 250 (Eng. trans. p. 250 n.); *La Méditerranée*, 1st ed., p. 707–708.

[279] Paul Bénichou, *Romances judéo-españoles de Marruecos*, Buenos Aires, 1946.

[280] H. Kellenbenz, *op. cit.*, p. 35 ff.

[281] A. Milano, *op. cit.*, p. 235.

harboured German Jews (*Todeschi*), some of whom, since there was not room for all, went to live in the old ghetto. These *Todeschi*, who had been accepted in the city at the time of the League of Cambrai, were poor Jews dealing in second-hand clothes and pawnbroking and they were to run the *Monte di Pietà* in Venice – 'li banchi della povertà'. Meanwhile certain Jews specializing in large-scale trade, Portuguese and Levantine, by turns detested and wooed by the Signoria, obtained a special status, probably after 1581.[282] But in 1633, all the Jews, including the *Ponentini*, were confined to the same ghettoes – hence the many social, religious and cultural conflicts within this artificial near-concentration-camp world.

Such differences could not however prevent the existence of a Jewish civilization in its own right, a civilization full of vitality and movement, and certainly not inert or 'fossilized', as Arnold Toynbee calls it.[283] It was on the contrary both vigilant and aggressive, swept from time to time by strange messianic outbursts, particularly in the early modern period when it was divided between, on the one hand, that rationalism which led some towards scepticism and atheism, well before Spinoza, and on the other, the propensity of the masses to irrational superstition and exaltation. All persecution tended to produce by reaction messianic movements, for example that of the so-called Messiahs David Rubeni and Diego Pires who caused a stir among the Portuguese Jews in the time of Charles V, between 1525 and 1531,[284] or the immense wave of popular feeling provoked by the Messianic propaganda of Sabbataï Zevi[285] in the East, Poland and even farther afield.

But even apart from these acute crises, it would be wrong to assume that the Jewish attitude was ordinarily peaceful and tolerant. There were unmistakable signs of activity, combative spirit and eagerness to proselytize. The ghetto may have been the prison within which the Jews were confined but it was also the citadel into which they withdrew to defend their faith and the continuity of the Talmud. A historian as sympathetic towards the Jewish cause as the great Lucio de Azevedo maintained that Jewish intolerance, at the beginning of the sixteenth century, was 'certainly greater than that of the Christians',[286] which is probably an exaggeration. But clearly intolerance there was. It was even rumoured – absurd though it seems – in about 1532, that the Jews had tried to convert Charles V to the Mosaic faith during his stay at Mantua![287]

[282] Cecil Roth, in *Studi in onore di Gino Luzzatto*, Milan, 1949; for example A.d.S., Venice, *Cinque Savii* 7, f°s 33–34. 15th December, 1609. On the three ghettoes and the obviously debatable origin of the word, details and arguments are set out in G. Tassini, *op. cit.*, p. 319–320; it is still not clear exactly how the three Jewish communities were divided among the three ghettoes, even after reading A. Milano, *op. cit.*, p. 281.

[283] Arnold Toynbee, *A Study of History*, abridged by D. C. Somervell, Vols. I–VI, p. 361, 380, 389, etc.

[284] J. Lucio de Azevedo, *op. cit.*, p. 68–73.

[285] L. Poliakov, *op. cit.*, II, p. 262 ff.

[286] *Op. cit.*, p. 39.

[287] F. Amadei, *Cronaca universale della città di Mantoa*, II, p. 548.

The ubiquity of Jewish communities. Willingly or unwillingly, the Jews were forced into the role of agents of cultural exchange. It could hardly have been otherwise. They were, or had been, everywhere; despite expulsion orders they did not always leave the forbidden land, and they might return. Officially, they were absent from England between 1290 and 1655, the date of their so-called 're-entry' under Cromwell; in fact London had its Jewish merchants from the beginning of the seventeenth century and perhaps earlier. France too in theory expelled all Jews in 1394 but they very soon reappeared (as *marranos* and outwardly Christian it is true) in Rouen, Nantes, Bordeaux, Bayonne, the natural stopping places for Portuguese *marranos* travelling to Antwerp and Amsterdam. Henri II, 'king of France, allowed the Jewish merchants of Mantua to enter the cities of his kingdom and to trade in the country. He also exempted them from taxes and when they went to present him with their thanks and homage, he showed himself benevolent towards them, that year',[288] no doubt in 1547. More remarkable, if not more important, was the rumour that circulated in spring 1597 in Paris and possibly in Nantes, where it was picked up by Spanish intelligence, to the effect that the king of France was thinking of 'bringing back the Jews whom the Most Christian King St. Louis had expelled'.[289] The rumour reappeared four years later in 1601. 'A leading Jew [of Portugal]', Philippe de Canaye the ambassador explained to Henri IV, 'told me that if Your Majesty would permit his nation to live in France, you would receive much profit from it, and would people your Kingdom with over 50,000 intelligent and hard working families.'[290] Towards 1610, among the Moriscos who entered France, usually on their way elsewhere, there were a notable number of Jews and Portuguese *marranos* who mingled with the exiles and 'probably settled under Christian masks in France and particularly in the Auvergne'.[291]

In the south of France, the Jews were few in number. Towards 1568–1570, they were driven out of the cities of Provence and were received amicably in Savoy.[292] In Marseilles, where municipal policy varied, there were only a few Jews at the beginning of the seventeenth century.[293] Jews expelled from Spain in 1492 settled in Languedoc, remained there and 'accustomed [the French] to trade with Barbary'.[294] As New Christians they became apothecaries and doctors at Montpellier: Felix Platter lodged in the house of one of them. In Avignon at the end of the century, when

[288] Joseph Ha Cohen, *op. cit.*, p. 127.

[289] A.N., K 1600, 4th April, 1597, *Relacion de algunas nuebas generales que se entendien de Nantes de Paris y otras partes desde 4 de abril 97:* '. . . quiere hazer benir los judios que hecho el cristianissimo Rey St Luis . . .'

[290] Quoted by L. Poliakov, *op. cit.*, II, p. 368, *Lettres et ambassades de Messire Philippe Champagne*, 1635, p. 62.

[291] Quoted by L. Poliakov, *op. cit.*, II, p. 367–368, from Francisque Michel, *Histoire des races maudites de la France et de l'Espagne*, 1847, p. 71 and 94.

[292] J. Ha Cohen, *op. cit.*, p. 160.

[293] H. Kellenbenz, *op. cit.*, p. 135.

[294] Jean Bodin, *Response . . .*, *op. cit.*, ed. H. Hauser, p. 14.

his brother Thomas was there, they numbered about 500, protected by the Pope, but did not have the right 'to buy either house, garden, field or meadow, within or without the town', and were reduced to the trades of tailor or old clothes dealer.[295]

Germany and Italy were of course too divided to be able to expel the Jews simultaneously from every region, and yet heaven knows they were harassed enough. One city would close its gates to them, another open them up. When Milan, after much hesitation, finally ordered the few 'Hebrews' in the city to leave in 1597, the latter went, as far as we can see, to Vercelli, Mantua, Modena, Verona, Padua 'and the surrounding localities'.[296] Their trek from door to door, even when unsuccessful, had farcical overtones: in Genoa for instance, from which city the Jews were solemnly expelled in 1516 – only to return in 1517.[297] The same kind of thing happened at Venice and Ragusa because they were always allowed back in the end: in May, 1515, Ragusa was stirred up by a Franciscan monk and drove out its Jews; the latter immediately organized a grain blockage in Apulia and Morea against the Republic of St Blaise (proving that they controlled the grain trade) and the city had to take them back; in 1545, the thought of expelling them again had hardly crossed the mind of the Ragusans before the sultan was calling them to order.[298] In 1550, it was the turn of Venice to think of expelling the Jews, but she immediately realized that they controlled and handled the bulk of her trade: wool, silk, sugar, spice – and that the Venetians themselves were often content merely to retail the merchandize of Jews, 'guadagnando le nostre solite provizioni', earning only the usual commission.[299] In fact Italy had taken in large numbers of Jews, following the mass expulsions from France, Spain and Portugal, particularly the papal states, where more Jews settled for preference than elsewhere. They became remarkably prosperous at Ancona: before their violent persecution at the hands of Paul IV in 1555 and 1556, they numbered 1770 heads of household and bought as much property as they wished, houses, vineyards, 'and bore no sign to distinguish them from Christians'.[300] In 1492, the expulsion of the Jews from Sicily affected over 40,000 persons,[301] we are told, the vast

[295] Thomas Platter, *Journal of a Younger Brother*, transl. by Seàn Jennett, London, 1963, p. 81.

[296] J. Ha Cohen, *op. cit.*, p. 200.

[297] *Ibid.*, p. 112–113.

[298] S. Razzi, *op. cit.*, p. 118–119 (1516); p. 159 (1545). NB the intervention of Sulaimān the Magnificent against the persecution of the Jews and *marranos* of Ancona, A. Milano, *op. cit.*, p. 253. C. Roth, *The House of Nasi, Doña Gracia*, Philadelphia, 1947, p. 135–174.

[299] Werner Sombart, *Die Juden und das Wirtschaftsleben*, 1922, English translation by M. Epstein, Glencoe Free Press, 1951, *The Jews and Modern Capitalism*, p. 17, based on the document published by David Kauffmann, 'Die Vertreibung der Marranen aus Venedig im Jahre 1550', in *The Jewish Quarterly Review*, 1901. On the expulsion order against the Marranos, Marciana, 2991 C. VII 4, f° 110 v° and 111; Museo Correr, Donà delle Rose, 46, f° 155, 8th July, 1550.

[300] Marciana, 6085, f° 32 v° ff: account of the persecutions of 1555 and 1556. Cf. also A. Milano, *op. cit.*, p. 247–253.

[301] L. Bianchini, *op. cit.*, I, p. 41. But not 160,000, A. Milano, *op. cit.*, p. 222.

bulk of whom were the humble artisans whose departure the island could not well afford. In Naples, by contrast, which did not fall under the control of the Catholic King until ten years later, the Jews were permitted to remain, in small numbers it is true but including such rich and active families as the Abravanel, until 1541.[302]

It may seem incongruous to compare these fugitive Jews with the vigilant bands of outlaws, but after all both Jews and brigands were able to take advantage of a complicated political map, whether in Italy or Germany. Germany had the convenient nearby refuge of Poland, towards which the wagons piled with the possessions of refugees could make their way; Italy offered convenient means of escape by sea and to the Levant. When, in 1571, there was talk of expelling the Jews from Venice, some were already aboard departing ships when the order was revoked.[303] Escape by sea was not without its risks of course. The master of a vessel might be tempted to sell his passengers and seize their belongings. In 1540, the captain of a Ragusan ship robbed his passengers, Jews fleeing from Naples, and abandoned them at Marseilles, where the king of France, François I, took pity on them and sent them in his own ships to the Levant.[304] In 1558, Jews escaping from Pesaro[305] made their way to Ragusa then took ship for the Levant. The ship's crew, possibly Ragusan, seized them and sold them as slaves in Apulia. In 1583, a Greek crew massacred 52 of their 53 Jewish passengers.[306]

Always in search of towns 'where their weary feet could rest',[307] the Jews finally and inevitably ended scattered everywhere. In 1514 there were Jews living in Cyprus, where the Rectors received orders from the Signoria of Venice not to authorize any of these Jews to wear the black instead of the yellow cap.[308] In Istanbul, twelve Cretan Jews are recorded in poor circumstances, and we learn that on their island, 'they number above 500'.[309] On another Venetian island, Corfu, they numbered 400, 'sparsi per la citta con le lor case conggionte con quelle di Christiani', scattered through the town, their houses mingled with those of Christians: it would be wise, continued our document, to separate them from each other for the satisfaction of both sides.[310] In fact the Jews of Corfu were always to enjoy certain advantages in their relations with the Venetian authorities.[311]

If one wished to pursue the dispersion of the Jews throughout the Greater Mediterranean and indeed the world, one could easily find them

302 A. Milano, *op. cit.*, p. 233.
303 J. Ha Cohen, *op. cit.*, p. 180.
304 *Ibid.*, p. 121.
305 *Ibid.*, p. 143.
306 A. Habanel and E. Eskenazi, *Fontes hebraici ad res oeconomicas socialesque terrarum balcanicarum saeculo XVI pertinentes*, Sofia, 1958, I, p. 71.
307 The expression, quoted in this case from Joseph Ha Cohen, was widely used.
308 Museo Correr, Donà delle Rose, 46, f° 55, 5th June, 1514.
309 Museo Correr, Donà delle Rose 21, f° 1, Constantinople, 5th March, 1561. See also, on the ghetto or 'zudeca' of Candia, A.d.S., Venice, Capi del Cons° dei X, Lettere, B^a 285, f° 74, Candia, 7th May, 1554.
310 Museo Correr, Donà delle Rose, 21, 1588.
311 A. Milano, *op. cit.*, p. 236, 281, 283 . . .

in Goa, Aden, in Persia, 'under the stick, in the shadow of which they pass their weary lives throughout the Levant', but this comment dates from 1660[312] when the wheel had turned and it would turn again. In 1693, a French document describes Portuguese and Italian Jews as having been settled in the Levant 'for forty years' and having placed themselves under the protection of the French consuls at Smyrna. They had also managed to enter Marseilles, where 'imperceptibly they had possessed themselves of a major part of the Levant trade, which obliged the late M. de Seignelay to expel them from Marseilles by royal ordinance'.[313] But they were soon handling the other end of the line, in the Levant. There were Jews in Madeira and so numerous were they on the island of São Tomé, that (these were obviously New Christians) they 'openly' practised Judaism;[314] they were among the earliest arrivals in America and the earliest martyrs, (in 1515 in Cuba), of the Spanish Inquisition,[315] which did not stop there; in 1543, Philip as regent of the kingdoms of Spain had expelled them – a purely theoretical gesture – from the Castilian Indies.[316] Jews were also numerous in North Africa as far south as the Sahara.

Judaism and capitalism. The Jew, originally, like the Armenian, a peasant, had many centuries before turned away from life on the land. Now he was invariably financier, supply-master, merchant, usurer, pawnbroker, doctor, artisan, tailor, weaver, even blacksmith. He was often very poor: sometimes an extremely modest pawnbroker. Amongst the poorest undoubtedly, were the Jewish women who sold haberdashery, handkerchiefs, napkins, and bed canopies,[317] in the markets of Turkey, and all the Jews scattered throughout the Balkans, whose disputes and occupations (usually modest) we can gather from the records of rabbinical decisions.[318] Pawnbrokers, even the very humblest, constituted as it were the bourgeoisie of these often impoverished communities. In Italy the number of such small moneylenders was high and their services were much appreciated in the countryside and small market towns. In September, 1573, the *podestà* of Capodistria[319] asked for a Jewish banker to be sent to the town, otherwise the inhabitants, victims of continually rising prices, would be obliged to go (as they were already) to usurers in Trieste who lent money at 30 and 40 per cent; this would never happen with a local

[312] *État de la Perse en 1660, par le P. Raphaël du Mans*, ed. Ch. Schefer, Paris, 1890, p. 46.

[313] A.N., A.E., B III 235, 1693.

[314] In Madeira as late as 1682, Abbé Prévost, *op. cit.*, III, p. 172; Lisbon, 14th February, 1632 '. . . the island (São Tomé) is so infested with New Christians, that they practise the Jewish rites almost openly', J. Cuvelier and L. Jadin, *L'ancien Congo d'après les archives romaines, 1518, 1640*, 1954, p. 498.

[315] Prologue by Fernando Ortiz to Lewis Hanke, *Las Casas . . .*, p. XXXVI.

[316] Jacob van Klaveren, *op. cit.*, p. 143.

[317] Pierre Belon, *op. cit.*, p. 182 and 182 v°.

[318] A. Habanel and E. Eškenasi, *op. cit.*, I, 1958 (sixteenth century); II, 1960 (seventeenth century).

[319] A.d.S., Venice, Senato Terra, 63, 20th September, 1573.

Jewish moneylender. In the following year, 1574, the *povera communità* of Castelfranco asked the Signoria of Venice, which granted the request on 6th April, to allow 'Josef ebreo di tener banco nella cittadina, col divieto però di poter prestare salvo che sopra beni mobili', to lend money on movables, that is to say personal belongings, only.[320] Similarly in 1575, the *communità* of Pordenone petitioned in turn 'for the sake of the many poor', for 'un ebreo a tener banco'.[321] All this does not necessarily imply that relations were always good between Jewish moneylenders and their Christian clients. In 1573, the *communità* of Cividale del Friul[322] had asked 'to be liberated from Hebrew voracity which is continually devouring and consuming the poor of this town'. A *monte di hebrei* was looted at Conegliano in July, 1607 by highwaymen, *fuorusciti*. The *capelletti* of the Signoria (seventeenth century *carabinieri*) gave chase and recovered the stolen goods (5000 ducats' worth of jewels and other pledges), killed four of the bandits whose heads were carried to Treviso, and brought back two prisoners alive.[323]

But as well as small-time moneylenders and usurers, there were the great Jewish merchant families, sometimes expelled only to be recalled, always in demand. We find them at Lisbon masquerading as new Christians, or, if they were rich, as perfect Christians, the Ximene, Caldeira and Evora. They might be innovators: for example Michael Rodríguez or Rodrigua, the Levantine Jew of Venice who conceived the idea of the port of Spalato;[324] they might be powerful like the rich Abravanel family. Samuel Abravanel and his relations for years controlled the fate of the Jews of Naples, lending money to the king, holding interests in the Madeira sugar trade, the Lanciano fairs and the grain trade.[325] Success on a colossal scale can be glimpsed in the unparalleled career of the Portuguese Mendes family, in particular of a nephew, Juan Mínguez or Miques, the Joseph Micas of Spanish *avisos* from the Levant.[326] A *marrano*, he

[320] *Ibid.*, 63, 6th April, 1574.

[321] *Ibid.*, 66, 1575.

[322] *Ibid.*, 60, 1573.

[323] A.d.S., Florence, Mediceo 3087, f° 348, 14th July, 1607.

[324] See above, Vol. I, p. 286–287.

[325] The *casa* of the *Abravaneles* was of Spanish origin. For its loans to the king see Simancas E° Napoles 1015, f° 101, 6th October, 1533; *ibid.*, f° 33; 1018, f° 21, 15th January, 1534, if usury is not practised by the Jews, it will be practised by Christians at triple rates, 'porque el fin de Ytalia como V. M. tiene mejor experimentado y conocido, es ynterese', *ibid.*, f° 58, 3rd October, 1534, 'at Naples there are 300 to 400 Jewish families'; 1017, f° 39, 28th March, 1534, new Christians arrested in Manfredonia, 'que debaxo de ser xpianos han bibido y biben como puros judíos'; 1018, f° 58, 3rd October, 1534: the city of Naples asks to keep its Jews, without them last year the poor would all have starved to death; 1031, f° 155, 25th August, 1540, anti-Jewish measures; 1033, f° 70, 19th June, 1541, their expulsion is fixed . . . A.d.S., Naples, Sommaria Partium, 242, f° 13 v°, 16th April, 1543, Samuel Abravanel has his factor, Gabriele Isaac, fetch 120 *carri* of grain from Termoli; *ibid.*, 120, f° 44, 8th June, 1526, one Simone Abravanel, 'a Jew living in Naples' was importing sugar from Madeira.

[326] L. Poliakov, *op. cit.*, II, p. 254 ff, for an excellent summary of this extraordinary man's career. The basic works are Cecil Roth's *The House of Nasi*, 2 vols., 1947 and *The Duke of Naxos*, 1948.

reverted to Judaism at Constantinople, where he became a sort of eastern Fugger, powerful almost until his death (1579), dreaming of becoming a 'King of the Jews' and founding a state in the Holy Land (he had surveyed the ruins of Tiberias), or 'King of Cyprus', finally contenting himself with the title conferred on him by the sultan, of Duke of Naxos, the name by which he is known to those historians, usually willing hagiographers, who have concerned themselves with him.

But even this outstanding success depended on the general economic situation. Historians writing about sixteenth-century Turkey have noted (rather late in the day perhaps) the triumph of Jewish merchants.[327] They, with the Greek merchants, were soon to be farming the fiscal revenues of the state and even the revenues of rich landowners and the network of their interests stretched over the whole empire. Pierre Belon, who observed them in about 1550, says of them: 'They have so taken hold of trade in merchandise in Turkey that the riches and revenue of the Turk are in their hands; for they put the highest price on the collection of tribute from the provinces, farming the salt-taxes, the taxes on shipped wine and other things in Turkey.' And, he concludes, 'since I have many times been obliged to use the services of the Jews and to frequent them, I have quickly learnt that theirs is the most intelligent of nations and the most malicious'.[328] Had it not been for such general success, careers like that of the Duke of Naxos would have been impossible, much as I imagine the fortunes after the Thirty Years' War of the German Jewish financiers, the *Hofjuden* or court Jews,[329] would have been inconceivable without the accumulation of riches following the Peace of Augsburg (1555) which paved the way for future revenge on German Jewry. Similarly, at the end of the sixteenth century, the network of Portuguese Jews who controlled the sugar and spice trades and possessed ample capital, furthered the success of Amsterdam. America too was completely covered with their network of relations.

That is not to say that all Jewish merchants were rich and trouble-free; nor that Judaism was, by any speculative vocation or ethical assumptions, responsible for what we now call the capitalism, or rather pre-capitalism of the sixteenth century; nor that 'Israel passes over Europe like the sun: at its coming new life bursts forth; at its going all falls into decay.'[330] It is rather that the Jews were able to adapt to the geography as well as the changing circumstances of the business world. If Israel was a 'sun' it was a sun teleguided from the ground. Jewish merchants went towards regions of growth and took advantage of their advance as much as they contributed to it. The services rendered were mutual. Capitalism can mean many things. It implies amongst others a system of calculation, an acquaintance with certain techniques relating to money and credit: even before the fall

[327] See above, p. 727 ff. [328] *Op. cit.*, p. 180 v° and 181.
[329] Werner Sombart, *op. cit.* (English version), p. 52 ff; L. Poliakov, *op. cit.*, I, p. 249 ff. (Eng. trans., p. 229 ff.)
[330] W. Sombart, *op. cit.*, p. 13.

of Jerusalem to the Crusaders in 1099, the Jews were already familiar with the *suftaya* (bill of exchange) and the *sakh* (cheque),[331] which were in common use in the Moslem world. This acquisition was maintained through all the forced migrations of Jewish communities.

But beyond this, capitalism, to be successful, presupposes a network, the organization of mutual confidence and cooperation throughout the world. The Revocation of the Edict of Nantes (1685) did not automatically lead to the success of Protestant banking, which had been inaugurated in the sixteenth century, but it ushered in a period of great prosperity for it, the Protestants possessing in France, Geneva, the Netherlands and England a network of intelligence and collaboration. The same had been true for centuries of the Jewish merchants. They formed the leading commercial network in the world, for they had representatives everywhere: in backward or under-developed regions where they were artisans, shopkeepers and pawnbrokers and in key cities where they participated in economic growth and booming trade. Their numbers might be very small: there were only 1424 Jews in Venice in 1586;[332] barely a hundred in Hamburg[333] at the beginning of the seventeenth century; 2000 at most in Amsterdam, 400 in Antwerp in 1570.[334] Giovanni Botero[335] does, at the end of the sixteenth century, mention 160,000 Jews in Constantinople and Salonica,[336] the latter city being the principal refuge of exiles, but he notes hardly 160 families at Valona, the same in Santa Mauro, 500 in Rhodes, 2500 altogether in Cairo, Alexandria, Tripoli in Syria, Aleppo and Ankara. Such figures are only moderately reliable. But we can say with confidence that where the population was densest, in Constantinople and Salonica for instance, difficulties could be expected and the exiles would have to be prepared to work at any trade, even the lowest-paid; these were the weavers and dyers of Salonica, Istanbul and elsewhere, the itinerant merchants at local fairs buying fleeces and hides. Smaller colonies, on the other hand, frequently consisted of opulent merchants, favoured by the concentration of several rich trades, often attracted by these trades and therefore recent arrivals.

In the thirteenth century, the Fairs of Champagne were the centre of western commerce. Into the fairs and out of them flowed every kind of merchandise. The Jews were there too, in the towns and villages of Champagne,[337] some of them concerned in the agricultural life of the region, others artisans, owning fields, vines, real estate or houses which

[331] I am of course referring to the *Geniza* of Cairo shortly to be published by A. Gothein.

[332] Daniele Beltrami, *Storia della popolazione di Venezia . . ., op. cit.*, p. 79.

[333] H. Kellenbenz, *Sephardim an der unteren Elbe*, 1958, p. 29.

[334] *Ibid.*, p. 139.

[335] Giovanni Botero, *op. cit.*, III, p. 111.

[336] Simancas, E° Napoles 1017, f° 42, viceroy of Naples to H.M., Naples, 26th April, 1534, Salonica, 'donde ay la mayor judería de Turquía'.

[337] Paul Bénichou, "Les Juifs en Champagne médiévale" in *Évidences*, November, 1951.

they bought and sold, but for the most part they were already merchants and moneylenders, 'lending, it appears, being more popular with them than trade', their loans being made to noblemen, particularly the Counts of Champagne, and to monasteries. Though attracted by the Champagne fairs and the prosperity surrounding them, the Jews (with a few exceptions) did not participate directly and certainly did not dominate them, but they did control certain of the approaches to them.

With the general recession of the sixteenth century, the only region not economically threatened in the West was Italy: Jewish merchants spread all over the country and a recent study[338] shows that they were colonizing the lower levels of usury, ousting their rivals from this elementary plane of commercial life.

In the fifteenth and sixteenth centuries, the major Mediterranean trade currents led to North Africa and the Levant. In 1509, when Spanish intervention provoked the massacre of Christian merchants by the crowd in Tlemcen, the Jews shared their fate.[339] They were also to be found in Bougie and in Tripoli which the Spaniards took over in 1510.[340] Again in Tlemcen, in 1541, when Spanish troops entered the city, 'the Jews who were there in great numbers were taken prisoner and sold as slaves by the conqueror. . . . Some of them were ransomed at Oran and Fez, others were taken as captives to Spain, where they were forced to deny the Eternal, the God of Israel.'[341] A few years earlier, a similar spectacle had marked the capture of Tunis by Charles V in 1535. The Jews 'were sold, both men and women', recounts the doctor Joseph Ha Cohen,[342] 'in the most diverse countries, but in Naples and Genoa, the Italian communities bought the freedom of many, may God remember them for it!'

In North Africa, according to Leo Africanus, the Jewish colonies were thriving and still belligerent at the beginning of the sixteenth century, and capable of resistance, thus managing to survive in the inhospitable Spanish *presidio* of Oran until 1668,[343] with a finger in every pie. An enquiry conducted in the Oran *presidio* in 1626[344] mentions the arrival of camel trains from the Sahara, one of them, coming from Tafilalet and Figuig, accompanied by 'Jews of war' – 'judíos de guerra' – in fact simple

[338] L. Poliakov, *Les Banchieri juifs et le Saint-Siège du XIIIe au XVIIe siècle*, 1965.

[339] H. Hefele, *op. cit.*, p. 321. The strength of the Jewish presence in North Africa explains the long survival of the Jewish colony in Oran under Spanish rule; Diego Suárez describes the Jewish quarter in the very centre of the city, with its synagogue and school; in 1667, the ghetto consisted of over 100 houses and 500 people: the Jews were expelled from Oran on the orders of Charles II on 31st March, 1669, according to J. Cazenave in *Bulletin de la Société de Géographie d'Alger*, 1929, p. 188.

[340] J. Ha Cohen, *op. cit.*, p. 110–111.

[341] *Ibid.*, p. 124. [342] *Ibid.*, p. 120.

[343] J. Caro Baroja, *op. cit.*, I, p. 217.

[344] 'Cargos y descargos del Marqués de Velada', answers to the charge of maladministration brought against Don Antonio Sancho Dávila y Toledo, marques de Velada, during his government of Oran, 1626–1628, f° 57 (P. de Gayangos, *Cat. Mss. in the Spanish language*, B.M., IV, 1893, p. 133).

merchants, for in Spain as in the Islamic countries, a distinction was made between the *Moros de paz*, subjects who lived near the citadel, and the still unsubdued *Moros de guerra*: in the same way there were Jews *de paz* and *de guerra*. But the presence of Jewish merchants on this ancient trade axis is in itself worth noting.

In the Levant, contemporary accounts all agree on the major role played by the Jewish merchants; they controlled the markets at Aleppo and (particularly Portuguese Jews) in Cairo, as moneylenders to whom the Christians often had recourse and in whose hands the entire caravan trade was clearly concentrated.

What else is there to say? In Venice, the Jewish presence was maintained in spite of tensions and quarrels followed by pacts or reconciliation. One expulsion certainly took place, that of the rich *marranos* in 1497,[345] following their speculation in the Sicilian wheat upon which Venice depended, but they were only a small fraction of the Jewish population and recent arrivals at that (who seem to have returned, since there was once more talk of expelling them in 1550[346] and we find them mentioned by name in Venice until the end of the century and even later). We have also seen that the Jews were present in Milan and the Milanese until 1597. In Rome, they led a rather cramped existence, but became prosperous at Ancona as long as Ancona thrived, that is until the first years of the seventeenth century; in Leghorn they were the architects of the Medici revival from its effective beginnings, that is after 1593.[347]

One would like above all to know what their situation was in Genoa, capital of world finance, but on this subject there is very little information. One thing is certain: there was hostility towards them. In Genoa, the jealousy felt by local artisans and doctors towards their Jewish rivals led to the expulsion of the community on 2nd April, 1550, the decree being 'proclaimed to the sound of trumpets, as it was', writes a witness, 'in the time of my father, Rabbi Jehoshua ha-Cohen' in 1516. The same witness, the physician Joseph Ha Cohen, went to live not far away, still in the territory of the *Dominante*, at Voltaggio, where he continued to practise medicine.[348] In 1559, there was a fresh outbreak of hostility from Genoa – or rather from one prominent Genoese citizen, Negron de Negri, 'that perverse man who was as a goad in the flesh' of the Jews;[349] he attempted, unsuccessfully as it turned out, to have them expelled from Piedmont. In June, 1567, the Genoese expelled them from the *Dominio* where they had been tolerated after the edict forbidding them to live within the city itself. Joseph Ha Cohen the doctor then left Voltaggio and moved 'to Castelleto, in the territory of Monferrato, where everyone received me with joy'.[350] It is frustrating to have no more precise information. Am I justified or not

[345] M. Sanudo, *op. cit.*, I, column 819, 13th November, 1497.

[346] Marciana 7991 C VII, 4, f°s 110 v° and 111, and Museo Correr, Donà della Rose 46, f° 155, 8th July, 1550.

[347] F. Braudel and R. Romano, *op. cit.*, pp. 26–27.

[348] J. Ha Cohen, *op. cit.*, p. 130–131. [349] *Ibid.*, p. 152. [350] *Ibid.*, p. 158.

in thinking that the wealthier Jewish merchants had access to the Piacenza fairs?

One last point to bear in mind is the spread of the *marranos* throughout the Mediterranean, preparing the way for the Dutch and marking the beginning of the age of Amsterdam in world history. In 1627, the Count Duke of Olivares introduced the Portuguese *marranos* to the vital business of the *asientos*, giving formal recognition to a new financial era which had in fact begun well before this date.[351] It was discernible by many signs. As early as 1605, there had been talk of granting 10,000 Jews permission to settle in Spain, to help order the finances of the Catholic King better than they had been under the rule of the Christian *asentistas*.[352] It would be easy enough to prolong the list and quote evidence of a Jewish presence in the seventeenth century in Marseilles, Leghorn, Smyrna, the three thriving cities of the Mediterranean; in Seville, Madrid and Lisbon, which were still important, and in Amsterdam and even London, where the rich merchant Antonio Fernández Carvajal, 'the great Jew', settled some time between 1630 and 1635.[353] I think the point has been sufficiently demonstrated.

Jews and the general economic situation. If a chronological table were drawn up of the persecutions, massacres, expulsions and forced conversions which make up the martyrology of Jewish history, a correlation would be discernible between changes in the immediate economic situation and the savagery of anti-Jewish measures. Persecution was always determined by, and accompanied, a worsening in the economic climate. It was not simply the hostility of their fellow men, whether princes or 'perverted' individuals (whose role will not of course be denied) which put an end to the happiness and prosperity of western Jewry in England (1290), Germany (1348–1375), Spain (the pogrom of Seville and forced conversions of 1391), or France (the definitive expulsion of the Jews from Paris in 1394). The chief culprit was the general recession of the western world. On this point it seems to me that no argument is possible. Similarly, to take the single example of the expulsion of the Jews from Spain (1492), this event 'of world-wide import', according to Werner Sombart[354] occurred late on in a long period of economic depression, which had begun with the reign of the Catholic Kings and lasted until at least 1509 and possibly 1520.

Just as the secular recession of 1350–1450 sent the Jews to Italy and its sheltered economy,[355] so the crisis of 1600–1650 finds them in the equally sheltered economic sector of the North Sea. The Protestant world saved them and showed them kindness and in return they saved and showed

[351] See above, Vol. I, p. 641–642.
[352] Espejo y Paz, *Las antiguas ferias de Medina del Campo*, 1912, p. 137.
[353] W. Sombart, *Krieg und Kapitalismus*, 1913, p. 147.
[354] *The Jews and Modern Capitalism*, p. 13.
[355] L. Poliakov, *Les Banchieri juifs* . . .

kindness to the Protestant world. After all, as Werner Sombart has remarked, Genoa was just as well placed as Hamburg or Amsterdam for access to the maritime routes to America, the Indies or China.[356]

But the parallelism between the economic situation and the vicissitudes of the Jewish people is evident not only in major events and over long periods, but even in minor crises, where it can be observed to the year, almost to the day. It was quite logical, to return to this minute example, that Ragusa should have contemplated expelling the Jews from the city in 1545, for the republic was going through a period of economic difficulty. Similar motives lay behind the measures so promptly taken by Venice against the Jews both within the city and on the mainland, during the long recession from 1559–1575, in particular during the war with Turkey, 1570–1573:[357] Levantine Jews were arrested; Jewish merchandise was confiscated, strict conditions were imposed on the maintenance of the Jewish colony in Venice (18th December, 1571), there was a proposal to expel the Jews from Brescia and from Venice itself; young Jews were seized in the Adriatic and sent to the galleys 'until the end of the war'. These were painful times for 'Jacob'.[358] And they were dictated, as far as one can see, by the economic situation. So too was the violent persecution of the Jews at Ferrara, in 1581, another item to be added to the already copious dossier on the pronounced cyclical crisis of 1580–1584.[359]

But when the long-term trend re-asserted itself between 1575 and 1595 and the sky lightened, an improvement appeared in the economic activity of the whole Mediterranean and in particular in that of the Jewish colonies, wherever they had taken root. In Rome, Sixtus V himself protected them (1585–1590)[360]. Thereafter it looks as if the part played by Jewish capitalism in maritime exchange grew unhindered. It was certainly the dominant force at Ancona,[361] but also at Ferrara,[362] if not at Venice itself. All these 'Portuguese' or 'Levantine' success stories; the

[356] *The Jews and Modern Capitalism*, p. 12.

[357] For the arrest at Venice of Turkish and Levantine Jewish merchants, on 5th March, 1570, Chronicle of Savina, Marciana, f° 326 v°; complaints by Jewish merchants at Constantinople, 16th December, 1570, A.d.S., Venice, Annali di Venezia, serie antica; the 24 points of the ruling to be observed by the Jews, A.d.S., Venice, Senato Terra 58, 18th December, 1571; on the same theme, Museo Correr, Cicogna, 1231, f° 16; the Jews expelled from Brescia, 4th September, 1572, A.d.S., Venice, Senato Terra 60; they were granted a reprieve until September, 1573, *ibid.*, 61, 8th March, 1573; for a ruling on the activities permitted and prohibited to the Jews, *ibid.*, 11th July, 1573; certificate of bankruptcy granted to the bank of 'Cervo hebreo'; whose failure must date back to 1565, *ibid.*, 20th June, 1573. Thereafter the climate was less tense. On the Jews who were expelled from Urbino and put in the galleys, J. Ha Cohen, *op. cit.*, p. 161.

[358] *Ibid.*, p. 174.

[359] Cecil Roth, *art. cit.*, p. 239.

[360] A. Milano, *op. cit.*, p. 257, J. Delumeau *op. cit.*, II, p. 854, 887–890.

[361] As goes without saying, but note the new exemptions granted to Levantine Jews at Ancona, a serious threat to Venice. A remarkable document, A.d.S., Venice, *Cinque Savii*, Busta 3, 10th August, 1597.

[362] J. Ha Cohen, *op. cit.*, p. 205, 1598.

liaison with the Moroccan Sūs and its sugar mills;[363] the creation of the port of Spalato;[364] the proposal made in March, 1587[365] by the influential Daniel Rodriga that a deposit account of 20,000 ducats be set up at Istanbul to be controlled by the *bailo* in return for an equivalent advance on the Venetian customs; or the suggestion, made in about 1589, that the Jews of Ferrara be received into the city:[366] such freedom of planning and action indicates a change in climate. The regime introduced for 'Levantine' and 'Ponentine' Jews (*levantini* and *ponentini*) in 1598, was a genuinely liberal one: they were given passes valid for ten years which would automatically be renewed on expiry unless they were renounced; the conditions were the same as those granted in 1589, ten years earlier. One small additional favour was granted: 'they may wear a black cap and the usual arms when travelling, but not in Venice'.[367] In fact, Venice now became, to Ferrara's loss, the major rendezvous for *marranos* in Italy, the point at which they made contact with Jews from Germany and the Levant, as appeared from one unmistakable sign: Venice had become an intellectual capital. *Marrano* literature, both Spanish and Portuguese, was produced by Venetian printing presses, until their role was eventually taken over by the printers of Amsterdam and Hamburg.[368]

So from Amsterdam to Lisbon, Venice and Istanbul, Jewish colonies were entering a period of success or at least of more comfortable circumstances. The habitual hunt for Jewish cargoes on board Mediterranean ships was by no means a fruitless activity nor an insignificant detail, but on the contrary the mark of a certain prosperity which excited the envy of a multitude of enemies. It was a hunt which had begun many years before. As early as 1552[369] and again in 1565,[370] Jewish protests had singled out for complaint the ships of the 'most evil monks' of Malta, that 'trap and net which catches booty stolen at the expense of Jews'.[371] By the end of the century, Tuscans, Sicilians, Neapolitans and Greeks from the islands had joined the pirate galleys;[372] perhaps the prizes had increased. There are other signs of this revival in Jewish fortunes, the reopening of commercial relations with Naples for instance. Following their expulsion in 1541 they had been allowed access only to the fairs of Lanciano and

[363] A.d.S., Venice, *Cinque Savii* 22, f° 52, 20th November, 1598; f° 73, 16th August, 1602, privilege and renewal of privilege of Rodrigo de Marchiana; *ibid.*, 138, f° 191, 22nd February, 1593, some Portuguese Jews proposed establishing commercial links with the Cabo da Guer in what is now the Bay of Agadir and these Jews were in fact the same Rodrigo di Marchiana and his brothers.

[364] See above, Vol. I, p. 286.

[365] A.d.S., Venice, *Cinque Savii*, 138, 18th March, 1587.

[366] Cecil Roth, *art. cit.*, p. 239.

[367] A.d.S., Venice, *Cinque Savii* 7, f° 30, 5th October, 1598.

[368] Hermann Kellenbenz, *op. cit.*, p. 43; see also Cecil Roth, *History of the Jews in Venice*, Jewish Communities Series, Philadelphia, 1930 and 'Les marranes à Venise' in *Revue des Études Juives*, 1931.

[369] J. Ha Cohen, *op. cit.*, p. 131.

[370] *Ibid.*, p. 172.

[371] *Ibid.*

[372] Cf. below, p. 879–880.

Lucera, it seems. But after 1590 there was talk of re-establishing their trading rights[373] and these were finally acquired in September, 1613.[374]

Historians already refer to the 'age' of the Fuggers and the 'age' of the Genoese: it is not entirely unrealistic in the present state of scholarship to talk of an 'age' of great Jewish merchants, beginning in the decade of the 1590s and lasting until 1621 or possibly even 1650. Their age was one of intellectual brilliance.

Understanding Spain. The destiny of the Jews cannot be studied outside the context of world history, in particular the history of capitalism. (It has been rather too quickly assumed that the Jews did not invent capitalism, which may well be true – can any single group claim to have done so? Certainly they participated wholeheartedly in its beginnings.) It may be helpful to approach this far from simple problem through the single but spectacular case of Spain. The destiny of the Jewish people both reflects and is reflected in the many-faceted mirror of Spanish history.

One major difficulty will be to prevent the emotions, vocabulary and polemic of our own age from intruding into this highly-charged debate; to refuse to be drawn by the simple language of moralists into the rigid separation of black and white, good and evil. I cannot consider Spain guilty of the murder of Israel. Has there been any civilization at any time in the past which has sacrificed its own existence to that of another? Certainly not Islam or Israel any more than anyone else. I say this dispassionately, for I am bound to share the feelings of twentieth century man; my sympathy lies with all those who are oppressed in their liberty, their persons, their possessions and their convictions. In the Spanish situation I am therefore naturally on the side of the Jews, the *conversos*, Protestants, *alumbrados* and Moriscos. But such feelings, which I cannot avoid, are irrelevant to the basic problem. To call sixteenth century Spain a 'totalitarian' or racist country strikes me as unreasonable. It has some harrowing scenes to offer, but then so do France, Germany, England, or Venice (from a reading of the judicial archives) at the same period.

Let me stress once more that the economic situation, a blind force in Spain, as much as in Turkey or the New World now entering international history, must take its share of the blame. When they expelled the Jews in 1492, Ferdinand and Isabella were not acting as individuals, in the aftermath of the fall of Granada, victory as always being a bad counsellor: their action was encouraged by the poor economic climate and the reluctance of certain wounds to heal. Civilizations, like economies, have their long-term history: they are prone to mass movements, carried as it were imperceptibly forward by the weight of history, sliding down a hidden slope so gradual that their movement is unaided and unheeded by man. And it is the fate of civilizations to 'divide'[375] themselves, to prune

[373] A.d.S., Naples, Sommaria Consultationum, 10, f°s 91–93, 30th March, 1590.

[374] *Ibid.*, 25, f°s 152 v° to 159, 8th September, 1613.

[375] The expression is Michel Foucault's, *Histoire de la folie à l'age classique*, 1961, p. iv.

their excess growth, shedding part of their heritage as they move forward. Every civilization is the heir to its own past and must choose between the possessions bequeathed by another generation. Some things must be left behind. No civilization has been forced to inflict so much change upon itself, to 'divide' itself or rather tear itself apart so much as the Iberian civilization in the age of its greatest glory, from the time of the Catholic Kings to Philip IV. I say *Iberian civilization* expressly. For this is a particular variety of western civilization, an outpost or promontory of it, at one time almost entirely washed over by foreign waters. During the 'extended' sixteenth century, the Peninsula, in order to reintegrate itself with Europe, turned itself into the Church Militant; it shed its two unwanted religions, the Moslem and the Hebrew. It refused to become either African or Oriental in a process which in some ways resembles that of modern de-colonization. Other destinies can be imagined for Iberia: it could have remained a bridge between Europe and Africa, in obedience to its geographical position and what was for centuries its historical vocation. It might have been possible – but a bridge implies two-way traffic. Europe conquered the Peninsula by way of the Pyrenees and by the Atlantic and Mediterranean shipping routes: along this frontier zone it defeated Islam with the victories of the Reconquest which were victories for Europe. As historians of the Peninsula will tell us, Claudio Sanchez Albornoz as well as Americo Castro, the 'Ultramontane' forces won the day, the reconquest of Spain by Europe going hand in hand with the purely Spanish reconquest of Moslem soil. The great discoveries later did the rest: they placed the Peninsula at the centre of the modern world, that is at the centre of European world conquest.

To say that Spain should not have become part of Europe is one point of view and it has been voiced.[376] But it is hard to see how she could have avoided it. Political considerations alone did not determine the expulsion of heterodoxies or create the Spanish Inquisition in 1478 and the Portuguese Inquisition in 1536; there was also popular pressure, the intolerance of the masses. To our eyes, the Inquisition seems abhorrent, less for the number of its victims,[377] which was relatively small, than for the methods it employed. But were the Inquisition, the Catholic Kings, the various rulers of Spain and Portugal really the major forces responsible for a combat urged by the profound desires of the multitude?

Before the nationalism of the nineteenth century, peoples felt truly united only by the bonds of religious belief; in other words by civilization. The massive cohesion of Spain in the fifteenth century was that of a people which had for centuries been the underdog in relation to another civilization, the weaker, the less intelligent, the less brilliant and the less rich, now suddenly liberated. Although the superior power at last, it had not yet acquired the internal confidence nor the reflexes of a superior power. It went on fighting. If the terrible Inquisition in the end claimed

[376] *La Méditerranée . . .*, 1st ed., p. 136, n. 1.
[377] Léon Poliakov, *op. cit.*, II, p. 204 to 217.

few victims, it was because there was little for it to get its teeth into. Spain was still subconsciously too fearful and too militant for heterodoxy to insinuate itself with ease. There was no place in Spain for Erasmianism or for the doubtful *converso* any more than for the Protestant.

In the context of this cultural conflict, the passionate and seductive arguments of Léon Poliakov do not entirely satisfy me. He sees only one side of the tragedy, the grievances of Israel, not recognizing those of the Spain of different periods, which were in no way illusory, fictitious or diabolical. A Christian Spain was struggling to be born. The glacier displaced by its emergence crushed the trees and houses in its path. And I prefer not to divert the debate to a moralizing level by saying that Spain was amply punished for her crimes, for the expulsion of 1492, the persecution inflicted on so many *conversos* and the angry measures taken against the Moriscos in 1609–1614. Some have said that these crimes and passions cost her her glory. But the most glorious age of Spain began precisely in 1492 and lasted undimmed until Rocroi (1643) or even 1650. The punishment, depending on which date one chooses, came at least forty years if not a century late. Nor can I accept that the expulsion of the Jews deprived Spain of a vigorous bourgeoisie. The truth is that a commercial bourgeoisie had never developed in Spain in the first place, as Felipe Ruiz Martín has shown, owing to the implantation there of a harmful international capitalism, that of the Genoese bankers and their equivalents. Another argument frequently heard is that the tragedy of *limpieza de sangre*, purity of blood, was to be the trial and scourge of Spain. No one would deny the trials it brought and their fearful sequels, but all western societies erected barriers in the seventeenth century, consecrated social privileges, without having the reasons attributed to Spain.

Let us accept rather that all civilizations move towards their destiny, whether willingly or unwillingly. If the train in which I am sitting moves off, the passenger in a train alongside has the sensation of moving in the opposite direction. Civilizations too may be carried past one another. Do they understand each other? I am not at all sure that they do. Spain was moving towards political unity, which could not be conceived, in the sixteenth century, as anything other than religious unity. Israel meanwhile was being carried towards the destiny of the *diaspora*, a single destiny in its way, but its theatre was the whole world, it spanned oceans and seas, new nations and ancient civilizations. The latter it disputed and defied. It was a modern destiny, ahead of its times. Even as lucid an observer as Francisco de Quevedo saw it as possessing diabolical features. The devil is always the Other, in this case the other civilization. *La Isla de los Monopantes* (1639) is a pamphlet directed against the Count Duke of Olivares and the *marrano* bankers of his entourage, and possibly not written by Quevedo himself. 'In Rouen', say the Jews of the *Island of Monopantes*, 'we hold the purse-strings of France against Spain and at the same time those of Spain against France; and in Spain, under a disguise which conceals our circumcision, we help the monarch [in this case

Philip IV] with the wealth we possess in Amsterdam, in the country of his mortal enemies. We do the same in Germany, Italy and Constantinople. We weave the blind web, the source of wars, helping everyone with money taken from the pocket of his greatest enemy, for our help is like that of the banker lending money at huge interest to a gambler who plays and loses, so that he will lose even more.'[378] In short a critique of capitalism.

From one civilization to another, every man gives his own version of the truth. The neighbour's version is never acceptable. The one thing of which we can be certain is that the destiny of Israel, its strength, its survival and its misfortunes are all the consequence of its remaining irreducible, refusing to be diluted, that is of being a civilization faithful to itself. Every civilization is its own heaven and hell.[379]

4. THE SPREAD OF CIVILIZATION

He who gives, dominates. The theory of the donor works not only at the level of individuals and societies but also for civilizations. That such giving may in the long run cause impoverishment is possible. But while it lasts it is a sign of superiority and this observation completes the central thesis of Part Two of this book: the Mediterranean remained, for a hundred years after Christopher Columbus and Vasco da Gama, the centre of the world, a strong and brilliant universe. How do we know? Because it was educating others, teaching them its own ways of life. And I would stress that it was the *whole* Mediterranean world; Moslem and Christian, which projected its light beyond its own shores. Even North African Islam, often treated by historians as a poor relation, spread its influence southwards towards the Saharan borders and through the desert as far as the *Bled es Sudan*. As for Turkish Islam, it illuminated a cultural area which it owned in part, from the Balkans to the lands of Araby, into the depths of Asia and as far as the Indian Ocean. The art of the Turkish Empire, of which the Sulaimānīye mosque is the crowning achievement, spread far afield, affirming its supremacy, and architecture was only one element in this vast cultural expansion.

Even more distinctive, to our eyes, is the penetrating influence of western Mediterranean civilization. It spread in fact against the current of world history, reaching out to northern Europe which was soon to become the centre of world power: Mediterranean, Latin culture was to Protestant Europe what Greece was to Rome. It rapidly crossed the Atlantic both in the sixteenth and seventeenth centuries, and with this geographical extension over the ocean, the Mediterranean sphere of influence was finally complete, embracing Hispano-Portuguese America, the most brilliant America of the time. To make identification even easier, a word given

[378] Passage quoted by L. Poliakov, *op. cit.*, II, p. 290.

[379] For an analogous line of thought, though inclined more towards a social explanation, see an original study shortly to be published, by Antonio José Saraiva, *L'Inquisition et la légende des Marranes*.

general currency by Burckhardt, the Baroque, conveniently designates the civilization of the Christian Mediterranean: wherever we find the Baroque we can recognize the mark of Mediterranean culture. The influence of the Renaissance – putting value judgements aside – cannot compare, for sheer mass and volume, with the enormous explosion of the Baroque. The Renaissance was the child of the Italian cities. The Baroque drew its strength both from the huge spiritual force of the Holy Roman Empire and from the huge temporal force of the Spanish Empire. With the Baroque a new light began to shine; since 1527 and 1530 and the tragic end of the great cities of the Renaissance, Florence and Rome, the tone had changed: new and more lurid colours now bathed the landscapes of western Europe.

Having said this, I hope the reader will appreciate my problem. This is a book about the Mediterranean: if I attempted to describe this enormous metamorphosis in all its aspects, I should be writing a book about the world. I therefore decided that a single demonstration would satisfy both the glory of the Mediterranean and the equilibrium of this book. Regretfully then, I have had to leave aside Islam, regretfully too Hispano-Portuguese America and the late but rare splendour of Ouro Preto in the mining heart of Brazil. The Baroque, the sprawling and extravagant Baroque, will be more than enough to occupy us, in the single sector, itself enormous, of western Europe.

The stages of the Baroque. Following Burckhardt, other German historians – H. Wölfflin, A. Riegl, A.-E. Brinckmann, W. Weisbach – brought the term Baroque into favour,[380] launching a vessel in which many were to sail. This initiative was conceived as a useful essay in classification, the distinction of an artistic period comparable for instance to a geological layer. To the sequence Roman, Gothic, Renaissance, we are invited to add a fourth term, the Baroque,[381] to be inserted just before the classical revival of French inspiration: the term does not correspond to an entirely clear and simple notion, for the Baroque can be described as a three or even four-tiered edifice.

Its origins are usually discerned in the *Pietà* sculpted by Michelangelo for St. Peter's between 1497 and 1499, and also in the *Stanze* of Raphael, the tumultuous movements of the *Fire in the Borgo* and the *Expulsion of Heliodorus from the Temple*, the St. Cecilia frescoes at Bologna which, according to Émile Mâle, already foreshadow the genius of the new

[380] On the word Baroque, see Pierre Charpentrat, 'De quelques acceptions du mot *Baroque*', in *Critique*, July, 1964.

[381] The origin of the word is obscure: possibly formal logic (from *baroco*, one of the terms of the sequence *barbaro, celarent, baroco*) according to L. Pfandl, *Geschichte der spanischen Literatur*, p. 214, n. 1 – or from the Spanish word *baruco* which is used in jewellery to describe an irregular pearl, according to G. Schnürer, *op. cit.*, p. 68 – or from the name of Federigo Barocci (Le Baroche in French) (1526– or 1528–1610) according to P. Lavedan, *Histoire de l'Art*, Clio, p. 302. It is not clear when the word entered historical vocabulary before Burckhardt made it popular.

age[382] and, it has been said, the 'language of gesture the Baroque was to make its own'.[383] One could also look for its origins in the cartoon of the *Battle of Anghiari* or (outside Italy this time) in some of Dürer's engravings: a strange gathering at the christening. One of the unquestionable fathers of the Baroque is said to be Correggio, the Correggio of the *Assumption of the Virgin* at Parma.[384] To be fully acceptable as a Baroque artist, he need only have manifested a little more disdain or indifference to the joys of the earth and the beauty of the Nude, that Nude on which Michelangelo expended such labour and love. On the other hand Michelangelo's taste for the grandiose, his pathos and *terribilità*, were, along with the *grazia* of Raphael and the movement and play of light in Correggio, the first gifts laid before the cradle of the infant Baroque. Thus endowed, the child grew quickly and had almost reached maturity when Correggio died in 1534, certainly by the time that Michelangelo, after seven years of exhausting labour had in 1541, finished the *Last Judgement* in which 'the terrors of the Middle Ages' were revived.[385]

The curtain fell suddenly on the splendours of the Renaissance after the Sack of Rome in 1527 and the capture of Florence in 1530. 'The fearful Sack of Rome'[386] appeared to contemporaries a divine punishment. It abruptly recalled the city to its Christian mission. While Clement VII held out in the Castel Sant'Angelo, the city was for months at the mercy of the soldiery and looting peasants. Nothing was spared. Raphael's pupils had scattered far and wide, Penni to Naples, Pierino del Vaga to Genoa, Giulio Romano to Mantua from which he would never want to return. 'So the pupils of Raphael had no pupils' concludes Stendhal rather hastily.[387] Thus the fragility of all artistic life, all life of the intellect, was once more revealed. The siege and capture of Florence 'a second judgement of God', the violent effect of which on economic life has been demonstrated by G. Parenti, repeated in 1530 the disaster of 1527. With this, 'something died, and died quickly'.[388] A new generation, of which Giuliano de' Medici predicted that it would be more Spartan than Athenian, was moving into position,[389] new fashions were triumphing. What had died was the Renaissance, perhaps Italy herself. What was triumphing was *la maniera*, imitation, emphasis, exaggeration: it inflated the work of those of Raphael's pupils who were still working and their academism had its followers.[390]

The wind of change was felt first in painting. This was the beginning of Mannerism, the definition and aims of which were outlined in 1557 by Lodovico Dolce in a formal apologia for *la maniera*. All Italy was intoxi-

[382] *L'art religieux après Le Concile de Trente* . . ., p. 188.
[383] Marcel Brion, *Michel Ange*, 1939, p. 149.
[384] G. Schnürer, *op. cit.*, p. 80.
[385] Pierre Lavedan, *op. cit.*, p. 293.
[386] Stendhal, *Promenades dans Rome*, ed. M. Lévy, 1858, II, p. 121.
[387] *Ibid.*, p. 121.
[388] Gonzague Truc, *Léon X*, p. 303.
[389] *Ibid.* [390] *Ibid.*

cated with it from about 1530–1540,[391] except for Venice, where there were a few *manieristi*, but where there was also for many years the indomitable Titian.

Mannerism was re-christened by the twentieth century as the Pre-Baroque, a long period typified by Tintoretto and ending with his death in 1590.[392] The last masterpiece of Mannerism was the great *Paradiso*, painted between 1589 and 1590 in the *Sala del Maggior Consiglio*, Venice. Almost immediately began the first phase of the Baroque, inaugurated, according to G. Schnürer, by Federico Barocci of Urbino, whose celebrated *Madonna del Popolo* is now in the Uffizi.[393] It was to be the reigning manner until 1630; but this was far from its end, for from the 'Italian Baroque' there immediately derived a vigorous art which was to flourish in Switzerland, southern Germany, Austria and Bohemia until the eighteenth and even nineteenth centuries, nourished by a fertile popular imagination which gave it the life it never possessed in its Italian period. It was here indeed, in the countries of central Europe, that the word Baroque (whatever its origins) began to be applied, in the eighteenth century to an art which was in fact already in decline. Hence, so German scholars have said, the equation Baroque = German; a false equation if one looks at the sources.

Begging the question. There is endless possibility for debate about this chronology and the assumptions behind it: clearly it extends and increases the significance of the Baroque. Equally one could spend a long time discussing what the Baroque is or is not: Gustav Schnürer has even called it a civilization in itself, the last oecumenical civilization proposed and imposed in Europe. The last? Again one could argue the point and in the first edition of this book, I was drawn into this tempting discussion. But such problems are rather different from the point we are at present considering, which is that whatever the precise nature of this civilization, it originated in the Mediterranean. The Mediterranean was the donor, the transmitter and therefore a superior force, whose teachings, way of life and tastes were adopted in lands far from its shores. It is this evidence of the vigour of that civilization, its resources and the reasons behind it, which must occupy us here.

Rome: centre for the diffusion of Mediterranean culture.[394] Rome was a great centre of cultural diffusion, by no means the only one, but certainly the most important. At the beginning of the sixteenth century it was still undistinguished, as Rabelais saw it on his first voyage in 1532 and as it is described in Marliani's *Topography* and various other guides. It was a

[391] G. Bihlmeyer, *op. cit.*, III, p. 131.

[392] Erich von der Bercken, *Die Gemälde des Jacopo Tintoretto*, Munich, 1942, 360 illustrations.

[393] G. Schnürer, *op. cit.*, p. 86–87.

[394] See Jean Delumeau's excellent book, *Vie économique et sociale de Rome dans la seconde moitié du XVIe siècle*, 1957, p. 246 ff.

small city, in the centre of a pastoral economy; strewn and ringed round with ancient monuments often half-destroyed, atrociously disfigured, more often still buried to their foundations under earth and rubble. The inhabited part of the city was characterized by brick houses, sordid narrow streets and vast vacant lots.

During the sixteenth century, the city was transformed, new life was breathed into it as palaces and churches rose from the ground. Its population grew, maintaining its level even in the seventeenth century, an age generally unfavourable to Mediterranean cities. Rome became a gigantic building site. Any artist could find work there, an army of architect-masons to begin with: Baldassare Peruzzi of Sienna (d. 1536), Sammicheli of Verona (d. 1549), Sansovino of Florence (d. 1570), Vignola (d. 1573) from northern Italy (the cradle of almost all the great Italian architects), Ligorio of Naples (d. 1580), Andrea Palladio of Vicenza (d. 1580), Pellegrini of Bologna (d. 1592). Olivieri, an exception, was a native of Rome (d. 1599). On the heels of these artisans, architects and stonecutters, pressed the army of painters necessary in an age of art when decorative painting reached its apogee. Domes and ceilings offered unlimited space to painters, while imposing upon them sometimes strictly defined themes. The sacred painting of the Baroque was the logical consequence of its architecture.

It was at this time that the basilica of St. Peter's was completed, and the Gesù built, between 1568 and 1575, by Giacomo Vignola, who died in 1573 without seeing his work completed. The first Jesuit church had now appeared; it was to serve many times as a model throughout Christendom. Every order would now wish to possess its own churches, in Rome and outside Rome, decorated in an individual style with the images of its particular patterns of worship. So there now sprang up, in the Eternal City and then all over the Christian world, the first churches with accolades and cupolas of sober geometry, of which the Val-de-Grâce is a later but still typical example in France.

The prodigious growth of Rome entailed vast expenditure. Stendhal correctly diagnosed the problem when he noted that 'only those countries which did not have to tremble for their authority were able to commission the great masterpieces of painting, sculpture and architecture of modern times'.[395] This brings us back to the history of the papal finances, which Clemens Bauer has re-oriented in a remarkable article:[396] it is now beyond dispute that the Popes drew enormous revenues from the papal states and also appealed successfully to public credit. Their religious policies and indeed their policy in general in Christendom, were pursued less at their own expense than at that of the national churches. The churches of France and Spain were abandoned to the covetousness and financial needs of His Most Christian Majesty and His Catholic Majesty respectively. The papal states, during the fifty years which concern us,

[395] Stendhal, *op. cit.*, II, p. 191.

[396] 'Die Epochen der Papstfinang' in *Hist. Zeitschr.*, 1928.

only rarely (in 1557 and during the three years of the Holy League) incurred heavy military expenses. So the papacy was able to allot a large budget to the fine arts. The invasion of the Mediterranean by American silver was to facilitate these luxurious investments. It was in the years after 1560–1570 that all the dreams of Leo X and Julius II were realized. Not only that but the religious orders, whose numbers were increased by the wave of Catholic piety, added their efforts to those of the Popes. Rome being also the capital of these little states within the state, their shop window so to speak, Jesuits, Dominicans, Carmelites and Franciscans all contributed their share of financial effort and artistic emulation and copied, outside Rome, the achievements of the capital. If there was an artistic and religious expansion of the Baroque, it was the work of these orders, above all that of St. Ignatius. It is for this reason that the adjective Jesuit seems to me in a way far more appropriate than that of Baroque to describe this expansion, notwithstanding the reservations which have been expressed concerning this claim.

I do not think it is necessary to retrace here the history of the widespread move towards monasticism, which preceded by many years the success of the Council of Trent, that first victory of the new generation. As early as 1517, the Oratory of Divine Love, founded in Genoa in the preceding century by Bernardino da Feltre, established itself at Rome. In the same year, Leo X agreed to the separation of the Observants from the Conventuals in the Franciscan order. From the ranks of the Reformed Franciscans were to emerge amongst others, the Capuchins in 1528. But it was not until 1540, the year of the founding of the Society of Jesus, that the movement gained impetus and could finally be said to be under way.

Three years earlier, in 1537, the Commission of Cardinals convoked by Paul III, had been pessimistic; it had even considered allowing corrupt congregations to die out in order to repopulate them later with new monks. Then in the decade of the 1540s the whole picture changed: the first round was played and won: the creation and reformation of monastic orders continued and the movement towards monastic renovation gathered pace. It moved even more quickly after the Council of Trent: the Oratorians were founded by St. Philip Neri in 1564; the Oblati of St. Charles Borromeo in 1578; the Congregation of Minor Clerks Regular founded by the Genoese Giovanni Agostino Adorno and St. Francesco Caracciolo in 1588 (the first establishment at Naples dates from 1589) and three years later, in 1592, the Fathers of the Christian Doctrine settled in Avignon.

Who can say what support the religious orders, often released, in order to do battle, from the old restrictions of choral life and monastic observance, 'vrais clers réguliers', brought to the papacy? Thanks to them the Church was saved; it was able, from Rome, to coordinate one of the most extraordinary revolutions from above in history. The battle it fought was waged with intelligence. The civilization it carried forward – whatever name we choose to give it – was a militant civilization; and its art was merely one more means to an end.

Baroque art then, often smacks of propaganda. In some respects it is an art done to order, with all the advantages and disadvantages that implies. Shrewd theologians and friars demanded of Rubens, Caracciolo, Domenichino, Ribera, Zurbaran or Murillo, the physical execution of pictures spiritually composed by themselves, turning them down if the execution appeared in any way deficient. In the fight against Protestantism, the enemy of decorated churches and images, the Church set out to build the most beautiful houses of God on earth, images of Paradise, portions of heaven. Art was a powerful means of combat and instruction; a means of stating, through the power of the image, the Immaculate Holiness of the Mother of God, the efficacious intervention of the saints, the reality and power of the Eucharistic sacrifice, the eminence of St. Peter, a means of arguing from the visions and ecstasies of the saints. Patiently compiled and transmitted, identical iconographical themes crossed and re-crossed Europe. If the Baroque exaggerates, if it is attracted by death and suffering, by martyrs depicted with unsparing realism, if it seems to have abandoned itself to a pessimistic view, to the Spanish *desengaño* of the seventeenth century, it is because this is an art which is preoccupied with convincing, because it desperately seeks the dramatic detail which will strike and hold the beholder's attention. It was intended for the use of the faithful, who were to be persuaded and gripped by it, who were to be taught by active demonstration, by an early version of *verismo*, the truth of certain contested notions, whether of Purgatory or of the immaculate Conception. It was a theatrical art and one conscious of its theatricality: had not the theatre itself provided the Jesuits with arms, notably in their conquest of Germany, in an age moreover when the theatre was establishing its rights everywhere, with strolling players and before long fixed stages?

So it was both a way of life and a way of belief which travelled northwards from the shores of the Mediterranean, towards the Rhine and the Danube as well as to Paris, the heart of France, where in the early years of the seventeenth century so many churches and convents were being built. It was a way of life and belief specifically Mediterranean: witness Jacob Burckhardt's description of Pius II processing through Viterbo 'surrounded by live tableaux representing the Last Supper, St. Michael battling with the Devil, the Resurrection of the Lord and the Virgin carried in triumph to heaven by Angels'.[397] One immediately thinks of Spanish processions with the *tratos* representing scenes from the Passion; no more than in Italy does this exclude *autos sacramentales*.[398] This then

[397] Quoted by G. Truc, *Léon X*, p. 123.

[398] M. La Torro y Badillo, *Representación de los autos sacramentales en el periodo de su mayor florecimiento* (1620 to 1681), Madrid, 1912. Ludwig Pfandl, *Geschichte der spanischen Literatur*, p. 124; Henri Merimée, *L'art dramatique à Valence depuis les origines jusqu'au commencement du XVIIe siècle*, 1913.

was a dramatic form of Christianity which northerners found astonishing. The manner of devotion and the flagellation practised by Spaniards shocked and scandalized the people of Flanders.[399] Baroque art, nourished on this southern religiosity, carried something of it to the North. A whole book could be written on the devotional practices imported to all parts of Europe, on the part played by men of the Mediterranean in the violent reclamation of the contested lands of the North which returned to the fold of the Roman Church. Remembering this, one can no longer talk of the decadence of the Mediterranean; unless decadence and the disintegration it implies can be credited with a powerful capacity for diffusing a dying civilization.

Another centre of cultural diffusion: Spain. Moving westwards from Vienna to Lyons then to Toulouse or Bayonne, one begins to feel the pull of another civilization – that of Spain. In Vienna and Munich, Roman and Italian (every kind of Italian) influence predominated. In France too, travellers, fashions and ideas from Rome and Italy were to be found, but stronger than either was the influence of Spain.

One of the problems of the Pyrenees was that they never allowed two-way traffic. Either France was the predominant cultural influence and everything travelled from north to south – as was the case from the eleventh and twelfth centuries until the fifteenth; or else Spain suddenly rose to eminence and reversed the flow from south to north, as happened in the sixteenth and seventeenth centuries. The traditional dialogue between France and Spain abruptly took a new turn and it was to change again in the eighteenth century. In the time of Cervantes, France hankered after the fashions and ideas of her neighbour beyond the Pyrenees, a country at once, mocked, reviled, feared and admired. Spain on the contrary broke off all contact, kept a watch over her frontiers, forbade her subjects in the Netherlands to study in France and recalled her medical students from Montpellier.[400]

It was a strange dialogue and as usual one without affection. Where, if not in the Netherlands, was the Spaniard so mocked as in France? A French translation appeared in 1608 of a satirical fantasy published in Middelburg by Simon Molard: *Emblèmes sur les actions, perfections et mœurs du Segnor espagnol.*[401] Poor *Segnor*! He is compared to every kind of beast, a devil in the house, a wolf at table, a swine in his bedchamber, a peacock in the street, a fox with women and more besides. 'So beware of the *Segnor* in all places' warns the pamphlet in conclusion. But the *Segnor* though ridiculed was envied and imitated as well. The influence of Spain was that of a strong nation, of an immense Empire 'on which the sun never set', of a civilization more refined than that of France.

[399] Georg Friederici, *op. cit.*, I, p. 469.

[400] On the last point, Francés de Alava to Philip II, Montpellier, 18th December, 1564, A.N., K 1502, B 18, no. 67, D.

[401] A. Morel Fatio, *Ambrosio de Salazar*, 1900, p. 52 ff.

Any man of culture in France was obliged to know Spanish and did: which helped several natives of the Peninsula, the *murciano* Ambrosio de Salazar for example, to achieve a successful career in France as teacher and grammarian in the time of Marie de Medici. The Castilian vocabulary colonized the French tongue and Brantôme was conspicuous among Hispanophiles,[402] habitually using a number of pseudo-Spanish terms (*blasonner*, *bourler*, *busquer fortune*, *habler* for *parler*, *tirer* for *lancer*, *treper* for *monter*, *care*, *garbe*, *marcher* '*à la soldade bizarrement*'[403]). It was fashionable to sprinkle one's conversation with Spanish words[404] which were as frequent at the time as Italianisms and the fashion required study, numerous teachers and imported books. Montaigne's father had read the *Epistolas familiares*, the *Libro Aureo de Marco Aurelio*, the *Relox des Principes* and the *Aviso de privados y doctrina de cortesanos*, the works of the celebrated bishop of Mondonedo, Antonio de Guevara.[405] Translations abounded. 'In Paris there was a veritable agency of translators from Castilian.'[406] Cervantes was much read. In 1617, his long book, *Los Trabajos de Persiles y Sigismunda*, was reprinted in Paris in Spanish, then translated into French.[407] And the picaresque novel had an even more devoted following. Soon would come the adaptations of Spanish comedies for the French stage. In England too, Italian and Spanish works were translated and incorporated into the country's intellectual heritage.

Alongside such literary influences came a host of minor cultural borrowings. The court of Louis XIII, said to be as Spanish as it was French, set the tone. Anything Spanish was in fashion. Women daubed their faces with 'Spanish white' and 'Spanish vermilion' – which did not necessarily come all the way from Spain. They doused themselves – and so did men – in perfumes, some of which came from Nice and Provence, but most of which, especially the more costly ones – forbidden to 'peasants'[408] – were imported from Spain and Italy. If Brantôme is to be believed, the women of those two countries 'have always been more excitingly and exquisitely perfumed than our great ladies in France'.[409] There was competition for the secrets of rare essences and beauty preparations at least as complicated as those of Molière's *précieuses*. A gallant would promise to buy his mistress gloves of 'Spanish leather' and indeed, although fine goods were already being produced in France and the reputation of French fashions and elegance was beginning to appear, Spanish gloves made from fine supple hides, the famous toiletwater of Córdoba, *guadameciles*, the gilded leather used for wall hangings, enjoyed the kind of prestige that the twentieth century was to give to the *article de Paris*. And they were just as expensive. When the wife of Simón Ruiz took it into her head to 'do

[402] A. Morel Fatio, *L'Espagne en France* in *Études sur l'Espagne*, I, Paris, 1895, 2nd ed., p. 30.

[403] *Ibid.*, p. 32.

[404] *Ibid.*, p. 40.

[405] *Ibid.*, p. 27; *Essais*, II, 1.

[406] *Ibid.*, p. 41.

[407] *Ibid.*

[408] Alfred Franklin, *La vie privée d'autrefois. Les magasins de nouveautés*, 1894–1898, II, p. 39. See also I, p. 183; II, p. 23–25, 75.

[409] IX, p. 253, quoted by Alfred Franklin, II, p. 39.

business' and sent 'perfumed gloves' from Spain to Florence, to be exchanged for Italian merchandise, her husband's partner, Baltasar Suárez, claimed that in this city of grave bourgeois, nobody would buy such a costly and frivolous article (three crowns a pair). But this was in 1584.[410] One wonders what the Florentines would have thought a couple of decades later.

In the field of literary imports, about which we know most, Spanish influence did not decline until the end of the reign of Louis XIII,[411] which brings us once more to the years 1630–1640, a turning point too in financial and economic history and a crucial date in the history of world wealth. The great age of Spanish influence was broadly the first half of the seventeenth century. The process had already begun in the sixteenth century, as France found that the geographical embrace of the Spanish Empire could not be endured with impunity. But it was not until the return of peace at the end of the century and in the first decades of the next, that the seeds which had fallen would bring forth plants and flowers. It was the return of peace which carried the 'triumphs' of the Baroque across Europe.

The supposed decline of the Mediterranean. If it had not for so long been thought that the energy of the Mediterranean was spent after the Renaissance, historians would perhaps have taken more trouble to study its influence at the end of the sixteenth and beginning of the seventeenth century. I would not wish to exaggerate its value, duration and effectiveness. And yet the cultural waves which the Baroque unfurled upon Europe were possibly more deep, full and uninterrupted than those even of the Renaissance. The Baroque was the product of two massive imperial civilizations, that of Rome and that of Spain. But how are we to chart their expansion, their tumultuous foreign adventures without the indispensable maps which no one has yet constructed? We have museum catalogues but no artistic atlases, histories of art and literature but none of civilization.

It is at least clear that yet again, the marginal regions of the Mediterranean, rather than its tumultuous centre, are the best place to view and possibly to decipher its destiny. Those Mediterranean influences which spill over its borders are sign enough of its forceful presence in the turmoil of exchange and conflict which makes up the life of the world. These influences underline, as the seventeenth century begins, the eminent position of the Mediterranean, the cradle of ancient civilization, in the building of the modern world on which it left so large a mark.

[410] F. Ruiz Martín, *Lettres de Florence . . .*, CXXI.

[411] A. Morel Fatio, *op. cit.*, I, p. 27. Let us note in passing the views expressed by Brémond, *op. cit.*, p. 310, not necessarily correctly, that tight-fitting clothes for men, which caused such scandal, came to France from Spain under Philip IV. On Spanish influence in England and in particular on Shakespeare, Ludwig Pfandl, *Geschichte der spanischen Literatur*, p. 98 and J. de Perrot in *Romanic Review*, V, 1914, p. 364.

CHAPTER VII

The Forms of War

War is not simply the antithesis of civilization.

Historians refer constantly to war without really knowing or seeking to know its true nature – or natures. We are as ignorant about war as the physicist is of the true nature of matter. We talk about it because we have to: it has never ceased to trouble the lives of men. Chroniclers give it first place in their narratives; contemporary observers are addicted to discussion of the responsibilities for and consequences of the wars they have witnessed.

While I am determined not to exaggerate the importance of battle history, I cannot allow myself to neglect the history of warfare, a powerful and persistent undercurrent of human life. During the fifty years with which we are concerned, war punctuated the year with its rhythms, opening and closing the gates of time. Even when the fighting was over, it exerted a hidden pressure, surviving under ground.

I shall not however claim to draw any philosophical conclusions about the 'nature' of war from an excursion into this dramatic field. Polemology, if indeed it is a science at all, is still in its infancy. It must learn to identify the long-term patterns, regular rhythms and relationships underlying the history of events: a stage we have not yet reached.

I. FORMAL WAR: NAVAL SQUADRONS AND FORTIFIED FRONTIERS

War in the Mediterranean: one thinks at once of the slim and powerful silhouettes of galleys, racing along the coast in summer, laid up in port during the winter. The documents abound with information about their voyages, their maintenance and their costly splendour. We have scores of expert estimates of what they cost in repairs, supplies, crews and money. Experience quickly showed how difficult it was to marshal them for major engagements, since they had to be accompanied, when sailing in large formation, by roundships carrying their bulky supplies. After a lengthy period of preparation, departure was sudden and the voyage usually rapid. Any point on the coast could be reached. But one should not exaggerate the harm a squadron of galleys could do. Any troops they disembarked were unlikely to go far inland. In 1535, Charles V captured the city of Tunis but went no further; in 1541, he made an unsuccessful attack on Algiers: his campaign took him only from Cape Matifou to the heights overlooking the town. The Turks fared no better: in 1565 their fleet arrived to besiege Malta and became immobilized there. In 1572, the aged García de Toledo advised Don John of Austria, after Lepanto,

25. THE SIEGE OF A FORT IN AFRICA, by Vicente Carducho. Possibly the Peñón de Vélez (1564). The elegant lines of different ship-types are beautifully rendered. Madrid, Academia di S. Fernando

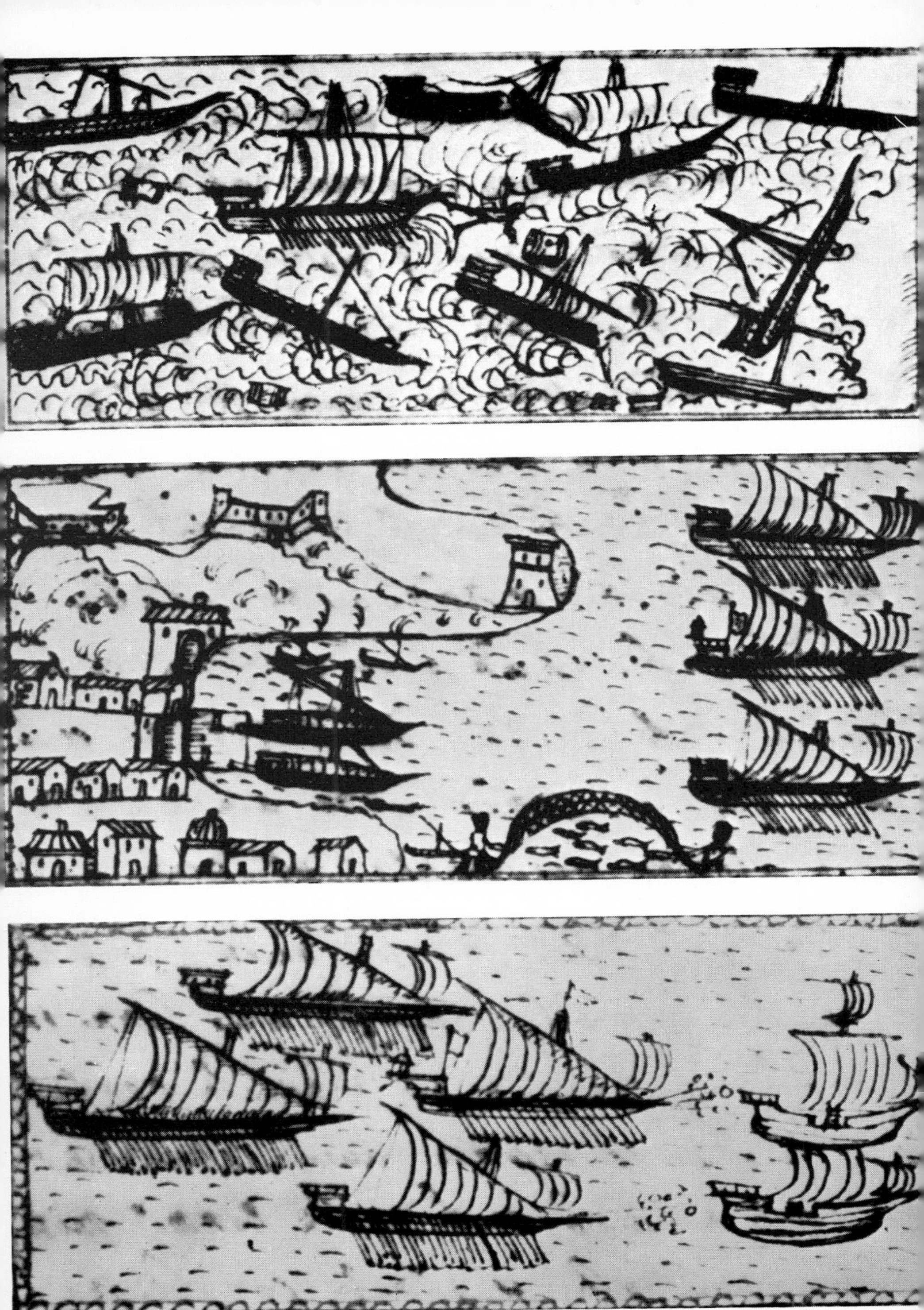

26, 27, & 28. GALLEYS IN A STORM, IN HARBOUR AND IN BATTLE. From sketches by a Tuscan prisoner, circa 1560. Manuscript in the Marciana Library, Venice

that if the victorious powers should launch an expedition to the Levant, it would be better to attack an island than the mainland.

One thinks too of the many armies which by the sixteenth century are remarkable for their greatly increased size. Moving them from place to place, even mustering them beforehand, presented enormous problems. It took months for the king of France to assemble mercenaries and cannon at Lyons in order to 'make a sudden sally over the mountains'.[1] In 1567, the Duke of Alva successfully moved his troops from Genoa to Brussels, but this was a peaceful manœuvre, not a series of encounters. It required all the vast resources of the Turkish Empire to transfer the sultan's armies from Istanbul to the Danube or to Armenia and then to wage a war so far from home. These were extraordinary and expensive feats of organization. When there was enemy opposition, any long march was ordinarily out of the question.

One remembers lastly the series of fortified positions, already looming large in the sixteenth century and soon, in the seventeenth, to become all-important. To the Turks and corsairs, Christendom presented a frontier bristling with defences, the product of the skill of its engineers and the strong arms of its labourers. These fortifications testify to the mentality of a whole civilization. *Limes* or Great Wall of China, a barrier invariably expresses a state of mind. That Christendom – and not Islam – should have surrounded itself with a string of forts is of some significance; indeed it is a major distinction to which we shall be returning.

But these images, familiar and essential though they are, cannot tell us everything about war in the Mediterranean. The scenes they present are those of official war. But no sooner was regular war suspended than a subterranean, unofficial conflict took its place – privateering on sea and brigandage on land – forms of war which had existed all along but which now increased to fill the gap like second growth and brushwood replacing a fallen forest. There are different 'levels' of warfare then, and it is only by studying the contrasts between them that sociologists and historians will make progress towards explaining them. The dialectic is essential.

War and technology. War has always been a matter of arms and techniques. Improved techniques can radically alter the course of events. Artillery for instance, transformed the conditions of war in the Mediterranean as it did everywhere else. The appearance, widespread adoption and modification – for it was always being modified – of artillery, constitute a series of technical revolutions. The problem is dating them. When and how did artillery reach the narrow decks of the galleys? When did it become the distinctive and formidable arm first of the galleasses or large galleys and later of the galleons and tall ships? When was it installed on the ramparts and gun-placements of fortresses and how did it keep up with armies on the move? Well before the victories of Sulaimān, Charles VIII's Italian

[1] A.d.S., Modena, Venezia 15, 77, VI, 104, J. Tebaldi to the Duke, Venice, 16th August, 1522.

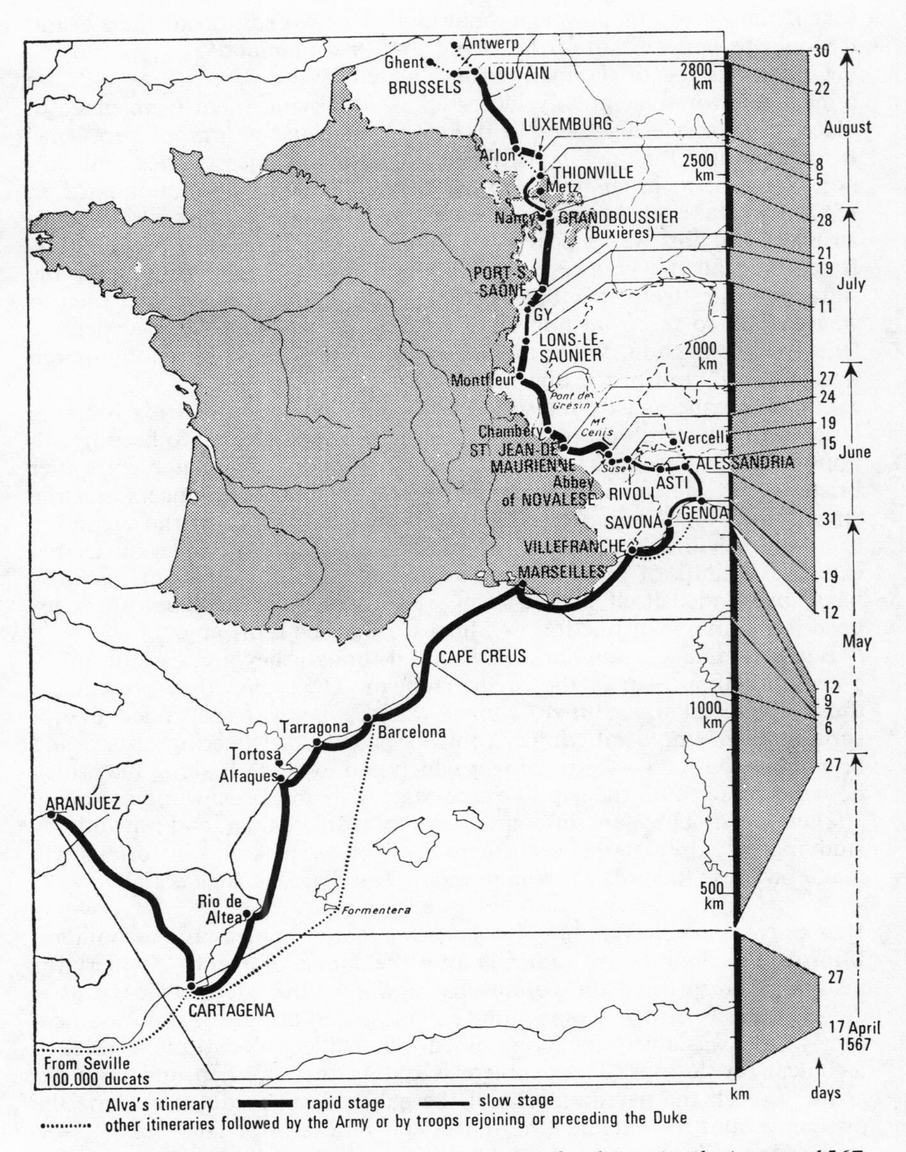

Fig. 63: *The Duke of Alva moves troops to Flanders, April–August, 1567*

To move an army almost 3000 km, without opposition it is true, was a feat in itself. Note the fast sea passages and the time taken to cross the Alps. Unwelcome on French territory, the army had to take the long way round. Measurements, research by J. J. Hémardinquer.

expedition in September, 1494, brought immediate and world-wide fame to field artillery. Several ages of gun manufacture succeeded each other – the age of bronze guns, iron guns, reinforced guns – and there were ages of *geographical* preponderance too, depending on the location of the munitions industry. The policies of Ferdinand the Catholic depended on the foundries of Málaga and Medina del Campo, the latter created in 1495, the former in 1499. Both were to decline quickly: the material they produced was to be worn out in Italy, immobilized in Africa or along the frontiers facing France.[2] A longer reign was that of the foundries of Milan and Ferrara.[3] Then the lead was taken by German and French foundries and, more important than either for supplying Spain and Portugal, the gunfounders of Flanders. From the first decades of the sixteenth century, the superiority of northern guns (and possibly of northern gunpowder too)[4] was becoming evident. These were all matters of consequence. When a hundred or so pieces of ordnance arrived at Málaga from Flanders,[5] the event was immediately noted in diplomatic correspondence. Similarly, the dispatch of forty pieces from Málaga to Messina was read by the Tuscan ambassador as a sign that an expedition was to be launched against Algiers, Tripoli or Barbary.[6] In 1567, Fourquevaux declared that 15,000 cannon balls would be sufficient to take Algiers.[7] This does not seem excessive if one is prepared to accept – as not all historians are – that Malta was saved in 1565 because the Duke of Florence had supplied it, the previous year, with 200 barrels of powder. Such at least was the opinion of a Spanish informant[8] and for us it is an opportunity to note the importance of Tuscany for the manufacture of gunpowder, cannon balls and match for arquebuses.

[2] Jose Arantegui y Sanz, *Apuntos históricos sobre la artilleria española en los siglos XIV y XV*, 1887; Jorge Vigon, *Historia de la artilleria española*, Vol. I, 1947. Was there a decline of the Málaga foundries? See Simancas E° 499, Cobre entregado al mayordomo de la artilleriá de Málaga, 1541–1543. On Málaga and its arsenal towards the middle of the century, Pedro de Medina, *op. cit.*, p. 156.

[3] Have Italian historians perhaps over-emphasized the decline of the foundries at Milan? Pieces were exported either by way of Genoa (chiefly arquebuses and side-arms, 30th August, 1561, Simancas E° 1126) or down the Po to Venice (artillery recorded on board a Portuguese vessel bound for Messina, Venice, 25th April, 1573, Simancas E° 1332).

[4] Cf. a curious document dated 1587 which I hope to publish on the English attack on Bahia, A.N., K. On the place of northern artillery in Spain, 1558, E. Albèri, *op. cit.*, VIII, p. 259.

[5] Nobili to the prince, Tuesday, 6th June, 1566, A.d.S., Florence, Mediceo 4897 *bis*. Other arms, notably arquebuses, arrived from Flanders too of course, witness the ship out of Flanders, carrying arms for the *presidios*, which was captured just inside the Straits of Gibraltar by the Algerine corsairs, the bishop of Limoges to the queen, 24th August, 1561, B.N., Paris, Fr. 16 103 '. . . since by God's will a good and stout ship come from Flanders to furnish all the forts in Barbary with arms has been attacked and taken after passing through the straits with loss of five or six thousand arquebuses, corselets, pistols and other kinds of offensive arms . . .'

[6] See previous note, Mediceo 4897a.

[7] *Op. cit.*, I, p. 167, 4th January, 1567.

[8] D. Francisco Sarmiento to the Grand Commander of Castile, Rome, 28th September, 1565, *CODOIN*, CI, p. 112–114. The opinion reported is that of the Grand Master of Malta.

It remains difficult to say precisely when these transformations occurred and had their effect. Occasional glimpses of developments are all we have. Similarly, while we can with confidence date from 1550[9] the appearance in the Venetian fleet of the great galleys or galleasses carrying artillery (they were undoubtedly technically responsible for the victory at Lepanto), we know next to nothing about the development in the Mediterranean of the armed galleons which we suddenly find being used by the Turks themselves towards the end of the century, on the route from Constantinople to Alexandria.[10] For although Christendom had a head start, technical advances passed from one shore of the sea to another, similar materials came to be used and the political impact of innovation was diminished. Artillery played as great a part in the Christian attack on Granada and North Africa as it did in the victories of the Turks in the Balkans, at the vital battle of Mohács for instance[11] or in Persia[12] or again in North Africa.

War and states. War is a waste of money. 'Les nerfs des batailles sont les pécunes', 'coine, is the sinews of war', as Rabelais was certainly not the first to say.

To make war or peace at the time of one's own choosing, never to *undergo* one or the other, was theoretically the privilege of the strong: but surprises were always possible. Opinions round every prince were divided and his own feelings might be precariously in the balance. The conflict was often incarnated in the eternal protagonists, the war party and the peace party, of which the court of Philip II, up to 1580, provides the classic example. For years the vital question was who would sway the Prudent King – the circle surrounding the spokesman of peace, Ruy Gómez (who remained a faction even after their leader's death), or the supporters of the Duke of Alva, the warmonger always prompt to favour aggression. Indeed what prince or political leader has not been constantly confronted with these two alternatives embodied in opposing factions? Richelieu himself, at the end of the dramatic year 1629, encountered opposition from his peace-loving Garde des Sceaux, Marillac.[13] The choice between the parties was often imposed by events and some 'man of the moment' was propelled to the forefront.

The expenses of war crippled states and many wars were unproductive. The inglorious and costly Irish wars ruined Elizabeth's finances towards the end of her brilliant reign and, more than any other single factor, prepared the way for the truce of 1604. The cost of war in the Mediterranean was so great that bankruptcy often followed, both in Spain and in Turkey.

[9] F. C. Lane, *op. cit.*, p. 31–32.

[10] E. Albèri, *op. cit.*, III, V (Matteo Zane), p. 104 (1594).

[11] Where Turkish success resulted from a concentration of artillery on the line of combat.

[12] The Persians feared Turkish artillery and arquebuses, J. Gassot, *op. cit.*, p. 23; '... for they make very little use of fire-arms ...'

[13] Georges Pagès, in *Rev. d'hist. mod.*, 1932, p. 114.

Philip II's expenditure was phenomenal. In 1571, it was estimated at Madrid that the maintenance of the allied fleet (belonging to Venice, the Pope and Spain) comprising 200 galleys, 100 roundships and 50,000 soldiers, would cost over four million ducats a year.[14] Floating cities, such war fleets devoured money and supplies. The annual upkeep of a galley equalled the cost of its original construction, about 6000 ducats[15] in 1560, and this figure later rose. Between 1534 and 1573, naval armaments tripled, even by conservative estimates. At the time of Lepanto, there were active in the Mediterranean between 500 and 600 galleys, both Christian and Moslem, that is (according to the figures given in notes)[16] between 150,000 and 200,000 men – oarsmen, soldiers and sailors, all flung into the hazards of shipboard life, or as García de Toledo said, to the mercy of the elements – earth, air, fire and water – for all four threatened the precarious existence of men at sea. One bill for provisions supplied to the fleet in Sicily (biscuit, wine, salt meat, rice, oil, salt, barley) comes to some 500,000 ducats.[17]

Regular fleet warfare then depended on the mobilization of large quantities of both money and men: ragged soldiers levied in Spain and given clothes on the march if at all; *Landsknechte* who had made their way to Italy on foot by way of Bolzano then queued up in La Spezia waiting for galleys, Italians, adventurers recruited or accepted to fill the gaps

[14] *Relatione fatta alla Maestá Cattolica in Madrid alli XV di luglio 1571* . . . B.N., Paris, Oc 1533, f°s 109 to 124.

[15] For the beginning of Philip II's reign see the fifty per cent *asiento* for the Tuscan galleys (Simancas E° 1446, f° 107), each galley receiving 250 ducats per month at eleven silver reals to the ducat. On the price of construction, Relacion de lo que han de costar las XV galeras que V.M. manda que se hagan en el reyno de Sicilia este año, 1564, Sim. E° 1128; the bill for 15 galleys is 95,000 *escudos* to the nearest round number, not counting the arms distributed to the men. This price is presented as very reasonable in the report. We learn that the hull of the galleys accounts for less than half the cost price, the rest is represented by sails, oars, yards, masts, rigging, chains, irons, receptacles, spades and other implements, casks, thread to sew the sails, tallow for the ship's bottom. Of the total cost of 95,000 crowns then, the hulls of the 15 galleys represent 37,500, the rigging 9000, the sails almost 20,000, yards and masts 3000, oars 2900, artillery 22,500. These figures omit the price of convicts and slaves which, along with the indispensable supplies of biscuit, constituted the major item of maintenance expense. On the 22 Sicilian galleys there were in May, 1576, 1102 convicts, 1517 slaves and 1205 volunteer oarsmen; in May, 1577 these figures had dropped to respectively 1207, 1440 and 661 (Simancas E° 1147) which works out at a total of 173 oarsmen to a galley in the first case and 143 in the second. Galleys were sometimes reinforced with extra oarsmen: the galley belonging to a grandson of Barbarossa (7th October, 1572, Serrano, *op. cit.*, II, p. 137) had 220 slaves at the oars. Besides oarsmen there were officers, crews and infantrymen. In August, 1570, total numbers on board the 20 Neapolitan galleys were 2940 men, or roughly 150 men to a galley. So including convicts, seamen and soldiers, each galley represented about 300 men. Naval war between 1571 and 1573 sent, aboard the 500 or 600 galleys both Christian and Moslem, 150,000 to 200,000 men to sea, not counting those immobilized on land, in ports and arsenals. For the study of prices see the admirable resources of the Archivio di Stato of Florence and notably: *Nota di quel bisogna per armar una galera atta a navicare*, Mediceo 2077, f° 128. See also Mediceo 2077, f° 60.

[16] See the preceding note.

[17] Simancas E° 1141.

caused by desertion and epidemics, and above all the long files of convicts trudging towards the ports. There were never enough of them to ply the red oars of the galleys – hence the constant need to press poor men,[18] capture slaves and recruit volunteers; Venice sought them as far afield as Bohemia. In Turkey and Egypt, forced levies exhausted the resources of the population. Willingly or unwillingly, massive numbers of men flocked to the coast.

When one remembers that land armies too were expensive – a Spanish *tercio* (about 5000 combatants) cost per campaign, in pay, provision and transport, 1,200,000 ducats, according to an estimate at the end of the century[19] – one readily understands the close connection between war with its prodigious expense and the revenues of a ruler. Through these revenues, war ultimately touched every human activity. But the rapid development and modernization of war made it too much for the old structures to handle; eventually it brought its own end. Peace was the result of chronic inadequacy, prolonged delays in paying the troops, insufficient armaments, of all the misfortunes which governments dreaded but could not escape.

War and civilizations. Every nation experienced these conflicts. But there are wars and wars. If we think in terms of civilizations, major participants in Mediterranean conflicts, we should make a distinction between 'internal' wars, within any one civilization, and 'external' wars, between two mutually hostile worlds, distinguishing in other words between on one hand the Crusades or *Jihāds* and on the other the intestinal conflicts of Christendom and Islam, for these great civilizations consumed themselves in an endless series of civil wars, fratricidal struggles between Protestant and Catholic, Sunnite and Shiite.

It is a distinction of major importance. In the first place, it provides a regular geographical delimitation: Christendom and Islam correspond to given areas with known frontiers, whether on land or sea. So much is obvious. But it also suggests an interesting chronology. As time passes, an age of 'external' wars is succeeded by an age of 'internal' wars. There is no clear dividing line, yet the transition is plain to see: it offers a new perspective on a confused period of history, illuminating it in a way which is neither artificial nor illusory. It is impossible to avoid the conviction that contrasting ideological patterns were first established and then replaced. In Christendom, where the documentary evidence is more plentiful, the Crusade, that is external war, was the dominant force until about 1570–1575. It might be supported with greater or smaller enthusiasm, and already there was a growing tendency towards evasion, excuses, luke-

[18] For example the treatment of the Spanish gipsies sent to the galleys, not for any crime but 'por la necessidad que havia de gente por el remo' . . ., Don John of Austria to Philip II, Cartagena, 17th April, 1575, Simancas E° 157, f° 11.

[19] Morel Fatio, *L'Espagne aux XVIe et XVIIe siècles*, *op. cit.*, p. 218 ff; Nicolas Sanchez-Albornoz, 'Gastos y alimentación de un ejército en el siglo XVI segun un presupuesto de la época', in *Cuadernos de Historia de España*, Buenos Aires, 1950.

warmness or downright refusal – by taxpayers on one hand and sceptics on the other. But the crusade had always had its fervent apologists and its critics. Dissident voices cannot conceal the fact that a wave of militant religion was flowing through Europe in the sixteenth century. In Spain this is only too evident. But it was equally true of France, despite the shifts and compromises of royal policy. It is easy to find, in Ronsard for instance, traces of the crusading spirit, tinged with Hellenism: let us save Greece 'the eye of the inhabited world' and work for Christ's kingdom. The sentiment persisted even in the northern countries which had passed or were about to pass over to Protestantism. *Türkenlieder* from the far-off battlefields of the south east, were being sung all over Germany. When he urged that Germany free herself from the yoke of Rome, Ulrich von Hütten demanded that the money thus recovered should be used to strengthen and extend the Reich at the expense of the Turk. Luther too was always actively in favour of a war against the lords of Constantinople; in Antwerp, there was often talk of marching against the Infidel and in England, where there was always concern over Catholic victories in the Mediterranean, there was at the same time rejoicing over the defeat of the Turk. Lepanto brought both joy and fear to English hearts.[20]

But Lepanto was the end of an era. The waning of the crusades had long been foreseeable. The brilliant victory of 1571 was an illusion: Don John of Austria, a crusader when the age of crusades was past, was as isolated from his contemporaries as his nephew, Dom Sebastian, the hero of Alcázarquivir. Their vision dated from other times. Part of the explanation at least is the strength of the Catholic reaction to the Reformation, at least after 1550, an ideological change of front. Mediterranean Christendom abandoned one war to fight another, as its religious fervour carried it in a new direction.

The change of heart became evident in Rome with the accession of Pope Gregory XIII (1572–1585) which was marked by an outbreak of renewed hostility towards Protestant Germany. This was now the major concern of the Supreme Pontiff, eclipsing the moribund Holy League which he had inherited and which collapsed in 1573 with the defection of the Venetians. Papal policy now looked northwards (conveniently for the success of Turco-Spanish negotiations). More than once, Philip II's advisers trembled for the consequences of the annual truces agreed with the sultan between 1578 and 1581. But the Vatican remained silent. Its aim was now to carry the struggle into the Protestant north and, to this end, to propel the king of Spain towards intervention in Irish affairs, thus touching a raw nerve in England: so the Prudent King appears in the unfamiliar role of follower, rather than standard-bearer, of the troops of the Counter-Reformation.

It is scarcely surprising then, that with the volte-face of the last third of

[20] See also the curious possible interpretations of Elizabeth's policy towards the sultan; she did not wish to appear on too close terms with the enemy of Christendom, W. A. R. Wood, *op. cit.*, p. 27.

the sixteenth century, the notion of the crusade against Islam should have lost much of its force. In 1581, we find the Spanish Church protesting – not against the abandonment of the Turkish wars, but against paying taxes to no purpose.

After 1600 however, as the Protestant wars declined in ferocity and peace slowly returned to the Christian countries of Europe, the idea of the crusade regained force and vigour on the shores of the Mediterranean, in France for instance on the occasion of the Turco-Imperial war of 1593–1606. 'After 1610', notes one historian,[21] 'the Turcophobia which reigned in public opinion, degenerated into a veritable mania.' An explosive mixture of plans and hopes now came together: until once again, a Protestant, 'internal' war put an end to them in 1618.

This overall interpretation seems virtually irrefutable, even though there is insufficient chronological evidence to tell us whether popular feeling followed or preceded – or as I prefer to think both followed and preceded – these volte-faces, provoking them, adding fuel to them and finally being consumed in the resulting explosion. But any explanation which takes account of only one belligerent is extremely unlikely to be satisfactory. We tend always to view the world through naïve western eyes. The other half of the Mediterranean had its own life to live, its own history to make. A recent study (the more remarkable for being so concise)[22] suggests that there were comparable phases, coincident *conjunctures* on the Turkish side too. The Christians abandoned the fight, tiring suddenly of the Mediterranean, but the Turks did precisely the same, at the same moment; they were still interested, it is true, in the Hungarian frontier and in naval war in the Mediterranean, but they were equally committed in the Red Sea, on the Indus and the Volga. With the years, the centre of gravity and the direction of the Turkish military effort shifted in accordance with the phases of a 'world' war (an idea frequently suggested to me in conversation by Frederic C. Lane). It is as if in this continuous history of violence, reaching from the Straits of Gibraltar and the canals of Holland to Syria and Turkestan, everything was interrelated. This was a history operating so to speak at the same voltage everywhere; its variations were electrically identical. At a certain point in time, Christian and Moslems clashed in the *Jihād* and Crusade, then turned their backs on one another, discovering internal conflicts. But this equation of confluent passions was also, as I shall try to show at the end of this second book,[23] the consequence of the slow rhythms of the economic conjuncture, identical throughout the known world which in the sixteenth century saw the beginning of its existence as a unit.

Defensive frontiers in the Balkans. To meet the Turkish threat, Mediterranean Christendom erected a chain of fortresses, now to be one of the

[21] L. Drapeyron, *art. cit.*, p. 134. On all these questions see G. de Vaumas, *op. cit.*, p. 92 ff.

[22] W. E. D. Allen, *Problems of Turkish Power in the XVIth Century*, London, 1963.

[23] See below, p. 897–898.

characteristic marks of its approach to war. As well as fighting, it was constantly extending its defensive and protective lines, encasing itself within a shell of armour. It was a policy both instinctive and unilateral: the Turks built neither very many nor very effective fortifications. Does this indicate a difference in levels of technology or in attitude: on one side confidence in the strength of the Janissaries, the *sipāhis* and the galleys, on the other a desire for security and even, in the major wars, a tendency to economize both strength and resources? Similarly, if the Christian powers maintained a large intelligence service in the Levant, it was not merely out of fear, but in order to obtain the most accurate estimate of the possibility of attack, so that the defence effort could be proportionate to it. The Turks will not come this year: so all extra troops are demobilized and all recruiting is cancelled. It is an absurd pastime, says Bandello,[24] racking one's brains to discover what the Turk or the Sophy will or will not do – he was right in the sense that the ingenious speech-makers to whom he was referring in reality knew nothing of the secret intentions of these mighty enemies, for all their perorations. But for princes it was a different matter: this 'pastime' often determined the extent of defence preparations.

So Mediterranean Christendom erected between itself and Islam a series of fortified fronts, military 'curtains', the defence lines behind which, conscious of its technical superiority, it felt secure. These lines stretched from Hungary to the Mediterranean coast, a series of fortified zones separating one civilization from another.

The Venetian 'limes'. On the edge of the western sea, Venice had long been keeping vigil. To meet the Turkish threat, a string of Venetian outposts and coastal watch-towers ran along the shores of Istria, Dalmatia and Albania, as far as the Ionian islands and beyond, to Crete and Cyprus: this last base, acquired by the Signoria in 1479, was to be hers until 1571. But Venice's straggling maritime empire, a parasite plant on eastern shores, had been broken through several times by Turkish attacks. To go no further back, the peace of 12th October, 1540[25] had deprived it of two precious positions on the Dalmatian coast, Nadino and Laurana [Vrana], of some *isolette* in the Aegean, Chios, Pátmos, Cesina, of some 'feudal' islands, Nios [Ios], the fief of the Pisani family, Stampalia [Astipália] the fief of the Quirini and Paros which belonged to the Venier family. Venice had also been forced to give up the important stations of Monemvasia and Napoli di Romania in Greece. Thirty-three years later, in the separate peace of 1573, completed by the difficult agreements of 1575,[26] she ceded more positions in Dalmatia, paid a war indemnity and renounced her claim to Cyprus, which had been a *de facto* Turkish possession since 1571. The Venetian Empire has often been compared to the

[24] *Op. cit.*, IX, p. 138.
[25] Giuseppe Cappelletti, *Storia della Repubblica di Venezia*, VIII, p. 302 ff.
[26] H. Kretschmayr, *op. cit.*, III, p. 74.

British; Venice's possessions at the end of the sixteenth century then would be the equivalent of the British Empire stripped of its holdings east of Suez. But the comparison is misleading: the frontier possessions of Venice were merely a string of tiny settlements in often archaic forts. The populations of the towns and the islands rarely exceeded a few thousand: in 1576, Zara had just over 7000 inhabitants,[27] Spalato just under 4000,[28] Cattaro only about a thousand after the epidemic of 1572, Cephalonia scarcely 20,000,[29] Zante 15,000,[30] Corfu 17,517.[31] Only Crete with its 200,000 inhabitants carried a certain weight and was indeed the major link in the revised chain. But the Greek island was known to be unreliable, as it was to prove in 1571 and again in 1669. The empire was insignificant, in demographic terms, compared with Venice and the Terraferma, the population of which was estimated about the same time as one and a half million.[32]

It was little short of a miracle then that the barrier held back the swirling tide of Turkish invasion. In 1539 after all, the Spanish had been unable ,to hold the bridgehead of Castelnuovo on the Balkan coast.[33] The remarkable solidity of the Venetian defences was a triumph of improvisation, the result of endlessly revised calculations: of the scrupulous maintenance of the fortresses, of the vigilance of the Arsenal, that mighty factory, of incessant patrols by roundships and galleys and, it should be added, of the loyalty of the frontier population, the high quality of the men who commanded in the name of the Signoria, and the courage of the deportees who were serving their sentence there, not to mention the efficiency of the artillery schools and the possibility of recruiting soldiers from the Albanian, Dalmatian and Greek populations of these troubled zones.

But at either extremity of the chain of defences, Venice encountered problems. The furthest point east was Cyprus, hard to defend and with an unreliable population. This island, like Rhodes, had the disadvantage of being close to Asia Minor and therefore vulnerable to Turkish attack; the defeat of 1571 forced the Venetians to withdraw to Crete, which had a narrow escape in 1572 and which the Signoria felt thereafter to be under constant threat from the ambitions of the conqueror. At the other end of the front, in the North, on the borders of Istria and Friuli, Venice's possessions were adjacent to Habsburg and almost to Turkish territory – a double menace, the more serious in that it threatened the Terraferma, Venice's very flesh and blood. Already between 1463 and 1479, the Turks had launched raids as far as Piave[34] and on the Habsburg side, the frontier,

[27] Relation of Andrea Giustiniano, 1576, B.N., Paris, Ital., 1220, f° 81.

[28] *Ibid.*, f° 69.

[29] *Ibid.*, f° 34 v° and 35.

[30] *Ibid.*, f° 25 v°.

[31] *Ibid.*, f° 39 ff.

[32] B.N., Paris, Ital., 427, f° 274, 1569.

[33] A. Morel Fatio, in *Mémoires de l'Académie des Inscriptions et Belles Lettres*, Vol. XXXIX, 1911, p. 12 ff, offprint. There had been an equally unsuccessful attempt, five years earlier, at Coron in the Morea.

[34] Fernand Grenard, *Grandeur et décadence de l'Asie*, p. 77.

although established *de facto* in 1518,[35] had not yet been accorded formal recognition. It was to meet all these perils that Venice was to build the massive and costly fortress of Palma at the end of the century.

The Venetian Empire, no more than a line of frontier posts, while it did not hold the Turkish Empire in a stranglehold was nevertheless something of an impediment. Venetians were fully aware of the extreme vulnerability of these outposts. Time after time, ambassadors and *baili* interceded in Constantinople, using persuasion or bribery, to protect them from the possibility of attack. And time after time, for political or commercial reasons, following border incidents over, say, a ship loading grain without a permit, an over-bold pirate vessel or an over-zealous Venetian patrol galley, tempers would rise and conflicts break out. In 1582, Sinān Pasha, in order to pick a quarrel with the Venetians, provocatively demanded the return of the islands 'which are the true feet joined to the body of the sultan's state'.[36]

Perhaps after all the Venetian line held because of its very weakness, because the Turks had already made breaches in it, doors and windows through which they could reach the West: Modon, which although poorly fortified, held out during the dramatic siege of 1572, and which Pierre Belon had, as early as 1550, described as 'the key to Turkey'; further north Navarino, which was fortified after 1573[37] and lastly Valona in Albania, which although unfortunately surrounded by a perennially restless region, was nevertheless an excellent base for expeditions to the western sea and Christendom. Can it be true that the gaps in the Venetian *limes*, by decreasing its effectiveness as a barrier, enabled it to survive longer?

On the Danube. North of the Balkans,[38] the Turkish Empire extended to and crossed the Danube, a vital but fragile frontier. It partly controlled the Danube provinces, although it was never to be unchallenged master of

[35] Carlo Schalk, *Rapporti commerciali tra Venezia e Vienna*, Venice, 1912, p. 5.

[36] X. de Salazar to H.M., V. 24th March, 1582, Simancas E° 1339.

[37] P. de Canaye, *op. cit.*, p. 181.

[38] References for this paragraph are to the vast and not always accessible literature on Hungary. A. Lefaivre, *Les Maggyars pendant la domination ottomane en Hongrie, 1526–1722*, Paris, 1902, 2 vols. More recent German books tend to be oriented by twentieth-century preoccupations, Rupert von Schumacher, *Des Reiches Hofzaun, Gesch. der deutschen Militärgrenze im Südosten*, Darmstadt, 1941; Roderich Gooss, *Die Sienbenbürger Sachsen in der Planung deutscher Südostpolitik*, 1941 (political and detailed); Friedrich von Cochenhausen, *Die Verteidigung Mitteleuropas*, Jena, 1940, biassed and summary; G. Müller, *Die Türkenherrschaft in Siebenbürgen*, 1923; Joh. Loserth, 'Steiermark und das Reich im letzten Viertel des 16. Jahrhunderts' in Zs. d. hist. Ver. f. *Steiermark*, 1927, concerning the mission of Friedrich von Herberstein who went to the Reich in 1594 to ask for help against the Turks. On religious life and the spread of Protestantism, there is an abundant bibliography, see K. Bihlmeyer's *Manuel*, Vol. III, p. 69; *Mémoires de Guillaume du Bellay*, *op. cit.*, II, p. 178. Hungarian soldiers fighting as light cavalry 'to whom the name Hussirer was sometimes given, were considered by the Germans little better than barbarians', G. Zeller, *Le siège de Metz*, Nancy, 1943, p. 15. On supplies for the Hungarian war, see Johannes Müller, *Zacharias Geizkofler*, *1560–1617*, Vienna, 1938.

the wooded mountains of Transylvania. To the west, it advanced along the longitudinal valleys of Croatia, beyond Zagreb to the strategic gorges of the Kulpa [Kupa], the upper Sava and the Drava, overlooked by those poor, mountainous, inaccessible regions, inhabited by few men, where the Dinaric bloc meets the towering mass of the Alps. So the Turkish frontier north of the Balkans became fixed fairly early both in the extreme west and the east: on both sides it was confined by geographical obstacles. There were human obstacles too: in the eastern lands of Moldavia and Wallachia, Tartar hordes periodically launched devastating and irresistible invasions. In the west, a German frontier was militarized, at least in the *Windischland*, between the middle reaches of the Sava and those of the Drava, under the command of the *Generalkapitän* of Laibach. The imperial order for its fortification was issued at Linz in 1538. In the Windisch March and later in Croatia, these military frontier installations were to grow up of their own accord under Charles V and Ferdinand. A ruling of 1542 laid down the administration of the whole area. As Nicholas Zrinyi was soon to write, in 1555, this was the breastplate, the *Vormauer* of Styria and therefore of the entire hereditary Austrian state. Was it not in fact this need for a common defence system, financed locally, which gradually cemented into a fairly recognizable unit the Austrian *Erbland*, previously a miscellaneous collection of small states and nations?[39] In 1578, the solid fortress of Karlstadt [Karlovac] rose up over the Kulpa; at about the same time, Hans Lenkovitch was given command over the Croatian and Slavonic frontier, the administration of which was redefined in the Bruck Edict (1578). Its most original feature was the settlement along the frontier of numerous Serbian peasants fleeing Turkish rule and territory. These peasants received land and privileges. They were grouped in large families, patriarchal and democratic communities, in which military and economic tasks were distributed by an Elder.

With the passing of time then, the organization of these military zones was strengthened; one might suppose that, as Busbecq remarks in a note,[40] if a frontier of this kind hardened along more or less stable lines, it was because for many years and at least until 1566, it was relatively untroubled. But peace and stability were only partial. For if resistance was possible at the edges, it was more risky in the central region of the frontier, on the vast bare Hungarian plains. Enough has already been said about the disasters of this tragic country, the terrible turmoil which was its lot after 1526, its quarrels and fratricidal divisions, its almost total subordination to Turkish rule in 1541. Incorporated into the Turkish Empire, Hungary left only a narrow frontier strip in Christian hands. Its plains and waterways were an invitation to invasion, and the Danube most of all. After the Turkish advance on Vienna in 1529, it was necessary, in order to defend what had become the chief rampart of the Germanic world, to multiply the artificial barriers along the roads and rivers; to create and

[39] F. von Cochenhausen, *op. cit.*, p. 86–87.
[40] *Lettres . . ., op. cit.*, Paris, 1748, II, p. 82 ff.

maintain a Danube fleet, about a hundred boats as estimated in 1532 by the *Generaloberst* of the Vienna Arsenal, Jeronimo de Zara. The *Salzamt* of Gmunden received orders to build these boats along with the barges it regularly provided for the transport of salt. They were known as *Nassarnschiffe* or *Nassadistenschiffe*; the word *nassades* is occasionally found in sixteenth-century French texts, but they eventually came to be called *Tscheiken*, from the Turkish *caïque*. Right up to the nineteenth century there were *Tscheiken* on the Danube, with *Tscheikisten* on board. In 1930, *Tscheiken* from the time of Prince Eugene were displayed at a retrospective exhibition at Klosterneuberg.

Towards the end of the sixteenth century, the location of the long Hungarian frontier became more permanent. It was never completely pacified. But despite constant border incidents, raids after captives or tribute, the line was more or less fixed. There gradually grew up an intricate network of forts, watch-towers, castles and fortresses, easy enough for a handful of marauders to slip through, but effective to prevent the advance of the regular armies against which it had been deployed. Here as elsewhere, in Croatia and Slavonia, peace had given an opportunity for construction, especially after 1568 and the truce of Adrianople, renewed in 1574–1576 and again in 1584. This comparatively untroubled period lasted unbroken until 1593, but twenty-five years of peace had been enough to make permanent the long and for many years indeterminate frontier. In 1567 evidently, it was still fragile: 'Certainly on that side', writes Chantonnay from Vienna,[41] 'Christendom is but poorly covered', especially, adds Fourquevaux, since the German soldiers in Hungary are particularly undistinguished. The Turks 'account them so many women and have beaten them as many times as they have come to blows'.[42] This was in 1567 but was even more true in 1593 when war broke out once more with the Turks. The Frenchman Jacques Bongars[43] who was to tour the frontier zone from Raab to Neutra in spring 1585, notes in his *Journal* the many precautions taken by the Christian defenders: in the Raab district alone, there are twelve fortresses with – in peace time – over 5000 foot soldiers and 300 cavalry garrisoned in them. At Komorn, as an extra precaution, there is a workshop actually manufacturing shot and powder inside the fort. Raids and skirmishes are everyday occurrences all along the *limes*.[44]

The central Mediterranean: along the coasts of Naples and Sicily. The coast of Naples and Sicily, along with Malta which provided a link with the Maghreb, was a military zone of a very different kind. Its strategic position derived from its position on the central axis of the sea. 'It was Italy's maritime front against the Turkish threat',[45] facing the

[41] *CODOIN*, CI, 7th June, 1567, p. 229.
[42] Fourquevaux, *op. cit.*, I, p. 239, 17th July, 1567.
[43] L. Anquez, *Henri IV et l'Allemagne*, p. XXI–XXIII.
[44] *Ibid.*, p. XXII.
[45] A. Renaudet, *L'Italie* . . ., p. 12.

watch-towers Italy possessed in Albania and Greece. Its function was at the same time to provide a naval base for the Spanish fleets, to offer resistance to Turkish armadas and to defend its own territory against pirate attacks.

Brindisi, Taranto, Augusta, Messina, Palermo and Naples were all possible rallying points for the Christian galleys. Brindisi and Taranto were possibly too far east, Palermo and Augusta looked more towards Africa than the Levant, and Naples was too far north. That left Messina. In times of danger, Messina was the crucial naval base of the western powers. Its commanding position on a narrow channel of water, its easy access to supplies of both Sicilian and foreign wheat, its proximity to Naples, all contributed to its suitability. From Naples it received men, sails, biscuit, hogsheads of wine, vinegar, fine powder, oars, match and rods for arquebuses, iron cannon balls. As for the city's position, we should not think of it in present-day strategic terms: in the years of Turkish supremacy, it was still possible for Moslem armadas to force a passage through the straits, a feat sometimes accomplished, though at great risk, by isolated galleys or pirate fleets. For even this narrow channel was immense, measured by the firing range of sixteenth-century artillery, and not easy to patrol.

At any time after the beginning of the sixteenth century, a traveller would have found both the coastal and the inland regions of Naples and Sicily strewn with fortresses and fortifications, many of them tumbledown and out of date. Only occasionally would there be provision for modern artillery in the way of platforms and cavaliers, only rarely would walls and ramparts be strengthened in anticipation of enemy gunfire, with vital installations above ground kept to a minimum. The demolition and reconstruction of these old-fashioned forts was to occupy several generations: Catania began in 1514[46] to add to its mediæval walls bastions capable of providing cross-fire. The work was not completed until 1617, after three-quarters of a century of effort and expenditure.

It was in about 1538 that major defence works began throughout the Mezzogiorno, on the initiative in Naples of Pietro de Toledo, in Sicily of Ferrante Gonzaga. For 1538 was the year of Prevesa and the year that Turkish fleets, invincible on sea, began to pound the coasts of Naples and Sicily. The anonymous *Vita di Pietro de Toledo*[47] indicates that it was at this time that the viceroy had work begun on the fortifications of Reggio, Castro, Otranto, Leuca, Gallipoli, Brindisi, Monopoli, Trani, Barlettta, Manfredonia and Vieste and also took steps to fortify Naples. From then on, it seems, watch-towers were being built along the Neapolitan coast. In 1567, 313 were built throughout the kingdom.[48] Pietro de Toledo's efforts in Naples were paralleled in Sicily between 1535 and 1543, by those

[46] Rosario Pennisi, 'Le Mura di Catania e loro fortificazioni nel 1621' in *Arch. st. per la Sicilia Orientale*, 1929, p. 110.

[47] *Arch. st. ital.*, Vol. IX, p. 34.

[48] Simancas E° 1056, f° 30.

of Ferrante Gonzaga.[49] He had 137 towers[50] built along the southern and eastern coasts, the former protected to some extent by the lie of the land, the latter exposed to Turkish attacks and soon to become a 'mere military frontier facing the Ottoman Empire'.[51] On this sensitive front, defence works had begun in 1532 at Syracuse.[52] This, as Ferrante Gonzaga himself said, in a report to the king,[53] was the only exposed coast of the island. The north coast was mountainous, the south, 'le più cattiva e più fluttuosa spiaggia di quei mari',[54] offered no harbour for an enemy fleet. But the east, with its low, fertile and easily accessible shores, was a different matter. It would therefore be necessary to fortify not only Syracuse but also Catania and Messina, which he found on his arrival in 1535 'abandonate et senza alcuno pensamente di defenderle'.[55] But they were still without defences when he left the island.

The transformation was not to be completed overnight – or even in the too brief reign of a viceroy. In Sicily, work continued under Ferrante's successors, as it did under those of Pietro de Toledo in Naples. It was a never-ending task,[56] forever being left off and begun again, as order succeeded counterorder. It was said in Naples that every viceroy faced with the construction of the kingdom's twenty fortresses (19 in 1594, to be precise) undid the work of his predecessor.[57] This is hardly fair in view of the difficulties they encountered. The administrators were held up by lack of funds, obliged to stop work in one place in order to begin somewhere else, to shore up installations which were already collapsing (the Sicilian watch-towers were completed in 1553 and had to be rebuilt between 1583 and 1594) to overhaul and modernize all the forts one by one. Then the line had to be extended westwards, evidence that the threat from the western side was increasing. Barbary corsairs and large Turkish expeditions before 1558 had been capturing Sicilian and Neopolitan positions from the rear; so defensive works had to be begun on the Tyrrhenian side, in Palermo,[58] Marsala,[59] Trapani,[60] Sorrento,[61] Naples[62] and Gaeta.

But the chief danger lay to the east and it was on this side that the defence system was concentrated. Take Naples in 1560: for a year, work

[49] G. Capasso, 'Il governo di D. Ferrante Gonzaga in Sicilia dal 1535 al 1543' in *Arch. st. sic.*, XXX and XXXI.

[50] G. La Mantia, 'La Sicilia e il suo dominio nell'Africa settentrionale dal secolo XI al XVI' in *Arch. st. sic.*, XLIV, p. 205, note.

[51] Hans Hochholzer, *art. cit.*, p. 287.

[52] L. Bianchini, *op. cit.*, I, p. 259–260.

[53] Milan, 31st July, 1546, B.N., Paris, Ital., 772, f° 164 ff.

[54] *Ibid.*, f° 164 v°.

[55] *Ibid.*

[56] Simancas E° 1050, f° 136, 3rd December, 1560 and E° 1052, f° 10.

[57] *Arch. st. it.*, Vol. IX, p. 248; Simancas E° 1051, f° 68.

[58] 2nd May, 1568, Simancas E° 1132; 1576, Simancas E° 1146, the question was still on the agenda.

[59] G. La Mantia, *art. cit.*, p. 224, note 2.

[60] L. Bianchini, *op. cit.*, I, p. 55.

[61] After being sacked by the Turks, 31st January, 1560, Simancas E° 1050, f° 14.

[62] 26th February, 1559, Simancas E° 1049, f° 91.

had been in progress to fortify Pescara,[63] the island of Brindisi and the large town of Taranto.[64] After some debate, it had been decided definitely to countermand the order first issued, then withdrawn, by the Duke of Alva (when he was viceroy of Naples in 1557) to dismantle the defence works of a series of small places on the Capo di Otranto and around Bari: Nolseta, Sovenazo, Vigella, Galignano, Nola, on condition that these little towns provide their own fortifications and protection, a detail which amply illustrates the difficulty of completing works and the imperfections of the defensive line. So, as summer approached, these various posts received reinforcements. The Naples militia provided 8000 to 10,000 men and might if necessary send 20,000. Since they had to cross the kingdom and be billeted among the population, there was relief that these were Neapolitan and not foreign troops.[65] So in May, 1560, 500 infantry were dispatched to Manfredonia, 700 to Barletta, 600 to Trani, 400 to Bisceglie, 300 to Monopoli, 1000 to Brindisi, plus three companies of Spaniards in the fort, 500 militiamen to Taranto, 800 to Otranto, 800 to Crotona. In addition, a thousand men-at-arms and 200 light horse were to be garrisoned in Apulia and 6000 Italians levied as reserves to be used in the event of attack.[66] As well as occupying the coast and strengthening these positions, as an extra precaution the authorities evacuated the *lugares abiertos*, the towns and villages exposed on the seaward side. In Sicily in 1573, since it was impossible to defend the whole island,[67] it was decided to protect only Messina, Augusta, Syracuse, Trapani and Milazzo, abandoning for the time being, since they were too vulnerable, Taormina, Catania, Terranova, Licata, Girgenti, Sciacca, Mazzara, Castellammare, Termini, Cefalu and Patti.

Such were the preoccupations of summer (at the approach of winter, the whole system was dismantled) of the viceroys of Naples and Sicily, until the 1580s and even later. At this time, since the Turkish threat was less imminent, the military presence came to be felt more of a burden, particularly in Sicily, where the cavalry (vital to the protection of this mountainous island) literally devoured the revenues of the kingdom. In all, considering the effort put into this sophisticated system of defences, the huge number of troops it employed, the complicated mechanism of dispatches, communications and signals it implied, it is hardly remarkable that the Turks received some harsh surprises when they met this flexible barrier. If 1538 roughly marks the beginning of this adaptable defence system, it was working to full capacity only after about 1558.[68] Its efficiency

[63] Fourquevaux knew about it, 29th December, 1565, Fourquevaux, *op. cit.*, I, p. 36.

[64] G. C. Speziale, *Storia militare di Taranto*, Bari, 1930.

[65] 10th January, 1560, Simancas E° 1050, f° 9; *Ordenanzas de la milicia de Napoles* (1563), printed, Simancas E° 1050, f° 54.

[66] Simancas E° 1050, f° 43 (18th May, 1560); similar arrangements were made in 1561, Simancas E° 1051, f° 52 (5th April, 1561).

[67] E. Albèri, *op. cit.*, II, V, p. 483.

[68] See for example the register of coastal garrisons at Naples in May, 1567, Simancas E° 1056, f° 67; in Sicily in 1583 or 1585, Simancas E° 1154.

was recognized by the Venetians. In 1583, a report by the *provveditore* of the fleet, Niccolò Suriano, declares: 'Not so long ago, the entire coast of Apulia, from the Capo di Santa Maria to Tronto, possessed very few watch towers. So the Turkish *fuste* were forever sailing up and down this coast, doing great damage both to shipping and to the territory and being content with their success here, did not penetrate to the heart of the Gulf. Now because of these towers, it seems that the people on the shores are well protected, and small boats can sail to and fro in full security, in the daytime. If an enemy ship appears, they can take refuge under the towers where they are quite safe, admirably defended by the artillery mounted there. So now the *fuste* came up past the mountain of Ancona, where they are sure to find good prizes without great risk.' Since off this part of the coast, Venetian vessels were being captured, rather than Spanish ships sailing to Naples, it is easy to understand the writer's concern and his conclusion that the Pope and the Dukes of Ferrara and Urbino ought to be building watch-towers like those in the Kingdom of Naples.[69] The labours of the Spanish viceroys were not in the end to be despised.

The defence of the coasts of Italy and Spain. The Naples-Sicily line, prolonged by the formidable Christian base at Malta to the Barbary coast, where the *presidio* of La Goletta was a Spanish possession until 1574, was not usually crossed by the Turkish fleets. Not that it was capable of stopping them; but once they had captured their prey, the Turks were rarely concerned to go farther afield. There was nothing to prevent them doing so however, any more than there was to prevent ships from moving between Turkey and Barbary. Meanwhile the pirates of Algiers were always active. So the Christian powers had to take serious steps to defend all coasts, equipping them with towers and fortresses.

Like the Sicilian defence works, these walls did not rise overnight: they would be started, then the site would be moved or modernization be called for. When and how? It is hard to say. In 1563,[70] it was realized that the old towers of Valencia would have to be replaced by new works allowing room for artillery placements. The immediate question at Barcelona was, who would pay, the king, the city or the *Lonja*?[71] In August, 1536, in Majorca,[72] watchers sighted enemy sails from the top of the *atalayas*. So there must have been towers on the island by this date. How long had they been there? In 1543, fortifications were begun at Alcudia, but what kind of fortifications were they? And when were the round watch-towers of Corsica built, as distinct from the square towers of village fortifications?[73] Was it in 1519–1520 that the coastguard patrols

[69] V. Lamansky, *op. cit.*, p. 600–601.
[70] 31st March, 1563, archive reference mislaid.
[71] A. de Capmany, *op. cit.*, IV, appendix p. 84, 20th July, 1556.
[72] 29th August, 1536, A.N., K 1690.
[73] P.B., 'Tours de guet et tours de défense. Constructeurs de tours', in *Petit Bastiais*, 19th June–14th July, 1937

were set up in Valencia on the model of the Holy *Hermandad*[74] with 'musters' and alarm systems? They cannot have amounted to much, since in 1559, Philip II in Brussels is expressing surprise that the fort at Alicante is held by only six men.[75] In 1576, plans were still being made for the fortification of Cartagena.[76] In Granada by contrast, in 1579, there was a coastal defence service, under the command of Sancho Davila, *Capitan general de la costa*,[77] perhaps because there were more specific reasons for taking precautions in this region. Similarly Sardinia was obliged to consider defence (detailed plans drawn up for the fortification of the island in 1574 still exist)[78] and built watch-towers under the administration of the viceroy Don Miguel de Moncada, in about 1587.[79] Fishermen off the island's coral reefs would take refuge under these towers and use artillery in their defence.[80]

Needless to say, these works were never completed. There was always more to be done to assure the protection of the *poveri naviganti*[81] and the inhabitants of the coasts. On the whole, these were defences on a much smaller scale than those described earlier. The Spanish coasts were often raided by corsairs, particularly the Barbary pirates, but had little to fear from the Constantinople fleet. There was quite a difference.

The coasts of North Africa. The North African defence network presents fewer problems to the historian than the others[82] – not that it is less complicated, but it is better documented. Although no more than a string of forts, the line of *presidios* was implicated in the histories of the regions it contained: along this frontier two civilizations met, hence the evidence from many sources which illuminates both the general and the particular history of the Spanish positions in North Africa. Established in the time of Ferdinand the Catholic, chiefly between 1509 and 1511, the *fronteras* had been deployed along the borders of an ancient but fragmented land, incapable of defending itself. Perhaps only the Aragonese preoccupation

[74] K. Häbler, *Gesch. Spaniens*, Vol. I, p. 26–27.

[75] 31st March, 1559, Simancas E° 137.

[76] *CODOIN*, II, p. 183.

[77] *CODOIN*, XXXI, p. 162, 165, 169. J. O. Asin, articles in *Boletin de la R. Academia Española*, 1928, XV, p. 347–395 and 496–542 and *Bulletin Hispanique*, XXXV, 1933, p. 450–453 and XXXIX, 1937, p. 244–245. See also Mariano Alcocer Martinez, *Castillos y fortalezas del antiguo reino de Granada*, Tangier, 1941; A. Gamir Sandoval, *Organización de la defensa de la costa del Reino de Granada desde su reconquista hasta finales del siglo XVI*, Granada, 1947.

[78] Relacion de todas las costas del Reyno de Cerdaña (undated), Simancas E° 327, an extremely important document, must date from after 1574.

[79] Francesco Corridore, *op. cit.*, p. 18.

[80] F. Podesta, *op. cit.*, p. 18.

[81] 20th March, 1579, A.d.S., Genoa, L. M. Spagna 8 2417.

[82] Fernand Braudel, 'Les Espagnols et l'Afrique du Nord' in *Revue Africaine*, 1928; 'Les Espagnols en Algérie' in *Histoire et Historiens de l'Algérie*, 1930. Since this article there has been only one general survey, Robert Ricard, 'Le Problème de l'occupation restreinte dans l'Afrique du Nord (XVe–XVIIIe siècle)' in *Annales d'histoire économique et sociale*, 1937, p. 426–437.

with the riches of Italy prevented Spain from moving in to capture the inland Maghreb. But the lost opportunity never presented itself again. In 1516, the Barbarossas settled in Algiers; in 1518, they placed themselves under the protection of the sultan; in 1529, their city liberated the small but troublesome fortress on the Peñon, which had been held by the Spaniards since 1510. Even before this date, Algiers had been the urban centre of the barren regions of the central Maghreb, sending out its swift marching columns, settling garrisons, drawing to the city all the trade of this vast intermediary zone. From now on a land held from within was to oppose and threaten the Spanish in North Africa. Charles V's two major expeditions, against Tunis in 1535 and against Mostaganem in 1558, did not alter the situation. And indeed after the failure at Mostaganem, which brought the collapse of grandiose projects of alliance with Morocco, a new age, the age of the *presidios*, had already begun.

Inaugurated in the reign of Philip II, it was an age marked by prudence and calculation rather than adventure. Grand schemes for African expeditions did not altogether vanish it is true. But there was much deliberation and little action now, except at points known – or thought – to be particularly weak. Such was the case of the expedition against Tripoli which ended in disaster at Djerba in 1560. Even this was undertaken less on the king's initiative than on that of the viceroy of Sicily, the Duke of Medina Celi, and of the Grand Master of Malta. The great expedition against the Peñon de Velez, mounted with over 100 galleys in 1564, was something of an anticlimax. The recapture of Tunis in 1573 by Don John of Austria and his obstinacy in clinging to his trophy in the face of opposition from his brother and his advisers, who desired only to see the position evacuated and destroyed, was merely a sudden outburst of megalomania, a brief revival of the spirit of Charles V's reign, of which there were several during the reign of the Prudent King.

Meanwhile the steady, discreet but in the long run effective labour of reinforcing and extending the *presidios* themselves was undertaken during the years 1560–1570. Mortar, lime, bricks, beams, planks, stones, baskets for earthworks, picks and shovels become the subject matter of all the letters from the *presidios*. Parallel to the authority of the captain of the position, the role and authority of that other figure, the *veedor* or quartermaster, was gradually increasing, as was that of the engineer-architect, a civilian, which did not always make for good relations. Giovanni Battista Antonelli for instance was in charge of the defence works here at Mers-el-Kebir[83] and another Italian, Il Fratino (whom Philip II was also to use in Navarre) moved the entire old *presidio* of Melilla, brick by brick, to the edge of the lagoon. Two of his drawings, preserved at Simancas, show the little settlement in its new site, a tiny cluster of houses round the church, with either side of it stretching the

[83] Juan Baptista Antoneli to Eraso, Mers-el-Kebir, 29th March, 1565, Simancas E° 486. His quarrel with F. de Valencia, F. de Valencia to the king, Mers-el-Kebir, 8th February, 1566, Simancas E° 486.

steep and endless coast. Il Fratino also worked at La Goletta,[84] where his relations with the governor, Alonso Pimentel, were stormy: their antagonism was that of men isolated from society, driven to almost murderous lengths and sending home mutual accusations.[85] The *presidio* grew nevertheless; around the original rectangle with its bastions of 'old Goletta', engravings of 1573 and 1574 show a string of new fortifications, completed in summer 1573.[86] There were also a windmill, powder magazines, cisterns and 'cavaliers' on which the powerful bronze artillery was mounted. For artillery was the strength and the *raison d'être* of the fortresses in Africa.

In the time of Philip II, the *presidios* expanded, forever building new fortifications, devouring constructional materials often carried over great distances (lime from Naples for Mers-el-Kebir for instance) and continually calling for more 'pioneers' – *gastadores*. Oran and its annex Mers-el-Kebir – which after 1580 was a model of its kind – hummed with activity. By the end of the century this was no mere fortress but a fortified zone, created at huge expense and with heroic effort. The soldier, like the humble *gastador*, had to handle pick and shovel. Diego Suárez, the soldier-chronicler of Oran, who had worked on the Escorial in his youth, could not find words to express the achievement. It was as good as the Escorial, he had to say in the end. But this exceptional masterpiece was only built in the final years of Philip II's reign, and in 1574 oddly enough it had been threatened with total destruction. The Spanish government was then on the verge of the second bankruptcy of the reign (1575). In Tunisia, Don John of Austria, who had just captured Tunis, was hanging on against instructions,[87] and his obstinacy led to the disaster of August–September, 1574, when the Turks succeeded in taking both La Goletta and Tunis. This double tragedy showed that the two fortresses, which had had to share supplies sent out from Spain, had finally been harmful to each other. In the light of this, it was logical to suppose that the double *presidio* of Oran – Mers-el-Kebir, linked only by a poor road of about a league, unsuitable for artillery, was also a mistake. The enquiry conducted on the spot by Prince Vespasiano Gonzaga, in December, 1574,[88] concluded that

[84] On the fortifications of La Goletta, Alonso Pimentel to the king, 29th May, 1566, Simancas E° 486; 9th June, 1565, *ibid.*; Luis Scriva to the king, 7th August, 1565, *ibid.*, the fortification 'va de tel arte que a bien menester remedio'; Philip to Figueroa, 5th November, 1565, Simancas E° 1394, he had decided to fortify La Goletta, borrowing 56,000 crowns from Adam Centurione; Fourquevaux, who knew the news, 24th December, 1565, announced the departure of Il Fratino and the carpenters, *op. cit.*, I, p. 10 and 19; *Lo que se ha hecho en la fortificacion de la Goleta; Instruction sopra il disegno della nova fabrica della Goleta*, 1566, Simancas E° 1130; Philip II to Don García de Toledo, Madrid, 16th February, 1567, orders to remit 50,000 crowns to Figueroa, to be sent immediately to La Goletta, Simancas E° 1056, f° 88; Fourquevaux, 30th September, 1567, *op. cit.*, I, p. 273.

[85] El Fratin to the king, La Goletta, 5th August, 1566, Simancas E° 486.

[86] 20th May, 1573, Simancas E° 1139.

[87] See below, Part III, Chapter IV.

[88] Vespasiano Gonzaga to Philip II, Oran, 23rd December, 1574, Simancas E° 78, on his return see B.N., Paris, Esp. 34, f° 145 v°; Mediceo 4906, f° 98; consultation of

Oran should be evacuated, the fort dismantled and rased to the ground, in order to concentrate the entire strength of the garrison at Mers-el-Kebir, which was better situated and had a good harbour. 'La Goletta' writes the author of the report, 'was lost the day we took Tunis.' As for fortifying Oran, all the engineers in the world would never be able to do it, short of building a great city there. But when the emergency was over, it was this 'great city'[89] which the Spaniards did in fact patiently hollow out of the rock, providing the stable environment in which there later flourished the 'Corte chica', or little Madrid, as it was called with some exaggeration in the eighteenth century.

The fall of the Tunisian bases in 1574, did not have the feared result. No subsequent disaster befell Sicily or Naples. It is true that the latter used the one arm left to them, the galley-fleets.[90] In 1576, the Marquis of Santa Cruz, with the galleys of Naples and Malta, led a punitive expedition along the shores of the Tunisian Sahel, ravaging the Kerkenna Islands, seizing the inhabitants and their livestock, burning the houses and leaving behind him damage amounting to over 20,000 ducats. The inhabitants of the coasts of the Sahel fled immediately and a reinforced galliot took the alarm to Constantinople.[91] Mobile squadrons had their advantages. The Spaniards now seem to have realized this and to have seen that the best way to defend the threatened coastline was to send out the galleys, instead of leaving them, as they had too often before the 1570s, lined up defensively at Messina, waiting for Turkish aggression. Many schemes of reconquest were advanced after the fall of Tunis. One of them, dating from 1581, stated as a general principle that naval strength was a prerequisite[92] – at least someone had seen the light.

The new method of defence – by aggression – was likely to be even more profitable than in the past as a result of the economic revival of the Maghreb. A Spanish report of 1581[93] describes Bône as a populous town, manufacturing quite good earthenware and exporting wool, butter, honey and wax; Bougie and Cherchel as export markets for the agricultural produce of their hinterland, which was not exhausted by the demands of the great markets of Algiers, as is proved by the fact that even nearer the city of the *re'īs*, in the estuary of the Oued el Harrach and off the point of Cape Matifou, boats would come to pick up wool, wheat and poultry for France, Valencia and Barcelona. These details concur with what Haëdo says about the activity of the Port of Algiers at about the

the Council of State, 23rd February, 1575, E° 78 (either the withdrawal to Mers-el-Kebir or the fortification of Arzeu).

[89] On the works at Oran and Mers-el-Kebir, Diego Suárez, *op. cit.*, p. 27–28 (over 30 years, the fortifications of Oran had cost a total of 3 million ducats), p. 148–149, 209 and 262.

[90] As is well observed by E. Pellissier de Raynaud, 'Expéditions et établissements des Espagnols en Barbarie' in *Exploration scient. de l'Algérie*, Vol. VI, 1844, in-8°, p. 3–120. Cf. also B.N., Paris, Ital., 127, f° 72.

[91] Relacion de lo que se hizo en la isla de los Querquenes, Simancas E° 1146.

[92] Relacion de todos los puertos de Berberia que deben de ganarse y fortificarse, Simancas E° 1339.

[93] *Ibid.*

same time in the 1580s. So, sailing the waters off the forbidding and inhospitable coasts of the Maghreb, there were now, as never before, many rich prizes to be had. Besides, was not this method more economical than the maintenance of the *presidios*? A financial report,[94] dating from some time between 1564 and 1568, sets out the expenses incurred by the *presidios*, from the Peñon de Velez, recovered in the West in 1564, to La Goletta (Tripoli had been lost in 1551 and Bougie captured by the Algerines in 1555, so they are not included). Total wages for the garrisons were as follows: the Peñon 12,000 ducats, Melilla 19,000, Oran and Mers-el-Kebir 90,000, La Goletta 88,000 – a grand total of 209,000 ducats.[95] Note the relatively high bill at La Goletta: the garrison of a thousand regular troops plus a thousand extraordinary troops, cost almost as much as the double *presidio* of Oran, manned at that time by 2700 soldiers and 90 light horse. This was because infantrymen at Oran were paid less (1000 maravedís a month) 'por ser la tierra muy barata', since the cost of living there was low.[96] In the west, only the garrison on the Peñon received the high wages of Italy.[97]

The figure of 200,000 ducats refers only to expenditure on personnel: there was much more besides, for instance the maintenance and construction of fortifications. Philip II sent 50,000 ducats for the building of the new Goletta in 1566 and another 50,000 two years later. These may not have been the only two payments. Then there was the heavy burden of supplying ammunition. A consignment[98] for La Goletta alone in 1565 consisted of 200 quintals of lead, 150 of match for arquebuses, 100 of fine gunpowder at 20 ducats a quintal, 1000 baskets for earth, 1000 shovels with handles, the total bill amounting to 4665 ducats, not including transportation. A similar cargo in 1560 had required eight galleys to transport it. Each *presidio* had its own building fund, from which the authorities might borrow if necessary and repay later. These accounts merit detailed study. They would enable us to calculate (quite apart from the original expense involved in capturing the positions, 500,000 ducats for the capture of the Peñon in 1564 for instance, not including the expenses of the fleet) the enormous sum required to maintain these tiny fortresses, always in need of repair, strengthening or extension, with their constant demand for food and supplies.

By way of comparison, at the same period, the guard on the Balearic islands (although they were seriously threatened) cost a mere 36,000 ducats, as did the guard on the coast from Cartagena to Cadiz. As for

[94] Relacion de lo que monta el sueldo de la gente de guerra que se entretiene en las fronteras de Africa, Simancas E° 486.

[95] B.N., Paris, Dupuy 22.

[96] Philip II to Peralte Arnalte, Escorial, 7th November, 1564, Simancas E° 144, f° 247.

[97] In 1525, total expenditure on the *presidios* was estimated at 77,000 ducats, E. Albèri, *op. cit.*, I, II, p. 43. In 1559, their maintenance is simply referred to as a great burden, without further details, E. Albèri, *op. cit.*, I, III, p. 345.

[98] Simancas E° 1054, f° 170.

the annual maintenance of a galley, it was at this time about 7000 ducats. The guard of the *presidios* immobilized, between 1564 and 1568, approximately 2500 regular troops (2850) and 2700 summer reinforcements (men transported there in spring and repatriated at the approach of winter, in theory that is, for delays in arrival and more frequently in departure, were common). 5000 men – more than His Catholic Majesty had in the whole Kingdom of Naples.[99] Without entering upon calculations and considerations worthy of the *speculativi* referred to by a Genoese agent, can one suggest that it might have been better to keep thirty galleys at sea than to hold on to the African *presidios*? What these figures unquestionably demonstrate is the magnitude of the sacrifice made by Spain for the Barbary coast.

The presidios*: only a second best.* Robert Ricard too[100] has questioned whether this solution, 'only second best', was not prolonged beyond the point of usefulness. Cortés on arriving in Mexico burnt his boats: he had to triumph or die. In North Africa, there was always the supply ship with its fresh water, fish, cloth or *garbanzos*. The administration was looking after you. Did the technical superiority of the Christian which allowed him to establish and then maintain *presidios* 'defended by cannon-fire', dispense him from more direct and perhaps more profitable effort? To some extent it must have. But the country was also defended by its immensity and its aridity. Here it was impossible to live like the *conquistadores* of America, who drove herds of cattle and pigs before them. As for sending settlers, it had been considered: in the time of Ferdinand the Catholic, there was a proposal to populate the towns with Castilian Moriscos; in 1543 to colonize Cape Bon.[101] But how were these deportees to live? In a Spain dazzled by the lure both of the New World and the good fare of Italy, where were the men to be found? There were also plans to make these strongholds economically viable, to create some kind of link with the vast interior, off which they would live. In the time of Ferdinand and later Charles V, there was actually some attempt at an economic policy[102] aiming at the development of these African positions

[99] Where the number varied: 2826, April, 1571, Simancas E° 1060, f° 128; 3297, 11th May, 1578, Simancas E° 1077.

[100] *Art. cit.*, note 82 above and *Bulletin Hispanique*, 1932, p. 347–349.

[101] Memorial de Rodrigo Cerbantes, Contador de la Goleta (about 1540), *Rev. Africaine*, 1928, p. 424.

[102] For instance the North African privileges granted to Catalan sailing ships; the pragmatic of 18th December, 1511, issued at Burgos and the new pragmatic granted by Queen Germaine in 1512; *Real executoria* issued at Logroño in the same year 1512, against the officers in Africa; the appointment of a Catalan consul at Tripoli; further protests at the Cortes of Monzon in 1537, against the African governors . . ., A. de Capmany, *op. cit.*, I, 2, p. 85–86, II, p. 320–322. But North African trade circuits bypassed the Christian and Tripolitanian positions, M. Sanudo, *Diarii*, XXVII, col. 25 (they were diverted to Misurata or Tadjoura); for the truth about Oran, *CODOIN*, XXV, p. 425, Karl J. von Hęfele, *op. cit.*, p. 321 (the massacre of Christian merchants at Tlemcen in 1509), caravans travelling to Bône in 1518, La Primaudie, *art. cit.*,

in hopes of making them a centre for Catalan shipping, and of obliging the Venetian galleys to put in there. But all in vain. The doubling of the customs duties in 1516[103] in all the Spanish ports in the Mediterranean did not have the effect of forcing the Venetian galleys to concentrate their African trade in Oran. The commercial currents of the Maghreb of themselves by-passed the Spanish *presidios* and preferred to use as export outlets, Tajura, La Misurata, Algiers and Bône, none of which was in Christian hands. The volume of traffic in these free ports gives some idea of the lack of success of the Spanish *fronteras*, just as in Morocco, at the end of the sixteenth century, the fortune of the Moroccan ports of Larache, Salé, Cabo da Guer (now Agadir) underlines the collapse of the Portuguese strongholds, which had long been prosperous trading-posts. Similarly, the trade even of Spain with North Africa[104] – which seems in any case to have concentrated more on Atlantic Morocco than on the Barbary coast – for all that it revived in the 1580s, bringing to the African coast cloth (woollens, silks, velvets, taffetas and peasant weaves), cochineal, salt, perfumes, lacquer, coral, saffron, thousands of caps, lined and unlined, from Córdoba and Toledo, cow and goat hides and even gold – all this trade (apart from a few crossings to Ceuta and Tangier) was transacted outside the *presidios*. They were virtually excluded from commercial routes. Under such conditions, the *presidios*, condemned to deal only with sutlers and camp-followers, neither prospered nor put out shoots. As grafts they had barely taken and the best they could hope for was to stay alive.

Life in the *presidios* must have been miserable. So near the water, rations rotted and men died of fever.[105] The soldiers were hungry all year round. For a long time, the only supplies came by sea. Later, but only at Oran, the surrounding countryside provided meat and grain, which had

p. 25. I think the best account at present available of Spanish policy towards Venetian trade between North Africa and Spain is H. Kretschmayr, *op. cit.*, II, p. 178. Spain apparently tried in 1516 to make this trade between Africa and the Iberian Peninsula pass through Oran, hence the doubling of customs tariffs in Spanish ports which was meant to hurt Venetian trade. In 1518, according to C. Manfroni, *op. cit.*, I, p. 38, Venice was trying to gain entry to Oran, an event which does not seem to fit what we know of the question. Later on, Charles V after taking Tunis was to operate an open-door policy, J. Dumont, *op. cit.*, IV, Part 2, p. 128, Jacques Mazzei, *Politica doganale differenziale*, 1930, p. 249, n. 1. There is still an immense amount of research to be done into these economic questions lying behind the Spanish 'crusade'. Cf. the valuable study by Robert Ricard, 'Contribution à l'étude du commerce génois au Maroc durant la période portugaise (1415–1550)', in *Ann. de l'Inst. d'Ét. Orientales*, Vol. III, 1937.

[103] G. Cappelletti, *Storia della Repubblica di Venezia*, VIII, p. 26–27.

[104] Besides Haëdo, *op. cit.*, p. 19, B.N., Paris, Esp. 60, f°s 112–113; 18th June, 1570, Simancas E° 334; *CODOIN*, XC, p. 504, Riba y Garcia, *op. cit.*, p. 293; Inquiry into the Barbary trade, 1565, Simancas E° 146; 1598, Simancas E° 178; 4th November, 1597, E° 179; 26th and 31st January, 1597, *ibid.*, 18th July, 1592, A.N., K 1708. In 1565, 30 ships left Cadiz for Morocco. In 1598, about 7000 dozen *bonetes* were exported.

[105] Pescara to the king, Palermo, 24th December, 1570, Simancas E° 1133, the hospitals of Palermo are crowded with sick from La Goletta.

become a regular supplement by the very end of the century.[106] Garrison life was in many ways similar to shipboard life, not without its hazards.

The supply station of Málaga, with its *proveedores*,[107] occasionally helped by the services of Cartagena, was responsible for supplying the western sector, Oran, Mers-el-Kebir, Melilla. There were certainly failures in service and prevarications, indeed it would be surprising to find the contrary. But one should not exaggerate these venial offences. The traffic passing through Málaga was very considerable. All supplies for Africa travelled from here: munitions, rations, construction materials, soldiers, convicts, labourers and prostitutes.[108] Supply and transport posed serious problems – wheat for instance had to be bought, then transported from the interior by files of pack donkeys,[109] which was costly for a start. To transfer it from the administration's granaries to the port and from the port to the *presidios* meant more work and more delay. The sea was infested with pirates. So it was only in winter, when pirates were few, that the risk would be taken of sending to Oran a *corchapin*, two or three boats, a *tartane* or perhaps a Marseilles or Venetian galleon,[110] placed under embargo, and requisitioned to transport supplies or munitions. On more than one occasion, the boat was seized by galliots from Tetouan or Algiers, and the Spanish would be lucky if they could buy it back from the corsairs when they anchored, as was their habit, off Cape Falcon. So pirates, quite as much as negligence by the administration, were responsible for the recurrent famines in the western *presidios*.

The fate of La Goletta was no different, despite its apparently fortunate location near the inexhaustible supplies of bread, wine, cheese and chick peas of Naples and Sicily. But those who succeeded in making the short crossing from Sicily could not always do so at their convenience. When Pimentel took command of La Goletta in 1569, the garrison was living off its reserves of cheese – without either bread or wine. The administration on the Italian side was partly to blame of course. Was it from here or from Spain that the garrison received 2000 pairs of shoes, of good Spanish leather, but in little girls' sizes?[111]

Furthermore, the internal organization of the *presidios* did not con-

[106] The Duke of Cardona to the king, Oran, 18th June, 1593, G.A.A. Series C 12, f° 81.

[107] Various letters from the *proveedores* are preserved at Simancas in the *legajos* E° 138, 144, 145: 7th, 21st, 28th January; 14th February, 6th March, 1559; E° 138, f°s 264, 265, 266, 276, 7th January, 14th September, 25th September, 29th November, 17th November, 31st December, 1564; E° 144, f°s 22, 91, 96, 278; E° 145, f°s 323 and 324. This Castilla series is not in order and the folios do not correspond to any numerical classification.

[108] They were not supposed to be shipped over of course, neither were infected soldiers or priests disguised as soldiers. La orden ql Señor Francisco de Cordoba . . ., Valladolid, 23rd June, 1559, Simancas E° 1201, f° 37. A Spanish courtesan at La Goletta and Tunis, Isabella de Luna, M. Bandello, *op. cit.*, VI, p. 336.

[109] Simancas E° 145, f°s 323 and 324, 25th September, 1564.

[110] R. de Portillo to the king, Mers-el-Kebir, 27th October, 1565, Simancas E° 486.

[111] About 1543, report by Rodrigo Cerbantes, G.G.A. Series C, file 3, no. 41.

tribute to their smooth running, as can be seen from the administration in 1564 of Mers-el-Kebir.[112] Soldiers were supplied with rations by the storekeepers at the price fixed on the shipping labels[113] and often on credit: this was the dangerous practice of allowing advances on their pay, by which soldiers could amass frightening debts, since they were always buying on credit from passing merchants. Sometimes, in times of difficulty, or through the complicity of the local authorities, prices would rise beyond all measure. To escape their intolerable burden of debts, the soldiers would desert and go over to Islam. The problem was aggravated by the fact that pay was lower in the *presidios* than in Italy – one more reason, when troops were being embarked for the *presidios*, not to tell them in advance the name of their destination and once they were there, never to repatriate them. Diego Suárez spent twenty-seven years in Oran, in spite of several attempts to escape as a stowaway in the galleys. Only the sick, and then not always, could come back from the hated coast to the hospitals of Sicily or Spain. The *presidios* then were virtually places of deportation. Nobles and rich men might be sent there as a punishment. The grandson of Columbus, Luis, arrested at Valladolid for trigamy, was condemned to ten years' exile: he arrived in Oran in 1563 and was to die there on 3rd February, 1573.[114]

For and against raids. Let us imagine the atmosphere in these garrisons. Each was the fief of its captain general, Melilla for many years that of the Medina Sidonia family, Oran that of the Alcaudete family; Tripoli was ceded in 1513 to Hugo de Moncada for life.[115] The governor reigned with his family and the lords who surrounded him. The favourite pastime of the rulers was the *razzia*, the planned sortie combining sport with work and, it must be admitted, strict necessity: it was the duty of the garrison to police the surrounding districts, protecting their inhabitants and dispersing intruders, collecting pledges, gathering information and requisitioning supplies. Necessity apart however, there was a certain temptation to play soldiers, to lay ambushes in the gardens of Tunis and kidnap unsuspecting farmers arriving to pick fruit or harvest a field of barley; or beyond the *sebka* at Oran, by turns glistening with salt or covered with water, to surprise a douar, the presence of which had been betrayed by hired spies. This was a more exciting, more dangerous and more profitable sport than hunting wild animals. Everyone had his share of the booty and the Captain General sometimes took the 'Quint' or royal fifth,[116] whether in grain, beasts or humans. Sometimes the soldiers

[112] Relacion de lo que han de guardar los officiales de la fortaleza de Melilla, 9th April, 1564, Simancas E° 486. [113] Diego Suárez, 28th July, 1571, B.N., Madrid, ch. 34.

[114] Alfredo Giannini, 'Il fondo italiano della Biblioteca Colombina di Seviglia' in *R. Instituto Orientale, Annali*, February, 1930, VIII, II. Other *desterrados*: Felipe de Borja, the natural brother of the Maestre de Montesa, Suárez, *op. cit.*, p. 147; the Duke of Veraguas, Almirante de las Indias, *ibid.*, p. 161; Don Gabriel de la Cueva, *ibid.*, p. 107 (1555). [115] G. La Mantia, *art. cit.*, p. 218.

[116] Diego Suárez, *Historia del Maestre ultimo de Montesa*, Madrid, 1889, p. 127.

themselves, tiring of their everyday fare, would go off in search of adventure, from a desire for fresh food, or money, or simply out of boredom. In many cases, such raids naturally prevented the establishment of vital good relations between the fortress and its hinterland, if, as they were intended to, they spread wide the terror of the name of Spain. Contemporary judgements are far from unanimous on this point. We must strike hard, says Diego Suárez, and at the same time be accommodating, increase the number of *Moros de paz*, the subdued populations who took shelter near the fortress and in turn protected it. 'Cuantos más moros más ganancias', writes the soldier-chronicler, repeating the old proverb that the greater the number of Moors, the greater the profit – in grain, everyday foods and livestock.[117] But was it possible to refrain from striking, terrorizing and therefore driving away the precious sources of supplies, without destroying what was by now the traditional way of life and pattern of defence of the *presidios*, the development, by persuasion or by force, of a zone of influence and protection as indispensable to the Spanish *presidio* as it was to the Portuguese *presidio* in Morocco? Without it, the fort would have suffocated.

Inevitably there were regrettable incidents, even serious mistakes. Orders on high authority came from Spain in 1564 to suspend the *razzias* in August and September: the local inhabitants, duly informed, hastened to bring grain and food to Oran. Whereupon the commander of Oran, Andrés Ponze, organized a *razzia* and returned with eleven prisoners. His exploit, measured in prices then current, may have represented a gain of possibly a thousand ducats, which was certainly not negligible. But Francisco de Valencia, the commanding officer at Mers-el-Kebir, refused to join the expedition. One may guess that Francisco was not over-fond of his colleagues in Oran. He refused – and wrote a report: such disobedience of orders has deprived Oran of its supply of wheat and barley – the local inhabitants have stopped coming to the *presidio* – is that a good thing? In more general terms, 'I must tell Your Majesty, the *razzias* organized here in the past, have, in my opinion, drawn the Turks into the Kingdom of Tlemcen'.[118]

[117] Diego Suárez, paragr. 471, G.G.A.; in favour of a truce, para. 469 and 470, *ibid.*, 481 and 482, but elsewhere, B.N., Madrid, ch. 34, we find him saying that the *razzias* serve a purpose, that it is by the terror that they inspire that the Spaniards control the flatlands, imposing *seguros* and sovereignty. One *razzia*, 13th to 16th November, 1571, brought in 350 prisoners and a great wealth of camels, goats and cattle. On the other hand, many *correrias* turned out badly and cost dear in men. Raids were more frequent in winter when the nights were long, Diego Suárez, *op. cit.*, p. 87, the twofold advantage of this policy of striking some and protecting others, p. 69; what the Moors brought to Oran, p. 50; what was occasionally contributed by the Kingdom of Tlemcen, p. 50 (wheat sometimes for export to Spain, Oran itself required 40,000 *fanegas* of wheat and 12,000 of barley a year); retired soldiers at Oran, p. 263; the techniques of raiding, p. 64 ff; sharing out the spoils, p. 125 ff, examples, p. 228–229, 260, 293. The system of dividing had changed after 1565, p. 90, and oddly enough in the soldier's favour.

[118] Francisco de Valencia to Philip II, Mers-el-Kebir, 8th February, 1565, Simancas E° 486.

This is rather a sweeping accusation. While the organized raids must be numbered among the causes of the difficult and isolated life of the *presidios*, they cannot explain the ultimate failure of Spain on North African soil, any more than the hunger of the ragged soldiers or the strange priests who ministered to their spiritual needs, like the Frenchman who passed himself off as a cleric in Melilla, although it is unlikely he ever took orders, and who led a charmed and usually half-intoxicated existence;[119] or any more than the bad faith of the native population, 'the greatest liars in the world' as a Spanish captain called them, 'the most treacherous in the world' as an Italian put it. Such reasons, which seemed obvious to contemporaries, dwindle in the light of history. The meagre use which Spain made of the African *presidios* is merely one more chapter in the policy of the Habsburgs, or rather of the Catholic world.

Defensive psychology. That a mighty civilization – Christendom – should have surrounded itself with defences against Islam, is both important and instructive. Islam, preferring wars of aggression, flinging masses of cavalry into the field, took no such precautions. It was, as Guillaume du Vair[120] said of the Turk, 'ever poised in the air' ready to strike at its enemies. Two different attitudes: how does one explain them? Émile Bourgeois[121] long ago remarked upon the indifference with which Christendom abandoned so much territory to the Turks, notably the Balkans and Constantinople, preoccupied as it was with trans-Atlantic expansion. What could be more logical than its attitude towards Islam, a policy of defence at least expense, using cannon and sophisticated fortifications: it was a way of turning one's back on the East.

If Islam, on the other hand, sought contact, and if necessary the desperate contact of war, it was because it wanted to continue the dialogue, or force it to continue. It needed access to the superior techniques of the enemy. Without them, it would be impossible for Islam to fulfil towards Asia the same role that Christendom fulfilled towards Islam itself. In this respect it is revealing to see the Turks, after experiencing western gunfire on the Carniola frontier, attempting, unsuccessfully as it happened, to train the *sipāhis* in the use of pistols against the Persians.[122] Even more conclusive is the evident connection between the nautical vocabulary of the Turks and that of the Christians: *kadrigha* (galley), *kalliotta* (galliot), *kalium* (galleon).[123] The easterners borrowed more than the word: by the end of the century, they were building mahonnas for the Black Sea on the model of western galleasses and, what was more, imitating the Christian galleons.[124] The Turks possessed about twenty of them, large cargo vessels, with a tonnage of about 1500 *botte*; in the last quarter of the century, they

[119] 12th February, 1559, Simancas E° 485; 2nd March, 1559, *ibid.*
[120] *Actions et traités*, 1606, p. 74, quoted by G. Atkinson, *op. cit.*, p. 369.
[121] *Manuel historique de politique étrangère*, Vol. I, Paris, 1892, p. 12.
[122] J. W. Zinkeisen, *op. cit.*, III, p. 173–174.
[123] J. von Hammer, *op. cit.*, VI, p. 184, n. l.
[124] E. Albèri, *op. cit.*, III, V, p. 404 (1594).

carried pilgrims, rice and sugar on the Egypt-Constantinople route[125] – and sometimes gold, although this was also transported overland.

The Turks did however build a *limes*: between themselves and the Persians, who were one degree poorer still.

2. PIRACY: A SUBSTITUTE FOR DECLARED WAR

By 1574, the age of war by armada, expeditionary force and heavy siege was practically over. It was revived to some extent after 1593, but effectively only on the Hungarian frontier, outside the Mediterranean. Did the end of official war mean peace? Not entirely, for by some apparently general law, warfare simply took on new forms, reappearing and spreading in this guise.

In France, the massive demobilization of armies which followed the peace of Cateau-Cambrésis contributed in no small measure to the outbreak of the Wars of Religion, disturbances far more serious in the long run than foreign wars. If Germany by contrast was quiet between 1555 and 1618, it was because her surplus of adventure-seeking troops had been sent abroad to Hungary, Italy and in particular to the Netherlands and France. The end of these foreign wars at the beginning of the seventeenth century was to be a mortal blow to her. The connection was perceived by Giovanni Botero who compared the French wars of his own time with peace in Spain, suggesting that France was paying the price for her inactivity abroad, while Spain reaped the benefit of being involved in every war in the world simultaneously.[126] The reward of making trouble for others was peace at home.

The suspension of major hostilities in the Mediterranean after 1574 was undoubtedly one cause of the subsequent series of political and social disturbances, including the increase in brigandage. On the water, the end of conflict between the great states brought to the forefront of the sea's history that secondary form of war, piracy.[127] Already a force to be reckoned with between 1550 and 1574, it expanded to fill any gaps left by the slackening of official war. From 1574–1580, it increased its activities even further, soon coming to dominate the now less spectacular history of the Mediterranean. The new capitals of warfare were not Constantinople, Madrid and Messina, but Algiers, Malta, Leghorn and Pisa. Upstarts had replaced the tired giants and international conflicts degenerated into a free-for-all.[128]

[125] *Ibid.*, p. 402. [126] *Op. cit.*, p. 127.

[127] On piracy, that vast and borderless subject, see the brilliant pages by Louis Dermigny, *La Chine et l'Occident. Le commerce à Canton au XVIIIe siècle, 1719–1833*, 1964, I, p. 92 ff, which give a picture of the 'great pirate belt' in the seventeenth century, from the West Indies to the Far East. The increase and ubiquity of piracy are related to the breaking up of the great empires: the Turkish and Spanish, the Empire of the Great Mogul and the decline of China under the Ming dynasty.

[128] The following pages draw on three basic works: Otto Eck, *Seeräuberei im Mittelmeer*, Munich and Berlin (1st ed. 1940, 2nd 1943), which I was not able to obtain

Piracy: an ancient and widespread industry. Piracy in the Mediterranean is as old as history. There are pirates in Boccaccio[129] and Cervantes[130] just as there are in Homer. Such antiquity may even have given it a more natural (dare one say a more human?) character than elsewhere. The equally troubled Atlantic was frequented in the sixteenth century by pirates certainly more cruel than those of the Mediterranean. Indeed in the Mediterranean, the words *piracy* and *pirates* were hardly in current usage before the beginning of the seventeenth century: *privateeering* and *privateers* or *corsairs* were the expressions commonly used and the distinction, which is perfectly clear in the legal sense, while it does not fundamentally change the elements of the problem, has its importance. Privateering is legitimate war, authorized either by a formal declaration of war or by letters of marque,[131] passports, commissions or instructions. Strange though it now appears to us, privateering had 'its own laws, rules, living customs and traditions'.[132] Drake's departure for the New World without any form of commission was considered illegal by many of his fellow countrymen.[133] In fact it would be wrong to suppose that there was not already in the sixteenth century a form of international law with its own conventions and some binding force. Islam and Christendom exchanged ambassadors, signed treaties and often respected their clauses. In the sense that the entire Mediterranean was an arena of constant conflict between two adjacent and warring civilizations, war was a permanent reality, excusing and justifying piracy; to justify it was to assimilate it to the neighbouring and in its way respectable category of privateering. The Spaniards in the sixteenth century use both terms: they speak of Barbary 'corsairs' in the Mediterranean and of French, English and Dutch 'pirates' in the Atlantic.[134] If the word piracy was extended in the seventeenth century to activities in the Mediterranean, it was because Spain now wished to stigmatize as dishonourable all robbery on the inland sea, recognizing that the privateering of the old days had degenerated into nothing more nor less than an underhand and disguised war waged by all the Christian powers against her trade, dominion and wealth. The word

until very recently (it is still not available in the Bibliothèque Nationale). Godfrey Fisher, *Barbary Legend. War, Trade and Piracy in North Africa*, Oxford, 1957, with its arguments in favour of the Barbaresques, makes it necessary to take another look at files which were once considered closed for good. And thirdly the book by Salvatore Boni, *I corsari barbareschi*, Turin, 1964, with its wealth of unpublished documents. Since very full bibliographies are appended to all three books, particularly the last mentioned, I shall not give full references here.

[129] 5th day, second novella.

[130] See in particular *Don Quixote*, the *Ilustre Fregona*, II, p. 55; *El amante liberal*, I, p. 100–101; *La española inglesa*, I, p. 249, 255.

[131] Few letters of marque were issued in the Mediterranean. One example, letters of marque and reprisal issued by Philip IV against the French, Madrid, 2nd August, 1625; B.N., Paris, Esp. 338, f° 313. On the ocean, letters of marque were more necessary since all pirate ships belonged nominally to Christian powers.

[132] S. Bono, *op. cit.*, *passim* and p. 12–13, 92 ff.

[133] G. Fisher, *op. cit.*, p. 140.

[134] *Ibid.*, *passim* and p. 84 and 139.

piracy was applied to the Algerine corsairs, according to one historian,[135] only after the capture of the Marmora by the Spanish (1614) when the corsairs of the town were driven to take refuge in Algiers. The word may have sailed in through the Straits of Gibraltar with the Atlantic ships; but this is only conjecture.

Privateering and piracy, the reader may think, came to much the same thing: similar cruelties, similar pressures determined the conduct of operations and the disposal of slaves or seized goods. All the same there was a difference: privateering was an ancient form of piracy native to the Mediterranean, with its own familiar customs, agreements and negotiations. While robbers and robbed were not actually accomplices before the event, like the popular figures of the *Commedia dell'Arte*, they were well used to methods of bargaining and reaching terms, hence the many networks of intermediaries (without the complicity of Leghorn and its open port, stolen goods would have rotted in the ports of Barbary). Hence too the many pitfalls and oversimplifications in wait for the unwary historian. Privateering in the sixteenth century was not the exclusive domain of any single group or sea-port; there was no single culprit. It was endemic. All, from the most wretched[136] to the most powerful, rich and poor alike, cities, lords and states, were caught up in a web of operations cast over the whole sea. In the past, western historians have encouraged us to see only the pirates of Islam, in particular the Barbary corsairs. The notorious fortune of Algiers tends to blind one to the rest. But this fortune was not unique; Malta and Leghorn were Christendom's Algiers, they too had their bagnios, their slave-markets and their sordid transactions. The fortune of Algiers itself calls for some serious reservations: who or what was behind its increased activity, particularly in the seventeenth century? We are indebted to Godfrey Fisher's excellent book *Barbary Legend* for opening our eyes. For it was not merely in Algiers that men hunted each other, threw their enemies into prison, sold or tortured them and became familiar with the miseries, horrors and gleams of sainthood of the 'concentration camp world': it was all over the Mediterranean.

Privateering often had little to do with either country or faith, but was merely a means of making a living. If the corsairs came home empty-handed there would be famine in Algiers.[137] Privateers in these circumstances took no heed of persons, nationalities or creeds, but became mere sea-robbers. The Uskoks of Segna and Fiume robbed Turks and Christians alike; the galleys and galleons of the *ponentini* (as western corsairs were called in the waters of the Levant) did just the same:[138] they seized anything that came their way, including Venetian or Marseilles vessels under

[135] C. Duro, according to G. Fisher, *op. cit.*, p. 138.

[136] S. Bono, *op. cit.*, p. 7, after A. Riggio: 'in Calabria, privateering by the Barbaresques had taken the form of an authentic class struggle'.

[137] D. de Haëdo, *op. cit.*, p. 116.

[138] Marin de Cavalli to the Doge, Pera, 8th September, 1559, A.d.S., Venice, Senato Secreta, Costantinopoli, 2/B, f° 186.

the pretext of confiscating any goods on board belonging to Jews or Turks. In vain both the Signoria and the Pope, protector of Ancona, protested, demanding that a flag should guarantee immunity to the cargo. The right of search, whether abused or not, was retained by Christian privateers. Turkish galleys invoked similar rights in order to seize Sicilian or Neapolitan cargoes. A legal fiction on both sides, the practice continued despite the severe blows which the Venetian galleys sometimes inflicted on corsairs of any nation.

When Ibiza was plundered in August, 1536, were the attackers French or Turkish?[139] How can one tell? In this case they were probably French since they carried off several sides of salt pork. Even amongst themselves, Christians and Moslems fought and looted. From Agde, during the summer of 1588, Montmorency's soldiers (who had not been paid, or so they said) began making pirate raids with a brigantine, capturing anything that sailed out of the gulf.[140] In 1590, corsairs from Cassis robbed two Provençal boats.[141] In 1593, a French ship, the *Jehan Baptiste*, probably from Brittany, carrying all the necessary certificates and passes issued by the Duke of Mercoeur and by the Spanish representatives at Nantes, Don Juan de Aguila, was nevertheless seized by Prince Doria, her cargo sold and her crew clapped in irons.[142] In 1596, French and above all Provençal *tartanes* were raiding the coasts of Naples and Sicily.[143] About twenty years earlier, during the summer of 1572,[144] a Marseilles freighter, the *Sainte-Marie et Saint-Jean*, master Antoine Banduf, returning from Alexandria with a rich cargo, became separated from the flotilla of other French ships by bad weather and met a Ragusan merchantman coming from Crete to fetch wheat from Sicily and take it to Valencia. The big ship captured the Marseilles boat and 'sent it to the bottom, drowning the said captain, his officers and mariners, having first looted and stolen the cargo'. Such were the hazards of life at sea. In 1566, the captain of a French vessel found himself in difficulties at Alicante – and to judge from the countless complaints of French sailors, the Spaniards could create powerful difficulties when they wanted to. But the captain was a bold man: he seized the men who boarded his ships and what was more scaled the walls of the town.[145] Anything was allowed – provided it succeeded. In 1575, a French ship took on board in Tripoli in Barbary a cargo of Moorish and Jewish passengers bound for Alexandria, 'people of all ages and both sexes' Without hesitation, the captain sailed straight to Naples where he sold his

[139] Bernard Pançalba, governor of the island, to the empress, Ibiza, 26th August, 1536, A.N., K 1690 (orig. Catalan, transl. into Castilian).

[140] Barcelona, 24th July, 1588, Simancas E° 336, f° 164.

[141] A. Com., Cassis, E E 7, 21st December, 1580.

[142] Henri IV to Philip III, Paris, February, 1600, Letters of Henri IV to Rochepot, p. 3–4.

[143] 25th December, 1596, Simancas E° 343.

[144] The consuls of Marseilles to the lords, dukes and governors of the city and Republic of Genoa, Marseilles, 20th April, 1574, A.d.S., Genoa, Francia, Lettere Consoli, 1 2618.

[145] Madrid, 28th March, 1566, A.N., K 1505, B 20, no. 91.

29. IN SIGHT OF TUNIS (1535). Tapestry after Vermeyen. In the distance, La Goletta and its channel, the lagoon of Tunis and the town. In the foreground, the galleys

30. ALGIERS IN 1563. A naive but fairly accurate sketch map: note the harbour mole, the rock at the harbour mouth, the lookout position ('cavalier, built by the French'), the Arsenal, the harbour mosque, the 'Soco' (Zoko or souk = market place), the Casbah and the ramparts. Archivo General de Simancas, E° 487, Mapas, planos y dibujos, VII-131

31. RAGUSA IN 1499–1501, from a painting by Nicolas Bozidarevitch. St. Blaise or Blasius, whose hands are visible in the foreground, is holding up a plan of the city. Note the harbour with its chain and moorings, and the city walls. Dominican Convent, Dubrovnik

passengers, and all their baggage.[146] Such mishaps were not infrequent: in 1592 for example, a certain Couture of Martigues having accepted some Turkish passengers at Rhodes, took them to Messina, instead of to their destination in Egypt.[147] And there was pure brigandage: during the summer of 1597, bandits armed several *leuti* and prowled the Genoese coast in search of prey.[148] What kind of history have we been taught, that these acts, familiar to seamen of all nationalities, should nevertheless seem so astonishing?

Privateering sponsored by cities. Like Monsieur Jourdain who had been speaking prose all his life, there must have been many seamen who sailed *more piratico* and would have been scandalized to hear themselves described as privateers, let alone pirates. When Sancho de Leyva, for instance, proposes in 1563 to set off with a few Sicilian galleys to the Barbary coast to fetch prisoners for the rowing benches, 'para ver si puede haver algunos sclavos',[149] what is one to call such an expedition? Fleets would often detach a few galleys to reconnoitre and pick up any victims who came their way. Piracy was simply another form of aggression, preying on men, ships, towns, villages, flocks; it meant eating the food of others in order to remain strong. In 1576, the Marquis of Santa Cruz went on a 'patrolling expedition' along the Tunisian coast – or as some would say more bluntly, on a pirate raid against the poor Kerkenna Islands.[150] Marauding was open to all: English merchantmen after 1580 were particularly notorious – having the reputation (invented by Mediterranean sailors) of being totally ruthless and unscrupulous. But privateering, bordering on piracy, was part of the tradition of the sea, *l'usanza del mare*.[151] The official navies of Mediterranean states harboured privateers, made a living from privateering and sometimes owed their origin to it. It was by privateering that Turkish might first revealed itself, off the coasts of Asia Minor in the fourteenth century.[152] As for the Turkish fleet itself, when it made an expedition to the western sea, what else was this but glorified 'piracy'?

Privateering – perhaps one should say 'true' privateering – was usually instigated by a city acting on its own authority or at any rate only marginally attached to a large state. This was as true in the age of Louis XIV as it was in the sixteenth century. When the Sun King could no longer maintain a regular battle fleet against England, he encouraged or allowed war by piracy; Saint-Malo and Dunkirk became belligerents in place of France.

[146] Henri III to Philip II, Paris, 30th September, 1575, A.N., K 1537, B 38, n° 113, Spanish copy.
[147] P. Grandchamp, *op. cit.*, I, p. 42.
[148] A.d.S., Florence, Mediceo 2845, Giulio Gotti to his brother, Genoa, 22nd August, 1597.
[149] 20th November, 1563, Simancas E° 1052, f° 44.
[150] Simancas E° 1146 or the same report, Simancas E° 1071, f° 78.
[151] S. Bono, *op. cit.*, p. 3.
[152] F. Grenard, *op. cit.*, p. 54; W. Heyd, *op. cit.*, II, p. 258.

Already in the sixteenth century, Dieppe and in particular La Rochelle, were centres of privateering, the latter operating virtually as a city-republic. The list of privateering centres in the Mediterranean is a roll-call of strategic cities: in Christendom, Valetta, Leghorn and Pisa, Naples, Messina, Palermo and Trapani, Malta, Palma de Majorca, Almeria, Valencia, Segna and Fiume; on the Moslem side, Valona, Durazzo, Tripoli in Barbary, Tunis-La Goletta, Bizerta, Algiers, Tetouan, Larache, Salé [Sallee].[153] From this list, three new towns stand out: Valetta, founded by the Knights of Malta in 1566; Leghorn, re-founded in a sense by Cosimo de' Medici; finally and above all the astonishing city of Algiers, the apotheosis of them all.

This was no longer the Berber town of the beginning of the century but a new city, sprung up American-fashion almost overnight, complete with its harbour mole, its lighthouse, its archaic but solid ramparts and beyond them the mighty defence works which completed its protection. Privateers found shelter and supplies here, as well as a fund of skilled labour: caulkers, gun-founders, carpenters, sails and oars, a busy market on which stolen goods could quickly be disposed of, men willing to sign up for adventures at sea, galley-slaves, not to mention the pleasures of a port of call without which the life of the corsair with its violent contrasts would have been incomplete. Alonso de Contreras, on his return from a pirate raid to Valetta, which was famous for other things besides duels and prayers, lost no time in spending all his gold with the *quiracas* in the brothels of the city. The *re'is* of Algiers, on returning from an expedition, would hold open house, either in their city residences or in their villas on the Sahel where the gardens were the most beautiful in the world.

Privateering required above all things a market for its spoils. Algiers could not become a pirate stronghold without also becoming an active commercial centre. By the time Haëdo turned his observant eye on the city, in 1580, the transformation had been accomplished. In order to have food and equipment and to sell the prizes of war, the city had to be open to foreign caravans and ships, to the boats bringing ransom for captives, Christian vessels – Marseillais and Catalan, Valencian, Corsican, Italian (from every corner of Italy), English or Dutch. It was necessary too to attract by the prospect of affluence *re'is* of every nation – Moslems or half-Moslems, sometimes northerners, in their galleys or nimble sailing ships.

A powerful city then, with freedom of movement was Algiers, the very place for a corsair capital. In the sixteenth century, all states, despite their disagreements, were deeply committed to the law of nations and supposed to respect it. But the cities of the corsairs had been known totally to disregard international law. They were worlds on the margins of society. Algiers, at the height of its prosperity, between 1580 and 1620, might or might not, depending on its own convenience, obey the orders

[153] R. Coindreau, *Les corsaires de Salé*, Paris, 1948.

of the sultan, and Istanbul was a long way off. On the Christian side, Malta was a rendezvous for adventurers with pretensions to autonomy. A revealing detail is the effort made in 1577–1578 by the Grand Duke of Tuscany, master of the Knights of St. Stephen,[154] to distinguish, when negotiating with the Turks, his own cause from that of the Knights: here was a prince at once exercising very real authority, yet at the same time protesting his powerlessness.

But the corsairs who operated from urban bases were not the only ones. There were lower levels of piracy, often little more than petty thieving. Besides the great predators, lesser scavengers prowled the seas, roaming among the Aegean islands or along the coasts of western Greece, in search of victims their own size. The mere sight of the watch-towers on the Apulian coast was enough to send them scurrying away from such dangerous haunts, back to the coasts and islands of the East. These were humble men with humble ambitions: to capture a fisherman perhaps or rob a granary, kidnap a few harvesters, steal some salt from the Turkish or Ragusan salt-pans at the mouth of the Narenta. It was pirates of this sort whom Pierre Belon[155] saw at work in the Aegean, 'three or four men, accustomed to the sea, boldly pursuing the life of adventure, poor men, with only a small barque or frigate or perhaps some ill-equipped brigantine: but they have their *bussolo* or mariner's compass to navigate by and also the means to wage war, that is a few small arms for firing at a little distance. Their rations are a sack of flour, some biscuit, a skin of oil, honey, a few bunches of garlic and onions and a little salt and on these they support themselves for a month. Thus armed they set out in search of prey. And if the winds keep them in port, they will drag their boat ashore and cover it with branches, chop wood with their axes, light a fire with their guns . . . and make a round loaf with their flour, which they cook in the same way as the Roman soldiers of antiquity on campaign.' From such humble beginnings too came the Caribbean buccaneers of the seventeenth century.[156]

Such minor carnivores did not always inflict the least damage, nor amass the smallest fortunes in the end. Privateering, like America, was the land of opportunity and the wind of prosperity was unpredictable. A humble shepherd boy could become 'king' of Algiers and many were the careers that began in rags to end in riches. When the Spaniards were trying in 1569 to bribe Euldj 'Alī, the Calabrian fisher-boy who had become 'king' of the Berber city and was soon to astonish the world by rebuilding the sultan's navy, they dangled a marquisate in front of him – the kind of offer they assumed would be irresistible to one of such humble birth.

[154] A.d.S., Florence, Mediceo 4274, 4279; Simancas E° 489, 1450, 1451; A.N., K 1672, no. 22; G. Vivoli, *op. cit.*, III, p. 155.

[155] *Op. cit.*, p. 86 v° ff.

[156] Alexandre O. Oexmelin, *Histoire des aventuriers flibustiers . . .*, Trévoux, 1775, Vol. I, p. 124–131.

The prizes. Without prizes, privateering would have died out. The gain was sometimes very small: if it were not for the traffic carrying salt from Corfu to Albania and gall-nuts on the return journey, the island would not have been plagued by Albanian pirates, the Venetian Senate was told in 1536.[157] A relationship can be established between robbers and robbed, varying as the victims defended themselves. Artillery made an early appearance aboard the galleys where it was to remain, albeit rather uncomfortably; then it was adopted by merchant vessels. By mid-century, the transition was complete.[158] In 1577, even the smallest vessels putting in to Seville had either bronze or iron guns, the number roughly corresponding to their tonnage.[159] The coasts were defended too, and more efficiently as time went by. The corsair might concentrate on sea-going vessels one year, on coasts the next: it was simply a question of supply and opportunity.

During the years 1560–1565, the Barbary corsairs ravaged the western sea. In those years, it would almost be true to say that the western Mediterranean was closed to shipping, as the chorus of Christian protests suggests (perhaps rather too insistently), as well as the fact that the Barbary pirates were now attacking even the coasts of Languedoc and Provence.[160] For now the very success of piracy was reducing the numbers of prizes and pirates could not live without booty, even among friends, unfortunately for the subjects of the French king. Algiers was still expanding at the beginning of the seventeenth century, but why? Her corsairs were now to be found venturing into the Levant (though perhaps less than has been claimed),[161] moving into the Adriatic, chasing the little boats of Marseilles and then, with the help of their northern recruits, passing through the Straits of Gibraltar, sailing out into the Atlantic, reaching English shores in 1631, attacking the clumsy Portuguese carracks and appearing as far off as Iceland, Newfoundland and the Baltic. Does this mean that their regular prey in the Mediterranean was growing scarce? Piracy in short, by its shifts and alterations, reacted in a characteristically direct and rapid fashion to the broader currents of Mediterranean life. The hunter had to follow the game. Its value as an indicator is unfortunately vitiated by the lack of statistical evidence. The descriptions, complaints, rumours and false reports we have are a slender basis for serious calculation.

The chronology of privateering. Certain key dates punctuate the history of privateering: 1508, 1522, 1538, 1571, 1580, 1600. In about 1500 (except in Venice), captives and convicts replaced the voluntary oarsmen who had

[157] V. Lamansky, *op. cit.*, p. 592, n. 1.
[158] Pierre Belon, *op. cit.*, p. 88 v°.
[159] See above, Vol. I, p. 301 ff.
[160] A. Com., Marseilles BB 40 f° 197 ff; 19th August, 1561, Sim. E° 13; E. Charrière, *op. cit.*, II, p. 659–661 (27th June, 1561), p. 799–803 (27th September, 1561); Bayonne, 28th June, 1565, A.N., K 1504, B 19, no. 34; Venice, 18th August, 1565, Simancas E° 1325; Charles IX to Fourquevaux, Orcamp, 20th August, 1566, Fourquevaux, *op. cit.*, p. 48–49.
[161] G. Fisher, *op. cit.*, p. 144.

until then almost exclusively manned the galleys.[162] 1522: the fall of Rhodes removed the last remaining barrier in the East to large-scale Moslem privateering.[163] 1538: Prevesa gave Islam that control of the sea of which the Christian victory at Lepanto was to deprive it in 1571. It is between these two dates (1538–1571) that the first great age of the Barbary corsairs can be situated, in particular during the decade from 1560 (after Djerba) to 1570, in those years when, apart from the siege of Malta, there were comparatively few major naval engagements. After 1580, both Christian and Moslem piracy, as the great armadas lay idle, increased at about the same rate. And finally after 1600, the Algerine corsairs, adopting completely new techniques, sailed out into the Atlantic.

Christian privateers. There had always been Christian pirates in the Mediterranean, even in the darkest hours. This piracy has not been fully recorded by historians, partly for psychological reasons, partly because it was carried out in very small ships, brigantines, frigates, *fregatillas*, barques, sometimes the tiniest fishing-smacks. The short distance separating Sicily or Spain from North Africa made it possible to proceed in such modest vessels; the meagreness of the booty made it necessary. The coast of the Maghreb, well defended by the Turks, was mountainous and deserted. In the old days, perhaps even in the fifteenth century, piracy might have been profitable in these waters. 'Non si puó corseggiare la riviera di Barberia come già si soleva' says a Venetian report in 1559.[164] What possible pickings were there off these shores by 1560? A few local inhabitants for slaves, a fishing boat, the odd brigantine laden with coarse woollen *baracans* or rancid butter. Game was scarce and the scavengers were correspondingly few, only occasionally receiving a mention in the documents of the time. One of these rare glimpses is Haëdo's account of the exploits of the Valencian Juan Canete,[165] master of a brigantine with fourteen oar-benches based at Majorca, an assiduous hunter off the coasts of Barbary who had been known to steal by night up to the very gates of Algiers and capture the people sleeping under the city walls. In the spring of 1550,[166] he ventured inside the port at night with the intention of setting light to the poorly guarded foists and galliots. The attempt was a failure. Nine years later, he was to be executed by his gaolers in the bagnio. In 1567, his plan was revived by another Valencian, Juan Gascon, who was employed with his brigantine in the Oran supply

[162] According to M. Sanudo, quoted by C. Manfroni, *op. cit.*, I, p. 37. The same true of France, royal letters of 1496, Alfred Spont, 'Les galères dans la Méditerranée de 1496 à 1518' in *Revue des Quest. hist.*, 1895; Alberto Tenenti, *Cristoforo da Canal. La marine vénitienne avant Lépante*, 1962, p. 78 ff. Venice had galleys manned by *condennati* only after 1542, *ibid.*, p. 82.

[163] Relacion de lo de Tremeti (1574). The Tremiti islands were key positions on the Adriatic coast of the Kingdom of Naples . . . Simancas E° 1333. 'Despues de la perdida de Rodas multiplicandose los cossarios en el mar Adriatico . . .'

[164] *Relazione di Soriano*, p. 54.

[165] *Op. cit.*, p. 158.

[166] And not in 1558 as C. Duro says, *op. cit.*, II, p. 16. Similar pirate raids were made in 1562 by one Francisco de Soto, based on Majorca, D. de Haëdo, *op. cit.*, p. 163 v°.

fleet and did a little privateering on his own account.[167] More successful than his predecessor, he entered the port and set fire to a few ships, but was later captured on the high seas by the *re'īs*.

Such occasional mentions tell us very little about the sailors of southern Spain. One has the impression that they must have become more active in 1580, since from then on they figure more frequently in the documents. They were probably never completely idle. When they emerge from the years of obscurity, we find them still using the same light vessels with their tall sails and still as audacious as ever. Take for example the account of the third voyage of one Juan Phelipe Romano, a sort of Scarlet Pimpernel for refugees from Algiers.[168] On 23rd May, 1595, he left the Grao of Valencia, probably aboard a Barbary brigantine captured the year before.[169] On 7th June, he lowered anchor near Algiers in a bay at the edge of a garden which was his rendezvous. The first night no one came, so he remained ashore and sent his companion back to the ship with orders to put back to sea and wait for a signal before returning. On the following night, the owner of the garden and his wife, with whom Romano had long ago made an agreement, arrived at the rendezvous. The refugee was a certain Juan Amador of Madrid, taken prisoner at Mostaganem in 1558 (that is about forty years before). In the interval he had embraced the Moslem faith, but now wished to return with his wife and a grand-daughter aged seven months. With him there embarked on the frigate the same night a 'princess', the *soldina*, daughter of Muṣṭafā, ten Christian captives and two black slaves belonging to her, as well as a young Morisco girl of twenty-two; one of the wives of Mami Re'īs, the daughter of a lieutenant in Majorca, who was also accompanied by slaves, four men and a woman, all Christians; a Portuguese master locksmith of Algiers, his wife and two children and finally some Christian slaves who were on the spot and took advantage of the chance to escape: a total of thirty-two passengers, whom Romano carried without incident to Valencia.

It makes a good story – but such *tours de force* were rare. On the whole these pirates operated in a very small way, the fishermen of Trapani for instance in their *luitelli*;[170] a certain amount of petty piracy was encouraged in 1614 (or perhaps earlier) for individual profit, by the Spanish government of Sardinia.[171] The only game of any significance in the western part of the sea was the Algerine pirate-vessel itself, but only the largest galleys of the squadron would risk taking on such a formidable foe. On the other hand, in about 1580, fishing boats in the Algiers region dared not

[167] Madrid, 13th June, 1567, Simancas E° 333.

[168] Relacion del tercero viaje q. ha hecho Juan Phelipe Romano a Argel (1595), Simancas E° 342.

[169] Viceroy of Valencia to Philip II, Valencia, 30th July, 1594, Simancas E° 341.

[170] Salomone Marino, in *A. st. sic.*, XXXVII, p. 18–19; a brigantine from Trapani privateering, 17th November, 1595, Simancas E° 1158.

[171] Amat di S. Filippo, *Misc. di storia italiana*, 1895, p. 49.

put further than half a league out to sea for fear of Christian frigates.[172]

The Levant was easily the most rewarding hunting-ground for Christian privateers. And to the Levant sailed a steady stream of well-manned galleys, brigantines, galleons, frigates[173] and swift sailing-ships well able to batter their way through the rough seas of the end of winter and the spring months. The reason was always the same: for the privateers the eastern Mediterranean meant rich prizes, to be found in the Aegean and even more on the Rhodes-Alexandria route, the route taken by pilgrims and by cargoes of spices, silks, wood, rice, wheat and sugar. The game was certainly plentiful, but the gamekeeper was vigilant: every year at the approach of spring, the Turks sent out their galley patrols, employed far more for the defence of ships than to guard the coast.

In the middle years of the century there were active in the Levant only the Maltese galleys, a few Tuscan galleys and the occasional sailing vessel like the galleon of the Genoese captain Cigala, put out of action in 1561;[174] with here and there a Sicilian ship, the galleon fitted in 1559 by the viceroy himself for example or the galliot fitted out the year before by the captain Joseph Santo.[175] The said captain, having captured off Alessio a Turkish vessel worth over 15,000 ducats, was obliged by bad weather to take refuge with the Venetians, who promptly confiscated the ship. It is only through this incident that we know of the boat's existence at all. In 1559, a Tuscan galley, the *Lupa*, and a galliot belonging to Andrea Doria both set off on a privateering expedition; the first was seized by the guard off Rhodes and the second after many adventures, limped exhausted into Cyprus and the clutches of the Venetians.[176] Irritation in the West at this quasi-Turkish behaviour on the part of the Signoria can easily be imagined. Why should the Venetians have the right, argued the Duke of Florence, to prevent a Christian ship from putting out against the Infidel, if the said Christian vessel did not enter one of their ports? 'Does not the sea belong to everybody?'[177] Poor Venetians: at the same time the Turks were reproaching them with failing to control the *ponentini*,[178] and Turkish reprisals, usually after due warning and usually effective, were a threat to all peaceful Christian travellers and merchants in the East.[179]

In these middle years of the century, the boldest western corsairs were the Knights of Malta, led by La Valette, in the years 1554–1555[180] and by Romegas in about 1560. In 1561, the latter captured 300 slaves and several rich cargoes at the mouth of the Nile;[181] in 1563, having set out with two

[172] D. de Haëdo, *op. cit.*, p. 44.

[173] Avviso from Const., October, 1568.

[174] D. de Haëdo, *op. cit.*, p. 160 v°; Pera, 9th April, 1561, A.d.S., Venice, Senato Secreta Costant., 3/C, Venice, 22nd March, 1561, Simancas E° 1324, f° 83.

[175] Venice, 27th September, 1559, Simancas E° 1323.

[176] A. de Herrera, *Historia general del mundo* . . ., Madrid, 1601, I, p. 15.

[177] *Ibid.*

[178] Pera, 13th July, 1560, A.d.S., Venice, Sena. Secreta Cost., 2/B f° 253.

[179] Busbecq, *Lettres*, *op. cit.*, II, p. 279, about 1556.

[180] J. B. E. Jurien de La Gravière, *Les chevaliers de Malte* . . ., 1887, I, p. 16–18.

[181] *Ibid.*, p. 63–64 and Simancas E° 1050, f° 27, 28th May, 1562.

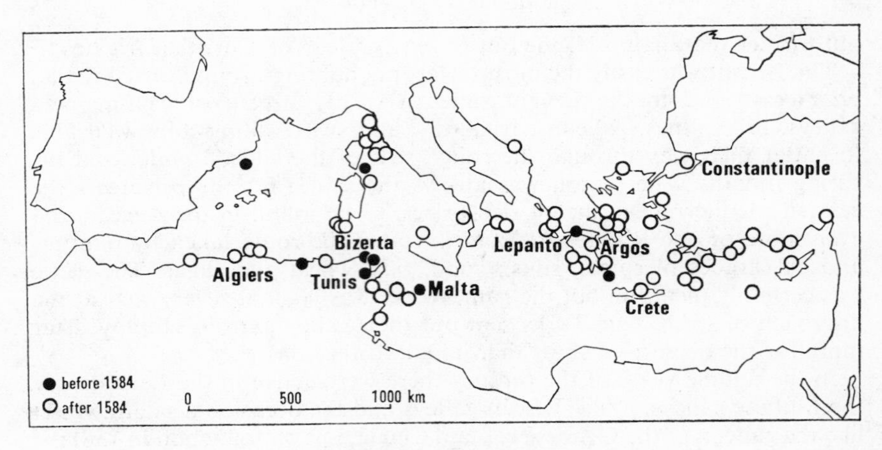

Fig. 64: *The privateers of Tuscany*

From G. G. Guarnieri, *op. cit.*, p. 36–37. The map represents the major exploits of the Florentine galleys of the Order of San Stefano between 1563 and 1688. Without wishing to read too much into this evidence, let us note in passing that before 1584, Tuscan attacks occurred in the western rather than the eastern part of the sea; after this date, they become general throughout the Mediterranean.

galleys,[182] he was seen sailing back to Cape Passaro[183] with over 500 slaves, black and white and, heaped on to two ships (the rest had been sunk) the cargoes of eight ships he had captured. These prizes, the letters add 'must have been very rich since they came from Alexandria'. In 1564, Romegas brought home three *corchapins* laden with oars, tow and munitions for Tripoli in Barbary, and a Turkish roundship of 1300 *salme* which had left Tripoli for Constantinople with a cargo of 113 black slaves. The ship was taken to Syracuse, the *corchapins* to Naples.[184]

Second place in those days went to the Florentines, who were presently to challenge the supremacy of the Knights of St. John. In 1562, Baccio Martelli[185] sailed to Rhodes, scoured the sea between Syria and the Barbary coast and captured a boat-load of Turks and Ethiopian Moors, the latter carrying gifts for the sultan: precious stones, a gold cross, conquered Christian standards and a *filza* of ritually amputated Christian noses. In 1564,[186] the Knights of St. Stephen made their first sortie *in forma di religione*, sailing to the Levant where they seized two rich Turkish vessels.

These were by no means the only prizes. But at this stage the Levant was not yet being plundered without mercy. A Venetian *avviso*, dating

[182] J. B. E. Jurien de La Gravière, *Les chevaliers de Malte . . .*, 1887, I, p. 64.
[183] Report from Messina, 1st June, 1563, Simancas E° 1052, f° 189.
[184] Per lre (lettere) di Messina, 7th May, 1564, Simancas E° 1383.
[185] G. Mecatti, *op. cit.*, II, p. 723.
[186] G. Vivoli, *op. cit.*, III, p. 53.

from early spring 1564, refers to twelve western galleys in the Aegean.[187] This was by no means negligible, but during the same period the Moslem foists and galliots shamelessly plundering the riches of the West could be counted in multiples of twenty or thirty. At this point honours were far from equal.

Christian piracy in the Levant. From about 1574 onwards however, the Levant was invaded by pirates from the West. The Knights of Malta now virtually abandoned the nearby Barbary coast to make expeditions into the eastern seas. There was a visible increase in the activity of the Florentine galleys. They always worked in teams of four or five fast and powerful ships. In 1574,[188] a voyage from Italy to Rhodes and Cyprus and back took them only twenty-nine days (leaving Messina on 7th August and arriving back at Catania on 5th September). They occasionally made a foray into western waters too. From time to time a galleon belonging to the Grand Duke would try its luck.[189] Guarnieri's picturesque book[190] celebrating these bloodthirsty feats does not tell the complete story of the restless voyages of the Knights of St. Stephen, from which there is so much to be learnt about the traffic in the Turkish seas, with its *gerbe*, *caramusali*, *passa cavalli*, barques, brigantines and heavy western ships. The extremely precise accounts of privateering expeditions in the Florentine Archives are full of lively detail. Here we can read how the Venetians come swooping down in their galleys like watchdogs between Cerigo and Cerigotto and force the galleys sailing under the red cross to turn about and head for Italy and the protection of darkness.[191] Elsewhere a series of long direct voyages made without incident is dismissed in a single line: the ships sail out under a fair wind and their destination soon appears on the horizon, a cape, a cluster of lights winking in the darkness or a flurry of sails, usually a sign that land is near. Other voyages are taken slowly, following the coast from one watering-place to the next, calling in at coves or anchoring off sandbanks. The prizes themselves are described briefly and without emotion: item a *caramusali*, the number of cannon shots fired by the *capitana* to shatter its yards or unmast it, casualties on our side, casualties among the enemy; then on to the cargo: Greeks, Turks, dried fish, sacks of rice, spices, carpets . . . and on to the next. A brief note explains the classic tricks of the trade: if one is bold enough to enter the Aegean 'alla turchesca, costeggiando la terra firma',[192] it

[187] Daniel Barbaro to the Doge, Pera, 28th March, 1564, A.d.S., Venice, Senato Secreta 4/D.

[188] See note 194.

[189] Silva to Philip II, V, 10th September, 1574, Simancas E° 1333.

[190] *Cavalieri di San Stefano . . .*, Pisa, 1928.

[191] NB the considerable importance for the Venetian patrols of the watch on Cerigo, from 1592 (E. Albèri, *op. cit.*, III, V, p. 430) the Venetian guard at Cerigo may have succeeded in protecting Turkish shipping. Cerigo, says Cigala, '. . . fanale e lanterna dell'Arcipelago e la lingua e la spia di tutti gli andamenti turcheschi . . .'

[192] Nota di vascelli presi (1575), A.d.S., Florence, Mediceo 2077, f° 536.

is sometimes possible to capture without a struggle passengers coming down to the quayside mistaking the Christian pirates for the galleys of the Grand Turk. We learn what is, among pirates, normal practice: any unwanted vessels are sunk after their cargo has been taken; the Genoese captain of a Venetian ship is tortured, with a heavy weight tied on his foot, until he admits to having on board some *robbe*, the property of Jews or Turks; a ransom[193] is fixed, say 1000 crowns payable in 250 lb. bales of silk at a crown a pound; or a captured ship is loaded to the brim with stolen rice or wheat, manned with a Greek crew and sent off to Sicily, with prayers to God and all his saints to bring it safe to port. The Greek crew of the previous victim (now at the bottom of the sea) is transferred to a plundered Turkish vessel – and if some Greek pope protests overmuch he is dumped at Malta without ceremony.[194]

In order to discover the whole truth about these ruthless expeditions the historian will have to unearth accurate accounts of battles and captures, calculate the profits and losses of this unique commercial enterprise and look closely at the equally specialized markets created by privateering, in particular the market in human beings which was the speciality of Malta, Messina and Leghorn. A register of captives intending to pay ransom (their birthplaces range from Fez to Persia and the Black Sea)[195] or a list of galley-slaves with their ages and place of origin gives an idea of the probable benefits of the operation to the pirate-knights of St. Stephen and their shrewd master. They can also be detected from the innumerable letters addressed to the Grand Duke by his rivals in Tripoli and Algiers:[196] would his highness be willing to release so-and-so in exchange for a person of his choice? Would he consent to hear the respectful petition addressed by the wife of Mami Arnaut to the Grand Duchess herself? And would he in any case deign to accept this horse as a gift?

In this respect too, times were changing. In 1599, the five galleys flying the red cross seized the citadel of Chios which they were to hold temporarily.[197] And in 1608, excelling themselves, the galleys of the Knights of St. Stephen captured all the Turkish ships off Rhodes carrying pilgrims to Mecca.[198] The reprisals envisaged at Constantinople had a half-hearted air about them. In 1609, there was some talk in the Divan of forbidding the pilgrimage to Jerusalem, in the hope of arousing indigna-

[193] Another example, 10th December, 1558, *Corpo. dipl. port.*, VIII, p. 78.

[194] All the details in this paragraph are taken from a report of 1574, A.d.S., Florence, Mediceo 2077, f°s 517 to 520 v° and a report of 1597, *ibid.*, f° 659 ff.

[195] *Nota delli schiavi* . . . (1579–1580), *ibid.*, f° 606 ff. List of killed or wounded galley-slaves, *ibid.*, f° 349.

[196] *Ibid.*, 4279, numerous missives from Algiers, from Muṣṭafā Agha, 15th April, 1585; from the wife of Arnaut Mami, 20th October, 1586; from Muḥammad Pasha, 'King' of Tripoli, June and July, 1587; from Arnaut Mami, 9th October, 1589; from Murād Bey, *capitan general de mar y tierra deste reyno de Argel*, 16th February, 1596, etc.

[197] The St. Stephen's galleys sailed under the red cross in the Levant, G. Vivoli, *op. cit.*, IV, p. 11. Capture of the fortress of Chios, G. Mecatti, *op. cit.*, II, p. 816.

[198] G. Vivoli, *op. cit.*, IV, p. 29–30.

tion in the Christian world against the pirates of Tuscany[199] – times had certainly changed by now. And the Knights of St. Stephen and St. John, the invaders of the Aegean as a document of 1591 calls them,[200] were not the only people to realize it. Other corsairs were forcing their way into the Levant: Sicilians, Neapolitans, even Barbary pirates,[201] not to mention the minuscule but not negligible scavengers of the Levant itself, often in league with the patrols to fleece what was left in the devastated Aegean. The Neapolitans (apart from a brief episode between 1575–1578) did not appear in large numbers before the end of the century,[202] if information given in Venice is correct. It was only then that the viceroys allowed ships to conduct privateering raids for public or private gain. It is not altogether surprising perhaps to find among these adventurers Alonso de Contreras, whose description of the plundering of the islands is particularly brutal, as well as two Provençal sea-captains who were credited in Paris with sinister intentions.[203]

From Sicily, on the other hand, a whole fleet of privateers was, even before 1574, beginning to launch raids on the Levant. The names of some are famous: Filippo Corona, Giovanni di Orta, Jacopo Calvo, Giulio Battista Corvaja and Pietro Corvaja, who were present at Lepanto, along with several others, notably the amazing Cesare Rizzo, an expert at reconnaissance in the Levant: from the great battle, in which he participated in his light *fregatina* decked in all her canvas, he was to bring home as a trophy to the Chapel of Santa Maria della Gracia in the parish of S. Nicolò Kalsa at Messina, a bell which the Turks 'havianu priso a l'isola di Cipro', in the previous year.[204] There were other names too: Pedro Lanza, a Greek from Corfu and a well-known chaser of frigates and galliots, Venetian ships and Venetian subjects, employed in 1576–1577 by Ribera, the governor of Bari and Otranto;[205] or Philip Cañadas, the notorious corsair who in 1588 commanded one of the privateering galliots of Pedro de Leyva, general of the Sicilian galleys,[206] another threat to Venetian shipping.

For the entire piratical world was, by the end of the century, eager to settle scores with the Republic of St. Mark. The galleys of the *Serenissima*

[199] Alonso de la Cueva to Philip III, Venice, 7th February, 1609, A.N., K 1679.

[200] C. 19th April, 1591, A.N., K 1675 '. . . para guardar el Arcipelago de la inbasion de Malta . . .'

[201] Barbary foists plunder Candia, H° Ferro to the Doge, Pera, 12th November, 1560, A.d.S., Venice, Sen° Secreta Cost., 2/B f° 291 v°; Simancas E° 1326, 12th August, 1567; A.N., K 1677, 7th July, 1600. For the seventeenth century, Paul Masson, *op. cit.*, p. 24, 33, 380.

[202] F° de Vera to Philip III, Venice, 10th July, 1601, A.N., K 1677.

[203] J. B. de Tassis to the Spanish ambassador at Genoa, Paris, 20th July, 1602, A.N., K 1630.

[204] Salomone Marino, in *Arch. Stor. Sic.*, XXXVII, p. 27.

[205] Relacion sobre lo del bergantin de Pedro Lanza . . ., Simancas E° 1336, 1577. Silva to Philip II, Venice, 20th November, 1577, *ibid.*

[206] Relacion que ha dado el embaxador de Venecia . . ., Simancas E° 1342. The document mentions two other galliots having been sent privateering by P. de Leyva on his own account, contrary to the king's orders.

patrolled in vain. There were many ways – if only by taxing her merchants in Taranto – of making Venice loosen her hold. Diplomatic protests in Florence or Madrid were rarely heeded. Venice did succeed in obtaining from Philip II a ban on privateering by Naples and Sicily. Naples obeyed, more or less; in Sicily, both private individuals and the viceroy himself continued their profitable trade. In any case, Philip's orders[207] dating from 1578, were taken to refer far more to the Turk, with whom negotiations were under way, than to Venice. The Venetians argued, to little or no avail, that seizing 'ropas de judios y de turcos' carried in Venetian ships harmed the trade of Venice and therefore in the long run that of Spain, with whom she had commercial ties, and furthermore that by such actions the pirates were molesting poor 'stateless' Jews who, although expelled from Spain, nevertheless considered themselves subjects of the Catholic King, as well as humble and peaceloving Turkish merchants.[208] Madrid usually viewed with equanimity the difficulties of Venice, whom she knew to be unkindly disposed towards Spain and whom she suspected of having made an illicit fortune out of the peace which Venetians were prepared to go to any lengths to preserve. In the Levant, even the Turks plundered Venetian ships, so much so that the all-round increase in piracy deserves careful consideration in relation to both Venice and Ragusa (Ragusan ships did not escape the right of search). The historian should ask himself whether the success of the western privateers does not explain in part why both Ragusa and Venice withdrew to the safer routes of the Adriatic, far from the seas and islands 'harassed' and 'starved' by the insolence of the 'vasselli christiani'.[209] In Venice, insurance rates tell their own story: for the Syria route, they went up to 20 per cent in 1611 and to 25 per cent in 1612.[210]

The first brilliant age of Algiers. In the western waters of the sea, Moslem privateering was equally prosperous and had been so for many years. It had several headquarters, but its fortunes are epitomized in the extraordinary career of Algiers.

Between 1560 and 1570, the western Mediterranean was infested with Barbary pirates, mostly from Algiers; some made their way to the Adriatic or to the coasts of Crete. The characteristic method of attack during these years was the regular assault in large groups if not in actual battle formation. In July, 1559, fourteen pirate vessels were sighted off Niebla in Andulasia;[211] two years later, fourteen galleys and galliots again appeared near Santi Pietri, off Seville.[212] In August, Jean Nicot reported '17

[207] Marcantonio Colonna to Philip II, Messina, 10th July, 1578, Simancas E° 1148.
[208] F^co de Vera to Philip III, Venice, 5th February, 1601, A.N., K 1677. An important and lengthy appeal to reason.
[209] V. Lamansky, *op. cit.*, p. 578 (1588), see also p. 592, 599, 601–602. Complicity of the Greek population.
[210] G. Berchet, *op. cit.*, p. 130 and 139.
[211] Simancas E° 138, 7th July, 1559.
[212] El Prior y los Consules de Sevilla to Philip II, 7th May, 1561, Simancas E° 140.

galères turcquesques' on the Portuguese Algarve.[213] At the same period, Dragut was operating off Sicily and in a single raid captured eight Sicilian galleys off Naples.[214] With a fleet of thirty-five sail, he blockaded Naples[215] in midsummer. Two years later, in September, 1563 (that is after the harvest) he was prowling off the coast of Sicily and was twice sighted at the Fossa di San Giovanni near Messina with twenty-eight ships.[216] In May 1563, twelve ships, four of them galleys, were reported off Gaeta.[217] In August, nine Algerine ships were sighted between Genoa and Savona;[218] in September there were thirteen of them off the Corsican coast.[219] Thirty-two appeared at the beginning of September on the Calabrian coast,[220] probably the same ships which were reported as a fleet of about thirty sail arriving off Naples one night and taking shelter near the island of Ponza.[221] Still in September, eight ships sailed past Pozzuoli, making for Gaeta[222] and twenty-five sails were sighted *sopra Santo Angelo in Ischia.*[223] In May, 1564, a fleet of forty-two sail appeared off Elba[224] (forty-five according to a French report[225]). Forty sails, this time off the coast of Languedoc on the track of the galleys bound for Italy, were reported by Fourquevaux in April, 1569.[226] A month later, twenty-five corsairs were seen sailing past the shores of Sicily, which they hardly bothered to trouble, being utterly engrossed in the pursuit of ships and boats.[227]

Such numbers explain the severe blows dealt by the corsairs, who on one occasion seized eight galleys and another time, off Málaga, captured twenty-eight Biscay ships (June, 1566).[228] In a single season they accounted for fifty ships in the Straits of Gibraltar and on the Atlantic coasts of Andalusia and the Algarve;[229] a raid into the kingdom of Granada furnished them with 4000 prisoners.[230] During this period, according to the Christians, the audacity of the corsairs knew no bounds.[231] Where once they had operated only by night, they now showed themselves in broad daylight. Their raids reached as far as the *Percheles*, the criminal quarter of Málaga.[232] The Cortes of Castile refer, in 1560, to the desolation and

[213] *Op. cit.*, p. 69.

[214] A.d.S., Naples, Farnesiane fasc. II, 2, f° 271, 28th June, 1561; Simancas E° 1126, 29th June, 1561; J. Nicot, *op. cit.*, p. 70, 17th August, 1561.

[215] The bishop of Limoges to the king, Madrid, 12th August, 1561, B.N., Paris Fr. 16103, f° 33 ff.

[216] Relacion de lo que ha hecho Dragut, 15–30th September, 1563, Simancas E° 1127.

[217] Simancas E° 1052, f° 182.

[218] Simancas E° 1392, 18th September, 1563.

[219] *Ibid.*

[220] Simancas E° 1052, f° 212.

[221] *Ibid.*, viceroy of Naples to G. Andrea Doria, 20th September, 1563.

[222] *Ibid.*, f° 214, 9th September, 1563.

[223] *Ibid.*, f° 217, 10th September, 1563.

[224] Simancas E° 1393, 24th May, 1564.

[225] Oysel to Charles IX, Rome, 4th May, 1564. E. Charrière, *op. cit.*, II, p. 755 note.

[226] *Op. cit.*, II, p. 69, 7th April.

[227] Simancas E° 1132, Pescara to Philip II, 18th June, 1569.

[228] Fourquevaux, *op. cit.*, I, p. 90.

[229] *Ibid.*, p. 122.

[230] *Ibid.*, p. 135.

[231] Simancas E° 1052, f° 184.

[232] Pedro de Salazar, *Hispania victrix*, 1570, p. 1 v°.

emptiness of the Peninsular coasts.[233] In 1563, when Philip II was at Valencia, 'all the talk', writes Saint-Sulpice,[234] 'is of tournaments, jousting, balls and other noble pastimes, while the Moors waste no time and even dare to capture vessels within a league of the city, stealing as much as they can carry'.

Valencia threatened, Naples under a blockade (in July, 1561, 500 men were unable to cross from Naples to Salerno because of corsairs),[235] Sicily and the Balearics surrounded – all these can be explained by geography, given the proximity of North Africa. But the corsairs were active as far north as the coasts of Languedoc, Provence and Liguria, which had until then seen little disturbance. Near Villefranche, in June, 1560, the Duke of Savoy himself barely escaped falling into their hands.[236] In the same month, June, 1560, Genoa's stocks of grain and wine ran low and prices rose: the boats which usually brought wines from Provence and Corsica dared not put to sea, for fear of the twenty-three pirate ships prowling the coast.[237] Nor were these isolated incidents: every summer the Genoese coast was plundered. In August, 1563, it was the turn of Celle and Albissola on the western riviera. All 'this trouble', wrote the Republic of Sauli, its ambassador in Spain, 'comes from the sea's being empty of galleys, there is not a single Christian ship to be seen'.[238] As a result, no shipping dared put out. In May of the following year, a memorandum from Marseilles, annotated by Philip II himself,[239] reports that fifty corsairs have put out from Algiers, thirty from Tripoli, sixteen from Bône and four from Velez (the Peñon which blocked the entry to this port was not captured by the Spanish until September, 1564). If this is to be taken literally, a hundred ships, galleys, galliots and foists were at large in the Mediterranean. The same source adds: 'It is raining Christians in Algiers.'

The second brilliant age of Algiers. Between 1580 and 1620, Algiers entered upon a second age of prosperity, as spectacular as the first and certainly more far-reaching. The corsair capital benefited both from the concentration of piracy and also from a technical revolution of decisive importance.

As it had towards the middle of the century, privateering was once again replacing fleet warfare. The southern islands were besieged for weeks, even months on end. 'The corsairs do much harm in this island', writes Marcantonio Colonna, viceroy of Sicily, in June, 1578, 'in the many coastal regions which are without towers.'[240] In 1579, some Barbary

[233] Quoted by C. Duro, *op. cit.*, II, p. 45–46.

[234] Quoted by H. Forneron, *op. cit.*, I, p. 351–352.

[235] 3rd July, 1561, Simancas E° 1051, f° 108.

[236] H. Forneron, *op. cit.*, I, p. 365; Campana, *op. cit.*, II, XII, p. 87 and v°; Pietro Egidi, *Emmanuele Filiberto*, II, p. 27 gives the date as 1st June, Campana gives 31st May. The raid was led by Euldj 'Alī. News reaches Spain, Maçuelo to Philip II, Toledo, 12th July, 1560, Simancas E° 139.

[237] Figueroa to Philip II, Genoa, 19th June, 1560, Simancas E° 139.

[238] A.d.S., Genoa, L.M. Spagna 3.2412.

[239] Avisos de Marsella, 2nd May, 1564, Simancas E° 1393.

[240] Marcantonio Colonna to Philip II, Messina, 26th June, 1578, Simancas E° 1148.

foists captured two of the galleys of the Sicilian fleet off Capri and the galleys of Naples were alerted in vain. As usual they were in dock, disarmed, without troops, their crew occupied in unloading goods from merchant vessels and other peaceful labours.[241] In 1582, the viceroy of Sicily was gloomy: 'the sea is crawling with pirates'.[242] The danger did not decrease as years went by. The fact that piracy was commonplace off the northern coasts of the sea is revealing. Not even far-away Catalonia (indeed it was particularly harassed), Provence or Marseilles was spared. On 11th February, 1584, the municipal council[243] is found discussing the ransoming of citizens of Marseilles who were prisoners in Algiers; on 17th March, 1585, the council decided[244] to 'inquire into the speediest means of ending the ravages wrought by the Barbary pirates on the Provençal coast'. Years went by without bringing any relief. During the winter of 1590, Marseilles decided to send an envoy to the king of Algiers to negotiate ransoms.[245] In Venice, which one would have expected to be protected by its position, the *procuratori sopra i capitoli*, on 3rd June, 1588, elected a consul for Algiers with particular responsibility for the interests of Venetian slaves.[246]

The corsairs were everywhere in these grim times. They had to be reckoned with in the Straits of Gibraltar and were to be met almost daily along the coast of Catalonia[247] and the Roman shores; they plundered the *madragues* of Andalusia as well as those of Sardinia.[248] As early as 1579, Haëdo notes with surprise: 'sixty-two ecclesiastics imprisoned in Algiers at one time – such a thing has never been seen before in Barbary'.[249] But it was to become a familiar enough sight in later years.

There is no shortage of explanations for this revival of prosperity in Algiers: in the first place it was a natural consequence of the general prosperity in the Mediterranean. As has already been remarked, when there were no merchant vessels, there were no pirates. This is one of the recurrent themes of Godfrey Fisher's book: the Mediterranean continued to experience commercial prosperity throughout and in spite of everything, until at least 1648.[250] One is bound to conclude that piracy cannot have had the disastrous results described or suggested by the chorus of contemporary accounts and complaints, since this prosperity endured despite the increased threat from corsairs. There was in fact a close connection between trade and piracy: when the former prospered, privateering paid off correspondingly. In short, privateering was a means of forcible

[241] E. Albèri, *op. cit.*, II, V, p. 469.

[242] To Philip II, Palermo, 6th June, 1582, Simancas E° 1150, '. . . el mar lleno de corsarios . . .'

[243] A. Communales Marseilles BB 46, f° 91 ff.

[244] *Ibid.*, f° 228 ff.

[245] *Ibid.*, BB 52, f°s 10 and 10 v° and f° 29.

[246] A.d.S., Venice, Cinque Savii 26.

[247] A. de Capmany, III, *op. cit.*, p. 226–227; IV, Appendix, p. 85; A.d.S., Florence, Mediceo 4903, Madrid, 3rd June, 1572.

[248] F. Corridore, *op. cit.*, p. 21. In Corsica at the end of the century, 61 villages were destroyed or burnt, Casanova, *Histoire de l'Église corse*, 1931, I, p. 102.

[249] *Op. cit.*, p. 153.

[250] *Op. cit.*, p. 158.

exchange throughout the Mediterranean. Another explanation[251] is the evident and growing lassitude of the major states. The Turks relinquished their hold on the seas of the Levant as Spain withdrew from those of the west. Gian Andrea Doria's expedition against Algiers[252] in 1601 was to be a mere gesture, no more. Above all, the dynamism of Algiers proved to be that of a new and rapidly growing city. With Leghorn, Smyrna and Marseilles it was one of the young powers of the sea. In Algiers, all life depended, needless to say, on the volume and success of pirate shipping, down to the pittance of the poorest muleteer in the city[253] or the cleanliness of the streets which was maintained by an army of slaves, even more of course the buildings under construction, costly mosques, rich men's villas, and aqueducts, the work apparently of Andalusian refugees. But the general standard of living was modest. Not all the janissaries made a fortune in commerce, although they often engaged in it. Privateering, the major industry, was the cohesive force of the city, creating a remarkable unanimity whether for the defence of the port or the exploitation of the sea, the hinterland or the masses of slaves. It was a disciplined city and the discipline was that of a rigorous judicial system, established and maintained by what was in effect an army quartered in urban barracks. All his life, Haëdo must have remembered the iron-studded shoes of the janissaries echoing through the streets of Algiers. The activities of the corsairs undoubtedly stimulated other sectors, revivifying and organizing them, drawing food and merchandise towards Algiers. Over a wide radius around the white city, as far as the mountains and distant plateaux, peace reigned. There followed for the town a period of rapid and abnormal growth, bringing both inward and outward changes to its social fabric.

Algiers in 1516–1538 was a city of Berbers and Andalusians, of renegade Greeks and of Turks, thrown together peli-mell. This was the period that saw the rise of the Barbarossas. Between 1560 and 1587, Algiers under Euldj 'Alī was becoming increasingly more Italian. After 1580–1590 and towards 1600 came the northerners, Englishmen and Dutchmen, one of whom was Simon Danser[254] (Dansa in the French and Italian documents) that is *der Tantzer*, the dancer – his real name was Simon Simonsen and he was a native of Dordrecht. The English consul in Algiers saw him arrive in 1609 aboard a ship 'of great force' built in Lübeck and manned by a mixed crew of Turkish, English and Dutch sailors, with about thirty prizes already to her credit that year.[255] Little is known with any certainty about his eventful life, his return to Christendom and Marseilles where he had a wife and children, his entry into the service of that city, his capture and probable execution years later in Tunis, on the orders of the Dey in February, 1616.[256] The fair-skinned invaders did not come empty-handed. They brought with them cargoes of sails, timber, pitch, gunpowder and

[251] See above, p. 703. [252] See below, p. 1234.

[253] O. Eck, *op. cit.*, p. 139 ff. For the whole of this paragraph see G. Fisher, *op. cit.*, *passim* and p. 96 ff.

[254] E. Mercier, *Histoire de l'Afrique septentrionale*, Paris, 1891, III, p. 189.

[255] G. Fisher, *op. cit.*, p. 174. [256] S. Bono, *op. cit.*, p. 361 and note 21.

cannon – and best of all their sailing-ships, the same ships which had for many years been sailing the Atlantic, running rings round the unwieldy galleons and carracks of the Iberians. Leghorn also welcomed the new arrivals. But Algiers put them to better use. Sailing ships now replaced the slim galleys and traditional galliots with their light tapered hulls which had been weighed down not with ballast, baggage and cannon, but with galley-slaves enduring agonies, rowing through rough seas if necessary, to preserve the advantage of speed over the heavy Christian galleys. Unbeatable galley-crews had been the main strength of the *re'is*. But now Algiers adopted the light sailing ships, which were also capable of speed and surprise.

In 1580, the Algiers fleet amounted to perhaps thirty-five galleys, twenty-five frigates and a certain number of brigantines and barques. Towards 1618, she probably possessed about a hundred sailing vessels of which the smallest had 18 to 20 guns. In 1623 (a rather more reliable figure furnished by Sir Thomas Roe, English commercial representative at the Golden Horn) the fleet consisted of seventy-five sail and several hundred small boats. From now on the Barbary pirates were concentrated almost exclusively at Algiers; once-fearsome Tripoli (in Italy in about 1580, the usual parting words to those putting out to sea had been 'May God preserve you from the galleys of Tripoli') by 1612 possessed only a couple of sailing ships; Tunis seven in 1625.[257] Was the same true of the ports in the West where, in 1610 and 1614, the Spaniards had taken Larache and the Marmora without a struggle?[258] Algiers in any case, was soon overflowing with wealth. A Portuguese prisoner[259] tells us that between 1621 and 1627 there were some twenty thousand captives in Algiers, a good half of whom were people 'of pure Christian stock', Portuguese, Flemish, Scottish, Hungarian, Danish, Irish, Slav, French, Spanish and Italian; the other half were heretics and idolaters – Syrians, Egyptians, even Japanese and Chinese, inhabitants of New Spain, Ethiopians. And every nation of course provided its crop of renegades: even allowing for the lack of precision in the account, it seems clear that the fabric of Algiers was now of many colours. Meanwhile the corsairs swarmed all over the sea, their city now of a size to dominate the entire Mediterranean. In 1624, the Algerines plundered Alexandretta, capturing two ships, a Frenchman and a Dutchman.[260] Even more significant, they sailed out of the Straits of Gibraltar, plundering Madeira in 1617, Iceland in 1627, reaching England, as we have already seen in 1631[261] and becoming, particularly in the 1630s, pirates of the Atlantic.[262] Moslem piracy had concluded an alliance

[257] *Ibid.*, p. 89.
[258] 20th November, A. Ballasteros y Beretta, *op. cit.*, IV, 1, p. 485.
[259] *Historia tragico-maritima, Nossa Senhora da Conceycão*, p. 38.
[260] H. Wätjen, *op. cit.*, p. 138, note 2; Paul Masson, *op. cit.*, p. 380.
[261] S. Bono, *op. cit.*, p. 178.
[262] J. Denucé, *op. cit.*, p. 20 and even earlier, Barbaresques (Turks) near Iceland in 1617, *ibid.*, p. 12.

with Atlantic piracy. According to some sources it was none other than the notorious Simon Danser (alias Simon Re'īs) who taught the mariners of Algiers, perhaps as early as 1601, how to slip through the difficult Straits of Gibraltar.[263]

Conclusion? This somewhat incomplete resumé of Algerine piracy does not lead to any self-evident conclusions. For my part, I am inclined to relate the activity of the corsairs to the general economic situation in the Mediterranean which was still far from unhealthy. Godfrey Fisher's illuminating book on the corsairs does not quarrel with this assumption, indeed the reverse. But he adds some nuances, not without justification. Fisher considers that the disruptive and to western minds wicked role played by Moslem pirates in general and the Barbary corsairs in particular has been much exaggerated. The opponents of Christianity acted in good faith quite as often as its defenders and servants. On this point, no impartial judge would disagree. But history has a habit of rejecting impartial judges. Less debatable, and here I am in complete agreement with my English colleague, is the suggestion that the activity of pirates in the Mediterranean as a whole has been greatly over-estimated. Too much attention has been paid to the protests and arguments of the inhabitants of Christian shores and historians have sometimes been rather hasty in drawing conclusions.

Piracy was not a visitation of God on the prosperity of the sea. To support his conclusions, Fisher argues that some of the figures require scrutiny: he considers a hundred a very high figure for the number of sailing ships in Algiers. The exact number and above all its annual variation, are indeed unknown. It seems fairly clear that these sailing vessels were small and sacrificed guns to speed;[264] they were often little more than marauders, robbing ships of the odd barrel of fish from Newfoundland or elsewhere. Their appearance off the English coast in 1631 was remarkable for its novelty rather than for the actual danger they represented.[265] Such wounds as they inflicted were often mere pinpricks.

Can we accept this interpretation? In some ways one is obliged to: in the past the problem has been too often dismissed unthinkingly and one-sidedly; Algiers was a phenomenon of world-wide international significance which cannot be considered as merely North African or even Moslem. On the other hand, documents read by other historians suggest a rather different story. A detailed study like Alberto Tenenti's[266] takes us back to the picture of a sea teeming with pirates who dealt savage blows. His survey of ships leaving or entering Venice between 1592 and 1609 obviously does not cover the whole Mediterranean. But since Venice had the doubtful privilege of being everyone's target, the survey has more than strictly local significance. Of the 250 to 300 ships plundered

[263] See above, Vol. I, p. 119.
[264] G. Fisher, *op. cit.*, p. 166–167. [265] *Ibid.*, p. 138.
[266] *Naufrages, corsaires et assurances maritimes à Venise, 1592–1609*, 1959.

during this brief space of time which can be marked on a chart, we are reasonably certain of the identity of the aggressor in 90 cases. Moslem corsairs carried off 44 prizes, northerners (Dutch and English) 24, Spanish 22. So Christian and Moslem piracy roughly balances out. To set beside these 250 to 300 captures, there are 350 shipwrecks, so man was almost as destructive as the elements.[267] If for the sake of argument we reckon Venetian traffic at about one-tenth of the total traffic of the sea, other things being equal, over the eighteen-year period from 1592 to 1609, something like 2500 to 3000 vessels were captured by pirates, that is 138 to 166 captures a year on average (not counting the raids on the coasts in search of men, merchandise and other property). These figures are not spectacular. But we should not lay too much emphasis on the foregoing low and extremely hypothetical estimates,[268] nor on the modest equipment of the pirate fleet. It was adequate to overcome any resistance in a sea where there was an abundance of small boats and little effective patrolling. And piracy, we should remember, was a matter of boarding and hand-to-hand fighting – using the sword and arquebus more than the cannon. If the small boats of the Uskoks were to be judged purely by their size and equipment, one would never suppose them to have been a threat to anyone – but they certainly were.

The essential point, without question, is the positive correlation between piracy and the economic health of the Mediterranean: and I would stress that it is positive: they rise and fall together. When piracy has little impact on peaceful trading, it is probably because prizes are hard to come by and possibly corresponds to a general falling-off in trade. Proof requires statistics and these are not available. We have no precise idea of the total number of pirate vessels, the volume of captured goods and the number of captives: the evidence *suggests* that all these figures were on the increase.

Ransoming prisoners. All over Christendom, institutions were being introduced to handle the ransoming of prisoners: rich men, as we know, arranged their own ransoms. In 1581, the papacy led the way: Gregory XIII created the *Opera Pia della Redenzione de' Schiavi* and attached it to the ancient and active *Arciconfraternità del Gonfalone* of Rome. The first ransoms were negotiated in 1583, the first mission arrived in Algiers in 1585.[269] In 1586, there was set up in Sicily the *Arciconfraternità della Redenzione dei Cattivi* with as headquarters the church of Santa Maria Nuova at Palermo. In fact this was merely a revival of a former institution

[267] *Ibid.*, p. 27 ff.

[268] The chief difficulty lies in estimating *relative* importance. I have tackled this problem in another work, *Capitalisme et civilisation matérielle, XVe–XVIIIe siècles*, Vol. I, Ch. 1. To follow closely a series of sixteenth-century figures means adapting to a different scale, on which everything depends.

[269] Salvatore Bono, 'Genovesi schiavi in Algeri barbaresca', in *Bollettino Ligustico*, 1953; 'La pirateria nel Mediterraneo, Romagnuoli schiavi dei Barbareschi', in *La Piê, Rassegna d'illustrazione romagnuola*, 1953.

which had already functioned in the fifteenth century.[270] On 29th October, 1597,[271] Genoa set up the diligent *Magistrato del Riscatto degli Schiavi* which was also the second incarnation of an organization dating back to 1403, the *Magistrato di Misericordia*. There was a need for administration and for tribunals on behalf of these prisoners, temporarily without rights of citizenship, who returned, when they did return at all, to incredibly complicated situations at home. After being absent too long, or having become renegades, they had left unresolved business behind them: their families had had to take steps to have the disappearance legally certified while the 'ministry of prisoners' for its part was intervening to protect the interests and effects of former captives. The long series of papers in Genoa is a magnificent mine of information for the researcher who is interested in the true history of these imprisonments rather than the picturesque narrative.

Saving prisoners was all very well, but even more important was the salvation of their souls. The religious orders devoted themselves seriously to this task. It meant slipping into Barbary under the plausible cover of arranging ransoms, reaching an agreement with the charitable organizations, obtaining a passage and the relevant funds from Rome, Spain, Genoa or elsewhere. Some idea of the difficulties inherent in such negotiations emerges from a letter written by the Capuchin friar Fra Ambrosio da Soncino, dated Marseilles, 7th December, 1600, and addressed to the *Magistrato del Riscatto* at Genoa. The Capuchins and Carmelites have divided the spiritual tasks between them, the latter taking Tetouan, the former Algiers. But he is distressed at the length of negotiations for a passage and 'the time necessary for the salvation of souls, for it is this and this alone that we desire'.[272]

The traffic in ransoms and the exchange of men and goods led to the establishment of new commercial circuits. The voyages of redemptionists increased in number and they carried aboard their ships either money or goods, all duly insured.[273] After 1579, it was all registered at the French Consulate at Algiers, as it had been in Tunis since 1574. Tabarka[274] was, towards 1600, another active headquarters for ransom, dealing with Tunis and Bizerta. The return of the liberated prisoners would be marked by grand ceremonies, processions and thanksgivings. As early as 1559[275] a convoy of released captives marched through the streets of Lisbon carrying on the end of sticks pieces of brown bread – the only food in the bagnios. Inevitably a network of communications was created by the capture, negotiation and release of prisoners. With piracy on both sides,

[270] G. La Mantia in *Archivio storico siciliano*, XLIV, p. 203.

[271] R. Russo, in *Archivio storico di Corsica*, 1931, p. 575–578. On ransoms, there is a vast amount of unpublished material.

[272] A.d.S., Genoa, M° del R° degli Schiavi, Atti, 659.

[273] *Ibid.*, 14th and 15th May, 1601, insurance on 2532 *lire* at 4 per cent (two insurers).

[274] *Ibid.*, many documents, for example Giacomo Sorli to Philip Lomellini, Tunis, 7th November, 1600.

[275] J. Nicot, *op. cit.*, p. 25, 21st September, 1559.

complicated bargaining situations could arise. A document from the French consulate at Tunis mentions a Sardinian priest, the slave of the wife of Mami Arnaut[276] who was himself a slave of the King of Spain. Such predicaments made exchanges possible, if not rapid.

And with overpopulation in the bagnios, escapes became more frequent. We have already noted the exploits of the frigate of Felipe Romano, the Valencian who operated a quasi-official escape route for prisoners from the bagnios of Algiers. Prisoners also arranged their own getaways and mass escapes were commonplace.[277] One day the refugees would steal a foist, another time a galley and sail out trusting to luck – one of the more cheering incidents in these unfortunate lives. The ease with which escapes were effected came largely from the growing numbers of that hybrid race, half-Christian, half-Moslem, living on the borders of the two worlds in a fraternal alliance which would have been even more evident if the states had not required appearances to be preserved. Fraternization could be the result of changing one's faith (not the noblest, but undoubtedly the most common cause) or of trade, whether in ransoms or in merchandise. In Constantinople, this was the special province of Italian renegades; in Algiers that of the seamen of Cape Corse, familiar both to the *re'īs* and to the bagnios, who fished for coral and sometimes transported wax, wool and hides; in Tunis it was virtually the monopoly of the French consuls, who were accused of securing the release of those they chose, even making sure, on payment of a bribe, that certain prisoners never returned.[278] And everywhere Jews acted as intermediaries.

These operations were all remunerative. Trading in Algiers brings a sure profit of 30 per cent, said a Genoese merchant under interrogation. And in Spain, several reminders had to be issued that it was unlawful to ship to Algiers certain prohibited goods, to buy stolen merchandise[279] or indeed any goods sold by corsairs. But the latter could always find buyers by looking to Italy and in particular to Leghorn. These links still existed in the seventeenth century. The capture of a Portuguese vessel in 1621 left in the hands of the *re'īs* of Algiers a collection of diamonds 'with which all Italy became rich' says the source.[280] The Turks, who did not know a great deal about precious stones, had sold them at low prices. But we have only brief glimpses of these daily, unspectacular transactions. Tunis, as much as, possibly more than Algiers, was a centre for clandestine trade: the Shanghai of the sixteenth century as a Sicilian historian has called it,[281] probably with some justification.

[276] P. Grandchamp, *op. cit.*, I, p. 43, 26th August, 1592.

[277] Relacion del tercer viaje que ha hecho J. Phelipe Romano a Argel (1594), Simancas E° 342.

[278] G. Atkinson, *op. cit.*, p. 133.

[279] For example the prohibition on the Valencians, 4th January, 1589, B.N., Esp. 60, f°s 441 and v°. Equally frequent are enumerations of *mercaderias no prohibidas*, 17th July, 1582, Simancas E° 329, I.

[280] *H. tragico-maritima*, *N. Senhora da Conceycão*, p. 19.

[281] Carmelo Trasselli, Noti preliminari sui *Ragusei in Sicilia*, unpublished article, p. 32 of typescript.

Fig. 65: *Christian prisoners on their way to Constantinople*
From a drawing by S. Schweigger, 1639.

One war replaces another. So when we say that war in the Mediterranean came to an end in 1574, we should make it clear which kind of war we mean. Regular war, maintained at great expense by the authoritarian expansion of major states, yes, that certainly came to an end. But the living materials of that war, the men who could no longer be kept in the war fleets by what had become inadequate rewards and wages (as a perceptive Venetian, the Captain of the Gulf, Filippo Pasqualigo, noted in 1588) were driven to a life of roving by the liquidation of international war. Sailors from the galleys, even sometimes the galleys themselves, deserting from the fleet, soldiers, or those who would normally have been soldiers, adventurers of large or small ambitions were all absorbed into the undeclared war which now raged on land and sea. One form of war was replacing another. Official war, sophisticated, modern and costly, now moved into northern Europe and the Mediterranean was left with its secondary, minor forms. Its societies, economies and civilizations had to adapt as best they could to what was on land guerrilla warfare, on sea warfare by piracy. And this war was to absorb much of their energy, regrets, bad consciences, vengeance and reprisals. Brigandage subsumed as it were the energies of a social war which never surfaced. Piracy consumed the passions that would in other times have gone into a crusade or *Jihād*; no one apart from madmen and saints was now interested in either of these.

With the general return of peace (1598, 1604, 1609) regular war died out in the North and the Atlantic; its spectre returned to haunt the Mediterranean with plans, threats and dreams. Was it to break out again? No: the abortive war launched by Osuna and Spain against Venice (1618–1619) provides a convenient test. It failed to spark off a wide conflict, proof perhaps that the Mediterranean was no longer able to support the burden, pay the terrible price; and yet its waters were still haunted by bloodshed.

Having come thus far, we are forced to a pessimistic conclusion. If the history of human aggression in the Mediterranean in the sixteenth century is neither fictitious nor illusory, war in its metamorphoses, revivals, Protean disguises and degenerate forms, reasserts its perennial nature: its red lines did not all break at once. *Bellum omnium pater*, the old adage was familiar to the men of the sixteenth century. War, the begetter of all things, the creature of all things, the river with a thousand sources, the sea without a shore: begetter of all things except peace, so ardently longed for, so rarely attained. Every age constructs its own war, its own types of war. In the Mediterranean, official hostilities were over after Lepanto. Major war had now move north and west to the Atlantic coasts – and was to stay there for centuries to come, where its true place was, where the heart of the world now beat. This shift, better than any argument, indicates and underlines the withdrawal of the Mediterranean from the centre of the stage. When in 1618, the first shots of the Thirty Years' War rang out, and nations went to battle once more, it was far from Mediterranean shores: the inland sea was no longer the troubled centre of the world.

CHAPTER VIII

By Way of Conclusion: Conjuncture and Conjunctures

If, after a series of chapters discussing the economics, politics, civilizations and wars of the Mediterranean, I now introduce the term *conjuncture*, my intention is not to bring together under a single heading all that has gone before, but to suggest possible new directions for research and some tentative explanatory hypotheses.

In the preceding pages, the reader has been constantly reminded of the inter-relationship between change and the near-permanent in history. If we now narrow our range to focus exclusively on the element of change, of movement, the picture alters dramatically: a mathematical parallel might be the transition, by eliminating one dimension, from solid geometry to the necessarily simpler field of plane geometry. In this case we are now faced with a narrative view of history, the episodic content of which – periods, crises, phases and turning-points – may tempt the historian to dramatize or to jump to convenient if sometimes fallacious explanations. For the economic conjuncture, the most obvious and familiar of those we have to deal with, very rapidly comes to tower above all the others, imposing upon them its own terminology and categories. Neo-materialism is an inviting path. How valid is it as an approach?

A word of warning. Our problem now is to imagine and locate the correlations between the rhythms of material life and the other diverse fluctuations of human existence. For there is no single conjuncture: we must visualize a series of overlapping histories, developing simultaneously. It would be too simple, too perfect, if this complex truth could be reduced to the rhythms of one dominant pattern. How clear, in any case, is that pattern itself? It is impossible to define even the economic conjuncture as a single movement given once and for all, complete with laws and consequences. François Simiand himself recognized at least two, when he spoke of the separate movements of the tides and the waves. But reality is not as simple as this relatively simple image. In the web of vibrations which makes up the economic world, the expert can without difficulty isolate tens, dozens of movements, distinguished by their length in time: the secular trend, 'longest of the long-term movements'; medium-term trends – the fifty-year Kondratieff cycle, the double or hypercycle, the intercycle;[1] and short-term fluctuations – inter-decennial movements and seasonal shifts. So, in the undifferentiated flow of economic life, several

[1] Gaston Imbert, *Des mouvements de longue durée Kondratieff*, 1959; see particularly p. 24 ff.

languages can be distinguished by the somewhat artificial process of analysis.

If then we propose to use the economy in order to locate the chain of causality stretching back into the past, we may be obliged to handle ten or twenty possible languages – and as many causal chains. History becomes many-stranded once more, bewilderingly complex and, who knows, in seeking to grasp all the different vibrations, waves of past time which ought ideally to accumulate like the divisions in the mechanism of a watch, the seconds, minutes, hours and days – perhaps we shall find the whole fabric slipping away between our fingers.

But rather than prolong a theoretical discussion, let us give it practical application. Here before us, we have the whole of the Mediterranean during the period I have sometimes described as the 'long' or extended sixteenth century, as this book has tried to reconstruct it. Putting aside for the moment our doubts and reservations, let us try to effect some measurements in terms of the secular trend and the medium-term economic conjecture, omitting for the time being short-term and seasonal fluctuations.

The secular trend. An economic upswing, beginning in about 1470, reached a peak, or slowed down for a while, during the years of record high prices 1590–1600, then continued after a fashion until 1650. These dates: 1470 (or 1450), 1590, 1595 or 1600, 1650, are only very approximate landmarks. The long upward movement is confirmed essentially by variations in grain prices which give us a clear and unequivocal series of figures. If the wage curve, say, or the production curve had been used as a basis for calculation one would no doubt find somewhat different chronologies, but they would ultimately have to be checked against the all-powerful grain curve.

It seems clear then that during the 'long' sixteenth century, a slow but powerful upsurge favoured the advance of the material economy and of everything dependent on it. It was the secret of the fundamental healthiness of the economy. 'In the sixteenth century,' Earl J. Hamilton once told me, 'every wound heals over.' There were always compensations: in industry, production might soar in one sector if it was declining in another; in the world of commerce, as soon as one type of capitalism was on the wane, it was succeeded by another.

This hidden resilience did not vanish overnight at the end of the sixteenth century – indeed the recession was slow to appear: not before the short but structural (in other words deeply felt) crisis of 1619–1623, according to Ruggiero Romano,[2] whose opinion is almost that of Carlo M. Cipolla;[3] possibly not even until the 1650s, as Emmanuel Le Roy

[2] Ruggiero Romano, *art. cit.*, in *Rivista storica italiana*, 1962.

[3] In particular in the article written in collaboration with Giuseppe Aleati, 'Il trend economico nello stato di Milano durante i secoli XVI e XVII: il caso di Pavia', in *B.S.P.S.P.*, 1950.

Ladurie,[4] René Baehrel,[5] Aldo de Maddalena[6] and Felipe Ruiz Martín[7] suggest, and as I am more and more inclined to think myself, within the limits of my observation. For on the downward path there were pauses, recoveries even in agriculture, which one would expect to have been the first affected. Felipe Ruiz Martín tells me:[8] 'the decline of Spanish agriculture after the crisis of 1582 was not as precipitous as is usually thought: within the general downward trend, there was a cyclical [i.e. short-term] recovery, between 1610 and 1615 and another in the 1630s. Disaster did not really strike until after the 1650s.'

There is clearly no simple solution to this much-debated question which is further complicated by the possible time-lag between economic trends in different parts of Europe; although on this last point too, I think it is an over-simplification merely to distinguish between northern Europe and the Mediterranean, the latter succumbing more quickly to the general decline of the seventeenth century. The debate is still open. As far as historians of the Mediterranean are concerned, we must once again rid ourselves of the persistent but false notion of its early decline. In the first edition of this book, I dated that decline about 1600 or even 1610–1620.[9] Today I would be quite willing to put it another thirty years later.

This having been said, it is interesting to note that the general estimates calculated long ago by professional economists also suggest as the terminal date of the long upward trend, the middle years of the seventeenth century, including within it that is the first fifty years of the century, although a certain deceleration of the growth rate was already visible.

On the point of take-off however, they are by no means agreed. We have the choice between Marie Kerhuel's estimate,[10] to which I personally incline (1470 or even 1450) and Jenny Griziotti Kretschmann's[11] (1510). Either date can be defended. The earlier, 1470, is deduced from the nominal price curves, the later from silver prices. Speaking for myself, I, like René Baehrel, prefer nominal price indexes as a basis for calculation – but that dispute need not concern us here.

Further light will no doubt be shed on the question by future historians using evidence of a rather different order. In Venice, which I have per-

[4] *Les paysans du Languedoc*.

[5] *Une croissance: la Basse Provence rurale* (*fin du XVI*[e] *siècle – 1789*), 1961. René Baehrel refers to the crisis of 1690; was it not already plain to see in about 1660? Cf. Emmanuel Le Roy Ladurie, 'Voies nouvelles pour l'histoire rurale (XVIe–XVIIIe siècles)', in *Études rurales*, 1964, p. 92–93.

[6] *Art. cit.*, in *Rivista int. di scienze econ.*, 1955.

[7] In a personal letter, 11th August, 1964.

[8] See preceding note.

[9] *La Méditerranée . . .*, 1st ed., p. 613, 1095, 1096–1097. 'I am not sure whether we cannot say that the period 1550 to 1580 was a B phase and that from 1580 to 1610 an A phase, the age of the final splendours of the Mediterranean.'

[10] *Les mouvements de longue durée des prix*, 1935. Thesis defended before the Faculty of Law at Rennes. Cf. resumé in Gaston Imbert, *op. cit.*, p. 20.

[11] *Il problema del trend secolare nelle fluttuazioni dei prezzi*, 1935: the author suggests that the long upward movement began in 1510 and ended in 1635 (in France) or 1650 (in England).

sonally studied in some detail, I am impressed, for example, by the scale of public building and decoration in the city after 1450: the replacement of the wooden bridges over the canals by stone bridges,[12] the digging of the great well near the church of Santa Maria di Brolio in August, 1445,[13] the construction in May, 1459 of a new loggia *in loco Rivoalto*,[14] where the weavers' shops were demolished to make way for the extension to the Doges' Palace. 'The city increases in beauty every day', notes a document of 1494,[15] 'let us hope people will respect its beauties.' In March, 1504,[16] indeed, order went out to remove from St. Mark's Square (which had boasted its magnificent clock tower since 1495[17]) the huts erected by the stone-masons, who had planted trees and vines alongside – 'et quod pejus est: è facta una latrina che ogniuno licensiosamente va lì a far spurtitie . . .'. Needless to say, this evidence does not prove anything one way or the other, either in Venice (where construction may have been carried out because or in spite of the economic climate) or in the Mediterranean as a whole. But it inclines me to classify the whole vigorous period from 1450 to 1650 as a unit, the 'long' sixteenth century, and therefore to agree with Jean Fourastié and his pupils[18] that the first wave of prosperity was independent of American bullion. To take a single city, in this case Venice, as an index, can be a fruitful exercise; it may even reveal a truer picture of the economic situation than we have from price curves. Similar views are held by Gilles Caster, who writes that 'energy returned to Toulouse in the years 1460–1470' and tells us that the same city, Toulouse, experienced a full century (1460–1560, the 'first' sixteenth century) of prosperity.[19] It would be useful to have confirmation from other sources.

That these two hundred years, 1450–1650, should form a coherent unit, at least in some respects, clearly demands some explanation. Whether cause or effect, we know for certain that these two centuries show a steady increase in population, varying somewhat from region to region and year to year, but never, as far as one can see, interrupted. It should be noted however, that the upward trend did not, as we have already pointed out, betoken a rise in living standards. During every period, until at least the eighteenth century, economic progress was inevitably at the expense of the ever-increasing masses, the victims of 'social massacres'.[20]

The constant pressure of the mounting secular trend clearly appears to

[12] A.d.S., Venice, Notatoio di Collegio 12, f° 32 v°, 18th November, 1475; 13, f° 17, 14th November, 1482; 14, f° 9, 10th February, 1490.

[13] *Ibid.*, 9, f° 26 v°, 12th August, 1445.

[14] A.d.S., Venice, Senato Terra, 4, f° 107 v°, 25th May, 1459.

[15] *Ibid.*, 12, f° 42 v°, 18th February, 1494.

[16] *Ibid.*, 15, f° 2, 4th March, 1504.

[17] *Ibid.*, 12, f° 115, 3rd November, 1495, the clock is *quasi fornito*, all that remains to be done is *fabricar il loco*.

[18] See above, Vol. I, p. 403 and note 247.

[19] Gilles Caster, *Le commerce du pastel et de l'épicerie à Toulouse, 1450–1561*, 1962, p. 381 and 383.

[20] Ernest Labrousse's expression.

have encouraged the growth first of territorial states and then of empires.[21] Its reversal was to create obvious difficulties for them. Economic growth, with all its vicissitudes, had favoured a comparatively open society. The aristocracy was, as we have seen, reinforced by the 'bourgeois' invasion, itself encouraged by a run of prosperity. The repeated evidence of such prosperity presupposes an upward turn in economic life. With the reversal of the secular trend, society presumably closed its doors, though here we have too little evidence to construct an acceptable chronology.

Intermediate-term fluctuations. Economic historians[22] are more or less in agreement concerning the following medium-term fluctuations between a series of low points: 1460, 1509, 1539, 1575, 1621 – and peaks: 1483, 1529, 1595, 1650, dates which are accurate to within a year or so. This gives us four successive waves, each with its rise and fall, the first lasting 49 years, the second 30, the third 36 and the last 46. The apparent regularity of this pattern conceals the fact that the ascending and descending phases of the third wave (1539–1575) are not as clearly defined as the others. The middle years of the true sixteenth century (1500–1600) were marked by a pause whose repercussions, although short-lived in Spain (from 1550 to 1559 or 1562 if we take Seville as the norm)[23] were longer lasting in France, England, the Netherlands and no doubt elsewhere as well. It is as if there were two distinct phases, the 'first' sixteenth century (the age of gold) and the 'second' sixteenth century (the age of silver),[24] with a difficult transition period.

Was it for this reason (amongst others) that there appeared during the sixteenth century in the extended sense, several successive types of capitalism (similar but not the same) and wage-patterns which veered from subsistence level to prosperity? Pierre Chaunu distinguishes two phases of capitalist expansion at Antwerp: 'It was the great hunger of 1470–1490 – the fall in living standards of the workers', he writes, 'which enabled the merchant class to lay the foundations of the city's prosperity. The apogee of Antwerp as a commercial centre coincides with the second age of hardship of the proletariat, 1520–1550. The collapse of Antwerp between 1566 and 1585 can be imputed not only to civil disturbances but also to what might, with a little exaggeration, be called the second age of affluence of the urban proletariat.'[25] His remarks, which are compatible with Earl J. Hamilton's classic thesis, may well find an application in the Mediterranean. I would distinguish between three successive types of capitalism in this region, though I am unable to link them to differential variations in

[21] See above, p. 657 ff.

[22] Gaston Imbert, *op. cit.*, p. 181 ff.

[23] Pierre Chaunu, *op. cit.*, *Conjoncture*, I, p. 255 ff. Recession concerning only American trade, *ibid.*, p. 149 ff.

[24] Frank Spooner, *op. cit.*, p. 8 ff.

[25] Pierre Chaunu, 'Sur le front de l'histoire des prix au XVIe siècle: de la mercuriale de Paris au port d'Anvers' in: *Annales E.S.C.*, 1961.

profits: until about 1530 a predominantly commercial capitalism; in the midcentury a form of industrial capitalism (directed by commerce) and, towards the end of the century, financial capitalism.[26] The 'affluence' – if we can call it that – of the Venetian wage-earner occurred at the end of the century.[27]

This very imperfect model is constructed from a certain amount of data only and naturally calls for discussion. The greatest problem, to my mind, is posed by the length of the mid-century stagnation, lasting from 1529 to 1575 according to the longest estimate, or possibly from 1539 to 1575. It seems clear, to me at any rate, that this period of stagnation coincided with the disappearance of northern ships from the Mediterranean.[28]

The bankruptcies of the Spanish Crown and economic fluctuations. The suspensions of payments by the Spanish Crown, which we have already discussed at some length,[29] fit adequately enough into this tentative schema for explanations to suggest themselves. The first of them (1557, 1560) occur near the peak of the third wave and the third, 1596, near the peak of the fourth wave; once more, the turning-point of an intercyclical boom opens the way to bankruptcy. These can be considered 'normal' bankruptcies, the result of external pressures and in their own way quite logical. As for those of 1575, 1607 and 1627, according to our model they must be abnormal, brought about not only by bleak economic conditions outside (which were certainly present) but also by internal pressures; they were deliberately planned, or at the very least willingly accepted, as we have already seen in the case of the key crisis of 1575, engineered by Philip II and his advisers in the hope of ridding themselves of the Genoese bankers at what seemed a propitious moment – a feat which was not in fact accomplished until fifty years later, with the bankruptcy of 1627. The suspension of payments in 1607 was the result of lavish spending by the Spanish treasury at the dawn of what was to be the Golden Age, first under Philip III and later Philip IV.[30]

There is then a distinction to be made between those bankruptcies which were intentional and those which were, in part at least, imposed by circumstances. The historian must take good care not to consider them as identical, despite the monotonous similarity of the warning signs.

War at home and abroad. Wars fit even more obligingly into our attempt at classification. We have already identified two types of war,[31] the internal (within Christendom or Islam) and the external (between two hostile civilizations). It is possible to claim that the *Jihād* or Crusade was almost invariably encouraged by an unfavourable economic situation. Civil

[26] See above, Vol. I, p. 318 ff and p. 341 ff.
[27] Domenico Sella, *art. cit.*, in *Annales E.S.C.*, 1957, p. 29–45.
[28] See Vol. I, p. 615–621.
[29] See Vol. I, p. 505–517 and below, p. 960 ff.
[30] See Vol. I, p. 514–515.
[31] See above, p. 842–844.

wars, in which Christian fought Christian and Moslem Moslem, were on the contrary usually preceded by a 'boom'; they come speedily to a halt when the economy takes a downward turn. In Christendom therefore, the major diplomatic treaties, the 'Ladies' Peace' (Paix des Dames) (1529), Cateau Cambrésis (1559), Vervins (1598) occur either at the very peak of an upward curve or close to one; the great battles between Turk and Christian on the other hand (Prevesa, 1538, Lepanto, 1571) occur where one would logically expect to find them, in periods of recession. I would not claim that this correlation is either perfect or inevitable. The capture of Belgrade by the Turks for instance, took place in 1521, Mohács in 1526, when according to our model, such confrontations should not have occurred. To cite another discrepancy, Charles VIII crossed the Alps in 1494, while the foregoing analysis suggests that the Italian wars should not have begun until 1509 (the year, I cannot resist pointing out, of Agnadello!) But if the timetable does not exactly fit the actions of France under Charles VIII and Louis XII, it seems to apply very well to Spain under Ferdinand and Isabella. The period 1483–1509 saw the conquest of Granada and the expeditions against North Africa. The latter were accelerated between 1509 and 1511 and came to a halt with the renewal of the so-called Italian wars.[32]

I would not wish to over-state my case, nor do I seek to suppress inconvenient evidence. It is a fact nevertheless that the Italian wars, although they did indeed break out in 1494, got off to a slow start. Similarly, while the years 1521–1526 undoubtedly opened up Hungary to the Turkish invasion, it has been suggested by some historians that Hungary was not truly overcome until later, that the lengthy conquest dragged out until 1541.

One might note, by contrast, that it was at the end of the sixteenth century, after 1595 – that is exactly when we should expect it – that anti-Turkish feeling came to a head: a crusade was planned, though never, it is true, put into execution. However war by privateer was waged throughout the Mediterranean with exceptional savagery, for reasons which must go beyond those of technique, economics or individual enterprise; passions played a part; in Spain, the expulsion of 300,000 Moriscos took place between 1609 and 1614, the outcome of one of the most brutal wars of the period; lastly in about 1621, a critical year, the war which had first flared up in Bohemia in 1618 was to find fresh fuel and ravage the heart of central Europe. The Thirty Years' War began punctually.

These coincidences teach us something. In fair weather, family quarrels came to the fore; when times were bad, war was declared against the Infidel. The rule seems to apply equally well to Islam. From Lepanto until the revival of the war against Germany in 1593, Turkey, turning towards Asia, flung herself into a fanatical war against Persia. From such observations, we can perhaps glimpse something of the psychological origins of major wars.

[32] Fernand Braudel, *art. cit.*, in *Revue Africaine*, 1928.

Within Christendom, outbreaks of anti-semitism appear to coincide with foreign wars. It was in times of economic depression that the Jews were persecuted throughout the Christian world.

Conjuncture and history. I make no claims for the infallibility of the foregoing analysis, any more than I would for any attempt to classify the known data of history by the use of explanatory grids derived from our notions of the many possible conjunctures.[33] Conjunctural analysis, even when it is pursued on several levels, cannot provide the total undisputed truth. It is however *one* of the necessary means of historical explanation and as such a useful formulation of the problem.

We have the problem of classifying on the one hand the economic conjunctures and on the other the non-economic conjunctures. The latter can be measured and situated according to their length in time: comparable, let us say, to the secular trend are long-term demographic movements, the changing dimensions of states and empires (the geographical conjuncture as it might be called), the presence or absence of social mobility in a given society, the intensity of industrial growth; parallel to the medium-term economic trend are rates of industrialization, the fluctuations of state finances and wars. A conjunctural scaffolding helps to construct a better house of history. But further research is essential and at this stage much prudence is called for. Classification will be no simple matter and should be approached with caution. The long-term trends of civilizations, their flowering in the traditional sense of the word, can still surprise and disconcert us. The Renaissance for instance, between 1480 and 1509, falls in a period of clear *cyclical* depression; the age of Lorenzo the Magnificent was one of economic stagnation.[34] The Golden Age in Spain and the splendours of the sixteenth century, even in Istanbul, all blossomed after the first great reversal of the secular trend. I have offered a possible explanation – but who shall say how valid it is? My suggestion would be that any economic recession leaves a certain amount of money lying idle in the coffers of the rich: the prodigal spending of this capital, for lack of investment openings, might produce a brilliant civilization lasting years or even decades.

This tentative answer may formulate the problem, but it does not resolve it – any more than the familiar observations we have all heard about the unexpected flowering of the Renaissance and Baroque and the troubled societies which give birth to them, of which they are, one might almost

[33] I am thinking in particular of several striking and provocative, though perhaps over-rich articles by Pierre Chaunu, 'Séville et la "Belgique", 1555–1648' in *Revue du Nord*, 1960; 'Le renversement de la tendance majeure des prix at des activités au XVIIe siècle. Problèmes de fait et de méthode', in *Studi in onore di Amintore Fanfani*, 1962; 'Minorités et conjoncture. L'expulsion des Morisques en 1609' in *Revue Historique*, 1961; and *art. cit.*, above, note 25. The pursuit of political events is something of a wild goose chase.

[34] Roberto Lopez and Harry A. Miskimin, 'The economic depression of the Renaissance' in *The Economic History Review*, XIV, no. 3, April, 1962, p. 115–126.

say, the morbid product. The Renaissance spelt the end of the city-state; with the Baroque, the great empires of the sixteenth century began to feel the cold wind at their back. Perhaps the extravagance of a civilization is a sign of its economic failure. Such problems take us well outside the narrow confines of the conjuncture, whether long or short term. But once more, it is a useful path by which to approach them.

Short-term crises. I have omitted from this discussion the question of short, inter-decennial crises, the history of which is becoming daily clearer. They were evidently both contagious and irresistible, as Ruggiero Romano has shown in his article on the international crisis of 1619–1623, from which I have frequently quoted. Did it have repercussions, as one might suppose, in Islam and in the New World? So far there is no evidence that it did. Another possible project would be to subject to close scrutiny, in the light of the recent work by Felipe Ruíz Martín, the short-term crisis of the years 1580–1584. This was the result, not merely, as I had originally supposed, of the swing of the pendulum which carried Spain and her rich treasury inevitably towards Portugal, but also of the cereal crisis which was now affecting the entire Iberian Peninsula obliging it to make huge cash payments to the countries of northern Europe, those 'necessary enemies' who were thus linked once again to the Peninsula. This great upheaval is visible in price movements in Spain, Venice, Florence and even France as well as in the level of trade. In Venice, the Tiepolo-Pisani bank failed. Detailed study of these short-term crises, violent disruptions of economic life, of their ramifications and above all their extremely variable character, ought to provide new landmarks in the evolution of the Mediterranean economy. The study in depth of the 'events' of economic life would be of great value to historians. It has yet to be undertaken. A major problem at the outset is our profound ignorance of the area under Turkish domination, where, from the evidence so far available, the economic conjuncture, in the sixteenth century at least, seems to have displayed certain similarities to that in the West.[35]

[35] See Vol. I, p. 518. Some of the little available evidence can be found in Ömer Lutfi Barkan and Traian Stoianovich, 'Factors in the decline of Ottoman society in the Balkans', in *Slavic Review*, 1962.

Part Three

Events, Politics and People

It was only after much hesitation that I decided to publish this third section, describing events in the Mediterranean during the fifty years of our study: it has strong affinities with frankly traditional historiography. Leopold von Ranke, if he were alive today, would find much that was familiar, both in subject matter and treatment, in the following pages. And yet it must be included, for there is more to history than the study of persistent structures and the slow progress of evolution. These permanent realities – the conservative societies, the economic systems trapped by the impossible, the enduring civilizations, all the legitimate ways of approaching history in depth we have surveyed in the preceding chapters, undoubtedly, to my mind, provide the essentials of man's past, or what we today in the twentieth century consider to be the essentials. But they cannot provide the total picture.

What is more, this method of reconstructing the past would have been most disappointing to Philip II's contemporaries. As spectators and actors on the sixteenth-century stage, in the Mediterranean and elsewhere, they felt, rightly or wrongly, that they were participating in a mighty drama which they regarded above all as one personal to them. Possibly, probably even, they were under an illusion. But this illusion, this feeling of being an eye-witness of a universal spectacle, helped to give meaning to their lives.

Events are the ephemera of history; they pass across its stage like fireflies, hardly glimpsed before they settle back into darkness and as often as not into oblivion. Every event, however brief, has to be sure a contribution to make, lights up some dark corner or even some wide vista of history. Nor is it only political history which benefits most, for every historical landscape – political, economic, social, even geographical – is illumined by the intermittent flare of the event. The preceding chapters have drawn heavily on this concrete evidence, without which we should often find it hard to see anything at all. I am by no means the sworn enemy of the event.

But our problem, as we approach this third section, is very different. Here we shall not simply be using the lantern provided by the history of events for research that goes beyond it, but we shall be asking ourselves, just as the good traditional historian must, whether if we pool all the rays of light, the result will be a valid form of history – a particular kind of history. I think the answer is yes on one condition: we must be aware that this history implies a choice between the events themselves, on at least two levels.

In the first place, this kind of history tends to recognize only 'important' events, building its hypotheses only on foundations which are solid or assumed to be so. The importance in question is obviously a matter of opinion. One definition of the important event is one which helps to explain, Taine's 'significant detail' – but it may often lead us off the subject and very far from the event itself. Another is the event with far-reaching consequences and repercussions as Henri Pirenne was fond of remarking. By this definition, to quote the remark made by a German historian,[1] the fall of Constantinople in 1453 was a non-event and Lepanto, the great Christian victory of 1571, had no consequences at all, as Voltaire took great pleasure in pointing out. (These two opinions are both, let me hasten to say, open to debate.) An event might also be defined as important if it seemed important to contemporaries, was one they used as a point of reference, a crucial landmark, even if its exact impact is exaggerated. To Frenchmen, the St. Bartholomew's Eve Massacre (24th August, 1572) was a traumatic turning-point in the history of their country and Michelet was later to give it the same passionate emphasis. The real turn of the tide however, in my view, came several years later, in 1575 or even 1580. Finally, one could say that any event which forms a link in a chain can be considered significant. But even 'serial' history is the result of a selection, made either by the historian or for him by the available documentary evidence.

Nowadays we have two fairly well established 'chains' to choose from, one built up by the research of the last twenty or thirty years – the chain of economic events and their short-term conjunctures; the other catalogued over the ages – the chain of political events in the wide sense, wars, diplomatic treaties, decisions and domestic upheavals. It is this second chain which, to the eyes of contemporary observers, took precedence over any other series of happenings. In the sixteenth century, when there was no shortage of chroniclers, when 'journalists' make their first appearance (the *fogliottanti* or the writers of *avvisi* in Rome and Venice) politics was the real game, from the point of view of all the spectators who passionately followed its course.

For us, there will always be two chains – not one. So even in the realms of traditional history, it would be difficult for us to tread exactly in Ranke's footsteps. In turn we should beware of assuming that these two chains preclude the existence of any others, or of falling into the trap of naïvely assuming that one can explain the other, when even now we can guess at further possible chains composed of data from social or cultural history and even from collective psychology.

In any case, by admitting that economic and political realities can be classified more easily, in the short or very short term, than other social manifestations, we imply the existence of some global order going beyond them and the need to continue the search for structure and category behind the event. When the first edition of this book was published,

[1] R. Busch-Zantner, *op. cit.*

André Piganiol wrote to me saying that I could have written it the other way round – beginning with events, then moving on from that spectacular and often misleading pageant to the structural features underlying it, and finally to the bedrock of history. Perhaps the metaphor of the hourglass, eternally reversible, is a fitting image for what I have left unsaid in this brief introduction.

CHAPTER I

1550–1559: War and Peace in Europe

1550–1559: these were grim times. War, which had been simmering under the surface for five or six years, broke out once more. Although the Mediterranean was not now the principal theatre of operations, more than once the region was violently shaken by the conflict. But what fighting there was was marginal to the main tide of history. Germany, Italy and the Netherlands were the great poles of attraction in Europe. Turkey was preoccupied above all with Persia. During these years then, the Mediterranean cannot be said to have had an autonomous history. Its fate was linked with that of its neighbours, near and distant. These links are in my view, crucial. When they were severed during the critical years 1558–1559, the Mediterranean was left to generate its own wars, an enterprise to which it devoted much energy.

I. THE ORIGINS OF THE WAR

1545–1550: peace in the Mediterranean. In 1550, the Mediterranean had been peaceful for several years. One after another its wars had ended. On 18th September, 1544,[1] the treaty of Crespy-en-Laonnois had been signed by Charles V and François I; it was a makeshift agreement, with bad faith on both sides, and the dynastic matches it arranged were soon to collapse; nevertheless it brought a lasting peace. One year later, on 10th November, 1545, after comparatively easy negotiations, Ferdinand reached a truce with the Turk.[2] The sultan insisted on humiliating terms, including payment of a tribute to the Porte. But this truce did more than any other single act to eliminate war from the Mediterranean, east and west. In 1545, France was able to withdraw from the sea twenty-five galleys which sailed out of the Straits of Gibraltar, under the command of Paulin de la Garde, to participate in an attempted landing on the Isle of Wight in the north.[3] These martial sounds in turn died down: in June, 1546, at Ardres, France made peace with England.[4]

[1] After the foray by the imperial troops which took them to Meaux, Ernest Lavisse, *Histoire de France*, V, 2, p. 116. The correct date is 18th September, Jean Dumont, *Corps universel diplomatique*, Amsterdam, 1726–1731, IV, 2, p. 280–287, and not 18th November, the date given in error by S. Romanin, *Storia documentata di Venezia*, Venice, 1853–1861, VI, p. 212.

[2] A.E. Esp. 224, Philip to Juan de la Vega, Madrid, 5th December, 1545, on the truce made between the king of the Romans and the sultan, minute, f° 342. On the renewal of the truce in 1547, B.N., Paris, Ital. 227.

[3] E. Lavisse, *op. cit.*, V, 2, p. 117; Georg Mentz, *Deutsche Geschichte, 1493–1618*, Tübingen, 1913, p. 227.

[4] *Ibid.*, p. 117, (8th June), Henri Hauser and Augustin Renaudet, *Les débuts de l'âge moderne*, 2nd ed., 1946, p. 468.

Peace had come as a result of financial pressures. But they were powerfully assisted by the providential disappearance of some of the great warriors of the first half of the century. Luther died on 18th February, 1546; July of the same year saw the end of the adventurous career of Barbarossa, formerly 'king' of Algiers and from 1533 until his death, Kapudan Pasha of the sultan's fleets.[5] On the night of 27th to 28th February, 1547,[6] it was the turn of the English king, Henry VIII, and on 31st March, François I died.[7] The coming to power of new men and their advisers meant a change in policy and ideas; in the interval there was peace.

In the Mediterranean, this calm came after a series of storms such as had not been experienced for centuries. For many years, despite the marauding of sea-robbers and continental wars, there had been order after a fashion in the Mediterranean. Since at least the twelfth century, it had been virtually a Christian preserve. With her merchants and soldiers in North Africa, with her island positions in the Levant and with her powerful fleets everywhere, Christendom had succeeded, to the greater prosperity of her trade and civilizations, in maintaining her rule against an inward-looking Islam confined to its continental territory. But this order had crumbled. Once certain crucial gates had been forced (with the fall of Rhodes in the East in 1522 and the liberation of Algiers in the West in 1529) the sea lay open to the Turkish fleet. Until then it had hardly ventured out, except on such isolated expeditions as the sack of Otranto in 1480. But between 1534 and 1540 to 1545, the situation was reversed in the course of a violent struggle: the Turks, allied to the Barbary corsairs and commanded by their famous leader Barbarossa, had gained supremacy over almost the entire Mediterranean.

This was a momentous event. The spectacular battles between the emperor and France or Germany have tended to push this war into the background of Charles V's career – quite wrongly, for with the beginnings of Turkish sea-power, with the *entente* between François I and Sulaimān (1535) and the forced alliance between Venice and Charles V during the years of the first League (1538–1540), the fate of the Mediterranean was at stake. And Christendom was on the losing side. The blame must lie with its own divisions; with Andrea Doria, sworn enemy of the Republic of Saint Mark and capable of all the treachery imputed to him; and with Charles V himself, who was neither willing nor able to back up Venice as a loyal ally. Habsburg diplomacy, resorting once more to petty methods, tried to induce Barbarossa to defect: he toyed with the idea throughout

[5] On his appointment to command of the Ottoman fleet, 1533, and the date of his death, see Charles-André Julien, *Histoire de l'Afrique du Nord*, Paris, 1931, p. 521. On his life, see the highly-coloured, fictionalized, but often very accurate biography, by Paul Achard, *La vie extraordinaire des frères Barberousse, corsaires et rois d'Alger*, Paris, 1939.

[6] O. de Selve, *op. cit.*, p. 95; S. Romanin, *op. cit.*, VI, p. 23.

[7] E. Lavisse, *op. cit.*, V, 2, p. 122; S. Romanin, VI, p. 222; O. de Selve, *op. cit.*, p. 124–126.

the interminable negotiations. Would he betray the Turks for a sizeable reward? And if so, what was his price? The entire North African coast as he demanded or only Bougie, Tripoli and Bône, as the Christians suggested?[8] In the end, all this secret activity came to nothing: on 27th September, 1538,[9] Doria's fleet abandoned the field of Prevesa to the galleys and foists of Barbarossa without putting up a fight.

The Christian defeat of 1538, it is sometimes said, bears no comparison with the punishment inflicted on the Turks in 1571; it was merely a setback, a loss of face. Maybe so, but its consequences lasted over a third of a century. In 1540, Venice left the League and agreed to pay the high price of a separate peace, negotiated with the aid of French diplomacy. Without the Venetian fleet, it was impossible for the western alliance to offer any resistance to the Turkish armada, soon to be reinforced by French galleys which took to plundering the Catalan coasts and the waters round the Balearics. The collective security of Mediterranean Christendom was gravely threatened; the Turks would now not merely attack, but go beyond Malta and the Sicilian channel. Christian seapower was reduced to a defensive, ineffectual but nevertheless costly strategy. It could now launch only privateering raids and a few hasty operations as winter approached, harassing the rearguard of the enemy fleet. The last major offensive of this kind, Charles V's expedition against Algiers, failed in 1541 before the city and its guardian 'saints'. The situation appeared in all its appalling clarity when the Turkish fleet, after capturing Nice, spent the winter of 1543 to 1544[10] in Toulon, to the bitter frustration of the co-religionists of the Most Christian King of France.

So after several centuries, the Moslems once more occupied the fertile coasts of the sea. As far as the Pillars of Hercules and even beyond them to the approaches of Seville with its rich American cargoes, the shipping of all Christian powers in the Mediterranean had to move in fear; unless, that is, they had come to terms with the Turk, as had his French allies in Marseilles, his Ragusan subjects, and the Venetian businessmen who favoured neutrality in all circumstances. And it was to the Moslem side that there flocked the adventurers of the seaports, renegades willing to work for the strongest master. It had the fastest ships, the largest and best trained galley-crews, and the most powerful new city in the Mediterranean, Algiers, headquarters of the Barbary corsairs.

Can we assume that this victory was planned in Constantinople with full awareness of its importance?[11] Turkish policy in 1545 would seem to suggest the contrary. The truce with the emperor can, if necessary, be

[8] C. Capasso, 'Barbarossa e Carlo V', in *Rivista storica ital.*, 1932, p. 169–209.

[9] *Ibid.*, p. 172 and note 1; C. Manfroni, *Storia della marina italiana*, Rome, 1896, p. 325 ff; Hermann Cardauns, *Von Nizza bis Crépy*, 1923, p. 24 and 29; C. Capasso, *Paolo III*, Messina, 1924, p. 452; Alberto Guglielmotti, *La guerra dei pirati e la marina pontificia dal 1500 al 1560*, Florence, 1876, Vol. II, p. 5 ff.

[10] E. Lavisse, *op. cit.*, V, 2, p. 112.

[11] N. Iorga, *Geschichte des osmanischen Reiches*, Gotha, 1908–1913, III, p. 76 ff. On Turkish policy in the West in general and on Asian complications, *ibid.*, p. 116 ff.

explained by the peace of Crespy: without the French diversion, it was impossible to overcome the imperial army. Sulaimān had therefore to postpone for the moment his plans for the conquest of the small strip of Hungary he did not yet possess. But on sea too, and this is more surprising, Turkey did not follow up her success. There was no major naval encounter until 1560. Was it because Barbarossa was no longer there; or was it rather because the Turks were obliged, in their war against Persia, to wage a difficult war, thousands of miles from Constantinople, across deserted, mountainous territory, where campaigning came to a halt in winter and where the army required enormous caravans of supplies? The Persian war of 1545, complicated by Sulaimān's dynastic quarrel with his rebel son Muṣṭafā,[12] plus what amounted to all-out war in the Red Sea and the Indian Ocean against the Portuguese (the second siege of Diu took place in 1546),[13] all diverted the powerful Turkish war machine away from the Mediterranean.

What was an ill wind for the Turks was a breathing-space for the Mediterranean cities. The more prudent among them, the Sicilian towns for instance,[14] took the opportunity to strengthen their fortifications. Their boats once more travelled the sea. Before long, some of the northern ships, which had almost disappeared from the Mediterranean in about 1535, began to make the trip south once more.[15] They mingled with Venetian and Florentine vessels returning from England which, on occasion, even dared to put into the ports on the Moroccan coast. Did the re-opening of communications between two shores, two civilizations, mean that peace had come?

The Africa affair. The trouble was that peace in the Mediterranean inevitably brought a revival in privateering. It is of course quite impossible to measure it statistically. But if all the known cases are catalogued, the number of references clearly shows that this miniature war expanded with impunity in the central regions of the Mediterranean. A book written by Pedro Salazar, in 1570,[16] describes the summer odyssey of some of these small marauders: two Turkish foists and a brigantine, belonging to the fleet attached to Dragut and based therefore in the Tunisian Sahel and the southern shore of Djerba. In June, 1550 – June was a good month for corsairs – these three ships were posted near Ischia at the entrance to the Gulf of Naples watching the rearguard of the Spanish fleet under Don García de Toledo, which had just moved off towards Sicily. They captured first – at no risk to themselves – a supply ship (the galleys were always followed by the quarter-master's roundships, which were difficult to defend). Then came a Christian frigate. Next, still off Naples, between the islands of Ventotene and Ponza, they attacked a boat full of pilgrims bound for Rome. The brigantine then left her companions

[12] *Ibid.*, p. 117.
[13] See Vol. I, p. 545, note 25.
[14] See above, p. 851.
[15] See Vol. I, p. 612–613.
[16] *Hispania victrix*, Medina del Campo, 1570.

and returned alone to Djerba. The two foists, making their way north, appeared at the mouth of the Tiber and then sailed towards the island of Elba. But one of them, having suffered some damage, returned to Bône and from there to Algiers, where she sold her spoils. The other carried on alone. Off Piombino she accompanied four of Dragut's galliots for a while, but soon let them sail for Spain, herself making for the coast of Corsica, where, however, her haul was meagre. She decided to turn back and sailed for Bizerta by way of the Sardinian coast and Bône. She arrived in Algiers in August. If we multiply this voyage by ten or twenty and remember the Christian privateers, who were also busy,[17] we shall have some idea of the place of privateering in Mediterranean life during the 1550s.

It posed nothing like the threat of the great armadas of course. Privateers were content to operate in a small way, keeping a respectful distance from cities, fortifications and war fleets. They hardly ever ventured near certain coasts. But others, the shores of Sicily and Naples for instance, were their favourite haunts and hunting-grounds for slaves. As important as slaves to the African privateers, was the wheat aboard ships leaving the *caricatori* on the south coast of Sicily; the *caricatori* themselves were sometimes attacked.

Of all the corsairs who preyed on Sicilian wheat, Dragut (Turghut) was the most dangerous. A Greek by birth, he was now about fifty years old and behind him lay a long and adventurous career including four years in the Genoese galleys, where he was still rowing at the beginning of 1544, when Barbarossa personally negotiated his ransom.[18] In 1550 he settled in Djerba.[19] It was here that he returned between voyages, spent his winters surrounded by his *re'īs* and recruited his crews. Little more than tolerated by the inhabitants of Djerba, he took advantage of a local dispute to seize, in 1550, the small town of Africa in the Tunisian Sahel. On its bare, narrow promontory, without trees or vineyards, lying north of Sfax approximately on a level with Kairouan, Africa had had its hour of splendour in the time of the Fatimids. Now in decline, a village rather than a town, to Dragut it represented, with its sheltered waters and tumble-down walls, a useful station on the way to Sicily and a home for himself while waiting for better things.

The takeover alarmed the authorities on the other side of the Sicilian channel. The viceroy of Naples, informed by express messenger from Genoa, immediately sent word of the capture of the little port, 'luogo forse di maggior importanzia che Algieri'[20] as it was called. Let us not assume too quickly that this was an exaggeration. What was threatened by Dragut's move, was not only the security of the coasts of Sicily, in-

[17] Charles Monchicourt, 'Épisodes de la carrière tunisienne de Dragut 1550–1551' in *Rev. Tun.*, 1917, on the exploits of Jean Moret (offprint p. 7 ff).

[18] *Ibid.*, p. 11. On the life of Dragut, see the study by the Turkish historian Ali Riza Seifi, *Dorghut Re'is*, 2nd ed., Constantinople, 1910 (an edition in Turco-Latin alphabet, 1932).

[19] *Ibid.*, p. 11.

[20] *Archivio storico ital.*, Vol. IX, p. 124 (24th March, 1550).

dispensable to the supply of the western Mediterranean, but also 'Tunisia', the decadent kingdom of the Hafsids, only partly controlled by the rulers at Tunis, whom Spain tolerated because she could (thanks to the *presidio* at La Goletta) protect them and if necessary, bring them to heel. Now it looked as if Tunisia, the Arab *Ifriqiya*, still rich and still coveted by the Sicilians, might be re-organized by the Turks into a stronger and more coherent state. Charles V had in 1535 travelled personally to snatch Tunis from Barbarossa, who had installed himself there the year before.[21] Could Christendom stand by and watch Dragut, who might one day receive direct Turkish support, seize a neighbouring position? The rapid growth of Algiers was fresh in men's minds. Africa might be just a beginning.

On 12th April (he must have been informed straight away), Charles V sent a letter to the sultan from Brussels, protesting at Dragut's movements. The *re'īs* had broken the truce. Malvezzi, who was on his way to Constantinople on behalf of Ferdinand, also received instructions from the emperor.[22]

Meanwhile, in April, Dragut was preparing for the hunting season. Having left Africa in the hands of a garrison of five hundred Turks, he was at Porto Farina on the 20th. A Sicilian dispatch reported the sighting of his fleet of thirty-five sail and added that he would be off in search of prey as soon as he had tallowed his ships' bottoms, and had favourable weather.[23] There was immediately much anxiety at Naples where the arrival of Doria's galleys was eagerly awaited. They did not actually arrive until some time later, on 7th May.[24] About ten days before this, on 29th April, a dispatch reported that Dragut had been sighted off Messina, lying in wait for the grain ships.[25] After this, his ships whether in groups or scattered like the three whose movements have been described, ranged up and down the Christian coasts. The lookouts did not always manage to give warning in time. On 7th May[26] all that was known in Naples about the corsair was that he had moved westwards, possibly to Spain.

Naturally some countering action had to be taken. Charles V's 'Kapudan Pasha', the aged prince Doria, arrived in Naples on 7th May with galleys which were poorly manned (they were at least a thousand oarsmen short), but quite capable of carrying out policing action. They carried two thousand foot soldiers.[27] When Doria left Naples on 11th,[28] he intended to seize Africa in Dragut's absence. But after encountering in his attack on the little port of Monastir, north of Africa, much more resistance than he had expected – if the defenders had been more expert, the entire

[21] F. Braudel, 'Les Espagnols et l'Afrique du Nord de 1492 à 1577' in *Revue Africaine*, 1928, p. 352 ff.

[22] Carl Lanz, *Correspondenz des Kaisers Karl V*, Leipzig, 1846, III, p. 3–4 (12th April, 1550).

[23] *Archivio storico italiano*, IX, p. 124 (20th April, 1550).

[24] *Ibid.*, p. 126–127.

[25] *Ibid.*, p. 125.

[26] *Ibid.*, p. 126–127.

[27] *Ibid.*, p. 127 (11th May, 1550).

[28] *Ibid.*

Spanish infantry would have been lost in the engagement[29] – he heeded the warning. Before tackling Africa, where he knew that cannon and arquebuses were waiting for him, he sent twenty-four galleys back to Naples with instructions to embark reinforcements, a thousand Spanish soldiers and heavy siege artillery. He also asked for a general to be appointed to the command of the expeditionary force: an experienced soldier, Juan de la Vega, viceroy of Sicily, was appointed on 3rd July.[30]

These measures were sufficient to throw Naples into a turmoil of preparations and excitement throughout the month of June. Some Franciscan friars joined the convoy 'con grandi crucifissi e con grande animo di far paura a quei cani'. Every man left 'with the highest resolve to fight or die'.[31] Morale, as we would say today, was high.

The siege began on 28th June.[32] It lasted almost three months. It was not until 10th September that, watched by Doria and his sailors, who stood by as spectators, the Spaniards, Italians and Knights of Malta finally took Africa.[33] It had been no simple matter: in the interval fresh reinforcements of 500 cavalry had been asked for and the bill sent by the *provveditore* of the Duke of Florence to Pisa, shows that the expeditionary force economized neither on cannon balls nor on gunpowder.[34]

It was only a minor victory. Dragut was no longer a danger. But the Sicilians were only to hold this remote position for a few years, making occasional conspiratorial contacts with the southern nomads – an easy, but not particularly productive undertaking.[35] Since the Knights of Malta were unwilling to be responsible for it, the little fortress was dismantled and its ramparts blown up[36] after a mysterious garrison mutiny. On 4th June, 1554,[37] the occupying troops were sent back to Sicily and then, since no units stood idle for long, used in the war of Siena.[38]

It seemed no more than a trifling engagement to the emperor who, in 1550 in Augsburg, had other things to worry about: his family affairs and the political and religious situation in Germany for a start. However, he wrote on 31st October, a long letter to the sultan,[39] complaining once more about the depredations of Dragut, which, he alleged, contravened the terms of the truce; and explaining why he had intervened. It is almost

[29] *Archivio storico italiano*, IX, p. 129–130 (10th June, 1550).

[30] *Ibid.*, p. 132 (5th July, 1550).

[31] *Ibid.*, p. 131 (16th June, 1550).

[32] The dates given in E. Mercier, *Histoire de l'Afrique septentrionale*, Paris, 1891, III, p. 72, are inaccurate.

[33] *Archivio storico ital.*, Vol. IX, p. 132, C. Monchicourt, *art. cit.*, p. 12.

[34] A.d.S., Florence, Mediceo 2077, f° 45.

[35] Agreement between governor of Africa and Sheik Sulaimān ben Saīd, 19th March, 1551, Sim. E° 1193.

[36] E. Pélissier de Raynaud, *Mémoires historiques et géographiques*, Paris, 1844, p. 83.

[37] Charles Monchicourt, 'Études Kairouanaises', Part I: 'Kairouan sous le Chabbîa' in *Revue Tunisienne*, 1932, p. 1–91 and 307–343; 1933, p. 285–319.

[38] Alphonse Rousseau, *Annales tunisiennes*, Algiers, 1864, p. 25, is wrong when he says the troops were evacuated to Spain. E. Pélissier de Raynaud, *op. cit.*, p. 83; Charles Féraud, *Annales Tripolitaines*, Paris, 1927, p. 56.

[39] C. Lanz, *op. cit.*, III, p. 9–11.

an apologetic letter. For never more than at this moment had the emperor been anxious to preserve peace with the Turk at all costs, failing which he would not have been free to act as he wished in Europe and Germany. To punish a corsair, an outlaw, did not necessarily in the sixteenth century mean offending the sultan. Every day the peace had to accommodate such actions and it had usually done so in the past. So Charles V did not consider the Africa affair particularly important. It was a miscalculation on his part, for the next year he had to face massive reprisals from the Turks. But other reasons, far weightier than the Africa episode, had their part to play here. Africa was merely a pretext.

Mühlberg and after. To see why, we must go back to the apparently peaceful years 1544, 1545, 1546, and to the great battle of Mühlberg on 20th April, 1547, which settled the fate of both Germany and Europe (if such an unstable fate can ever be called settled) and consequently that of the Mediterranean. For the emperor this was his hour of triumph, more significant even than Pavia. Germany belonged to him at last: in the past, what Charles had almost always lacked was the regular support of the German world. Not only was it a triumph, it was a miracle: difficulties had melted away around him as if to help him execute a plan so long cherished. On 18th September, 1544, the war with France had ended. In December, 1545[40] the Council of Trent had met again and the Church had scored an important success. In November came the truce with the Turk. In June, 1545,[41] the papacy at last made a formal alliance with the emperor, the precious confirmation of an alliance which had existed *de facto* for years against the German Protestants, but which had not prevented Rome from disapproving Charles's Fabian policy towards the powerful Schmalkaldic League, nor Charles from being obliged to act prudently in view of the unpredictable power at Rome, which displayed alternately hostility and sympathy towards him. Now the prospects were much brighter after the negotiations by Cardinal Farnese at the Diet of Worms in March, 1545.[42] The support of Rome meant troops and money – over three thousand ducats – not to mention half of the Church's revenue in Spain, the *mezzo frutti*, as they were called in Rome. So it was a financial triumph too.[43]

However it was some time before the emperor made up his mind to strike the first blow, probably because his chancellery was overwhelmed by a mass of administrative documents and because of the usual problem of military supplies. In September, 1545,[44] Juan de la Vega, the then

[40] S. Romanin, *op. cit.*, VI, p. 214; 13th December, 1545, P. Richard, *Histoire des Conciles*, Paris, 1930, Vol. IX, 1, p. 222.

[41] P. Richard, *op. cit.*, IX, 1, p. 214.

[42] *Ibid.*, p. 209 ff.

[43] *Ibid.*, p. 214 and Buschbell, 'Die Sendung des Pedro Marquina . . .', in *Span. Forsch. der Görresgesellschaft*, Münster, 1928, I, 10, p. 311 ff. On the concessions of 1547, J. J. Döllinger, *Dokumente zur Geschichte Karls V . . .*, Regensburg, 1862, p. 72 ff.

[44] Quoted by Buschbell, *art. cit.*, p. 316.

imperial ambassador in Rome, fretted as he saw the campaign season coming to an end. It seemed the ideal time for intervention, with the neutrality or even partial complicity of France, and on the Turkish side, if not neutrality, at least inaction. In September, Juan de la Vega confided to his secretary, Pedro de Marquina, that he was sending to court a long speech to be read to the emperor – a speech full of utopian visions. If victorious, Charles V should make the empire a hereditary state, 'y quittar aquella cirimonia de election de manera que viniesse hereditario el imperio como los otros estados'. Then the Pope, the emperor and the king of France could form an alliance with a view to the conquest of England and the recovery of Hungary from the Turks. France would regain Boulogne in compensation for Milan. The Duke of Orléans would receive, with the hand of one of Ferdinand's daughters, reconquered Hungary as a dowry. A fragile amalgam of dreams, projects and visions, this document gives some idea of the special atmosphere at the courts of both Pope and emperor. In the sixteenth-century world, so divided against itself, certain circles were unbelievably haunted by the notion of a return to world unity and by the old dreams of crusade. Charles V himself can hardly be understood outside this current of thought.

I do not intend however in this study of one world, the Mediterranean, to focus attention on another, the Germanic, vital though it was in these mid-century years. My aim here is to show how, after being prepared over long years by German and extra-German circumstances, chief among which was the pacification of the Mediterranean itself, war broke out in Germany; how this war brought the triumph of the emperor, but how at the same time this triumph brought his enemies closer together – by their combined efforts, they were to swing the balance of power in Europe against him. What interests us here is that the war, apparently confined to Germany, spread gradually to neighbouring parts of Europe and then to the Mediterranean. This is the chain never revealed, although perfectly visible when one looks for it, that leads from the distant battlefield of Mühlberg in April, 1547, to the fresh outbreak of war in the Mediterranean, three years later.

What exactly did this victory in the mists of the Elbe on 24th April, 1547 bring to the emperor? In the first place, undoubtedly, it brought prestige, so unexpected was it and so rapid as to surprise the victor himself. Not that the war had been brilliantly managed by any means: secrecy had not been total, troops had been assembled slowly and the transport of heavy artillery, without any escort, could easily have been intercepted.[45] But the Protestants, divided amongst themselves, panic-stricken by the last-minute treachery of Maurice of Saxony, left their leaders and thousands of their men in the hands of the enemy. The retreat became a rout.[46] Charles V had rid himself at one stroke of what 'had for fifteen

[45] S. Romanin, *op. cit.*, VI, p. 221, according to the report made by Lorenzo Contarini in 1548.

[46] Georg Mentz, *op. cit.*, p. 209.

years been his greatest torment', the Schmalkaldic League, the organization of Protestant German princes who refused to bow to Rome or to the emperor's will.[47]

Now that Germany was subdued, Charles V intended to introduce a new political and religious order, which led him to the controversial Interim of Augsburg (1548) and the equally vexed question of the imperial succession. The latter concerns us more nearly than the former. For the emperor tried to ensure for his son, Philip of Spain, the future rule of Germany, in other words to join the German inheritance to the Burgundian and Spanish inheritance – against the clearly expressed will of the German people. From 1546 onwards, a Protestant slogan had been: 'Kein Walsch soll uns regieren, dazu auch kein Spaniol.'[48] Nor were Catholic Germans of any other mind. In September, 1550, the elector of Trèves said openly 'che non vuol che Spagnuoli commandino alla Germania'.[49] In November of the same year the cardinal of Augsburg gave vent to his ill humour at Spanish insolence and declared that Germany could only be ruled by a German prince.[50] 'Many princes, rather than elect Philip, announce that they would prefer to make peace with the Turk', wrote the Venetians in February, 1551.[51]

To disregard these feelings was madness. But after Mühlberg, what could the victor not do? Only a few free towns were still offering resistance and how long could they hold out? No help could be expected from outside: the Turks had renewed the truce with the imperial army for another five years (19th June, 1547).[52] France indeed had shown some slight signs of interest, but François I had died before Mühlberg and the new king was already occupied in the North, at least he intended to be: the war between France and England over Boulogne had broken out again in 1548.[53] In Rome serious problems had arisen for the emperor, singularly revealing of the papal position. But these problems were not insurmountable and besides Paul III was to die on 10th November, 1549.[54] The Habsburgs could do as they pleased in Germany. What it pleased them to do in fact was quarrel.

For many years the Habsburgs had surrounded the emperor with a family circle of great loyalty; indeed without it, Charles V's empire is almost unthinkable. But when the time for inheritance came, as in any other family, this loyalty suffered under the strain. The problem of the succession had already arisen before Mühlberg, in 1546 and probably

[47] G. de Leva, *Storia documentata di Carlo V . . .*, Venice, 1863–1881, III, p. 320 ff.

[48] Joseph Lortz, *Die Reformation in Deutschland*, Fribourg-en-Brisgau, 1941, II, p. 264, note 1.

[49] Domenico Morosino and Fco Badoer to the Doge, Augsburg, 15th September, 1550, G. Turba, *Venetianische Despechen*, 1, 2, p. 451 ff.

[50] *Ibid.*, p. 478, Augsburg, 30th November, 1550.

[51] *Ibid.*, p. 509, Augsburg, 15th February, 1551.

[52] B.N., Paris, Ital. 227, S. Romanin, *op. cit.*, VI, p. 214.

[53] In March, 1548, cf., Germaine Ganier, *La politique du Connétable Anne de Montmorency* (diplôme of École des Hautes Études, Le Havre), 1957.

[54] P. Richard, *op. cit.*, IX, 1, p. 439.

earlier too. It cropped up again in 1547, when the Diet met at Augsburg, still crowded with soldiers. The emperor himself brought it to mind constantly by his preoccupation with his own death, the *meditatio mortis* which inspired his many wills.

For he was already, at 47, an old man. In those days any soldier who had endured the hard life of campaigning was worn out at fifty. The long life of Anne de Montmorency for instance astonished his contemporaries. Henry VIII and François I, Charles V's coevals, had both died in the year of Mühlberg, the first aged 56, the second 53. The emperor was besides crippled with gout and at death's door, or so ambassadors reported from time to time. They confidently predicted the approaching end of the old man, 'who is seen for days on end in a dark humour, one hand paralyzed, one leg tucked under him, refusing to give audiences and spending his time taking clocks and watches apart and putting them together again'.[55]

And yet this man was still possessed by one passionate desire: to pass on to his son Philip, the inheritance intact. It was a dream dictated both by politics and by affection, for he loved this orderly, thoughtful and respectful son, the disciple he had schooled himself, both personally and from a distance. Master now of Germany and of Europe, his first action was to call his son to his side. Philip, who had been ruling in Spain since 1542, left Valladolid on 2nd October, 1548, leaving in his place his cousin Maximilian, son of Ferdinand. He was twenty-one years old and it was his first tour of Europe, described to us in all its ceremonial detail by a scrupulous, if not picturesque chronicler.[56] With Philip travelled the flower of the Spanish nobility, fathers and sons.[57] To carry them from the little Catalan port of Rosas to Genoa, the old admiral Doria's entire fleet was brought into service; musicians played on the galleys with their many-coloured oars and their shining gilded prows. On land, there was a series of triumphal arches, festivities, speeches and banquets, all the way to Brussels where the heir to the world was reunited with his father on 1st April, 1549. Charles immediately had him recognized as heir apparent to the Netherlands, an unusual step since the latter were still nominally under the authority of the Holy Roman Empire. However the young prince was 'inaugurated' as Count of Flanders and Duke of Brabant. He was exhibited to the towns of both the south and the north, which one after another from spring to autumn, 1549, had to mount an official welcome. Next came the voyage to Germany, which was to bring more bitter quarrelling than ever before about the succession.

In Augsburg, where the Diet had been convoked, the Habsburgs in August, 1550 opened their family council. Amid smiles and formal compliments a debate began which was to continue hardly unbroken for

[55] A detail frequently mentioned: Fernand Hayward, *Histoire de la Maison de Savoie*, 1941, II, p. 12.

[56] Juan Christoval Calvete de Estrella, *El felicisimo viaje del . . . Principe Don Felipe*, Antwerp, 1552.

[57] L. Pfandl, *Philippe II*, *op. cit.*, p. 170.

six months. Charles V was opposed by the ambitions of his brother, or rather his brother's family, the 'Ferdinandians', of whom the most determined was Ferdinand's eldest son, Maximilian, at that time king of Bohemia, the emperor's nephew and son-in-law. In fact it was Charles himself who had engineered the rise of Ferdinand's family. In 1516, at the time of the Spanish succession, Ferdinand had made way for his elder brother in spite of possible intrigues. His reward had quickly followed: by the treaty of 1522, he had received the Austrian Erbland intact. Nine years later in January, 1531, he had been promoted King of the Romans and in this capacity had governed Germany during the long absences of his brother. The 'apanaged' house had made its own additions, in 1526 annexing Bohemia, the fortress of middle Europe, and Hungary, or rather what had been left of Hungary by the Turks. In 1550, it was well placed. As Germany wanted neither to be subdued to foreign rule nor to embrace Catholicism, and therefore rejected the Spanish dominion which personified both, it turned naturally to the house of Vienna. It was Maximilian, not Philip, whom Germany wished to see succeed Ferdinand.

Charles had one ally: his sister, Mary of Hungary, passionately devoted to his family, governor of the Netherlands since 1531. Perhaps the succession project was her handiwork.[58] She it was at any rate who undertook to persuade Ferdinand. Was he not obliged to her as much as to Charles? In 1526, after Mohács, when her husband, Louis of Hungary, had been killed, she had helped Ferdinand to obtain the dead man's inheritance. In September, she went to Augsburg and lectured the unwilling Ferdinand, patiently and insistently. When she returned to the Netherlands, she left a far less tense situation behind. It is true that if the matter was dropped for the moment, it was in order to wait for Maximilian. As soon as he arrived, the debates began again and soon became acrimonious. These were strange conversations, held in French in memory of the Burgundian forefathers of the *Chartreuse* of Dijon; the Habsburgs, quarrelling like any ordinary heirs in the notary's office, were arguing over Germany and Europe.

When Maximilian arrived, the whole tone of the meetings changed and his indiscretions made public a debate which had until then been conducted in private. The journals of the ambassadors are full of sensational details. Charles V was angry and near to despair: 'I swear to you I can bear no more of this, or I shall die', he wrote[59] to his sister in December, 1550. Nothing had ever affected him so much as the attitude of his royal brother, not all his troubles with 'the king of France now dead', nor the current 'blustering' of the French Constable, Montmorency. Upon hearing which, Mary returned in January. This time all attempts at conciliation were in vain; so Charles V decided to impose his will by the *Diktat* of 9th March, 1551,[60] the text of which was written, rather

[58] L. Pfandl, *op. cit.*, p. 161. [59] C. Lanz, *op. cit.*, III, p. 20.

[60] F. Auguste Minet, *Charles Quint, son abdication et sa mort*, Paris, 1868, p. 39 and note 1.

mysteriously, in the emperor's own chambers, by the bishop of Arras. The imperial title was reserved for Philip, but at some future date, for the golden crown would be inherited first by his uncle, while Philip succeeded to the title of King of the Romans; on the death of Ferdinand, Philip would become emperor and Maximilian King of the Romans. Philip was also shortly to receive a promise that he would be invested with the 'feudal' authority possessed by the emperor in Italy, with the title of Imperial Vicar of the Italian territories.[61]

But this ruling was to remain a dead letter.[62] Rebuked and threatened, Ferdinand's party knew that they could count on better days. Maximilian could pose as the protector of the Lutherans, the friend of Maurice of Saxony, when he was not wooing the king of France. And this, argues Ludwig Pfandl, not entirely convincingly,[63] was the true reason for Charles V's obstinacy: he did not wish the empire to fall into the hands of such an unreliable character, a semi-heretic. The solution he did choose however was most unlikely to find approval. Hardly had the talks at Augsburg ended before pamphlets and placards appeared to warn the emperor. It has often been said that young Philip was to blame for the scheme's failure and it is probably true that this serious-minded, rather distant boy, ignorant both of the language and customs of a country once described by a contemporary[64] as fonder of drink than of Lutheran doctrine, had little to offer personally to help tip the scales. But it is impossible to see how he could have won anyway. The Augsburg ruling had already been condemned in advance by Germany and by Europe.

Most of all by Germany: how could any power hope to hold it with foreign regiments, made up of Italians and Spaniards, extravagant southerners against whom popular feeling, violent from the start, was running high? Nor could these regiments be maintained indefinitely: armies cost money. Their departure from Germany in August, 1551[65] meant a notable depletion of the force that had won Mühlberg. The emperor had few friends in Germany. Even the Catholic towns of the South did not flock readily to his side. They valued their freedom and, more than freedom, peace. As for the princes, they could not be relied upon, the more so in that the fragmented and ungovernable German world could always offer pretexts for the surrounding countries of Europe to intervene. And Europe did not want an imperial victory any more than Germany did.

Thus, in Germany and around Germany, the threat of war slowly grew. Slowly, for it took time to make alliances, raise troops and arrange for supplies. Diplomats were able to give warning far in advance of the ominous preparations under way.

[61] Meeting of 6th October, 1551, Simancas, Capitulaciones con la casa de Austria, 4.

[62] Can one even say with Ranke that it was one of the triumphs of Austrian diplomacy?

[63] L. Pfandl, *Philippe II*, *op. cit.*, p. 159.

[64] The Venetian Mocenigo, in 1548, L. Pfandl, *op. cit.*, p. 199.

[65] Charles V to Ferdinand, Munich, 15th August, 1551, C. Lanz, *op. cit.*, III, p. 68–71.

On this occasion, Simon Renard, imperial ambassador at the French court, was the best informed witness, for France played the leading role in the planned offensive. She had had her hands free since ending the English war by the treaty of 24th March, 1550.[66] Even before this date, Simon Renard had been worried, with reason, by her diplomatic moves. The king of France had tried for instance to persuade the Turk to break the truce before its time lim ithad expired (letter dated 17th January, 1550).[67] At the same time he had his agents at Bremen and entertained Spanish exiles at his court; it was even said that he planned to attack in the direction of Fuenterrabia (letter from Philip to Renard, 27th January).[68] *Mañas de Franceses*, wrote Philip.[69] But French diplomatic correspondence confirms that the rumour was correct. At the centre of these comings and goings was the unpredictable policy and person of the Constable, Montmorency, a man of prudence but also at times violent and outspoken. Certainly this was a far cry from the 'collaboration' of 1540.[70]

From the moment that the obstacle of the English war was removed, French counter-action was organized with energy and efficiency. Simon Renard took note of all the echoes that reached him. On 2nd April, French envoys were sent to Turkey and to Algiers; the troops no longer needed at the fort of Boulogne were sent to Piedmont;[71] on the 25th, the Venetians greeted with unconcealed delight the news of the peace between France and England;[72] it appeared to them a guarantee that France would not give back Piedmont and would continue to act, both in the North and in Italy, as a counterbalance to Spanish power. On the same day, 25th April, a French agent was sent to the Sharif; the latter had been worrying Spain lately by his incursions into the Oran region and by rumours that he was planning some action against the Peninsula itself.[73] The French government agent, so it was said, went to offer him the assistance of the French fleet, now not needed against England. The target for attack was thought to be the Kingdom of Granada.

Of course one never knew with the French. 'Sire,' wrote Simon Renard, also on 25th April, 'everything ordered or discussed in this kingdom is so often changed and altered, that it is very difficult to discover and report what they are really doing.' Talking too much – the traditional French vice – was after all, like the silence and secrecy of the Spanish, just another way of concealing the truth. And yet, concluded Renard a few months later: 'The king of France does not trust the emperor and in order to

[66] A.N., 1489; W. Oncken, *op. cit.*, XII (Portuguese edition), p. 1047; S. Romanin, *op. cit.*, VI, p. 224.

[67] A.N., K 1489.

[68] *Ibid.*

[69] To Simon Renard, 27th January, 1550, *ibid.*

[70] I refer to Mlle. Ganier's study cited above.

[71] A.N., K 1489, copy.

[72] *Ibid.*, Poissy, 25th April, 1550. Deciphered, Spanish translation.

[73] On the invasion of the Oran region by the latter, dispatch dated 17th August, 1550, *Alxarife passa en Argel con un gruesso exercito por conquistar . . .*, *ibid.*

thwart his designs he is negotiating with the Germans, the Swiss, the Moors and the Infidels'[74] and also, he added on 1st September, with the outlaws of Naples, *los foraxidos*, with the Duke of Albret and the Moroccan Sharif.[75] On 6th December, Fuenterrabia is mentioned again as being the point the French king would attack, 'knowing that Fuenterrabia is the key to Spain'.[76] The Venetians asked nothing better than that war should break out once more between France and Spain and it looked as if France had made up her mind. 'What incites them to it, is the intelligence and connections they have in Germany.' At the first hostile gesture, Germany will rise up – had Maurice of Saxony not hinted as much at the Diet? And there was encouragement from the Grand Turk who promised to come with 'such a great fleet that he could drive Your Majesty from Barbary, Sicily and Naples and deliver to the French all the cities he had conquered'. Simon Renard had heard wind of these projects from various sources and confirmation had come from a certain Demetico, a Greek working as an interpreter (of Arabic and no doubt Turkish) and living in Paris. An ambassador from the 'king' of Algiers to the king of France had arrived, it seems, the very day that Henri II entered Blois. He was received by the king and the Constable. 'They spoke of the victory which Your Majesty won this year at Africa.' The latest information was that the Turk would break the truce on the pretext that fortifications were being built in Hungary, contrary to the terms of the past conventions.

The following year, Simon Renard[77] gave another series of extremely detailed and probably quite accurate reports, concerning Fuenterrabia, the German towns, Italy and Barbary where a Knight of Malta had reported the arrival of sails and oars from Marseilles. Then the warning signs increased: the French ambassador returned to Constantinople on 12th April, a sure sign that something important was about to happen. On 27th May, Montluc sailed for Italy and 40 galleys were being fitted at Marseilles. In the end war broke out over Parma, where Pope Julius III had attacked the Farnese; behind the Farnese was the king of France; behind the Pope was the empire. The first shots, fired by intermediaries, were muffled and indistinct; but this was the beginning of the great war, whose rumblings had grown louder all over Europe and which was at last breaking out: on 15th July, it was learned at Augsburg that the Turkish fleet had arrived off the coast of Naples.[78]

2. WAR IN THE MEDITERRANEAN AND OUTSIDE THE MEDITERRANEAN

For the Turks struck the first blow. How could they let the Christians settle in securely on the African coast, from Tripoli, now held by the

[74] A.N., K 1489, copy, Simon Renard to the king and queen of Bohemia, 31st August, 1550.
[75] *Ibid.*
[76] *Ibid.*
[77] Under the same number, A.N., K 1489.
[78] Fano to Julius III, 15th July, 1551, *Nunt.-Berichte aus Deutschland*, Berlin, 1901, I, 12, p. 44 ff.

Knights of Malta,[79] to Africa and La Goletta, along that vital line which could completely disrupt, or at any rate seriously hinder, their communications with the West? Dragut was not strong enough alone to stand up to Andrea Doria's fleets. He was only able to escape them, at Djerba in April, 1551, by the desperate stratagem of cutting a canal through the salt-flats in the southern part of the island.[80] Dragut was in danger of being driven out of North Africa altogether. In addition, the Knights of Malta were said to be considering abandoning their rocky and infertile island and transferring their base to Africa and Tripoli, where they could govern the nearby seas. Were the Turks going to give them time to build a new, impregnable fortress of Rhodes[81] at the very gateway to Barbary?

The fall of Tripoli: 14th August, 1551. Progress was so slow however, that the Turks were able to wait for a suitable pretext according to the best diplomatic rules. The emperor had built fortifications, contravening the terms of the truce, along the Hungarian frontier. He was plotting something in Transylvania.[82] He had attacked Dragut, the sultan's ally. In February, 1551, an emissary from the Turk, a Ragusan (who travelled overland from Constantinople to Augsburg) arrived to see the emperor. The latter must demolish the fort of Zoenok and give back the town of Africa, or there would be war.[83] Oddly enough, Sinān Pasha, arriving off the Messina light, with his entire fleet waiting in the Fossa di San Giovanni, made a further demand for the restoration of Africa, in a letter to the viceroy.[84] The request was of course refused. But no one knew what the Turkish fleet would do next. Would it sail to Malta, Africa or Tripoli? Or would it sail on westwards to rendezvous with the French galleys? What would the French do then, wondered Charles V at Augsburg.[85]

The fleet, after a pretence of assault, reached the port of Malta on 18th July,[86] attempted a landing, then made for the island of Gozo, which was violently sacked, five or six thousand prisoners being taken by the Turks.[87] On 30th July, it set sail for the African coast. In Malta as in Tripoli, in

[79] Since 1530. The city had been captured in 1510 by Pedro Navarro, F. Braudel, *art. cit.*, in *Revue Africaine*, 1928, p. 223.

[80] C. Monchicourt, 'Épisodes de la carrière tunisienne de Dragut', in *Revue Tunisienne*, 1917, p. 317–324.

[81] Giacomo Bosio, *I cavalieri gerosolimitani a Tripoli negli anni 1530–31*, ed. S. Aurigemma, 1937, p. 129.

[82] J. W. Zinkeisen, *op. cit.*, II, p. 869.

[83] G. Bosio, *op. cit.*, p. 164.

[84] G. Turba, *Venetianische Depeschen*, 12, p. 507, Augsburg, 10th February, 1551.

[85] Nuncio to Julius III, Augsburg, 15th July, 1551, 'hora si starà aspettando dove ella batta, benché si crede che habbia a fare la impresa de Affrica. Sua Maesta aspetta parimente con sommo desiderio veder quel che Francia farà con questa armata. . . .' *Nunt.-Berichte aus Deutschland*, I, 12, p. 44 ff.

[86] G. Bosio, *op. cit.*, p. 164.

[87] E. Rossi, *Il dominio degli Spagnuoli e dei Cavalieri di Malta a Tripoli*, Airoldi, 1937, p. 70; 6000 according to Charles Féraud, *Ann. trip.*, p. 40; 5000 according to C. Monchicourt, 'Dragut, amiral turc' in *Revue tun.*, 1930, off-print, p. 5; 6000, Giovanni Francesco Bela, *Melite illustrata*, quoted by Julius Beloch, *op. cit.*, 1, p. 165.

the first week of August, there was still hope that it might be a false alarm. The French ambassador in Turkey, d'Aramon, had arrived in Naples on 1st August, en route for Constantinople. A rumour circulated that he was there to take the armada west with him for the winter. But the expeditionary force was soon landing at Zuara to the west and Tadjoura to the east of Tripoli.

Tripoli, captured in July, 1510 by the Spanish, had been handed over by them to the Knights of Malta in 1530. It was not much of a place: a small native settlement, with a population of Arabs serving the Christians, surrounded by an inferior wall, strengthened in places with towers, but built essentially of packed earth. Facing the port was an old-fashioned fort with four corner-towers and walls built partly of stone, but again mostly of earth. Finally, commanding with its cannon the entry to the harbour (a very large and deep harbour which could accommodate vessels of 1200 *salme*) was a small fortress, built on a spit of land, leading to the islands which guarded the mouth of the harbour to the west, the *castillegio* or *Bordj el Mandrik*, as the Arabs called it; a rather inferior construction as a result of the lack of stone or wood in this arid region and also, it was said, on account of the avarice of the Grand Master, Jean d'Olmedes. Inside the position, under the command of Fra Gaspar de Vallier, marshal of the *langue d'Auvergne*, who was to prove a mediocre leader, were thirty knights and 630 Calabrian and Sicilian mercenaries, last-minute recruits of less than outstanding quality.[88]

So the siege was unremarkable, in spite of the short time the Turks had left before winter. The assailants were able to land and obtain supplies with ease, dig trenches and line up three batteries of twelve pieces in front of the fort; the besieged soldiers mutinied and forced their leaders to capitulate. The negotiations were brief. The Turks insisted that the town's fortifications be left to them intact. In exchange, on the intervention of the French ambassador, who had joined the Turkish fleet, the Knights were given their lives and liberty. They returned to Malta, a rather sheepish band aboard the galleys of the ambassador, while the soldiers, as a punishment for their disobedience, were left in the hands of the enemy.[89]

Such at any rate is the account given by Bosio, the 'bourgeois' of Malta, who from his own position of safety, gives as his source the arguments used at the trial of the responsible commanders, which subsequently took place in Malta. Apparently blame was heaped on the heads of the captive soldiers, who were not there to defend themselves. But at the time many rumours circulated about the affair. Not to speak of the

[88] For details see C. Féraud, *op. cit.*, esp. p. 40 on the avarice of Olmedes, E. Rossi and G. Bosio, *op. cit.*

[89] For an account of the siege, besides the works already mentioned, Salomone Marino, 'I siciliani nelle guerre contro l'Infedeli nel secolo XVI', in *A. storico Siciliano*, XXXVII, p. 1–29; C. Manfroni, *op. cit.*, p. 43–44; Jean Chesneau, *Voyage de Monsieur d'Aramon dans le Levant*, 1887, p. 52; Nicolas de Nicolay, *Navig. et pérégrinations . . .*, 1576, p. 44.

serious accusations levelled against the ambassador d'Aramon, did the French Knight Gaspar de Vallier betray the position, as the historian Salomone Marino has claimed? Or should the Grand Master himself, the Spaniard Olmedes, who had been guilty at the very least of lack of foresight, bear responsibility?

Wherever the blame lies, what mattered now was that with Tripoli, the Turk possessed a valuable military position and a link with the Barbary states. The traditional port for the African interior, the town was restored to its former glory. While the Christians had occupied the town, the Sahara trade had been diverted to Tadjoura, near Tripoli, a town ruled by Murād Agha, the rough leader whom the victory of 1551 placed at the head of the Pashalik of Tripoli. Once more, gold dust and slaves travelled to this city 'rich in gold'.

The Turkish raid also gave the signal for the general war which had been under preparation in Europe. French provocation continued, while the Imperials, as a precautionary measure, in August seized all French vessels in the Netherlands.[90] The nuncio, insulted by Henri II and the Constable, announced to anyone who would listen, that war was on the way.[91] Troops were being levied in Gascony,[92] and the 30,000 men and 7000 horses of the Duke of Guise were crossing the frontiers of Bar and Burgundy. They would take some time to reach Italy however, where the situation at Parma and Mirandola was not going too well for the French, reported Renard.[93] But the Marseilles galleys had apparently received instructions to join the Turkish armada.[94]

These threats put pressure on the emperor. There were financial pressures too, and serious ones, at this difficult time when he was menaced on every side. Fearing for Sicily, he gave orders in August that the Spanish and Italian troops in Würtemburg be sent there. Few acts were to be more important than this apparently simple step. Charles V, when he left Germany, agreed that his brother should occupy, but only at his own expense and if he thought it worthwhile, the positions from which the troops were withdrawn. It so happened that Ferdinand was at this time much preoccupied by the Hungarian frontier, where the war was also spreading and where, although supported by Transylvania, which had rallied for the moment to the Habsburg cause, he was finding it difficult to offer resistance to the devastating raids made by the Beglerbeg of Rumelia, Muḥammad Sokolli.[95]

By removing the occupying troops, Charles V was directly encouraging the German rising of 1552. Did he overestimate the Turkish danger? If so, he was not the only one. The viceroy of Valencia, Tomás de Villanueva,

[90] Simon Renard to Charles V, Orléans, 5th August, 1551, A.N., K 1489.
[91] *Ibid.*
[92] Simon Renard to Philip, Orléans, 5th August, 1551, A.N., K 1489.
[93] Cf. note 90.
[94] Simon Renard to H.R.H., Blois, 11th April, 1551, A.N., K 1489.
[95] J. W. Zinkeisen, *op. cit.*, II, p. 869.

warned Philip on 15th August of the possibility of a landing.[96] Villegaignon in Malta on 24th August pleaded with Anne de Montmorency: 'unless it please the king and yourself to intercede with the Grand Turk that he leave us in peace, we are in danger of being undone'.[97]

As if the king of France cared about saving Malta! He had other things to think about. Hostilities had begun at Parma unofficially and were spreading all over Europe. Only an official declaration was lacking and the French king took the initiative. As an ally of the Duke of Parma, he began by breaking with the Pope on 1st September, the day of the re-opening of the Council of Trent. On the 12th, he sent home the imperial ambassador, Simon Renard[98] and recalled his own ambassadors.[99] In an open act of war, Brissac had already taken without difficulty the small positions of Chieri and San Damiano.[100] Even earlier, in August, Paulin de la Garde, the commander of the French galleys, had seized fifteen Spanish vessels off the coast of Italy[101] and returned to the attack in the same month in Barcelona harbour, where French sailors sailed away with four large roundships, a newly-launched galley and a frigate belonging to Prince Doria.[102] From then on, war measures by France increased: she sent troops to Italy, began fitting out ships in Brittany[103] and seized ships belonging to imperial subjects in French ports.[104] Finally October saw the beginnings of the ultimate alliance of the king of France with the Protestant princes of Germany.[105]

Did the dénouement surprise the imperial party? Apparently not (despite what Fuester[106] says), at least according to the letters of Mary of Hungary in the Netherlands, concerned but lucid and, as usual, in favour of bold measures, for instance winning over England and thus obtaining a safe anchorage for imperial ships. 'It is even said', she suggests,[107] 'that the said kingdom is easily conquerable and particularly at this time, being divided and very poor.' In any case, it would be politic now to simulate affection and trust towards the sons of Ferdinand, and for the moment to let the question of the succession drop. The Germans might receive some comfort from such moves and be incited to help His Majesty. If the war

[96] Valencia, 15th August, 1551, *Colección de documentos ineditos* (*CODOIN*), V, 117.

[97] Malta, 24th August, 1551, Guillaume Ribier, *Lettres et mémoires d'État*, Paris, 1666, p. 387–389.

[98] M. Tridon, *Simon Renard, ses ambassades, ses négociations, sa lutte avec le cardinal Granvelle*, Besançon, 1882, p. 54.

[99] *Ibid.*, p. 55 and 65, Henri II's ambassadors were the bishop of Marillac and the abbé of Bassefontaine.

[100] S. Romanin, *op. cit.*, VI, p. 225.

[101] Anthony of Bourbon to M. d'Humières, Coucy, 8th September, 1551, *Lettres d'Antoine de Bourbon*, ed. Marquis of Rochambeau, 1887, p. 26 and note 2.

[102] Philip to Simon Renard, Toro, 27th September, 1551, A.N., K 1489, min.

[103] *Ibid.*

[104] Avisos del embassador de Francia, September, 1551, A.N., K 1489.

[105] W. Oncken, *op. cit.*, XII, p. 1064, 3rd and 5th October, 1551.

[106] Eduard Fueter, *Geschichte des europäischen Staatensystems*, Münich, 1919, p. 321.

[107] Mary of Hungary to the bishop of Arras, 5th October, 1551, C. Lanz, *op. cit.*, III, p. 81–82.

is won, it will be easy to see that the empire falls into the right hands: but it has to be won first. Mary of Hungary seems to have foreseen everything, how the French would gain the support of Protestant England and 'sow dissension' in Germany, as well as the alliance of the said French with Duke 'Mauris' of Saxony. As for him, she suggests, could he not be offered a commission on the Turkish front, since there was a Turkish front again,[108] in Hungary? It would be one way of getting him out of the way or, if he refused, of forcing him to reveal his game.

Charles V consistently failed to understand this game. It was his only tactical error. As for the rest, he had no illusions and, gouty as he was, travelled to Innsbrück to be in a position to keep a close watch on Italy. So once more he prepared to go to war with the Most Christian King of France.[109]

1552: the flames of war. It was in the next year, 1552, that all the accumulated tensions were released in one vast explosion; the fires which burned all over Europe were part of a single conflagration, although they broke out in so many places, either successively or simultaneously that their essential relatedness has sometimes been obscured. Throughout almost the whole continent, 1552 unleashed a chain of wars.

First there was an internal German war, the so-called *Fürstenrevolution*, although it was more than a 'princes' revolution'; it was also a religious,[110] even a social war. It ended disastrously for Charles V. Driven from Innsbruck, he had to flee on 19th April before the troops of Duke Maurice and lost Germany as quickly as he had won it in 1547. His 'tyranny' as Bucer called it, collapsed in a few months, between the beginning of February and 1st August, 1552, which, with the treaty of Passau, brought the restoration of German liberties and, temporarily, peace between the emperor and Germany.

In the west, Germany was concerned in a foreign war, which falls into two phases. In the first phase, the French king, in accordance with his agreements with the German Protestants, ratified by the treaty of Chambord, 15th January, 1552, began his 'march on the Rhine': he occupied Toul and Metz on 10th April,[111] arrived on the banks of the Rhine in May and then, when his German allies began talks with the emperor, prudently retreated westwards. Verdun was occupied by the retreating troops in June.[112] In the second phase, the emperor, after coming to terms with

[108] Where things were going badly for the imperial army, Fco. Badoer to the Doge, Vienna, 22nd October, 1551, G. Turba, *Venetianische Depeschen*, *op. cit.*, I, 2, p. 518 ff. Temesvar was threatened by the Turks.

[109] Camaiani to Julius III, Brixen, 28th October, 1551, *Nunt.-Ber. aus Deutschland*, Series I, 12, p. 91 ff. Fano to Montepulciano, Innsbruck, 6th November, 1551, *ibid.*, p. 97 ff, 14th December, 1551, *ibid.*, p. 111.

[110] Charles V to Philip, Villach, 9th June, 1552, J. J. Döllinger, *op. cit.*, p. 200 ff.

[111] E. Lavisse, V, 2, p. 149; G. Zeller, *La réunion de Metz à la France, 1552–1648*, 2 vols., Paris–Strasbourg, 1927, I, p. 35–36, 285–289, 305–306.

[112] E. Lavisse, V, 2, p. 150.

Germany, crossed it from south to north, with reassembled troops, in order to recapture Metz: the siege, which began on 19th October, ended on 1st January, 1553 with the defeat and withdrawal of the emperor.[113] A third German war, in the east this time, was being waged on the Hungarian frontier against the Turks. A particularly arduous war, it went badly for Ferdinand, who did not receive assistance from the German princes, led by Maurice of Saxony, until the end of the year. On 30th July, 1552, Temesvar was taken by the Turks.[114]

On the frontiers of Luxemburg and the Netherlands, there was a series of minor hostilities, insignificant by comparison.

In Italy, the war was sporadic: skirmishes, sieges, guerrilla warfare in the mountains of Piedmont, with truces from time to time. The treaty of 29th April put an end to the war between the king of France and Pope Julius III,[115] but almost immediately, to restore the balance, Siena rose up on 26th July, to cries of *Francia, Francia*, expelled the imperials and proclaimed her independence. This turned out to be quite a serious episode, since it interrupted the Spanish lines of communication, and was not over until after the fall of Siena in April, 1555, to the imperials and Cosimo de' Medici.[116]

In addition to these continental wars, there were naval operations in the Mediterranean. They make no sense without reference to the whole complex of war, of which they were only a detail and by no means the most important from the military point of view. In 1552, these operations consisted of no more than the movements of the Turkish armada, which sailed by the normal route to Messina and on 5th August defeated Andrea Doria's fleet between the island of Ponza and Terracina[117] – and of the French galleys under Paulin de la Garde, which had received orders to join the Turkish admiral.

However the Turkish navy did not, despite strong pressure from France, press on to the west. The viceroy of Valencia informed Philip of Spain that the Levantine armada had entered Majorca on 13th August, 1552, but this time, as in the previous summer, invasion was only a scare.[118] Perhaps Sinān Pasha was in a hurry to return to the East, for personal reasons or because of the war against Persia. At any rate, he did not wait for the French galleys and they, like the Turkish fleet in Toulon in 1543, had to winter far from home, in Chios.[119] We have a documentary

[113] G. Zeller, *Le siège de Metz par Charles-Quint, oct.–déc. 1552*, Nancy, 1943.

[114] J. W. Zinkeisen, *op. cit.*, II, 873.

[115] The agreement was accepted by Charles V at Innsbruck, 10th May, 1552, Simancas, *Patronato Real*, no. 1527.

[116] S. Romanin, *op. cit.*, VI, p. 226, Henri Hauser, *Préponderance espagnole*, 2nd ed., 1940, p. 475.

[117] For all these dates, C. Monchicourt, *art. cit.*, offprint, p. 6, references to E. Charrière, *op. cit.*, II, p. 167, 169, 179–181, 182 note, 200, 201. On the defeat at Ponza, Édouard Petit, *André Doria, un amiral condottiere au XVIe siècle*, 1887, p. 321. In the night following the defeat at Terracina, the Turks captured seven galleys crowded with troops, C. Manfroni, *op. cit.*, III, p. 382.

[118] *CODOIN*, V, p. 123.

[119] C. Monchicourt, *art. cit.*, p. 7.

record of their arrival off the coast of Naples, near Reggio where landing parties supplied themselves at little cost with local produce, killing cows and pigs and cutting down trees in the orchards for their wood. Two cabin-boys who deserted at this point, one Italian and one from Nice, reported that the galleys were on their way to make the Turkish fleet turn back and capture Naples or Salerno. Does this incident have any connection with the conspiracy of the Prince of Salerno, Don Ferrante Sanseverino, who was in fact aboard the French fleet, a conspiracy with which Venice would have nothing to do and which seems to have failed as a result of the belated arrival of the fleet?[120] Once again, we may note how obsessed French policy was with Naples. Perhaps, if the Turkish fleet had not been so anxious to return home, it might have won a substantial victory. Neither Genoa nor Naples could have resisted the combined efforts of the two allies and Andrea Doria would not have had time to put to sea again to supply or reinforce the threatened cities.

But the Turks had no such long-term aims. For them, the operations of the fleet were purely looting expeditions: when the holds were full, the fleet returned to the Levant – perhaps even, as an obstinate but uncontrollable rumour was soon to suggest, carrying large bribes from the Spanish or Genoese.

So the major political problems of the dramatic year 1552 were not in the Mediterranean. They had nothing to do either with Dragut, Sinān Pasha, or with the now aged Andrea Doria. The men who hold the keys to this year are the emperor, Henri II or the enigmatic Maurice of Saxony. Indifferent to matters of religion, realistic, some would say amoral, few characters are as mysterious as Duke Maurice. He it was who led the reaction against Charles V, made him pay for Mühlberg by forcing him to flee over the Brenner to Villach. Then he stopped, in his moment of victory, just when he held, as it was said, a bill of exchange on Italy. Why? Because his soldiers would not have followed him? Because he would rather not be a French satellite? Was he perhaps that rarest of creatures, a clear-sighted and thoughtful statesman, anxious to put an end to war in Germany, or was he merely in league with the party of Ferdinand, in that year of ups and downs, and realized the difficulties of the Germans in the East against Islam? All these solutions have been suggested at one time or another and if it is difficult for us to choose between them, it is because this strange character suddenly departed this world, taking with him the key to the answer, his own life.[121]

As for the old emperor, was he really, as Eduard Fueter suggests, the victim of mistakes by his own diplomatic service; or, as I prefer to think, of his own stubbornness? Perhaps he thought he would be able to save the

[120] Relacion de la viaje de las galeras de Francia despues del ultimo aviso, undated (Thursday, 25th August or 25th September, 1552), A.N., K 1489. On the refusal of Venice, S. Romanin, *op. cit.*, VI, p. 226, documents in V. Lamansky, *op. cit.*, The difficulty of resistance at Genoa or Naples, C. Manfroni, *op. cit.*, III, p. 382–383.

[121] 11th July, 1553, W. Oncken, Portuguese edition, *op. cit.*, XII, p. 1084.

situation without fighting, since for eight whole years he had managed to avoid outright conflict on the French front; without fighting – in other words, without spending money. His financial difficulties at this time were very grave; it was only after the flight from Innsbruck that the Habsburg Empire decided on an all-out effort. Perhaps as Richard Ehrenberg suggests, Charles V really was saved, in June 1552, by the 400,000 ducats lent him by Anton Fugger.[122] This money enabled him to speak with firmness at Passau. The substantial aid from Florence (in the form of a 200,000 ducat loan), Naples (800,000 ducats) and above all from Spain, gave new vigour to the great imperial machine.[123] And after 1552 Spanish shipments of silver outside the Peninsula to Genoa and above all Antwerp, reached major proportions for the first time.[124] Charles V has been accused of improvidence. But how could he have foreseen the damaging near-treachery of the Ferdinandians? His greatest error was to insist on staying near Augsburg, instead of travelling (as he attempted to do, but too late) to the Netherlands, which was the seat of his power, his stronghold and even now, his treasury.[125] It was only from there, as he should have known since 1544, that a blow could be struck against France.

The policy of Henri II has also given rise to much conjecture. Was the journey to Germany a reversal of Valois policy, wonders Henri Hauser, only to answer at once in the negative.[126] Henri II returned almost immediately to his obsession with Italy, the insistent mirage. The nuncio Santa Croce was to write at the beginning of 1553, 'The Most Christian King is entirely occupied with Italian matters'.[127] The German expedition then was a mere accident. In fact, the king had hardly any choice in the matter; his aim was to resist the mighty Habsburg Empire and, therefore, to strike it at the same time as its other enemies and at the most vulnerable point. Events carried him first in one direction, then another. In 1552 they carried him eastwards and with him the historian of the Mediterranean, inasmuch as Spanish power was taking up positions in Germany and making the Netherlands both its source of supply and at times, its vital strategic base.

Corsica becomes French and England Spanish. The following year, the Mediterranean and its dependent regions continued to be peripheral to international affairs. What activity was there? An expedition by the Algerine corsairs as far as Gibraltar; a campaign by the Turkish armada, rather late in the year, which reached Corsica in conjunction with the

[122] Richard Ehrenberg, *Das Zietalter der Fugger*, Jena, 1896, I, p. 152–154.
[123] G. Turba, *Venetianische Depeschen*, I, 2, p. 526, Innsbruck, 13th May, 1552.
[124] See Vol. I, p. 480 ff.
[125] G. Zeller, *L'organisation défensive des frontières du Nord et de l'Est au XVII*e *siècle*, Nancy–Paris–Strasbourg, 1928, p. 4.
[126] *Le préponderance espagnole*, p. 475.
[127] Quoted by Henri Hauser, see note 126.

French galleys; and finally at the end of the summer, the occupation of Corsica by French troops and the *fuorusciti* of the island,[128] three operations which although spectacular, were less important than they appeared.

Salīh Re'īs,[129] a 'Moor' born in Alexandria in Egypt, and brought up under Barbarossa, had been since 1522 the seventh 'king' of Algiers. Arriving in his city in April, he first brought to heel the tribal chiefs of Touggourt and Ouargla who were refusing to pay tribute: from these triumphal raids, he returned laden with gold and with the promise that the tribute, which consisted of a few dozen black women from the depths of Africa, would be paid every year. The winter of 1552–1553 in Algiers was devoted to the thorough outfitting of a fleet, and at the beginning of June Salīh Re'īs sailed out with 40 vessels, galleys, galliots and brigantines – all well armed. The season's first encounter however, at Majorca, was a humiliating setback; and afterwards, along the Spanish coast where the patrols had been alerted in time, the corsairs had no luck. It was only in the straits that they had the good fortune to seize five Portuguese caravels, which, as it happened, were carrying a governor of Vélez, pretender to the Sharif's throne, returning from the Peninsula, where, with a party of followers, he had been seeking to advance his cause. The whole flotilla, caravels, Portuguese, Moroccans and all, were captured and transported to Vélez, where Salīh Re'īs, says Haëdo, presented the booty to the Sharif, both as a token of friendship and neighbourliness and to dissuade him from launching more of his frequent raids on the Oran region. Nevertheless three months later, fresh border incidents were reported near Tlemcen and the master of Algiers had to spend his winter once more preparing an expedition, this time against Morocco. He had taken the precaution of bringing back to Algiers with him the pretender, Ba Hassun.

Salīh Re'īs had probably returned to his base by the time the Turkish fleet, commanded by Dragut (and accompanied by Paulin de la Garde and the French galleys) arrived off the Italian coast. Intrigues and possibly the complicity of the Vizier Rustem Pasha, who had been bribed by the imperials,[130] had held up the departure of the fleet, which was less strong than that of the previous year and had a new commander, Dragut having replaced Sinān Pasha. Moreover instead of proceeding straight to the coast of the Tuscan Maremma, the Turkish ships wasted time plundering: the island of Pantelleria was sacked in August, then the grain port at Licata on the Sicilian coast. Talks between Dragut and the Tunisians (the king of Tunis had just quarrelled with the Spaniards at La Goletta) also held up the fleet between Sicily and Africa. Thanks to these delays, Andrea Doria had time, while stationing most of his ships at Genoa, to supply Christian positions and to place along the Italian coast enough fast ships to keep a close watch on the movements of the enemy.

[128] Henri Joly, *La corse française au XVIe siècle*, Lyons, 1942, p. 55.
[129] D. de Haëdo, *Epitome de los Reyes de Argel*, f° 66 v° ff.
[130] C. Lanz, *op. cit.*, III, p. 576, G. de Ribier, *op. cit.*, II, p. 436.

The latter did not arrive in the Tyrrhenian Sea until 3rd August.[131] A few days later, the ships attacked Elba, sacking Capoliveri, Rio Marina, Marciana and Porto Longone. But the main objective, Cosmopolis – more familiar perhaps as Portoferraio – resisted their attacks. It was only then, after considering an attack on Piombino, that the fleet helped transport French troops from the Sienese Maremma to Corsica.

The French commanders had been holding a council of war at Castiglione della Pescara[132] and the proposal of Marshal de Termes, commander of the French forces at Parma, had won the day. With the support of Paulin de la Garde and the Corsican exiles (chief among whom was Sampiero Corso) he had decided, without explicit orders from the king, to invade the island. The invasion was accomplished with a minimum of difficulty. Bastia was taken on 24th August, and the Baron de la Garde arrived at Saint-Florent on the 26th. Next it was the turn of Corte, inland, where the Genoese merchants of Bastia had taken refuge. Finally, at the beginning of September, Bonifacio surrendered after an attempt at resistance:[133] this town with Calvi was the place of greatest Genoese presence on the island. Laden with spoils and, what was more, a promise of money extorted from Paulin de la Garde by Dragut himself, the Turks refused to prolong the blockade of Calvi, the last position on the island occupied by the Genoese, and went home. On 1st October,[134] their fleet passed through the Straits of Messina. In December it was back in Constantinople.

Had a heaven-sent opportunity to destroy the House of Austria been lost? Certainly the Turks did not strike with all the force of which they were capable.[135] There may have been corruption – but other reasons could well explain their less than total effort, which must have been planned from the start, since only sixty galleys left Constantinople that year. In the East, there was still the Persian war. In the same year, 1553, a London merchant[136] in Aleppo, Anthony Jenkinson, saw Sulaimān the Magnificent enter the city with his sumptuous train, en route for Persia: 6000 light horse, 16,000 janissaries, 1000 pages of honour, 'all dressed in gold', accompanied the sultan, who wore a gold-embroidered robe and an enormous turban of silk and linen and rode on a white horse. There were over 300,000 men in the army and 200,000 camels followed as transport. Small wonder then that war in the Mediterranean that year was limited.

However, thanks to the Turks, the French now had a foothold in Corsica. By the end of the summer, the island was in their hands. News of the invasion stupefied the Genoese government, astounded Cosimo de' Medici and the imperial forces and provoked a flood of censure from

[131] C. Manfroni, *op. cit.*, III, p. 386.

[132] Paul de Termes to Montmorency, Castiglione della Pescara, 23rd August, 1553, B.N., Paris, Fr. 20 642, f° 165, copy, quoted by H. Joly, *op. cit.*, p. 55.

[133] J. Chesneau, *Le Voyage de Monsieur d'Aramon*, *op. cit.*, p. 161.

[134] H. Joly, *op. cit.*, p. 53. The detour on the outward journey was made in order to spare the Kingdom of Naples, considered as being in a way French territory.

[135] *Ibid.*, p. 385, C. Monchicourt, *art. cit.*

[136] R. Hakluyt, *The principal navigations* . . ., II, p. 112.

the papacy. The conquest had been very rapid. Sampiero Corso and the exiles, assisted by the inhabitants of the island, had done almost all the work single-handed. Rightly or wrongly, justifiably or unjustifiably, Corsica hated the Genoese, hated them whether as foreign masters, as usurious merchants of the towns, or as penniless immigrants who came over, as to all colonial countries, to seek a new living. Was this true of the entire population? It was certainly true of the noble families, whom the Genoese had forced into submission, but equally of the mass of the people, exasperated by poor harvests and the economic crisis and whose normal life had been disturbed by the new agricultural methods, introduced by the colonists. To all these people, the Genoese occupation was an 'assassinio perpetuo'.[137]

For all that, in this island where population outran resources, war only aggravated distress. French, Genoese, Turks, Algerines, sturdy German soldiers, Italian or Spanish mercenaries in the pay of the *Dominante* and, it must be added, Sampiero's men themselves – all this mass of soldiers had to live. They looted, spoiled crops and burned villages. Corsica had the misfortune to possess an external significance far greater than its intrinsic value, and in the war between the Valois and the Habsburgs it was vital to communications. More than in Parma, more even than in Siena, the French presence in Corsica hampered internal communication between the imperial forces and their allies. 'All ships sailing from Cartagena, Valencia, Barcelona (or Málaga or Alicante for that matter) to Genoa, Leghorn or Naples, inevitably passed within sight of the Corsican coast: this was even more true in the sixteenth century, for since Barbary pirates infested that part of the sea between Sardinia and the African shore, the normal shipping lines ran round Cape Corse or went through the Straits of Bonifacio. Moreover, the small tonnage of vessels in those days prevented them from making direct crossings, and ships sailing from Spain to Italy naturally put in to Corsican ports.'[138] Contemporaries were immediately aware, whether they greeted the news with joy or dismay, of the importance of the conquest of this 'bridle of Italy' as Sampiero Corso called it.[139]

Imperial retaliation was not long in coming. As soon as bad weather put an end to the summer alliance of French and Turks and restored the normal balance between the western fleets, the situation was reversed. Genoa and Tuscany now had the advantage of proximity, while the island, without the support of the French galleys which had returned to Marseilles[140] and what was more, directly threatened by the Genoese still holding out in Calvi, was protected only by the rebels and by 5000 old soldiers. Henri II does in fact seem to have begun indirect negotiations with Genoa in November.[141] But the latter, egged on by the Duke of

[137] Tommaseo, *Proemio alle lettere di Pasquale Paoli*, p. CLIII, quoted by H. Joly, *op. cit.*, p. 28.

[138] H. Joly, *op. cit.*, p. 8.

[139] *Ibid.*, p. 9.

[140] *Ibid.*, p. 71 and 72.

[141] *Ibid.*, p. 117.

Florence, was already appealing to the emperor[142] and had raised 800,000 ducats and levied 15,000 men. The expeditionary force sent by Doria left the city on 9th, lingered off the Genoese coast and finally reached Cape Corse on 15th. On 16th it entered the Gulf of Saint-Florent, whose garrison surrendered on 17th February following.[143] A painful war was beginning.

The year 1553 then was a fairly eventful one in the Mediterranean. But alongside the vast war which raged throughout Europe, these clashes in the south were of minor significance. The great drama of the year was the English succession. On 3rd July, 1553,[144] Edward VI died; with him there disappeared an officially Protestant England, benevolently disposed towards France and certainly hostile to the Habsburgs, so that in the Netherlands, thanks were given to God simultaneously for the death of Maurice of Saxony (11th July) and for the accession of Mary Tudor.[145] This accession however in a much-divided kingdom was particularly difficult and immediately presented the problem, in itself no simple matter, of the queen's marriage. After overcoming many obstacles and at the last moment defeating the candidature of a rather mature suitor, Don Luis of Portugal, Mary's own uncle,[146] the young Prince Philip won her hand. This success 'which was attended with much jealousy'[147] was the work of the emperor, of Granvelle, and in particular it was the masterpiece of the ambassador, Simon Renard. The marriage contract was signed on 12th July and published throughout the kingdom two days later.[148]

At the very moment when the Habsburg Empire was under severe attack, its fortunes were restored by this unexpected success. In the Netherlands, ceasing to rely on a divided Germany which he now abandoned as if on purpose to divide itself even further, the emperor turned for support to England. He concentrated his forces near the North Sea, the Mediterranean of the North, which was now virtually under his control, as were the great shipping routes leading into it from the Atlantic. He made the Netherlands his impregnable fortress.[149] For the Most Christian King of France, the outlook during the winter of 1553–1554 was therefore rather gloomy. Would he even be able, as the Venetian ambassador hoped, to prevent the Spanish prince from reaching his new kingdom (an opportunity did present itself to Villegaignon, but he did not take it)[150] and in any case what would he have gained by it? In

[142] H. Joly, *op. cit.*, p. 14, note 1.

[143] 17th, H. Joly, *op. cit.*, p. 106, not 27th, C. Manfroni, *op. cit.*, III, p. 389.

[144] W. Oncken, *op. cit.*, XII, p. 1086, 6th July.

[145] Da Mula to the Doge, Brussels, 29th July, 1553, G. Turba, *Venetianische Depeschen*, 1, 2, p. 617. On the recognition of Mary Tudor as Queen of England, *Reconocimiento de Maria Tudor por Reina d'Inglaterra*, Simancas E° 505–506, f° 7.

[146] Enrique Pacheco y de Leiva, 'Grave error politico de Carlos I' in *Rev. de Archivos, Bibl. y Museos*, 1921, p. 60–84.

[147] Granvelle to Renard, 14th January, 1553, quoted by M. Tridon, *op. cit.*, p. 85.

[148] M. Tridon, *op. cit.*, p. 84. The result had been achieved in November, 1553, Charles V to the queen of Portugal, Brussels, 21st November, 1553, in E. Pacheco *art. cit.*, p. 279–280.

[149] W. Oncken, *op. cit.*, XII, p. 1086.

[150] Ch. de la Roncière, *Histoire de la marine française*, 1934, III, p. 491–492.

Germany the contract was an equally harsh blow. 'The German princes', wrote the Venetian ambassador at the imperial court on 30th December, 1553,[151] 'are still afraid that the prince of Spain, now approaching closer to Germany, by gaining England, will be able with the assistance of this new kingdom and thanks to the divisions among the Germans, to try once more to appropriate by force the "coadjutorship" of the empire which he tried to obtain before by negotiation.'

So even before it was concluded, the English marriage was throwing its weight into the balance of power.[152] The only consolation for the emperor's enemies was that the marriage had not yet been consummated, that the island kingdom was convulsed with grave internal dissensions, which the French by their appeals to the people of London were seeking to aggravate even further.[153] There was even some talk in February, 1554, of transporting the queen to safety in Calais.[154] It was not then the support of England that Charles had acquired, merely the sympathy of a queen who was not uncontested ruler in her own country, who was uncertain what means she had at her command and was not assured even of Spanish assistance, which could be intercepted by the French in the Channel,[155] a queen in fact who had even less money than the emperor himself or his son.

The several abdications of Charles V: 1554–1556. It so happened that money, or the lack of it, was crucial during this phase of the war. On the emperor's side, there were constant problems with the Fuggers, the Schetz and other money lenders of Augsburg, Antwerp or Genoa.[156] The French king for his part was able to obtain a loan on the exchange at Lyons: 1553 was to be the year of the 'Grand Party'. But all the money borrowed had to be repaid, and in order to do so, more taxes had to be levied. Throughout the kingdom there was unusual discontent, the origins of which went even further back.

Already in 1547,[157] the Constable had had to quell peasant disturbances in Guyenne caused by high taxes. In April, 1552[158] reports reaching Spain suggested that while France was not short of grain or bread, discontent

[151] Da Mula to the Doge, Brussels, 30th December, 1553, G. Turba, *op. cit.*, I, 2, p. 640.

[152] Charles V to Philip, 1st January, 1554, A.E. Esp. 229, f° 79. On the attitude of Soranzo in England, da Mula, 2nd March, 1554, G. Turba, *op. cit.*, I, 2, p. 645, note 2.

[153] The Constable to the Cardinal of Paris (at Rome), Paris, 3rd February, 1554, A.N., K 1489 (copy in Italian). Simon Renard to Charles V, London, 29th January, 1554, A.E. Esp. 229, f° 79; the same to the same, 8th February, 1554, f° 80; 19th February, 1554, March, 1554, *ibid.*; *CODOIN*, III, p. 458.

[154] E. Lavisse, *op. cit.*, V, 2, p. 158.

[155] They stationed troops near Calais to this effect, the Constable to the Cardinal of Paris, Paris, 3rd February, 1554, Italian copy, A.N. K 1489.

[156] Charles V to Philip, Brussels, 13th March, 1554, A.E. Esp. 229, f° 81; 21st March, 1554, f° 82; 1st April, 1554, f° 83, 3rd April, 1554, f° 84. Da Mula to the Doge, Brussels, 20th May, 1554, G. Turba, *op. cit.*, I, 2, p. 648 ff.

[157] E. Lavisse, *op. cit.*, V, 2, p. 137.

[158] Avisos de Francia, Nantes, 26th June, 1552, A.N. K 1489

was widespread because of the taxes which did not spare even the monasteries, nor the hospices of Saint-Antoine or Saint-Lazare. The war which broke out again in 1552 brought ruin to the shopkeepers and peasants, humble people who had much to fear from the exactions of the nobility. 'Does not every gentleman', went on the report, 'take whatever he needs when he finds it? These men are like Moors without masters.' True this is a Spanish report and as such to be treated with caution, but in April, 1554, a dispatch from France to Tuscany[159] also reports the weariness of the population, the poor condition of the army, the inability of the king, for lack of money, to levy Swiss mercenaries, the rising taxes (once again), the melting down of private silver, the sale of letters of nobility, the contributions demanded from the clergy. Indeed in all the Christian countries, France, Spain, Italy and Germany, war-weariness was widespread. In August the Pope would try to turn it to account in his bid for peace.[160]

The Turkish Empire itself, with its armies engaged in Persia, was no better off. In 1555, Codignac, the French ambassador, had to travel to the headquarters of the army fighting the Sophy to ask the sultan in person to send a fleet.[161]

Perhaps what has in the past been taken to be the result of intrigue and calculation, was often no more than lack of funds. Throughout these two years, 1554–1555, the war was conducted half-heartedly everywhere: a siege war along the frontiers of the Netherlands and on the borders of Piedmont where Brissac[162] took by surprise the fortified town of Casale in June, 1555; a minor naval war in the Mediterranean, where the Turkish fleet put in only a brief appearance: in 1554, under Dragut, it lingered far too long at Durazzo, at least to the minds of the French, who, together with the galliots of Algiers, were meanwhile attempting to intervene in Corsica and on the coast of the Tuscan Maremma.[163] They met little opposition particularly since a certain number of Spanish galleys had been sent to the Atlantic to accompany Philip to England. But Dragut arrived late and had hardly sailed once up the coast of Naples, before turning homewards again. French agents accused him of treason[164] and afterwards sought by every means to have him removed from his command. It is quite possible after all, that Dragut had received money from the imperial side. But the following year, he was only second in command, under the orders of a new admiral, Piāle Pasha, young and inexperienced. And yet, despite the king of France, who had called for a 'strong and determined' war effort,[165] the Turkish fleet did no more than stand by, without participating in the siege of Calvi which was able, thanks to

[159] Avisos de Francia, 3rd April, 1554, A.d.S., Florence, Mediceo 424, f° 5, quoted by H. Joly, *op. cit.*, p. 119.

[160] H. Joly, *op. cit.*, p. 118.

[161] C. Manfroni, *op. cit.*, III, p. 392 and references to E. Charrière, *op. cit.*

[162] H. Joly, *op. cit.*, p. 122.

[163] It was in the course of these operations that Leone Strozzi perished.

[164] C. Manfroni, *op. cit.*, III, p. 391.

[165] *Ibid.*, p. 392; E. Charrière, *Négociations . . .*, II, p. 351.

32. BARBAROSSA, after Capriolo, *Ritratti di cento capitani illustri*, Rome, 1596, f° 113 v°

33. CHARLES V, from the Arras Collection

supplies from Genoa, to resist the French. It stood by equally calmly in August, at the siege of Bastia which had been recaptured by the enemy during the previous year. Finally after a few fruitless attacks on the Tuscan coast and islands, pleading lack of supplies and poor weather,[166] it turned round and sailed home. Does this record not suggest that the fleet had, as in the previous year, received instructions to act with restraint?

The lack of means of the large states allowed the smaller ones to act to more effect than usual. We have already seen what energy Genoa put into her Corsican war: in 1554–1555, she managed to drive the French from the greater part of the island.[167] Cosimo de' Medici too made sterling efforts: although poorly backed up at sea by Andrea Doria, who was in the first place prudent by nature and in the second did not, being Genoese, view Tuscan expansion with a wholly favourable eye, Cosimo nevertheless forced the French in Siena to capitulate on 21st April, 1555: a few months later, he recaptured Orbitello on the coast of the Maremma. Now there only remained the 'republic' of Montalcino in the Apennines, refuge of the Sienese patriots and of a few Frenchmen.[168] But at the end of 1555, Cosimo. attacked it, beginning his operation in the Val di Chiana.[169]

The state of Algiers alone warrants a longer mention, during this two-year period, than the Ottoman fleet. In 1554,[170] Salīh Re'īs, taking his army by sea to the 'new' port of Melilla and then overland to Taza and Fez, which he entered in triumph, led an astonishingly rapid raid against Morocco. The Moroccan cavalry was unable to resist the Turkish arquebuses. But the victorious expedition led to nothing, for the protégé of the Algerines (the same Ba Hassun who had been captured the year before), now installed by them as ruler in Fez, was soon to be killed by the former Sharif, who re-entered the city once the invaders, laden with spoils and carrying large sums of gold showered upon them by their grateful protégé, had ridden away on the horses and mules of the Moroccans. All that remained to the Algerines from this excursion was the rocky islet, the Peñon de Vélez, of which more later.[171]

The following year, 1555, their activity was directed eastwards against Bougie or rather against the Spanish *presidio* there, for it was no longer a true town. Inside the former boundaries of the native settlement, the *presidio* was a small, triangular, fortified zone with a fort at each corner;

[166] Marquis de Sarria to Princess Joanna, Rome, 22nd November, 1555; J. J. Döllinger, *op. cit.*, p. 214–216.

[167] During the winter the Genoese fleet came out of hiding. Of the 12 galleys under his command, Gian Andrea Doria, whose career was just beginning, lost nine in January, 1556 in a gale (the *libeccio*) off the Corsican coast, C. Manfroni, *op. cit.*, III, p. 394.

[168] Lucien Romier, *Les origines politiques des guerres de religion*, Paris, 1914, II, p. 393–440.

[169] Coggiola, 'Ascanio della Corna', p. 114, note 1, December, 1555.

[170] D. de Haëdo, *Epitome . . .*, *op. cit.*, f°s 68 and 68 v°.

[171] See below, p. 1000–1001.

the imperial castle, a rectangular building reminiscent of the origina fortress at La Goletta; the great Castle and the smaller Castle of the Sea, old Moorish buildings facing the coast.[172] Inside these ramparts lived a hundred or so men and a few dozen horses. For supplies, the garrison had to rely on raids as much as on the supply-ships. It was on a foraging expedition that the old governor, Luis Peralta, died in an ambush, leaving his son Alonso to succeed to the office.[173] In June, 1555, Salīh Re'īs left Algiers with several thousand soldiers, including renegade *ex-copeteros*, meanwhile sending by sea a small supply fleet with provisions and artillery: two galleys, a barque, a French *saëte* requisitioned at Algiers – a very small flotilla, most of the corsairs' vessels having left to join Leone Strozzi's fleet. But these means proved adequate: the fort was unable to resist the artillery and its defenders fled to the nearby town, which was virtually indefensible. Alonso de Peralta surrendered shortly afterwards against a promise that he and forty of his companions of his own choice, should be guaranteed their lives and given a passage back to Spain on the French *saëte*. The defeat had widespread repercussions in Spain.[174] In Valencia, Catalonia and Castille, there was talk of an expedition in reprisal, the movement being led by Siliceo,[175] Archbishop of Toledo. Then the tumult died down, as it tends to do, notes Luis Cabrera, in these affairs of honour and reputation, when large sums of money are required. The expedition was postponed on the pretext that the emperor was not in the kingdom; but resentment remained such that Alonso de Peralta was arrested as soon as he set foot in Spain, and tried and beheaded at Valladolid on 4th May, 1556.[176] Was he so very guilty? As soon as Bougie was attacked, he had sent a call for help in good time to Spain; from Spain orders were dispatched to the Duke of Alva, at that time viceroy of Naples, but so slowly that by the time Andrea Doria, alerted by the duke, was in Naples with his galleys ready to sail, news of the capitulation had already arrived.[177]

While lesser powers were settling their own differences, the game of international diplomacy among the major powers was being pursued as usual. The death of Pope Julius III on 22nd March, 1555,[178] had robbed Charles V of a valuable ally. The king of France inherited what he had lost, when after the reign of Marcellus II, which lasted only a few weeks,[179] Paul IV was elected on 23rd May, 1555,[180] the very day when Franco-

[172] Paule Wintzer, 'Bougie, place forte espagnole', in *B. Soc. géogr. d'Alger*, 1932, p. 185–222, esp. p. 204 ff and 221.

[173] Diego Suárez, *Hist. del maestre ultimo que fue de Montesa . . .*, Madrid, 1889, p. 106–107.

[174] Luis de Cabrera, *Felipe II, Rey de España*, Madrid, 1877, I, p. 42.

[175] Peticiones del Cardenal de Toledo para la jornada de Argel y Bugia y Conquista de Africa, Simancas E° 511–513.

[176] Paule Wintzer, *art. cit.*, p. 221. In his favour see Diego Suárez, *op. cit.*, p. 107.

[177] The Duke of Alva to Princess Joanna, 29th March, 1556, Simancas E° 1049, f° 11.

[178] G. Mecatti, *Storia cronologica della Città di Firenze*, *op. cit.*, II, p. 697.

[179] Coggiola, 'Ascanio . . .', p. 97.

[180] H. Joly, *op. cit.*, p. 122; S. Romanin, *op. cit.*, VI, p. 230.

Imperial talks in view of peace opened at Marcq.[181] No signs appeared at first of the new Pope's violent hostility to the emperor, but it was sufficient in itself to threaten the peace which seemed to be approaching in the North. And indeed a secret treaty (the existence of which was nevertheless known both in Venice and Brussels), dated 13th October, 1555, assured the French of the formal alliance of the Pope, if the plans for peace should come to nothing.[182]

Within the empire itself, equally important changes were being felt. Philip had arrived in England without incident in 1554[183] and diplomatic correspondence was ablaze with speculation: did the queen love him or not? Would they have children? (As early as 1555 people were saying they would not.) At the same time it was learnt that Charles V was handing over to his son, now king of England, the kingdoms of Sicily and Naples and the Duchy of Milan.[184] No doubt the gesture was meant to add weight to the prospects of the young bridegroom, in the same way that Ferdinand had named his son Maximilian, king of Bohemia in 1551: these were matters of prestige and protocol. However the surrender of these titles – as the will drawn up by Charles in the same year, 1554, attests – already foreshadowed the abdication of the emperor, or rather the abdications in the plural. One normally thinks only of the affecting and tearful scene at Brussels, the abdication of the Netherlands, when Charles announced for the first time before the States-General, on 25th October, 1555, his intention of withdrawing from the world.[185] In fact he had already given up Sicily and Naples and the Milanese; in January, 1556 while he was out of the country he quietly abdicated the Spanish throne.[186] It was not until 1558, shortly before his death, that he yielded the imperial gold crown, his final abdication, postponed at the urgent request of Ferdinand himself,[187] who was disturbed by the prospect of a risky election, and possibly also at

[181] H. Joly, *op. cit.*, p. 120.

[182] Simancas P° Real, n° 1538, 13th October, 1555, Coggiola, *art. cit.*, p. 246.

[183] Philip to Princess Joanna, Windsor, 9th August, 1554, A.E. Esp. 229, f° 84. *Viaje de Felipe II (sic) a Inglaterra quando en 1555 fué a casar con la Reina Dª Maria*, *CODOIN*, I, p. 564.

[184] It is difficult to give precise dates. On 25th July, 1554, the minute of Charles V's resignation of the kingdom of Naples was presented to Philip by the regent Figueroa (Simancas E° 3636, 25th July, 1554, G. Mecatti, *op. cit.*, II, p. 693). On 2nd October of the same year, Julius III conceded the investiture of the kingdoms of Naples and Sicily to Philip (Simancas E° 3638, 23rd October, 1554) then on 18th November, the Pope conceded to him as a fief the kingdoms of Sicily and Jerusalem (Simancas E° 1533, Rome, 18th November, 1554). On Naples see Lodovico Bianchini, *Della Storia delle Finanze del Regno di Napoli*, 1839, p. 52–53. The abdication of Charles V from the throne of Sicily, according to Bianchini, took place on 16th January, 1556, but the abdication was made in the name of 'Carolus et Joana reges Castelle' and must therefore date from before the death of Joanna the Mad in 1555.

[185] For an abridged account see Charles Bratli, *Philippe II, roi d'Espagne*, Paris, 1912, p. 87 ff or L. Pfandl, *op. cit.*, p. 272 ff.

[186] *Renuncia de Carlos V en favor de Felipe II de los reinos de Castilla*, Simancas E° 511–513.

[187] For instance Ferdinand to Philip II, Vienna, 24th May, 1556, *CODOIN*, II, p. 421 or Charles V to Ferdinand, Brussels, 8th August, 1556, *ibid.*, p. 707–709.

the request of Philip who felt the need for his father's supporting presence in the Netherlands and Italy.

It is perhaps wrong to dismiss these abdications, as historians have done since Mignet and Gachard, as the fruits of a personal, private conflict. There is also the war-torn climate of these years 1554–1556 to consider. Charles V may have wished to spare his son the dangers of succeeding during the confusion following his death. If he voluntarily renounced one of his most cherished projects, handing over to Ferdinand the great German ship of state, it was because in the years since 1552 and 1553, he had realized how impossible it was to steer. He had himself left the helm when he charged Ferdinand in 1555 with the trouble and responsibility of negotiating the Peace of Augsburg, the peace which kept Germany free of strife for the rest of the century, but which Charles himself hated with all his heart. In any case Germany, unreliable at the best of times, could be replaced by England, the dowry Philip had received in marriage, in the imperial balance of power. To abandon Germany was perhaps the only way to be rid of war and the colossal expenditure it had brought him.

For whatever reason, and the event was to prove vital to the Mediterranean world, Philip's empire detached itself from the German bloc. The last tie was broken in July, 1558, when Philip II claimed the Imperial Vicariate in Italy, which he had been promised by the convention of 1551.[188] His ambassador to Ferdinand received from the latter on 22nd July, 1558 an eloquent reply: '. . . having examined the matter you wish to bring to my attention on behalf of the Most Serene King of England and Spain, our very dear and beloved nephew, on the subject of the title of Imperial Lieutenant in Italy . . ., you may say for our part to His Highness that as we bear in mind having made the promise to him, we have every desire to keep it . . .'. But it was a delicate matter. '. . . His Highness must remember that when my lord the Emperor and myself and others known to his Highness discussed making him coadjutor of the Empire, together with my son, King Maximilian, I pointed out to them that inconveniences, troubles and tumults within the Empire might follow such a step: and that such a course could not be successful. However, out of respect for the Emperor and in obedience to his will, we had to do what was done; and not long afterwards it was realized that I had been a better prophet than we wished, for, as soon as they were informed of our design, Duke Maurice and the princes rose up in arms . . .'.[189]

And now, asked Ferdinand, was it worth running a similar risk over the vicariate, reinforcing the accusation that the Habsburgs wished to make the empire *hereditario*? To combat the major powers of Germany would be most unwise, 'given the present problems and necessities by which we are beset, his Highness by France and the Turk, myself by the Turk and the rebels in Hungary, not to speak of religious matters and other problems, of which there is no lack', not to mention either the difficulties which will be raised by the Pope, sworn enemy of the House of Habsburg.

[188] Cf. above, p. 915–916. [189] *CODOIN*, XCVIII, p. 24.

'Moreover', Ferdinand went on, 'there is another drawback: in order to fulfil these functions, his Highness will be obliged to reside in Italy: it was on this implicit condition that the promise was made, and it was never our intention that it could possibly be exercised from Flanders, England or Spain. . . .' We may hasten through Ferdinand's prose, which sounds more ironic to our ears no doubt than it did to Philip, to hear his conclusion: 'On these conditions then, we give our solemn word, as of now, that the moment His Highness goes to Italy, we will send him our patents in the due form . . .'; a promise he was never called upon to keep: Philip II was soon to find himself king only of Spain.

It was probably his surrender of Germany, implicit in the first abdications of the emperor, which above all contributed to peace in Europe. The talks which had opened in Marcq and, contrary to widespread belief, had proceeded uninterrupted,[190] ended finally in the truce of Vaucelles, hastily thrown together on 5th June, 1556,[191] thanks to the good offices of the queen of England and just in time before the campaign season began.

This truce did not solve anything; it merely sanctioned the status quo. But it put an end to the fighting, in other words to the spending of vast sums of money, which was what everyone wanted: 'Lack of money is felt everywhere in these times', complained Ferdinand, who was hoping, through the good offices of the French, to reach a truce with the Turk,[192] while Charles V, now that the horizon was calmer, was thinking of departing for Spain,[193] finally taking his farewell of the world and power and leaving Philip behind in the Netherlands. Thus the empire would survive in more or less its present form, with Brussels as the political and military capital, and Antwerp as the financial capital. It was a comforting thought: from Brussels it was easy to keep an eye on Europe and govern it. But would Europe let itself be governed?

3. THE RETURN OF WAR: THE INITIATIVE STILL COMES FROM THE NORTH

The Truce of Vaucelles is broken. It is difficult to understand why the truce of Vaucelles was broken. Given the exhaustion of both sides, it should have contented all parties, at least for a while: France was allowed to keep her conquests, notably Savoy and Piedmont; the Habsburgs once more looked like masters of the world. They held Sicily, Naples, Piacenza, Milan – in other words they controlled Italy, for Piedmont, in

[190] As evidenced by H. Joly, *op. cit.*, p. 126, who contradicts Francis Decrue de Stoutz, *Anne de Montmorency*, Paris, 1899, II, p. I.

[191] A. d'Aubigné, *Histoire universelle*, Paris, 1886, I, p. 125; E. Lavisse, *op. cit.*, V, 2, p. 160 says 15th February, but the king of France made the truce public on 13th (13th February, 1556, A.N., K 1489), F. Hayward, *op. cit.*, II, 18.

[192] Ferdinand to Charles V, Vienna, 22nd May, 1556, C. Lanz, *op. cit.*, III, p. 69, 702.

[193] He was to disembark at Laredo on 6th October, 1556, L. P. Gachard, *Retraite et mort de Charles Quint*, Brussels, 1854, p. 137.

the sixteenth century could hardly be called part of Italy. It was apparently a perfect opportunity for the papacy to transform the truce into a general peace. This was the traditional role of the papacy[194] and Paul IV felt obliged at least to pay lip-service to it; he ordered official rejoicings,[195] visited the signatories and even claimed responsibility for the agreement in front of the Venetian ambassador Navagero,[196] but he deceived no one, least of all the Venetians.

In fact, news of the truce had been received at Rome as a great shock.[197] The rumour immediately circulated that it had been negotiated against the Pope's wishes and in spite of all his efforts.[198] It was thanks to him at any rate that it was broken. That one man should have revived, alone and with such speed, a still smouldering war, gives one to reflect on the role of individuals in the dramatic episodes of history. In Paul IV, old in years (born in 1477, he was 79 when he acceded to St. Peter's throne) but a man of astonishing passion, vigorous health and impressive piety (he was to found the Theatine Order) the Church had found an intransigent warrior who revived against Charles V the conflict which the death of Paul III had interrupted in 1549: Rome's eternal hatred of Caesar, of the man who had ordered the sack of Rome in 1527, the man who had let the Protestants triumph in Germany and accepted the Peace of Augsburg.

This then was one aspect of the antipathy of Paul IV for Charles V, the antipathy of a Pope, not to be underestimated. But there was another: that of the Neapolitan, the head of the Francophile Carafa family, who hated Charles V as the master of Naples and the enemy of his kinsmen, a hatred nourished on bitter grievances and desires for vengeance. Old enough to remember Italy when it was free, he also resented the emperor as the foreigner, the invader, the representative of the Spanish, 'those heretics, schismatics, cursed of God, a race of Jews and Moors, the dregs of the world'.[199] That the idea of Italian liberty was dear to him is attested by the following words (addressed to the Venetian ambassador, after the setback to papal policy): 'You will repent, my dear lords of Venice and all you others, who would not seize the opportunity to rid yourselves of this scourge. . . . French and Spaniards, both are barbarians and it would be better that they stayed at home.'[200]

Paul IV was a man who acted according to the impulses of his own heart and mind. A preacher and theologian, he lived more in the world of dreams and ideas than in the real world around him. 'He is a man who

[194] Philip II to Princess Joanna, London, 13th April, 1557, A.E. Esp. 232, f° 232.

[195] Badoero to the Senate, Brussels, 7th March, 1556, Coggiola, *art. cit.*, p. 108, note.

[196] Navagero to the Senate, Rome, 21st February, 1556, Coggiola, *art. cit.*, p. 232–233.

[197] Badoero to the Senate, Brussels, 1st March, 1556, Coggiola, *art. cit.*, p. 108, note. [198] *Ibid.*

[199] Report of Bernard Navagero, 1558, E. Albèri, *Relazioni . . .*, II, 3, p. 389.

[200] Ernesto Pontieri, 'Il papato e la sua funzione morale e politica in Italia durante la preponderanza spagnuola', in *Archivio storico italiano*, 1938, Vol. II, p. 72.

considers the affairs of state only in general terms, like a philosopher', Marillac said of him.[201]

Putting all this together one can better understand the policy adopted by the Supreme Pontiff and its explosive force during the years 1556 and 1557. Even so, that is not the whole story, for the Pope was not alone: he had not one, but several policies, and he was not responsible for all of them. Around him were his relatives and advisers, including one formidable character, Cardinal Carlo Carafa, a strange man, as passionate as the Pope but without his more noble qualities: greedy, insatiable, short-tempered, aggressive and not over-scrupulous; negotiating with the imperial court as well as with the French and prepared to go to considerable lengths in such dealings.

Having arrived at the French court in June, 1556 as legate *a latere*, he had left with formal promises of intervention[202] from the 'pacific' Montmorency; even Coligny had allowed himself to be drawn into his schemes.[203] A few months passed and in October and November, in the course of the talks which began between the Pope and the Duke of Alva and which were to result on 18th November in a truce of forty days,[204] Cardinal Carafa found himself in direct contact with the Duke of Alva, who had reached Ostia. The results of the meeting were somewhat unexpected: the Carafa family asked the Spaniards not only for the positions still held by the French in Tuscany, but for Siena too. The papers of Della Casa contain a curious *Discorso al Card. Caraffa per impetrare dalla M. dell' Imperatore lo stato e dominio di Siena*[205] and the Spanish archives contain a memorandum of 22nd January, 1557 detailing the *condiciones con que S. M. tendra por bien de dar al conde de Montorio* – the cardinal's brother – *el Estado de Sena para la efectuation del accordio que se trata con S.S.*[206] It was the same Carafa who went to Venice to try to persuade the Signoria to enter the struggle against the Habsburgs and to share in the possible division of their possessions in Italy. The Venetians refused, not wishing, as they said to 'come home with a handful of flies' or as we might say, of dust.

This is the man whom some historians have declared to be the faithful interpreter of the policy and thought of Paul IV; others disagree. The truth is not easily discovered.

What we do know is that Paul IV very soon showed unmistakable

[201] E. Lavisse, *op. cit.*, V, 2, p. 163.

[202] Henri II to Ottavio Farnese, Fontainebleau, 29th June, 1556, Coggiola, *art. cit.*, p. 256–257; F. Decrue, *Anne de Montmorency*, II, p. 186.

[203] H. Patry, 'Coligny et la Papauté en 1556–1557' in *Bul. de la Soc. d'histoire du protestantisme français*, Vol. 41, 1902, p. 577–585.

[204] The Duke of Alva returned to Ostia on 14th November: lo que refiere un hombre que fue a Francia estos dias a entenderlo que alla se hazia (December, 1556), A.N., K 1490. The truce was signed on 18th November (Simancas, Patronato Real, nº 1580) and extended on 27th December, 1556, *ibid.*, no. 1591.

[205] *Opere*, Milan, 1806, p. 119–131, quoted by Coggiola, p. 225 ff.

[206] Same date, Philip II to Cardinal Carafa, Simancas *Patronato Real*, no. 1614.

signs of his ill-will towards the Habsburgs.[207] It was even said that he was prepared to convoke a Council to strip the emperor of his dignities. So the chief problem for the Habsburgs was to find out what the king of France intended to do: if he remained neutral, the Pope could quite well be managed; otherwise, war would break out again, even if the French king intended to remain in a state of 'covert war', as it was to be called in the seventeenth century. After July, doubt was no longer possible, the peace talks between Ruy Gómez and the Constable had broken down, the Constable being dissatisfied with the terms concerning prisoners of war and the increased ransom he was being asked for his son. At Brussels there were no illusions: 'In order to have the shadow of a pretext, they are waiting for the Duke of Alva to do something that offends the Pope.'[208]

Let me stress once again: that the Carafa policy should have produced such striking results, so quickly, is quite amazing. But the French might fear that by failing to take Rome's part, they were making the odds favour their enemy. As it was, they resorted to indirect means, supporting the Pope without breaking the truce. In fact it was probably precisely because it was so rapid that papal intervention was so effective. The passions inflamed by the previous conflict had not been soothed; the French were still thinking about Naples and Milan while the emperor – who was already considered as having left the world – flew into a passion against Paul IV. He had all the dispatches read to him, decided in June to postpone his journey to Spain and, spurred on perhaps by the memories of his violent clash with Rome long ago, instructed the Duke of Alva to retaliate to the preparations being made by the Supreme Pontiff. This against the advice of Philip, who wished to avoid a rupture at all costs. The impending struggle would be one brought about by old men, still fighting battles of long ago and seeking fresh pretexts to inflame old passions.

Saint-Quentin. So true was this that against all apparent logic the war which had been revived in Italy, and because of Italy, was not actually fought in the Mediterranean region at all. One reason, it is true, may have been the inactivity of the great Turkish fleets, since France without the support of her powerful ally could attempt no major offensive in the southern seas. Only a few Turkish galleys put out in 1556 and they merely joined up with several privateers and Hasan Corso for a brief siege of Oran,[209] while in 1557, the decisive year of the war, the Turks did not organize the equivalent of even this minor diversion.

[207] In the case of the Colonna family for example, whom he stripped of their estates, the Colonna being well known as supporters of Spain. Cf. his relations with Naples which were never cordial.

[208] Lo que contienen dos cartas del embaxador en Francia de 9 y 13 de julio 1556, A.N., K 1489.

[209] D. de Haëdo, *op. cit.*, f° 69 v° and 70; Jean Cazenave, 'Un Corse roi d'Alger (Hasan Corso)' in *Rev. Afrique Latine*, 1923, p. 397–404; Socorro de Oran, Simancas E° 511–513.

In December, 1556, François, Duke of Guise, had crossed the Alps with a large army: 12,000 foot, 400 men-at-arms and 800 light horse.[210] It was rumoured that he had even more troops.[211] What use could be made of this army and the Italian troops levied[212] by France's only transalpine ally, the Duke of Ferrara, nominally in command of the French forces in Italy, though in practice he left it in the hands of his son-in-law, François of Guise? The most sensible thing would probably have been to attack the Milanese. But being an ambitious man, dreaming of conquests and crowns, perhaps of that of Naples for himself, the Duke of Guise was unlikely to remain long deaf to the appeals of Paul IV, who had renounced the truce signed with the Spanish in November and extended in December, 1556, and was prodigal with promises. Simon Renard reported on 12th January that the Pope was determined to use all his 'popery', and the revenues of the Church, to pursue the war.[213] He apparently planned to return to the French Bologna and Perugia, from which cities they could inflict maximum annoyance on the Duke of Florence. It is easy to see why Guise marched his army to Rome. But on arrival, he spent a whole month in diplomatic intrigues attacking the Kingdom of Naples only on 5th April, and then without much success. In May he was obliged to assume a defensive position and in August he received orders to return to France.

Thus abandoned, the Pope was forced to sue, this time for a permanent settlement. The peace, negotiated with exemplary moderation by the Duke of Alva, was made public on 14th September.[214] The news was celebrated with great rejoicing of which two examples will suffice: in Palermo in September 'si ficiro li luminarii per la pace fatta fra la Santità del Papa Paolo quarto con la Maestà del Re Filippo Seconde, nostro Re';[215] and on 18th November[216] at Valladolid, church bells were rung, processions held and Te Deum sung.

I need not stress the importance of this peace between Spain and the papacy; it marks a turning point in western history, Rome's submission to the Habsburg yoke or perhaps it would be more accurate (for under Paul IV this submission was never total, to mention only the difficulties he raised over recognizing the newly-elected emperor in 1558) to call it

[210] E. Lavisse, *op. cit.*, V, 2, p. 167.

[211] Un hombre que se envio a Francia y bolvio a Perpiñan a los XXV de enero ha referido lo siguiente – 28th January, 1557 – A.N., K 1490. 30,000 foot, 10,000 cavalry in Piedmont; marginal note reads 'todo es mentira'. Simon Renard was better informed, Simon Renard to Philip II, 12th January, 1557, A.N., K 1490 (a total of 12,000 men).

[212] Simon Renard to Philip II, 12th January, 1557, A.N., K 1490.

[213] *Ibid.*

[214] Cavi, 14th September, 1557. Capitulación publica sobre la paz entre Felipe II y Paulo IV ortogada entre el duque de Alba y el Cardinal Caraffa. Simancas *Patronato Real*, nº 1626. Secret clauses on the fortifications of Paliano, *ibid.*, no. 1625.

[215] Palmerino B. Com. Palermo Qq D 84. The date given, 11th September, must surely be wrong.

[216] Juan Vásquez to Charles V, Valladolid, 18th November, 1557, L. P. Gachard, *La retraite* . . . I, doc. nº CXXI.

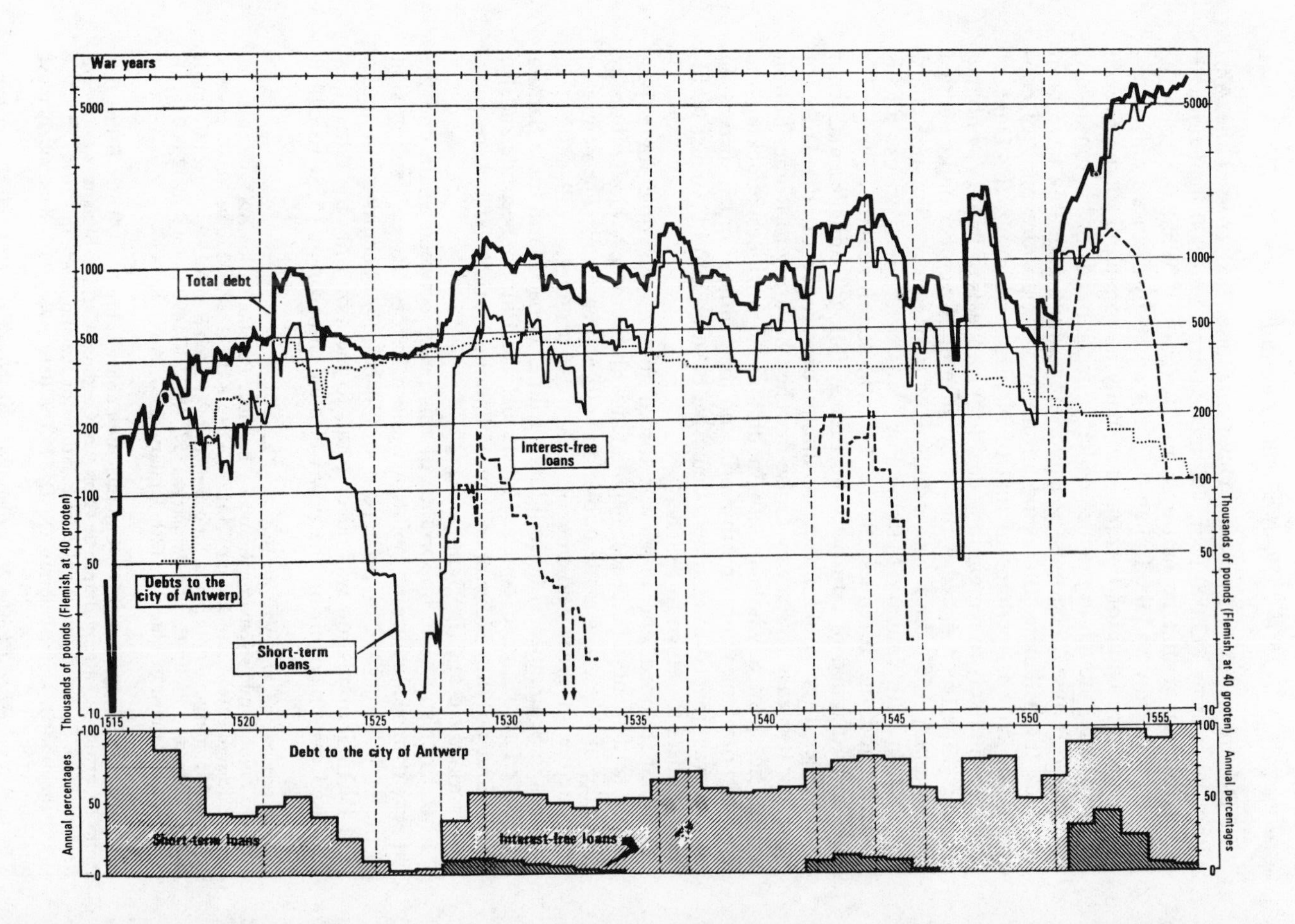

War years
Thousands of pounds (Flemish, at 40 grooten)
Annual percentages
5000
1000
500
200
100
50
10
0
Total debt
Debts to the city of Antwerp
Short-term loans
Interest-free loans
Debt to the city of Antwerp
1515
1520
1525
1530
1535
1540
1545
1550
1555

the alliance of Rome and Spain. It was an alliance which was to last until 1580–1590, for the greater protection of Catholicism and the Church,[217] for the triumph of the Counter-Reformation, which but for this alliance of temporal and spiritual forces would never have been assured.

The Duke of Guise, already back in the Milanese,[218] now had to cross the Alps again, at the news of the disaster of Saint-Quentin (10th August, 1557). Coligny had managed to slip into the fortress, the day after it had been invested by the Spanish. The army led by the Constable to relieve the town was surprised and scattered along the banks of the Somme by the bulk of the enemy forces on 10th August. The battle was followed by a massacre and the wholesale capture of prisoners, including the Constable himself. In the rear of the troops, Philip received hourly reports of the victory. 'At eleven in the evening', he wrote to his father, 'a messenger arrived from the battlefield and told us of the rout of the enemy and the capture of the Constable; at one o'clock, by another messenger, we had confirmation of the defeat, but not of the Constable's capture. I came here [to Beaurevoir] this morning, to be on the spot tomorrow. A member of the household of my cousin [Emmanuel Philibert] reports that he has seen the Constable and the prisoners of whom Your Majesty shall see the list.'[219]

Saint-Quentin taken, the king of France disarmed – what could Spain not do to his kingdom? 'Provided, that is,' noted Philip II, 'that we do not run out of money, "si no falta el dinero".'[220] The fateful word had been spoken. And indeed the situation of the Spanish treasury was catastrophic. The decree of 1st January, 1557 had announced the bankruptcy of the Spanish state. Any major project would be difficult to carry out, unless it was decided to risk everything in one last desperate gamble, by marching on Paris, against all the rules, as Emmanuel Philibert wanted to

(Opposite) Fig. 66: *Loans obtained by Charles V and Philip II from Antwerp financiers, 1515–1556*

From Fernand Braudel, 'Les emprunts de Charles V sur la place d'Anvers', in *Charles Quint et son temps* (CNRS), 1959.

There were three kinds of debt: to the city of Antwerp; to local merchants (short-term loans) and to prominent individuals (interest-free loans). The percentages are given in the second diagram – short-term loans became the dominant pattern. The oscillations of this enormous floating debt follow the fortunes of war: shaded areas represent war years. Each one leads to an immediate rise in the total debt. The war against the German Protestants is recorded in its two phases. The logarithmic scale adopted here disguises the sharp rise in the 1550's from about 500,000 pounds to 5 million pounds: the reign of Philip II was beginning. To complete the picture we need similar information concerning Medina del Campo at the very least.

---

[217] Paul Herre, *Papsttum und Papstwahl im Zeitalter Philipps II*, Leipzig, 1907.

[218] Philip II to Charles V, Beaurevoir, 11th August, 1557, autog. A.N., K 1490. See the same file for several documents on the battle of Saint-Quentin.

[219] Philip II to Charles V, see preceding note.

[220] *Ibid.*

do and as Charles V, when he heard of the victory in his retreat, was to wish too. Who knows what would have happened if they had had their way? What did happen was that by wasting time besieging small places like Ham, Le Catelet, Saint-Quentin and Noyon (Saint-Quentin having resisted after the defeat of the relieving army) the imperial forces threw away the advantage their victory had given them.

The king of France had time therefore, to prepare counter-measures, to assemble troops and to wait for Guise to arrive. Oddly enough, in the financial centres of Europe, the defeated king's credit was better than that of his conqueror. In the dead of winter, on 31st December, Guise attacked Calais and captured the town on 6th January. The English had lost their ancient stronghold in France through over-confidence and perhaps too through not accepting Spanish reinforcements in time. At any rate, the situation was now restored from the French point of view. True, on 13th July Marshal de Termes was beaten at Gravelines and the defeat was a serious one, because of the intervention of the English fleet, but the Duke of Guise[221] had at the end of June, taken Thionville; Metz was now threatened and again a victory balanced a defeat.

In the same year, 1558, there arrived in the Mediterranean a mighty armada from the East, in answer to French requests.[222] It appeared off the coast of Naples in the first week of June; on the 7th it was sighted at Squillace,[223] a small Calabrian port; on the 13th it was 'in le bocche di Napoli'[224] and sailing on rapidly[225] past its usual stopping-places. It was able to surprise Sorrento and Massa, for the inhabitants, although warned by special messenger, had not supposed the danger to be imminent. On 26th June, pillaging as it went, it was off Procida; from there it set sail for the western Mediterranean.[226] Failing to find the French galleys in the Gulf of Genoa, it went on to the Balearics, where Piāle Pasha seized the little town of Ciudadella on Minorca[227] and threw Valencia into a state of alarm, with fears of a rising among the Moriscos.[228] The French then per-

[221] Philip II to the Count of Feria, 29th June, 1558, *CODOIN*, LXXXVII, p. 68.

[222] Cesaréo Fernández Duro, *Armada española*, Madrid, 1895–1903, II, p. 9 ff. The Doge and governors of Genoa to Jacomo de Negro, ambassador in Spain, Genoa, 23rd May, 1558, A.d.S., Genoa, Inghilterra, I, 2273. On the role of the French ambassador De la Vigne, Piero to the Duke of Florence, Venice, 22nd January, 1558, Mediceo 2974, f° 124. The fleet arrived earlier than usual. Dispatch from Constantinople, 10th April, 1558, Simancas E° 1049, f° 40.

[223] Pedro de Urries, governor of Calabria, to the viceroy of Naples, 7th June, 1558, Simancas E° 1049, f° 43. It was captured on 13th, and Reggio plundered. C. Manfroni, *op. cit.*, III, p. 401.

[224] Instruction data Mag^co^ Fran^co^ Coste misso ad classem Turchorum pro rebus publicis, Genoa, 20th June, 1558, minute, A.d.S., Genoa, Costantinopoli 1558–1565, 1–2169. C. Manfroni, *op. cit.*, III, p. 401, note refers I think, to another copy of this instruction.

[225] It sailed past Torre del Greco, Cardinal de Siguenza to H.R.H., Rome, 16th June, 1558, Simancas E° 1889, f° 142, A.E. Esp. 290, f° 27.

[226] Don Juan Manrique to H.R.H., Naples, 26th June, 1558, Simancas E° 1049, f°41.

[227] C. Fernández Duro, *Armada española* . . ., II, p. 11.

[228] *Ibid.*, p. 12.

suaded it to return to Toulon and Nice, but on arrival, Piāle Pasha refused to take any action against Bastia. His refusal had several causes: the news of Gravelines, the epidemic which was decimating his oarsmen, forcing him to tow some of the galleys, but above all, and this we know for a fact, the Turkish admiral had been richly bribed by Genoa.

He went home, followed at a respectful distance by the galleys in the service of Spain, paying no heed to the protests of the French. The raid although costly to Christendom, had not been of major significance in the war.

So with the Roman problem settled since September 1557, the two enemies were able once again to negotiate for peace. In a word, they were back to the 1556 situation, with two differences: on 21st September, 1558, Charles V died at Yuste and Philip II's presence in Spain thus became more necessary than at any time in the past (as we shall see); then on 17th November,[229] Mary Tudor herself died, bringing to an end the union between England and the Spanish Empire so dangerous to France. England was once more in the throes of a succession crisis. All eyes once more were on the North.

The treaty of Cateau-Cambrésis. The English question may have had more influence than is generally suspected on the negotiations which were to end with the treaty of Cateau-Cambrésis on 2nd and 3rd April, 1559.

There can be no doubt that financial exhaustion forced the adversaries to a settlement. Besides, it had already been demonstrated that neither side could expect a military victory. On the French side there were also weighty domestic problems. If one takes literally all the diplomatic reports coming out of the kingdom, it would be difficult to find a more discontented country, a more impoverished aristocracy, or a population more distressed than in France. The description may be exaggerated, but it is not substantially incorrect. The country was also seething with Protestant ferment, against which Henri II's government was resolved to take violent action: of the two signatories, he was undoubtedly the more 'Catholic', the more determined to strike down heresy. To do so, he needed peace. Finally there were the dynastic feuds so conspicuous during the reign of Henri II, the political battles between the houses of Guise and Montmorency, which were soon to add fuel to the wars of religion, themselves often no more than struggles for power. Venetian reports stressed 'that if there is peace, the Constable is the most important man in France; if there is war he is a prisoner, deprived of all importance'[230] as was indeed only too evident.

These points have all been made before: long ago by Alphonse de Ruble,[231] as well as in the brilliant studies by Lucien Romier.[232] But we

[229] G. Turba, *op. cit.*, I, 3, p. 81, note 3.

[230] Marin de Cavali to the Doge, Pera, 16th December, 1558, A.d.S., Venice, Senato Secreta, Cost., Filza 2 B, f° 102.

[231] *Le traité de Cateau-Cambrésis*, 1889.

[232] *Les Origines politiques des guerres de religion*, Paris, 2 vols., 1913–1914.

may still add a little to their accounts. The treaty of Cateau-Cambrésis has been considered by French historians, as it was by certain contemporary figures (Brissac,[233] the man responsible for French Piedmont in particular) as a disaster for France. It is perhaps worth examining the other point of view. The chief advantages France derived from this peace were two marriages: the marriage of Emmanuel Philibert to Marguerite, and that of Philip II to Elisabeth of France, the little girl who was to be the 'queen of peace' in Spain. We tend nowadays to underestimate such advantages. Yet it is undeniable that sixteenth-century politics were above all dynastic: marriages were important transactions, the fruit often of lengthy calculations and the occasion too for much duplicity, prevarication and deception. The Spanish marriage was a brilliant success for France if only because it eliminated the possibility of another match. Elizabeth of England could have been Philip II's wife if she had wanted to: the offer was put to her in all sincerity in October, 1558, but Elizabeth declined.[234] The marriage with a French princess, apart from its other benefits, was a guarantee against yet another alliance between England and the Spanish empire.

On the debit side, the treaty confirmed that France had abandoned all claim to Italy and given up Savoy and Piedmont, provinces which had close ties with the kingdom and had been easily assimilated, thus creating a barrier against any future intervention by France in the Peninsula; finally, by surrendering Corsica in spite of formal promises to the contrary, France lost a major strategic base in the Mediterranean. But France had only given up part of Corsica – the rest was not hers to give. And the treaty also left her with five strongholds in Piedmont, including Turin. The immediate future was thus safeguarded. The positions were given up on 2nd November, 1562, it is true.[235] But even after this date, France had a bridge-head on the other side of the Alps. Hence the anger of the Duke of Nevers[236] when he learnt in September, 1574, that Henri III, passing through Turin, had presented 'Monsieur de Savoie' with the two fortresses of Pignerol (Pinerolo) and Savillian (Savigliano), given to France in 1562 as compensation. Henri III's only possession beyond the Alps were now the undefendable towns and villages of the marquisate of Saluzzo. 'And it must give me much distress', added the duke, 'fearing that the universe will be greatly astonished to see that hardly has Your Majesty entered upon his kingdom, before he is seeking to dismember it and, what is more, close the door for ever on [going] to Italy, after seeing with his own eyes the beauty of that land.' As for poor Italy, 'unhappy to see all means of delivery fading away . . . [she] will have just occasion to

[233] Guy de Brémond d'Ars, *Le père de Mme de Rambouillet, Jean de Vivonne, sa vie et ses ambassades*, Paris, 1884, p. 14; Lucien Romier, *Origines . . ., op. cit.*, II, Bk. V, Chap. II, p. 83–86; B.N., OC 1534, f° 93 etc.

[234] Elizabeth to Philip II, Westminster, 3rd October, 1558, A.N., K 1491, B 10, n° 110 (in Latin).

[235] Baron A. Ruble, *op. cit.*, p. 55.

[236] To Henri III, 25th September, 1574, copy, Simancas E° 1241.

bewail her misery, seeing that she will be submitted to Spanish power for all time'. If it was still possible to 'close the door' on Italy in 1574, fifteen years after the treaty of Cateau-Cambrésis, it was perhaps because the sacrifices of 1559 were not as great as is sometimes supposed.

Unfortunately it was not only Italy which was sacrificed, but Savoy and above all Piedmont, a state partly integrated into France, linked with the Swiss cantons, reaching the sea by the narrow corridor of Nice and Villefranche, and touching the plains of northern Italy on the other side of the mountains. This was certainly not an indissoluble part of Italy, but a land apart, with its own customs and traditions, even to the eyes of an Italian like Bandello,[237] who can hardly be suspected of partiality in this case. France under Henri II seems to have given up Piedmont rather unthinkingly, in the haste to reach a settlement, with a clear lack of awareness of its possibilities and also with unforgivable cruelty. It also coldly abandoned the Sienese to Cosimo de' Medici and Corsica to Genoa. The Sienese *fuorusciti* in vain tried to buy their freedom from Philip II with gold.

However the treaty of 1559 concealed a calculation by France. For did not Henri II's zealous pursuit of heresy, both inside and outside his own territories, form part of a manœuvre against England? When Mary Tudor died in November, another Mary, Mary Stuart, married to the French Dauphin on 24th April, 1558,[238] had from the dynastic point of view, a strong claim to the English throne. The more so since at the same time Elizabeth was progressing prudently, but visibly, towards Protestantism. Her moves were causing concern at Rome. Philip II on the other hand, tried hard to prevent the excommunication of the young queen: it would open the path to a French invasion, the notion of which was no secret: even poets spoke of it, Ronsard in April, 1559 in an ode of celebration addressed to Henri II, and before that du Bellay, just after the death of Mary Tudor, in an extremely explicit sonnet.[239]

Nothing better demonstrates the importance of the North and the English affair than a long memorandum placed before Philip II in June, 1559,[240] which disturbed him enough to make him postpone his trip to Spain. This unsigned document, which Philip sent to his sister Joanna, regent in his absence of the Spanish kingdoms, was no doubt the work of the non-Spanish advisers of the prince. It stated the case, in thirty-four clauses, for the king's continued presence in Flanders, heart of the North. The French were planning an invasion of England. 'If England were lost, it cannot convincingly be denied, though some would dispute it, that these lands of Flanders themselves would be in imminent danger. And the loss

[237] *Op. cit.*, VII, p. 198, 205. [238] A. d'Aubigné, *op. cit.*, I, p. 41.

[239] 'Ils veulent que par vous la France et l'Angleterre/Changent en longue paix l'héréditaire guerre.'

[240] Apuntamientos para embiar a España (undated, May–June, 1559), Simancas E° 137, f°s 95–97. A copy of this important document is to be found in A.E., Esp., 290. On the meeting of the principal persons 'di qsti paesi' and their desire, on account of the 'garbuglio' of England and Scotland, that the king spend the winter in Flanders, Minerboti to the duke, 2nd July, 1559, A.d.S., Florence, Mediceo 4029.

of England, within a short space of time, is considered certain for many reasons.' There were the claims of the French Dauphin, the weakness of the English kingdom, its divisions, the poor state of its defences, the English Catholics' need for a protector and the facility with which France, using her navy and Scotland as a base, could launch an invasion, not to mention the fact that the Pope could deprive the present queen of her crown. Clearly the king could not consider supporting the Church's enemies in England, for moral reasons; and if he were to do so, he would find himself opposed by 'the majority of the island's population' (a claim which tells us that in the Netherlands at least, England was thought to be a predominantly Catholic country). Could he stand by and watch the king of France carry off such a momentous undertaking? The latter would in all likelihood formally proclaim and maintain the peace under his own name, meanwhile entrusting the expedition to the Dauphin, in other words pursuing his own ends despite the treaty he had signed. But if Philip were to remain in person in the Netherlands, the French king would not dare to attack.

Staff papers should never be taken literally; but there is evidence that this plan was more than a project in the air. If Philip II did not wish to go through France on his way to Spain, if he evaded the blandishments being prepared for him, it was no doubt to avoid being implicated in some scheme of this nature. The Duke of Alva represented Philip at the marriage ceremony at Notre-Dame in Paris. 'The French', he wrote in code to the king, 'seek to display great amity towards Your Majesty in their conversation. Those who surround the king cannot say three words without two of them mentioning the love and friendship the Christian King bears Your Majesty and the aid he will bring him in all his enterprises. It is perhaps the truth, for it corresponds to reason. It is also possible that they are offering to participate in Your Majesty's enterprises in the hope of preventing him from causing the failure of any of their own.'[241] Thus suspicions are voiced at the very moment when the king of France, as first earnest of an *entente cordiale*, was offering to send his galleys on the expedition which Philip was believed, in France at any rate, to be planning against Algiers. The suspicion is given plainer utterance in another letter from the duke[242] in which he expresses surprise tinged with scorn that everyone at the French court, down to the merest equerry, knows the content of the secret debates of the Conseil d'État, tells anyone who will listen that France and Spain could between them impose their laws on Christendom and that 'if Your Majesty were to help the Most Christian King in his enterprise against England, he might aid Your Majesty to become master of Italy'.[243] However, he says in substance in a letter written in July and countersigned by Ruy Gómez, the French should not be allowed to establish themselves in England. To become involved in their schemes would be both risky and dangerous,

[241] The Duke of Alva to Philip II, Paris, 26th June, 1559, A.N., K 1492, B 10, f° 43 a. [242] *Ibid.*, June, 1559, f° 44. [243] *Ibid.*

'bearing in mind what happened at Naples'. 'It is our firm opinion that Your Majesty should let it be known, as of now and clearly, even if it is not his intention . . . that out Lord Prince (Don Carlos) will come to the Netherlands as soon as Your Majesty leaves, so that both French and English will know that Your Majesty is not leaving this position unprotected.'[244]

Elizabeth for her part was disturbed by French activity in the ports of Normandy and was preparing to take action both in Scotland and against France. The conspiracy of Amboise of 1560 had religious and social significance, but it was not unconnected with foreign affairs.[245] It is true that by now the France of Henri II had given way to a much weaker regime. The king who signed the Treaty of Cateau-Cambrésis had died as the result of an accident on 10th July, 1559,[246] and his death, with the troubles it brought in its wake, made it impossible, at least temporarily, for France to take a major part in international events.

This was an unfortunate blow for France. But if we are to weigh up the results of the treaty of 1559, we should include, to offset the familiar list of losses so often detailed by historians – the surrender of Italy and loss of Corsica – the hope of gaining England, a hope which was for a brief space close to being realized, but which the future was to dash to the ground.

Philip II's return to Spain. Philip II had never liked the North. As early as 1555 he had had ideas of leaving his father in Flanders and returning to Spain.[247] Mary of Hungary[248] was beside herself with indignation – were the 'fogs' of the North suitable for old men and the southern sun for the young? In 1558, Philip, who had not changed his views, thought of asking his aunt, who had accompanied the emperor to Spain in the autumn of 1556, to take his place in the Netherlands. But Mary of Hungary, after finally agreeing,[249] died in 1558. It was not until 1559, four months after the signing of the treaty of Cateau-Cambrésis, and one month after the death of his father-in-law, Henri II, that Philip was able to make the trip.

Biographers and historians have paid little attention to his decision.[250] The writer of the second part of Mariana's history[251] does not even mention

[244] Paris, 11th July, 1559, *ibid.*, f° 49.

[245] J. Dureng, 'La complicité de l'Angleterre dans le complot d'Amboise", in *Rev. hist. mod.*, Vol. VI, p. 248 ff; Lucien Romier, *La conjuration d'Amboise*, 3rd ed., p. 73; E. Charrière, *op. cit.*, II, p. 595.

[246] Ruy Gómez and the Duke of Alva to Philip II, Paris, 8th July, 1559, A.N., K 1492, f° 48, Henri II is beyond hope.

[247] L. P. Gachard, *op. cit.*, I, p. 122 ff, 27th May, 1555.

[248] *Ibid.*, p. 124, the queen of Hungary to the bishop of Arras, 29th May, 1556.

[249] *Ibid.*, I, p. XLI ff; p. 341–352; II, p. CXXXVII ff, p. 390.

[250] *Historiae de rebus Hispaniae* . . ., Vol. I (the only one published) of the later series, by Father Manuel José de Medrano, Madrid, 1741.

[251] Let me add that this incident has been inaccurately narrated and that the chronology given is usually incorrect. Philip II embarked at Flushing on 25th August and disembarked on 8th September at Laredo. According to Campana, the king sailed on 27th, while Gregorio Leti gives the sailing date as 26th . . . Modern historians, from Robertson and Prescott on, have reproduced this traditional version.

it and his narrative jumps without explanation from the Netherlands to Spain. Yet with this journey Philip II's personal empire, a unit which was to remain stable over the years, finally emerged from the inheritance of Charles V. At the same time a new European order was being established. In 1558, two essential positions had been lost, without a struggle, by the new sovereign: the death of Mary Tudor and his father's abdication from the empire had deprived Philip of England and the empire. Of the two events, one, as we have seen, was inevitable anyhow: against the combined antagonism of Protestant Germany, Ferdinand and Maximilian, Philip II had little chance of success. But at almost the very moment when Germany was coalescing into a closed and foreign bloc against him, a pure accident, the unforeseen death of Mary Tudor in November, destroyed the alliance of England and Spain and put an end to the vision of an Anglo-Flemish state centred round the North Sea.

One has only to think of what Philip II might have been – ruler over both England and the German empire – to appreciate the immense significance of these events. The title of emperor, even stripped of all substance, would have avoided the irritating disputes about precedence: it would have reinforced Spanish authority in Italy and conferred upon the war against the Turks, both on the plains of Hungary and in the Mediterranean, a single command. Moreover, with the support or neutrality of England, the war of the Netherlands would have gone very differently, and the struggle for control of the Atlantic, which dominated the second part of the century, would not have ended in disaster. But above all, who could fail to see that the centre of gravity of Philip II's empire was being shifted from the North to the South by force of circumstances? The Peace of Cateau-Cambrésis, by confirming the Spanish presence in Italy, helped to make southern Europe the focus of the foreign policy of the Catholic King at the expense of other more urgent and possibly more fruitful endeavours.

Philip II's return visit to Spain in August-September, 1559 put the finishing touches to this process. From now on, Philip was to remain in the Peninsula, a prisoner so to speak in Spain. True, contrary to the legend which pictures him walled up in the Escorial, he travelled a great deal even after this,[252] but always within the Peninsula.

Gounon-Loubens,[253] whose book though written long ago is still useful, criticizes Philip for not moving his capital from Madrid to Lisbon after the conquest of Portugal, for not realizing the importance of the Atlantic. At first sight it looks as if the departure from Brussels in spring, 1559 was an error of the same sort. Philip II quite deliberately placed himself, for the entire duration of his reign, outside the centre of Europe. His policies were forced to contend with the unfavourable arithmetic of distances: it is an easy matter to demonstrate statistically that news reached Brussels long before Madrid, whether it came from Milan, Naples or Venice, not

[252] For a summary of his travels see C. Bratli, *op. cit.*, p. 188, note 280 and p. 101–102.
[253] *Essai sur l'administration de la Castille au XVIe siècle*, 1860, p. 43–44.

to mention Germany, England or France. Spain and Spain alone became the heart of Philip's dominions, and from this powerful heart flowed now swiftly, now sluggishly, the vital force behind his policies. It was from Spain that the king was now to view events and judge them; in the moral climate of Spain that his policy was to be shaped; with Spanish interests that his entourage was preoccupied, and by Spaniards that he was surrounded.

For the king's return had its effect on the composition of his circle of advisers. Similarly in Charles V's day, the emperor's visits to his various dominions, even when they were of short duration, had caused different men to move into or out of favour and prominence. In 1546, speaking of Perrenot, Bernardo Navagero, the Venetian ambassador, noted in passing:[254] '. . . as long as the emperor was outside Spain and stayed in Germany or Flanders, his credit rose remarkably'. On leaving the Netherlands, Philip cut himself off from his Flemish and Burgundian advisers: the separation had a certain effect as is demonstrated by the case of Granvelle, Perrenot's son. True, the bishop of Arras, who had travelled all over Charles V's empire, remained in the Netherlands in an enviable position: he was Philip II's accredited representative at the court of Margaret of Parma. But his situation no longer compared with the position he had occupied in the councils under the emperor and under Philip before the latter's departure in 1559. For twenty years then, he remained far from his sovereign. The importance of the ultimate meeting of the two men is well known; a period of imperialist expansion followed the arrival of Granvelle in Madrid in 1579.[255]

By returning to Spain, Philip delivered himself up for many years into the hands of the Spanish advisers. He gained the inestimable affection of his Peninsular kingdoms. After the interminable wanderings of the emperor, his continuous presence was received by Spain as a blessing, which stirred the country 'to its inmost parts'.[256] 'Vast and numerous though the king's dominions may be', wrote the Duke of Feria in 1595, 'I doubt whether he reigns anywhere more completely than in the hearts of Spain.'[257]

There certainly seems to have been nothing improvised about the voyage itself, continually being planned and as often postponed. Much has been made of Philip II's personal feelings: loving the Netherlands as little as they loved him, 'disgusted with this place of sojourn', he is said to have hastened to leave them never to return.[258] This is rather a bold assumption: only his haste is known for certain. The French ambassador, Sébastien de l'Aubespine, wrote to his royal master[259] from Ghent on

[254] E. Albèri, *Relazioni*, I, 1, p. 293 ff, July, 1546.

[255] M. Philippson, *Ein Ministerium unter Philipp II. Kardinal Granvella am spanischen Hofe, 1579–1586*, 1895.

[256] Cf. the article by C. Pérez-Bustamante, 'Las instrucciones de Felipe II a Juan Bautista de Tassis' in *Rev. de la Biblioteca, Archivo y Museo*, Vol. V, 1928, p. 241–258.

[257] Simancas E° 343.

[258] Louis Paris, *op. cit.*, p. 42, note 1.

[259] *Ibid.*, p. 42.

27th July: 'It is unbelievable how this prince makes haste and urges on all speed in order not to be delayed or prevented from leaving.' Elizabeth's ambassador reported the rumour, whispered in Spanish circles, that the king would never return to the Netherlands, and Margaret of Parma spoke of 'His Majesty's desire to arrive in Spain'. But this desire had serious roots. Philip's Spanish advisers at Brussels had encouraged it since 1555, against the 'Burgundian' party, that of Granvelle, Courteville, Egmont and the Prince of Orange. They had personal reasons too, no doubt: they wanted to return to their homes, their own ways and their own interests, some perhaps to take advantage of the massive sale of crown lands then taking place in their native country. But they were also thinking of the interests of Spain.

The prolonged absence of the ruler had led to the gradual slackening of the governmental machine. The Spanish dominions had three capitals and three governments: Brussels, from which the king conducted wars and controlled the strings of diplomacy; the monastery at Yuste, where Charles V had very soon, despite his original resolution, taken up the reigns of government again; and Valladolid, where Princess Joanna heard the advice of the councils and shouldered the essential burden of administration. The division of powers between the three capitals was most unequal and communication between them, in spite of the large number of couriers, imperfect. It is the subject of constant complaint in all official correspondence and this lack of coordination was to produce rapid consequences. All decisions made at Valladolid had to be ratified by the sovereign – the delays caused by this incredible detour can be imagined. Spain was virtually ungoverned. The death of Charles V at Yuste in September, 1558, made matters even worse, Princess Joanna being clearly unequal to the situation.

It was in the euphoric atmosphere of victory that Philip left Brussels. Representatives from all over Italy flocked round him, offering money and presenting petitions, Cosimo de' Medici asking to keep Siena, the Grand Master of Malta asking for the orders necessary for an expedition against Tripoli, the Republic of Genoa seeking detailed rulings on the recovery of Corsica, the Farnese seeking to drive out the Duchess of Lorraine and to ensure that the government of the Netherlands was safely in the hands of Margaret of Parma. Amid the receptions and Te Deums, Philip II distributed his final favours to the lords of Flanders and defined the powers of the new governor. On 11th August he was at Flushing. There he waited two weeks for a favourable wind, spending the time visiting island after island, castle after castle. Finally, on 25th August, the royal fleet set sail.

We have a very detailed account of the voyage in the *Journal* of Jean de Vandenesse,[260] with further information in several letters written to

[260] Here is the brief account given by Jean de Vandenesse: '. . . on Thursday, St. Bartholomew's day', he wrote, '23rd August, His Majesty supped at the said Sobourg [Soubourg]; and after supper came to Flushing. And at about eleven in the evening

Margaret of Parma by Ardinghelli,[261] the tutor of the young Alexander Farnese: a hostage of Spanish policy, whom his mother had agreed to allow to be brought up in Spain, he accompanied the king on this voyage. Let us point out in passing that the traditional account of this voyage – to be read in Watson, Prescott or Bratli – of Philip's romantic landing

went aboard his ship, remaining at anchor until late on Friday when he set sail. On the said day, at about nine o'clock in the morning, the princes and lords of the Netherlands took their leave of the King and of all his company; which they did not do without regret, sighs and tears, pitiful to see, seeing their natural King abandoning them. . . . And about midday there arrived the Duchess of Parma, accompanied by the prince her son and several other lords, to take leave of His Majesty. At vesper hour, His Majesty set sail and passing with moderately fair winds through the channels and dangers of the banks in sight of Dunkirk, Calais and Dover, sailed to the channel near the isle of Vicq [Wight]. On entering the sea of Spain, we were becalmed and were fifteen days at sea. And on the eighth September, the day of our Lady, His Majesty and several other ships entered the port of Laredo, where His Majesty disembarked and went to hear mass in the church and slept there that night, which was a Friday, and it was impossible that day to disembark as much as we should. The hookers which are bulky vessels and some other ships were unable to enter port so soon. And on the Saturday, His Majesty left Laredo about one hour after midday to go to Colibre [Colindres] which is half a league further inland than Laredo. And that same hour there arose such a violent storm both on sea and land that the ships which were at anchor in the harbour, could not help but capsize and perish and it was pitiful to see the loss of ships, passengers and baggage. And the others were forced to trust to fortune on the sea. On land trees were uprooted and tiles blown from the roofs of the houses and it lasted all that day and all night . . .' in L. P. Gachard and Piot, *Collections des voyages des souverains des Pays-Bas*, 1876–1882, IV, p. 68 ff.

[261] Here is a summary of Ardinghelli's version: Ardinghelli followed Philip II's movements in Zeeland and kept in touch with him. On 23rd August, he contacted Margaret of Parma so that she could take her leave of Philip. Having embarked on 25th, he took advantage on the way of passing ships to send news of the prince's health. On 26th August, between Dover and Calais, he reports that all is well and that pilots have been taken on aboard to guide the ships safely through the sandbanks. Philip II will not dismiss these precious pilots, he writes on 27th, before the isle of Wight. The king himself is perhaps responsible for the slow progress: a fair wind rose, but His Majesty did not want to be separated from the hookers, otherwise the other ships could have been thirty leagues ahead. 'The voyage is bound to be successful', he writes, 'all the dangerous places are past and today week, we hope to be in Spain.' A small Spanish boat carried another letter from him, dated 31st. The weather continues fine. 'We shall be out of the Channel tonight. . . .' Then there is a gap in the correspondence until 8th September, on which day Ardinghelli writes: 'Praise be to God who brought us all safe and sound to this port of Laredo. After leaving the English channel, the weather was so changeable that it misled the mariners more than once and we were hindered sometimes by calm seas and sometimes by contrary winds, but thank God there were no storms. Last night a north wind began to blow and has brought us to land this night to our great joy. . .'. From Laredo again (he did not leave the port until 14th) Ardinghelli writes on 10th: 'Last Saturday, (9th September) in the middle of the day there sprang up a storm on the sea so terrible that we thanked God we were ashore. The ships in the harbour survived only with much difficulty . . . three of them capsized but no men or goods, were lost. The hookers which were coming behind, must have been in great danger and until now we have no news of them and are therefore in much fear on their behalf. . . .' However on 13th, the 'Flanders fleet' arrived, 'without having suffered any loss in the storm'. They were joyfully welcomed: on board were the servants and belongings of the lords accompanying Philip II. The Ardinghelli letters are in the Farnese archives in Naples, Spagna fascio 2, from f° 186 to f° 251.

at Laredo, is wrong from start to finish. The king did not arrive alone, at risk of his life in a tiny boat, while behind him in the ocean, his entire fleet was sinking, carrying his treasure, his thousand noble attendants and their precious belongings to the bottom of the sea. There was indeed a storm, which tossed the heavy hookers following the convoy, but a letter written by Philip II himself, on 26th September, 1559, says that only one ship is missing.[262] As for the king, he was already ashore and probably had been for at least twenty-four hours. The whole incident was invented, perhaps by Gregorio Leti, who gives a detailed account of this so-called disaster, 'a veritable forecast of all the disgraces and misfortunes which thereafter befell the king'.[263]

4. SPAIN IN MID-CENTURY

What kind of Spain did the king find when he landed? One thing is certain: it was a Spain eager to welcome him back.

For years his return had been clamoured for: by the Regent and her councils since 1555;[264] by the Cortes of Castile, which met in 1558;[265] by Charles V himself, who esteemed it indispensable and by all the officials of the Peninsula. In Francisco Osorio's correspondence,[266] almost every page contains an allusion to the return of the king, which would settle everything, he says, when the news is bad, and make things even better, when the news is good. When at last the great tidings reached him, 'the joy and contentment caused by the peace and the longed-for return of Your Majesty to these kingdoms are so great', he writes on 17th May, 1559, 'that it is beyond my power to express them!'[267]

The situation in Spain was serious indeed. Although the country had been spared the direct experience of war, it had provided an endless supply of men, ships and money – especially money. Socially, economically and politically, the country was plunged into chaos, racked by deep discontents aggravated even further by a religious crisis which was taken very seriously.

The Protestant scare. In 1558,[268] there had been discovered in Seville, Valladolid and one or two smaller centres, several 'Protestant' com-

[262] Philip II to Chantonnay, 26th September, 1559 (not 1560 as it is described on the listing), A.N., K 1493, B 11, f° 100 (minute) '. . . of the ships accompanying the armada on which I came to these kingdoms, only one is missing and has not yet arrived. It is owned by one Franscisco de Bolivar of Santander. It was carrying the possessions of the regents of my council for Italy as well as of some of my secretaries and other servants, as you will see from the attached memorandum.' There were reports that the ship had reached La Rochelle. On the lost ship, see L. P. Gachard, *Retraite. . ., op. cit.*, II, p. LVII.

[263] G. Leti, *Vita del Catolico re Filippo II*, 1679, I, p. 135.

[264] L. P. Gachard, *Retraite . . ., op. cit.*, I, p. 122 ff.

[265] *Actas de las Cortes de Castilla*, 1558, I.

[266] *CODOIN*, XXVII. [267] *Ibid.*, p. 202.

[268] L. P. Gachard, *Retraite . . ., op. cit.*, II, p. 401 ff, but see in particular the standard authorities, E. Schäfer and Marcel Bataillon; E. Albèri, *op. cit.*, I, III p. 401–402.

munities: we may as well call them that, although the expression is misleading; after all that is what they were thought to be. The news was received by Charles V, as it was by his son, with consternation, so much so that it has sometimes been suggested that the voyage in 1559 was connected with this outbreak of 'Protestantism'. The second auto-da-fé, on the Plaza Mayor of Valladolid, in fact took place one month after the landing at Laredo.[269] The Danish historian Bratli is merely following tradition when he writes that Philip II having received the bad news from Seville and Valladolid, 'longed only for the moment when he could set foot in Spain once more'.[270]

Did the spectacular repression organized by the Inquisition really mean that the Protestant movement in Spain was sufficiently widespread to constitute a danger? Marcel Bataillon in *Érasme et l'Espagne*[271] paints a very different picture, showing that the so-called 'Protestants' of 1558 were essentially the heirs to a spiritual tradition with roots far back in Spanish religious life and owing nothing to Lutheran doctrine. The spiritual flames rising from Valladolid and Seville, when looked at closely, are multicoloured, like fires fed by the dust of several different minerals: some of them very rare and precious. Who can say what was contributed to this blaze by the *conversos*, Augustín Cazalla or Constantino for example – with their mystical Jewish traditions; or how much was illuminism, that strange metal produced exclusively in Spain, the purified version of which was her major mystical strain; how much of the alloy was made up of Erasmian notions of a religion of the spirit, oriented towards the inner life? The years 1520 to 1530 brought to the Peninsula, at that time still open to the spiritual influences of the outside world, first the ideas of the Erasmians, then those of the Valdesians. Twenty years later, these ideas were still alive, transposed but recognizable: and while a smattering of Lutheran thought may have combined with them, it is certainly not true that there was an organized Protestant cult in Spain, a dissident religion like that of the French Huguenots. The Spanish heresy, if it was opposed in some respects to Catholic tradition, tended more towards attempting to save not only the spirit, but the Church and its institutions, in short to preserve orthodoxy. Such at any rate, were its hopes.

So why the repression of 1559, if nothing or virtually nothing new had been introduced to the centres of the new religious thought? According to Marcel Bataillon,[272] the new element was the method of repression: Catholic intransigence had become aware of itself, eager to attack as a proof of self-confidence, to inspire terror by example. Gone now was the irenical policy of the emperor, and the uncertainty of a strained situation

[269] Juan Ortega y Rubio, *Historia de Valladolid*, 1881, II, p. 57 (first auto-da-fé); p. 58 (second auto-da-fé); p. 64: half the victims had been saved until the king's arrival. [270] C. Bratli, *op. cit.*, p. 93.

[271] P. 555 ff. See review by Lucien Febvre, 'Une conquête de l'histoire: l'Espagne d'Érasme', in *Ann. d'hist. soc.*, Vol. XI, 1939, p. 28–42.

[272] *Op. cit.*, p. 533 ff. Should one also take into account the economic depression, which was a poor counsellor? See above, p. 897–898.

which obscured all the dividing lines and confused positions. Protestant intransigence had made things clearer. After 1555, after the success of the reformed church in Germany and the abdication of Charles V, the two sides took up unequivocal positions; merciless repression was introduced first in Italy, then in Spain, independently of each other as it happened; the Spanish Inquisition was an autonomous body and the relations between Philip II and Paul IV were far from amicable, although both men were drawn in the same direction. The situation developed rapidly. The Spain to which Philip returned had already gone over to the Counter-Reformation and repression; this was the doing not of the king himself, but of his times, of the events occurring all over Christendom, of the rise of Geneva and the response of Rome, a vast spiritual conflict into which Philip II was drawn, without having in any way created it. However, it was no accident that the king was present on the Plaza Mayor on 8th October, stressing by his presence the exemplary punishment of the 'Lutherans'.

We should not after all underestimate Philip's anxiety: having witnessed the course of events in Germany and France, he had much to fear in 1558. But by 1559, even before his return, he had realized that the problem was less serious than he had thought, as his correspondence with the Infanta Joanna shows after this date, by its lack of concern for and infrequent mention of the religious question. On 26th June,[273] acknowledging the arrival of the long description of the first auto-da-fé in May, he expresses his hope that this way 'a remedy will be found for the great evil that has been sown'. His tone is calm. 'Tan gran mal como estava sembrado': but the harvest had never been allowed to ripen.

The persecution meant the end of heresy in Spain. Perhaps this easy victory (after all many movements have resisted persecution) can be explained by the fact that Erasmianism or Protestantism were foreign transplants in Spain – grafts which had 'taken', had even put out shoots and flowers, but for how long? In the history of civilization, fifty years is a very short time. The ground was unsuitable and the tree did not lend itself to grafting. In the end all that remained of this 'Protestantism' was that part which could be assimilated into the Spanish mystical tradition, the refuge of personal prayer, the direction taken by St. Teresa and St. John of the Cross.

Moreover, the movement had never been popular, indeed the reverse. The Archbishop of Toledo declared in May, 1558 that the people did not appear to have been contaminated.[274] The prisoners had had to be transferred to Valladolid after dark, for fear that the masses and the children might stone them, so great was public indignation against them.[275] The

[273] Simancas E° 137, f° 123 and 124.

[274] Luis Quijada to Philip II, 1st May, 1558, ed. J. J. Döllinger, *op. cit.*, p. 243.

[275] Memorandum from the Archbishop of Seville to Charles V, 2nd June, 1558, published by L. P. Gachard in *Retraite . . .*, *op. cit.*, II, p. 417–425: 'God be praised', writes Vásquez to Charles V on 5th July, 1558, 'the harm is less than was thought', *ibid.*, p. 447–449.

accused were only a small group of men, an elite of humanists and mystics, an elite too of Spanish noblemen, whom the Grand Inquisitor, unlike his predecessors, did not spare in 1558.

This must be the explanation of the report spread by the Venetians, quite erroneously it seems, that 'under cover of religion, several uprisings had been organized with the connivance of the great lords'.[276] An even more explicit account appears in a letter written by the bishop of Dax from Venice, in March, 1559,[277] in which he reports that 'in the last few days, there have been rumours at St. Mark's which have lately been confirmed, that in Spain four of the greatest princes of the kingdom have risen in favour of the Lutheran heresy, in which they are so determined . . . that they constrain all their vassals by force to embrace their cause; if King Philip does not take immediate steps to remedy this, he will be in danger of being the weaker party'. But then Venice, like Rome, was a breeding-ground for false rumours: the cardinal of Rambouillet wrote one day to Charles IX, that 'the news which is sent from here [Rome] to Venice and from Venice here, is taken no more seriously in Italy than palace gossip in France'. It does not seem likely that Spanish 'Protestantism' had political repercussions. But the confusion was made possible by the existence in Spain, alongside religious dissidence, of political discontent which also caused her rulers much anxiety.

Political discontent. Spain under Philip II is usually spoken of as a unified country. The term requires some qualification. Centralization certainly increased during this long and authoritarian reign, but popular liberties in 1559 were continually being nibbled away, the laws remained unchanged and the memory persisted of past rebellions. Royal authority was neither unlimited nor unopposed. It had to contend with the *fueros*, with the fabulous wealth of the clergy, the independence of an opulent aristocracy, the sometimes open defiance of the Moriscos and the disobedience of government officials. The years 1556–1559 seem to have witnessed a serious diminution of respect for the state, what might be called a crisis of insubordination.

This was no open rebellion, but a wave of discontent and disaffection visible in small details of the kind collected by the historian Llorente,[278] which are meaningful only when they are laid side by side. When Charles V 'bowed down with laurels, years and infirmities', landed in Laredo in 1556, he found only a few gentlemen waiting to meet him; the old emperor was hurt and surprised.[279] A little later, Charles V's sisters, Elizabeth and Mary, queens of France and Hungary, were travelling in the Peninsula; on the journey from Jarandilla to Badajoz, several noblemen, having been summoned to attend them, did not reply to the convocation and did not trouble to excuse themselves.[280] A little while before, the

[276] Report by Marcantonio da Mula, E. Albèri, *Relazioni* . . ., I, 3, p. 402 ff.
[277] 6th and 11th March, 1559, E. Charrière, *Négociations* . . ., *op. cit.*, II, p. 563.
[278] 'La primera crisis de hacienda en tiempo de Felipe II' in *Revista de España*, I, 1868, p. 317–361. [279] *Ibid.* [280] *Ibid.*

two queens wishing to stay at Guadalajara, had asked the Duke of Infantado to allow them the use of his palace, the very same building in which Philip II's third marriage was to be celebrated.[281] The duke had refused, to the great indignation of the two ladies and the emperor, who did not however, despite their complaints, wish to force the hand of the duke, a prominent noble from whom he had received many services. In January, 1558, the *corregidor* of Plasencia had decided to put into effect certain directives at Cuacos, a village near Yuste where the emperor was represented by an *alguazil*; a dispute arose between the two men and the *corregidor* ended it by having the *alguazil* arrested and imprisoned.[282]

Taking advantage of government deficiencies, and of the incompetence of jurists and ministers, every individual tried to obtain some extra privilege. In October, 1559[283] Philip II, preoccupied by his financial deficit, was anxious to discover abuses to be suppressed and economies to be made. An old adviser, the alcalde of the Chancellery of Valladolid, the *licenciado* Palomares, sent him a curious letter detailing the excessive pretensions of the grand nobility in judicial matters. He reminded the king that on the occasion of his voyage to Germany in 1548–1550, seven or eight of the grandees, having met at the Convent of San Pablo in Valladolid, had requested for all *títulos* the privilege of being judged only by the sovereign. They had also demanded that when criminal cases originating on lands owned by nobles were heard by the royal tribunal, any fines paid should be the property of the landlord. In support of their claim, they cited a law of Guadalajara, dating from the reign of John I and which, according to Palomares, was either apocryphal or ambiguous. In 1556 (the date is significant: the king was absent again, having left for England in 1554), there was another meeting at San Pablo; the same claims were put forward, only to be rejected by the Infanta Joanna. The lords had then resorted to less direct methods; to the deeds of sale of domanial lands, principally those sold in 1559, those who drafted the contracts introduced clauses potentially threatening crown authority, again citing the Guadalajara law. It was a royal official, the *licenciado* Juan de Vargas, who first slipped these crucial words into the text of a sale to his own advantage, with the aim of retaining for himself the proceeds of criminal justice on the lands he was buying. His example was naturally followed by others 'Certain of your servants and councillors of State have enacted similar sales', adds Palomares, 'and Your Majesty should be on the watch for it.' Even the highest officials were open to temptation.

The actions of the lords and purchasers of *lugares de vasallos* are an index of the inadequacy of the state, a measure of the failings and weaknesses which encouraged such inroads on its authority. It is not surprising then that the obstacles encountered by royal authority in normal times should have increased in size. The towns, when threatened with the

[281] L. P. Gachard, *La Retraite . . .*, *op. cit.*, I, p. 206–207, 7th November, 1557, and II, p. 278–279, 15th November, 1557.

[282] *Ibid.*, I, p. 240–242, 5th January, 1558.

[283] Simancas, E° 137.

loss of their rights of jurisdiction, defended themselves stoutly, sent personal deputies to the king, and very often had their way. Similarly it was often the very officers of the *Contratación* who helped the merchants of Seville to evade government measures. In the spring of 1557, the government had seized all the silver which the Indies fleet was carrying for individuals: 'and whereas out of the seven or eight million ducats that arrived, we had succeeded in claiming five million, the merchants have been so cunning that only 500,000 is left now', complained Charles V indignantly![284] Only the personal intervention of the emperor was sufficient to set in motion the judicial machinery against the offenders. In the autumn of the same year,[285] it was judged more prudent to send a squadron under Alvaro de Bazan to meet the Indies fleet; arriving at San Lucar on 7th September, the squadron seized the money and transported it to Santander from where it was sent on to the Netherlands. To such means was royal authority now reduced.

There were times when it dared not even intervene, for instance when the viceroy of Aragon, the Duke of Francavila, had a 'demonstrator' garrotted, ignoring the rights of jurisdiction of the *fuero*; his action unleashed a riot, the Cortes met independently without the royal convocation, the viceroy was forced to flee to the Aljaferia and the government of Valladolid, when it was informed, disavowed him.[286] It could not afford to alienate Aragon, especially during a war against France! Similarly the inquisitors of Valencia who examined – it was their daily bread – the cases against the *Tagarinos*, the local Moriscos, received instructions to proceed with caution. A letter to the Council of the Inquisition, dated 4th June, 1557,[287] reads as follows: 'You wrote to us . . ., on 4th September last, . . . that as these are dangerous times, we should for the present suspend hearings of cases against the *Tagarinos*.'

Understandably, armed with such advice, government officials were timorous and hesitant to act even under orders. The inquisitor Arteaga, writing on 28th February to the *Suprema*,[288] reports that the alguazil of the Holy Office of Barcelona has come to ask him to execute at Valencia the judgements made by that tribunal. 'If I have not apprehended the persons indicated in the letter of requisition, it was in order to avoid the scandal and the great disturbance which might result in this city, given the present times and since the persons incriminated are for the most part prominent officials of this place.' *The present times* – difficult times and a period of sore trials for monarchical rule.

[284] L. P. Gachard, *op. cit.*, 1, p. 137–139, 1st April, 1557; p. 148–149, 12th May, 1557; on these points and on the punishment of the 'oficiales', A.E. Esp., 296, 8th and 9th June, 1557; on the diversion of a ship laden with precious metal to Portugal, L. P. Gachard, *op. cit.*, I, p. 142–144.

[285] *Ibid.*, I, p. 172, Martin de Gaztelu to Juan Vásquez, 18th September, 1557.

[286] Juan A. Llorente, 'La primera crisis . . .', *art. cit.*

[287] A.H.N., Valencia Inquisition, Libro I.

[288] *Ibid.*, concerning demands from Barcelona that the judgements should be executed.

Financial difficulties. The sovereign was far from having a free hand; his every action was determined by the greatest of the worries which had brought him back to Spain: financial insolvency.

The vast deficit in the imperial finances which he had inherited was such that with the first payments made necessary by the renewal of hostilities, his credit collapsed. On 1st January, 1557, the bankruptcy was made official.[289] But was it a true bankruptcy? The first of Philip's famous decrees was merely the consolidation of the floating debt. The royal treasury was fed by loans or advances made at high rates and on onerous conditions by the merchants who alone (in view of the dispersed nature of the Spanish Empire and the presence of the sovereign in the Netherlands) were able to mobilize on its behalf, revenues far off both in time and space. The treasury paid them high interest rates and reimbursed them at the regular fairs. The debts of the state, then, were represented by a mass of documents of different kinds. The decree did not annul the debts, but specified that they would be paid back in *juros*, perpetual or life annuities, which carried an interest of 5 per cent in principle. 1st January, 1557 was named as the initial date of these repayments.

The bankers first protested then accepted it, the Fuggers putting up livelier resistance than the others. The decree evidently injured the bankers: not only was the interest on their loans reduced, but their capital was also immobilized. They could still sell these annuities – and several of them did – but their action led to a dramatic fall in prices, to the disadvantage of the sellers. This is the reason why by the time the Fuggers capitulated,[290] the *juros* had fallen to 50 or 40 per cent of their nominal value. The forced exchange of short-term debts at high interest (12 and 13 per cent for perpetual annuities at 5 per cent, although bringing great financial loss to the creditors, was nevertheless not technically bankruptcy.

The expedient enabled the state to survive somehow, until the Treaty of Cateau-Cambrésis, but it did not remove all its difficulties. The Genoese bankers, the only ones who would still make loans to the Catholic King, exacted stiffer terms than in the past. If any proof is needed, let us consider the two *asientos* signed at Valladolid in 1558. In the first, Niccolò Grimaldi, a Genoese banker[291] advanced one million in gold to the king: 'the said Niccolò Grimaldi undertakes to pay in Flanders 800,000 crowns at 72 *grooten* (Flemish groats) per crown, and in the following manner: 300,000 when the first ships arrive from Peru, 250,000 at the end of November and the remaining 250,000 crowns at the end of December of this year 1558. Another 200,000 crowns he undertakes to pay at Milan, at 11 reals to the crown, in the course of November and December of this year, half in each month.' The king in return gave certain guarantees:

[289] On this subject see the classic studies by K. Haebler and R. Ehrenberg and above, Vol. I, p. 500 ff.

[290] For the text of the *asiento* with the Fuggers, A.d.S., Naples, Carte Farnesiane, fasc. 1634.

[291] B.N., Paris, Fr. 15 875, f° 476 and 476 v°.

'H.M. will repay the said million in Spain, at 400 *maravedís* per crown and in the following manner: 300,000 immediately, from the money which is at Laredo, 300,000 from the gold and silver arriving by the first ships from Peru and in the event that the payments are not made in October of this year, the said Grimaldi will not be obliged to make his payments at the end of November and December, either in Flanders or in Milan; 300,000 crowns from the *servicios* of Castile in 1559 and bills of exchange without interest will be delivered to him; the remaining 166,666 crowns of the 400 million *maravedís* to be payable in annuities at 10 per cent. He will be repaid 540,000 crowns of outstanding debts as follows: 110,000 in annuities at 10 per cent, 135,000 at 12 per cent, 170,000 at 14 per cent and 25,000 assigned on the mines. The interest on this sum will be reckoned up to the end of 1556 at 14 per cent and for the year 1557 at 8 per cent. He is also granted permission to export one million in gold from Spain.'

These figures reveal exceptionally harsh terms. The anonymous French commentary which accompanies them, states that 'this Genoese merchant is lending nothing of his own, as one can see and yet for the favour he is doing king Philip by letting him have money on the exchange at Antwerp and Milan on the same conditions as he is paid here, he is making 50 *maravedís* for each crown, since he is paid at a rate of 400, when it is worth only 350, that is 15 per cent, and he will gain practically as much in Flanders, since he is lending at 72 groats to the crown when it is worth 78'. The commentator wonders why the king of Spain ever signed such an agreement. If he had money at Laredo, would it not be simpler to have it sent directly? He can see only two possible advantages, the avoidance of the risks of sea transport and the reduction of the rates of interest on old debts. Accumulated debt weighed heavily on Philip's policy towards his moneylenders.

Equally disadvantageous was the loan of 600,000 crowns agreed in the same year 1558 by Constantino Gentile, another Genoese merchant.[292] It was to be repaid as follows: 125,000 crowns immediately at Seville; the same again at Seville in July, 1558; and 350,000 crowns assigned on the *servicios* of Castile. Add to this the extravagant advantages of exchange rates and the consolidation of 1,400,000 crowns of old debts: the same anonymous commentator has no difficulty showing who benefited most from the deal.

That in both examples quoted here the entire burden should fall on Castile (as in the curious *asiento* negotiated with the Fuggers on 1st January, 1557, of which there is a copy in the Farnese archives[293] in Naples), should cause no surprise. This was invariably the case for loans negotiated during these difficult years: all of them were advanced on the strength of the ordinary and extraordinary taxes of Castile and the precious metals brought by the Indies fleets. Philip II's credit depended in the

[292] B.N., Paris, Fr. 15 875, f° 478 to 479.
[293] A.d.S., Naples, Carte Farnesiane, fasc. 1634.

last analysis on the credit of Spain. And this was already deeply undermined.

Indeed the country had been unscrupulously exploited. During the war against Paul IV, money had had to be literally extorted from prelates who had only given it because they were weary of the struggle. Then as necessity knows no laws, as much as possible of the money brought home by the fleets from the Indies was seized. This money had been either destined for merchants in Seville or carried on the persons of individuals returning from the Indies. The expropriations, repeated in 1556, 1557, and 1558, left very unpleasant memories. It was not until 1559 that Philip decided to repay the capital he had seized, but two-thirds of the reimbursement was in *juros*. The joy with which his move was greeted among the merchants is sufficient indication that despite the incomplete nature of the restitution, most people had despaired of receiving anything at all.[294]

After Cateau-Cambrésis, Philip II seems to have felt some remorse for his actions,'. . . it seems reasonable to us', he says, 'not to take anything either from merchant passengers [on the Indies fleets] or from individuals, but on the contrary to let them receive freely all the money that is sent to them'.[295] It was rather late in the day. Ten years later, when it was whispered that the government had decided to return to its old methods, many people preferred to stay in America rather than risk having their wealth seized if they returned to Spain.[296]

As for the normal sources of money, apart from taxation in Castile, they all seem to have been drawn on heavily in advance. Others had to be found, hence various financial expedients, a list of which can be found in a letter from the Infanta Joanna to the king, dated 26th July, 1557:[297] the sale of titles of nobility, the granting of patents of legitimacy to the sons of ecclesiastics, the creation of municipal offices, sale of crown lands and rights of jurisdiction. These sales brought more disturbance than anything else to the Spanish kingdoms. It seems clear that they benefited the high nobility, but we know very little about them: they require careful study, as do the sales of church lands after 1570. The towns were the earliest victims, for crown lands were often in reality municipal lands which now fell into the hands of the aristocracy. But many villages took advantage of the opportunity to buy themselves, thereby liberating themselves from the jurisdiction of the towns. Thus Simancas freed itself from the jurisdiction of Valladolid.

The creation of new municipal offices was another way of forcing the towns to contribute, for the state received the proceeds from the sale of

[294] It was only moderate joy of course, and at first there was even some discontent. One third of the debt was cancelled and the rest converted into *juros* at 20 per cent, Philip to Princess Joanna, 26th June, 1559, Simancas E° 137, f° 121.

[295] Philip II to Princess Joanna, Brussels, 26th June, 1559, Simancas E° 137, f° 123 and 124.

[296] Simancas E° 137, 13th July, 1559.

[297] Manuel Danvila, *El poder civil en España*, Madrid, 1885, V, p. 364 ff.

office and the town theareafter paid the salary.[298] The complaints of the latter are understandable.[299] In defence of their money, they did not hesitate to send representatives as far as Flanders. Philip II could not remain deaf to all their requests. He intervened to stop several agreements which had almost been concluded and finally forbade the sale of petty office. But once more, these sensible measures came far too late. Many injustices had already been perpetrated, as can be seen from the letter referred to earlier, of the *licenciado* Palomares. Of some of these, the usurpation of domanial lands in Granada for example, little or no record remains.[300] In 1559, the state of the treasury was desperate. Philip had made peace with France, but until the peace was signed, he had had to maintain an army in the field; then he had had to disband it, which could only be done by giving out back pay. For lack of money, the army could not be demobilized and the debt rose ever higher in a vicious circle. Philip asked Spain for seventeen hundred thousand crowns in March[301] but the Regent was able to raise only two loans, one for 800,000 and the other for 300,000, her last effort endangering incidentally the credit of the factor Francisco López del Campo, the official in charge of treasury payments in Spain. Already, to save him, the closing date for the fair at Villálon had been postponed until June: 'The factor', wrote Joanna to her brother on 13th July, 1559,[302] 'was preparing to go there and as far as possible fulfil his obligation, concerning the sum of *maravedís* set out for Your Majesty in the accompanying memorandum. The principal source on which he relied was the silver expected from the Indies on the fleet, which has just arrived and which we have learnt today, brought nothing either for Your Majesty or for anyone else.' It was the viceroy of New Spain, according to the municipal officials of Seville[303] who had decided that no bullion should be carried by the fleet for fear of privateers.

Under the circumstances, it was impossible to meet the June deadline: 'the Villalón fair[304] will be prolonged until the feast of St. James', continues the Infanta, 'in order to seek a remedy meanwhile, for the Council of Finance has decided that it is imperative to repay at the fair, even if we have to borrow against the *servicios* [of Castile] for the year 1561, which have not yet been granted, or on some other security. Any rate of interest or other disadvantage is greatly to be preferred to the ruin of the factor's credit. For it is only thanks to him that Your Majesty has been served and

[298] M. Danvila, *op. cit.*, V, p. 346 ff.

[299] Complaints received from Burgos (10th February, 1559), Seville (Simancas E° 137), Guadalajara (B.N., Paris, Esp. 278 f° 13 to 14, 5th November, 1557).

[300] All I know about the inquiry into the usurped lands of Granada, is the name of the official who conducted it, Dr. Sanctiago, 'oydor de Valladolid' and this from a letter from Philip II to Princess Joanna, 29th July, 1559, Simancas E° 518, f° 20 and 21. It is merely mentioned in another letter from Philip II, 27th April, 1559, Simancas E° 137, f° 139.

[301] See Princess Joanna's reply on this subject, 27th April, 1559, Simancas E° 137 f° 139; M. Danvila, *op. cit.*; V, p. 372.

[302] Simancas E° 137.

[303] See next note.

[304] 13th July, 1559, Simancas E° 137.

provided with money up to the present, and will be again, so long as we can meet our creditors at the fair. The resources at our command are a few sales of vassals, but Your Majesty has restricted these, particularly in the case of Seville, just when a sale worth 150,000 ducats was being negotiated with the Duke of Alcalá, who wishes to acquire 1500 vassals . . .'. At the same time, the Regent sent economic experts, Dr. Velasco amongst others, to Philip to tell him exactly how things stood,[305] fearing lest her brother should still be under any illusion.

In the Netherlands, Philip could see no solution. 'By staying here', he wrote on 24th June,[306] 'all I shall achieve will be to lose myself and these states [the Netherlands]. . . . The best thing is for us all to seek some remedy . . . and if I cannot find one here, I shall seek one in Spain', a clear enough indication of his purpose. Philip had doubts about the efficiency of his sister Joanna, whose mind was much taken with her liberal charity, her religious devotion and her ambitious dreams – her charity was limited by her brother and her dreams were centred on the Infant Don Carlos, whom, it was said, she wished to marry in order to maintain herself in the first rank. Perhaps the king remembered the journey made to the Peninsula by Ruy Gómez in 1557.[307] Where the favourite had succeeded, maybe the sovereign could try his luck. Salvation could be sought only in Spain and by the sovereign himself. When contrary winds forced him to wait a fortnight in the islands of Zeeland, the king showed signs of distress, not because of personal discomfort but, as he wrote on 24th August to the bishop of Arras, 'because I see that along with my arrival in Spain, I am postponing the moment when I may find the necessary remedy both for this country and that'.[308]

These details help us to understand the dramatic letter written by Philip II to Granvelle on 27th December, 1559,[309] by which time the true situation of Spain was no secret to him. 'Believe me', wrote the sovereign, 'I had every intention of providing the Netherlands with all the things I know to be necessary. . . . But I give you my word that I found here a situation worse than that I left behind, so that it is absolutely impossible for me to help you or even, in this country, to satisfy such trifling needs that you would be astonished to see them. I confess to you that when I was in Flanders, I never believed the situation could be so bad here and I have found no remedy as yet, except for the dowry money[310] as you will see from the letter I have sent my sister [Margaret of Parma].' The dis-

[305] We have several references concerning the mission of this man (Velasco, not de Lasco as he is called in Granvelle's papers, *Papiers d'État*, *op. cit.*, V, p. 454). The mission is mentioned in a letter from Philip II to the princess, Brussels, 18th June, 1559, Simancas E° 137 and 20th May, *ibid.*, f° 116.

[306] Granvelle, *op. cit.*, V, p. 606.

[307] L. P. Gachard, *La Retraite* . . ., *op. cit.*, II, p. LIII–LIV; M. Danvila, *op. cit.*, V, p. 351 (1557).

[308] Granvelle, *Papiers d'état*, V, p. 641–644.

[309] *Ibid.*, Toledo, 27th December, 1559, p. 672.

[310] Of the new queen of Spain.

34. PHILIP II, CIRCA 1555. Anonymous drawing, B.N., Paris

35. PHILIP II. Detail from the Gloria, *The Burial of the Count of Orgaz*, by El Greco, Toledo, 1586

illusion shown in this letter is unmistakably genuine. There was nothing left in Spain, because too much had been taken from her; and thus through lack of foresight, the sources of the empire's wealth had apparently dried up for the time being. Hence Philip II's belated return to moderation which we noted earlier. Was it because he recognized the necessity of maintaining these sources at their normal level that he determined to remain in Spain for the rest of his life?

In 1570 there opened in Córdoba the session of the Cortes of Castile, which was to close the following year at Madrid. At the opening sitting, Eraso speaking on behalf of the king, gave a brief history of the years which had passed since the previous meeting of the assembly in 1566: 'The king as you know has resided throughout these years in Spain, although there were grave and pressing reasons for his leaving in order to attend in person to matters in others of his States, as was made clear to the *Reino* in the last Cortes. But His Majesty knows how necessary his presence is in these kingdoms . . . not only for their own good and advantage, but also in order to provide . . . for the needs of the other States, for these kingdoms, of all his dominions, are the vital centre, the head and principal part. And considering also the great love he bears you, His Majesty so ordered matters that while remedies to instant necessity were to be sought, his absence from Spain might be avoided.'[311]

Reading between the lines of this official speech, we see that Philip could not leave Spain, the heart of his states and their treasure, without immense risk. Brussels was an admirable political capital, true, but politics were not everything. Valladolid was the financial capital of the Spanish Empire; the *asientos* were arranged there and the rhythm of the fairs of Castile, outside its gates, fixed their closing-dates. It was essential that it should be so, that the master of the empire, concentrating in his person the chief burden of state expenditure, should be present in the country which received the silver of America. However, all this became clear to the king only after his return to Spain. The orders that he had previously sent from the Netherlands to his governors of Spain undoubtedly revealed such ignorance – ignorance admitted by Philip himself in his letter to Granvelle – that his correspondents more than once found them quite ludicrous. We know this from Philip himself: in the margin of a letter from the Infanta Joanna[312] in which she says that after the meeting of the councillors of Valladolid she has to inform him that everyone feels very differently from him; deems it impossible to send any money; and considers that he should come to Spain instead, the king wrote (having been informed by some malicious tongue apparently) 'se han hart reydo de mi, they made much mock of me'. Who were they? The councillors and the

[311] The reference is to Philip II's proposed journey to the Netherlands (1566–1568), *Actas de las Cortes de Castilla*, III, p. 15–24.

[312] Marginal notes made by Philip II on the letter sent to him by the princess on 14th July, 1559, Simancas E° 137, f° 229. The text was verified at my request by Don Miguel Bordonau, the then head archivist at Simancas.

princess, all better informed of realities in the Peninsula and anxious for his return.

So Philip II returned to Spain to learn that the situation was even graver than he had imagined. It is therefore difficult to see by what aberration this exhausted country was to persist in the war in the Mediterranean, allowing a conflict to develop which could have been quelled and which, as we shall see, instead grew more serious. But was the Prudent King really responsible?

CHAPTER II

The Last Six Years of Turkish Supremacy: 1559–1565

From the treaty of Cateau-Cambrésis (April, 1559) to the siege of Malta (May to September, 1565) Mediterranean history forms a coherent whole. For six entire years it was independent of the major events in eastern and northern Europe; freed for the moment from their other responsibilities, the giants who shared the sea between them – Turk and Spaniard – returned to their single combat. As yet it was not an all-out struggle. Did either side really seek a fight to the death? Were they not both victims of short-term initiatives and objectives which in the end carried them further than they had intended? One might imagine so from the indeterminate policy of Spain, apparently the product of events rather than of a resolute decision. Turkish policy too, during the last years of the great Sulaimān's reign, bears similar marks. The single major change in the West was the creation of a powerful navy in the service of Spain. But the question was whether Spain would use this navy efficiently and indeed whether it was adequate to ensure control of the sea.

1. WAR AGAINST THE TURK: A SPANISH FOLLY?

The war continued in the Mediterranean, just when it was ending in western Europe: in Germany with the internal peace of Augsburg, between the Spanish Empire and the papacy by the agreements of September, 1557, and between France and Spain by the treaty of Cateau-Cambrésis. Everywhere peace was returning: everywhere that is except the Mediterranean. Here the war persisted, albeit veering between violent clashes of arms and prolonged periods of inactivity. Both motives and events in this war are obscure. The economic recession of 1559–1575 cannot have been entirely responsible.

The breakdown of Turco-Spanish negotiations. Between the two major powers, Spain and Turkey, war was by no means inevitable, or at least so Philip II's advisers thought in 1558. A truce with Turkey lasting several years appeared to be the necessary condition for less restricted action in the West. On 21st May, 1558, Philip sent the bishop of Aguila to his uncle Ferdinand, with very precise instructions.[1] He had received from the emperor a letter, dated 2nd January, informing him that talks with the Turks were being held at Vienna, that it had been decided to pay the

[1] *Instrucción de lo que vos el Reverendo padre obispo del Aguila habéis de decir a la Majestad del Serenissimo Rey é Emperador, nuestro muy caro y muy amado tío donde de presente os enviamos*, Brussels, 21st May, 1558, *CODOIN*, XCVIII, p. 6–10.

arrears of the annual tribute to the Porte (stipulated in the treaty of 1547 and not paid since 1550) and even to consent to an increase. Philip II expressed his approval, 'knowing as I do today that there is small hope of Christendom being able to oppose with all the necessary force a power as mighty as that of the Turk, I cannot but agree with the prudent recommendations of your own subjects, the Hungarians, Bohemians and Austrians, supported by the German electors . . .'. It so happened that an intermediary 'who has experience and intelligence of the Turkish court', had undertaken, only a few days before, to obtain a promise of peace with Spain from the sultan, if the king should wish it. 'For several particular reasons', the king continued, 'I did not wish such negotiations *to be proposed on my behalf*. Neither did I wish to burn all the bridges between us, keeping it in mind that the Turk, fearing my forces in the Mediterranean, might, if he knew that I could be persuaded to allow myself to be included in the peace under discussion with Your Majesty, possibly moderate and soften his terms.' A typical example of Spanish diplomacy, haughty, determined to respect 'pundonor', but not unwilling to take an indirect path if necessary. Philip II, reluctant to make the first overtures to the Turk, is less scrupulous when he thinks he may be able to use Vienna as a go-between.[2]

During the first months of 1559, the king was still envisaging negotiations as a possibility. We know of the existence for example of a memorandum dated 5th March, listing the possible conditions for a truce of ten or twelve years with the Grand Turk.[3] In a letter written on the 6th to the secretary of his ambassador at Venice, Garci Hernández, the king says that he has 'chosen Nicolò Secco to go to the Grand Turk along with Francisco de Franchis and to discuss the truce which the latter has, as you know, set under way'.[4] The said Nicolò Secco had been to the court of the emperor and would have to see the Duke of Sessa next to receive his instructions. On the 6th too, Nicolas Cid, the paymaster-general of the army in Lombardy, received orders to pay 2000 crowns to Secco, 'para el gasto de cierto camino que ha de hazer a mi servicio', and 5000 to Garci Hernández, who knew to whom he had to pass it on. The instructions sent to Nicolò Secco[5] on the same day give us even more details. The whole

[2] As for the person 'who has experience and intelligence of the Turkish court', and who was instructed informally to explore the ground in Constantinople, he must have been Francisco de Franchis Tortorino, a Genoese attached to the Mahonna (or Maona; a trading consortium) of Chios. Sent on a mission by Genoa following the corruptive dealings with Piãle Pasha during the summer of 1558, he no doubt offered his services at the same time to Philip II. A handsome register in the Genoese Archives gives a detailed description of the voyage of Francisco de Franchis (Costantinopoli, 1.2169) and consular letters mention his presence at Naples and Messina where, incidentally, he had some trouble with the Spanish authorities, before moving to Venice, A.d.S, Genoa, Napoli, Lettere Consoli, 2,2635; Gregorio Leti, *op. cit.*, I, p. 302, mentions his mission; he was accompanied by one Nicolò Gritti.

[3] *CODOIN*, XCVIII, p. 53–54.

[4] Brussels, 6th March, 1559, Simancas E° 485.

[5] *Instruction del Rey a Nicolò Secco para tratar con el Turco*, Brussels, 6th March, 1559, Simancas E° 485.

affair seems to have been the brainchild of Francisco de Franchis Tortorino, who was now passing through Venice on his way to Constantinople for the second time, again on behalf of the Republic of Genoa. He had long discussions with the ambassador, Vargas, concerning the possible cooperation of Rustem Pasha (who was Grand Vizier at the time) and the gifts it would be suitable to offer him.

Nicolò Secco, who had previously been ambassador to Turkey, was to join Franchis at Venice and to travel with him to Ragusa. From there, Franchis was to continue alone and would only call Secco in the event of the imminent ratification of the proposed truce. So once more, as in 1558, Philip II was not inclined to rush things. Secco was authorized to sign a truce of ten, twelve or even fifteen years, with the Turks. For as long as the truce should last, eight to ten thousand crowns would be paid annually to Rustem Pasha. And if it is at all possible, adds Philip, to secure through the good offices of Rustem a promise that the Turkish fleet will not come out this summer, 'it would be permissible to offer him twelve to fifteen thousand crowns, payable without delay and all together, either at Venice or Constantinople, as he pleases'.

I have gone into such detail over this affair merely to establish the facts about the negotiations and the intentions of Philip II before Cateau-Cambrésis. For once the peace had been signed, things were quite different. On 8th April, 1559,[6] Philip explains his change of mind in a long letter to the Duke of Sessa: 'You have seen', he writes, 'what I told you about . . . the truce with the Turk . . . and the dispatches I sent you in order that Nicolò Secco should handle the matter. Since then I have been informed by the emperor that a three-year truce has been negotiated by his ambassadors, between himself and the Turk. The latter would on no account accept that I should be included in the said truce. My main objective . . . had been to discover whether, by this means, some benefit might be obtained for the emperor and his affairs; it seems to me that that objective has been achieved. Furthermore, we have recently made peace with the king of France, so it may be supposed that the Turk, deprived of his ally and having no friendly port [in the West] for his fleet, will not send it against Christendom. In addition to all this, there is his great age, his reported desire for rest and the trouble caused him by the discord, ill-feeling and rivalry between his sons.'[7] The conclusion is therefore: 'cancel the voyage by Francisco de Franchis and Nicolò Secco, since it will be impossible to attempt or do anything at the moment without great loss of our authority'. The original Spanish indeed expresses it more strongly: 'sin gran desautoridad nuestra'. And this was the real motive: in order not to lose face, Philip II, who was now free from troubles in the West, did not pursue his peace moves. His attitude could not fail to produce consequences.

And indeed as early as June, Philip was acquiescing in the projects of the Knights of Malta and the viceroy of Sicily against Tripoli. Writing to

[6] Brussels, 8th April, 1559, Simancas E° 1210.

[7] Selīm and Bāyazīd.

the Duke of Florence to ask for the help of his galleys, he says, 'Since it has pleased the Lord our God that peace should be signed with the Most Christian King of France, it seemed to me that it would be to the greater glory of God and of service to all Christendom that the galleys for which I am paying in Italy should not lie idle for the rest of the summer, but be employed to destroy corsairs and to assure freedom of shipping. . . . I have therefore authorized an expedition against Tripoli.'[8] An expedition against Tripoli meant an expedition against Dragut, who had been *beglerbeg* of the town since 1556. But had it not been proved, in 1550, that any move against Dragut provoked a riposte from the Turks?

Philip II's face-saving policy was thus immediately responsible for the reopening of the conflict – and the more responsible in that the situation lent itself to peace-making. 'Affairs are going badly for the Turk', notes the Duke of Sessa on 4th December, 1559, 'because of the dissension between his sons';[9] and the level-headed Duke of Alcalá, viceroy of Naples, wrote to Philip on 10th January, 1560, just when the expedition against Tripoli had emptied his kingdom of a great many troops: 'I would remind Your Majesty that it is an appropriate moment to negotiate some truce with the Turk, both because of the conflict between his sons and because of the great need for peace felt by Your Majesty's dominions. Here in Naples, everyone thinks it is most necessary.'[10]

Not only did Philip refuse to sue for peace on his own behalf, but he also intervened with the emperor to dissuade him from concluding the agreement which had very nearly been reached. If we are to believe the Venetian ambassador, Giacomo Soranzo,[11] the articles of the peace had not yet been returned to Vienna at the end of October, and Philip II, who had been consulted during the interval, advised the emperor not to accept them; he even offered to harass the sultan in the Mediterranean, promised the emperor troops and money and suggested that he might, through the king of Portugal, make approaches to Bāyazīd and the Sophy; in short he neglected no argument. Would it not be more profitable for Ferdinand to seize Transylvania than to make peace with the sultan? His advice did not apparently fall on deaf ears.[12]

The naval supremacy of the Turks. Philip had some reason and perhaps some excuse for his actions.

One reason was that following the signature of the treaty of Cateau-Cambrésis, Henri II had demobilized his Mediterranean fleet. There was to be practically no military shipping in the ports of the South of France for the rest of the century and even longer. An extra guarantee was thus

[8] 15th June, 1559, Simancas E° 1124, f° 295.

[9] Summary of the letters of the Duke of Sessa, 1st, 4th and 7th December, 1559 (4th Dec.), Simancas E° 1210, f° 142.

[10] 10th January, 1560, Simancas E° 1050, f° 9.

[11] To the Doge, Vienna, 25th October, 1559, G. Turba, *op. cit.*, I, 3, p. 108 ff.

[12] Soranzo to the Doge, 22nd November, 1559, *ibid.*, p. 120 ff.

provided for the peace which applied in the Mediterranean as well as to Europe,[13] and the Spaniards regained freedom of movement.

Philip II also had the excuse that he had not yet taken the full measure of Turkish might. He had hardly any experience of Ottoman seapower, for Prevesa had not been considered a major naval encounter even at the time and as far as he was concerned, it was ancient history; on land, Spaniards had only participated as individuals in the war on the Hungarian frontier, when they had participated at all. Only twice had the Spanish (whom Charles V had landed in 1534 at Coron and in 1538 at Castelnuovo, where they had led the usual life of the *presidios*, with periodic alerts and sorties) met in battle with Barbarossa, and then he had dislodged them in 1534 and 1539. But from such remote and unequal battles, what lessons could be learnt? It was only in 1560 in Djerba and in 1565 in Malta, that the Spanish infantry was to take the full measure of its enemy.

Moreover, in Turkey, the dispute between the sultan's sons had unleashed a number of breaches of discipline, provincial disturbances and even social conflicts. The French ambassador, De La Vigne, writes to the bishop of Dax, in July, 1559[14] that the slaves all favour the rebellious son of Sulaimān, Bāyazīd. That the latter should have been defeated by the favourite, Selīm, was no solution since Bāyazīd had fled to Persia and a domestic war, not totally subdued, threatened to develop into a foreign war. The Turks, writes De La Vigne in September,[15] 'find themselves more embarrassed than ever they were before by domestic matters'. Philip II may very well have thought that this was not the moment to talk peace with them but on the contrary to defeat them in war.[16]

The summer of 1559 seemed to prove him right: that year, the Turkish fleet did not venture beyond the Albanian coast and, rather the worse for wear, turned for home in the autumn without making any attacks on Christian powers. Philip II undoubtedly relied rather too heavily on the assumption that it could not threaten the West without the complicity of France. Without this aid, he reasoned, it would have to be content with a few raids during the fine season. So in spite of its numerical inferiority the Spanish fleet might risk an action, either at the end of the fine weather, in winter, or in the spring, before the enemy emerged. The main thing was to avoid being surprised, particularly when the proposed action was to take place in the central region of the sea.

For here, Spain faced a double danger: the Barbary corsairs on one hand, between Tripoli and Salé, and the Turks on the other. The two groups were independent and always separated during the winter, but combined forces during the fine weather. The Barbary corsairs were

[13] The king to the viceroy of Sicily, Brussels, 4th April, 1559, Simancas E° 1124, f° 304.

[14] E. Charrière, *op. cit.*, II, p. 596, note.

[15] *Ibid.*, p. 603.

[16] Marin de Cavalli to the Doge, Pera, 18th March, 1559, A.d.S. Venice, Senato Secreta, Costant. Filza 2/B.

comfortably installed in the western Mediterranean and their reign was growing in prosperity: in the central Maghreb, Algiers was expanding and collecting around it an empire which formed a direct threat to Spain. Needless to say, this empire was not a model of political coherence. There were large pockets of dissidence, the mountains of Kabylia for instance; but the major routes were under control. We have already seen how, in 1552, Salīh Re'īs, seventh king of Algiers, had led an expedition as far as Ouargla; and in 1553 another as far as Fez. Fez had been recaptured and the Sharif had even briefly held Tlemcen in 1557. Pursued by the Turks, he had been forced to retreat towards his capital, but at a short distance from the town – thanks to his numerous cavalry and to the 'Elches', Moriscos who had taken refuge in Morocco and who were particularly useful for their skill at handling the arquebus – he had stopped the advance of troops of Hasan Pasha, Barbarossa's son. In the west, the Algerian-Moroccan frontier had finally proved easier to cross than to move. But in the east, the state of Algiers had succeeded in dislodging the *presidio* at Bougie in 1555. Finally, in 1558, it had won an immense victory at Oran.

Since the beginning of the century, in 1509, the Spaniards had played a waiting game at Oran, succeeding several times in annexing Tlemcen. This war of nerves practised consciously by Count Martín de Alcaudete, had come to an end in 1551 however, from the day when a Turkish garrison managed to instal itself permanently in Tlemcen. It was a constant thorn in the flesh of the *presidio* and in an attempt to remedy this, as well as to restore the morale of his garrison, with troops levied in part on his own estates in Andalusia, old Don Martín, *El Viejo*, as he was called to distinguish him from his son, organized an expedition against Mostaganem, 12 leagues east of Oran. To deprive the Turks of Mostaganem would mean breaking their communications with Tlemcen, for through this port they received all the supplies and artillery necessary for their operations on the western front. Competently led, such an expedition could not fail to succeed against a position as poorly fortified as Mostaganem. But time was wasted training the new recruits on sorties into the countryside around Oran which alerted the whole of North Africa. Then *El Viejo* proceeded to put his plans into operation with exceeding slowness and caution. On 26th August, surprised by forces from Algiers and native troops, he succumbed to superior numbers and over 12,000 Spaniards fell into the hands of the conqueror. The houses of Algiers were full of these new slaves and the following year many of them forswore their faith to go and fight in Lesser Kabylia in the army of Hasan Pasha.[17]

These details show the strength and determination with which the new Turkish state was carving out a place for itself in the Maghreb. Even more evident was its growing seapower, east as far as Sicily, north as far as Sardinia and west as far as Gibraltar: 'The Turks have lately sailed

[17] D. de Haëdo, *op. cit.*, p. 73, 74. On Spanish policy in North Africa, see my article in *Revue Africaine*, 1928; Jean Cazenave, *Les sources de l'histoire d'Oran*, 1933.

with fourteen or fifteen galleys to the Algarve', writes Nicot, French ambassador to Lisbon, on 4th September, 1559,[18] 'and have wrought ravage among the people. By the time I arrived, they had withdrawn . . .'. They had done more damage in Castile, and in 'Caliz'[19] had raised 'a white flag, putting up for ransom all their victims and there every captive was bought back'. These were 'Turks' of the Barbary variety.

But the state of Algiers, although the most powerful of the Barbary states, was not the only one. In the eastern Maghreb, the 'kingdom' of Tripoli was developing on the Algiers pattern, particularly since Dragut had taken control of it in 1556. There was this difference, that the state of Tripoli could only sustain itself at the expense of a hinterland desperately poor and difficult to subdue, especially in the region of 'Darrien' [Gharian], the inhabitants of which could easily interrupt the routes which brought slaves and gold from the Sudan. Confined on the landward side, Tripoli turned to the sea: her major income came from this side, from the coast of Sicily, so close and ready at hand. And beyond Sicily, Dragut was threatening the material existence of the western Mediterranean, 'hasta Cataluña y Valencia que morian de hambre', up to and including Catalonia and Valencia which were dying of hunger, as the Duke of Medina Celi,[20] viceroy of Sicily and the major promoter of the expedition against Tripoli, wrote in June, 1559.

The Djerba expedition.[21] This expedition took a different direction from the one originally intended, and finally headed for Djerba but only after several twists of fate which we shall set out here as briefly as possible.

If the final decision to launch the expedition was not reached until 15th June, 1559, the date of the orders and instruction dispatched from Brussels,[22] the notion had been suggested earlier and Philip II was not solely responsible. All the evidence points to the leading roles played by the Duke of Medina Celi, viceroy of Sicily, and the Grand Master of Malta, Jean de La Valette. Bound by the ties of close friendship,[23] they both had cause to resent the terrible corsair of Tripoli, and Jean de La Valette, who had been a remarkable governor of Tripoli[24] in the days when it was held by the Knights, was moved both by the nostalgia of a

[18] Jean Nicot, *Correspondance* . . ., ed. E. Falgairolle, p. 7.

[19] Cadiz.

[20] To the king, 20th June, 1559, Simancas E° 485.

[21] For most of the details in this section, the reader is referred to Charles Monchicourt's *L'expédition espagnole contre l'île de Djerba*, Paris, 1913, a work of exemplary erudition. My own notes refer for the most part to sources not mentioned in this book.

[22] To the viceroy of Sicily, same date, Simancas E° 1124, f° 300; instructions to the commander Guimeran, same date, *ibid.*, f°s 278, 279; to the Grand Master of Malta, same date, *ibid.*, f° 302, etc.

[23] Don Lorenzo van der Hammeny León, *Don Felipe El Prudente* . . ., Madrid, 1625, f° 146 v°.

[24] Jean de La Valette, of the 'langue' of Provence, Grand Master of the Order from 1557 to 1568. He had been governor of Tripoli from 1546 to 1549. Cf. the extracts from Bosio, *I Cavalieri gerosolimitani a Tripoli*, ed. S. Aurigemma, 1937, p. 271–272.

one-time 'African' and the ambitions of a head of state. For if Tripoli was recaptured, it would undoubtedly revert to his Order. As for Juan de la Cerda, Duke of Medina Celi, apart from Sicilian interests, he was driven by the urge to repeat, perhaps even more brilliantly, the success won by his predecessor, Juan de la Vega, in the Africa affair of 1550. Circumstances appeared to favour the design. Tripoli was poorly fortified, with a garrison of scarcely 500 Turkish soldiers. Dragut, constantly obliged to take action in the hinterland, was at open war with the 'king' of Kairouan, the Shābbiyya emir whose troops, according to a dispatch from La Goletta, had already defeated Dragut and whose moral authority was high, 'quasi come il Papa tra Cristiani', claims Campana,[25] with some exaggeration. Finally, one was always sure to receive some assistance from the nomadic 'Moors', who had suffered rather too much at the hands of the Turks to be benevolently disposed towards them. The Duke of Medina Celi had contacts with them (and even, through a certain Jafar Catania, found some allies among Dragut's followers). He himself recognized, however, that despite the letters and the professed sentiments of these sheikhs, it would be imprudent to rely on them.[26]

It was one of the Knights of Malta, the commander Guimeran, who was to present the Tripoli project to the king at Brussels. The matter soon progressed beyond the preliminary stage. On 8th May, 1559, Philip II asked for a report from the viceroy; but the report had not yet left Sicily when the king made up his mind,[27] notifying the Duke of Medina Celi and appointing him to the command of the expedition, in a letter dated 15th June, in which he explains his motives: the peace with France; the desirability of removing Italy's troublesome neighbour; the disarray of Dragut's followers on their return from the mountains of Gharian, surrounded and practically besieged as they were by hostile Moors; and finally the ease with which such an expedition could be launched in summer, before the corsair had securely fortified his position. In his instructions for the commander Guimeran, which were dispatched the same day, the king mentions yet another favourable factor: according to all reports, the Turks were not preparing to send out a large fleet that year. The king placed the Italian galleys at Medina Celi's disposal, ordering the Spanish galleys on the contrary to return home, where they had instructions to protect the coasts against marauders. When their commander, Juan de Mendoza, later refused to join the expedition, he was therefore merely obeying orders.[28]

[25] *Op. cit.*, p. 82–83.

[26] The Duke of Medina Celi to Philip II, 20th July, 1559, Simancas E° 1, f° 204.

[27] The decision is dated 15th June, the report 20th.

[28] C. Monchicourt, *op. cit.*, p. 93, gives the impression that Don Juan was acting on his own initiative. R. B. Merriman, *op. cit.*, IV, p. 102, suggests that Mendoza may have received orders; his hypothesis is proved by Philip's letter (see previous references, notes 22 and 8). On this point and on the lack of protection of the Spanish coast, see also L'Aubespine to the king, 20th July, 1559, E. Charrière, *op. cit.*, II, p. 600, note; L. Paris, *Négociations sous François II*, p. 24; C. Duro, *op. cit.*, II, p. 46.

The expedition was to be launched then using only the Italian section of Philip II's navy, the galleys of Sicily and Naples, galleys hired from the Genoese, the Tuscans, the Sicilians, the Duke of Monaco and the allied fleet of the Pope and the Knights of St. John. It was not difficult to assemble these ships, left without occupation since peace had been signed with France, in the usual convenient port of Messina. It was rather more difficult to collect adequate supplies and above all the indispensable troops. Originally, Philip II had envisaged embarking 8000 Spaniards, of whom 5000 would come from the garrisons in Milan and Naples and 2000 from the kingdom of Sicily. Together with the thousand men promised by Guimeran as Malta's contribution, this would surely be sufficient.[29] But even before he had news of the king's decision, Medina Celi, in his memorandum of 20th June, was asking for about twenty thousand men, although he considered two artillery batteries sufficient, considering the vulnerability of the position. These figures illustrate the conflict from the start between the royal project – a swift punitive expedition carried out in the summer months – and the more massive operation the viceroy was hoping to mount. So the king, when it proved difficult to withdraw Spanish troops from Lombardy (the French positions in Piedmont had not yet been handed over) gave orders on 14th July[30] that they should be replaced by the 2000 Italians whom the Duke of Alcalá had just sent from Naples to Messina, aboard galleys intended for the expedition. His chief concern was that the ships should lose no time in sailing back to Genoa to pick up troops and, in Philip's own words,[31] 'that the enterprise be executed in what remains of the fine season'. Speed, in the king's view, was the first priority.

But Medina Celi asked for more men, obliging Philip to repeat on 7th August[32] the order to send the Spaniards in Lombardy, 'con la mayor brevedad' to Sicily. However at this point, the Duke of Sessa found a fresh argument, in Henri II's death, for not sending his men.[33] Every order meant a further delay as letters slowly made their way between Ghent, Naples, Milan and Messina. On 10th August, Gian Andrea Doria wrote to the king[34] that he was sending Alvaro de Sande with one galley to Genoa; he would then travel to Milan and try to persuade the Duke of Sessa to send not only the Spaniards, but also two thousand Germans and

[29] N.B. the curious remarks by A. de Herrera, *op. cit.*, I, p. 14; after 1559, there was the problem everywhere of finding employment for the disbanded armies. The planned expedition was one way of removing from Spanish Italy the soldiers 'left over from the war in Piedmont and for whom there could be no better use than action against the Infidel'.

[30] Philip II to Guimeran, Ghent, 14th July, 1559, Simancas E° 1124, f° 331.

[31] To the viceroy of Sicily, Ghent, 14th July, 1559, Simancas E° 1124, f° 321.

[32] To Guimeran, Ghent, 7th August, 1559, Simancas E° 1124, f° 330.

[33] Figueroa to Princess Joanna, Genoa, 7th August, 1559, Simancas E° 1388, f°s 162–163.

[34] Gian Andrea Doria to Philip II, Messina, 10th August, 1559, Simancas E° 1124, f° 335, in Italian. In later years, Gian Andrea Doria almost always corresponded with the king in Spanish.

two thousand Italians, to be levied in Lombardy. This would raise fresh problems of galley transport, not to mention the necessary convoy of supply ships. On 11th August, the Duke of Sessa in Milan finally agreed,[35] since the handing over of the French positions to the Dukes of Savoy and Mantua had taken place as arranged. However it was a whole month before the Spanish, German and Italian infantry troops, promised one after another, finally arrived in Genoa.[36] On 14th September, Figueroa, the Spanish ambassador in the city announced their embarkation aboard a few roundships and eleven galleys: 'They are all magnificent and reliable troops. If they are not prevented by the weather, they will leave in full strength, without losing a moment'. But it was already 14th September!

In Naples too there were delays and problems. Gian Andrea Doria wrote on 14th September[37] that the galleys of the Order of Malta had left for Naples to fetch the Italian infantry which the knights had levied there. He himself had sent galleys to Taranto to fetch five Italian companies which the viceroy of Naples had agreed to contribute to the expeditionary force, and the ships were to go on to Otranto to pick up powder and shot. But only the day before he had received a letter from the viceroy who was now unwilling to let him have the said infantry, since he had received 'certain news of the arrival of the Turkish fleet, about 80 sail, at Valona, where it has embarked 1500 *sipāhis*.'[38] Doria was immediately concerned for his galleys: 'please God that they return safe and sound . . .'. Meanwhile time was slipping by and Philip II grew alarmed: 'I am extremely anxious about the success of the expedition', he writes on 8th October,[39] 'it being so late in the season.' From Syracuse, where the fleet had assembled, Don Sancho de Leyva, who was in command of the Sicilian galleys, wrote on 30th November: 'I have not hesitated to say to the Duke of Medina Celi, and several times, that the major chance of success of this expedition lay in speed and that delay was its worst enemy . . . and yet troops and supplies have been searched out from every part of Italy.'[40]

In my view it is essential to bear in mind the long delays in these preparations.[41] When the fleet finally left Syracuse on 1st December, during a spell of fine weather,[42] it included forty-seven galleys, four galliots, three galleons (in all fifty-four warships and thirty-six supply

[35] The Duke of Sessa to the king, Milan, 11th August, 1559, Simancas E° 1210, f° 203.

[36] Figueroa to Philip II, Genoa, 14th September, 1559, Simancas E° 1388.

[37] Gian Andrea Doria to Philip II, Messina, 14th September, 1559, Simancas E° 1124, f° 336.

[38] *Ibid.*

[39] Philip II to the Duke of Medina Celi, Valladolid, 8th October, 1559, Simancas E° 1124, f° 325–326.

[40] To Philip II, Simancas E° 1124, f° 270.

[41] Countless documents testify to these delays, see in particular Simancas E° 1049, f°s 185, 188, 189, 225, 227, 251, 272.

[42] Gio. Lomellino to the Signoria of Genoa, Messina, 10th December, 1559, A.d.S. Genoa, Lettere Consoli, Napoli–Messina, 1–2634.

vessels).[43] On board were ten to twelve thousand men,[44] an army larger than that which had attacked the town of Africa in 1550 and inferior only to the expeditions led personally by Charles V against Tunis and Algiers. Its size alone then is sufficient to explain the delays in marshalling it, but the arrival of the Ottoman fleet at Valona in August delayed it even further.[45] Herrera claims that if the Turkish armada, numbering almost a hundred sail, did not venture any further west, it was because of the respect felt for the galleys assembled at Messina.[46] A more truthful version might be that both fleets remained at a wary distance. It was not until the Turks had turned back to the east in October, that the viceroy of Naples had finally agreed to release the last companies necessary for the expedition, which he was holding near Taranto.[47] The Christian fleet had then sailed from Messina to Syracuse.

By now of course, any hope of surprise had vanished. All Europe knew about the expedition; so did the Turks and the corsairs. Dragut was preparing to defend himself. A French ship which left Marseilles on 25th November, carried at least as far as Milo news of the armada lying at Messina,[48] one of whose patrol ships had been captured by Dragut in the autumn.[49] Venetian diplomatic correspondence[50] buzzed with information, much of it correct, and the Turks began to assemble hastily in Constantinople a fleet which would consist, it was said of 250 sail. The opinion of Maximilian in Vienna was 'that the expedition has been noised abroad so far ahead that the Turks have been provided both with the motive and the time to mount a fleet of this size'.[51]

Was it to recapture some element of surprise that the fleet set sail in the unlikely month of December? As every seaman knew, it was madness to depart at that time of year. But the duke was a soldier, not a sailor: he insisted on his plan against all advice and the fleet left Messina. Almost immediately it encountered bad weather. The only solution was to make for Malta. The sailors must have prevailed here, for the armada was kept in Malta by poor weather for 10 weeks, until 10th February, 1560. During the long wait, epidemics decimated the expeditionary force which lost two thousand men without a shot being fired.

The galleys and roundships left Malta at different times, with a rendezvous arranged near Zuara. The roundships arrived late: the galleys were there by 16th February, after a detour to the Kerkenna Islands and Djerba, during which they had captured two ships carrying oil, barracans

[43] C. Monchicourt, *op. cit.*, p. 88. [44] *Ibid.*, p. 92.

[45] Gio. Lomellino to the Signoria of Genoa, Messina, 24th August, 1559, see note 42 above.

[46] *Op. cit.*, I, p. 15.

[47] Figueroa to Philip II, Genoa, 27th October, 1559, Simancas E° 1338, f° 16.

[48] Marin de Cavalli to the Doge, Pera, 29th January (1560), A.d.S. Venice, Senato Secreta, Cost. 2/B, f° 222 v°.

[49] C. Monchicourt, *op. cit.*, p. 100.

[50] 31st January, 1560, for instance, *C.S.P.*, VII, p. 150.

[51] Giacomo Soranzo to the Doge, Vienna, 3rd February, 1560, G. Turba, *op. cit.*, I, p. 134.

[a cloth of camel hair, used by the Turks for cloaks] and spices[52] and failed to take two galliots which fled with Euldj 'Alī to raise the alarm at Constantinople. Above all their movements alerted Dragut, who was in Djerba, and gave him time to get to Tripoli, where as one might imagine, there was much anxiety. This was how the situation appeared to the Venetian *bailo* in Constantinople, at the other end of the Mediterranean: four of Dragut's galleys had arrived at the Turkish capital. 'It is said that they carry, besides slaves, many rich gifts from Dragut, a sign that he is in desperate straits. He asks for immediate assistance, saying that he has only 1500 Turks under his command: all the corsairs who were wintering at Tripoli with about 15 ships, left at once without permission as soon as they heard of the approach of the Spanish fleet . . .'.[53]

If at this point the armada had attacked Tripoli, it might have had some chance of taking the city. It had already made a mistake in failing to catch Dragut at Djerba, for if the corsair had been prevented from leaving the island, the four hundred Turks in the garrison at Tripoli would have been forced to surrender to the Spanish, as the Duke of Medina Celi later admitted.[54] But on the sandbanks of Palo, near Zuara, the armada was once again immobilized by bad weather during the second half of February; further delays, further sickness and loss of men. On 2nd March it set sail, but now for Djerba, no doubt because it had been learnt that Dragut was back in Tripoli: if not the city then it would take the island, with its palms, olive-trees and sheep – the island of oil and wool. Troops were landed on 7th March, without incident. At the beginning of April, the Genoese consul, Lomellino, was able to announce from Messina (where the news had just arrived): 'l'armada nostra' – a meaningful expression – 'has taken Djerba'.[55]

By this date the Duke of Medina Celi had indeed established with much solemnity and paternalism, the government of the king of Spain in his new possession. He had invested a sheikh of his choice; he was seeing that the Djerbans were not molested and obliging his soldiers to pay for everything they took from the inhabitants. Moreover supplies were sent by the Hafsid ruler of Tunis and the Shābbiyya of Kairouan. Meanwhile on the north coast of the island, work was begun on the construction of a fortress, an extremely difficult task since there was an acute shortage of wood, stone and lime. The inhabitants of the island gave little effective assistance beyond providing camel-trains. So it was the army which, although racked by fevers, had to deploy its own men for this hard labour. Meanwhile the more practical ships' masters were buying up oil, horses, camels, or hides, wool and 'barracans'.

News was reaching Naples and Sicily not only from North Africa but

[52] Messina, 3rd April, 1560, A.d.S. Genoa, Lettere Consoli, Napoli–Messina, 1-2634.

[53] The *bailo* to the Doge, Pera, 30th March, 1560, A.d.S. Venice, Senato Secreta, Cost. 2/B.

[54] In his notes on the *Memorial de D. Alvaro*, C. Monchicourt, *op. cit.*, p. 100, note 2.

[55] 3rd April, 1560, A.d.S., Genoa, Lettere Consoli . . ., 1–2634.

from the Levant: and it was bad news. The viceroy of Naples was informed, at the beginning of April, that the Turkish fleet was preparing to sail much earlier than usual. He asked the king to order more warships to Messina, in particular the Spanish galleys. They would not be sufficient to engage the Turk, but at least might make it more difficult for him to land troops and artillery. He also wrote to Medina Celi, asking him to send back with the galleys the infantry he had lent him, directing it to Taranto.[56] On 21st he confided his fears to the king: if he did not receive his troops back, he would be obliged to levy Italians, which would mean more expense. He argued therefore for the return of all or part of the expeditionary force, adding 'I have informed [the Duke of Medina Celi] that in my opinion, it is foolish to wait for the Turkish armada to arrive while Your Majesty's fleet is still held up building a fort on Djerba'. A few days later, he learnt from a traveller returning from Constantinople, that the Turkish armada had left to take aid to Tripoli.[57] On 13th May,[58] he was informed that it had already left Modon. Immediately he sent word by land to Sicily and by frigate to the occupation army on Djerba. To the king he wrote, 'I consider that Your Majesty's armada is in no little peril . . .'. The dispatch which arrived on the 14th carried the news that the Turkish fleet had been sighted off Zante heading for Barbary.[59] But by then, everything was over at Djerba.

Piāle Pasha's fleet had indeed travelled as quickly as news of its arrival. By 8th May, it was between Malta and Gozo. It had sailed straight there with astonishing speed. To travel from Constantinople to Djerba in twenty days was something of a record. The duke, who had been expecting the Turks some time in June, saw them arrive on 11th May. He had been warned one day in advance by a frigate from Malta. No one in Djerba envisaged fighting. It seemed to everyone, as Cirini later recalled, 'that a retreat in good order would be better than a brave struggle'.[60] Should this attitude be attributed to some kind of inferiority complex, to the lack of sang-froid of the leaders, or to the desire of most of them to carry to safety the cargoes they had heaped up in their holds during their sojourn on the island? It was these cargoes which the *visitador* Quiroga later alleged lay at the origin of the disaster: had it not been for them, he said, for the anxiety to load as much booty as possible before leaving, the instructions given by the viceroy of Naples would have been followed and the Turkish fleet, arriving at Djerba would have found that the position had been evacuated a few days earlier.[61]

However, flight itself was no easy matter. Not wishing to abandon the Italian and German infantry, who were still ashore, the duke lost precious

[56] The viceroy of Naples to Philip II, 4th April, 1560, Simancas E° 1050, f° 28; to the Duke of Medina Celi, 20th April, *ibid.*, f° 32; to the king, 21st April, f° 32.

[57] To the king, 5th May, 1560, Simancas E° 1050, f° 36.

[58] *Ibid.*, f° 39.

[59] To the king, 16th May, 1560, *ibid.*, f° 40.

[60] P. 32 and 32 v°, quoted by C. Monchicourt, *op. cit.*, p. 109.

[61] The *visitador* Quiroga to the king, Naples, 3rd June, 1560, Simancas E° 1050 f° 63.

time in the night of the 10th to 11th. The next day, when the Turkish fleet attacked, there was immediate and utter panic.[62] Everything was sacrificed in the frantic desire to get away, including the famous cargoes – bales of wool, jars of oil, horses and camels – all thrown overboard with anything else that might make the ships heavier. Cigala, used to the life of a pirate in the Levant, was one of the very few who dared to confront the enemy. He held him off for a while and finally escaped. But of the 48 galleys and galliots composing the Christian armada at the beginning of the encounter, 28 were lost, not counting the ships which fell into enemy hands. Such a total débâcle had rarely been seen in Mediterranean waters.

The news spread rapidly to Sicily, Naples, Genoa and Spain and indeed throughout Europe. At two in the morning of 18th May, five galleys which had escaped from the battle scene arrived in Naples, three belonging to Antonio Doria, one to Bendinelli Sauli, one to Stefano de' Mari. They carried the bad news with many horrifying details. It is worth noting that these first arrivals were, by no coincidence, hired galleys, or as they were called, the galleys of *asentistas*, individuals who had made *asientos* or contracts with the king of Spain and who were therefore concerned above all with safeguarding their capital. Almost simultaneously other fugitives began to arrive aboard frigates or even smaller craft. Among the fortunate men who had slipped through the Turkish clutches was Gian Andrea Doria, commander of the fleet, the viceroy himself and some of his close associates, 'miraculously escaped to Malta and from Malta to Messina'.[63]

Meanwhile, several thousand men were left in the fort with abundant supplies – enough for a year it was said. What was to be done about them? From La Goletta – which did not receive news of the disaster until 26th May,[64] possibly by way of Sicily – Alonso de la Cueva wrote to the king on 30th: despite the pleas addressed to him by the viceroy of Naples, he did not think there was any possibility (he was quite right, the individual concerned was a lackey of the Turks) of using the king of Tunis, vassal of His Majesty to relieve the fort on Djerba. If the fort had been built not on the site of the old castle but at Rocchetta, where the fleet had first landed, the besieged would have had a deep harbour and drinking water; it would have been possible to assist them; as it was . . .

For a while, the Duke of Alcalá continued to cast about him wildly, envisaging measure after measure. Then he calmed down at the news of the survival of his colleague, the Duke of Medina Celi.[65] The latter moreover brought him a section of the Italian infantry who had escaped the disaster, to help protect Naples, until the Spanish infantry lost at Djerba had been replaced by fresh levies from Spain.[66]

[62] As described by Machiavelli's son, C. Monchicourt, *op. cit.*, p. 111.

[63] The Signoria of Genoa to Sauli, Genoa, 19th May, 1560, A.d.S. Genoa, L.M. Spagna, 3.2412.

[64] To the king, 30th May, 1560, Simancas E° 485.

[65] The Duke of Alcalá to Philip II, Naples, 31st May, 1560, Simancas E° 1050, f° 56.

[66] Alcalá to the king, 1st June, 1560, *ibid.*, f° 60.

As for Philip II, he heard the news on about 2nd June, through Genoa.[67] He was informed that 30 galleys and 32 boats had been lost,[68] that only 17 galleys had arrived safe in port – figures which were very close to the truth. The king, after deliberating the matter with the Duke of Alva, Antonio de Toledo, Juan de Manrique and Gutierre López de Padilla, decided to send immediately to Messina some person of authority to replace the viceroy, whose fate was still unknown, and to have 5000 foot soldiers levied in Calabria and sent to Sicily, along with artillery and munitions from the reserves in Naples.[69] It was even rumoured that Philip II was about to ask the king of France for the support of his fleet.[70] On 3rd June he appointed to the government of Sicily Don García de Toledo, the viceroy of Catalonia.

He was thus preparing for the relief of the fort, where it was believed that the Duke of Medina Celi was still besieged. On 8th June, reassuring news was received from Sicily: all the more reason, cried the king with exaltation, [71]to think of the men in the fort; to save those who had served the crown was a sacred duty. He intended to assemble up to 64 galleys at Messina, and had ordered an embargo to be laid on thirty large and well-armed roundships. Italians levied locally, the Spaniards from Lombardy (who were to be replaced by 3000 'alemanes altos', men from southern Germany) a total of 14,000 foot soldiers, were supposed to board the relieving fleet, under the command of Don García de Toledo. Finally a large quantity of grain was to be sent to Genoa for the manufacture of biscuit.

Everything was ready. Then on 13th June,[72] Philip received a letter from Don García de Toledo informing him that the viceroy of Sicily was safe.[73] On 15th the king abruptly countermanded his orders, alleging that according to all reports, the besieged had sufficient supplies for eight months, while the Turkish fleet only had enough for two and would not therefore prolong the siege.[74] All the preparations were cancelled. But some time elapsed before the new orders reached their destinations, time during which the agitation aroused by the Djerba affair continued to rage. Old Admiral Doria sent his advice: he considered it imprudent to launch a direct attack without a sufficient fleet of galleys. It would be better to attempt a diversionary raid in the Levant. The Signoria of Genoa offered four galleys for the relief of the fort. The lord of Piombino placed one galley at the disposition of the king of Spain: if it was not wanted,

[67] By a letter from Figueroa and dispatches from Cardinal Cigala and the Genoese ambassadors, Philip II to the viceroy of Naples, 2nd June, 1560, Simancas E° 1050, f° 63. On the size of the loss, Gresham, 16th June, 1560, mentions the figure of 65 ships, Ms. Record Office, n° 194.

[68] Tiepolo to the Doge, Toledo, 2nd June, 1560, *C.S.P. Venetian*, VII, p. 212–213.

[69] *Ibid.* [70] *Ibid.*

[71] Philip II to the Duke of Alcalá, Toledo, 8th June, 1560, Simancas E° 1050, f° 69.

[72] Barcelona, 9th June, 1560, Simancas E° 327.

[73] Don García de Toledo to Philip II, Barcelona, 12th June, 1560, Simancas E° 327.

[74] Philip II to Don García de Toledo, Toledo, 15th June, 1560, Simancas E° 327. Don García's reply, dated 23rd June, *ibid.*

he said, he would send it 'buscar su ventura'.[75] The Duke of Savoy announced that he had three, one in good order, the second with only its oarsmen and the third with nothing at all, but he was waiting for four galleys expected from the king of France.[76] Estefano de Mari had just bought two galleys from cardinal Vitelli, and he was prepared to hire them out to the king of Spain. Domenico Cigala, a Genoese living in Venice and formerly in the service of the emperor, offered to go himself to Turkey and Persia.[77] In Sicily, the Duke of Medina Celi threw himself energetically into preparations: in July through his intervention, seven galleys were being built for Palermo, Messina and the *Regia Corte*.[78] Earlier, in April, six had already been launched thus compensating in advance for those lost at Djerba.[79]

Once more, the incident illustrates the state of Franco-Spanish relations: the request for the French galleys was never clearly formulated in Philip's name. As Michiel told the Doge of Venice, 22nd June, 1560, the Spanish king feared a refusal more than he needed an acceptance.[80] Suspicion and accusations kept two states apart. Philip II had just sent back the French servants of the young queen. He had not changed his position concerning England. Equally striking is the equivocation on the French side, although the unrest beginning to be felt throughout the kingdom, the importance of which was somewhat exaggerated, was pushing the government of the Guises towards collaboration with Spain. The French ambassador in Spain informed Tiepolo, who repeated it to the Doge, on 25th June,[81] that he had offered the Spanish the galleys of Marseilles and some troops, but this was on 25th June, ten days after Philip II had cancelled the operation. The Duke of Alva made it quite clear in September: 'lately, at the time of the defeat at Djerba, we never dared ask [the French] for galleys in the rescue operation that Your Majesty was then preparing, for having several times made indirect approaches, I never found the response sufficiently encouraging to advise Your Majesty to propose it to them. When the time was past . . . and it was clear that their assistance would no longer be of any use, the ambassador came to tell me that if we needed any galleys, they would be made ready.'[82] France employed a hesitant policy, or perhaps it would be fairer to say that both sides continued to respect the political lines of the past, finding it difficult to break free of long-standing attitudes. Was not the

[75] Summary of Figueroa's letters of 3rd, 5th, 10th and 12th July, 1560, Simancas E° 1389. [76] *Ibid.* [77] *Ibid.*

[78] The Duke of Medina Celi to the king, 9th July, 1560, C. Monchicourt, *op. cit.*, p. 237.

[79] The Duke of Alcalá to Philip II, Naples, 9th October, 1560, Simancas E° 1050, f° 137.

[80] Michiel to the Doge, Chartres, 22nd June, 1560, *C.S.P. Venetian*, VII, p. 228.

[81] *C.S.P. Venetian*, VII, p. 229. The Duke of Alva to Philip II, Alva, 19th September, 1560, original in Simancas E° 139, a copy exists.

[82] B.N., Paris, Esp. 161, f°s 15 to 21. See E. Charrière, *op. cit.*, II, p. 621–623, on rumours of Franco-Spanish collaboration.

king of France still in touch with the sultan, whose friendship he did not care to lose,[83] as well as with the Algerines, who had sent emissaries to him and to whom arms were shipped from Marseilles?[84] At the same time Henry, king of Navarre, or as the Spanish called him, 'Monsieur de Vendôme', although not in power in France, and indeed persecuted by the Guises, was employed in plotting with the Sharif of Morocco.[85]

To return to Djerba, we have seen what a stir was caused by this little incident: in a few days its echoes sounded all over Europe. Even in Vienna, where a short time before there had been eagerness to cross swords with the Turk, the disaster gave Ferdinand and his advisers food for thought.[86] It is difficult not to believe that Philip II's prestige suffered in the affair, although the letters written by the Spanish ambassador at Vienna claimed that the counter-measures taken by his master had increased his reputation even more than a victory at Tripoli would have!

From this point of view, was the king's brutal decision to abandon the besieged men the best solution? If the sailors at Djerba had conducted themselves with reprehensible cowardice, the troops ashore, under the command of Alvaro de Sande, an experienced soldier, had performed their duty honourably. Although surrounded, he had not lost contact with the outside world; he was still able on 11th July, to send a letter to the viceroy of Sicily.[87] Perhaps there really was reason to believe that the Turkish fleet would abandon the blockade for lack of supplies at the end of the summer. The viceroy of Naples had understood that it would do so if no relief expedition was mounted. He informed the governor of La Goletta on 26th June (that is before he knew that Philip II had cancelled the expedition), imagining that it would be no bad thing to drop a few calculated indiscretions to the Turks, that preparations for the relief expedition were taking time.[88] It is certainly a fact that the Turkish leaders were showing little enthusiasm to attack. Time was passing, their losses were heavy. In July, Nassuf Agha, an intimate of Piāle Pasha, arrived in Constantinople and made no secret of his feeling that the fort would not be taken.[89] Meanwhile disquieting news was arriving from Persia: the Sophy, it was reported, was dead and his successor loved Bāyazīd as a brother.[90] There arrived in Genoa on 15th July (and heaven knows from where or when or how he came) a self-styled ambassador from Bāyazīd, whom Figueroa received and made much of before letting

[83] The king to the bishop of Limoges, 16th September, 1560, L. Paris, *op. cit.*, p. 523–530.

[84] Chantonnay to Philip II, 2nd February, 1560, A.N., K 1493, f° 39; L. Romier, *La conjuration d'Amboise*, 1923, p. 123. The queen of Spain to Catherine de Medici, September, 1560, L. Paris, *Négociations sous le règne de François II*, p. 510.

[85] Chantonnay to Philip II, Melun, 31st August, 1560, A.N., K 1493, f° 83; 3rd September, 1560, L. Paris, *op. cit.*, p. 506–509.

[86] 3rd July, 1560, *CODOIN*, XCVIII, p. 155–158.

[87] Simancas E° 1389.

[88] Simancas E° 1050, f° 84.

[89] 13th July, 1560, E. Charrière, *op. cit.*, II, p. 616–618.

[90] Constantinople, 17th and 27th July, 1560, *ibid.*, 618–621.

him go on to Nice aboard a brigantine.[91] It was not until he reached Spain that he was recognized as an impostor.

All these hopes were to be dashed. The Turks, if they did not launch an all-out attack on the fort, nevertheless seized all the neighbouring wells, reducing the besieged to the water in their cisterns, which soon ran dry in the heat of July. Alvaro de Sande then attempted a sortie, in the course of which he was taken prisoner, on 29th July. Two days later, the fort surrendered. This at least was the explanation given by the prisoner, when he wrote to the Duke of Medina Celi on 6th August,[92] laying the blame incidentally on his soldiers: 'if I had found in these men the qualities I have previously met in other troops under my command, we should have won the greatest victory seen for many years'. This was rather an inflated claim to make for an unsuccessful sortie and one is tempted to agree with Busbecq's opinion of him a little while later in Turkey: 'stout, impeded by his weight' – and inclined to be easily dispirited. It is possible then to advance another explanation of the second disaster of Djerba, as Duro does[93] – that the fault lay in the command. The most convenient accusation though is the one made in the letter sent to the king by Don Sancho de Leyva, from the depths of his prison in 1561, in which he calls to task those responsible for launching the expedition:[94] 'the double disaster is a judgement of God. If any more expeditions are to be sent, let blasphemers be closely watched, and the command be given this time to a man who is a good loyal Catholic' – outpourings typical of a prisoner mourning over the causes of his imprisonment, but also, for this was a lucid and experienced man whom we shall later find after his release, in command of the galleys of Naples, outpourings typical of a sixteenth-century Catholic.

After the surrender of the fort, the Turkish armada was free for action once more. Gian Andrea Doria, who was crossing from Malta to Africa with reinforcements for La Goletta, turned back as soon as he heard, abandoning his planned expedition against Tripoli.[95] And it was the victorious fleet which eventually put into the town before setting sail for Gozo,[96] where it arrived on 13th August;[97] from there it set out on a plundering expedition, sailing up the Sicilian coast and taking Augusta,[98] looting and burning the villages and hamlets on the coast of the Abruzzi.[99] But a dispatch sent on 4th September reported that the ships had careened at

[91] Figueroa to the king, Genoa, 26th July, 1560, Simancas E° 1389. The imposture discovered, Sauli to the Republic of Genoa, Toledo, 14th December, 1560, A.d.S. Genoa, Lettere Ministri, Spagna 2.2411.

[92] 6th August, 1560, Simancas E° 445, copy.

[93] *Op. cit.*, II, p. 36.

[94] B.N., Madrid, Ms 11085, 9th April, 1561.

[95] C. Monchicourt, *op. cit.*, p. 133; Gian Andrea Doria arrived in Malta on 8th August, 1560, G. Andrea Doria to Philip II, 8th August, 1560, Simancas E° 1125. He had been on the point of attacking Tripoli when news of the fall of the fort reached him. Doria to the king, 9th September, 1560, *ibid.*

[96] 18th August, 1560, Simancas E° 1050, f° 120.

[97] C. Monchicourt, *op. cit.*, p. 134.

[98] *Ibid.*

[99] G. Hernández to Philip II, Venice, 21st August, 1560, Simancas E° 1325.

Prevesa[100] where Piāle Pasha, having received orders to return to Constantinople, left the *sipāhis* (they later returned home overland) and set sail on 1st September for Navarino. A number of dispatches confirmed these reports and the viceroy of Naples prepared to disband the troops still positioned at Crotona and Otranto.[101] On 1st October, Piāle Pasha made his triumphal entry into Constantinople on the green-painted flag-ship, followed by 15 galleys painted bright red and the rest of the fleet amid salvoes and the huzzas of the crowd and to the deafening sounds of drums and trumpets. Busbecq has left a description of the arrival, of the long procession of captives[102] and of the festivities throughout the city, where the Christians were, for a time, harshly treated.

The event justified such enthusiasm: Islam had won the battle for the domination of the central Mediterranean.[103] Tripoli, where Turkish rule had had such a narrow escape, was now more firmly held than ever. Christendom was in anguish: hardly had the Turkish armada left the shores of Italy, than the Christian powers were thinking with apprehension of the calamities which it would bring when it returned – the next year. The Duke of Monteleone and the viceroy of Naples both talked of an expedition which the Grand Turk was intending to launch against La Goletta in 1561.[104] When it was learned in Vienna, on 28th December,[105] that the Turk was fitting 120 galleys, imagination already saw them sailing for La Goletta. This obsession was nourished by the audacity of the corsairs, following the Islamic victory. Despite the winter, they sailed up as far as Tuscany.[106] All the Italian and Spanish coasts were on the alert.[107] It was said at Constantinople that the Spaniards were so panic-stricken by the defeat they had suffered at Djerba, that they were on the point of evacuating Oran.[108]

That stage had not yet been reached. But it is true that the double disaster at Djerba had led to some useful re-thinking. From high-placed officials downwards, everyone hastened to send advice to Madrid and the advice was usually that the king could not protect the shores of his Mediterranean possessions without a strong fleet. It was also essential, said the Duke of Alva,[109] to reinforce the garrisons in Italy (which were evidently weak: we have seen how difficult it was to move even a few men from them); their small size and their depletion at the time of Djerba had undoubtedly contributed to the 'bubbling up' of intrigues in Italy which had

[100] Corfu, 4th September, 1560, Simancas E° 1050, f° 129.

[101] The viceroy of Naples to Philip II, Simancas E° 1050, f° 128.

[102] *The Turkish Letters*, p. 173 ff.

[103] R. B. Merriman, *op. cit.*, IV, p. 107.

[104] Monteleone to the king, 30th August, 1560, Simancas E° 1050, f° 121. The Duke of Alcalá to the king, Naples, 3rd September, 1560, *ibid.*, f° 124.

[105] The Count of Luna to the king, 28th December, 1560, *CODOIN*, XCVIII, p. 189–192.

[106] Florence, 10th July, 1560, Simancas E° 1446.

[107] G. Hernández to Philip II, Venice, 20th July, 1560, Simancas E° 1324, f° 47.

[108] H° Ferro to the Doge, Pera, 12th November, 1560, A.d.S. Venice, Senato Secreta, Cost. 2/B, f°s 290–291.

[109] 19th September, 1560, B.N., Esp. 161, f°s 15 to 21.

been particularly evident since the autumn.[110] But above all it was essential to be strong on sea.

Not everyone saw this; some were still preoccupied exclusively with defensive measures on land, the Duke of Alcalá for example, who was anxious to fortify the islands of Minorca and Ibiza, which he knew to be poorly protected.[111] Others were more lucid. In a rather emotional letter written on 9th July, 1560, the Duke of Medina Celi says: 'We must draw strength from our weaknesses; let Your Majesty sell us all, and myself first, if only he can become lord of the seas. Only thus will he have peace and tranquillity and will his subjects be defended, but if he does not, then all will go ill for us.'[112] 'Señor del mar' – a phrase which recurs several times under the pen of the viceroy of Sardinia, Alvaro de Madrigal, whether he is beseeching the king to become lord of the seas, or congratulating him for his steps in that direction, 'for this is what is most needed for the peace of Christendom and the preservation of his States'.[113] Similar wishes were expressed to the king in the same year, 1560, by Doctor Juan de Sepulveda[114] and by the eccentric Doctor Buschia, a little-known Spanish agent in Ragusa, one of those informers who, being paid by the word, often passed on the gossip of the taverns.[115]

Diplomatic dreams followed their usual course, but their final object was the same. All turned their eyes to Venice: in the present plight of Christendom, Venice alone – as we shall see – could help the western powers regain supremacy on sea. Knowing the sturdy self-interest of the Venetians, one may smile: such requests amounted to a notice to shut up shop. But pens continued to scratch regardless of such embarrassing difficulties. In Vienna, on 8th October, the count of Luna thought 'that it would be most helpful to Your Majesty to return to the League which the Venetians [once] formed with the Emperor, my lord, may God rest his soul . . .'.[116] And it seems that in Rome the idea of a league against the Turks, including Venice, was indeed brought up during the conversations between the new Pope, Pius IV, and Don Juan de Zuñiga who with his brother, the commander of Castile, at that time represented Philip II's interests at Rome. 'I have replied in another letter', writes Philip to the two brothers, 'to what you, Don Juan de Zuñiga, wrote to me and you will find in it my opinion of your conversation with His Holiness concerning a league with the Venetians against the Turks. Here, separately, I wish to inform you both of the suggestions concerning the same matter, made

[110] 9th October, 1560, Simancas E° 1850, f° 139, J. de Mendoza to the king, Palamos, 1st September, 1560, Sim. E° 327.

[111] 26th August, 1560, Simancas E° 1058, f° 118.

[112] C. Monchicourt, p. 237.

[113] Cagliari, 25th August, 1560, Simancas E° 327.

[114] *CODOIN*, VIII, p. 560.

[115] On doctor Buschia, see some of his letters in A.N., Paris, series K, 1493, B 11, f° 111 (20th, 28th, 30th September, 4th, 8th and 13th October, 1560). On inventive informers in the Levant, Granvelle to Philip II, Naples, 31st January, 1572, Simancas E° 1061.

[116] *CODOIN*, XCVIII, p. 182.

(in the name of the Duke of Urbino and by way of Ruy Gómez) by the Count of Landriano, who offers to negotiate such an alliance if I should wish it. To this offer I replied that since it was a matter so beneficial to the glory of God and the good of Christendom, I should be most glad of it. Upon which, news reached me of the death of the Doge of the republic, the predecessor of the present Doge [so the matter must have been discussed before 17th August, 1559]. I therefore suspended the talks for several days. But the Count of Landriano told me that the duke would take the matter up again.'[117]

In a way then the Djerba disaster was a salutary shock. It placed Philip II's empire face to face with its Mediterranean responsibilities. It forced it into some action. Djerba and the year 1560 marked the apogee of Ottoman power in the Mediterranean. After this year it was to decline – not by its own fault, but as a result of the massive labour of naval construction which was to begin that year from Palermo and Messina along the entire western coast of Italy and the whole Mediterranean coast of Spain.

2. SPAIN'S RECOVERY

The rally of Spain's fighting forces could hardly have occurred if the Turks had not unexpectedly and inexplicably allowed them a breathing space. The Turkish armada did not leave port in any strength in 1561 – nor in 1562, 1563 or 1564. For four years running, Christendom escaped with a scare. Four years running, the same scenario was enacted. During the winter months, there was much talk in dispatches of the Turks' military preparations: they were planning to launch a large army, attack La Goletta and Sardinia . . . In summer there would be an exaggerated scare which would fade quickly as the threat was seen to be imaginary. So military defence programmes planned in winter were not always fully executed, funds were withdrawn, troops demobilized, transports cut down and levies neglected. Spanish policy in the Mediterranean seemed to have a twofold rhythm. From the mass of documents which survive, it is extremely easy to chart its phases.

The years 1561 to 1564. Would the Turkish fleet invade in 1561? With the bitter memory of Djerba fresh in all minds, everyone feared so, in the cold winter of 1560–1561, when food shortage struck in one place,[118] plague in another.[119] A Frenchman happened to be returning from Constantinople with a Ragusan; from their conversation on the voyage, the Ragusan declared, on arriving in his native city in January, 1561, that according to his companion, the Turkish army had returned from Persia and that

[117] Philip II to the Grand Commander and to Don Juan de Zuñiga, Madrid, 23rd October, 1560, Simancas E° 1324, f° 48.

[118] Dolu to the cardinal of Lorraine, Constantinople, 5th March, 1561, E. Charrière, *op. cit.*, II, p. 652–653, high prices, poverty and plague.

[119] *Ibid.*

the fleet to be launched that year would be 'importantissima'.[120] The viceroy of Naples, after studying all the dispatches passed on to him, concluded on 5th January that the fleet would sail earlier than usual. In twenty or thirty days' time he would be taking steps to put the marine defences on the alert. Would he be sent the Spanish reinforcements he had been promised; would La Goletta be prepared in time?[121] A month later we find the viceroy of Sicily claiming that Sicily, Oran and La Goletta would be the principal targets of the enemy attack (11th February).[122] Not to mention the sorties of the Algerine corsairs, which were by contrast only too real and such a menace that Philip II on 28th February refused the viceroy of Majorca permission to leave the island.[123] Dispatches from Corfu, dated 30th March (and received at Naples on 2nd May) were still predicting a Turkish fleet of a hundred galleys[124] and Antonio Doria, who sailed to La Goletta in April, feared an encounter with the corsairs who were thought to be blockading the *presidio* while waiting to rendezvous with the Turkish armada.[125] The first dispatch from Constantinople which reported that there would only be a small fleet that year, confining itself to the defence of the coast of the Levant, is dated 9th April, 1561.[126] This news, which was amply confirmed by subsequent events,[127] cannot have reached Naples before June at the earliest and until then Christendom remained on the alert. A series of demands by La Goletta for water cisterns and artillery[128] were met between April and June. And in May, the viceroy of Naples was asking the Pope to authorize Marcantonio Colonna to participate if necessary in the defence of Naples.[129] Such moves give sufficient indication, I think, of the workings of this tremendous political and military machine, more regular in its functioning than historians usually give it credit for.

If we follow events through the eyes of the Neapolitans, it was not until the beginning of August that tension really relaxed, with the demobilization of the Italians posted to coastal defence.[130] In Spain the alert was over at the beginning of September: 'now that the fine season and fears of a Turkish attack are ended', writes the bishop of Limoges from Madrid on 5th September.[131] Leaving Constantinople in June, the Turkish

[120] Ragusa, 2nd January, 1561, Simancas E° 1051, f° 11.

[121] The viceroy of Naples to Philip II, 6th January, 1561, Simancas E° 1051, f° 12.

[122] To the king, Trapani, Simancas E° 1126.

[123] Philip II to the viceroy of Majorca, Aranjuez, 28th February, 1561, Simancas E° 328.

[124] Corfu, 30th March, 1561, Simancas E° 1051, f° 51.

[125] *Relación que haze Antonio Doria*, 18th April, 1561, Simancas E° 1051, f° 62.

[126] Constantinople, 9th April, 1561, Simancas E° 1051, f° 54.

[127] 12th, 14th April, 1561, Simancas E° 1051, f° 55; Liesma, 16th April, 1561, *ibid.*, f° 56.

[128] Alonso de la Cueva to the viceroy of Naples, La Goletta, 17th April, 1561, E° 1051, f° 57.

[129] The viceroy of Naples to Marcantonio Colonna, Naples, 9th May, 1561, Simancas E° 1051, f° 78.

[130] Viceroy of Naples to Philip II, 9th August, 1561, Simancas E° 1051, f° 119.

[131] Madrid, 5th September, 1561, B.N., Paris, Fr. 16103, f° 44 ff.

fleet consisting of about fifty galleys, was content merely to sail to Modon and back. It had left Modon by the beginning of July and on 19th August set sail from Zante for Constantinople.[132] Why such a limited effort?

Contemporary documents offer no more than impression and conjecture. Was it because of Persia yet again, as suggested by a letter written by Boistaillé in Venice to Catherine de Medici, on 7th June, 1561?[133] An earlier letter, dated 11th May,[134] had hinted as much: 'King "Phelippes" . . . has no other nor more sure way to contain the Grand Signior in these kingdoms than by this bridle. And he may be sure that the Grand Signior would not otherwise have let him off so lightly this year as he has done, sending only forty galleys out from Constantinople.' Let us remark in passing that Venice was better informed than Naples on 9th May; on 8th June the viceroy of Sicily still did not know whether the threat from the Turk was serious or not.[135] However, Boistaillé did not consider that affairs in Persia were sufficient to explain the abnormally short tour of the armada, which had turned for home by July.[136] Had Piāle Pasha died as it was rumoured, he was wondering on 11th July? Piāle Pasha was not dead – but Rustem was and 'Alī Pasha had succeeded to the post of Grand Vizier;[137] rivalry between the sultan's ministers may perhaps have had something to do with the fleet's movements.[138] Many rumours continued to circulate, including even the suggestion that the galleys were needed for action in the Black Sea.[139]

The report of the Spanish ambassador at Vienna, 14th September,[140] gives more details: the Turk has been unable to reach agreement with the Sophy; being much angered by this, he has ordered war to be declared on the Persians. It is said that he intends to spend the winter at Aleppo preparing for the coming campaign. But it is also said, and here the count of Luna's information is worthy of remark, that 'he will not dare to leave Constantinople, not only because he cannot count on the loyalty of his son Selīm, but also because he fears, such is the popularity of Bāyazīd among his subjects, that a rebellion may break out in these parts and make it impossible for him to return'. We may guess at the background to the war against Bāyazīd suggested by this document and should not therefore underestimate its social aspects. Turkey was, it seems, threatened not to say paralysed by this conflict, to the very heart of the state. To these explanations advanced by Christian observers, we might perhaps add a hypothesis of our own: the year 1561 seems, throughout the Turkish

[132] 'Lo que se entiende de Levante . . . de Corfu', 10th August, 1561, Simancas E° 1051, f° 120.

[133] E. Charrière, *op. cit.*, II, p. 657–658. [134] *Ibid.*, p. 653–654.

[135] Viceroy of Sicily to Philip II, Messina, 8th June, 1561, Simancas E° 1126. Viceroy of Naples to the king, Naples, 7th July, 1561.

[136] E. Charrière, *op. cit.*, II, p. 661.

[137] H° Ferro to the Doge, Pera, 10th July, 1561, A.d.S. Venice, Senato Secreta Cost. 3/C. Rustem Pasha died on 9th July.

[138] The bishop of Limoges to the king, Madrid, 5th September, 1561, B.N., Paris, Fr. 16103, f° 44 ff.

[139] E. Charrière, *op. cit.*, II, p. 657–658. [140] *CODOIN*, XCII, p. 240–244.

Empire, to have been a year of poor harvests, of disputes with the Venetians over grain and of outbreaks of epidemic. All this must have contributed to the final result.

In 1562, the news from Constantinople was less disturbing. The only slightly sensational reports concerned an ambassador from the king of Tunis, tearing his garments before the Grand Turk,[141] or (but this was of interest only to Genoa) the voyage made by Sampiero Corso to Constantinople, by way of Algiers.[142] Defence preparations, which began later than the previous year, ended sooner. The Turkish fleet attempted no attack and the alert was pronounced over by the end of May in Naples,[143] and in the first fortnight of June at Madrid.[144] Strangely enough, but it was evidently the sequel of the exaggerated and groundless fears of the summer before, this time the absence of an attack was regarded as quite natural, and no one took great pains to discover some reason for it. The viceroy of Naples simply writes that the sultan will not send his fleet against La Goletta, 'either because of the dispute between his sons, or because he wants to keep it at home, or because he knows that the position is strongly defended'.[145]

There must have been good and pressing reasons anyway, because in the same year the Turks renewed the truce with the emperor which had been suspended since 1558.[146] On this occasion Alvaro de Sande, Don Sancho de Leyva and Don Berenguer de Requesens were freed in return for ransoms.[147] Evidently the sultan wished to concentrate his full attention on the East, since the *de facto* peace which he could choose to force on Christendom at sea was not enough: he also desired to have his land armies released from the western front.

During the next winter, Christendom grew accustomed to peace, while officially still taking precautions. There was of course some talk of the threat to La Goletta and Sardinia. But as early as January, 1563, disputes over the grain which Venice was as usual trying to obtain from the Aegean, indicated that Turkish granaries were far from full.[148] It was also learnt at an early date that Sampiero Corso's voyage had come to nothing. Significantly, it was Philip II himself in the Escorial who prudently sounded the alarm[149] and ordered supplies to be sent to La Goletta, the

[141] 28th May, 1562, Simancas E° 1052, f° 27.

[142] Sampiero did not actually arrive in Constantinople until January, 1563. See the many documents in A.d.S. Genoa, Spagna, 3.2412 and Costantinopoli, 1.2169.

[143] Viceroy of Naples to Marcantonio Colonna, 24th May, 1562, Simancas E° 1051, f° 87.

[144] Philip II to the viceroy of Naples, 14th June, 1562, Simancas E° 1051, f° 96.

[145] See note 143.

[146] Daniel Barbarigo to the Doge, Pera, 5th August, 1562, A.d.S. Venice, Senato Secreta 3/C; Venice, 20th August, 1562, an eight-year truce was agreed, *CODOIN*, XCVII, p. 369–372, C. Monchicourt, *op. cit.*, p. 142.

[147] Constantinople, 30th August, 1562, E. Charrière, *op. cit.*, II, p. 702–707.

[148] 6th–17th January, 1563, *ibid.*, p. 716–719.

[149] Philip II to the Dukes of Savoy and Florence, Escorial, 8th March, 1563, Simancas E° 1393.

tiny garrison which swallowed so many men and supplies. As early as the beginning of June, Naples was certain there would be no armada. An informant who had left Constantinople on 29th April, reached the city on 5th June with good news which nobody questioned.[150] All subsequent reports confirmed that the Turks had merely launched a certain number of galleys without arming them and had only sent out the few that were needed for the defence of the Aegean.

In 1564, there was little change. In January there was some talk of Turkish preparations, which alarmed even the Venetians.[151] But on 12th February it was learnt from Constantinople that there would be no armada.[152] At about the same time the Duke of Alcalá was in the process of arranging for a thousand men to be sent to La Goletta, but, he said, this was in spite of the information he had received and not because of it.[153] All was calm. Sampiero Corso was able to profit by the lull to negotiate through intermediaries, with the Spanish ambassador in Paris, Francés de Alava.[154] He complained about the Genoese rule, reminded the ambassador that Corsica belonged by rights to the Crown of Aragon and that the Corsicans were subjects of the Catholic King. Francés de Alava concluded that in any case the two Corsican captains with whom he was in contact on this occasion, were familiar with the situation in the Levant and might be of some service to the king.

True, there was another alarm at the beginning of May. Ruy Gómez was to mention it to the French ambassador.[155] But the crisis was over before the end of the month.[156] Detailed dispatches from Constantinople[157] dated 27th May and 6th June, explain why, in spite of the protests of the *re'īs* who were eager to put to sea, the armada could not be sent out: sixty galleys at present being made seaworthy, will be launched, but so far nothing has been done about finding oarsmen and rations, so they will not be ready before 10th or 15th July at the earliest. Then tallowing of the keels and the embarkation of the *sipāhis* from the garrisons will take them up to August. So in all probability they will not sail to the west at all. In mid-June, Philip II therefore quite sensibly decided to turn his fleet on the Barbary corsairs.[158] Troops from Naples were now being moved not to Messina and La Goletta but towards Genoa and Spain, to Málaga to be precise.[159] There was still a shadow of uncertainty in August: on 2nd, Sauli reported to the Signoria of Genoa[160] that Madrid had spread

[150] Simancas E° 1052, f° 169.

[151] Narbonne, 2nd January, 1564, Edmond Cabié, *Ambassade en Espagne de Jean Ébrard, Seigneur de Saint-Sulpice*, Albi, 1903, p. 212.

[152] Constantinople, 12th February, 1564, Simancas E° 1053, f° 19.

[153] Viceroy of Naples to Philip II, Naples, 17th February, 1564, Simancas E° 1053, f° 22.

[154] To Philip II, Paris, 17th March, 1564, A.N., K 1501, no. 48 G.

[155] Saint-Sulpice to the king, 11th March, 1564, E. Cabié, *op. cit.*, p. 262–263.

[156] *Ibid.*, p. 269, 29th May, 1564.

[157] Simancas E° 1053, f° 54.

[158] Early July, 1564, E. Cabié, *op. cit.*, p. 270.

[159] *Ibid.*, p. 279.

[160] A.d.S. Genoa, L.M. Spagna, 3.2412.

word of the approach of the Turkish armada, but that this was a rumour considered to be 'per vanitā' and 'that therefore less anxiety was felt about the situation in Corsica' (for Corsica had just risen up under the leadership of Sampiero Corso). This is the last we hear of the Turkish armada that year, 1564, when Christendom, not unduly perturbed about the East, was preoccupied above all with events in the western sea, with Corsica in the first place, but also with the victorious expedition led by Don García de Toledo against the tiny Peñón de Velez on the Moroccan coast.

Within a few weeks and months, the guessing-game began again, with the first frosts of winter. On 29th December, 1564, Maximilian was talking to the Venetian ambassador at Vienna, Leonardo Contarini.[161] It was thought that a large Turkish armada would sail as soon as fine weather returned 'and what will you Venetians do then, with Cyprus in the front line and likely to take the fancy of the Turks?' 'Venice will fortify it', replied the ambassador, asking in return about the Corsican situation where Genoa's problems were likely to be increased by the arrival of a Turkish fleet. 'Sampiero Corso may have no declared allies', said the emperor, 'but he has had secret contacts with a certain prince, so secret that nobody at all knows about them.' A typical winter conversation in front of the map of Europe, but future events were to confirm them in at least one respect: 1565, unlike the past four years, would hear the sound of battle.

A double enemy: the corsairs and the winter seas, 1561–1564. The Turks had presented the Spanish Empire with the unexpected gift of four years' peace. Those years were turned to good account. In the first place action was taken against the corsairs. They had not eclipsed themselves at the same time as Turkish warships and quite naturally, every year, the Spanish navy was tempted to turn against them the forces which the threat from Turkey had made it assemble, but which were left idle when the crisis never materialized.

Philip II's new fleet was forged in this painful struggle against expert enemies who were both difficult to track down in the vast expanses of the sea and difficult to harm in their African lairs.

They dealt the Spaniards some very heavy blows. In July, 1561, the entire Sicilian squadron of seven galleys fell into an ambush laid by Dragut off the Lipari islands.[162] It had been placed under the command of the Catalan Knight of Malta, Guimeran, the same Guimeran 'who was held in much esteem at Saint-Quentin', as the bishop of Limoges wrote to the French king, 'However he was reckoned to be a better commander on land than on sea, where his apprenticeship was interrupted by Dragut, and many good men were lost with him, as you have no doubt, Sire,

[161] To the Doge, G. Turba, *op. cit.*, I, 3, p. 289–290.

[162] Diego Suárez, quoted by General Didier, *Histoire d'Oran*, 1927, VI, p. 99, note 5.

already heard by now from Italy'.[163] Amongst the losses, he adds, they 'want to conceal from us another ship lost between Naples and Sicily and said to be carrying three of the old companies from Flanders lately transported to Italy'. Taking advantage of the recall of the king's galleys to the coast of Spain, Dragut, 'with thirty-five vessels, has held the kingdom of Naples in such a noose, that a fortnight ago a messenger arrived on foot from the marquis of Tariffa, the governor of the said Naples', begging Philip 'to send back the said galleys for those of Malta, of Sicily, and other neighbouring ports are so harassed and confined by the said Dragut that not one of them can pass from one place to another'. Fortunately, adds the bishop, the Turkish armada has not arrived, 'which is in faith a blessing from God, as any man can see since a handful of pirates and brigands holds this prince, from the Straits of Gibraltar to Sicily, in such servitude, that the Infidels descend wherever they please on his lands, if not on his forts'.

The true history of this reign of terror from Gibraltar to Sicily, can be discovered only by investigating the archives of the towns of the Balearics, Andalusia and Valencia, where there seems to have been a connection between unrest among the Moriscos and outbursts of aggression by the corsairs. Haëdo suggests as much on several occasions, when he talks of activity at the corsair base at Cherchel, inhabited almost entirely by Morisco fugitives who were still in touch with their friends and relatives on the Spanish coast.[164]

The corsairs certainly captured many prizes in the summer of 1561. As it drew to an end, the time for Christian revenge approached and reports begin to tell a different story: in September the Spaniards were thought to be planning to take Monastir.[165] In this season, the corsairs were returning home one by one, taking refuge from the stormy sea. The government of Spain, however, did not intend to lay up its ships: it was still militarily at a disadvantage. The prince of Melfi, who had been appointed to the command of the Spanish galleys on the death of Andrea Doria, protested against orders given by landlubbers who did not know what it was like to be tossed in a galley by the winter waves of the Mediterranean – who did not realize the wear and damage which storms could inflict on the narrow oarships.[166]

He was right of course. But when it was a matter of acting to maintain vital communications, the weaker party was automatically forced to operate in bad weather when the seas were empty and the enemy safe in harbour. The viceroy of Sicily, still the Duke of Medina Celi, sharply reminded the prince of Melfi that the king had given the command for the galleys to proceed to Messina; orders were orders.[167] So the movements planned were to take place. La Goletta received supplies of munitions in

[163] 24th August, 1561, B.N., Paris, Fr. 16103.

[164] C. Duro, *op. cit.*, II, p. 44. See above, Part II, Chapter 4.

[165] The bishop of Limoges to the king, 5th September, 1561, B.N., Paris, Fr. 16103, f° 44 ff.

[166] Simancas E° 1051, f° 131.

[167] *Ibid.*, f° 139.

October[168] and at the beginning of November the Spanish fleet was again at Trapani. That it was no longer at Messina shows that it had retreated inside its own lines, while the prince of Melfi had shown no desire to move outside them in the first place. The weather was terrible in any case[169] and it was no doubt at this time (the document bears no precise date) that a convoy of galleys had to abandon a voyage to La Goletta. Orders were meaningless in the face of a tumultuous sea. The viceroy of Sicily finally decided to send a large roundship with two thousand *salme* of wheat to supply the needs of the abnormally large garrison.[170] There was talk in January of re-embarking the superfluous Spanish troops in the *presidio*, commanded by Juan de Romero,[171] but fetching them back presented the same problem as supplying them. The prince of Melfi, if he had dared, would have found it easy to demonstrate that the policy of keeping the Spanish galleys on duty, so costly in itself, had not produced many results that winter.

In spring, privateering by the corsairs began again with renewed vigour. On 1st March, 1562, a letter from La Goletta[172] reported that Dragut had left in search of grain. In April, the vessels of Algiers tried to storm Tabarka.[173] Juan de Mendoza meanwhile managed between May and June to lead a large convoy of roundships to La Goletta under the escort of about twenty galleys,[174] without meeting any enemy vessels and almost without mishap. In the same month of May, pirate-ships from Algiers were at Marseilles.[175] They claimed to have captured en route a Ragusan roundship coming from Alexandria carrying goods belonging to Florentine merchants, and a Venetian roundship laden with malmsey wines. There was also mention of a 'town' they had captured, near Porto Maurizio, where they had taken fifty-six prisoners. 'They came to Marseilles to renew their stocks of biscuit and other rations and to begin their attacks again. In the night, they secretly loaded thirty-six barrels of gunpowder and saltpetre.' Then we lose trace of the corsairs. They must have been very persistent on the northern shores of the sea since in June we find Juan de Mendoza patrolling the coast of Naples up to the mouth of the Tiber, with thirty-two galleys, at the request of the Pope.[176] And in July, there arrived in Algiers, along with Sampiero Corso, an ambassador from France charged with the mission

[168] Simancas E° 1051, f° 49.

[169] Viceroy of Sicily to Philip II, Palermo, 8th November, 1561, Simancas E° 1126.

[170] Viceroy to Philip (undated in my file), *ibid.*

[171] Bishop of Limoges to the queen, Madrid, 3rd January, 1562, B.N., Paris, Fr. 16103, f° 129 v°. In June, 1562, the Spaniards were still in La Goletta. Report of voyage by J. de Mendoza, Simancas E° 1052, f° 33.

[172] Alonso de la Cueva to the viceroy of Sicily, 1st March, 1562, Simancas E° 1127.

[173] Figueroa to Philip II, Genoa, 9th May, 1562, Simancas E° 1391.

[174] Report of voyage by Juan de Mendoza . . ., Simancas E° 1052, f° 33; Don J. de Mendoza returned to Palermo on 9th May, 1562, Simancas E° 1127.

[175] *Per lettere di Marsiglia*, 21st May, A.d.S. Genoa, L.M. Spagna, 3.2412.

[176] Viceroy of Naples to Philip II, 4th July, 1562, Simancas E° 1052, f° 45.

of claiming compensation for the damage caused by the sea-robbers of the city.[177]

In September, the Spaniards began to hit back. It was announced in Barcelona that three corsair galliots had been taken at Ponza; and likewise a few foists (unconfirmed), at Tortosa.[178] The Spaniards scored another success with a new supply expedition to the insatiable La Goletta, led by Gian Andrea Doria in September.[179] Juan de Mendoza had returned towards the Spanish coast with the fleets of Sicily and Spain as well as a few private galleys, to patrol the coast and to carry supplies and men to Oran.[180] But surprised by a strong easterly gale in the harbour of Málaga, the twenty-eight galleys were obliged to take refuge to the leeward in Herradura Bay. The *Instructions Nautiques*[181] describe this silted mooring, with depths of twenty to thirty metres, as dangerous in an onshore wind. Hardly had the galleys taken refuge there, when they were surprised by a gale blowing from the south.[182] It was a total disaster: twenty-five of the twenty-eight galleys were lost and between 2500 and 5000 men perished. Only a fraction of the fittings could be salvaged from the wrecks.

The news reached Gaeta on 8th November, 1562 and was passed on to Naples.[183] The catastrophe coming so soon after Djerba, aroused considerable emotion.[184] But Philip II's government was able to turn weakness to strength: on 12th December, 1562,[185] in a special session of the Cortes, an extraordinary subsidy was requested for the defence of the African frontiers, and for the fitting of new galleys.[186] Spain's maritime recovery, although substantially set back, was pursued with even more vigour than before. What had been lost at Herradura was the naval protection of the Peninsular coastline and of the *presidio* of Oran, the only position worthy of the name, according to the bishop of Limoges, that Spain possessed in Africa. The great offensive launched the following year against Oran by the corsairs of Algiers, was undoubtedly a consequence of the Herradura tragedy.

It was an assault in the grand manner, quite different from that led by Hasan Corso in 1556. The siege lasted two months, from the first days of April[187] until 8th June, 1563. The Spanish garrison had been alerted in

[177] Algiers, 12th July, 1562, A.d.S. Genoa, L.M. Spagna, 3.2412.

[178] Sauli to the Signoria of Genoa, Barcelona, 13th September, 1562, *ibid.*

[179] To the king, La Goletta, 30th September, 1562, Simancas E° 486.

[180] Saint-Sulpice to the king, 26th October, 1562, E. Cabié, *op. cit.*, p. 90.

[181] N° 345, p. 83.

[182] *Relación de como se perdieron las galeras en la Herradura*, 1562, Simancas E° 444, f° 217; C. Duro (*op. cit.*, II, p. 47 ff) does not seem to have consulted the sources.

[183] J. de Figueroa to the viceroy of Naples, Gaeta, 8th November, 1562, Simancas E° 1052, f° 67.

[184] C. Duro, *op. cit.*, II, p. 48.

[185] Agostinho Gavy de Mendonça, *Historia do famoso cerco que o xarife pos a fortaleza de Mazagão no ano de 1562*, Lisbon, 1607.

[186] C. Duro, *op. cit.*, II, p. 49.

[187] On 3rd or 4th, according to the traditional versions, perhaps not until 8th. At this date, the forces of Algiers were still two leagues from the town on the landward side. Philip II to Figueroa, Segovia, 18th April, 1563, Simancas E° 1392; *Lo que ha*

advance, on 20th March, when it had learnt of the arrival at Mazagran of 4000 'tiradores que so los que van delante del campo del Rey de Argel'. The spies added that had it not been for the rains, the king of Algiers himself would have been there at the same time as them. He was thought to be reaching Mostaganem on Friday, 26th March, at the same time as forty ships, including two caravels and a merchant *naveta* which had happened to be at the mole in Algiers and were subsequently laden by the corsairs with powder, cannon balls, 'bestiones de madera' and biscuit. Four galleys arrived carrying the artillery and finally, ten large galleys (were they the same ten, captured from the Christians at Djerba, which Hasan Pasha had brought back from Constantinople to Algiers?)[188] were being sent in two squadrons to the coast of Spain to investigate the possibility of relief being sent from the Peninsula.[189]

Armed with this information, the two sons of the Count of Alcaudete who commanded the two *presidios* of Oran, Martín the elder and Alonso, were able to sound the alarm before the double forces of Algiers, on land and sea, were upon them. They had to defend both Oran itself and, on the other side of the harbour of Mers-el-Kebir, the small fort on a peninsula, which commanded the moorings in the harbour. The Algerines, after some hesitation, concentrated their attack on Mers-el-Kebir, and in particular on the little fortress of San Salvador, which had just been built on the heights overlooking Mers-el-Kebir from the landward side; then after the fall of San Salvador on 8th May following a siege of twenty-three days, on Mers-el-Kebir itself and its small garrison of a few hundred men. The defenders succeeded however, despite the long bombardment from 8th to 22nd May, in repelling the first assault on the 22nd, inflicting severe losses on their enemy. The Algerines decided to attack the fort from another direction, from 22nd May to 2nd June, then attempted an assault simultaneously from the old and the new battery emplacements, while the forward guns of their ships battered the position from the seaward side. The attack was unsuccessful and resulted in the Turks evacuating eight galliot-loads of wounded men to Algiers.[190]

So Mers-el-Kebir had held out. It is true that the proximity of the Spanish coast provided both positions with valuable help. Through the Algerine blockade slipped a number of fast galleys and even more small boats, whose pilots, Gaspar Fernández, Alonso Fernández and others, were the real saviours of the besieged garrisons, bringing them rations, munitions and reinforcements. Between 1st May and 4th June, over two hundred gentlemen crossed over by this means from Spain to Oran. In Cartagena, the Marquis de los Vélez, whose name was much feared in the

---

passado en el campo de Oran y Almarçaquibir . . ., Toledo, 1563; Document, B.N., Paris, Oi 69.

[188] D. de Haëdo, *op. cit.*, p. 75 v°.

[189] Summary of letters of the Count of Alcaudete, March, 1563, Simancas E° 486.

[190] *Relación de lo que se entiende de Oran por cartas del Conde de Alcaudete de dos de junio 1563 rescibidas a cinco del mismo*, Simancas E° 486.

Moslem world, held open house for them, so liberally that the towns-people could find no fresh meat or fish in the market.[191] However the situation in Mers-el-Kebir was bad: the exhausted garrison had practically nothing to eat, 'sino algunae çeçinas de jumentos y animales nunca usados', except a little dried meat from donkeys and other animals not usually eaten. The relief fleet arrived only just in time on 8th June, and put the Turkish 'dogs' to flight.

That it should have been there at all, only two months after the beginning of the siege is a miracle, when one knows that the galleys which accomplished the feat almost all came from Italy. The interesting thing, to the historian that is, about the siege, which made much stir in Spain (both Cervantes and Lope de Vega made it the subject of a play) is not so much the heroic conduct of Don Martín and his men at Mers-el-Kebir, as the rapidity with which help was organized. It is one of the very few good examples of Spanish dispatch.

On 3rd April before the siege had even begun, Philip II having received the spies' reports, had sent an express courier to his ambassador at Genoa, Figueroa, requesting him to send the galleys of Gian Andrea Doria, Marco Centurione, Cardinal Borromeo and the dukes of Savoy and Tuscany, to rendezvous in the first place at the port of Rosas. 'Let haste be made to save every possible hour', wrote the king, 'for until the moment I see them here, I cannot help but feel justifiable anxiety.'[192] These orders, received at Messina on 23rd April[193] meant the recall to Spain of all the Italian warships, except for the galleys of Sicily and of the Knights of Malta. 'Lo que más importa', wrote Philip to Don García de Toledo on 25th April, 'es la venida de las galeras de Italia.'[194]

Similar views were held in Italy. The viceroy of Naples in a letter to Figueroa dated 25th April,[195] said that he had been told about the siege of Oran by letters sent as early as 28th March (that is before the king's orders of 3rd April) and knowing that the Turkish armada would not come out that year, he immediately realized 'that it would be most valuable to His Majesty's service if on the twenty-two galleys which Gian Andrea Doria could provide and four others from the kingdom, twenty-six in all, the admiral could embark two thousand Spanish soldiers and sail by way of Sardinia, Minorca and Ibiza to Cartagena [that is directly, without stopping at the Rosas rendezvous] where he would await orders from His Majesty'. On the same day,[196] Doria announced to the king that he would soon be at Cartagena. Philip II, receiving his letter on 17th May at Madrid,[197] wrote his answer the same day informing Doria that biscuit would be made at Cartagena in preparation for the arrival of the galleys

[191] *Lo que ha passado* . . ., B.N., Paris, Oi 69.
[192] The king repeats this plea for urgency in his letter of 18th April, Simancas E° 1392.
[193] Viceroy of Sicily to Philip II, Messina, 23rd April, 1563, Simancas E° 1127.
[194] Madrid, 25th April, 1563, Simancas E° 330.
[195] Simancas E° 1052, f° 156.
[196] This letter is quoted from the king's reply, see next note.
[197] Madrid, Simancas E° 1392.

and that some would be brought from Barcelona and Málaga as well. He added that for various reasons – the possible late arrival of the Italian fleet, the general feeling that this was a Spanish domestic affair, and the fact that on its return from Oran the armada would have to be divided into two squadrons, one of which would return under Doria to Italy to protect the coast from corsairs – for all these reasons, he had chosen as leader of the expedition, Don Francisco de Mendoza, captain general of the Spanish galleys.

So by the beginning of June there were assembled at Cartagena forty-two galleys, of which four were Spanish. Eight remained in port (four belonging to the Duke of Savoy, four to Genoa). The other thirty-four pounced on Oran on 8th, but succeeded in capturing only three *redondos* (roundships), a dozen barques and a French *saëte* (found to be laden with lead, munitions and coats of mail). All the great oarships had had time to flee.[198] Perhaps it was because, according to a dispatch from Bône dated 3rd June retransmitted through Marseilles, the Algiers fleet had been preparing to leave in any case.[199] The operation was nevertheless considered to be a great success. The king wrote as much on 17th June[200] to the viceroy of Naples, whom Diego Suárez in his valuable chronicles of Oran praises as the architect of the victory. No one would contradict this. But among those deserving praise, Philip II himself should be included and congratulations on the victory could be extended to the whole Spanish military machine, which for once worked perfectly, possibly because of the bitter experience of the past and because the sphere of action was limited and close to home.[201]

In Madrid however there were even grander ambitions. Hardly had the fleet returned to Cartagena, than it received orders from the king to take the Peñón de Vélez by surprise. Francisco de Mendoza, who was ill, left Sancho de Leyva in command of the operation, plans for which had been drawn up by the governor of Melilla. But the Turkish garrison on the little island was alerted by the sound of the oars and the army which landed in front of Vélez lacked determination. Instead of pressing on, bombarding the Turks and storming the position, a majority of the chiefs of staff decided to re-embark the men and postpone the attack. The fleet returned to Málaga in the first week of August.[202] The corsairs, when informed of the unsuccessful expedition, redoubled their raids on the Spanish coast. They even went as far as the Canary Islands, which they had never done before. Meanwhile the Spanish galleys completed their supply trips to Oran, taking over at the end of August the 20,000 ducats needed for the

[198] Madrid, Simancas E° 1392.

[199] Simancas E° 1392.

[200] The existence of this letter is assumed from the viceroy's letter of reply, 23rd July, 1563, Simancas E° 1052, f° 207.

[201] R. B. Merriman, *op. cit.*, IV, p. 110 refers to this as a superhuman effort – an exaggeration surely.

[202] On 2nd according to Salazar, 6th according to Cabrera, Duro, *op. cit.*, II, p. 55–59.

soldiers' pay. A few days later they had crossed the straits and were in the Puerto de Santa María, the seaport of Seville.[203]

The year's achievements were not negligible, taken overall. But the following year, 1564, Spain accomplished more: she considered the time was now ripe to take the offensive, probably because of the greater security in the East and the generally stable political situation. Perhaps too the appointment on 10th February, 1564 of García de Toledo to the post of Captain General of the Sea played its part. But above all, Spain was beginning to feel her strength. One indication of the change of mood was the request made (before Don García's appointment) by Sancho de Leyva, commander of the Neapolitan galleys: in January he asked the king's permission to take five of his own galleys, one belonging to Stefano de' Mari, and any of the Sicilian galleys that were available, to pursue the foists and galliots of the corsairs along the Barbary coast, in order to capture enough prisoners to man the benches of the new ships being fitted.[204] Very early that spring, along with the ritual problem of supplying La Goletta and Oran, there was talk in high places of making another attempt at the Peñón de Vélez. The official decision was taken in April.[205]

The whole operation was a masterpiece of methodical and painstaking organization to which a mass of unpublished documents[206] in the archives, bears witness. Everything was so well under control that Philip was able to announce to the French ambassador[207] on 12th June that the war fleet would be sent against Africa. The preliminary phase was over and Don García was engaged in assembling troops and galleys in Italy before sending them to Spain and finally to Africa.[208] On the 14th, he made a triumphal entry[209] into Naples with thirty-three galleys.[210] Once again, Philip II concerned himself minutely with the movements of the fleet; he ordered all his services to comply with any requests made by Don García and 'to hasten the operation, for with the wind sitting as it does now, he will be arriving very soon. Have inquiries made whether more soldiers will be required; the Duke of Alcalá writes that he can only provide 1200 men, who will be arriving under the command of Captain Carillo de Quesada'.[211]

Don García sailed to Spain by way of Genoa, following the roundabout

[203] Gómez Verdugo to Francisco de Eraso, 29th August, 1563, Simancas E° 143 f° 117.

[204] Sancho de Leyva to the king, Naples, 13th January, 1564, Simancas E° 1053, f° 8. But we know that Sancho de Leyva in fact sailed for La Goletta, viceroy to the king, Naples, 17th February, 1564, Simancas E° 1053, f° 22.

[205] Philip II to Don García de Toledo, Valencia, April, 1564, *CODOIN*, XXVII, p. 398.

[206] Down to the instructions for the building of shallops issued to the *proveedores* of Málaga, *CODOIN*, XXVII, p. 410, 17th May, 1564.

[207] 12th June, 1564, E. Cabié, *op. cit.*, p. 270.

[208] Don García de Toledo to Philip II, Naples, 15th June, 1564, Simancas E° 1053, f° 64.

[209] Don Juan de Çapata to Eraso, 15th June, 1564, *ibid.*, f° 63.

[210] Viceroy of Naples to Philip II, 15th June, 1564, *ibid.*, f° 60.

[211] Marginal note in Philip's hand on the letter sent to him by Don García de Toledo, Naples, 16th June, 1564, Simancas E° 1053, f° 65.

course along the north shore, instead of the short cut through the islands taken by Gian Andrea Doria the year before. The first rendezvous of the fleet was at Palamos, on the Catalan coast, where it was joined on 6th June by the Spanish galleys under Álvaro de Bazán, whose illustrious career was just beginning. Gian Andrea Doria, with 22 galleys, moored there on the 26th.[212] Next to arrive were the galleys and ships of Pagan Doria, who had waited at La Spezia to pick up German troops. On 15th August the fleet was at Málaga.[213] Don García left it briefly, to go to Cadiz to meet the Portuguese galleys promised for the expedition. His appearance spread panic from Estepona and Marbella to Gibraltar, along a coast so used to pirate-raids that it automatically assumed that he was an enemy. Then, rather slowly, the fleet finally assembled in the neighbouring ports of Marbella and Málaga. By the end of August it numbered 90 to 100 galleys,[214] plus a certain number of caravels, galleons and brigantines, a total of 150 sail, and 16,000 soldiers: an ostentatious and futile display of strength as it was described rather maliciously at Venice.[215] It had at any rate served to drive away the corsairs: three of their galleys and an armed galleon were captured and six or eight others were pursued and very nearly caught.

On 31st August, after a voyage of three days, the fleet arrived off the Peñón. As in 1563, the town had been abandoned by its inhabitants. In the harbour three Catalan ships were on fire, having been captured by the active pirates of Vélez, who had indeed left on a pirate raid with Kara Muṣṭafa, so little did they expect a Christian attack on their town. Don García nevertheless proceeded with caution employing almost excessive means. A large and strongly organized bridgehead protected the operation against the tiny fortified island from the landward side. Against all expectations, after a few days' bombardment, the garrison abandoned the rock on 6th September. The Spanish fortified it, left cannon, men and supplies there, then evacuated the bridgehead, after razing to the ground the walls of the town of Vélez. It was at this point that on 11th September there were several serious brushes with the inhabitants.[216]

The whole affair amounted to a great deal of publicity and expense over what was in the end a very minor operation. No doubt it was useful in demonstrating dramatically to the papacy that the subsidies provided by the Church for the struggle against the Infidel had not been given in vain. 'El Papa esta a la mira', as Philip II said.[217] Contemporary observers all

[212] J. B. E. Jurien de La Gravière, *Les Chevaliers de Malte*, Paris, 1887, I, p. 98.

[213] *Ibid.*, p. 99.

[214] Duro's figures. On 29th August, Saint-Sulpice mentions 62 galleys, with reference to Cadiz alone (E. Cabié, *op. cit.*, p. 291–292). 70 or so galleys, it was thought in France, 13th August, 1564, A.N., K 1502, no. 296.

[215] J. B. E. Jurien de La Gravière, *op. cit.*, I, p. 111, note 1.

[216] Don García de Toledo to the king, Málaga, 16th September, 1564, *CODOIN*, XXVII, p. 527.

[217] Philip II to Figueroa, 3rd August, 1564, Simancas E° 1393 and not E° 931, printed in error in F. Braudel, *Rev. Afr.*, 1928, p. 395, note 1.

stress the spectacular aspect of the expedition. There were strategic aims too: the organization under a new commander of the Hispanic fleet and the desire to crush the small but aggressive pirate centre of Vélez, too close for comfort to the Spanish coast and the Seville shipping routes. From now on (as during the years 1508 to 1525) a Spanish garrison stood guard on the little island. García de Toledo did not leave before all the necessary arrangements had been made. But he left in some haste then, for he was needed elsewhere: in Corsica, the Signoria of Genoa, faced with the outbreak of Sampiero Corso's rebellion, was clamouring for help.

The Corsican uprising. The Corsican revolt had been coming for a long time. The treaty of Cateau-Cambrésis had been a sore blow to the islanders. Between 1559 and 1584, Sampiero Corso had exhausted himself in making passionate representations to various rulers, without however achieving anything positive. But he had only to land on 12th June, 1564 in the bay of Valinco with a small band of men for the island to rise up in arms – it was ready to explode at the first spark. Sampiero immediately advanced on Corte and captured it. One of the most tragic wars the island had ever seen was about to begin: prisoners massacred, villages burned and crops ruined – Corsica was to witness all these and more.

It came as no real surprise to Genoa. Whatever may have been said officially, the Signoria had known for a long time how restless the island was, and how hostile to its overlord. Genoese agents had kept a close and very well-informed watch on the voyages of Sampiero Corso and his intrigues in France, in Algiers, Tuscany and Turkey. Genoese intelligence knew he was in Marseilles and that he was in possession of an armed galley. His landing had therefore been foreseen; but the rapid consequences of his attack, the almost immediate effect of the propaganda he spread and the number of people who flocked to his side may have been rather unexpected.

Who was behind Sampiero, everyone wanted to know when news of his success spread. The king of France, who had lent him the galley with which he landed? The Turkish corsairs?[218] Perhaps the Duke of Florence, it was soon whispered.[219] Sampiero probably did have backing from all these powerful sources – but only in an indirect and guarded way. The major strength of the revolt came from the poverty of the Corsican mountain regions, from the oppressed population of the island, harassed by usurers and the tax collectors of the *Dominante*. Genoa, needless to say, did not admit anything of the sort: it was in her interest to stress the role of foreign backing in order to obtain the intervention of Philip II and she lost no time in doing so, particularly emphasizing French complicity which was indeed only too evident. 'The Corsican affair', wrote Figueroa on 7th July, 'has more foundation than it was first thought. Sampiero has

[218] Figueroa to the king, Genoa, 27th June, 1564, Simancas E° 1393.

[219] Notably in Venice, that centre for news, true and false, and speculation, G. Hernández to Philip II, Venice, 12th September, 1564, Simancas E° 1325.

roused the population and a good part of the island has rallied to him. News has been received that in Provence, Monsieur de Carces is levying seven bands of infantry to send to him, although the French say they are for coastal defence'. Meanwhile Genoese merchants in Lyons[220] passed word to the Signoria of the reactions and activities of the French.

Philip II, on receiving the news, agreed with Don García de Toledo, who was in favour of sailing to Corsica with thirty galleys while Gian Andrea Doria and Ibarra continued to load supplies and embark German troops. So Don García was asked in a letter written on July 18th, whereever he might be when he received the order, to make for the island. We cannot, wrote the king, let Sampiero, who already controls Istria and is threatening Ajaccio, gain control of the whole island, Sampiero, an 'aficionado' of France who would turn Corsica into 'una scala para los Turcos moros enemigos de nuestra santa fe cathólica'.[221] France, he wrote to his ambassador in the French capital, is behaving inexcusably.[222] 'I cannot believe that it is with the approval of the king and queen that the said Corso has undertaken his course of the action, nor even that it is with their knowledge, since the affair is one so incompatible with our friendship and fraternity and so contrary to the observance of the peace. And yet there are so many obvious and unmistakable signs that they cannot seriously insist they they knew nothing about it'.

Unfortunately for Philip and even more unfortunately for Genoa, the order of 18th July directing the galleys to Corsica did not reach Don García until he was already on the Spanish coast and all set for the expedition against Vélez. Should he be sent back again? It would have meant losing time and compromising the expedition against the Peñon; explaining the position to Figueroa, the king added that he had been 'warned that the Pope is keeping a sharp watch to see that the money he provided for fitting galleys is really to be used in an action against the Infidel'.[223] For all these reasons, Philip II allowed the voyage to Gibraltar and Morocco to go ahead. Not until the end of autumn would the Spanish be able to consider Corsica.

So the action at the Peñón gave Sampiero and his partisans a valuable breathing-space. The news spread by Genoa became more and more alarming. Figueroa, in a letter dated 5th August, 1564,[224] talked of increasing French involvement, of frigates coming and going between the island and Provence, of secret meetings held by the Fieschi (Genoese outlaws) and the Corsicans in the house of Thomas Corso (otherwise known as Thomas Lenche, the founder of the Bastion de France) 'who is the habitual provider of oars, powder, sails and other contraband goods to Algiers'. Catherine de Medici however disclaimed all responsibility

[220] Hernández to the king, *ibid.*
[221] Philip II to Don García de Toledo, 18th July, 1564, Simancas E° 1393.
[222] 2nd August, 1564, A.N., K 1502.
[223] Philip II to Figueroa, 3rd August, 1564, Simancas E° 1393.
[224] Simancas E° 1393.

for the affair and even offered herself as a mediator. If galleys were being prepared at Marseilles, she said, it was only in preparation for the forthcoming visit of the king. She even confided to Francés de Alava, through the cardinal de Rambouillet, that the fact that Don García de Toledo's fleet, made up of so many galleys, had gone past the French ports without even requesting 'refreshment' had made her suspicious![225] Her assurances did not prevent Genoa from continuing to accuse France[226] and from worrying about the ten galleys which the Marquis d'Elbeuf was keeping in readiness at Marseilles.

Meanwhile the Spanish war machine, undamaged by the Peñón action, could hope to repeat in Corsica the success it had won in Africa. On 31st August, 1564, the Duke of Alva wrote to Figueroa that Don García de Toledo, having finished his work at Vélez, would leave only a score of galleys in Spain and would sail immediately to Corsica. Philip II assured Figueroa at the same time that there would be no opposition from France to the expedition which was being prepared more or less officially and which was indeed common knowledge. The ambassador of the Duke of Florence sent word of it to his master on 22nd September, 1564[227] and Philip II himself did the same the next day.[228]

The Genoese however were not satisfied; the preparations seemed to them to be taking too long. On the 24th, the ambassador Sauli claimed to know nothing about the armada which was supposed to be at Cartagena.[229] On 9th October his irritation became clear: 'If the armada is long in coming, let your Illustrious Lordships blame the excessive phlegm and slow nature of the lords of this country and not my negligence, for I have not ceased to press our cause both with His Majesty and his ministers'.[230] His reproaches are perhaps a little unjust. In August, Lorenzo Suárez de Figueroa, the son of the Spanish ambassador at Genoa, had been sent to Milan to levy 1500 Italians for action in Corsica. On the 26th they were embarked in three roundships which were prevented from reaching the island only by bad weather. Lorenzo was their colonel.

The Genoese might be excused their impatience however. Sampiero had routed the troops of Estefano Doria[231] and, as time went by, they feared that some autumn arrangement might be agreed to their disadvantage. Philip II himself said that an agreement with Sampiero was desirable, in order to avoid the expense of a war which might be prolonged by the harsh nature of the Corsican terrain.[232] When the Genoese were

[225] Don Francés de Alava to Philip II, 13th August, 1564, A.N., K 1502, n. 96.

[226] Nuevas de Francia . . ., received 3rd September, 1564, Simancas E° 351.

[227] Garces to the Duke of Florence, Madrid, 22nd September, 1564, A.d.S. Florence, Mediceo 4897, f° 36 v°.

[228] Philip II to the Duke of Florence, Madrid, 23rd September, 1564, Simancas E° 1446, f° 112.

[229] Sauli to the Signoria, Madrid, 24th September, 1564, A.d.S. Genoa, L.M. Spagna, 3.2412.

[230] 9th October, *ibid.*

[231] Philip II to Figueroa, Madrid, 25th October, 1564, Simancas E° 1393.

[232] *Ibid.*

approached on this subject, they took it very badly. In the end, García de Toledo arrived on 25th October at Savona.[233] But the summer was over and he had no wish to risk his fleet. He proposed to send twenty galleys with the infantry levied in Spain and Piedmont, whereas the Genoese wanted a massive show of strength by the whole armada at Porto Vecchio.[234] They did not get it. Francés de Alava wrote from Arles on 20th November that if the Genoese had not subdued the rebellion before winter, the best course of action would be to negotiate with the rebels as had already been suggested to him from the French side.[235] But Genoa would not hear of it and Philip II for his part refused French mediation.[236]

So winter came, and still the war went on. Assistance from outside continued to arrive in the island, from France again (not necessarily with the admitted knowledge of the king and queen)[237] but also from Leghorn, the source of frigates carrying munitions and money.[238] Sampiero even had contacts with His Holiness.[239] So the war took a bad turn for the Genoese.[240] Would the twenty galleys and the Spanish troops whom Gian Andrea Doria had taken to Bastia[241] be sufficient to alter its course? The bad weather did not only hinder naval operations (Don García was unable to get more than 25 miles from Genoa on 14th December),[242] but it also made action on land more difficult. It was learnt on 25th November that the expeditionary force which had left Bastia to take help to besieged Corte, had had to turn back because of bad weather and sickness. This retreat was hardly compensated by the recapture of Porto Vecchio, which Gian Andrea Doria took almost without a fight in mid-December, or by the capture of a village or so in the Balagne hills. Genoa now held only a few positions on the coast and in the interior, while the rest of the island gradually passed over to the insurgents. Cooped up in the *presidio*s, the soldiers of the *Dominante* had more to fear from epidemics and lack of supplies than from the enemy.

Peace in Europe. Sampiero's rebellion was to last a long time, but being confined to an island, it had little effect on European life in general. This is a point worth remembering, for if the Hispanic world was able to pause for breath and subsequently to restore a shaky situation, it was through the providential coincidence of peace on the Turkish front and a European truce which may after all have been no more than the natural consequence

[233] Figueroa to Philip II, Genoa, 27th October, 1564, Simancas E° 1393.

[234] Figueroa to Philip, 8th November, Simancas E° 1054, f° 21.

[235] A.N., K 1502, B 18, no. 51a.

[236] Philip II to Francés de Alava, 31st December, 1564, A.N., K 1502, B 18, no. 77.

[237] Figueroa to Francés de Alava, Genoa, 1st December, 1564, A.N., K 1502, B 18, no. 60.

[238] *Ibid.* Aboard one of the frigates was a Corsican, a friend of Sampiero's, named Piovanelo, who was captured en route by the Barbary corsairs.

[239] *Ibid.*

[240] *Ibid.*

[241] Figueroa to Philip II, 3rd December, 1564, Simancas E° 1393.

[242] Figueroa to Philip, 21st December, 1564, *ibid.*

of the exhausting wars of Charles V. From 1552 to 1559, these wars had absorbed all the resources of the European states and more besides. They were followed by a series of financial collapses in Spain, France and, by repercussion throughout Europe; hence major war was for the time being impossible in this paralyzed world which had been for years the scene of conflict.

The break-up of Charles's empire moreover, had brought a period of relative calm. Germany under the 'Ferdinandians' had recovered its autonomy and Europe forgot its fear of a Habsburg universal monarchy. Spanish imperialism was not yet a threat: that was not to come until 1580. In place of great wars, there now appeared only local quarrels fed by newly-released energies. The internal troubles in France were the direct consequence of the demobilization of the army, along with the lack of occupation of a minor nobility which was poorer than at the beginning of the century and which the royal power could no longer use in Italy.

Only one major conflict had survived the treaty of Cateau-Cambrésis: the opposition between France under the Valois and England. It was an old quarrel, going back to at least 1558 and the Dauphin's marriage. French foreign policy was thus preoccupied with the north, not the Mediterranean. But the two adversaries, both racked by political and religious strife within, were incapable of a real confrontation, and all the more driven to insults and complaints against each other, made either to the Pope or Philip II. The latter made no contribution to an agreement and did not take sides, seeing in this providential northern dispute a source of tranquillity for himself. It would take us too far from our subject to delve into this policy of bad faith[243] and unadulterated *raison d'état*—which was in the end to prove rather short-sighted. France certainly drew no benefit from it, but Spain in pursuing this course saved or helped to save Elizabeth's fragile kingdom. How was Philip II to know that it would grow so strong so quickly?

To keep France in check seemed at the time an essential requirement for peace in Spain. The task proved an easy one: the year 1560 saw the beginning of the reign of Catherine de Medici and trouble soon followed. It gave Philip who was determined to protect his dominions from the Protestant contagion, an opportunity to offer troops. His offer gave him a lasting hold over the neighbouring kingdom. He even thought it worthwhile buying himself allies in France, following the good old traditions of Habsburg policy and indeed the diplomatic habits of the century. The Spaniards were thus drawn into prolonged negotiations with Antoine de Bourbon. Who was deceiving whom? This file is one we should not disturb, were it not that it leads back to the Mediterranean, to Sardinia first and then to Tunis.

[243] We may appreciate the following argument for example: (the king to Chantonnay, Madrid, 10th November, 1562, A.N., K 1496, B 14, no. 126) Philip II had told Saint-Sulpice that he could not declare himself hostile to the queen of England, 'por causa de las antiguas alianças'.

The negotiations began in 1561,[244] as soon as Antoine de Bourbon rose to an eminence more apparent than real perhaps, with the title of Lieutenant General of the Kingdom. The man whom Spain called 'Monsieur de Vendôme' was in fact king of Navarre, the Spanish part of which Philip II was occupying without any legal right. To recover this territory beyond the mountains, or at any rate to encourage conspiracy inside it,[245] to meddle in Spanish affairs beyond the barrier of the Pyrenees, was a temptation which no king of Navarre since 1511 – the date of the Spanish conquest – was able to resist, not even the future Henri IV. However there was another course: if the recovery of Spanish Navarre was out of the question, it might be possible to obtain compensation, and upon this course Monsieur de Vendôme boldly embarked. He demanded the kingdom of Sardinia and made his claims known as far afield as Rome.[246] One of his agents, who figures in Spanish documents as Bermejo or Vermejo (but this is only a pseudonym for the sake of secrecy) was received at Madrid in 1562[247] by Ruy Gómez and the Duke of Alva, neither of whom were satisfied with the services rendered by Antoine de Bourbon with his visible penchant for heretics. Speculating on the ambitions of the said 'Vendôme', the two ministers proposed to Vermejo that his master should receive the kingdom of Tunis, with a promise to help him conquer it. But said the agent, what exactly does this kingdom consist of? 'I told him', declared the Duke of Alva, 'that no one was better placed than myself to inform him, for the emperor . . . had certain intentions towards this kingdom . . . and had spoken of it to me in great detail'. There follows an idyllic description of the kingdom of Tunis, whose fame is such that 'few are ignorant of it', a meeting-place for 'all the merchandise travelling from the Levant to the West and from the West to the Levant'; a fertile land which brings forth in abundance wheat, oil, wool and livestock; a string of ports which are both excellent and easy to defend. . . . Nothing like the poor kingdom of Sardinia, which in any case had its own laws and which the king had no authority to abrogate.

We do not know how the king of Navarre received this tempting proposal. But we do know that Catherine de Medici was disturbed by the negotiations between Bourbon and Spain and also by a rumour that the kingdom of Sardinia was likely to be ceded which circulated in Genoa in September, 1562.[248] 'News has arrived in this city' wrote Figueroa on the 9th, 'that Don Juan de Mendoza has taken possession of the kingdom of

[244] The conversations had already begun in September: the bishop of Limoges to Catherine de Medici, Madrid, 24th September, 1561, B.N., Paris, Fr. 15875, f° 194; Chantonnay to Philip II, Saint-Cloud, 21st November, 1561, A.N., K 1494, B 12, no. 111; Chantonnay to Philip II, Poissy, 28th November, 1561, *ibid.*, no. 115.

[245] G. Soranzo to the Doge, Vienna, 25th December, 1561, a conspiracy in favour of the king of Navarre had been discovered at Pamplona, G. Turba, *op. cit.*, I, p. 95 ff.

[246] Morone to the Duke of Alva, Rome, 2nd October, 1561, Joseph Susta, *Die Römische Curie und das Konzil von Trient unter Pius IV*, Vienna, 1904, I, p. 259.

[247] The Duke of Alva to Chantonnay, Madrid, 18th January, 1562, A.N., K 1496, B 14, no. 38.

[248] Figueroa to Philip II, 9th October, 1562, Simancas E° 1391.

Sardinia and will turn it over on your orders to Monsieur de Vendôme, a rumour which hardly seems worthy of credit'. The negotiations were brutally interrupted: Monsieur de Vendôme, wounded under the walls of Rouen, was to die of his wounds. Philip II was very soon informed that 'the doctors and surgeons hold out no hope for him'[249] and had letters of condolence made out in advance, with the date left blank.

It is a minor incident, but it demonstrates how France was surrounded by a web of watchful, smooth-tongued diplomacy, a little slow to act, Machiavellian if it could be done without expense, haughty and attached to protocol, and always active if not always as effectively as it supposed. For if Europe posed no great problems for the Spanish empire, was it entirely the happy result of the activities of Ruy Gómez and the subtlety of the Duke of Alva? Was it perhaps because from time to time, France could be left 'le bec dans l'eau' as the bishop of Limoges wrote? Because Philip II was the only adult male ruler (*dixit* Limoges again) in a Europe where most of the thrones belonged to children or inexperienced women? Was it not also because Europe was thoroughly exhausted? One thing is certain: by contrast with a Turkey embroiled in a war far from Mediterranean shores, Spain had freedom of movement, was not threatened or worried by Europe, at least for the time being. And Spain was able to put that time to good use.

A few figures on the maritime recovery of Spain. It is extremely difficult to demonstrate statistically the real state of naval armaments in the sixteenth century. In the first place, which ships should be counted? Alongside the galleys, galliots and foists (*fustes*) we should also take account of the auxiliary fleet of roundships, used as supply vessels but also as warships if necessary, for they carried guns. At the end of 1563 and beginning of 1564, the Spanish government commandeered about a hundred shallops and *zabras* belonging to fishermen on the Biscay and Cantabrian coasts, small boats of about 70 tons, equipped with voluntary oarsmen and guns. This auxiliary fleet was organized in Catalonia by Álvaro de Bazán: we do not know under what conditions or with what aim. It appears that these ships, of small tonnage and specifically designed for the ocean, were involved in naval encounters in the Mediterranean merely as supply ships. Spanish strategic command did not realize the great value these light ocean-going sailing-ships might have – and were to have later on.

If we stick exclusively to warships, we must mention in addition to the all-powerful galleys those lesser forms of the galley, the foist and the galliot. It is true that they were used mainly by the corsairs. The chief difficulty really is that Philip's fleet was in fact a combination of various fleets, a coalition of four squadrons: those of Naples, Spain, Sicily and the Genoese galleys hired by Spain (consisting mainly of Gian Andrea Doria's ships). They were occasionally joined by the galleys of Monaco, Savoy,

[249] Saint-Sulpice to Catherine de Medici, Madrid, 25th November, 1562, B.N., Paris, Fr. 15877, f° 386.

Tuscany and the Knights of Malta, which does not make our calculations any easier.

In order to measure the full strength of the Spanish navy, I have tried to estimate for each of the years 1560–1564, the number of galleys assembled either at Messina or elsewhere, but preferably at Messina, since this represents the fleet effectively mobilized.

In 1560, the year of Djerba, the Christian fleet numbered 154 warships, of which 47 were galleys and four galliots,[250] yielding a ratio of one galley to every three other warships. To these 47 galleys, we must add the Spanish squadron, which had been called to coastal defence duties and did not take part in the expedition, and about ten galleys belonging to the Knights, Tuscany, Genoa and Savoy. The measures taken when the king was considering relieving the fort of Djerba, enable us to calculate the strength of this reserve. On 8th June, 1560, Philip II,[251] estimating his possible galley strength, thought that there might be a total of 64;[252] his figure can be taken to be accurate, but it includes of course the twenty galleys that had escaped from Djerba. So if we add the other 44 to the 47 on the expedition – a total of 91 – we know the total number of warships on which Spain could rely directly or indirectly just after Cateau-Cambrésis. It is a very considerable number, but the Djerba disaster cut it to 64. The loss was the more serious in that most of the captured vessels were taken into enemy service: in 1562, the ten great galleys which brought Hasan Pasha from Algiers were part of the spoils of Djerba.

The reaction of the Italian arsenals was swift. In Sicily, new taxes were levied for naval construction.[253] In Naples, the six galleys lost at Djerba had been replaced by 9th October.[254] The only problem, but it was a serious one, was finding oarsmen. At the same time, Cosimo de' Medici was intensifying his naval efforts as was the Duke of Savoy. Letters from Figueroa in July, 1560 indicated that Philip II might find galleys to hire in Genoa.[255] Gian Andrea Doria meanwhile was reconstituting his own fleet and in January, 1561 bought two galleys from Cardinal Santa Fiore.[256]

[250] C. Monchicourt, *op. cit.*, p. 88.

[251] Philip II to the viceroy of Naples, Toledo, 8th June, 1560, Simancas E° 1059, f° 69.

[252] A Genoese estimate (*Conto che si fa delle galere che S. Mta Cattca potrà metere insieme*, A.d.S. Genoa, L.M. Spagna, 2.2411, 1560) provides an interesting breakdown of the figures: Spanish galleys (20); Genoese (6); belonging to Prince Doria, not counting those at Djerba (6); owned by the Duke of Florence (3); the Duke of Savoy (2); the Count of Nicolera (1); the king of Portugal (4); Paolo Santa Fiore (2), 'delle salve' (23), total 67. A Sicilian document dated 1560 (Simancas E° 1125) gives the total number as 74, with the following breakdown: galleys owned by the Pope (2); by Spain (20); Prince Doria (10); Genoa (8); Knights of Malta (5); the Duke of Florence (7); the Duke of Savoy (6); Antonio Doria (4); Cigala (2); Cardinal Vitelli (3); Paolo Sforza (2); Naples (3); Bendinelli Sauli (1); Stefano de Mari (1).

[253] L. Bianchini, *op. cit.*, I, p. 54.

[254] Viceroy of Naples to the king, Simancas E° 1050, f° 137.

[255] Summary of Figueroa's letters to the king of 3rd, 5th, 10th and 12th June, 1560, Simancas E° 1389.

[256] Viceroy of Naples to Philip II, Naples, 12th January, 1561, Simancas E° 1051, f° 17.

Shipbuilding meant above all finding money. It was on this occasion that Philip II asked Rome not only for the *cruzada*, which he was granted,[257] but for the 'subsidy' as well. He obtained it in January, 1561: 300,000 gold ducats a year for five years.[258] He decided it was not enough. In April, 1562, after lengthy negotiations and through the complaisance of Pius IV, the subsidy was raised to 420,000 ducats, for ten years instead of five, and (to the vehement indignation of the Spanish clergy) backdated to 1560.[259] According to calculations by Paolo Tiepolo the *subsidio* and *cruzada* combined brought Philip 750,000 ducats in 1563, not to mention other revenues collected both inside and outside Spain with permission from the Holy See: 1,970,000 ducats a year according to a Roman memorandum of 1565.[260]

Once the question of money was solved, there still remained technical problems. Philip II had at his disposal all the shipyards and manpower in the West, apart from those of Provence. But, at least during the year 1561, he did not throw himself into this task as energetically as he might have. The money from the Spanish Church was not immediately available, or else it was used to stop up the enormous gaps in the Spanish budget: above all, the king and his advisers did not wish to foot the bill for rearmament undertaken by the Italian 'potentates'. If military effort there was to be, it was of course for the good and protection of the whole of Christendom. It was only fair that the 'potentates' should henceforth make the same sacrifices as Spain and pay their own bills. So in March, 1561,[261] the Spanish government asked for the help of the Portuguese galleys against the Barbary corsairs. And when on 1st April the Marquis of Favara was sent to Italy with the task of negotiating the mobilization of all the galleys belonging to its allies, he specifically stated that his government did not wish to take on any galleys 'a sueldo'. From the lord of Piombino, the Republic of Genoa, the Duke of Savoy, the Duke and Duchess of Mantua and the Duke of Florence, Spain sought favours, pleading that she had very few galleys left and that those still under construction in her realms would not be ready for a while.[262] A letter from the Genoese ambassador in Spain said that of all those who had offered galleys 'a sueldo', only Marco Centurione had been given a contract, for four or five galleys for the year 1562.[263] However Gian Andrea Doria, the major provider of hired galleys had received 100,000 crowns payable at the October fair, of the 130,000 owed to him for the remaining equipment of his ships.[264] If one bears in mind the slowness with which new galleys were being launched and fitted, and the loss in June of the seven Sicilian galleys

[257] L. von Pastor, *op. cit.*, XVI, p. 256, note 1.

[258] *Ibid.* [259] *Ibid.*, p. 257. [260] *Ibid.*

[261] Tiepolo to the Doge, Toledo, 26th March, 1561, *C.S.P. Venetian*, VII, p. 305.

[262] Instructions of Fernando de Sylva, Marquis of La Favara, 1st April, 1561, Simancas E° 1126.

[263] Sauli to the Signoria of Genoa, Toledo, 27th April, 1561, A.d.S. Genoa, L.M. Spagna, 22411.

[264] Tiepolo to the Doge, 26th April, 1561, *C.S.P. Venetian*, VII, p. 310.

captured by Dragut,[265] we may conclude that the Spanish fleet did not in 1561 make good the losses of the previous year. The prince of Melfi was able to assemble for his autumn campaign only fifty-four galleys.[266]

It was not until the year 1561 that a really concentrated effort got under way in Spain. Even the Barcelona arsenal was brought back into service. Spain's neighbours were sufficiently disturbed by these activities for Catherine de Medici to send Mgr. Dozances on a special mission to her son-in-law, for the sole purpose of eliminating any possible misunderstandings.[267] That was in December. The same winter, the Duke of Joyeuse on express instructions from the king, moved his companies to the Spanish border. Although, he wrote, I do not believe there is any real danger on these frontiers. What was certain was that 'for two months now the said king of Spain has had the shipyards in Barcelona working most diligently to finish several galleys and other seagoing vessels and has had, and is still having, a great quantity of biscuit made. Common report has it that it is intended for an expedition against Arger [Algiers] this summer and I know, Sire, for a truth that the king of Spain is much besought by all his people to make war on Algiers, because of the great subjection in which the king of the said Algiers now holds the Spanish, for they cannot trade by sea without great risk'.[268] A month later, on 17th January, 1562, the bishop of Limoges gave a similar report on the galleys, 'which are being built and made ready on every side with all diligence, and in Catalonia and the neighbouring kingdoms, another four thousand feet of timber have been cut to meet their needs, not to mention the galleys being built in Naples and Sicily, and they have brought master craftsmen and workers from Genoa and some from our own Provence'.[269]

But the ship-building proceeded slowly: the wood had to be seasoned before it could be used, so this activity could not produce immediate results. And this year as before, Philip II did not wish to mobilize on his behalf all the available ships of the western Mediterranean. An official document of 14th June, 1562 provides for no more than 56 galleys to be made available to supreme command, 32 to operate under the orders of Don Juan de Mendoza and 24 under Doria.[270] However the detailed register

[265] The Duke of Medina Celi to the viceroy of Naples, 30th June, 1561, Simancas E° 1051, f° 100, copy.

[266] The bishop of Limoges to the king, Madrid, 5th September, 1561, B.N., Paris, Fr. 16103, f° 44 ff, copy, and also his letter of 12th August mentioned earlier.

[267] *Los puntos en que han hablado a S. M. Mos. Dosance y el embax^{or} Limoges*, Madrid, 10th December, 1561, A.N., K 1495, B. 13, n° 96.

[268] Joyeuse to the king, Narbonne, 28th December, 1561, B.N., Paris, Fr. 15875, f° 460.

[269] Memoranda from bishop of Limoges, 27th January, 1562, B.N., Paris, Fr. 16103, f° 144 v°, copy.

[270] Philip II to the viceroy of Naples, 14th June, 1562, Simancas E° 1052, f° 96. The composition of the squadrons was as follows: (a) Don Juan de Mendoza's squadron, 12 Spanish galleys (of which four were detached to be at the disposal of the Prior and Consuls of Seville); 6 from Naples; 6 owned by Antonio Doria; 4 owned by Count Federico Borromeo, 2 by Estefano Doria, 2 by Bendinelli Sauli; (b) Gian

shows that the convoy was not to include either the Sicilian galleys or those of the Pope, of Tuscany, of Genoa, or those belonging to individuals such as the Duke of Monaco and the lord of Piombino. It would be difficult to give exact figures for these unemployed galleys; going by past years, one might suppose them to have numbered between twenty and thirty. So the total naval strength at the disposal of the Spaniards was perhaps 80 to 90 galleys: the losses at Djerba had barely been made good. At this point there came the fresh disaster of Herradura: 25 galleys lost, and Spanish naval strength was brutally reduced to a level such as had not been seen for years; the efforts of an entire year had gone for nothing.

Major disasters call for major remedies. On 12th December, 1562, Philip II convoked the Cortes of Castile at Madrid. The 'proposicion' read at the opening session – what, as Duro says, would today be called the speech from the throne – outlined the reasons, relating both to the Mediterranean and the Atlantic, why a large fleet was necessary:[271] the conclusion, I hardly need add, was a request for more taxation.

These measures concerned the future. In 1563, the construction of new ships could only partly compensate for the losses of the Spanish fleet. When the campaigning season came round again, Philip called once more on his Italian allies, the Duke of Savoy, the Republic of Genoa and the Duke of Florence. On 8th March, he thought he had at his disposal about 70 galleys[272] which he intended as in 1562 to divide, sending half to Spain and half to Italy. All his plans were upset by the siege of Oran. It was only with some difficulty that he was able to muster the 34 galleys which saved the besieged garrison. For there was always a rather wide margin between the number of galleys which could be mobilized for a foreign expedition and the total complement of the fleet, a certain proportion of ships being kept for coastal protection.

The king did not see the fruits of his labours until 1564. In September Don García de Toledo was able to assemble 90 to 102 galleys between the coasts of Spain and Africa (to cite only the extreme figures given by contemporaries). Even 90 was a considerable jump. It is true that the new commander of the Spanish navy, believing the information he had received about the Turks, had made the bold decision to assemble all the available galleys at a single point in the western sea, leaving behind hardly any reserves or coastguards. Sampiero Corso's landing – by coincidence? – took place behind the back of the mighty fleet sailing westwards. It is also true that this time the king had not scrupled to call on all his allies for assistance whether paid for or not: the Vélez fleet was not the navy of the king of Spain, but that of all western Christendom except France. Amongst others it included ten galleys belonging to the Duke of Savoy, seven from

Andrea Doria's squadron: 12 galleys owned by the said Gian Andrea, as agreed by the recent *asiento*; 4 owned by the Knights of Malta; 4 owned by Marco Centurione; 2 by the Duke of Terranova, and 2 by Cigala.

[271] C. Duro, *op. cit.*, II, p. 49.

[272] Philip II to the dukes of Savoy and Florence, San Lorenzo, 8th March, 1563, Simancas E° 1392.

the Duke of Florence and eight from the king of Portugal.[273] If one adds the mercenary ships, about thirty 'allied' vessels sailed with Philip's fleet.

However some new ships had left the yards. The Naples squadron which in January had consisted of four galleys in service, two launched but not yet fitted, two completed in the arsenal and four still on the stocks,[274] by June consisted of 11 galleys in service,[275] a twelfth only needing oarsmen,[276] four others launched and four still under construction, a total of 20, of which eleven were in service. After the slow start, it seems that progress was rapid. At the end of 1564, the Spanish arsenals were working at full stretch. The Barcelona yards were given particular attention by Don García himself, a former viceroy of Catalonia, and the first results were encouraging: in spite of the losses, the 1559 total was not merely reached but exceeded.

Don García de Toledo. Was this salutary reaction the result of a conscious and consistent policy, indicating that Philip II had a clear vision of his interests and responsibilities in the Mediterranean? Perhaps it was only the imminent peril, Djerba, and a series of unfortunate accidents which propelled Philip towards an effort he had not previously contemplated. He would himself, it appears, have been content to accept the small-scale hostilities of the years 1561–1564 for many years to come, having no wish to take the initiative in bold projects and heavy expense. He had neither the breadth of vision, nor the fanaticism of the determined crusader. As far as he was concerned, the eastern horizon stopped at Sicily and Naples. It is even likely that when Maximilian, now elected emperor, opened talks at Constantinople in 1564, in view of an extension of the truce of 1562, which was again under discussion because of Ferdinand's death, Philip once more, as in 1558, tried to have some part in the negotiations. Hammer mentions in this context, among the papers preserved at Vienna, a report by the 'internuncio', the imperial agent at Constantinople, Albert Wyss, dated 22nd December, 1564.[277]

So behind Don García de Toledo there was apparently no fixed policy, none of the conditions which in a few years were to produce or at any rate, make possible the glorious career of Don John of Austria. Perhaps he also lacked what youth and his temperament conferred upon Don John – a taste for adventure. In 1564, Don García was an old man, crippled with gout and rheumatism. Yet it was he who succeeded in shaping the Spanish fleet, turning it into an effective and powerful weapon.

The son of Don Pedro de Toledo, the famous viceroy of Naples who had governed the kingdom with a firm hand and contributed handsomely to the embellishment of its capital, Don García seems to have inherited

[273] C. Duro, *op. cit.*, III, p. 67.

[274] Sancho de Leyva to Philip II, Naples, 13th January, 1564, Simancas E° 1053, f° 8.

[275] Viceroy of Naples to the king, 15th June, 1564, Simancas E° 1053, f° 60.

[276] 29th June, 1564, *ibid.*, f° 73.

[277] J. von Hammer, *op. cit.*, VI, p. 118.

his father's sense of grandeur, his taste for doing things on a large scale. Marquis of Villefranca on the death of his older brother, he had begun his service in 1539 with two galleys of his own under the command of Prince Doria. At twenty-one he was appointed to the command of the Naples squadron, a favour granted to his father, but which gave him heavy responsibilities at an early age. He saw active service at Tunis, Algiers, Sfax, Kelibia and Mahdia, in Greece, at Nice, during the war of Siena and in Corsica. For health reasons, at least so he said, he resigned his post on 25th April, 1558, and was made viceroy and captain general of Catalonia and Roussillon. It was here, after the alert of 1560 during which there was some talk of giving him the fleet and kingdom of Sicily, that he received his appointment to the post of *Capitan General de la Mar*, dated 10th February, 1564.[278] On 7th October of the same year,[279] at his own request and as a recompense for the victory at the Peñón, he was made viceroy of Sicily. He thus combined with his naval command, rule over the island which he wanted to transform into an arsenal and magazine.

Evidently then this was a man of some strategic vision. He was aware of the value of the services he rendered ('peleo por su servicio', I fight for the king's service, he wrote),[280] and the knowledge that he was doing his duty gave him the courage to make precise demands and to speak his mind clearly. 'It is impossible to describe or imagine the condition in which I found the fleet', he wrote to Eraso from Málaga on 17th August, 1564, on assuming active command. At the same time he was writing to the king: 'Your Majesty must know that it is indispensable that I have a firm hand with the fleet, in view of its present condition, if I am to do my job properly and safeguard his finances. I know that I will gain little by being unpopular, but I confess that I cannot shut my eyes to embezzlement and maladministration, in an area under my authority'.[281]

Honest and demanding,[282] thoughtful for the future and orderly, thus he appears in his correspondence. But he was also a man of lucid mind, capable of sharp observation and tactful diplomacy. The letter he wrote to Philip II from Gaeta on 14th December, 1564,[283] shows an intelligent appreciation of the problems in Spain's relations with the papacy. During Don García's interview with the Pope, Pius IV had wandered from topic to topic, complaining once again about the Spanish, about the persons sent to him by Philip II, the terms those persons had used in speaking to him, the count of Luna and Vargas, the king's attitude to the Council.

[278] C. Duro, *op. cit.*, III, p. 61, note 2 and p. 62, note 1.
[279] *Ibid.*, p. 64, note 3.
[280] Don García de Toledo to Eraso, Málaga, 17th August, 1564, *CODOIN*, XXVII, p. 452, quoted by C. Duro, *op. cit.*, III, p. 65–66.
[281] 22nd August, 1564, quoted by C. Duro, *op. cit.*, III, p. 66.
[282] Concerning the Naples galleys for instance, G. de Toledo to the viceroy of Naples, 23rd January, 1565, Simancas E° 1054, f° 52.
[283] Don García de Toledo to Philip II, Gaeta, 14th December, 1564, *CODOIN*, CI, p. 93–105.

For four hours by the clock, Don García had sat and listened without replying to the complaints and without mentioning the object of his visit. Two days later when the storm was over, he began to outline the naval achievements of the past year. The Pope replied, meaningfully, that he was pleased to see some result at last from the subsidy he had been paying for so long. His interlocutor then moved on to technical matters: a fleet cannot be built in a day, the uninterrupted labour of the past years can only be revealed to the world by massive assemblies in the coming year. But the Pope was unconvinced: he spoke of nothing less than an expedition against Algiers: what was the Peñón de Velez by comparison? The meaning of a sentence quoted earlier from Philip's correspondence now becomes clear: 'El Papa esta a la mira', the Pope has his eye on us. The Pope was indeed watching Spain and with a far from approving gaze.

3. MALTA: A TRIAL OF STRENGTH (18TH MAY – 8TH SEPTEMBER, 1565)

At the risk of lapsing into sensationalism, I am tempted to say that Malta or rather the sudden arrival in Malta of the Turkish fleet in May, 1565, hit Europe like a hurricane. But the hurricane – whose consequences made it one of the great events of the century – came only partly as a surprise to the governments concerned. How could the sultan have fitted out and perfected his mighty military machine without some echo reaching Europe? At the end of 1564, in Vienna, where they were always well-informed of Turkish moves, Maximilian told the Venetian ambassador that a large fleet would leave Constantinople 'a tempo nuevo'. Philip was preparing his ships but was Cyprus not in danger too?[284] The prediction game was already beginning.

Was it a surprise? At the beginning of January, Don García wrote to the king from Naples[285] that it was essential to wind up the Corsican affair before April, that is before the arrival of the Turkish fleet. Spain had to be free in the West in order to be ready to resist in the East what was very soon realized to be a major attack. On 20th January, Pétrémol wrote to Catherine de Medici from Constantinople that the Turkish armada would probably aim for Malta, but this was merely hearsay and he knew no more about it.[286] The name of Malta naturally came to mind whenever there was any question of a Turkish assault. At the end of January, Don García de Toledo considered going both to Malta and to La Goletta, for these two places, along with Sicily, which was too large to be under serious threat, were the foremost bastions of Christianity facing the East, and the positions the Turk was likeliest to attack.

[284] Leonardo Contarini to the Doge, Venice, 29th December, 1564, G. Turba, *op. cit.*, I, 3, p. 289.

[285] Don G. de Toledo to the king, Naples, 7th January, 1565, *CODOIN*, XXVII, p. 558.

[286] E. Charrière, *op. cit.*, II, p. 774–776.

All through the winter and into spring, alarming reports came in. According to dispatches sent on 10th February,[287] work was proceeding 'a furia' in the Turkish arsenal; by mid-April 140 galleys, 10 mahonnas (or great galleasses), 20 roundships and 15 *caramusalis* would certainly be ready for action. Beside these alarms, it hardly seemed to matter that Álvaro de Bazán, in command of the Spanish galleys, had succeeded in blocking the Rio de Tetuan (now Oued-Martine) by sinking several transports at the mouth of the river;[288] or that the corsairs had captured three ships out from Málaga and were proposing to sell them back for a ransom off Cape Falcon as usual.[289] Even the sensational interview at Bayonne was not sufficient distraction[290] and the new ships (eight galley hulls were launched at Barcelona and three galliots at Málaga)[291] were not enough to set minds at rest. For over everything hovered the distressing reality, the certainty, confirmed with every fresh report, that the expected armada was a mighty one and would be backed up by the ships of the corsairs, both in the Levant and the western sea. It is possible that Hasan Pasha in Algiers had indeed, as Haëdo says, been informed of the proposed attack on Malta as early as winter, 1564. All the look-out posts, whether in Constantinople or nearer home in Corfu and Ragusa, agreed. A dispatch dated 8th April, 1564, from Ragusa, announced that Piāle Pasha's first twenty galleys had left the straits on 20th March,[292] adding that general rumour spoke of Malta as the fleet's target, but nothing was known with any certainty.[293]

The Spanish government for its part feared an attack on La Goletta[294] and on 22nd March measures had been taken to levy four thousand foot soldiers in Spain, some of whom were to be sent to Corsica and some to serve as infantry troops aboard the galleys. Philip II dispatched warnings all round: 'The Turkish fleet will be coming with more galleys than in past years', he wrote on 7th April to the Prior and Consuls of Seville,[295]

[287] Constantinople, 10th February, 1565, Simancas E° 1054, f° 64.

[288] Álvaro de Bazán to Philip II, Oran, 10th March, 1565, Simancas E° 486, see E. Cat, *Mission bibliographique en Espagne*, 1891, p. 122–126.

[289] Rodrigo Portillo to the king, Mers-el-Kébir, 13th March, 1565, Simancas E° 485.

[290] Viceroy of Naples to Philip II, 14th March, 1565, Simancas E° 1054, f° 70.

[291] Francavila to the king, Barcelona, 19th March, 1565, Simancas E° 332, Philip II to the *proveedores* of Málaga, Madrid, 30th March, 1565, Simancas E° 145.

[292] Constantinople, 20th March, Corfu, 29th March, Ragusa, 8th April, 1565, Simancas E° 1054, f° 71; on 22nd, according to Jurien de La Gravière, *op. cit.*, I, p. 169.

[293] In Madrid on 6th April, the Tuscan ambassador Garces, gave Philip the dispatches from the Levant received via Florence: they reported on the size but not the intentions of the fleet. Garces to the Duke of Florence, Madrid, 6th April, 1565, A.d.S. Florence, Mediceo 1897, f° 88. Similarly, Pétrémol, in his letter to Du Ferrier, 7th April, 1565, E. Charrière, *op. cit.*, II, p. 783–785, reports the departure of the bulk of the fleet from Constantinople on 30th, but does not know whether it was heading for Malta or La Goletta. The date 30th March is also given by a dispatch from Constantinople, 8th April, 1565, Simancas E° 1054, f° 85.

[294] Philip II to the Dean of Cartagena (Alberto Clavijo, *proveedor* of Málaga), Madrid, 22nd March, 1565, Simancas E° 145.

[295] Aranjuez, 7th April, 1565, Simancas E° 145.

informing them of the orders given to Álvaro de Bazán: he was to go to Cartagena to embark Spanish troops for Corsica, then to return to Majorca and maintain his watch for corsairs. In Naples, the viceroy considered on 8th April that in view of the seriousness of the threat, he would levy between 10,000 and 12,000 men and would travel in person to Apulia.[296] As for the rumour that the Turks were planning to attack Piombino with the connivance of the Duke of Florence, he did not believe it.[297]

With the usual time-lag, news of the progress of the Turkish armada began to arrive in the West. On 17th April, 40 galleys were reported in the straits of Negropont; thirty more joined them there on the 19th; the rest of the fleet, 150 sail, was at Chios.[298] So the ships had taken two weeks and in some cases longer, to reach the Aegean. On the way they had completed their supplies, notably biscuit, and had taken troops on board. Dragut had insisted that the armada sail early and was said to have asked for fifty galleys in order to prevent Philip's navy from concentrating its forces. At Corfu it was rumoured that the fleet was bound for Malta, but the writer of the dispatch was cautious about believing it: 'in view of the preparations', he wrote, 'it is thought most likely that it is making for La Goletta'.[299] In May it arrived at Navarino;[300] by the 18th it was at Malta.[301]

Once more, the Turkish fleet had sailed with all speed to obtain maximum benefit from surprise. On 17th May, Carlos de Aragona had sent a short dispatch by special messenger from Syracuse to Don García de Toledo: 'at one in the morning, the guard at Casibile fired thirty salvoes. For them to fire so many it must, I fear, be the Turkish fleet.'[302] The news was soon confirmed: on the 17th the Turkish fleet had been 'discovered' off Cape Passero and the viceroy of Naples informed the king on 22nd, in a letter accompanying a detailed dispatch from Don García.[303] The king received the first definite reports on 6th June.[304]

Although they had been warned of the danger, those reponsible for defence, the Spaniards and the Grand Master, were surprised by the speed with which it approached, the Grand Master in particular: he had hesitated to spend money and to begin the necessary demolition work on the island of Malta. There were delays in the arrival of supplies and reinforcements, and five galleys belonging to the Knights, in excellent condition, were blockaded in the harbour and unable to give the Christian fleet the slightest assistance.[305]

[296] Viceroy of Naples to Philip II, Naples, 8th April, 1565, Simancas E° 1054, f° 80.
[297] Viceroy of Naples to Philip II, Naples, 8th April, 1565, *ibid.*, f° 81.
[298] *Ibid.*, f° 94, dispatch from Corfu, 30th April, 1565.
[299] *Ibid.*
[300] J. B. E. Jurien de La Gravière, *op. cit.*, I, p. 172.
[301] *Ibid.*
[302] Simancas E° 1125.
[303] Simancas E° 1054, f° 106.
[304] *Recidiba a VI de junio*, note on document quoted in note 303.
[305] C. Duro, *op. cit.*, III, p. 76 ff.

The resistance of the Knights. But the Grand Master, Jean de La Valette Parisot, and his knights defended themselves with great courage and in the end saved the day.

Arriving before the island on 18th May, the Turkish fleet immediately made use of the large bay of Marsa Scirocco on the south-eastern shore, one of the best harbours in Malta after the bay of Marsamuscetto, which was later to form the harbour of Valletta. 3000 men were landed in the night of 18th to 19th and 20,000 the next day. Overwhelmed, the island was occupied without difficulty. The Knights were left holding only the small fort of St. Elmo which commanded access to Marsamuscetto and the Old City, the Borgo, a large fortified area, as well as the strong forts of St. Michael and St. Angelo. Naval considerations obliged the Turks to begin their siege on 24th May with the weakest position, St. Elmo, in the hope that by capturing it they would have full control over the port. The battery began on 31st May. But the position was not captured until 23rd June after an extremely violent bombardment. Not one of the defenders escaped. But their determined resistance had saved Malta. It had given the island the vital time it needed to prepare for the assault and to finish the defence works at the Borgo and Fort St. Michael, designed by the Knights' military architect M° Evangelista. It had also enabled the Spanish to catch up. Only fortuitous circumstances prevented Juan de Cardona, the commander of the Sicilian galleys, from bringing help to Malta before the fall of St. Elmo. His little detachment of 600 men landed however at a still opportune moment on 20th June and was able to reach the Old City (Citta Vecchia), evidence that the besieging forces did not exercise perfect control over land or sea.

With St. Elmo taken, the Turks concentrated their efforts, both on land and sea, on the extensive but in part makeshift defence of Fort St. Michael. Artillery bombardment, assault forces, mines, attacks from the sea – no effort was spared and none of these measures got the better of the defenders. The almost miraculous salvation of the position was finally assured on 7th August by the intervention of the Grand Master in person and by a cavalry sortie from the Old City: hurling itself upon the Turkish rearguard, it spread utter panic. One month later, on 7th September, the Turkish army had made no progress whatever. Its ranks had been thinned by repeated assaults, it was suffering from epidemics and even from food shortage. Reinforcements and supplies from Constantinople failed to arrive. Both besiegers and besieged had in fact reached the same pitch of exhaustion. And at this point Don García de Toledo acted.

The relief of Malta. Historians have blamed Don García for his delay. But have they always examined thoroughly the conditions under which he had to operate? The loss of Malta would undoubtedly have been a disaster for Christendom.[306] But the loss of the Spanish fleet, just after it had been built at such cost, would have exposed Spain to inescapable

[306] P. Herre, *op. cit.*, p. 53; H. Kretschmayr, *op. cit.*, III, p. 48.

danger.[307] In addition, let us not forget that when it came to a struggle between the western Mediterranean and the Levant, the eastern sea was the more navigable; during the manœuvres of Spanish fleets, the Gulf of Lions presented a very different kind of obstacle from the Aegean with its islands. A commander seeking to concentrate his fleet in a hurry had to contend not only with distance but also with all the many patrol, transport and supply duties the navy was called upon to perform in a sea where corsairs threatened every shore. Trips had to be made to Genoa, Leghorn, Civitavecchia and Naples to load supplies, money and troops. And in Corsica the rebellion was still in full swing and gaining terrain.

Some idea of these problems can be gained by studying the voyages of the Spanish squadron under Álvaro de Bazán.[308] At the beginning of May it was at Málaga, loading cannon and munitions destined for Oran. From Oran, it returned to Cartagena and the nineteen galleys and two roundships took on board 1500 men whom they carried to Mers-el-Kébir. Not until 27th June was it at Barcelona;[309] on 6th July it was at Genoa, on the 21st at Naples and in each port some trifling task held it up. All over the western Mediterranean there were a thousand similar movements: levies of troops, processions of convicts, the freighting of transport ships and the dispatch of funds. It all took time. It had taken until August or September in 1564 to assemble the fleet for the Peñón expedition. This year it looked as if it would take equally long. On 25th June, two days after the fall of St. Elmo, Don García still had only 25 galleys. By the end of August he had about a hundred, counting the good with the bad. In the circumstances was he right to wait? Was he right to refuse to risk his ships in small detachments?

When most of the ships were at the rendezvous, a council of war was held at Messina, at the beginning of August,[310] on the best strategy to adopt. Bold spirits recommended sending a landing party with sixty reinforced galleys; the more prudent and experienced, the 'practical sailors', advised moving to Syracuse to await developments. Ten days later, with the arrival of Gian Andrea Doria, Don García at last had his full complement of galleys. Whereupon he decided abruptly, without consulting anyone, to carry a landing party to the island with his reinforced galleys. On 26th August, the relief fleet left Sicily. Bad weather drove it to the western coast of the island, to Favignana. From there it sailed to Trapani, where about a thousand soldiers took the opportunity to desert. A fair wind brought it back to Lampedusa and finally to Gozo, north of Malta. The squall which had surprised the fleet on its departure

[307] J. B. E. Jurien de La Gravière, *op. cit.*, II, p. 140.

[308] In May, Álvaro de Bazán had nineteen galleys under his command, Tello to Philip II, Seville, 29th May, 1565, Simancas E° 145, f° 284. His squadron was increased and he arrived in Naples with 42 galleys.

[309] Jurien de La Gravière, *op. cit.*, II, p. 167.

[310] *Ibid.*, p. 172 ff.

had most opportunely emptied the 'channel' of Malta of all shipping, but it was impossible for the Christian galleys to rendezvous at the right time off Gozo. In the end Don García, discouraged, returned to Sicily on 5th September. This bungled departure brought him much criticism, scorn and mockery not to mention the harsh judgements of historians. But the very next day, on the insistence of Gian Andrea Doria, the fleet put back to sea; on the night of the 7th it crossed the straits between Gozo and Malta and found itself standing off the bay of Friuli in fairly rough seas. Wishing to avoid the dangers of a night landing, Don García gave orders to await daybreak; the landing was effected in an orderly fashion, in an hour and a half, on the beach at Mellieha Bay. The fleet then returned to Sicily.

The landing party, under the leadership of Álvaro de Sande and Ascanio de la Corna, progressed slowly at first, weighed down by the baggage which had to be carried by the men since there were no pack animals. It made its way painfully to the outskirts of the Old City, where it was billeted in large storehouses outside the walls. Should they press on further? The Grand Master thought not. For the Turks had abandoned their positions, evacuated the fort of St. Elmo and were re-embarking. In the circumstances it would be better not to send out an expeditionary force, already burdened with sickness, towards the Turkish positions which would be full of detritus and corpses, to avoid the risk of an outbreak of plague. However, having been informed by a Spanish deserter, a Morisco, of the modest size of the Christian landing-party (5000 men), the Turkish leaders attempted a counter-offensive. Bringing back on shore several thousand men, they sent them to the interior of the island and into the Old City, where they were slaughtered in the narrow twisting streets of the town; those who escaped hurried back to Piāle Pasha's galleys which sailed for the Levant, the bulk of the fleet making for Zante. On 12th September, the last Turkish sail had disappeared over the horizon. On hearing the news, Don García, who had just embarked a fresh expeditionary force on to his sixty reinforced galleys at Messina, decided to land them all at Syracuse. What good would more troops be in an island devastated by war and short of supplies? On the 14th he entered the port of Malta with his fleet in order to take off the Neapolitan and Sicilian infantry and set off at once for the Levant, in the hope of capturing at least some of the Turkish roundships in the rear of the armada. He reached Cerigo[311] on 23rd and lay in wait there almost a week, but was prevented from attaining his objective by bad weather. On 7th October he was back at Messina.[312]

News of the victory spread fast. It had reached Naples by 13th

[311] Por cartas del Duque de Seminara de Otranto a 29 de 7bre, 1565, Simancas E° 1054, f° 207. On 22nd, Don García was between Zante and Modon, before the uninhabited island of Strafaria, having left Cerigo, a Venetian possession, with the intention of waiting for the Turkish fleet 'la qual forçosamente havia de pasar por alli'.

[312] Jurien de La Gravière, *op. cit.*, II, p. 224.

September[313] and Rome by the 19th.[314] On 6th October, or maybe before,[315] it caused consternation at Constantinople. Christians 'could not walk in the streets of the city for fear of the stones which were hurled at them by the Turks, who were universally in mourning, one for a brother, another for a son, husband or friend'.[316] Meanwhile in the West, Christian rejoicings were as heartfelt as Christian fears had been earlier in the year. As late as 22nd September, 1565, little optimism[317] had been entertained at Madrid. Here is the enthusiastic testimony of the Sieur de Bourdeilles (alias Brantôme), who with so many others had arrived at Messina too late to be shipped to Malta: 'A hundred thousand years from now, the great king Philip of Spain will still be worthy of praise and renown, and worthy that all Christendom should pray as many years for the salvation of his soul, if God have not yet already given him a seat in Paradise for having so nobly delivered so many gentlemen in Malta, which was about to follow Rhodes into enemy hands.'[318] At Rome, where there had been such panic earlier in the summer, at the news of the Turkish fleet, the heroism of the Knights received its full due and thanks were offered to God for his intervention, but no gratitude was expressed towards the Spanish; on the contrary. The tone was set by the Pope who did not pardon them for their delays, nor for the difficulties they had given him since his accession. Cardinal Pacheco, on receiving the news of the victory had requested an audience with the Pope, which turned out to be an exceedingly unpleasant occasion. On the cardinal's suggesting that this might be a good moment to grant the king the *quinquenio*, it was, he wrote 'as if I had opened fire on him with an arquebus'. 'Send him the *quinquenio*?' he said in the end: 'He will be lucky if he gets it when he asks me for it.' In his public audience a few minutes later, the Pope succeeded in speaking of the victory without mentioning the king of Spain, his admiral, or his troops, attributing the entire victory to God and the Knights of St. John.[319]

The role of Spain and Philip II. And yet the merits of Philip and his admiral seem beyond doubt to us. Jurien de la Gravière, who frequently compares Malta to Sebastopol, gives a more just appreciation of the event than other historians. Even Vertot ('My siege is finished!') criticizes Don García for his prudence and delay without considering any of the arithmetic

[313] The Duke of Alcalá to Philip II, Naples, 12th September, 1565, Simancas E° 1054, f° 194.

[314] Pedro d'Avila to G. Pérez, Rome, 22nd September, 1565, J. J. Döllinger, p. 629. Cardinal Pacheco had sent a courier off at midnight to carry news of the victory to the king, Cardinal Pacheco to Philip, 23rd September, 1565, *CODOIN*, CI, p. 106–107.

[315] Constantinople, 6th October, 1565, Simancas E° 1054, f° 210; Pétrémol to Charles IX, Constantinople, 7th October, 1565, E. Charrière, *op. cit.*, II, p. 804–805.

[316] See note 315.

[317] Garces to the Duke of Florence, Madrid, 22nd September, 1565, original in Spanish, A.d.S. Florence, Mediceo 4897, f° 148.

[318] Quoted by Jurien de La Gravière, *op. cit.*, II, p. 201.

[319] Cardinal Pacheco to Philip II, Rome, 23rd September, 1565, *CODOIN*, CI, p. 106–107.

behind this delay. Manfroni, in his *Storia della marina italiana*, gives the entire credit to the Italians; the Spanish were apparently not worth mentioning. Historians have only too often absorbed the national prejudices of the early chroniclers.

Without any question, the victory at Malta marked a new phase of the Spanish recovery, a recovery which was not achieved by accident and which had been pursued diligently throughout the year 1565. Fourquevaux, arriving at Madrid at the end of the year as representative of the king of France, wrote on 21st November that 40 galleys were being built at Barcelona, twenty at Naples and twelve in Sicily.[320] Probably (he adds and we hear the voice of the governor of Narbonne) the king of France will be asked permission to take from the forest of Quillan a number of 'rames à galoche', oars for the Barcelona galleys. The enormous effort of the king of Spain inspired others: the Duke of Florence was thus busy constructing a new fleet.

For the Turkish peril was not considered to have vanished overnight with the retreat from Malta. In fact it appeared more threatening than ever at the end of the year. The sultan was increasing his shipbuilding and on 25th September there was talk in Constantinople (where, it is true, news had not yet arrived of the failure of the armada) of further major offensives, in particular against Apulia.[321] News of the 'rout' of the fleet, as the French ambassador wrote, merely strengthened the desire for revenge. In spite of difficulties in obtaining wood, there was talk of building a hundred boats in the arsenal and the sultan himself had even mentioned a figure of five hundred. 'He has given orders', says a dispatch dated 19th October, 'that fifty thousand oarsmen and fifty thousand *assupirs* be in a state of readiness by mid-March in Anatolia, Egypt and Greece.' Malta, Sicily and Apulia were to be the target of this attack. Fourquevaux writing from Madrid on 3rd November says that it is feared that the Turk might 'make an almighty effort on land and sea if he does not die of anger from his army being repulsed at Malta'.[322] On 21st November,[323] Madrid heard from Vienna that the next year the sultan intended to pit his entire military strength against Philip II, including the janissaries and his guard. Dispatches dated 12th December also announced that Sulaimān had declared war on the emperor and would lead an army of two hundred thousand men against him,[324] but this was understood to be a gesture made by the sultan against the advice of his councillors. The West was convinced that the Turkish fleet would be sent against Malta under the same commanders as in 1565, for once the sultan allowed the island time to construct defences he would never be able to capture it. It was thought therefore that the sultan and the emperor would come to terms.

[320] Fourquevaux, *op. cit.*, I, p. 10–14.
[321] Constantinople, 25th September, 1565, Simancas E° 1054, f° 205.
[322] Fourquevaux, *op. cit.*, I, p. 6.
[323] *Ibid.*, p. 13.
[324] Constantinople, 16th December, 1565, Simancas E° 1055, f° 14.

These reports were taken very seriously by the Spanish government. On 5th November, 1565, Philip gave orders to fortify La Goletta: he had decided, he wrote to Figueroa, to grant it the necessary 56,000 ducats.[325] The decision seems to have been a definite one since he asked Adamo Centurione to arrange a *cambio* for this sum. In obedience to his orders, there began to rise up, around the old fortress a new Goletta ('Goletta la Nueva', facing 'Goletta la Vieja'). Moreover apart from Álvaro de Bazán's twelve galleys which were recalled to Spain, the king left his entire fleet in Sicily.[326] Had not the Grand Master threatened to abandon the island if he did not receive help? At the end of December, the king of Spain helped him to the tune of 50,000 ducats (30,000 in cash and 20,000 in supplies and ammunition) as well as 6000 infantrymen, at least according to the Tuscan agent.[327] It is thought that the Turk can only attack Malta or La Goletta, writes Fourquevaux on 6th January, 1566. If he strikes at Malta, the Catholic King will send 3000 Germans, 5000 Spaniards and Italians who will dig in on the mountain of St. Elmo, for the Borgo is beyond repair. If they attack La Goletta, the king will send 12,000 men to encamp around the fortress.

All these measures, though praiseworthy in themselves, did not however add up to a singleminded policy with any claim to direct the course of events. There was at Madrid, it is true, some vague plan of a league against the Turk; Philip was said to be interested in an alliance with Venice – but how seriously? After all the Venetians had rejoiced when they heard of the fall of St. Elmo.[328] As good honest traders they considered the Knights of St. John the fly in the ointment of East-West commercial relations, and never failed to tell the Turks what was afoot in the West. So when Fourquevaux came asking information from his colleague, the Venetian ambassador, the latter immediately reassured him: the Signoria had no intention whatever of signing an alliance with the king of Spain.

The same was true of a common policy between France and Spain: mere talk and nothing else. The much trumpeted meeting at Bayonne did not mark a turning-point in history as contemporaries and, later, historians believed. On the French side of the Pyrenees lay a restless kingdom, deeply disturbed, with treason already evident. At its head was an anxious woman and on the throne a boy-king. Catherine decided to show her son's face throughout the kingdom – something like a modern propaganda-tour: it was a successful tour as it happened, but it took time. When the travellers arrived in the South, it looked a suitable moment to arrange a possible meeting between the rulers of France and Spain. It matters little who thought of the idea first (possibly Montluc, who was to some extent in league with Spain). Philip II in any case declined to make a personal

[325] Philip II to Figueroa, 5th November, 1565, Simancas E° 1394.

[326] Fourquevaux to the king, 21st November, 1565, Fourquevaux, *op. cit.*, I, p. 10–14.

[327] A.d.S. Florence, Mediceo 4897 a, 29th December, 1565, Fourquevaux, *op. cit.*, I, 36, 25,000 crowns and 3000 Spanish troops.

[328] Garci Hernández to Philip II, Venice, 26th July, 1565, Simancas E° 1325.

appearance and it was only at his wife's request that he agreed in January, 1565 to allow her a brief reunion with her family. The fact that he saw fit and perhaps deemed it politic to wait to be begged for his consent, does not necessarily mean that the conversation left him indifferent.[329]

For beyond the Pyrenees, the great Hispanic world was still peaceful, but burdened increasingly with imperial responsibilities and a treasury burdened with debts. In Philip the strengths and weaknesses of the empire were incarnate. At his side his third wife, Elisabeth of Valois, known in Spain as Isabella, 'Reina de la Paz' might have a role to play. She was still almost a child, hardly a woman; not the unhappy bride one sometimes reads about. She appears to have adapted fairly quickly to life in Spain and at Bayonne in any case she played to perfection the part she had been taught. Francés de Alava, the Spanish ambassador at the French court, writing on 1st July to Philip II, says of the young queen: 'I assure Your Majesty with the deepest sincerity that Her Majesty won over the hearts of all right-thinking persons, especially when she spoke of religion and the sentiments of brotherhood and friendship which Your Majesty entertains and will continue to entertain towards the king of France.'[330] We may take his word for it.

Having left home on 8th April,[331] the young queen arrived on 10th June at Saint-Jean-de-Luz[332] where she was met by her mother. They entered Bayonne together on the 14th. Elisabeth stayed there until 2nd July, a little longer than had been intended.[333] The family reunion was an opportunity for the two governments to exchange guarantees, to talk of marriages (always the major talking point of royal meetings in the sixteenth century) and finally to take leave of each other empty-handed, each side less convinced than ever of the sincerity of the other. It was a non-event in history – to us that is, but not of course to the participants and contemporary observers.

Certainly Philip II did not consider it so: he had sent the Duke of Alva and Don Juan Manrique with the queen as observers and advisers. The imposing figure of the duke dominates the talks as they are described by chroniclers and historians. What Spain wanted was that France should be immobilized, engrossed in her own domestic and foreign problems. This was neither a friendly nor a diabolical policy: it was almost a necessity for the Spanish Empire, grouped as it was around France and automatically feeling the repercussions of any movements in that country, particularly in the Netherlands, which was so evidently vulnerable since

[329] Saint-Sulpice, 22nd January, 1565, E. Cabié, *op. cit.*, p. 338; Philip II to Figueroa, 3rd February, 1565; Garces to the Duke of Florence, A.d.S. Florence, Mediceo 4899, f° 64.

[330] Bayonne, 1st July, 1565, A.N., K 1504, B 19, n° 46.

[331] Luis Cabrera de Córdoba, *op. cit.*, I, p. 423, gives the dates of 8th and 14th June.

[332] The Duke of Alva and Don Juan Manrique to the king, Saint-Jean-de-Luz, 11th June, 1565, A.N., K 1504, B 19.

[333] The Duke of Alva and Manrique to the king, Bayonne, 28th and 29th June, 1565, *ibid.*, no. 37 (summary).

the events of 1564. But this was a great deal to ask of France in the name of religion, which once again provided a convenient cloak. Nothing was offered in exchange. How could the queen mother abandon her policy of tolerance for a tactic which, being only too evidently of Spanish origin, could only further divide and weaken the kingdom of France?

Despite the official smiles and festivities then, these profound differences were evident. There were even, before and during the interview, a certain number of hitches. Thus on 7th February, when Catherine de Medici had already sent orders to Bayonne that provision be made and apartments prepared 'a la española' for the queen of Spain and her ladies, Francés de Alava reported from Toulouse that there was a rumour to the effect that the French sovereigns were to be accompanied – could it be true? – by the heretical 'Madame de Vendôme', Jeanne d'Albret. These words in the report were underlined by Philip, who wrote in the margin: 'si tal es, yo no dexare ir a la Reyna, if so I shall not let the queen go'.[334] And he immediately[335] informed the French ambassador[336] that he did not wish either the queen of Navarre or the prince of Condé to be present at the interview. Another incident occurred in June, just before the arrival of the queen of Spain, when Francés de Alava learnt that, horror of horrors, a Turkish ambassador had landed at Marseilles. Catherine de Medici, taxed with this by the ambassador, defended herself as best she could. She hastily dispatched to her son-in-law an envoy, M. de Lansac, who arrived at Aranjuez on the very day, 10th June, 1565, that the queen of Spain was greeting her mother at Saint-Jean-de-Luz. The excuse carried by Lansac was as follows: the king and the queen of France do not know what brings this ambassador to their country and they have sent the Baron de La Garde to meet him and find out. If his mission should relate to anything contrary to the interest of the king of Spain, he will of course be refused an audience. 'I replied', wrote Philip to Francés de Alava, 'that I was sure this was so, but pointed out that many people could not but be astonished that this envoy should arrive at the very moment when the Turk has sent out a fleet against me. That nevertheless I was confident that the ambassador would receive a reply which . . . made plain to all the great love there is between my person and the king of France.'[337]

It may have been only a small thing but it did not lessen Spanish suspicions. The Turkish ambassador cut short his visit to the queen mother and left on 27th June. By now the talks were in full session, and the queen hastened to inform the Duke of Alva that she had spoken with the Turk only of the depredations committed in Provence,[338] upon which the Turk

[334] F. de Alava to Philip II, Toulouse, 7th February, 1565, A.N., K 1503, B 19, no. 33a. Autograph note in margin by Philip II.

[335] Bearing in mind the delays of distance.

[336] Saint-Sulpice to Catherine de Medici, 16th March, 1565, E. Cabié, p. 357–358.

[337] Aranjuez, 12th June, 1565, A.N., K 1504, B 19, no. 11.

[338] It is quite possible that these robberies were invented, H. Forneron, *Hist. de Philippe II*, I, p. 322.

had promised that reparation would be made, on condition that a French envoy be sent to the sultan. So the French are thinking of sending an embassy to Turkey, concluded the duke. Since the Turkish armada is already in the West, he retorted to the queen, 'there can evidently be no question of sending anyone to Constantinople. And next year the king of Spain's fleet will be so mighty that the sultan's will be able to do very little damage.'[339]

It seems then that at Bayonne, the Spaniards considered it self-evident that France would abandon her traditional friendship with the Turk and that they were seeking to draw her into a league against both heresy and the sultan. Definite overtures were made a few months later. These talks, Fourquevaux told the queen, seem 'to be intended to drive you into an alliance of great consequence'. The Spaniards were making use of the desires expressed by Catherine at Bayonne. The queen had spoken of marriages, which might perhaps end in a league. The Spaniards spoke principally of a league, beginning 'by the tail', as Fourquevaux put it.[340] Consider the dangers of consenting to such a league, warned the ambassador, when the 'Turk is at such peace with Her Majesty, and the French are more welcome in his ports and lands than they are in such places in the lands and kingdoms of the said king, and France being moreover so placed that she can expect little harm from the Turkish forces. If we are to break the peace with the said Turk and sacrifice all the trade and traffic of your subjects, His Majesty [Philip] ought to consent to any request Your Majesty might make of him.' What Catherine was in fact asking, was a series of advantageous marriages for her children and Fourquevaux considered it very unlikely that they would come about, notably that of the Duke of Orleans with Philip II's sister, Joanna, who did not seem in favour of it. Similarly the marriage of her daughter Marguerite de Valois to Don Carlos. Spanish diplomacy was only pretending to consider these matches; it was a way if not to control then at least to restrain the French government.

There was nothing very grand about Spanish designs however. Madrid was using the notion of a universal Catholic policy as a screen. The only policy sincerely pursued was a Spanish one (a policy common to all Catholicism could in any case only come from Rome, where Pius IV had just died). Spain did not even visualize a grand Mediterranean policy at this stage: that would have presupposed fervour, passion, interest, financial security and freedom of movement which were not, or not yet, possessed by the Prudent King. He felt himself to be surrounded by threats: in the Mediterranean of course, but also from the Protestant privateers in the Atlantic; from France on the frontier of the Netherlands and from the Netherlands themselves, where disturbances were beginning to menace the Spanish resources centred at the great meeting-place of Antwerp. In December, 1565 there first emerged the rumour, which

[339] See above, note 333.

[340] *Op. cit.*, I, p. 20, 25th December, 1565.

was to circulate and reappear for years, that Philip intended to go to Flanders.[341]

In fact circumstances conspired to prevent Philip II from pursuing any grand design at least more than momentarily. During the first ten years of his reign he could deal only with the most pressing matters, the most imminent dangers; and he had to do so at the least expense, to avoid mortgaging the future. We are still far from the imperialist follies of the end of the reign when Philip II was to depart radically from the image of the Prudent King.

[341] Don Francés de Alava to Philip II, 13th December, 1565, aut. A.N., K 1504, B 19, no. 95.

CHAPTER III

Origins of the Holy League: 1566–1570

Between 1566 and 1570, events moved quickly. Perhaps this was the logical consequence of the comparatively calm period harshly and suddenly terminated by the drama of Malta in autumn 1565.

But uncertainty persisted. Would the newly-found force of the Spanish Empire find expression in vigorous projects and campaigns in the Mediterranean, or would it be directed towards the Netherlands, that other pole of attraction in Philip II's states? These hesitations contributed to the far from settled political climate. Who would decide in the end? Men or circumstances, the latter sometimes in strange combination? The West, or the Turkish East, still 'poised in the air' ready to pounce on Christendom?

I. NETHERLANDS OR MEDITERRANEAN?

The election of Pius V. On 8th January, 1566, an unexpected vote placed on the papal throne Cardinal Ghislieri, better known to his contemporaries as the 'cardinal of Alessándria'. Out of recognition to Charles Borromeo and his party to whom he owed his election, he took the name of Pius V, honouring a predecessor who had not however particularly liked him. Pius IV, Pius V: between the two men the contrast was striking. Son of a rich and powerful Milanese family, the first was a politician, a jurist and a man of the Renaissance. Pius V had been a shepherd-boy. He was one of the countless sons of the poor in whom the Church, in the age of Counter-Reformation, often found her most zealous servants. Indeed as the century drew on, it was the poor who increasingly set the tone within the Church – the poor, or the upstarts (as Alfonso de Ferrara scornfully called them, after seeking unsuccessfully in 1566 to secure the election of his uncle, Cardinal Ippolito d'Este). Pius V was indeed one of these 'upstarts', not a 'princely' pope, not a man familiar with the ways of the world and prepared to make those compromises without which 'the world' would not go round. He had the passion, rigour and intransigence of the poor, and on occasions their extreme harshness and refusal to forgive. He was certainly not a Renaissance pope: their time was past. One historian has compared him to a medieval pope; another suggests, and I am inclined to agree, that he was something of an Old Testament pope.[1]

Born on 17th January, 1504 at Bosco,[2] near Alessándria, it was only by chance that he received any schooling. At fourteen, he entered the Dominican monastery at Voghera. He took the vows of the order in 1521

[1] Jean Héritier, *Catherine de Médicis*, 1940, p. 439.

[2] For these and further biographical details see Pastor, *op. cit.*, XVII, p. 37–59.

at the monastery of Vigevano and was ordained priest seven years later, after studying at Bologna and Genoa. From then on he led a rigorously simple life as Fra Michele of Alessándria, the humblest of all the Dominicans, steadfastly poor and travelling, when he travelled at all, on foot, his scrip upon his back. Honours came to him but always burdened him and they were accompanied by harsh duties. Prior, then *provveditore*, by 1550 he was inquisitor of the diocese of Como, at a sensitive point on the defensive frontier of Catholicism. He fought doggedly for the faith here. Not surprisingly, his confiscation of bundles of heretical books in the year 1550 brought some wrath down upon his head. But it also brought a trip to Rome and his first contact with the cardinals of the Inquisition, notably Cardinal Carafa, who took an interest in him from this time. Carafa's backing obtained him the post of commissioner-general of the Inquisition under Julius III. On the accession of Paul IV on 4th September, 1556, he became bishop of Sutri and Nepi, but in order to keep him at his side, the Pope appointed him Prefect of the Palace of the Inquisition. On 15th March, 1557, he promoted him cardinal. The future Pius V was indeed a man after Paul IV's own heart; he had the same intransigence, the same impassioned violence and will of iron. Naturally his relations with Paul IV's successor were less cordial: Pius IV was too 'worldly', too willing to compromise, too ready to please to get on with the 'cardinal of Alessándria'. The man who became Pope in 1566 should have taken the name Paul V.

This old man, with his bald head and flowing white beard, once described as an ascetic who was nothing but skin and bone,[3] possessed despite his fragile appearance incredible vitality and capacity for action. He never took any rest even during the terrible days of the sirocco in Rome; he ate sparingly: 'at midday, soup with bread, a couple of eggs and half a glass of wine; in the evening a vegetable soup, salad, some shellfish and a cooked fruit. Meat appeared on the table only twice a week.'[4] In November, 1566, when visiting the defence works on the coast, he was seen walking on foot, as in the old days, beside his litter.[5] It was his virtue which had brought him the votes of the Sacred College, not his intrigues or those of princes who for once did not intervene in the election.[6] In 1565, Requesens had written to Philip II: 'he is a theologian and a good man, of exemplary life and great religious zeal. In my view, he is the cardinal we need as Pope in the present times.'[7]

[3] Report by Cusano, 26th January, 1566, Vienna Archives, quoted by Pastor, *op. cit.*, XVII, p. 42, and note 2. [4] *Ibid.*, p. 44. [5] *Ibid.*, p. 45.

[6] There was only a hint of intervention on his behalf by Philip II, L. Wahrmund, *Das Ausschliessungsrecht (jus exclusiva) der kath. Staaten Österreich, Frankreich und Spanien bei den Papstwahlen*, Vienna, 1888, p. 26. Requesens to Philip II, Rome, 7th January, 1566: *Ha sido ayudado y favorecido por parte de V. M.*; Luciano Serrano, *Correspondencia diplomática entre España y la Santa Sede*, Madrid, 1914, I, p. 77 and note 2.

[7] Requesens to Philip II, 5th January, 1565, in J. J. Döllinger, *op. cit.*, I, p. 571–578, quoted by Pastor, *op. cit.*, XVII, p. 11 and 59.

Once on St. Peter's throne, Pius V remained true to his character and became a legend in his lifetime. Before the first year of his pontificate was out, Requesens had many times reiterated his opinion that the Church had not had a better leader for three centuries, that he was a saint. We find the same judgement from Granvelle's pen.[8] It is impossible to approach Pius without referring to his extraordinary personality. The slightest text in his hand gives a strange impression of violence and presence. He lived in a supernatural element, lost in his devotions, and the fact that he was not of this world, enmeshed in the petty calculations of politics, made Pius V a great historical force, unpredictable and dangerous. An imperial councillor wrote as early as 1567: 'We should like it even better if the present Holy Father were no longer with us, however great, inexpressible, unparalleled and extraordinary his holiness may be.'[9] To some people, evidently, this holiness was not entirely welcome.

Intransigent and visionary, Pius V had a highly developed sense of the conflicts between Christendom and the Infidels and Heretics. His dream was to wage these mighty battles and to bring a speedy solution to the quarrels that divided Christendom against itself. He very soon adopted Pius II's former plan to unite the Christian princes against the Turk. One of his first gestures was to ask Philip II to forget the dispute with France over precedence at Rome, a dispute which during the reign of Pius IV had led to the recall of Requesens.[10] Such quarrels only served to drive the Most Christian King into an alliance with the Turk.

Another of his first actions was to contribute to Spain's naval strength. The bargaining that had always accompanied any concession of ecclesiastical revenues to Spain, the favours that had to be bestowed on the relations and favourites of the Pope, the accessory expenses and the time it could take, are a familiar story. It so happened that the five-year subsidy granted towards the Spanish galleys by Pius IV lapsed at the very time of the new Pope's election: he renewed it immediately, without question. On 11th January, 1566, four days after the accession of Pius V, Requesens, in a letter to Gonzalo Pérez, expressed his unreserved delight at this *quinquenio* which had not cost the king a single *maravedi*. 'Last time, it cost us 15,000 ducats in income from the vassals of the kingdom of Naples, and 12,000 ducats' pension in Spain for the Pope's nephews, not to mention the large sums spent on sending ministers to handle the negotiations.'[11] The new pontificate evidently intended to turn over a new leaf. In Pius V the Church had undoubtedly found an energetic leader, committed to a new crusade. As it happened, the events of the year 1566 could not help but create a climate favourable for crusades.

[8] *Ibid.*, p. 48–49.

[9] Zasius to Archduke Albert V of Bavaria, F. Hartlaub, *Don Juan d'Austria*, Berlin, 1940, p. 35; V. Bibl, *Maximilian II. der rätselhafte Kaiser*, 1929, p. 303.

[10] Pius V to Philip II, Rome, 24th January, 1566, L. Serrano, *op. cit.*, I, p. 111.

[11] Rome, 11th January, 1566, *ibid.*, I, p. 90.

The Turks in Hungary and in the Adriatic. In November and December, 1565, disturbing news arrived from the Levant. The Venetian ambassador, admitted to a public audience at the doors of the conclave on 30th December, 1565, begged the cardinals, in view of the bad tidings, to make all speed in choosing the new Pope.[12] 'Common report' spoke of an armada more powerful even than that which had besieged Malta.

These dispatches in November and December explain the large-scale operations discussed in military headquarters. Philip II reminded Chantonnay on 16th January that in view of news that the Turkish fleet would be more numerous and more powerful than that of 1565, he had decided to strengthen the defences of the two most threatened positions: he was sending to Malta, to join the forces of the Knights themselves, 1000 Spaniards from the old companies, 2000 Germans and 3000 Italians; to La Goletta, where the new fortress was not yet ready, he would send 5000 trained Spaniards, 4000 Italians and 3000 Germans, a total of 12,000 men who could be encamped for lack of space within the fortress, on the 'mountains' surrounding it, where there were abundant supplies of water.[13] These plans involved the multiple details in which Philip's painstaking, bureaucratic machine excelled. Its work, not concealed this time, but on the contrary deliberately publicized, was duly reported by all the foreign ambassadors at Madrid.[14] Orders were given loud and clear; appointment succeeded appointment; Ascanio della Corna received command of the Germans who were to go to Malta; Don Hernando de Toledo, the son of the Duke of Alva, was appointed to La Goletta; Don Alvaro de Sande to Oran.[15] On 26th January, Fourquevaux mentions a convoy of 2000 Spaniards from the garrisons of Naples bound for Oran. He even writes that the Spaniards are actually hoping for a Turkish attack on Apulia or Sicily, since they are assured that in such a situation, 'all Christendom will immediately rush to the rescue'. A month later, the same writer reports that the king of Spain is said to have offered four Italian cities to the Venetians in order to draw them into an alliance against the Turk.[16]

Fourquevaux was also suspicious of the widespread publicity surrounding Spanish rearmament. Could the figures given him by the Duke of Alva be exaggerated? His suspicions were unjustified however, for the same figures are to be found in orders and communications sent by the king.[17] It still remains to be explained why the Spaniards, normally so

[12] Requesens to the king, Rome, 30th December, 1565, L. Serrano, *op. cit.*, I, p. 67.

[13] Philip II to Chantonnay, Madrid, 16th January, 1566, *CODOIN*, CI, p. 119–123; to F. de Alava, Madrid, 16th January, 1566, A.N., K 1505, B 20, no. 65; Nobili to the prince, Madrid, 18th January, 1566, A.d.S. Florence, Mediceo 4897a; Nobili to the prince, 21st January, 1566, *ibid.*

[14] The Tuscan representative for instance, 15th January, 1566, see preceding note.

[15] Letters sent by Nobili on 18th January and 16th February, 1566; 17th January, 1566, Fourquevaux, *op. cit.*, I, p. 47; 22nd January, p. 47–48; 11th February, p. 52.

[16] I, p. 61.

[17] Viceroy of Naples to Philip II, 23rd January, 1566, Simancas E° 1055, f° 11. Instructions to Prior Don Antonio de Toledo, 18th February, 1566, Simancas E° 1131.

reticent about such matters, chose to make their preparations very conspicuous. Was it in order to conceal preparations of a different kind? They kept very quiet about the naval constructions under way in both Naples and Barcelona and, in the former city at least, hampered by a shortage of convicts.[18]

Meanwhile news was arriving from Constantinople which, if true, rendered most of these precautions unnecessary. A report dated 10th January declared that the fleet was indeed planning to put to sea, but that it would be a smaller one than at Malta, for the Turks were short both of oarsmen and munitions. It was thought that about a hundred galleys might sail under Piāle Pasha, without any major objective, but possibly considering a raid as far west as Genoa to divide the Hispanic fleet. In addition, and this was the more sensational information, all the signs confirmed that the aged Sulaimān intended to go in person to Hungary and from there to march on Vienna.[19] In fact war had already broken out again on the long Balkan frontier, in 1565. In vain Maximilian had dispatched agents and letters to stop it and to encourage a return to the truce agreed in 1562; any return to the truce was very far from the mind of the sultan, who was engaged upon major military preparations: there was talk of 200,000 Turks and 40,000 Tartars. The Turkish leaders were making ready for the campaign, ruining themselves by purchases of camels and horses, which were already 'carissimos'. Another telling sign was that the old *sandjak-beg* of Rhodes, 'Alī Portuc, guardian of the archipelago, was on his way, with his galleys, to the Danube, where he was charged with the mission of building boats and tackle to help the army cross the river.[20]

The plans for naval attacks upon the West were not however abandoned. On 27th February, the oarsmen were reported to have arrived at Constantinople,[21] a sign that the galleys were ready. They were said to be planning to leave on about 1st April. But all sources agreed that there would be fewer than a hundred.[22] And if there was to be a war in Hungary, less danger could automatically be expected in the Mediterranean.[23]

[18] Letter from viceroy of Naples, 23rd January, 1566 (see preceding note). García de Toledo to the viceroy of Naples, Naples, 2nd February, 1566, Simancas E° 1055, f° 24; viceroy of Naples to Philip II, Naples, 16th April, 1566, Simancas E° 1055, f° 103, concerning the fifteen galleys to be built at Naples for the kingdom of Sicily. Could eight of them not be built at Genoa?

[19] Simancas E° 1055, f° 7.

[20] *Ibid.* '. . . por el Danubio a hazer fabricar barcones y a hazer xarcias para pasar los exercitos'.

[21] Dispatch from Constantinople, 27th February, 1566, Simancas E° 1055, f° 53.

[22] *Ibid.*, dispatch from Corfu, 28th February, in letters from Cesare de Palma and Annibale Protótico, Simancas E° 1055, f° 49; this dispatch refers to 80 or 90 galleys only, and these were not intended to leave Turkish waters. Dispatch from Chios, 1st March, 1566, Simancas E° 1055, f° 59; dispatch from Corfu, 16th March, 1566, Simancas E° 1055, f°s 67 and 68; only a dispatch from Constantinople, 15th March, 1566 (Lepanto, 25th March), Simancas E° 1055, f° 54, refers to 130 galleys.

[23] It looked more than ever as if a Hungarian war was likely, with a march on Vienna, dispatches from Ragusa, 26th February, 1566, Simancas E° 1055, f° 61. On 15th

Genoa, which had always, possibly because of the large number of Genoese renegades, been equipped with the best intelligence network in the Levant, was informed, in a letter dated 9th February, 1566, that the Turkish armada was planning to enter the 'Gulf of Venice', first attacking Fiume and collecting what would undoubtedly be very considerable booty there, then opening a passage to carry support to the Grand Turk's army in Hungary.[24] It would come no further west unless it heard that the Spanish fleet was not massed for action.

There was a general feeling of relief. The people of Malta, writes Requesens to the king on 18th April, feel that the terrible danger looming over them is quite dispersed.[25] Philip II seemed in May on the point of countermanding the large troop movements ordered in winter[26] and the viceroy of Naples, anxious to economize, asked permission on 20th April to dismiss the Germans as soon as he had recovered 1500 of the Spaniards of Naples, who had been lent to Don García de Toledo and who were either in Sicily or at La Goletta.[27]

Meanwhile the Turkish fleet had left Constantinople on 30th March, with 106 galleys according to some sources while others said 90, including the ten galleys of Alexandria.[28] It was evidently in no hurry to cross the Aegean, occupying itself first with the elimination (without a battle as it happened) of the Genoese presence on the island of Chios, merely exiling at this stage, the *signori mahonesi* with their wives and children, to Kaffa on the Black Sea.[29] At Corfu on 10th May, the fleet was expected to be entering the *Golfo* soon,[30] but it was not in fact sighted in the straits until 10th July.[31] On the 11th, it was at Valona[32] and passed quickly on to Durazzo, then to the bay of Bocche di Cattaro and to Castelnuovo, where it probably arrived on the 23rd.[33]

On receiving this intelligence, the Grand Master of Malta and Don

---

March, there was a report from Constantinople of troop movements towards Sofia (Simancas E° 1055, f° 64). From Corfu on the 16th came reports of the massing of *sipāhis* at Adrianople and the casting of many cannons in Constantinople (*Por carta de Corfu de 16 de março 1566*, Simancas E° 1055, f°s 67 and 68).

[24] B. Ferrero to the Republic of Genoa, Constantinople, 9th February, 1566, A.d.S. Genoa, Costantinopoli 22170. Confirmed by letters from Corfu of 16th March, Simancas E° 1055, f°s 67–68.

[25] L. Serrano, *op. cit.*, I, p. 184.

[26] 29th May, 1566, Fourquevaux, *op. cit.*, I, p. 64.

[27] 20th April, 1566, Simancas E° 1055, f° 104.

[28] Dispatch from Ragusa, 27th April, 1566, Simancas E° 1055, f° 13. Dispatch from Corfu, 3rd May, 1566, Simancas E° 1055, f° 124.

[29] Letters from Constantinople, 16th July, 1566, A.d.S., Genoa, Constantinople 22170.

[30] *Avvisi venuti con la reggia fregata da Levante dall'ysola de Corfu de dove partete alli 10 di maggio 1566*, Simancas E° 1055.

[31] Corfu, 11th July; the dispatch arrived at Otranto on the 12th, Simancas E° 1055, f° 155.

[32] Report by master of a galleon, Simancas E° 1055, f° 163.

[33] G. Hernández to the king, Venice, 1st August, 1566, Simancas E° 1365. Marquis of Caparso to the Duke of Alcalá, Bari, 24th July, 1566, Simancas E° 1055, f° 180.

García de Toledo decided to take off the unnecessary troops, the season being too far advanced for the Turkish fleet to contemplate attacking the island.[34] Eighteen galleys arrived to take off the German soldiers, and the marquis of Pescara, appointed several months earlier to the general command of the troops sent to the island by the Spanish king, left his post, having nothing more to do. Far from causing alarm, the entry of the Turkish fleet to the Adriatic seems to have made the Spaniards heave a sigh of relief. After all, the *Golfo* was the Venetians' affair. It was up to them now to arm themselves, negotiate and take precautions. What danger could it present to the Spanish? The Neapolitan coast was on the alert, well defended, and its inhabitants had fled several leagues inland.

According to Venetian intelligence, the Turkish fleet which arrived at Cattaro on about 21st July numbered 140 sail, of which 120 were galleys, galliots or foists. On the 22nd, Piāle Pasha had sailed with three galleys to Ragusa, where he collected the tribute of the Republic of St. Blaise.[35] A few days later, the armada began attacks on the infertile coasts of the Abruzzi.[36] On 29th July, it landed 6000 to 7000 men near Francavilla, seized the town which had been deserted by its inhabitants and burned it. From Francavilla, one galley accompanied by two sloops set out to reconnoitre the waters of Pescara, but a few cannon shots from the town's defences sent the reconnaissance vessels packing: the armada made for Ortona a Mare. Here again, it found the town abandoned, and burned it along with several coastal villages. On 5th August, the Turks made a foray eight miles inland, to the hamlet of Serra Capriola in the province of Capitanata. It was a bad move: they found themselves unexpectedly faced with a vigorous defence which forced them to retreat in disorder. On the evening of the 6th, the fleet appeared off Vasto, with 80 galleys, but vanished again into the night. On the 10th, news reached Naples in letters from the governor of Capitanata, that bad weather had driven four Turkish galleys on to the coast near Fortore.[37] The crews had of course fled, but orders had been given to recover any artillery and rigging, then to fire the hulls to forestall any attempt by the Turks to refloat them. All was in order, the report added, if the enemy should attack the coasts of the kingdom again. Meanwhile the authorities congratulated themselves on the little damage inflicted by its attacks so far. The viceroy, on learning of the departure of the fleet from Chios, had ordered full-scale evacuation of all non-defended points of the coast and the Turks had consequently found the shore deserted. They had taken the grand total of three prisoners. On every previous occasion that the Turkish armada had

[34] Grand Master to Philip II, Malta, 25th July, 1566, Simancas E° 1131. Philip II to G. Andrea Doria, Segovia, 11th August, 1566, Simancas E° 1395.

[35] Garci Hernández to the king, Venice, 1st August, 1566, Simancas E° 1325.

[36] Letter from the governor of the Abruzzi, 1st August, 1566, Simancas E° 1055, f° 165; *Relación de lo que la armada del Turco ha hecho en el Reyno de Napoles desde que fue descubierta hasta los seys de agosto 1566*, Simancas E° 1325.

[37] *Por cartas de D. de Mendoza*, 7th August, 1566, from San Juan Redondo, Simancas E° 1055, f° 171.

appeared off the kingdom, it had made off with at least 5000 to 6000 captives, even when the galleys of the Catholic King had been in pursuit. As for material damage, that was less than had at first been thought.[38]

And it appeared that the Turkish fleet had already turned for home. On 13th August it was greasing the keels at Castelnuovo; then it made for Lepanto, with its oarsmen in poor condition and decimated by sickness. Shortly afterwards it reached Prevesa and from there it was said to be heading home.[39] The West was somewhat surprised therefore to see it returning to Albania, the 'Cimara' as it was called,[40] in September. It sailed up the coast as far as Valona. Was this merely to punish the Albanian rebels,[41] the viceroy wondered, without undue anxiety, for the coasts of Naples, where Spanish troops had now relieved the Germans, had been maintained on the alert. The new danger had vanished of its own accord before winter returned.

Such was the naval campaign of 1566: for both sides, for the Turks with a limited campaign in the Adriatic sea, and for the Spanish who were content to wait, it was an uneventful season. The Spaniards made no attempt to attack Tunis or Algiers as they had briefly let it be understood they might.[42] They were content to relax unmolested during a summer which for a change brought danger to other powers. Venice had shown too little solicitude for others in the past to inspire much pity now, and she appeared to be the only target of enemy attack. The Turks had come to her very doorstep in the Adriatic, against all past conventions: Venice was extremely alarmed and took swift action. In July, a hundred galleys were launched[43] and it was perhaps this firm attitude which halted the Turks' progress northwards. However that may be, Venice was most disturbed and her anxiety was shared by all Italy, including the Pope who even agreed to support the Venetian requests. At the end of July and beginning of August, he asked Cardinal Alessandrino to write, and himself wrote to Don García de Toledo urging him to take his entire fleet to Brindisi, for the Venetians had said that they were fitting a hundred galleys, and that if they were joined by Don García's ships, they could attack the Turkish fleet.[44] Don García replied on 7th August,[45] swearing to the Pope that

[38] Viceroy of Naples to Philip II, Naples, 16th August, 1566, Simancas E° 1055, f° 177; Fourquevaux, *op. cit.*, I, p. 110–111, 123.

[39] *Por cartas de J. Daça castellano de Veste* (Vasto?), 6th August, 1566, Simancas E° 1055, f° 169. *Lo que se entiende de la armada por carta de Bari de los 19 de agosto 1566*, Simancas E° 1055, f° 178; viceroy of Naples to Philip II, Naples, 5th September, 1566, E° 1055, f° 190.

[40] Viceroy of Naples to Philip II, Naples, 14th September, 1566, Sim. E° 1055, f° 197.

[41] Viceroy of Naples to Philip II, Naples, 27th September, 1566, *ibid.*, f° 200.

[42] Fourquevaux, *op. cit.*, I, p. 84–85, 6th May, 1566; Nobili to the prince, Madrid, 6th May, 1566, A.d.S. Florence, Mediceo 4897a, '. . . et molti dicono che Sua Maesta vuol andar sopra . . . Argeri et dicono che in consiglio [Philip II] ha parlato di voler andar in persone benché questo io no lo credo . . .', Nobili to the prince, 7th May, 1566, *ibid.*

[43] G. Cappelletti, *op. cit.*, VIII, p. 373.

[44] Viceroy of Naples to Philip II, Naples, 6th August, 1566, Simancas E° 1055, f° 168.

[45] 7th August, 1566, Simancas E° 1055, f° 170.

he would defend the States of the Church as he would those of the king of Spain, but without agreeing to the audacious plan. There is little doubt that the Turkish fleet, caught between the twin jaws of the Venetian and Spanish forces, would not have escaped. But Venice can only have intended to put up a fight for a brief period, when she was actually threatened; and further south the cautious, prudent and ailing Don García had no orders to move into the attack. The Pope was probably the only man who considered this a magnificent occasion to destroy Sulaimān's fleet.

The war in the Adriatic, limited as its effects eventually proved to be, must have seemed a dramatic development at the time, because of the general upheaval sweeping through Europe. In France (take Brantôme for example) it was a time when young men were seeking action, adventure and travel: one might set off to fight in Hungary with the young Duke of Guise; another go to Naples; a third, like Montluc's son, go in search of adventure on the Atlantic and die during the capture of Madeira.[46] No one seemed to want to stay at home. Even Philip II talked of travelling. Everywhere war was showing its face, from the Netherlands which virtually rose up in arms in August, to the Adriatic or the Black Sea where the dark red line of a furious continental struggle prolonged over vast distances the war between the fleets in the Adriatic.

War breaks out again in Hungary. The death of the emperor Ferdinand (25th July, 1564) had served the Turks as a pretext for demanding back payments of the tribute and for calling the 1562 truce into question. The payment was made on 4th February, 1565,[47] and in return the truce was confirmed for another eight years. But Maximilian, who had not abandoned his plan to attack Transylvania, had assembled a large army and taken Tokaj and Szerencs. And to interfere in Transylvania or to hinder Turkish action in any way in that area was bound to inflame past discords and lead to a 'covert' war which as usual took the form of a series of surprise attacks and sieges. The long Hungarian frontier was more troubled than ever in 1565. Maximilian, who found he had disturbed a hornets' nest in Transylvania, made vain efforts to bring about peace, sincere only to some extent, for he desired peace but did not intend to withdraw. Moreover, he was faced with the powerful enmity of the Grand Vizier Muḥammad Sokolli and the sultan himself was anxious to wipe out the disgrace of Malta by some brilliant victory. At this point, Arslan, the pasha of Buda who from his forward position had never ceased to call for war, urging that Christian Hungary was desperately short of troops, led

[46] Edmond Falgairolle, *Une expédition française à l'île de Madère en 1566*, 1895; Nobili to the prince, Madrid, 6th October, 1566, A.d.S. Florence, Mediceo 4897a; *Calendar of State Papers, Venetian*, VII, 12th November, 1566, p. 386; Montluc disavowed by Charles IX, the king to Fourquevaux, Saint-Maur-des-Fossés, 14th November, 1566, p. 59–60; the Portuguese the aggressors, 29th November, 1566, Fourquevaux, *op. cit.*, I, p. 144.

[47] Josef von Hammer-Purgstall, *Histoire de l'Empire ottoman*, Paris, 1835–1843 (translated from the German), VI, p. 206.

the way by himself attacking the small fort of Palotta on 9th June, 1566; his move was a little hasty for just as he was about to storm the position, the Imperials delivered it and went on to carry off Veszprem and Tata, massacring everyone in the town indiscriminately, friend and foe, Turk and Hungarian alike.[48]

Thus the war of Hungary began once more. One cannot really say it was a surprise. At Vienna, everyone was well aware that Turkish reaction was inevitable. The German Diet had granted the exceptional aid, for one year, of 25 *Römermonate*, plus eight for each of the next three years.[49] The Spanish ambassador in London mentioned in connection with this aid, on 29th April, 1566, a figure of 20,000 foot soldiers and 4000 cavalry for three years.[50] Meanwhile Maximilian obtained from the Pope and Philip II assistance both in money and troops; the documents differ on the actual amount, but it was undoubtedly considerable. The Tuscan agent at Madrid talked, on 23rd March, 1566, of 6000 Spanish troops and 10,000 crowns a month (which had in fact been paid by Philip II since 1565).[51] These sums were to be paid through the Fuggers and the Genoese bankers.[52] A month or two later (6th June) he was speaking of 12,000 crowns a month, not counting a down payment of 300,000 crowns.[53]

So the emperor had both the time and means to prepare for war. By the summer, he had assembled a rather motley group of 40,000 soldiers,[54] near Vienna; these enabled him to do little more than remain on the defensive: but such was in any case his intention. In view of the immense distance between Constantinople and Buda, he assumed that the enormous Turkish army would not arrive very soon: the march was reckoned to take 90 *giornate*. There would therefore be little time in which to fight, for in October the Turks would be halted by the cold and by problems of supply, which could not but be considerable for a large army in a virtually empty countryside. This at least is how the emperor put it to Leonardo Contarini, the Venetian ambassador,[55] with a certain amount of *bragadoccio* apparently, since the same ambassador echoed him in passing on some manifestly exaggerated numbers on 20th June, reckoning the imperial army at 50,000 infantry and 20,000 cavalry, as well as a large Danube fleet.[56]

In reality, Maximilian's army seems to have been very little different from the one Busbecq had seen and judged so severely in 1562. Fourquevaux was not mistaken in thinking that the war could not redound to his

[48] Hammer, *op. cit.*, VI, p. 215.

[49] Georg Mentz, *Deutsche Geschichte*, Tübingen, 1913, p. 278.

[50] G. de Silva to the king, London, 29th April, 1566, *CODOIN*, LXXXIX, p. 308.

[51] Nobili to the prince, Madrid, 23rd March, 1566, A.d.S. Florence, Mediceo 4897a.

[52] Leonardo Contarini to the Doge, Augsburg, 30th March, 1566, in G. Turba, *op. cit.*, I, 3, p. 313.

[53] Nobili to the prince, Madrid, 6th June, 1566, A.d.S. Florence, Mediceo 4897a.

[54] Georg Mentz, *op. cit.*, p. 278.

[55] Leonardo Contarini to the Doge, Augsburg, 1st June, 1566, in G. Turba, *op. cit.*, I, 3, p. 320.

[56] Contarini to the Doge, 20th June, 1566, *ibid.*, p. 324.

advantage and in hoping 'that the Grand Signior of the Turks persists and perseveres in his enterprise of Hungary; for otherwise the vermin of Germany is greatly to be feared, if there should be peace on that front'.[57] Unfortunately for the France of the wars of religion, so often ravaged by the 'reîtres' (*Reiter* = horseman), peace was to be re-established in Hungary in 1568 and to last until 1593.

Against Maximilian's army, an enormous Turkish force, divided into several bodies, was moving into Hungary, 300,000 men according to intelligence received by Charles IX, armed after their usual fashion 'with such extreme quantities of artillery and other munitions that it is frightful to behold'.[58] The sultan had left Constantinople on 1st May,[59] in greater state than on any of his twelve previous campaigns. He travelled in a carriage, his health no longer permitting him to ride on horseback, along the great military and trade route from Constantinople to Belgrade via Adrianople, Sofia and Niš. Bumpy stretches of the road had been levelled as much as possible in advance of the imperial carriage, while along the road the Turks organized the energetic pursuit of the innumerable bandits who attacked the army and, more often, its supplies. For their benefit, a few gibbets were always erected near the camps. On the other side of Belgrade the major problem was not the negotiation with the Transylvanian leader but the crossing of the rivers, the Sava at Sabac,[60] the Danube near Vukovar[61] and the Drava at Esseg (Osijek) on 18th and 19th July.[62] On each occasion a bridge had to be built by the army over the high waters of the rivers (the Danube in particular), and not without some difficulty. Beyond Esseg an incident, the successful raid by an imperial captain, diverted the march of the Turkish army to the fort of Szigetvár, at some distance from Pecs, commanded by the said captain, Count Nicholas Zriny. The sultan and his forces arrived before the marshes surrounding the town on 5th August; on 8th September the position was captured.

But Turkish operations, hardly begun, were already doomed; three days before this victory, during the night of 5th to 6th September, Sulaimān the Magnificent died, 'either of old age, the effect of dysentery or an attack of apoplexy' says Hammer.[63] Whatever he died of, it is from this moment that many a historian dates the 'decline of the Ottoman Empire'.[64] It is an exercise in pin-pointing which for once carries some

[57] 18th August, 1566, *op. cit.*, I, p. 109.

[58] Charles IX to Fourquevaux, Orcamp, 20th August, 1566, C. Douais, *Lettres . . . à M. de Fourquevaux*, p. 49.

[59] J. von Hammer, *op. cit.*, VI, p. 216.

[60] *Ibid.*, p. 219, says the Danube at Sabac.

[61] *Ibid.*, p. 223.

[62] *Ibid.*, p. 224.

[63] *Ibid.*, p. 231; F. Hartlaub, *op. cit.*, p. 23; there are numerous unpublished dispatches on the subject from Constantinople, 23rd September, 1566, Simancas E° 1055, f° 198; viceroy of Naples to Philip II, 5th November, 1566, *ibid.*, f° 215.

[64] For instance N. Iorga; Paul Herre, *Europäische Politik im Cyprischen Krieg*, Leipzig, 1902, p. 8.

weight, for this empire, so dependent on its leader, was now to pass from the hands of the Magnificent, the Lawmaker, as the Turks called him, to those of the weak Selīm, the 'son of the Jewess', fonder of Cyprus wines than of battle campaigns. The Grand Vizier, Muhammad Sokolli, concealed the sultan's death, giving Selīm time to hurry from Kutahya to Constantinople to take unchallenged possession of the vacant throne. The war went on into the winter in a desultory fashion, with successes on both sides. That is, successes were announced on both sides. On 1st September, 1566 for instance, the Archduke Charles announced from Gorizia, on the fringe of the main theatre of operations,[65] that a successful raid had been carried out by the captain of Croatia, who was returning with prisoners and cattle stolen in Bosnia. The news was immediately passed on to Spain through Genoa. Was this the same action which was reported in Paris as a major battle between the Archduke Charles and the Turks in which the Duke of Ferrara was said to have perished?[66]

In fact the war was practically over. When winter came, the Turks withdrew and the imperial army disbanded of its own accord without having to be 'demobilized'. In Paris in December, it was rumoured that a truce would be agreed with the Turks before the end of the Diet.[67] As for Philip II, he had very prudently, as early as September, requested that there should be retained for his own use 'all that the emperor disbands if the Turk withdraws'.[68] We know why he took this precaution: Philip had just witnessed another chasm open before his eyes in the same year, 1566; this time in the Netherlands.

The Netherlands in 1556.[69] It is not the purpose of this book to study the deep-seated and complex origins of the war of the Netherlands – political, social, economic (if one remembers the great hunger of 1565),[70] religious and cultural origins – nor to repeat after other historians how inevitable that war was. We are here concerned simply with its effect upon the foreign policy of Philip II, which it diverted forcibly to the North and away from the Mediterranean, only a short time after the dramatic siege of Malta. As long as the Netherlands were 'Spanish' in name alone (until perhaps 1544, the date when they first became the fortress against France that they were to remain for a hundred years; or perhaps even until 1555, the year of Charles V's abdication) until this date then, the Netherlands had been

[65] *De Guricia, primero de set^bre 1566*, Simancas E° 1395.

[66] F. de Alava to the king, Paris, 10th November, 1566, A.N., K 1506, B 20, no. 76.

[67] Alava to the king, Paris, 8th December, 1566, *ibid.*, no. 88 'estar aqui algo sospechosos del Emperador paresçiendoles que hara treguas con el Turco antes que se acabe la dieta . . .'.

[68] Memorandum from Saint-Sulpice, 27th September, 1566, Fourquevaux, *op. cit.*, I, p. 134.

[69] See Pierre Chaunu's article (a little obscure from the very wealth of information) 'Séville et la "Belgique"', in *Revue du Nord*, April–June, 1960, which attempts to situate the affair of the Netherlands in the imperial history of Spain and the international situation as viewed from Seville.

[70] Léon Van der Essen, *Alexandre Farnèse*, Brussels, 1933, I, p. 125–126.

virtually left to their own devices, free to play their role of a crossroads open to Germany, France and England. This was a free country, with its own rights, political safeguards and financial privileges: a second Italy, highly urbanized, 'industrialized', dependent upon the outside world and for that reason and several others, difficult to govern. It had remained more of a landed society however than is sometimes realized and therefore retained a powerful aristocracy, with the houses of Orange, Montmorency (the French family was a younger branch) and Egmont. This aristocracy, mindful of its privileges and interests and eager to govern, was in close contact with the far-away party quarrels at the Spanish court, in particular with the peace party, led by Ruy Gómez, after 1559. This opens up horizons whose importance must be revealed if the complete history of the troubles in the Netherlands is ever written.

Inevitably, if only by their geographical situation in the heart of the northern nations, the 'pays de par deçà' did not escape the many currents of the Reformation. Its ideas reached them by land and sea. In its Lutheran form, the Reformation very quickly reached the 'flamingants' (Flemish-speaking) regions; towards mid-century it was suggesting its own solution of religious tolerance: a religious truce, an early version of the edict of Nantes.[71] But soon along the routes from the South, provider of wheat and wine, from France, the Reformation won its biggest victories and this time it was Calvinism that triumphed – the 'Romish' Reformation, militant and aggressive,[72] setting up its synods, active cells which had not been foreseen and would not have been tolerated by the peace of Augsburg. Infiltrating first the French-speaking regions, it quickly spread beyond them to the entire Netherlands. It therefore contributed to opening them up to the south. Freed from Germany politically, the Netherlands now achieved spiritual liberation, turning towards divided France. Persecution of the Lutherans became difficult as their numbers decreased. Moreover England was too near for the Netherlands to escape its influence and its determined policy, despite the persistence of certain rivalries. England offered a refuge to Flemish refugees, even the most humble – the workers who settled in Norwich for example – and this assistance wove links from one shore to the other of the North Sea.

Obviously we must make a distinction between the different currents of unrest in the Netherlands. They were of different origins: there was popular agitation, predominantly religious and often social in origin, and an aristocratic movement. The latter, essentially political in origin, manifested itself in the recall of Granvelle in 1564, and by the Compromise of the palace of the Count of Culemborch in April, 1566; it preceded by four months the rioting in the second half of August, a popular and iconoclastic rising which led to the sack of churches and the smashing of images, and spread with alarming rapidity from Tournai to Antwerp right across the Netherlands. These were two completely different

[71] Georges Pagès, 'Les paix de religion et l'Édit de Nantes', in *Rev. d'hist. mod.*, 1936, p. 393–413. [72] W. Platzhoff, *op. cit.*, p. 20.

movements. Margaret of Parma skilfully turned them against each other, encouraging the nobles, except for William of Orange and Brederode who fled to Germany, to oppose the people and the towns. She thus succeeded in re-establishing if not her authority, at least some order, without expense, without bringing in troops and with undoubted skill.

But this policy had its limits and, both in Spain and outside Spain, its enemies: in Rome, Granvelle, who did not remain inactive, in Spain, the Duke of Alva with all his party. It was of course true that Margaret's success compromised the authority of Philip II and the defence of Catholicism; had she not accepted, albeit unofficially, that the reformed religion could continue to be practised in those places where it was already prevalent before the troubles broke out in August? This was a considerable concession. For in Spain, Philip II, as the famous letters from the Wood of Segovia in 1565 prove, was unwilling to yield on any matter of substance. True, in order to gain time he had agreed to certain minor acts of clemency, a 'general pardon' (but only on political grounds, not for religious offences), very minor concessions whose only aim was to prevent the regent's credit being ruined. Plenty of evidence on the other hand indicates his real desire to deal severely with the rebels. The revitalized strength of the Hispanic empire in the Mediterranean, the influx of silver on the Indies fleets – these Philip II was determined to use in the North. Intransigence and lack of comprehension; Philip would often be reproached for these in the future. For it was clearly a hopeless task to try to regulate traffic at this crossroads of Europe; it was foolishness to try to enclose the 'pays d'en bas' when they made their living from the entire world, and were indispensable to the life of the rest of Europe already pressing at their gates and ready if necessary to break them down. To transform this country into the military camp it became between 1556 and 1561; to endow it with a separate religious administration as was attempted with the creation of the new bishoprics; to try to prevent students travelling to Paris, to mention only some of the remedies which had been applied previously to no effect, were so many futile measures. But to turn it into another Spain was an error more serious than all these. And yet this is what Philip II, prisoner of Spain, dreamed of in 1566.

Before he had heard news of the popular disturbances of the second half of August, he was writing to Requesens, his ambassador at Rome: 'You may assure His Holiness that rather than suffer the slightest thing to prejudice the true religion and the service of God, I would lose all my States, I would lose my life a hundred times over if I could, for I am not and will not be the ruler of heretics.'[73] In Rome too, the situation was imperfectly understood. Pius V had advised Philip to intervene energetically and in person: 'The plague of heresy', he wrote to the king on 24th February, 1556,[74] 'is spreading so fast in France and Burgundy, that I

[73] L. Serrano, *op. cit.*, I, no. 122, p. 316, 12th August, quoted by B. de Meester, *Le Saint-Siège et les troubles des Pays-Bas, 1566–1572*, Louvain, 1934, p. 20–21.

[74] *Ibid.*, I, p. 131.

think it cannot now be remedied but by a voyage by Your Catholic Majesty.' This famous voyage is mentioned by Fourquevaux as early as 9th April[75] and he returns to it on every occasion – when reporting the convocation to Spain of Francisco de Ibarra, commissioner-general of war, supplies and munitions, a specialist in transport, a summons which might have many consequences, 'for there is not a man in Spain to touch him in such matters';[76] and again in connection with the expedition which was being prepared supposedly against Algiers, but which might very well be intended for the Netherlands. And indeed the expedition against Algiers failed to materialize: in August the rumour naturally reappeared in ever greater detail that Philip was planning to go to Flanders.[77]

As early as 18th August, Fourquevaux knew that there were plans to move northward the large Spanish force already mobilized in the Mediterranean,[78] that is 5000 or 6000 Spaniards from Naples and Sicily and 7000 to 8000 Italians, the best troops from the Mediterranean fronts and garrisons. The Duke of Alva's great enterprise was to culminate with the Naples *tercio* lodged in Ghent, the Lombardy *tercio* in Liège and the Sicilian *tercio* in Brussels.[79] It meant disarming the Mediterranean both directly and indirectly: directly by moving out tried and tested troops and indirectly by the expense such movements implied. Had not the figure of three millions in gold been mentioned to Philip in the Council of State?

In addition this policy necessitated an extremely careful strategy in the North: towards France, where the Protestants[80] were preparing to help their brothers in religion; towards Germany, where the Prince of Orange and his brother Louis of Nassau[81] were succeeding in levying troops, despite an edict by the emperor forbidding them to; and towards England. All three were dangerous territories, where Spain could trust no one, individual, party or sovereign.[82] Philip II's attention, like that of his best advisers and servants, was inevitably diverted from the South to the North.

Naturally such measures seemed even more necessary after the August troubles. Friend and foe alike (including the Duke of Alva, perhaps in all sincerity) expressed surprise at the slow reaction of Spain.[83] It was not

[75] *Ibid.*, I, p. 67.

[76] 30th April, *ibid.*, I, p. 84.

[77] *Ibid.*, I, p. 104; Nobili to the prince, Madrid, 11th August, 1566, A.d.S. Florence, Mediceo 4897a.

[78] 18th August, 1566, *op. cit.*, I, p. 109.

[79] H. Forneron, *op. cit.*, II, p. 230.

[80] Philip II to F. de Alava, Segovia, 3rd October, 1566, A.N., K 1506.

[81] F. de Alava to Philip II, Paris, 21st September, 1566, A.N., K 1506, no. 57.

[82] On Philip's suspicions of the emperor, Fourquevaux to the queen, Madrid, 2nd November, 1566, Fourquevaux, *op. cit.*, III, p. 25–26. That Philip distrusted Catherine de Medici goes without saying.

[83] Dispatch from Saint-Sulpice, 21st September, 1566, Fourquevaux, *op. cit.*, I, p. 133.

until 25th September that Saint-Sulpice heard that Alva was to leave for Flanders, preceding his sovereign,[84] and at the same time that the most compromised of the trouble-makers – or at any rate those most anxious to escape Spanish retribution – had chosen to leave their dangerous homeland, some of them fleeing to the no less dangerous land of France. Thus it was that a stream of Flemish Anabaptists from Ghent and Antwerp came to settle in Dieppe.[85]

By now as it happened (perhaps as a result of Margaret's policy) the situation in the Netherlands was settling down in favour of the Catholic King. Or so it appeared. The king referred on 30th November, 1566[86] to the 'apparent improvement of affairs in Flanders', but added that it was not so great that the measures now being executed could be relaxed on any point.

This was on 30th November, it is true, several months after the riots. But could the Spanish government, faced as usual with the problems of distance and with its many responsibilities, have acted any more quickly? After being surprised in the early spring by the activities of the 'Beggars', or as they were then called, the 'remonstrants', and surprised again in August by the popular wave of iconoclasm, could it have prepared an immediate response? Alva probably gave the real reason for the delay in a conversation with Fourquevaux in early December, 1566: it was 'the assaults made by the Turks on Christendom and certain other considerations' which had prevented Spain from 'remedying the excesses of her subjects in the Netherlands'.[87] The Turkish alarm in the Mediterranean was not over until the end of August; until then how could Spain possibly release the experienced Spanish companies who were to play the leading role in Alva's expedition?

And, *vice versa*, events in the Netherlands no longer permitted Philip II to act imprudently in the Mediterranean. This double burden, this double set of problems, helps to explain the apparently hesitant policy of the king of Spain. It explains his repeated refusal to follow the advice of the Pope who was constantly suggesting to him that he adopt a firm and effective policy in one place or the other, without understanding that in neither problem area could Philip afford the luxury of an all-out engagement.

First of all Pius V tried to entice Philip into a league against the Turk, an ancient dream which Piāle Pasha's campaign in the Italian waters of the Adriatic had revived during the summer of 1566. On 23rd December, the nuncio in Spain reported to Cardinal Alessandrino what he had been told by Alva: 'His Majesty greatly approved the holy zeal and excellent in-

[84] Fourquevaux, *op. cit.*, I, p. 134.

[85] F. de Alava to the king, Paris, 4th October, 1566, received on the 20th, A.N., K 1506, no. 62.

[86] Philip II to F. de Alava, Aranjuez, 30th November, 1566, A.N., K 1506, no. 87.

[87] Fourquevaux to the king, Madrid, 9th December, 1566, Fourquevaux, *op. cit.*, I, p. 147.

tentions of His Holiness; . . . and he was also rather in favour of a league and union', but for the time being, it would be 'useless, for such enterprises can only be undertaken when the princes involved have a full complement of reliable forces and when they have confidence one in another; whereas at the present time those forces are divided, diminished and set at odds by mutual suspicions'. Moreover, the king of Spain is obliged for the moment 'to lead an urgent and necessary enterprise against his own subjects' in Flanders.[88] In this pessimistic survey of the international horizon with special attention to the stormclouds, ending predictably with the refusal of a request, Alva's usual method of argument is easily recognizable. But his point of view appears to have been shared by Philip II who refused for a long time to choose between the Netherlands and the Mediterranean.[89]

It was a difficult choice to make, for Spain could hardly refuse to fight the Turks; self-defence was essential. But it was another thing to attack them. At the end of the year 1566, the Spanish government had little desire to aggravate what was not exactly peace in the Mediterranean but at least no more than intermittent war with frequent periods of calm. Neither did it wish to disturb the useful, not to say indispensable pseudo-peace in Europe.

Spain therefore hesitated to join Rome in a spectacular alliance which the Protestant world was unlikely to accept without demur: it would be the perfect pretext for trouble in the Netherlands along all the open frontiers of this great market-place. In Germany, England and France (with the party of Coligny and Condé) enemies were poised to attack; only the slightest provocation or pretext was needed. Hence Philip's concern not to turn the situation in the Netherlands into a religious confrontation. In vain Pius V exhorted him to crusade openly against heresy. Whatever his personal feelings, for the time being Philip wished to play the role of the sovereign leading his subjects back to useful submission, using against them merely the imprescriptible rights of any sovereign over his subjects. As Alva explained to Fourquevaux on 9th December,[90] it was merely a question of 'bringing troublesome subjects to heel. . . . There being no longer any question of religion in the dispute, but only lack of respect towards His Majesty, with outrageous flouting of his authority and orders: a thing tolerable to no prince who would reign and maintain his states in peace.'

It is a clear but unconvincing argument. The massive preparations made by Spain, 'reasonable' as they might seem to Alva, spread great anxiety throughout Europe. Was this not, under cover of an expedition against Flanders, an operation whose real target was France? Fourquevaux for one thought so,[91] and many were ready to side with him in France

[88] L. Serrano, *op. cit.*, I, p. 425–426.

[89] Paul Herre, *op. cit.*, p. 41, Castagna to Alessandrino, Madrid, 13th January, 1567; Fourquevaux, *op. cit.*, I, p. 172 ff, 18th January, 1567.

[90] *Ibid.*, I, p. 147.

[91] 4th January, 1567, *ibid.*, I, p. 160.

where feeling ran high (and was deliberately encouraged to by the Protestants) against the Spaniards who had just massacred the French colonists in Florida. Elizabeth of England's anxiety was no less acute, concealed though it was behind excessive politeness. In October, the queen had officially rejoiced at the reported success of the imperial troops against the Turks.[92] 'If only she means it!' exclaimed the sceptical Spanish ambassador at London, Guzmán de Silva. On 10th October, on learning that Philip was to travel to Flanders by way of Italy, she expressed well-publicized disappointment: if only he had chosen the Atlantic route, she might have had the pleasure of entertaining him![93] Was a great army to accompany the king? she could only wish it were even greater to punish such ungrateful subjects.[94] Her extravagant protests convinced no-one and did not prevent the queen from later expressing fears of an anti-Protestant alliance composed of the emperor, the Pope and His Catholic Majesty. Even Venice was disturbed by the Spanish troop movements and thought it necessary to put Bergamo in a state of alert.[95] As for the German princes they had a thousand reasons, both political and religious, to be worried. In May, 1567, before the arrival of the Duke of Alva,[96] they began to take precautions: ambassadors from the Elector of Saxony, the Duke of Wurttemberg, the Margrave of Brandenburg and the Landgrave of Hesse arrived in the Netherlands, in order to ask protection for the Lutherans (*di lege Martinista*), who had not participated in the Calvinist rebellion.

But this is not the place to study the sudden extension of Spanish foreign policy in the year 1566 – to study it as it should be studied, in a total European context, in that climate of rising tension and religious fervour which aggravated the confessional struggles of the century and dictated the course of events; a climate which, more than the intransigence or inefficiency of Philip II, his so-called imprudence and his only too real lack of comprehension, was the true origin of the troubles in the Netherlands.

1567–1568: the Mediterranean eclipsed by the Netherlands. In 1567 and 1568, the Mediterranean became a secondary sphere of Spanish activity, both because Spain's eyes were elsewhere and because this preoccupation coincided with a general trend towards disarmament in the Mediterranean. As far as Spain is concerned the reason is clear: her vital resources, her money and her attention were engaged elsewhere. As for the Ottoman Empire, the precise reason is still obscure. No doubt it was beset with

[92] G. de Silva to the king, London, 5th October, 1566, *CODOIN*, LXXXIX, p. 381.
[93] De Silva to the king, London, 10th December, 1566, *ibid.*, p. 416.
[94] De Silva to the king, London, 18th January, 1567, *ibid.*, p. 427.
[95] Antwerp, 24th May, 1567, Van Houtte, 'Un journal manuscrit intéressant 1557–1648: les Avvisi' in *Bulletin de la Comm. Royale d'histoire*, I, 1926, LXXXIX (4th Bulletin), p. 375.
[96] Fourquevaux, *op. cit.*, I, p. 166, 4th January, 1567.

problems on the Persian front,[97] but these were less grave than dispatches from Constantinople suggested. No doubt it was also distracted by the continuing war in Hungary, but this conflict, waged throughout the fine season of 1567, without much energy (the only spectacular incident was the raid, launched by the Tartars, not the Turks, on the Austrian frontier, during which they are said, how truly we do not know, to have taken 90,000 Christian prisoners) was to culminate in a new eight-year truce signed on 17th February, 1568, but negotiated the year before.[98] No doubt too, the Turks had some difficulties in Albania[99] but these were perennial difficulties, of secondary importance. As for those they encountered in Egypt and the Red Sea[100] they did not, at least not until 1569, affect the essential life of the empire. Should we explain the Turkish lack of action by the great losses suffered during the 1566 campaign in Hungary, or by the accession of Selīm II, who had little taste for warlike expeditions? So his contemporaries thought[101] and historians have tended to follow

[97] Dispatches from Constantinople, 10th March, 1567, Simancas E° 1056, f° 23; Chefalonia, 24th, 26th March, 5th, 10th April, 1567, *ibid.*, f° 34; *Copia de capitulo de carta que scrive Baltasar Protótico de la Chefalonia a 10 de avril a D. García de Toledo*, *ibid.*, f° 38; Protótico to García de Toledo, 12th April, 1567, f° 36; *Memoria de lo que yo Juan Dorta he entendido de los que governian en Gorfo* (sic) *es lo siguiente* (24th April, 1567), f° 45, Chefalonia, 21st April, f° 47; Corfu, 28th April, *ibid.*, dispatch from Constantinople, 17th May, 1567, f° 60; Fourquevaux to the king, Madrid, 2nd August, 1567, Fourquevaux, *op. cit.*, I, p. 248, 'the Sophy is doing nothing, but is sending an ambassador to the said Turk'; a dispatch from Constantinople, 8th January, 1568, Simancas E° 1056, f° 126, says that the writer will believe in the arrival of Persian ambassadors when he sees it.

[98] J. von Hammer, *op. cit.*, VI, p. 313–317; Chantonnay to Philip II, Vienna, 4th June, 1567, *CODOIN*, CI, p. 151–152; G. Hernández to Philip II, Venice, 26th January, 1567, Simancas E° 1326. The truce which had been outlined well before, was an eight-year not a three-year agreement as Fourquevaux described it (before the event it is true) on 30th June, 1567, Fourquevaux, *op. cit.*, I, p. 219. The reduction in hostilities did not prevent the emperor from requesting Spanish subsidies, 24th August, 1567, *ibid.*, I, p. 255. There was a prolonged skirmish on the Transylvanian front, Chantonnay to Philip II, Vienna, 30th August, 1567, *CODOIN*, CI, p. 263. On imperial anxieties concerning the outcome of the negotiations, 20th December, 1567, Fourquevaux, *op. cit.*, I, p. 311; Nobili to the prince, 25th December, 1567, A.d.S. Florence, Mediceo 4898, f° 153. *Lo que se escrive de C. por cartas de VII de março 1568*, Simancas E° 1056, f° 135, the truce, it is said will cover a period of ten years and include the Most Christian King of France, the king of Poland and Venice.

[99] Constantinople, 10th March, 1567, Simancas E° 1056, f° 23, Albanian risings in the region of Valona and Sopotico; Corfu, 28th April, 1567, *ibid.*, f° 47; viceroy of Naples to Philip II, 11th May, 1567, Simancas E° 1056, f° 57, the Albanian rebels are in Zulati, Procunati, Lopoze, Tribizoti . . .; *De la Cimarra a XII de Junio 1567*, *ibid.*, f° 6; Corfu, 26th December, 1567, *ibid.*, f° 131.

[100] Revolt in the Yemen, Constantinople, 10th March, 1567, see previous note; *Copia de un capitulo de carta que scrive de Venecia J. López en XXI de hen° 1568*, Simancas E° 1066, f° 130; Constantinople, 17th November, 1568, Simancas E° 1057, f° 2.

[101] '. . . Que el Turco attiende solo a dar se plazer y buen tiempo y a comer y a bever dexando todo el govierno en manos del primer Baxa', Constantinople, 10th March, 1567, Simancas E° 1056 '. . . que el Turco continua sus plazeres y el presume mucho que por ventura sera la destrución deste imperio que los Dios lo permita', Constantinople, 17th May, 1567, Simancas E° 1056, f° 60.

them. And there may well be something in it, although one should not forget that behind this 'unworthy successor' to Sulaimān, as F. Hartlaub calls him,[102] the first of the 'do-nothing sultans' as Ranke termed him,[103] was the active figure of Muḥammad Sokolli, the astonishing Grand Vizier who was himself worthy of the great age of Sulaimān. Possibly these two quiet years, 1567 and 1568, conceal a secret plan to strike at Venice and to isolate her in advance. In autumn 1567, dispatches reported the building of a fort in Karamania, opposite Cyprus, as well as the building of roads in the interior of the region. It was already being concluded from these reports that an attack on the island was likely. Was it in order to be free to attack Venice that Selīm and his councillors signed the truce of 1568 with the emperor?

It is also certain that Turkey suffered the insidious but damaging blows of a succession of poor harvests. In February, 1566, Venice applied to Philip II for grain: from this detail alone we are able to date the first difficulties in the East.[104] A dispatch 'worthy of credit', suggested that in the month of April, people were dying of hunger in Egypt and Syria.[105] This economic situation may well explain the simultaneous disturbances in the Arab world. The next harvest, that of 1566, was particularly bad in the eastern half of the Mediterranean, throughout Greece, from Constantinople to Albania.[106] So it is not suprising if 1567 was also a difficult year. Haëdo reports great scarcity in Algiers which continued into the next year, 1568, for the 1567 harvest does not seem to have been sufficient to mend matters.[107] In November, an agent of the viceroy of Naples reported a terrible shortage of bread in Constantinople.[108] And plague, that almost inseparable companion of hunger, made its appearance at the same time.[109] In March, 1568 a dispatch reporting the conclusion of the truce with the emperor, says in so many words that it was signed, 'because of the tumults among the Moors and the scarcity of food, particularly barley. . .'.[110] So the 1567 harvest must, at best, have been mediocre. It was not until after that of 1568 that a letter finally gave an optimistic bulletin: 'In Constantinople the public health situation is good and there is plenty of food, despite the scarcity of barley'.[111] During these years

[102] *Op. cit.*, p. 24.

[103] *Op. cit.*, p. 54. See also Vicomte A. de la Jonquière, *Histoire de l'empire ottoman*, Vol. I, 1914, p. 204.

[104] Garces to the prince, Madrid, 13th February, 1566, A.d.S. Florence, Mediceo 4897, f° 116.

[105] Constantinople, 30th April, 1566, Simancas E° 1395.

[106] Simancas E° 1055, f° 77.

[107] *Ibid.*, 76 v°.

[108] *Relación de lo que refiere uno de los hombres que el Marqués de Capurso embio a Constantinopla por orden del Duque de Alcalá por octubre passado* (1567), Simancas E° 1056, f° not noted.

[109] Corfu, 26th December, 1567, dispatches arriving at Lecce on 12th January, 1568, Simancas E° 1056, f° 131.

[110] Constantinople, 5th March, 1567, Simancas E° 1058, f° 133.

[111] Constantinople, 17th November, 1568, Simancas E° 1057, f° 2.

1567–1568, the food situation was scarcely any better in the western. Mediterranean.[112]

For whatever reason then, both Turk and Spaniard spent these years spying upon one another, determined not to go to war but correspondingly interested in spreading as many false reports as possible and successfully alarming each other, one side believing that the enemy fleet was preparing to attack now La Goletta, now Malta, now Ragusa and Apulia, Cyprus and Corfu;[113] the other fearing a raid on Tripoli,[114] Tunis or Algiers.[115] The fears were short-lived: the intelligence services of both sides were sufficiently organized for this shadow-boxing – or war of nerves – not to deceive anyone for long. But it obliged both adversaries to take precautions which weighed heavily on the Mediterranean as a whole.

In 1567 for instance, the familiar pattern repeated itself. The viceroy of Naples alerted all coasts and had strategic positions occupied[116] while Messina and Sicily were subject to the usual naval bustle of summer.[117] The Turks for their part brought out the fleet in 1568. This was a purely defensive measure for the fleet turned about after reaching Valona;[118] but even the approach of the hundred or so galleys was sufficient to trigger all the defence mechanisms of the eastern coast of Italy.

Did Philip II at a time when he badly needed a maximum of military strength elsewhere, consider these costly precautions an unnecessary luxury? At any rate the idea of a truce with the Turk surfaced once more in Spain and after the fever of rearmament and the reign of the soldier, diplomacy came into its own again. It is perhaps worth underlining the pragmatism and lack of preconceived notions in Spanish policy on at least four occasions (1558–1559, 1563–1564; 1567; 1575–1581) and no doubt at other times unknown to us: a policy very different from that traditionally allotted to her by history.

Once more there was no question of visible official negotiations. (Philip

[112] Famine in Catalonia, 3rd May, 1566, Simancas E° 366; shortage at Genoa, Figueroa to Philip II, Genoa, 26th March, 1566, Simancas E° 1395; in Aragon, 13th February, 1567, Fourquevaux, *op. cit.*, III, p. 36; Sicily exported large quantities of grain in 1567–1568, 209,518 *salme*, but only 93,337 in 1568–1569, *Relationi delli fromenti estratti del regno de Sicilia*, 1570, Simancas E° 1124; the large export figure for 1567–1568 must surely be linked to increased demand in the Mediterranean.

[113] Constantinople, 10th August, 1567, dispatches arriving at Lecce on 20th October. Simancas E° 1056, f° 80; Constantinople, 20th October, 1567 (Ital.), *ibid.*, f° 91; Constantinople, 16th May (1568), *ibid.*, f° 151. For rumours of an impending attack on Ragusa, dispatch from Corfu, 12th June, 1568, *ibid.*, f° 16.

[114] The Duke of Terranova, president of the kingdom of Sicily, to Antonio Pérez, Palermo, 30th September, 1568, Simancas E° 1132; Pescara to the king, Palermo, 18th October, 1568, *ibid.*

[115] Against Tunis, 21st May, 1567, against Algiers and Djerba, 15th February, 1567, Fourquevaux, *op. cit.*, I, p. 180, 15th March, *ibid.*, p. 190. Note that after January, 1567, the Spaniards were virtually certain that no Turkish Armada would sail, Lecce, 29th January, 1567, Simancas E° 1056, f° 17.

[116] Viceroy of Naples to the king, 11th May, 1567, Simancas E° 1056, f° 57.

[117] But less busy than usual with so many galleys detached for duties in the west.

[118] Chantonnay to the king, Vienna, 12th August, 1568, *CODOIN*, p. 469–479.

II was still receiving the previous subsidies from Rome which were supposed to pay for his wars against the Turk; it was essential if the negotiations should fail, not to be compromised by them). In January, 1567, Titian introduced the Spanish agent in Venice, Garci Hernández, to a Turkish ambassador who was passing through the city.[119] This Turk claimed that the emperor would undoubtedly obtain the truce he was seeking from the sultan and that the Catholic King could perfectly well be included in it if he so desired. He entered into some rather strange details concerning sums of money still not paid for Alvaro de Sande's ransom, offered his services at Constantinople and said he had written to the emperor on the matter but had not received a reply. He even implied that in 1566, had it not been for the blunders of Michael Cernovitch, he would have brought the talks to a successful conclusion, and that at that time the king of Spain was included in the planned truce, 'as Your Majesty must know', wrote Garci Hernández. Finally this ambassador, one Alban Bey, a dragoman of the Grand Turk, offered his services for any other negotiations and indicated the home of a merchant of Pera, Domenigo Cayano, as a convenient postal address. Minor actors and minor matters, but they all add up. In May, 1567, in Paris this time,[120] Francés de Alava also made contact with a Turkish envoy, who had come to France to discuss in particular the claims of Micas the Jew, favourite of Selīm II and a very important person at Constantinople, where he played a role not unlike that of the Fuggers though on a smaller scale and was graced with the title of Duke of Naxos; we have already had occasion to mention this unusual character. It was in his name, and while continuing to handle his interests, that the Turkish agent offered his services to Spain, proposing notably to use his influence to obtain a truce between the Catholic King and the sultan. Ruy Gómez, added the envoy, must know many secrets concerning these matters – an interesting piece of information. Since all news eventually leaks out (albeit with many distortions in the process), Fourquevaux reported a curious rumour circulating in Madrid at this time: the Turk was said to have sent his head dragoman to the king of France, asking him to 'mediate a truce with the Catholic King'.[121]

If several Turks thus offered their services – for a price – to the king of Spain, it was no doubt because they were better informed than Fourquevaux of his real intentions. For Philip II had already begun negotiations. Chantonnay, his ambassador, had already left for Vienna with extremely precise instructions: once again, he was to try to obtain a truce without actually asking for it. On 23rd May, 1567, he wrote to his master that the

[119] Garci Hernández to the king, Venice, 25th January, 1567, Simancas E° 1326. Was this really Titian the painter?

[120] To the king, 6th May, 1567, A.N., K 1508, B 21, no. 6, passages deciphered between the lines.

[121] Fourquevaux to the queen, Madrid, 3rd May, 1563, Fourquevaux, *op. cit.*, III, p. 42, 8th May, I, p. 351. According to the writer, Philip did not wish to be included in the Turco-imperial truce.

emperor had sent to Constantinople the bishop of Agriá, who in the past, during Sulaimān's reign, had acted as ambassador for Ferdinand – to negotiate on his behalf in Turkey. Chantonnay had given him a copy of the paper entrusted to him by Philip containing the conditions on which his sovereign 'would consent to negotiate some agreement with the Turk'. The whole affair was of course to be presented as a project suggested by the emperor, not by Philip.[122] A little later, we find Philip congratulating his ambassador for handling the mission so tactfully.[123]

The result was that in December that year, the imperial ambassadors, in obedience to their instructions and in order to further their own affairs, that is to say the conclusion of the truce, offered to include the Catholic King in it, broaching the subject with Muḥammad Sokolli in the Croat tongue for the purpose of secrecy.[124] But the Grand Vizier remained unmoved: if Philip wanted a truce why did he not send an ambassador? Negotiations did not stop here however, but were diverted into the eager hands of Joseph Micas.[125] As late as June, 1568, Chantonnay was able to write to Philip II that he had refused to receive a Turkish ambassador who had come to Vienna just after the signing of the truce and who had wished to visit him 'a su posada'.[126] Philip II replied on 18th July,[127] approving his ambassador's actions and asking him to maintain the negotiations on the pre-arranged terms. So negotiations continued after this, but we know little about their subsequent course, nor do we know exactly why they failed.

Very probably, Philip was not sufficiently anxious for peace to pay the high price demanded. Total suspension of hostilities in the Mediterranean would certainly have eliminated a continual source of expense and harassment. But 'hostilities' were not at present such as to present any very great danger. Even if they had, the Spanish navy was now in a fit state to meet any emergency. Philip had seventy galleys in Spain, not counting those still under construction at Barcelona, a hundred in all, according to Fourquevaux.[128] To these of course could be added the very considerable strength of the Italian squadrons. It might not be enough to go out and attack a mighty Turkish fleet, but it was certainly enough to prevent one from acting as it pleased. And it was also sufficient to keep in check the foists and galliots of the corsairs: the Spanish fleet was able in 1567 and 1568 to continue undisturbed in its mission of clearing the straits of all sea-robbers,[129] particularly those of Algiers who had in 1566

[122] Chantonnay to the king, Vienna, 23rd May, 1567, *CODOIN*, CI, p. 213–219.
[123] Philip II to Chantonnay, Madrid, 26th September, 1567, *ibid.*, p. 280–281.
[124] Muḥammad Sokolli was born in Trebigni, near Ragusa.
[125] Chantonnay to Philip II, Vienna, 28th February, 1568, *CODOIN*, CI, p. 378–379.
[126] Chantonnay to Philip, Vienna, 12th June, 1568, *ibid.*, p. 432–436.
[127] Philip to Chantonnay, Madrid, 18th July, 1568, *ibid.*, p. 450.
[128] Madrid, 24th August, 1568, Fourquevaux, *op. cit.*, I, p. 256.
[129] Nobili to the prince, Madrid, June, 1567, A.d.S. Florence, Mediceo 4898, f° 246; G. Andrea Doria to the king, Cadiz, 26th June, 1567, Simancas E° 149. Arrival of G. A. Doria at Málaga, 11th July, the *proveedores* to Philip II, 12th July, 1567, Simancas E° 149, f° 197.

brazenly looted the coast of Andalusia as far west as Seville.[130]

Unchallenged masters of their coastal waters, the Spaniards were able to use Mediterranean routes for the concentration of forces destined for Flanders.[131] These troop movements, which had begun well before 1567, gave rise in the new year to a series of sea voyages. The Neapolitan infantry embarked in January.[132] The old Spanish companies were assembled at Milan shortly afterwards,[133] to the discomfort and indeed molestation of the townspeople with whom they were billeted. The transfer of these troops across Europe was no small problem for Spanish diplomacy which was anxious not to arouse fears and to obtain proper legal safe-conducts well in advance and over the whole route. Not surprisingly it met refusals – from the king of France for a start[134] – and ill-will towards the army was almost unanimous.[135]

We may add the problem of supply[136] and of maritime transport.[137] Ships had to be found and the sea-voyages which were undertaken despite the dangers of the winter season did not always end safely. On 9th February at Málaga, twenty-nine ships laden with munitions, supplies and pieces of artillery capsized and sank.[138] As for the Duke of Alva, after waiting some time – perhaps as Fourquevaux suggests[139] deliberately stalling to see what the Turks would do – he embarked at Cartagena on 27th April[140] with several companies of 'bizognes', raw recruits. At Genoa, where he found himself at the beginning of August,[141] he received

[130] Fourquevaux to the king, 2nd March, 1567, Fourquevaux, *op. cit.*, I, p. 187–192.

[131] Fourquevaux to the king, 4th January, 1567, *ibid.*, I, p. 160; viceroy of Naples to the king, 8th January, 1567, Simancas E° 1056, f° 11.

[132] See previous note, 8th January, 1567.

[133] 13th February, 1567, Fourquevaux, *op. cit.*, p. 177.

[134] 18th January, 1567, *ibid.*, I, p. 169; Charles IX to Fourquevaux, 25th February, 1567, *ibid.*, p. 83.

[135] Difficulties with the Swiss cantons, 18th April, 1567, A.N., K 150, no. 98. Venetian fears, Garci Hernández to the king, Venice, 13th April, 1567, Simancas E° 1326. Will there be some move against Geneva on the way? Madrid, 15th April, 1567, Fourquevaux, *op. cit.*, I, p. 202–203. Fears of the duke of Lorraine, F. de Alava to Philip II, Paris, 17th April, 1567, A.N., K 1507, no. 104.

[136] One small example: three ships are sent to Málaga, one carrying rice and beans, Diego López de Aguilera to Philip II, Simancas E° 149, f° 205.

[137] 30th June, 1567, Fourquevaux, *op. cit.*, I, p. 220.

[138] 15th February, 1567, *ibid.*, p. 180; Nobili to the prince, Madrid, 15th February, 1567, A.d.S. Florence, Mediceo 4897a, 'Hieri venne avviso che nella spiaggia di Málaga combattute da grave tempestà sono affondate 27 tra navi et barche cariche di biscotti, d'armi, di munitioni et d'alchuni pochi soldati. . . .'

[139] Fourquevaux to the king, Madrid, 15th April, 1567, Fourquevaux, *op. cit.*, I, p. 202. The Turk will not come, 'que no aura armada este año', the viceroy of Naples informed Philip II, Pozzuoli, 4th April, 1567, Simancas E° 1056, f° 31.

[140] He embarked on the evening of the 26th, Alva to Philip, Cartagena, 26th April, 1567, *CODOIN*, IV, p. 351. Alva to the king, Cartagena, 27th April, *ibid.*, p. 354. So he cannot have left on the 17th as Fourquevaux's published letters suggest, *op. cit.*, I, p. 209. Nobili to the prince, Madrid, 3rd May, 1567, A.d.S. Florence, Mediceo 4898, f°s 50 and 50 v°; Nobili to the prince, 4th May, 1567, A.d.S. Florence, *ibid.*, 4897a; *ibid.*, 12th May, 1567, 4898, f°s 58 and 59 v°, 'Parti il Ducca d'Alva di Cartagena alli 27 del passato et sperase che a questa hora sia arrivato in Italia'.

[141] Figueroa to Philip II, Genoa, 8th August, 1567, Simancas E° 1390.

a warm welcome, but was also besieged with warnings and with grievances, particularly concerning Corsica where the assassination of Sampiero Corso on 17th January, 1567[142] had not brought peace[143] and where French intervention continued.[144]

Some historians have found Spain's military preparations extremely slow. They have not fully taken the measure of these troop movements (alongside the regular soldiers travelled valets, wives, plus veritable battalions of prostitutes) not to mention the transport of supplies organized at the same time, pressing into service numerous roundships which could be used to carry recruits as well as the sacks of beans and rice and the inevitable biscuit. It was the largest exercise in troop movement that the century had yet seen. At one end of the chain, in Andalusia, the recruiting sergeant's drum was still beating, as the first Spanish troops, after their long overland march, were setting foot in the Netherlands, hundreds of leagues away from the Peninsula. But was it not folly to place the chief interest of the Spanish monarchy so far away, to embark on a conflict so far from home?

There may have been a last-minute hesitation on Spain's part in May, 1567. Alva was sailing towards the Italian coast. 'After his departure', writes the Tuscan agent Nobili, on 12th May, 'the Gentlemen of the Council, seeing that affairs in Flanders seemed to be going so well for His Majesty, held many conversations debating whether the Duke should continue to Flanders, or whether it would not be better to proceed against Algiers or Tripoli'.[145] The debate concluded as follows: in favour of recalling Alva, four votes out of eight; the other four considered that his voyage was necessary. Nobili adds 'The latter opinion seems to have carried the day'. It could hardly have been otherwise. Could Ruy Gómez and his friends (who were not perhaps altogether sorry to see Alva sail away over the horizon) have stopped the mighty military machine once it had been set rolling? It provides an interesting glimpse of moments of indecision inside Spanish governmental circles in response to outside events.

It appears that in so far as the Netherlands had hoped for a peaceful solution – that is a measure of freedom of conscience in exchange for money[146] – Ruy Gómez was the man on whom they counted. In January, 1567, it was rumoured in Madrid that the departure was imminent not of the Duke of Alva, but of Ruy Gómez, who would go to bring peace

[142] A. de Ruble, *Le traité de Cateau-Cambrésis*, *op. cit.*, p. 82.

[143] Adamo Centurione even proposed ceding Corsica to the Catholic King, Figueroa to Philip, Genoa, 8th August, 1567, Simancas E° 1390. Figueroa, to whom the proposal was submitted, immediately rejected it and reported to Spain.

[144] Figueroa to Philip, Genoa, 15th May, 1568, Simancas E° 1390.

[145] Nobili to the prince, Madrid, 12th May, 1567, A.d.S. Florence, Mediceo 4898; on the restoration of the situation in the Netherlands, F. de Alava to Philip, Paris, 9th April, 1567, A.N., K 1507, no. 99; Alava to Philip, Paris, 5th May, 1567, *ibid.*, K 1508, B 21, no. 16b; he writes of Margaret of Parma: 'significale quan enteramente quedan ya en obediencia de V. M^d^ todos aquellos estados sea dios loado. . . .'

[146] 4th January, 1567, Fourquevaux, *op. cit.*, I, p. 156.

without force, 'for all the states of the said country are calling for him'.[147] This suggests that the old connection remarked in 1559, between Ruy Gómez and his party and the great lords of the Netherlands still operated six years later. Alva favoured intervention. Ruy Gómez, if we are to believe the nuncio,[148] preferred the idea of a league against the Turk. But although the two rivals were officially and even publicly reconciled in March, and although Ruy Gómez refused to let himself be outmanoeuvred by Fourquevaux (who when he went to him with complaints about Alva, met a polite and insincerely sympathetic reception),[149] the rivalry persisted between the two parties, in great as in small affairs.

However if the king tolerated their conflict and rivalry, he kept control over them. In March, 1567, without consulting anyone at all, he made his own nominations to vacant commanderies and benefices 'by which [the party leaders] were much mortified'. Decisions of policy were still taken by the Catholic King.[150]

Finally at any rate the Spanish empire intervened in Flanders in 1567 in such strength that its neighbours remained worried for months.[151] In France after the *coup* at Meaux in September, civil war had broken out again. It was to culminate in the battle of Saint-Denis on 10th November[152] but soon afterwards died down. In the Treaty of Longjumeau (23rd March, 1568) the prince of Condé may perhaps have sacrificed the interests of the mass of his followers to those of the Protestant nobility alone;[153] but, as Philip saw, he was thus freeing his hands for action in the Netherlands.[154] Already the disturbances in France had considerably hampered Spanish communications, with couriers having to take either the Atlantic or the Mediterranean route, one from San Sebastián, the other from Barcelona and both desperately slow.[155] In Germany, the Protestants were alert and beginning to agitate, as could be seen from the

[147] Fourquevaux, *op. cit.*, I, p. 165; Nobili to the duke, 7th December, 1566, A.d.S. Florence, Mediceo 4898, f° 8, gives the same report earlier. But Ruy Gómez had been scheming.

[148] Castagna to Alessandrino, Madrid, 7th January, 1567, quoted by Paul Herre, *op. cit.*, p. 41. Ruy Gómez even, one day, sounded out the French ambassador on the subject of a possible French action against the Turk, then dropped the subject, Fourquevaux to the king, Madrid, 15th April, 1567, Fourquevaux, *op. cit.*, I, p. 204.

[149] Fourquevaux, 18th January, 1567, *ibid.*, I, p. 170.

[150] Sigismondo de Cavalli to the Signoria of Venice, Madrid, 7th May, 1568, *C.S.P., Venetian*, VIII, p. 423–424.

[151] 23rd January, 1567, Fourquevaux, *op. cit.*, I, p. 183. N.B. F. de Alava's remark to Philip II (Paris, 23rd April, 1567, A.N., K 1507, no. 106) concerning the king of France and his councillors: 'Va les cada dia cresciendo el temor de la passada de V. Md'.

[152] Charles IX to Fourquevaux, Paris, 14th November, 1567, C. Douais, *Lettres de Charles IX à M. de Fourquevaux*, p. 129 ff., news of the victory of Saint-Denis. Departure of the Flemish gentlemen who were in the Protestant camp. Francés de Alava to the king, Paris, 23rd October, 1567, A.N., K 1508, B 21, no. 81.

[153] E. Lavisse, *Histoire de France*, VI, I, p. 100.

[154] R. B. Merriman, *op. cit.*, IV, p. 289.

[155] Nobili to the duke, Madrid, 30th October, 1567, A.d.S. Florence, Mediceo 4898, f° 122.

Gröningen rising.[156] In England, Elizabeth was playing for time, continuing to send polite messages but also complaints. In June, 1567, Cecil pointed out to Guzmán de Silva that there were rumours of a league against the Protestants as well as some plan to support the queen of Scots. And the surest proof of these dealings was that the emperor, in order to be free to participate in such a league had just agreed a truce with the Turks on conditions extremely disadvantageous to himself and by which the members of the Privy Council were shocked.[157] Finally England was beginning to use a weapon which gradually carried more and more weight: in the Atlantic, war was beginning in earnest between British privateers and the Spanish galleons.

Anti-Spanish reaction was quietly crystallizing in faraway northern Europe, circling warily round the massive force Philip II had sent there and which was not perhaps such a magnificent weapon as Spanish diplomacy liked to think. To start with, having been brought from a very great distance, it cost a great deal of money. When desertions began to thin its ranks, and the recruiter's drum had to sound once more in Andalusia and elsewhere to fill these gaps, further heavy expenses and further deplorable delays were the result. And the proud and magnificent army in Flanders, having no sea protection, was at the mercy of any attacker who deprived it of the great ocean supply route taken by the Biscayan *zabras*.

There was a moment however when it was believed that a large fleet might be assembled. Philip II, who had actually announced his departure for Flanders and decided not to go by Genoa, had serious, or at any rate visible preparations made on the Cantabrian coast at Santander. Menéndez d'Aviles had returned from Florida in time, it seemed, to take command of the royal fleet. Then suddenly everything was cancelled. Historians will no doubt always wonder if this was some 'Castilian' subtlety, some prolonged deception to keep Europe on tenterhooks and also to wring money out of the Cortes of Castile, even perhaps to conceal as long as possible the reasons for sending the duke of Alva[158] to the north (by far the most tempting explanation). At all events, once again, not a whisper of the royal intentions leaked out. Philip II was certainly not a man who revealed his mind. Even in his own court, everyone was in the dark in 1567. Fourquevaux, excusing himself for knowing so little, wrote one day that even the members of the inner Council 'do not know where they are, this monarch never makes public his intentions until the very last minute'.[159] So we shall probably never know whether Philip II's voyage

[156] Madrid, 15th April, 1567, Fourquevaux, *op. cit.*, I, p. 201; Requesens to the king, Rome, 19th April, 1567, L. Serrano, *op. cit.*, II, p. 90.

[157] G. de Silva to the king, 21st June, 1567, A.E. Esp. 270, f° 175, articles (in Italian) of the league against the Protestants.

[158] L. Van der Essen, *op. cit.*, I, p. 151, references to Campana and Strada.

[159] Fourquevaux to the queen, 23rd February, 1567, Fourquevaux, *op. cit.*, III, p. 58; Granvelle to the king, Rome, 14th March, 1567, *Correspondance*, I, p. 294. On receiving bad news from the Netherlands, 'all at once, overnight', the king resolved to go to Italy. 24th May, 1567, Fourquevaux, I, p. 192; in Italy, it was announced that

was a trick or not. At any rate it was one of the major subjects of political speculation in 1567 and 1568. France was so convinced of it that Catherine de Medici actually thought of arranging a meeting at Boulogne.[160] But there is no formal evidence that Philip did not sincerely intend to go. Clearly his presence in the North – accompanied by a fleet – could have had a decisive effect upon the courts of events. But after the arrival of Alva in Brussels in August[161] and the restoration of order in the Nether-

---

the king would not go to Flanders, Granvelle to the king, 15th April, 1567, *Correspondance*, II, p. 382. Was the voyage a screen for an expedition against Algiers, Nobili to the prince, Madrid, 4th May, 1567, A.d.S. Florence, Mediceo 4897a; the departure is certain but the princes of Bohemia have made no preparations, Nobili to the prince, 18th June, 1567, *ibid.*, 4898, f°s 62 and 62 v°, but the same day Nobili, *ibid.*, f° 67 v°, announces their departure. Preparations for the voyage, Nobili, *ibid.*, 4897a, 26th June; Fourquevaux to the queen, Madrid, 30th June, 1567, Fourquevaux, *op. cit.*, I, p. 228, orders for departure have been given by Philip II: 'I do not know whether this is some Castilian subtlety, but that is how it is'. Philip II to Granvelle, Madrid, 12th July, 1567, *CODOIN*, IV, p. 373, writes of his firm intention to leave; Nobili to the prince, Madrid, 17th July, 1567, A.d.S. Florence, Mediceo 4898, f° 77 '. . . che l'andata di sua Mta in Flandra riscalda assai et tutti questi grandi lo dicono absolutamente . . .'. Fourquevaux to the king, end of July, 1567, Fourquevaux, *op. cit.*, I, p. 241, it is not known which sea-route Philip will take, Nobili to the prince, Madrid, 24th July, 1567, Mediceo 4898, f°s 75 and 75 v°, Pedro Meléndez has arrived from Florida just in time to command the royal fleet; Francés de Alava to Catherine de Medici, Paris, 8th August, 1567, A.N., K 1508, B 21, no. 42, the regent of the Netherlands is sending ships ahead of the royal fleet; Granvelle to the king, Rome, 17th August, 1567, is pleased to hear of the proposed journey: 'although in Spain and in Flanders and even more here, there are many people who refuse to believe in Your Majesty's journey and prate after their own fantasies . . .'. Time passed: Fourquevaux to the queen, 21st August, 1567, 'and for the said king to embark in September would be the act of a madman who wants to drown himself and all his men'; G. Correr to the Doge, Compiègne, 4th September, 1567, *C.S.P. Venetian*, VII, p. 403, according to letters from the merchants, the Catholic King 'is not likely to arrive'. G. de Silva to the king, London, 6th September, 1567, *CODOIN*, LXXXIX, p. 541, according to the French ambassador, Philip II is more likely to travel by way of Santander than Corunna. If the voyage does not take place there will be an expedition against Algiers. Circular letter from Philip II to the dukes of Florence, Ferrara, Urbino, Mantua, Madrid, 22nd September, 1567, he will not travel to Flanders by the western passage, 'por sus enfermedades y por ser el camino tan trabajoso que non ha sido possible, para estar ya a la boca del invierno . . .'. Fourquevaux to the king, Madrid, 23rd September, 1567, *op. cit.*, I, p. 367, the journey has been postponed until spring. Viceroy of Naples to the king, Naples, 30th September, 1567, Simancas E° 1056, f° 96, has received a letter informing him that the journey is postponed until the spring. Nobili to the prince, 17th November, 1567, Mediceo 4898, f° 128, all men of judgement are in favour of the journey; Nobili to the prince, *ibid.*, f° 137, 27th November, 1567, Don Diego de Córdoba has told him, in the king's antechamber, that Philip II intends to go to Flanders with sufficient troops to fill the gaps in the Spanish companies. The cardinal of Lorraine to Philip II, Reims, 16th January, 1568, A.N., K 1509, B 22, no. 3a, urges him to go to the Netherlands and continue his repression of the heretics; F. de Alava to the king, Paris, 27th March, 1568, A.N., K 1509, B 22, no. 35a, 'Todo lo de aquellos estados esta quietissimo desseando la venida de V. M^d^ para el remedio dellos'.

[160] E. Haury, 'Projet d'entrevue de Catherine de Médicis et de Philippe II devant Boulogne', in *Bulletin de la Société acad. de Boulogne-sur-Mer*, VI (extract in B.N., Paris, in-8°, Lb 33/543); Fourquevaux to the queen, Madrid, 24th August, 1567, Fourquevaux, *op. cit.*, I, p. 254.

[161] 24th August, Van Houtte, *art. cit.*, p. 376.

lands, was it necessary, now that a royal victory appeared to have been achieved, for Philip himself to set out for the North 'in the jaws of winter'? He was repeatedly assured, in 1568 too at least until the spring, that 'lo de Flandes esta quietissimo'.[162] And, a further deterrent, 1568 was the year of Don Carlos's tragic death, a more moving and cruel episode in real life than in any of the romantic legends.[163] With his son imprisoned in January (he was to die on 24th July following) could the father very well leave?[164]

It matters little in the end: for the tragedy of Spanish policy in the Netherlands was the result not of the abortive journey or journeys of Philip II (to think so, comments Ludwig Pfandl,[165] is grossly to overestimate royal magnetism) but of the voyage planned, conceived, premeditated and carried out – by the Duke of Alva.

2. THE WAR OF GRANADA: A TURNING-POINT

From the end of the year 1568 – in the dead of winter, unusually enough – and with increasing frequency in 1569, wars began to break out one after another all round the Mediterranean, some far away, but some very near to its shores; their flames were bright and of varying duration, but all bore witness to the increasingly tragic situation.

The rising tide of war. Far from the Mediterranean, the war of the Netherlands had begun: it was war in earnest, no longer to be dismissed as mere 'troubles'. In this conflict, Spain and through her the whole Mediterranean world was deeply implicated. The arrival of the Duke of Alva in August, 1567 had seen the beginning of a reign of terror which for a time reduced the opposition to silence. But in April, 1568, resistance began: the first hopeless attacks by Villiers and Louis of Nassau[166] were followed in July by William of Orange's equally unsuccessful campaign which ended in near-ridicule in November on the borders of Picardy.[167]

But if supremacy on land now lay with Alva, who retained it for a long time (at least until April, 1572 and the Brill rising) war on the sea was a

[162] Cf. end of note 159; F. de Alava to the king, 1st March, 1568, A.N., K 1509, B 22, f° 26, already reports that according to the Duke of Alva, the Netherlands are quiet.

[163] As I said earlier, Viktor Bibl's book, *Der Tod des Don Carlos* is not to be relied on.

[164] Fourquevaux to the king, Madrid, 26th July, 1568, Fourquevaux, *op. cit.*, I, p. 371. The death of Don Carlos relieved the king of much worry 'and he will be able to leave his kingdom whenever he wishes, without fear of sedition during his absence'.

[165] *Philippe II*, (French trans.), p. 128.

[166] H. Pirenne, *Histoire de Belgique*, Brussels, 1911, IV, p. 13. At this period, the activities of the prince of Orange did not greatly disturb the Spanish king, Philip II to F. de Alava, Aranjuez, 13th May, 1568, A.N., K 1511, B 22, no. 31.

[167] H. Pirenne, *ibid.*, p. 14, 15. Alva's letter to Philip reporting the victory, Cateau-Cambrésis, 23rd November, 1568, several copies extant, B.N., Paris, Esp. 361; Gachard, *Correspondance de Philippe II*, II, p. 49; *CODOIN*, IV, p. 506. On the prince of Orange's campaign in France in 1569, some unpublished documents are edited by Kervyn de Lettenhove, *Com. Royale d'Histoire*, 4th Series, 1886, XIII, p. 66–74.

different matter. In 1568 the battle of the Atlantic between Spaniards and Protestants first got under way.[168] It degenerated into an economic war, undercover and underhand, between England and Spain. Blows were struck on both sides. The British Isles were deprived of their normal supplies of Spanish oil, used in processing of wool.[169] In return, the great silver route of the Spanish Empire was interrupted with all the consequences that entailed, and the Biscayan *zabras*, with their cargoes of precious metals, were seized. The Duke of Alva did not immediately grasp the significance of this battle for the Channel and the Atlantic. Like many another politician incapable of reading the signs of the future, he did not realize that the Spanish Netherlands had more to fear from England than from either France or Germany.

Others were not so blind nor did they hesitate to speak. Pius V who was always ready to strike down heresy at the first sign (he excommunicated Elizabeth in February, 1569) proclaimed with vehemence in a letter sent on 8th July, 1568[170] that now or never was the time for the 'enterprise of England'. The ambassador Guerau de Spes, who was implicated in intrigues in the island on behalf of Mary Stuart and in the rising of the northern earls, organizing them as best he could, may have erred by optimism, paying too much attention to a small sector of Spanish policy, and failing to appreciate the whole. But he was not perhaps mistaken in thinking that in 1569 Spain held three cards which might be played against England: the situation in Scotland, that in Ireland,[171] and the imminent rebellion of the great Catholic nobles in the north. Philip II was tempted by this bold strategy, but Alva opposed it and won over his sovereign, pleading lack of funds and the European situation. Alva, far from being the great man of history books, was both narrow-minded and politically short-sighted, able to fight battles at close range only. He delayed the general pardon beyond all measure; allowed the queen of Scotland to flee to England and let Scotland turn Protestant; allowed the northern earls to rise in vain,[172] their rebellion being quickly and easily crushed by

[168] Van Houtte, *art. cit.*, p. 385, 386, 16th April, 1569; rumours of war with England, M. de Gomiécourt to Zayas, Paris, 2nd August, 1569, war with England said to be approaching fast: 'dize por ciertos avisos q la Reyna d'Ingleterra arma para Normandia es de temer que no sera para lo de Flandes'; Van Houtte, *art. cit.*, p. 388. On this serious crisis of 1569 see above, Part II, Chapter 2.

[169] G. de Spes to the king, London, 1st June, 1569, *CODOIN*, XC, p. 254. de Spes to the king, *ibid.*, p. 276, London, 5th August, 1569: 'Es la falta de açeite aqui tan importante que de simiente de rabanos sacan açeite para adereszar la lana . . .'.

[170] B.N., Madrid, MS 1750, f°s 281–283. On the whole English question, see O. de Törne's informative *Don Juan d'Autriche*, I, Helsingfors (Helsinki), 1915.

[171] On troubles in Ireland in the regions near Scotland, *CODOIN*, XC, p. 171. 8th January, 1570.

[172] Rome, 3rd November, 1569, L. Serrano, *op. cit.*, p. 186. *Proyecto del Papa sobre la sublevación de Inglaterra contra Isabel: piensa Çuñiga que interviniendo el Rey en la empresa se lograria la concesión de la cruzada*, Simancas E° 106, 5th November, 1569; *C.S.P.*, *Venetian*, VII, p. 479; London, 24th December, 1569, *CODOIN*, XC, p. 316; 26th December, p. 317. *Conquista de Inglaterra y comissión alli de consejero d'Assonleville*, Simancas E° 541. *Estado de los negocios de Inglaterra*, *ibid.* Alva's opinion is

Elizabeth; finally, instead of striking at England while it was still insecure, he negotiated and quibbled when time was not on his side. The epithet of prudent should have been applied in 1569 not to Philip, but to Alva, at a time when distance and events made him the master of the hour; Alva, who was unable to see that far away from the narrow terrorized zone which he thought he controlled, the war of the future was already showing its true face.

In France, where the peace of Longjumeau had not lasted long, the third war of religion had begun in August, 1568, with the flight of Coligny and Condé towards the Loire;[173] a move linked beyond any doubt with Spanish involvement in Flanders. But the withdrawal of the Huguenot leaders to the south, hardly brought the war any nearer the Mediterranean – where it had been feared during July that the German 'reistres', idle since Longjumeau, might with the help of the Protestants, have turned and pushed on towards Italy.[174] The hypothesis was received in Spain with scepticism, and the fresh outbreak of hostilities in France soon made it irrelevant. The third civil war, so violent and cruel, only briefly concerned the Mediterranean region, in spring 1570, when during his headlong progress after the year of Jarnac and Montcontour, Coligny made his way from the Garonne valley to the Mediterranean south and the Rhône valley.[175] There was also an alert in Spain in the summer of 1569 when the admiral reached Guyenne,[176] causing the Catholics to ask for Spanish support in the Pyrenees such as they had in the past received on the Flanders side.[177] Even without such support the disproportion of forces between Catholics and Protestants was already flagrant. But was victory the simple matter it seemed to Francés de Alava, who took the view that the court had only to be sufficiently determined for the Protestants and in particular the admiral to be crushed without any difficulty? The Spanish ambassador blamed the king's advisers, Montmorency, Morvilliers, Limoges, Lansac, Vieilleville[178] – as if courage did not count, as if distance did not count and as if the persecuted sect in France did not rely on powerful assistance from Protestants abroad.

So much for European wars. But in the East too, and again far from Mediterranean shores, war was raging on the vast borders of Turkey,

expressed in his letter to Don Juan de Zuñiga, Simancas E° 913, copy in A.E. Esp. 295, f°s 186 to 188: it is impossible to conquer England while still facing the hostility of Germany and France. [173] E. Lavisse, *op. cit.*, VI, 1, p. 106 ff.

[174] The duke of Alcalá to the Pope, 24th July, 1568, Simancas E° 1856, f° 17.

[175] E. Lavisse, *op. cit.*, VI, 1, p. 111.

[176] Francés de Alava to the king, Paris, 9th June, 1569, A.N., K 1514, B 26, n° 122.

[177] Francés de Alava to the Duke of Alva, Orleans, 19th June, 1569, A.N., K 1513, B 25, no. 54. J. Manrique inspects the frontier of Navarre, J. de Samaniega to the Duke of Parma, Madrid, 12th July, 1569, A.d.S. Naples, Farnesiane, Spagna 5/1, f° 272; the same information in Castagna to Alessandrino, 13th July, 1569, L. Serrano, *op. cit.*, I, p. 112. Fears for Navarre, but chiefly only if the civil wars in France should end, 19th August, 1569, Fourquevaux, II, p. 110; 17th September, 1569, *ibid.*, p. 117.

[178] Francés de Alava to the king, Tours, 29th October, 1569, A.N., K 1512, B 24, no. 139.

from the approaches of the Black Sea to the coasts of the Red Sea, that great peripheral war which was to do more in 1569 than any other factor to paralyze any possible action by Selīm and his fleets in the Mediterranean. It was a strange war too, indirectly challenging the hostile mass of Persia which had become a great turntable between Asia, particularly central Asia, India and the Indian Ocean.

The first area of conflict was the southern part of what is now Russia. The Turks, in alliance with the Tartars of Crimea and relying both upon them and upon massive levies of Rumanian peasants whom they intended to use as a labour force were trying to get back from the Russians Kazan and Astrakhan[179] which the latter had taken in 1551 and 1553. Without taking literally the figures given by European informants, we may suppose that in view of the immensity of the battle area, this war required large numbers of troops, massive transports and a gigantic assembly of supplies and munitions at Azov, the operational base.

Was the aim of this expedition merely to strike at the Muscovites, to punish them (as ultra-formalist Turkish diplomacy put it, anxious as usual to justify its actions) and bring succour to a vassal, the Khan of Crimea who had been unjustly attacked by the Russians? In fact both Turks and Russians in the sixteenth century had suffered equally from depredations by these nomads and would not have been entirely averse to combining forces against the buffer-state.[180] Nor does the expedition give the impression of forming part of a great Turkish design to open up the way to central Asia. There remains the more simple explanation that this was a long-distance military manoeuvre against the Persians. The Turks planned to dig a canal from the Don to the Volga, thus linking the Caspian to the Black Sea and preparing the way for their galleys to reach the inland shores of Persia. The Sophy realized this evident danger;[181] consequently, in circumstances about which we know nothing, he did his best to incite the peoples and princes of the Caucasus against the Turks. The fact that the grand design was thwarted in 1570 when the Russians seized the aggressors' war materiel and artillery, does not detract from the scale of the undertaking.[182]

At the same time, yet another war was becoming more noticeable, one

[179] Constantinople, 14th March, 1569, E. Charrière, *op. cit.*, III, p. 57–61. Const., 26th March, 1569, Simancas E° 1057, f° 45, Bª Ferraro to the Signoria of Genoa, Const., 11th June, 1569, A.d.S. Genoa, Cost. 2/2170, C. 16th November, 1569, Simancas E° 105, f° 3. Thomas de Cornoça to the king, Venice, 9th December, 1569, Simancas E° 1326: the Turkish armada is being prepared slowly, for the sultan needs his forces on the Muscovy, Arabian and Persian fronts. Const., 10th December, 1569, Simancas E° 1058, f° 6.

[180] Less however than W. Platzhoff suggests (*op. cit.*, p. 32).

[181] Const., 28th January, 1569, Simancas E° 1057, f° 27: '... que no queria [the new king of Persia] que el fiume [*sic*] Volga se cortasse porque dello le vernia mucho daño por poder yr despues con varcas hasta sus estados a saqueallos ...'.

[182] Dispatches from Constantinople, received from Prague via Venice, Prague, 28th January, 1570, A.N., K 1515, B 27, no. 21. For the problem of the Don-Volga canal and an account of the winter campaign see J. von Hammer, *op. cit.*, VI, p. 338 ff.

which had in fact begun two years earlier, and covered all the Arab countries, from Egypt to Syria, the sorest spot being the revolt of the Yemen in the south.[183] 'The Arab countries' between them included the entire zone crossed by the great trade routes of the Levant. This revolt cost the Turks almost two millions in gold every year according to contemporary estimates,[184] without counting the great expense and difficulties of a foreign war.

The most persuasive evidence for the intensity of these two wars behind Turkish lines is the inactivity of the sultan on the Mediterranean front, his almost total absence from the sea,[185] where the Christians, at least in the west and centre, could do much as they pleased. In August and September, the galleys of Doria and Juan de Cardona were able to hunt down corsairs in the sea near Sicily unmolested and with some success.[186] But more needs to be discovered about the concealed workings of Ottoman history before we can evaluate the policy of the Grand Turk with any degree of precision. For suddenly, at the end of 1569, the Constantinople arsenal shook itself out of the torpor of the four preceding years with massive preparations for the expeditions against Cyprus.[187]

[183] F. de Alava to the king, Paris, 4th February, 1569, A.N., K 1514, B 26, no. 41, 'el Turco esta muy embaraçado y empachado porque los Arabios prosperan y el Sofi los fomenta . . .'. A dispatch from Alexandria, 14th April, 1569, Simancas E° 1325, gives many details of the war in the Yemen, presenting it as the result entirely of the exactions of Turkish governors, a wealth of detail so great as to be almost unintelligible, as is the account by Hammer (*op. cit.*, VI, p. 342 ff) which for some reason reproduces all the information in the document. Const., 11th June, 1569, see above, note 179: the Turks will probably restore order in the Yemen, Const., 16th October, 1569, E. Charrière, *op. cit.*, II, p. 82–99. Thomas Cornoça to the king, Venice, 29th September, 1569, Simancas E° 1326, Aden recaptured by the Turks? On the plans for driving a canal through the Suez isthmus after the submission of Arabia, Hammer, *op. cit.*, VI, p. 341.

[184] Const., 16th October, 1569, E. Charrière, *op. cit.*, III, p. 82–90, 800,000 ducats for the Yemen, a million for Syria.

[185] Viceroy of Naples to the king, 14th January, 1569, Simancas E° 1057, f° 18, there will be no Turkish armada this year. F. de Alava to Zayas, Paris, 15th January, 1569, A.N., K 1514, B 26, no. 23, conflicting rumours, the Turks will attack either La Goletta or Alexandria; viceroy of Naples to the king, Naples, 19th January, 1569, Simancas E° 1057, f° 2, no armada at all. Th. de Cornoça to the king, Venice, 24th January, 1569, Simancas E° 1326, there will be no armada because of Arabia; Constantinople, 28th January, 1569, see note 181 above, no armada; 14th March, 1569, E. Charrière, *op. cit.*, III, p. 57–61. J. López to the king, Venice, 2nd July, 1569, Simancas E° 1326: 'Venecianos se han resuelto de embiar las galeaças en Alexandria y Suria ques señal que no saldra armada por este año . . .'.

[186] Pescara to the king, Messina, 31st August, 1569 and 2nd September, 1569, Simancas E° 1132.

[187] Bª Ferraro to the Signoria of Genoa. Const., 23rd July, 1569, A.d.S. Genoa, Cost. 2/2170. Const., 24th and 29th May (dispatches received at Naples on 18th July, Simancas E° 1057, f° 59). Const., 7th August, 1569, Simancas E° 1057, f° 72: 'que los aparatos maritímos continuan a gran furia y casi a la descubierta'; Const., 18th September, 1569, Simancas E° 1057, f° 76; Const., 29th September and 2nd October, 1569, Simancas E° 1057, f° 9; Const., 16th November, 1569, Simancas E° 1058, f° 3; Constantinople, 26th December, 1569, Simancas E° 1058, f° 8: 'quest'appar[a]to di quest'armata sia per Spagna . . .'. On these armaments, see the report of the *bailo* Barbaro in the Vienna Archives, Hammer, *op. cit.*, VI, p. 336, notes 1 and 2.

The original plan for this expedition probably dates back several years, at least to the fortification of Karamania in 1567. Perhaps it was because of this, partly at least, that the sultan so readily agreed to a truce with the emperor. Perhaps too it was for this reason that he wanted to settle his internal problems before embarking on such a major undertaking. Hence the energy with which in 1569 he pursued the wars described above. It is said that Joseph Micas, to whom the sultan had promised Cyprus, had already incorporated its emblem into his coat of arms. It is also said that the fire which conveniently devastated the Venetian arsenal on 13th September, 1569, was the work of the same Micas, or rather of agents in his pay.[188]

What is certain is that the Turkish attack on Venice had been in the air for some time. No doubt this was the reason for rumours of an alliance between Spain and the Signoria[189] which Venice prudently fostered (the centre of such rumours being Rome). Venice of course possessed a formidable fleet, but compared to the massive Ottoman power she was slight and vulnerable. If attacked by this monster – and she had always, heaven knows, made every concession to avoid such a disaster – she could only hold out if supported by the rest of Christendom, by Italy and Spain, in other words by Philip II, who ruled both.

The beginning of the war of Granada. Into this world of rising tension, where the arms race was gathering speed, the war of Granada erupted with a violence out of all proportion to the real importance of what was, at the beginning at any rate, a second- or even third-rate military operation. But it is impossible to describe the extent to which this 'staggering' news roused hopes and calculations in the outside world, or to which it exalted passions and transformed the climate in Spain.

The facts are well known. It all began with a minor incident: on the night of Christmas Day, 1568, a handful of Moriscos reached Granada and penetrated the city, calling out loudly upon all who wanted to defend the religion of Muḥammad to follow them.[190] Where sixty men had gone in, a thousand emerged.[191] 'It is becoming clear', wrote the nuncio at Madrid, 'that this was not such a great event as will perhaps be claimed'.[192] And indeed the Albaïcin, the great Moorish quarter, had remained quiet. But there is no proof that had it not been for the snow blocking the mountain passes, Granada would not have been invaded by a consider-

[188] P. Herre, *op. cit.*, p. 15. Julian López to the king, Venice, 15th September, 1569, Simancas E° 1326. By way of equivalent cf. the plan to fire the Arsenal at Constantinople by one Julio Cesar Carrachialo, viceroy of Naples to the king, 23rd March, 1569, Simancas E° 1057, f° 43.

[189] Summary of letters from the viceroy of Naples dated 22nd, 26th, 29th, 31st August, 1569, Simancas E° 1056, f° 192.

[190] Madrid, 5th January, 1569, L. Serrano, *op. cit.*, III, p. 23–24.

[191] Sauli to the Signoria of Genoa, 5th January, 1569, A.d.S. Genoa, Lettere Ministri Spagna 4, 2413.

[192] See above, note 190.

36. SIXTUS V. Venetian School, Pinacoteca Vaticana

37. DON JOHN OF AUSTRIA. Anonymous portrait, Musée de Versailles

able force.[193] The attack might then have had a far greater impact and inflamed the whole city. Its lack of success obliged the rebels to take refuge in the mountains where other Moriscos soon joined them, for the most part inhabitants of other parts of the kingdom rather than of its capital.[194] They were estimated at about 4000. 'Some say more, some say less', wrote Sauli to the Republic of Genoa. 'So far I have not been able to discover the truth. It is thought that it will be but a straw blaze for the rising has occurred in the bad season and immense preparations are afoot against the rebels. From Córdoba, Ubeda and Baeza and other places, large numbers of horsemen and footsoldiers are on their way'.[195] According to some reports, the Moriscos had fortified Orgiva, a village owned by the Duke of Sessa, but without artillery what could be done? 'They say', Sauli went on, 'that among them are 300 Turks, but from other sources it has been learnt that these are but eight or nine survivors from a galley wrecked upon these shores'. His report is on the optimistic side; these were early days and of course the author of these lines was not unbiased. Sauli, like all Genoese (or for that matter Tuscan) agents, called the Spaniards 'our side'. Fourquevaux, as we may imagine, was both more objective and less optimistic.

For this war of religion, a war between hostile civilizations, spread spontaneously and rapidly over a land well-primed by hatred and poverty.[196] By January, Almeria was blockaded by the rebels.[197] In February the Duke of Sessa, whose domains, towns, villages and vassals were all in Granada, estimated the rebels at 150,000 of whom 45,000 were in a position to bear arms.[198] In March, the rebellion moved down from the mountains into the plains.[199] And the links between the rebels and Algiers were no longer in doubt to anyone.[200]

In fact the rebellion had been taken seriously by the authorities from the start, possibly because the south of Spain had been inconsiderately deprived of manpower by Alva's expedition. Here more than anywhere else recruitment had been heavy. And Spain was unaccustomed to war at home. She was not prepared for the fight. The first precaution was to send money for immediate levies to the Marquis of Mondéjar, Captain General of Granada and to the Marquis of Los Vélez, the governor of Murcia. At the same time the Spanish galleys were alerted to prevent the possible arrival of help from North Africa.[201]

The Spanish government tried to prevent news of the rebellion leaking

[193] Pedro de Medina, *op. cit.*, p. 147 v°.

[194] See above, note 191.

[195] *Ibid.*

[196] Hurtado de Mendoza, *op. cit.*, p. 71.

[197] 13th January, 1569, Fourquevaux, *op. cit.*, II, p. 45.

[198] 28th February, 1569, *ibid.*, II, p. 56.

[199] Dispatches from Spain, 20th March, 1569, *ibid.*, II, p. 62.

[200] Sauli to the Signoria of Genoa, Madrid, 14th April, 1569, A.d.S. Genoa, Lettere Ministri, Spagna, 1/2413.

[201] Philip II to Requesens, 15th January, 1569, Simancas E° 910, orders to come to the coast of Spain with 24 or 28 galleys.

out: 'sera bien de tener secreto lo de Granada', wrote Philip on the margin of a letter to the viceroy of Naples.[202] How can we, came the reply from Naples, on 19th February,[203] when news of the troubles is already being broadcast by Genoa and Rome? Of course it soon made its way to Constantinople, where the rebels had themselves sent an appeal for assistance.[204] From the very heart of Spain, all the way to Turkey, an uninterrupted chain of intelligence operated, not to mention the itinerant and fugitive Moriscos who were indefatigable walkers, travellers and intermediaries: they had their agents and spokesmen in North Africa as well as in Constantinople. And it is by no means sure that the Turks did not have a hand in the Morisco rising as they had in the fire at the Venetian arsenal. In June, 1568, Don John held a long conversation in Barcelona with a Greek captain who offered to organize an uprising in Morea. If such an interview could take place, why should there not have been some similar meeting in 1565, 1566, 1567 or 1568 between a 'Tagarin' or a 'Mudejar' and one of the commanders of the Turkish fleet?

In any case, as soon as the news had travelled over the waters of the Mediterranean and the landmasses of Europe, there were at least two Morisco wars: the real one, fought out in the highlands of the Sierra Nevada, an incoherent, disappointing conflict, a mountain war, full of surprises, difficulties and sometimes atrocious cruelty; and the other, the 'war of Granada' as the outside world saw it through the most contradictory reports destined to arouse every kind of passion in Europe and also in the Levant, eastern curiosity, satisfied by a closely woven net of intelligence and espionage, the counterpart of that other network which we can more easily trace from East to West (because it concerned western Europe, and western Europe kept its documents in order).

Whatever the true dimensions of the war, Spain was certainly deeply disturbed by this domestic conflict. In January, 1569, the rising was the subject of every conversation at court,[205] 'the most striking news here at the moment', as Fourquevaux said.[206] 'The alarm is great, throughout the whole kingdom', continues the ambassador,[207] taking the opportunity for a little moralizing: it is indeed a sign of the times he reflects when subjects rebel against their legitimate sovereigns, as they have against Charles IX of France, against Mary Stuart in Scotland and against the Catholic

[202] Simancas E° 1057, f° 105, Madrid, 20th January, 1569.

[203] Viceroy of Naples to the king, Naples, 19th February, 1569, Simancas E° 1057, f° 36.

[204] Bª Ferrero to the Signoria of Genoa, Constantinople, 23rd July, 1569, A.d.S. Genoa, Cost. 2/2170: 'li mori di Granata etiam scriven qui al grand Sre et a tutti li principalli suplichando che se li manda socorso de arme solo che de gente sono assai', and ask that the armada should be sent in 1570; the regions near the Straits of Gibraltar are poorly guarded and easy to capture. A. de Herrera, *Libro de agricultura . . ., op. cit.*, 1598, reported in his first dialogue that the revolt of Granada was even known in Constantinople, where it was at first taken to be a false report: a curious item of public opinion (*segun se dize*) to say the least, 341 v°.

[205] 13th January, 1569, Fourquevaux, *op. cit.*, II, p. 47–48.

[206] *Ibid.*, p. 45.

[207] *Ibid.*, p. 46.

King in Flanders. 'The world today is given to sedition and subjects to rebellion in divers places.' Charles IX replied with questionable sincerity that he hoped the leaders of sedition would soon be punished along with 'all those who like them have taken up arms to trouble the state of their king and sovereign'.[208] How distressed was he really at the situation which for Spain was becoming all-absorbing?

Retaliatory measures were indeed taking time to arrange. It was impossible to act swiftly in the wild barren mountains where a column of soldiers might starve to death – and actually did on occasions. It was impossible to blockade the long coastline of the rebellious province with its innumerable bays offering access to the ships of Algiers and Barbary bringing men, munitions and weapons (taking Christian prisoners as payment: a man for a musket),[209] artillery[210] and rations, rice, grain or flour. It was a coastline where the local nobleman – not the king – was the master and where smuggling and privateering had their established routes and practices.[211] Both on land and sea, the war went badly for the authorities to begin with. Mondéjar was an excellent commander, but behind his back the future Cardinal Deza spoke against him and paralysed his actions, pushing to the fore the incompetent marquis of Los Vélez. The inefficiency of the repression merely served to extend this cruel war which was already gathering momentum.

None of which prevented Philip II putting a bold face on matters. Of the tumult in Granada, he 'appears to take little note', observes Fourquevaux.[212] He claims that the Turks have other things to worry about, that help for the rebels from Algiers is impossible with his galleys on guard; all that is required to restore order is the action of the 'Christian communities', that is to say of the Andalusian militia. The official optimism was not shared abroad, where Spanish representatives struggled manfully against unkind rumours of disaster. In London, Guerau de Spes was particularly bitter. In May it was even 'published openly' that other Spanish kingdoms had risen too against His Majesty: 'The people of this country do not realise the loyalty of the Spaniards.'[213]

It was true that the Spaniards could not crush these growing rumours with news of any clear military victory won on a named battlefield with published lists of the dead and captured. All engagements were untidy affairs, fought with small numbers, in this minor but cruel war where manhunts beyond all official control were launched by both sides.[214]

[208] Charles IX to Fourquevaux, Metz, 14th March, 1569, C. Douais, *Lettres de Charles IX à M. de Fourquevaux*, p. 206.

[209] H. Forneron, *op. cit.*, II, p. 161.

[210] 23rd January, 1569, Fourquevaux, *op. cit.*, II, p. 51.

[211] *Avisos sobre cosas tocantes al Reyno de Granada*, 1569, Simancas E° 151, f° 83.

[212] Fourquevaux, *op. cit.*, II, p. 56.

[213] Guerau de Spes to the king, London, 2nd April, 1569, *CODOIN*, XC, p. 219, de Spes to the king, 9th May, 1569, *ibid.*, p. 228. Quotation from second letter.

[214] Like American frontier wars, G. Friederici, *Der Charakter*, *op. cit.*, I, p. 463. In the extraordinary letter from the rebel king, Muhammad Ibn Umaiya, to Don

It was a war fought by patrol and did not lend itself to triumphant communiqués. The storm which battered and dispersed the galleys of the Grand Commander of Castile off Marseilles on 8th April, as they were returning to the Spanish coast, appears as an event of some importance in this context, although it was by no means the total catastrophe that Spain's enemies were pleased to announce.[215]

Badly begun and badly managed, with mistakes at every level, the Granada war was getting nowhere and costing money. The appointment of Don John of Austria to overall command in April, did not at first change anything. Experience had shown that the militia was not sufficient, that troops taken from the *tercios* of Naples[216] and Lombardy[217] would have to be brought from Italy and more levied in Catalonia.[218] The reinforcements took time to arrive and the situation was not clearly reversed until January, 1570, when Don John, free at last to act, decided to strike the first major blow – a whole year after the insurrection had begun.

What had been accomplished in the meantime? Very little apart from the encouragement of hopes – notably that famine alone would bring the rebels to heel.[219] And Granada itself had been held, Don John having express orders not to leave the city until the end of the year.[220] This has been interpreted by some commentators as a stab in the back for Don John from his half-brother, Philip II: it is unnecessary to invent personal reasons for a policy sufficiently explained by extra-personal factors. Such an interpretation also disregards the very real fears of the authorities, voiced by Francés de Alava: 'God grant that before the dog [i.e. the sultan, whose naval preparations had been reported] can arm himself, the rebels of the Alpujarras will have been chastised.'[221] There was also the fear that the sedition might spread outside the kingdom and that the

John of Austria, Ferreyra, 23rd July, 1569, Archives of Gov.-Gen. Algeria, Register no. 1686, f° 175–179, the king of the insurgents says that every day he receives (as his personal share) six to ten Christian prisoners.

[215] Michel Orvieto to Margaret of Parma, Madrid, 1st April, 1569, A.d.S. Naples, Farnesiane, Spagna, fasc. 5/1, f° 242, reports the movement of the galleys towards Spain. Pescara to the Duke of Alcalá, 17th April, 1569, Simancas E° 1057, f° 53: on hearing of the disaster he has decided to send to Spain the squadron of Juan de Cardona. *Récit du succez et journée que le Grand Commandeur de Castille a eu allant avec vingt-cinq galères contre les Mores*, Lyons, Benoist-Rigaud 8° doc., 14 p., B.N., Paris, Oi 69 (1569), a pro-Spanish report; Fourquevaux, *op. cit.*, II, p. 75, 4th May, 1569: if it were not for the galley patrols, the Moors would flee to North Africa.

[216] H. Forneron, *op. cit.*, II, p. 178, these troops were probably commanded by Alvaro de Bazán, see next note. Sauli to the Signoria of Genoa, Madrid, 20th May, 1569, A.d.S. Genoa, Lettere Ministri, Spagna 4/2413; Philip II to Don John of Austria, Aranjuez, 20th May, 1569, *CODOIN*, XXVIII, p. 10.

[217] J. de Samaniega to the Duke of Parma, Madrid, 18th May, 1569, A.d.S. Naples, Farnesiane, Spagna, fasc. 5/1, f°s 256–257.

[218] *Ibid.*, f° 274, Samaniega to Parma, Madrid, 25th June, 1569.

[219] Sauli to the Signoria of Genoa, see note 216.

[220] Philip to Don John of Austria, Aranjuez, 20th May, 1569, *CODOIN*, XXVIII, p. 10. Granada was the only position protected, Castagna to Alessandrino, Madrid, 13th July, 1569, L. Serrano, *op. cit.*, III, p. 111.

[221] To Zayas, Paris, 2nd August, 1569, A.N., K 1511, B 24, no. 35.

Moriscos in Aragon 'might take leave of their senses as they have in Granada'.[222] Were they to do so, the Spanish army would not be facing 30,000 men (the presumed number of rebels at the beginning of August) but at least 100,000.[223]

It was not a question of 'stabbing Don John in the back', but rather of properly assessing the situation. It had taken time, quite some time for Philip to realize that the 'success' reported by a courier[224] and apparently decisive, was nothing of the sort, for the Moriscos could flee 'like deer' from the Spanish arquebuses[225] and be not a bit the worse for it; time to see that the rebellion had to be taken seriously, that the situation as it presented itself in the autumn (lowlands and cities in the hands of the Christians, mountains held by the rebels),[226] threatened to drag on dangerously long in view of the extensive preparations visibly under way at Constantinople.

This realization had come rather late in the day. But once again, Philip II was paralysed by his own apprehensions. The threat from the Turkish fleet obviously made it necessary to have done with Granada quickly, but it was a threat facing Italy quite as much as Spain, and in October, Juan de Zuñiga asked the king to send reinforcements to the Spanish army in Italy,[227] the land which had been drained of men for the benefit of Flanders and would have to be drained once more (in December) for Granada. Money, yet more money, was badly needed too and the vast expenditure in Flanders had already absorbed so much. Only when on the one hand a more definite threat of foreign intervention materialized and on the other when the war took a clear turn for the worse, did Philip resign himself to the necessary steps.

On 26th October, the nuncio was told officially that if the war continued through the winter, if it should spread to other Morisco regions and if the Turks should intervene, then Spain herself might fall into Moslem hands. The nuncio may well have thought that this admission was intended to wring further concessions from the Holy Father, notably the *cruzada*, but it undoubtedly corresponded to a very real anxiety.[228] Madrid was well aware of the contacts between the Moriscos and the rest of the Moslem world. During the autumn, it was learnt not only that the Morisco ambassadors on their way back from Constantinople, had been received at Algiers, where they had been promised thousands of arquebuses,[229] but also that three Jews, rich merchants who had come to the French court because of their interests in the moneys owing to Joseph Micas, had

[222] Madrid, 6th August, 1569, Fourquevaux, *op. cit.*, II, p. 101–102.
[223] *Ibid.*, p. 109, 19th August, 1569.
[224] *Ibid.*, p. 107.
[225] *Ibid.*, 19th August, 1569, p. 111.
[226] *Ibid.*, 17th September, 1569, p. 117–118.
[227] To the king, Rome, 14th October, 1569, L. Serrano, *op. cit.*, III, p. 163.
[228] Madrid, 26th October, 1569, *ibid.*, p. 180.
[229] 31st October, 1569, Fourquevaux, *op. cit.*, II, p. 128–129. C. Pereyra suggests (*Imperio español*, p. 168) that the English may have assisted the Moriscos. . . .

reported that the armada would sail in 1570 'a dar color y ayuda a los moros de Granada'. The request for military aid had been made to the Turks by the Morisco envoys, as well as on behalf of the kings of Morocco, Fez and 'three or four others of Barbary'.[230] This news coincided with reports reaching Madrid almost simultaneously that the Sharif was preparing a military expedition against the *presidios* in Morocco: taken together they engendered fears that a concerted Moslem invasion of Spain was about to take place.[231]

By now it was recognized by everyone, by the nuncio,[232] by the Tuscan agent Nobili,[233] and by Philip II himself, in his correspondence with Don John,[234] that the war in Granada was going badly. In any case the signs were clear. In December all the Moriscos were expelled from the city of Granada, an extreme and desperate measure[235] which makes one think that the information received in the Netherlands in the last days of the year was correct: that is that the Moors, sometimes armed with arquebuses, were making numerous incursions into the kingdom, so that neither at Granada, nor in Seville did the townspeople 'dare to put their noses out of doors'.[236] It was time to act. On 26th December, Philip decided, in order to be closer to the scene of operations, to hold the Cortes of Castile at Córdoba.[237]

One consequence of Granada: Euldj 'Alī takes Tunis. Philip's problems in Granada were to cost one of the Barbary 'kings' his throne. Charles V's protégé, Mawlāy Ḥasan, who had been replaced on the throne of Tunis by the emperor as a puppet against the Turks, had been overthrown by his own son, Mawlāy Hamīda. The new king, caught between Spaniards, Turks and his 'subjects', that is the people of Tunis and the Arabs, both nomadic and sedentary, of the South, had governed as best he could, which was not very well. Perhaps as a remark by Haëdo suggests,[238] he relied on the people as allies against the Tunisian lords, but at all events he appears to have betrayed and alienated practically everybody. After twenty years on the throne then, he had not a few enemies within his kingdom and his authority was more fragile than that of any previous

[230] F. de Alava to the king, Tours, 9th December, 1569, A.N., K 1513, B 25, no. 138. Grandchamp de Grantrye to Catherine de Medici, Const., 16th October, 1569, E. Charrière, *op. cit.*, III, p. 94, Toulon might possibly be placed at the disposal of the Turks.

[231] Dispatches from Spain, 19th December, 1569, Fourquevaux, *op. cit.*, II, p. 165.

[232] Madrid, 2nd October, 1569, L. Serrano, *op. cit.*, III, p. 161.

[233] Nobili to the prince, Barcelona, 4th December, 1569, A.d.S. Florence, Mediceo 4898, f° 550.

[234] Madrid, 26th November, 1569, *CODOIN*, XXVII, p. 38.

[235] Fourquevaux, *op. cit.*, II, p. 165.

[236] Ferrals to Charles IX, Brussels, 29th December, 1569, B.N., Paris, Fr. 16123, f° 297 v°, '. . . de sorte que ceulx de Grenade et moins ceulx de Séville n'ausent mectre le nez dehors'.

[237] Philip to Guerau de Spes, Madrid, 26th December, 1569, *CODOIN*, XC, p. 318.

[238] *Op. cit.*, p. 78.

ruler of Tunis. The victim was ripe for the kill and the war of Granada made sure that he fell into Algerian hands.

For although the people of Algiers had helped the Moriscos in their rebellion in 1569 – out of self-interest, for money and also out of religious zeal (the central depot for the many weapons donated for the rebel cause was a mosque in Algiers) – there is little indication that Euldj 'Alī, king of Algiers since 1568, was prepared to run any great risks for them. As Haëdo remarks, he had taken greater pains over the defence of his city than over Granada: perhaps on orders from Constantinople, but it seems much more probable that he was under pressure from Spanish agents: we have at least one set of instructions, given to a certain J. B. Gonguzza delle Castelle, who was sent to Algiers in 1569.[239] In any case, a large-scale attempt to help the Moriscos would have meant challenging Spain's maritime defences, a potentially costly operation. Perhaps too, Euldj 'Alī did not particularly wish to see Spain's economic blockade against him prolonged unduly.[240]

But above all why should he trouble to help others when the war of Granada offered such a heaven-sent opportunity to execute a project dear to the heart of every ruler of Algiers: the conquest of the entire North African Peninsula? Memories were revived of Barbarossa's attack of 1534 and the expedition led by Salīh Re'īs in another direction against Fez in 1554. Proof of the efficiency of Spanish intelligence and of its network in Algiers was the fact that Madrid knew all about Euldj 'Alī's plans, the return of the Morisco ambassadors from Constantinople,[241] the 'king's' problems with Deli Hasan, ruler of Biskra,[242] and finally the planned capture of Tunis. Indeed, on 8th October a Spanish captive in Algiers, Captain Hierónimo de Mendoza, said he had heard from a very reliable source that Euldj 'Alī was preparing to attack Mawlāy Hamīda. A second letter sent on 29th October confirmed the first and Philip II gave orders that Don Alonso Pimentel, the governor of La Goletta, be immediately informed.[243]

Meanwhile Euldj 'Alī had already left Algiers some time in October[244] with 4000 or 5000 janissaries taking the overland route through Constantine and Bône;[245] the season being too far advanced for safe shipping, he did not send an accompanying fleet. As he passed through Greater

[239] Simancas E° 487.

[240] Trade to Algiers was in fact forbidden. A French merchant arrested at Valencia, 1569, Simancas E° 487; no ships in Málaga because of the war. Inquisition of Granada to C° S° of the Inquisition, A.H.N., 2604, 17th March, 1570.

[241] Madrid, 31st October, 1569, Fourquevaux, *op. cit.*, II, p. 128–129.

[242] Dispatches from Spain, 19th December, 1569, *ibid.*, p. 165.

[243] Jerónimo de Mendoza to the Count of Benavente, Algiers, 8th October, 1569, Simancas E° 333; Sauli to the Signoria of Genoa, Córdoba, 26th February, 1570, A.d.S. Genoa, L.M., Spagna, 4/2513; Mendoza to Benavente again, 29th October, Simancas E° 487. Philip II wrote on the margin of this report: 'sera bien embiar con este coreo a don A° Pimentel'.

[244] *Memorias del Cautivo*, p. 2, September, not October.

[245] D. de Haëdo, *op. cit.*, 78 v°.

and Lesser Kabylia, many volunteers joined his army, notably several thousand horsemen, with whom he emerged on to the plain of Béjà, two short days' march from Tunis. Mawlāy Hamīda's soldiers scattered without fighting and the fallen king took refuge in his city but, feeling insufficiently secure there, made for the Spanish fortress of La Goletta with a few followers and what was left of his treasure. At the end of December, according to Haëdo (but a dispatch from Algiers gives the more likely date of 19th January),[246] Euldj 'Alī entered Tunis without a struggle. The Calabrian, well received by the people of Tunis,[247] occupied the royal palace and instituted his own rule in the city, ordering, threatening and punishing; in March[248] he returned to Algiers, leaving in Tunis a large garrison which was maintained at the expense of the citizens, under the command of one of his lieutenants, the Sardinian renegade Cayto Ramadan.[249]

But had it not been for the major alarm in Granada, had a Christian fleet been massed at Messina, the operation would have been extremely hazardous. Would Spain accept the new situation?[250]

Granada and the war of Cyprus. The capture of Tunis in January, 1570, was the consequence of an imbalance created by the war of Granada. This imbalance was also a factor in the war of Cyprus, the great event of 1570 and in the eventual formation of a league between Rome, Venice and Spain, the direct consequence of the Turkish attack.

For the Morisco war lasted throughout 1570, at least until 30th November, on which date Don John left Granada if not totally, at least very largely pacified. All along, the war had continued to be both difficult and costly. But with the new year it had taken a new turn. There is no doubt that the young Don John of Austria – he was twenty-three – was already a gifted leader of spirit and courage. The king had placed plentiful means at his disposal and the royal presence at Córdoba as from January, which shortened delays in communications and obliged the executors of orders to display more energy, presented definite advantages, not always appreciated perhaps by the courtiers and diplomats, who were obliged to live

[246] Dispatches from Algiers, 22nd February, 1570, Simancas E° 487; Palmerini, B. Com. Palermo, Qq D. 84, dates the event in 1569, but is often unreliable. News of the capture of Tunis did not reach Rome until the night of 8th–9th February, 1570. The bishop of Le Mans to the king, Rome, 13th February, 1570, B.N., Paris, Fr. 17.989, f°s 147 and 147 v°. The news reached Constantinople on 2nd April, 1570, Const., 7th April, 1570, Simancas E° 1058, f° 41. *Memorias del Cautivo*, p. 5, capture of Tunis dated January, 1570.

[247] Dispatches from Algiers, see previous note, the Calabrian was well received by the inhabitants of Tunis ('fue muy bien recebido de todos ellos').

[248] Haëdo says February; dispatches from Algiers, 1st–6th April, 1570, Simancas E° 487, report that he left Tunis on 10th March.

[249] *Memorias del Cautivo*, p. 5.

[250] Fourquevaux to the king, Córdoba, 22nd April, 1570, Fourquevaux, *op. cit.*, II, p. 216.

in the rear of the army and to grapple daily with the minor worries of food and lodging.[251]

The situation did not change overnight. The first major engagement, the siege of the little Morisco town of Galera, perched on a high and inaccessible hilltop to which the artillery could not easily do any damage[252] was not a success. The garrison fought with superhuman energy and the attackers themselves had to show uncommon courage to capture the position, after a terrible massacre. The fallen town opened up the road to the mountains: the victorious army set out on its march. But from the heights of the Sierra de Serón, Morisco forces poured down upon it and unconquerable panic made the victors of the previous day turn tail.[253] It was in the course of this near-disaster that Don John's tutor, Luis Quijada, met his death.[254]

This guerrilla war was bound to demoralize the regular soldier, to drive him with every fresh event towards cruelty, cowardice or despair. Don John speaks himself in March[255] of the demoralization and indiscipline of his troops.[256] Hardly had they been assembled than they deserted. The lure of plunder unleashed a spontaneous private war which spread like leprosy reaching even regions untouched by official war. Every city in Spain was crammed with Morisco slaves for sale: boatloads were sent off to Italy. But in the mountains rebel forces still numbered about 25,000, of whom 4000 were Turkish or Barbary troops. They still had abundant supplies (wheat cost 10 reals per staro and other cereals only 4, notes an informant):[257] they had figs and raisins and could count on rations from the brigantines and foists of Algiers.[258] Above all they were buoyed up by the hope of Turkish intervention.[259] Undoubtedly they were only protected from the superior forces of their adversary by the mountains. But that was no small obstacle. The two columns into which Spanish forces were divided, one under Don John himself, the other under the Duke of Sessa, made very slow progress. On 27th March, Sauli, giving

[251] Nobili to the prince, Madrid, 18th January, 1570, A.d.S. Florence, Mediceo 4849, f°s 10 and 11 v°: 'quivi dicono ch'è gran penuria di tutte le cose onde non è molto approbata questa gita como non necessaria', f° 9 v°; '. . . che noi andiamo in una provincia penuriosa di tutt'i viveri per cagione del mal recolto et della guerra de Mori: staremo a ridosso d'un esercito che già patisce infinitamente . . .'.

[252] Madrid, 3rd February, 1570, Fourquevaux, *op. cit.*, II, p. 190.

[253] Sauli to the Signoria of Genoa, Córdoba, 26th February, 1570, A.d.S. Genoa, L.M. Spagna, 4/2413.

[254] Don John to Philip II, Caniles, 19th February, 1570, *CODOIN*, XXVIII, p. 49; Don John to Philip, Caniles, 25th February, 1570, A.E. Esp. 236, f° 13.

[255] To Philip II, Tijola, 12th March, 1570, *CODOIN*, XXVIII, p. 79.

[256] Don John to Philip, 30th March, 1570, 6th May, 1570, *ibid.*, p. 83 and 89.

[257] Was Nobili's information (note 252) correct? Some wheat at any rate was sent to the rebels from Algiers, dispatches from Algiers, 1st and 6th April, Simancas E° 487.

[258] Dietrichstein to Maximilian II, Seville, 17th May, 1570, P. Herre, *op. cit.*, p.113, note 1. The king of Algiers had apparently promised to send five ships of supplies and munitions.

[259] Nobili to the prince, Córdoba, 27th March, 1570, A.d.S. Florence, Mediceo 4899, f°s 59 ff.

details of the latest progress, concluded a letter full of 'very good news' with the words: 'e con questo, li Mori restano del tutto esclusi della pianura', the Moors have been driven entirely out of the plain. . . .[260]

Trying to follow the true course of events, as the diplomatic correspondents in Madrid had to, is a maddening task: good and bad news by turns is reported with an apparent lack of consequence which in the end completely baffled the best observers and made Nobili write (roughly translated), 'The war with the Moors is like a course of hot and cold baths . . .'.[261] In May, 1570 for instance, quite near Seville, about ten thousand Moriscos, vassals of the Duke of Medina Sidonia and the Duke of Arcos, rose up: this was bad news. But it was soon learnt, with joyful surprise, that the rebels had not joined the insurrection in the mountains: the king had had the excellent idea of sending their overlords who had calmed them and persuaded them for the most part to go home. They had risen up only because the Spaniards, taking advantage of the proximity of the theatre of war, had been carrying them off in order to sell them as prisoners of war and had been stealing their goods and their wives.[262] A similar incident occurred in March, in a village of Valencia where a revolt broke out but collapsed again almost immediately. As these two examples indicate, the war was by no means confined to a limited area. But the chief danger for the Spanish government was in the Alpujarras, in and wild and untameable world where the dissidents had taken refuge. Mountain warfare, even in the twentieth century, has never been easy. This war of 1570, wrote Fourquevaux, 'consumes and burns Spain with a slow flame'.[263]

What arms could not achieve, diplomacy, or what would perhaps in a later age have been called the department of Moorish affairs, finally obtained. The first rebel leader had been assassinated, a fate also suffered by his successor. The captain general of the rebellion, Hernando el Habaquí, arrived in Don John's camp on 20th May, kissed the prince's hands and laid down his sword. A peace treaty was signed. The Moriscos obtained a pardon and permission to wear national costume but had to surrender within ten days and lay down their arms at specified places. Barbary citizens were allowed to leave for North Africa without hindrance.[264] As can be seen the terms of the truce were quite generous, which led to talk of the 'natural clemency of His Majesty'[265] but they are

[260] Sauli to the Signoria of Genoa, Córdoba, 27th March, 1570, A.d.S. Genoa, L.M. Spagna, 4/2413.

[261] Nobili to the prince, Seville, 16th May, 1570, Florence, Mediceo 4899, f° 94 v°.

[262] Nobili to the prince, *ibid.*, f° 95 v°; Juan de Samaniega to the Duke of Parma, Córdoba, 18th March, 1570, A.d.S. Naples, Carte Farnesiane, Spagna, fasc. 3/2, f° 356.

[263] Fourquevaux to the king, Seville, 22nd May, 1570, Fourquevaux, *op. cit.*, II, p. 222.

[264] Nobili to the prince, Córdoba, 25th May, 1570, A.d.S. Florence, 4899, f° 166 v° and ransoming of captives, P. Herre, *op. cit.*, p. 118.

[265] Zayas to F. de Alava, Ubeda, 4th June, 1570, A.N., K 1517, B 28, no. 70.

above all an indication of his anxiety to be free at any costs from this harmful struggle. Was it really peace at last?

By 15th June, 30,000 Moors had laid down their arms: both roundships and oarships had been placed at the disposal of the Turks who were returning to Africa[266] and an ultimatum was issued for the surrender of the remaining rebels, expiring on 24th June, the feast of John the Baptist.[267] But even by 17th June, the Inquisitors were beginning to complain: in Granada, Moriscos claiming to have repented were parading about with their arms and describing their exploits in public quite freely, boasting of the numbers of Christians they had killed and of all they had done 'to offend our holy Catholic faith'.[268] Many of them have surrendered, says another letter, but none of them has yet come to confess his sins to the tribunal of the Inquisition.[269] Was the capitulation merely a stratagem? Small foists from Larache continued to ship arms, in such numbers that Sancho de Leyva's galleys had already seized four or five of these little craft.[270] And while the embarkation of the Africans was delayed,[271] the Christian soldiers, who had not been paid, broke ranks with the usual consequences.[272] None of this exactly contributed to inspire the rebels with a healthy respect for or fear of the Spanish forces.

In fact sporadic war continued in the mountains with dangerous raids being launched against isolated Christians.[273] 'The Moors of Granada are gradually being subdued', wrote Sauli,[274] 'although there are still not a few who persist in their rebellion. 400 Turks have left for Barbary, but the Barbary Moors remain.' Two or three thousand Moriscos were still in the hills with their 'king' and declared that they would submit only if they were granted leave to stay in the Alpujarras, a concession to which the royal government would not agree. 'To subdue them', writes Sauli, 'I ventured the opinion that a campaign lasting at least a year would be necessary, for the said Moors have harvested much grain and have sown millet and other cereals. Our troops cannot prevent them from harvesting these as they have not sufficient numbers for such an operation.'[275]

In the circumstances, few could have any illusions about the 'pacification' of the kingdom. Don John said once more in a letter written on 14th August that there would be peace only when the Moriscos had been expelled.[276] But evacuating a whole kingdom was rather different from

[266] Dispatches from Granada, 16th June, 1570, Fourquevaux, *op. cit.*, II, p. 227.

[267] *Ibid.*, p. 226.

[268] The Inquisitors of Granada to the Supreme Council of the Inquisition, 17th June, 1570, A.H.N., 2604. [269] *Ibid.*, 9th July, 1570.

[270] Madrid, 29th June, 1570, Fourquevaux, *op. cit.*, II, p. 241, 'laden with rice, wheat and flour', *ibid.*, 11th July.

[271] Don John of Austria to Philip II, 2nd July, 1570, *CODOIN*, XXVIII, p. 110.

[272] *Ibid.*, p. 111.

[273] Fourquevaux, *op. cit.*, II, p. 241–242.

[274] Madrid, 13th July, 1570, A.d.S. Genoa, L.M. Spagna, 4.2413.

[275] *Ibid.*, 5th August, 1570.

[276] Don John of Austria to Philip II, 14th August, 1570, *CODOIN*, XXVIII, p. 126. Don John to Ruy Gómez, same date, *ibid.*, p. 128.

evacuating a town, an exercise carried out the year before at Granada. So at Madrid the official version was that the war was over. All the ambassadors conveyed this message back home,[277] while at the very same time, Don John of Austria, who was still on the spot, was discussing various ways of attempting to subdue the Moriscos of Ronda and to penetrate the Alpujarras.[278] In September, the chief items on his agenda were the destruction of the rebels' vines and orchards,[279] the pursuit of deserters – another headache – and at the same time the recruiting of fresh troops. For the war was dragging on, a fire that refused to be extinguished.[280] The rebel leader now no longer sought direct confrontations, but was content to slip from hill to hill. All the forts built on the mountains, all the garrisons quartered in them were powerless to prevent the insurgents slipping through the net and making premeditated attacks on the Christians.[281]

It was at this point that the Spanish government decided upon massive deportation. There can be no doubt that this measure was decisive. Zayas wrote on 13th October to Francés de Alava: 'the Granada affair can be considered by Your Lordship as virtually concluded, as is fitting to the service and reputation of His Majesty'.[282] This time it was true. At the beginning of November, Don John announced that the Málaga region and the sierras of Bentomiz and Ronda had been pacified.[283] In the meantime the deportations had been carried out. They affected 50,000 persons, perhaps more, and essentially depopulated the lowlands. That the operation was pitiful to see we may judge from the much-quoted words of Don John himself, who was nevertheless in favour of the expulsion. 'It was the saddest sight in the world', he wrote on 5th November to Ruy Gómez, 'for at the moment of departure there was so much rain, wind and snow that the poor folk clung together lamenting. One cannot deny that the spectacle of the depopulation of a kingdom is the most pitiful anyone can imagine. Well it is over now.'[284] The Medici envoy to Madrid, the knight Nobili, whose correspondence I have so often cited, clearly saw the significance of this inhuman but effective measure and wrote to the Grand Duke: 'the Granada affair is now finished and I would sum it up in a word: the Moors who had surrendered and those of the lowlands were keeping the war alive, for it was they who secretly provided the rebels with food'.[285] These were the people who were deported.

[277] Juan de Samaniega to the Duke of Parma, Madrid, 20th August, 1570, A.d.S. Naples, Carte Farnesiane, fasc. 5/1, f° 394.

[278] To Ruy Gómez de Silva, Guadix, 29th August, 1570, *CODOIN*, XXVIII, p. 133.

[279] Madrid, 20th September, 1570, Fourquevaux, *op. cit.*, II, p. 268.

[280] Dispatches from Spain, September, 1570, *ibid.*, II, p. 262–263; Madrid, 11th October, 1570, *ibid.*, II, p. 280.

[281] *Ibid.*, p. 277.

[282] A.N., K 1516, B 28, no. 7.

[283] To Philip II, 9th November, 1570, *CODOIN*, XXVIII, p. 140.

[284] Don John of Austria to Ruy Gómez, 5th November, 1570, 'Al fin, Señor, este es hecho', quoted by H. Forneron, *op. cit.*, II, p. 189–190; and by O. de Törne, *op. cit.*, I, p. 201.

[285] To the grand duke, Madrid, 22nd January, 1571, A.d.S. Florence, Mediceo 4903.

There now remained of the rebellion in the kingdom of Granada only a few thousand Moors living like brigands[286] and like them operating in bands. But were there not as many if not more in the Catalan Pyrenees? It may not have been perfect peace, but it was a situation approaching the norm, with the usual margin for inadequate law enforcement. The Moriscos had been resettled in Castile while to the now pacified kingdom Old Christians moved in to colonize the good land of Granada: costly though the war had been, the Christians at any rate gained something from it.[287]

On 30th November, Don John left Granada, the scene of his first military apprenticeship, never to see it again. On 13th December he was in Madrid,[288] where another task awaited him, the result perhaps of the war he had just ended. For if the Turks were at last attacking Cyprus (the cherished and already venerable project of their army chiefs of staff) was it not amongst other reasons because Spain seemed to be absorbed at the other end of the sea in her domestic conflict?

The early stages of the war of Cyprus.[289] We can for once see the workings of Turkish policy, in this winter of 1569–1570, with rare if not perfect clarity. For Turkey in the sixteenth century is historically speaking uncharted territory. We western historians can only see it from outside through the reports of the official and unofficial agents of western powers. But in 1569–1570 the Grand Vizier, Muḥammad Sokolli was intimate with the Venetian *bailo* and his personal policy was not identical with that of his government. The difference between these two lines of conduct allows us to penetrate further than usual into the heart of the Turkish Empire. Or so historians have said. Are they right?

The new emperor Selīm, as everyone knew when he succeeded to the throne, was anything but warlike. But custom demanded that the beginning of his reign be marked by a brilliant victory, the profits of which would enable him to construct and endow the mosques traditionally required of a new sovereign. We have already noted, though without venturing any definite explanation, the semi-inactivity of the years 1567, 1568 and 1569. In 1569, at the moment when the war of Granada first broke out, the demands of the Russian campaign and the large-scale operations in the Red Sea precluded any action in the West. But since the Moriscos had not yet laid down their arms by the autumn, the problem

[286] Sauli to the Signoria of Genoa, Madrid, 11th January, 1571, A.d.S. Genoa, L.M., Spagna, 4/2413, talked of 2500 Moors living as *bandoleri*.

[287] A. de Herrera, *op. cit.*, p. 341 ff.

[288] A. de Fouché-Delbosc, 'Conseils d'un Milanais à Don Juan d'Autriche', in *Revue Hispanique*, 1901, p. 60, note a.

[289] There is a superabundance of sources; see for example references in Hammer, *op. cit.*, Vol. VI; Paolo Paruta, *Hist. venetiana*, 2nd part, *Guerra di Cipro*; Uberto Foglietta, *De sacro foedere in Selimum*, libri IV, Genoa, 1587; Giampietro Contarini, *Historia delle cose successe dal principio della guerra mossa da Selim Ottomano a' Venetiani*, Venice, 1576.

of supporting them became urgent. Should the Turks send an armada to the Spanish coast? In the Peninsula such a move was indeed feared. But the Turkish fleet would require a base on either the French or the Barbary coast. The request for the use of Toulon was made quite openly, so openly indeed that one may ask whether the intention was not so much to obtain use of the port as to frighten Spain, which learnt of the application and reacted predictably. Did the Turks ever seriously consider helping the Moriscos?

In the first place, was direct assistance to the Granada rebels technically possible, at such long distance, given the evident need for winter quarters for the galleys? Paul Herre[290] in his learned work on the war of Cyprus thought that it was, but only in so far as he believed it was the policy sincerely desired and energetically pursued by the Grand Vizier, an imperial policy worthy of Sulaimān the Magnificent himself and proved by numerous documents. But what are these documents? The letters of the *bailo* who reports on his conversations with the Vizier. It is not entirely out of the question that the Vizier was deliberately misleading his interlocutor. The trust he apparently had in him, the confidences, favours and conversations which did not cease even with the definite rupture between Venice and Turkey, all have something very 'Balkan' and oriental about them. Above all did they not serve the interests of the sultan's general policy? To distract Venice from the coming peril, and to remain on good terms with her (for diplomacy could always be useful) – may not necessarily have been the aim of the deception, if deception there was; but I certainly do not believe Sokolli's utterances to be as sincere as Herre considers them, concerned as he is, along with many other historians, to demonstrate the decadence of Turkey; illustrating it by measuring the distance between the 'eagle's policy' of Sokolli (leave Venice alone but assist the Moriscos) and the mediocre short-sighted policy of Selīm (strike Venice at the outer edge of her empire, in Cyprus, where she is known to be vulnerable).

This version is to my mind a somewhat oversimplified explanation, based on very little evidence, of Turkish policy, so elusive at the centres of decision. (For example as early as 1563, Ragusan intelligence reported plans for the capture of Cyprus, by none other than Sulaimān the Magnificent.) Born a Bosnian, taken from his Christian parents at an early age, slowly climbing the ladder of the Ottoman administration, in 1555 vizier of the Divan, ten years later Grand Vizier and Selīm's son-in-law, Sokolli had grown up in a court where survival was difficult, under an implacable and greatly feared master. What an apprenticeship in self-control and dissimulation he must have served! He was said to be the friend of Venice: he certainly performed services for her – in return for very handsome rewards. Such actions had never committed a minister in the court of the Grand Turk. He was said to be peace-loving, indeed pacifist: something of an exaggeration for what he wanted was a *pax*

[290] Paul Herre, *Europäische Politik im cyprischen Krieg, 1570–1573*, Leipzig, 1902.

turcica, burdensome upon the conquered, but glorious for the Sublime Porte. Moreover, if Sokolli's policy really was the one attributed to him, how could he have retained control of affairs, staying at the helm while the boat had embarked on a different course? It has been said that his policy was pursued discreetly, hinted at only, tempered if necessary, and abandoned at awkward moments. But where is the evidence of such suppleness? Even if it had existed, it is almost impossible to believe that this duplicity – the betrayal or at any rate undermining of his sovereign's plans – could have allowed him to resist all his rivals, the vizier Lala Muṣṭafa, the sultan's former tutor, the admiral Piāle Pasha, a great intriguer, and finally the influential Jew Micas – when we are aware of constant bitter rivalries and dissensions in the Seray. In passing we may wonder what to think of the last mentioned of these adversaries, that shadowy figure, a dishonest creditor (at least in his claims on France), a man who would in modern times be called a born spy, if not a stage villain. On him too, the West – but only the West, alas – possesses letters and documents: Micas was in contact with the Grand Duke of Tuscany, with Genoa, with Spain, perhaps with Portugal. Was he a traitor? Or, to extend our hypothesis, was he too acting on orders, playing a calculated role (not forgetting to turn it to personal profit) in a policy which was more concerted than historians have allowed? Paul Herre merely dismisses him as an individual 'whose hands were anything but clean',[291] acting solely out of spite or for gain: out of spite against Venice, which had sequestered some of his wife's property; for gain since as of 1569[292] he was said to be anxious to become king of Cyprus, and to settle a colony of his co-religionists there. He may well have had such motives – but that is not the whole story. Without wishing certainly to paint him as white as his biographer and devoted admirer does,[293] let us admit that we are ill-equipped to judge him, for lack of evidence, and more generally that it is risky in the extreme to approach a major chapter in Turkish history along the rather doubtful paths of biography and anecdote.

If hypothesis there must be, there is nothing to disprove the existence of a Turkish policy centred undeviatingly and from an early date on an objective – Cyprus: a policy which dictated certain actions in which every actor had his role to play; one to soothe the fears of Venice, another to handle relations with Spain or France. The Turks put up smoke screens on every side. They did not deliberately discourage the Moriscos but did not nudge the Barbaresques out of their sluggish semi-support for them; they blamed the latter for running risks at Tunis,[294] but practised on their own account a very similar policy. For although far from eager to aid the waiting Moriscos either directly or indirectly by an attack on the Spanish

[291] *Op. cit.*, p. 13.

[292] E. Charrière, *op. cit.*, III, p. 87–88; Iorga, *op. cit.*, III, p. 141.

[293] J. Reznik, *Le duc Joseph de Naxos, contribution à l'histoire juive du XVIe siècle*, Paris, 1936.

[294] Constantinople, 7th April, 1570, Simancas E° 1058, f° 41.

fleet, the Turks sought to profit from the help the Moriscos unwittingly gave them, to pursue their own ends without risk.

Determined not to let any opportunity slip by, they were also seeking in the same year to renew the Franco-Turkish alliance which had cooled considerably with the rapprochement between France and Spain in 1567–1568. This must be the reason behind the request concerning Toulon in 1569, which was a way of testing the ground, as well as for the extravagant voyage to Constantinople of the no less extravagant Claude Du Bourg. Arriving at an abnormally delicate moment in France's eastern policy, and complicating matters further with his 'folles imaginations' his 'absurd fancies',[295] he quickly quarrelled with the official ambassador, tried to overthrow the banker Micas and to win the favour of the Grand Vizier Sokolli; in the meantime, he was negotiating on behalf of Genoa and, on his way back through Venice, during the winter of 1568–1569 when he was accompanied by a Turkish ambassador-extraordinary, he nevertheless carried in his pocket the 'renewal' of the Capitulations. For Turkey was at this time extremely anxious to see once more in the West a France following traditional policy and favouring Turkish interests. The usual picture presented – with unconscious western arrogance – by historians is of France attracting the Turks and exploiting them for her own ends. The other side of the coin shows Turkey soliciting France and luring her towards the East. In 1569 and 1570 for instance, there was some talk of the crown of Poland for the Duke of Anjou and the prince of Transylvania as a bridegroom for Marguerite de Valois.[296]

But how, in a France torn by the third civil war, could the Franco-Turkish *entente* be revived? The French king, committed now to the Catholic side, could not make such a volte-face. The best evidence for this is that he gave his consent to the arrest in Venice both of Du Bourg and the Turk accompanying him.[297] Du Bourg wrote letter upon letter, promising the earth – he had such things to tell! He was finally allowed to leave Venice – but only to be arrested once more at Mirandola. Not that that prevented the king of France from boasting that poor Du Bourg had acted as mediator between the Signoria and the Turks![298]

However, after the fall of the Guises and of the intransigent Catholic party, civil war was to halt for a while. The armistice of 14th July was soon followed by the conciliatory Edict of Saint-Germain on 8th August.[299]

[295] Charles IX to Du Ferrier, 6th October, 1571, B.N., Paris, Fr. 16170, f° 32 v° ff.

[296] Paul Herre, *op. cit.*, p. 25 and 146.

[297] Madrid, 10th March, 1570, Fourquevaux, *op. cit.*, II, p. 202, the *chā'ūsh* was said to be intending to ask the king of France to have supplies of food made ready on the coasts of Provence and Languedoc, for the Turkish armada. 'There is much talk about it.' Salazar to H.M., Venice, 5th December, 1570, A.N., K 1672, G 1, no. 159, Claude Du Bourg is still being held at Mirandola, there is talk of setting him free; on the extraordinary career of Du Bourg, see above, Vol. I, p. 374–375.

[298] Instructions from Charles IX to Paul de Foix, 12th April, 1570, B.N., Paris, Fr. 16080, f° 166, quoted by P. Herre, *op. cit.*, p. 161.

[299] E. Lavisse, VI, 1, p. 113; Philip II to F. de Alava, Madrid, 3rd September, 1570, A.N., K 1517, B 28, no. 89: 'pernicioso concierto y pazes que esse pobre Rey ha

Catherine then turned towards the Protestants, to such a degree that before long there was talk of 'infernal marriages' – (Alava's expression)[300] – that of the Duke of Anjou and Elizabeth, and of Henri of Navarre and Marguerite of Valois. On the diplomatic level, France was now proceeding towards a grand anti-Spanish alignment, the Italian-born queen not merely shifting her allegiance between the two parties within France, but also between the two great powers on whom they relied so heavily. There was a sudden clarification of the French situation which is almost always incorrectly explained solely in terms of personal squabbles.

Meanwhile what of the Turks?

Never perhaps had there been a winter so tempestuous, so wild and hostile to the rapid circulation of news, as that winter of 1569–1570.[301] The bad weather, adding more than usual to distances, strengthened the wall of silence behind which the Turks operated. It had been known however for months that they were furiously re-arming[302] and the evidence suggested that a major attack was on the way – but where? Malta, La Goletta, Cyprus? While speculation still continued in the West, the Turks had already struck Venice wherever they could find a vulnerable

hecho con sus rebeldes que me ha causado la pena y sentimiento que podeis considerar viendo la poca cuenta que se ha tenido con lo que tocava al servicio y honora de Dios. . . .'

[300] To the king, Paris, 2nd September, 1570, A.N., K 1517, B 28, no. 87.

[301] In February and on into March, the galleys were unable to leave Naples harbour, the weather being 'the worst we have had in this season for many years', viceroy of Naples to the king, Naples, 11th March, 1570, Simancas E° 1058, f° 34.

[302] Dispatches from Corfu, received 10th January, 1570, Simancas E° 1058, f° 13, reported that great preparations were under way for an expedition against Malta: 20 mahonnas were under construction at Nicomedia, 22 large culverins were being cast at Constantinople, 10,000 oarsmen rounded up in Anatolia, the fleet will number 250 sail, of which 175 are galleys, 'et si murmura anchora per la Goleta'. Constantinople, 21st January, 1570, Simancas E° 1058, f° 19: Cyprus is commonly reported to be the destination of the fleet; there is a risk of looting by Algerine corsairs along the coasts of Dalmatia. Algiers, 26th January, 1570 (by Valencia), Simancas E° 334, the armada will make for Cyprus or La Goletta. Bishop of Le Mans to Charles IX, Rome, 2nd(?) February, 1570, B.N., Paris, Fr. 17989, f°s 145 v° and 146: the Turkish armada will sail against Malta or more probably La Goletta. Chantonnay to the king, Prague, 15th February, 1570, *CODOIN*, CIII, p. 450–453: the emperor has heard from a spy that the Turk will attack Cyprus and will not carry aid to the Moriscos. On hearing the news concerning Cyprus, Maximilian is talking of a possible alliance between himself, the king of Spain, Poland and the Reich. Thomas de Cornoça to the king, Venice, 25th February, 1570, Simancas E° 1327: the Turks intend to attack Cyprus. Castagna to Alessandrino, Córdoba, 11th and 22nd March, 1570, quoted by P. Herre, *op. cit.*, p. 61–62, note 1: the Turkish fleet will enter Spanish waters. Dispatches from Corfu, received in Naples on 31st March, 1570, Simancas E° 1058, f° 30, the Turkish armada will attack Cyprus. Pescara to the king, Palermo, 12th June, 1570, on the mission of the Knight of Malta Barelli, who had been sent to the East in search of information, 'mayormente presuponiendo que algun intento avia para dar calor a las cosas de Granada . . .', Simancas E° 1133. The said Barelli recommended to Miques, viceroy of Naples to the Duke of Albuquerque, Naples, 24th June, 1570. Constantinople, 18th May, Simancas E° 1058, f° 66: against Cyprus. Viceroy of Naples to the king, Naples, same date, *ibid.*, f° 64.

point. Because of the winter, the news took some time even to reach Venice although the city was on the alert.

Right up to the last moment, Venice had refused, against all the evidence, to believe in her bad luck. Bad luck was the word for it: for Venice, so often mocked for her prudence, Venice the courtesan who slept with the Turk, could hardly be accused of bad management. She was the victim not of her policy, but of her scattered and sprawling empire, a mere string of coastal stations, of her economy, which forced her, as it was to force free-trading England in the nineteenth century, to depend on the outside world, to derive her livelihood from her import and export trade. Venice's policy could never under any circumstances be that of the great empires, Turkish and Spanish, of which she was in a way the frontier, empires that were rich in manpower, revenues and space. The policy of the Republic, constantly under review, is therefore only comprehensible in the light of *raison d'état*. By the second half of the sixteenth century however, this sensible prudence had outlived its usefulness, for the world had moved on, the tide had turned against Venice, against her political formula and indeed her whole way of life.

In 1570, she had been at peace for thirty years: it had been a prosperous peace but it had affected her political structure and weakened her defences, more than either she or anyone else suspected. Her system of fortifications in particular, so formidable in the past but now outdated, the disorganization of her military administration reported every day with despair by army leaders, and above all of her fleet, left her at a much lower level of preparedness than a few decades earlier.[303] Having grown used to the peace which, although constantly threatened over the years was always saved at the last moment, familiar with the many intrigues of the Balkan world, Venice had come to consider a large deployment of means unnecessary. She no longer took seriously the policy or the hostile gestures of the Turks.

At first the Turks tried to intimidate her. They cherished the hope that Venice would yield without a struggle, that the first hostile act would lead to immediate negotiations. So Venetian merchants were arrested in January and their goods sequestered.[304] In the Morea, the measure seems to have been applied in mid-February. Ships suffered the same fate: two Venetian cargo vessels at Constantinople, already fully laden, were emptied of their cargoes and requisitioned for the fleet. But this is very far from the 'seizure of Venetian ships' described as a massive operation by some historians. In winter, ships in a position to be seized before hearing of the measure cannot have been many in number. Nor did the measure in itself seem particularly alarming. The ships had received promises of indemnity and their requisitioning was a fairly normal event in the sixteenth century. If Venice took fright – as she naturally did, arming herself as a result – it was at the news that the sultan was about to lead

[303] Alberto Tenenti, *Cristoforo da Canal, la marine vénitienne avant Lépante*, 1962, see esp. p. 175 ff.

[304] Paul Herre, *op. cit.*, p. 16, gives the date as 13th January.

his army through Anatolia and Caramania and that he had positioned seven hundred more janissaries at Castelnuovo.[305] This did not prevent – indeed the reverse – the Venetian government at Cattaro from sending gifts to the son of Muṣṭafā Pasha, who was passing through Castelnuovo.[306] It had of course been public knowledge in Constantinople since 1st February that the Turks intended to demand the unconditional handing over of Cyprus, in the name (even at this early date!) of the historical rights of Turkey. But the news, although it appears in Grandchamp's correspondence on 9th February,[307] did not reach Venice until March, by which time the war had begun.

Wishing no doubt to follow up their diplomatic initiative with a more precise threat, the Turks had indeed attacked the Venetian possessions. On Monday, 27th February, bad weather drove on to the Adriatic coast near Pescara, the small Venetian boat of a certain Bomino di Chioggia, who had left Zara on Sunday the 26th, and had been driven at great speed out of his way, south of Ancona, by a terrible *temporal*. The master, who was well known to a number of merchants, reported that 25,000 to 30,000 Turks had unexpectedly attacked Zara and that the merest chance had allowed two galleys, thanks to their cannon, to repulse the assailants and sound the alarm.[308] The figures cannot be verified, but it was the harsh truth that an attack had been launched on the long, winding line of Dalmatian outposts, containing 60,000 souls (according to a report of 1576). Upon these posts the Turks hurled themselves, causing much havoc.[309]

Finally official news of the Turkish demands reached Venice.[310] The *chā'ūsh* Ubat, who left Constantinople on 1st February,[311] passed through Ragusa towards mid-March[312] and was admitted to an audience with the Venetian Senate on the 27th. But he was not allowed to make a formal statement of his errand (the agenda had been prepared in advance) and was instead obliged to listen to stern words from the Venetians. The Turkish demand was rejected by 199 votes out of 220.[313] For Venice had

[305] Thomas de Cornoça to the king, Venice, 26th February, 1570, Simancas E° 1327.

[306] J. López to the king, Venice, 11th March, 1570, Simancas E° 1327. No further letters from the *bailo*; his arrest; Turkish invasion of Dalmatia; war certain. López to the king, 16th March, 1570, *ibid.*, they are said to be sending someone from Constantinople to Venice to demand the cession of Cyprus.

[307] To the king, E. Charrière, *op. cit.*, III, p. 101–104.

[308] [Doctor] *Morcat* (of the Consejo de Capuana) *que al presente esta por governor en las provincias de Abruço al duque de Alcala*, Civita de Chiete, 28th February, 1570, Simancas E° 1058, f° 37.

[309] J. López to the king, Venice, 1st April, 1570, Simancas E° 1327, Turkish 'correrias' around Zara. Cavalli to the Senate, Córdoba, 1st April, 1570, rumours that Zara has fallen (P. Herre, *op. cit.*, p. 85, note 2); 9 (?) castles taken by the Turks, Sokolli to Charles IX, Constantinople, 16th November, 1570, E. Charrière, *op. cit.*, III, p. 137.

[310] See note 306.

[311] P. Herre, *op. cit.*, p. 16.

[312] L. Voinovitch, *Depeschen des Francesco Gondola, 1570–1573*, Vienna, 1907, p. 21.

[313] *Ibid.*

resolved to fight. In mid-March, she had sent an ambassador extraordinary to Philip II[314] and, alerted by the Venetians, Pius V had sent an envoy, Luis de Torres, to the Spanish king; his intervention was to be decisive.[315]

Venice made plain her warlike intentions. She fitted and launched her reserve fleet, manned her galleasses, provided Fausto's galleon with magnificent artillery, levied troops, accepting any offered by the cities of the mainland, sent to Cyprus an expeditionary force which the Turks failed to intercept, dispatched several thousand men to Dalmatia and sent an engineer to Zara.[316] All this deployment of force was carried out with much fanfare before the spring, but not necessarily with definite intent to make use of it. The Turks did not land on the southern tip of Cyprus until July. Until then, Venice was content to abide by the rules: above all not to appear intimidated, to meet threat with threat, and violence with violence. At the announcement of the confiscation of the goods and persons of her nationals, she replied in kind.[317] But when the Turks yielded on this point, she was quick to do likewise.[318] It is true that the new doge elected at Venice on 5th May, Pietro Loredano, represented the war party.[319] But the peace party was far from being silenced. Behind the unanimity of 27th March, a strategic front which no doubt seemed politic to the pacifists themselves, there was ample scope for splits, divisions and about-turns. If the Turks had withdrawn, Venice would promptly have forgotten her armaments, the interests of Christendom and her dear sister in religion, Spain.

Spain at the beginning of 1570 was both anxious and handicapped: anxious about the massive scale of Turkish armament and handicapped by the war of Granada, which was still engaging a large part of her land and sea forces. She was also handicapped by the events in the North, before which she was a helpless bystander, neutral in a sense, since the Duke of Alva had decided against large-scale military intervention; but it was an extremely costly neutrality.

Deprived of the free use of most of their galleys, which were now busy patrolling the coast of Granada, and having left Italy dangerously unprotected, the Spaniards' reaction to the news from the Levant was very similar to the pattern for alerts before Malta: all positions in Naples, Sicily and North Africa were put in a state of readiness. This too was an expensive operation, but it could not be avoided: there was no reason to think that the Turks would spare Spanish possessions. The dispatches insisted that they would not attack but could dispatches always be relied

[314] F. de Alava to the king, Angers, 4th April, 1570, A.N., K 1517, B 28, no. 59.

[315] R. B. Merriman, *op. cit.*, IV, p. 126–127 and the book by A. Dragonetti de Torres, *La lega di Lepanto nel carteggio diplomático inedito di Don Luys de Torres*, Turin, 1931.

[316] Juan López to the king, 11th March, 16th March, 31st March, 1570 (for Fausto's galleon), 9th June, 1570, Simancas E° 1327.

[317] Paul Herre, *op. cit.*, p. 27–28.

[318] *Ibid.*

[319] H. Kretschmayr, *Geschichte von Venedig*, *op. cit.*, III, p. 54.

on? Nor was there unanimity among the dispatch-writers. It was not difficult to introduce murky water into the channels of information. It has been claimed that Venice often used her notorious intelligence service as a weapon to alarm the rest of Christendom and to sustain the psychosis of the Turkish peril. It is certainly the case that in the sixteenth century only a very limited credence was placed on the reports she spread. It was just as easy for the Venetians to organize deception from the enemy citadel, Constantinople, as from Venice, and if need be to instigate dispatches intended to find their way to the Catholic King himself. Be that as it may, on 12th March the viceroy of Naples wrote: 'I have received a letter from Constantinople, dated 22nd January, from one of our most reliable agents. . . . It confirms me in my opinion, to wit that in spite of what we have heard of the opening of hostilities in Dalmatia, the great armada now under preparation is not intended merely for the Venetians'.[320]

So the Spaniards assembled as many galleys as they could muster in southern Italy and for lack of the usual Spanish troops, levied Germans for Milan and Naples and Italians for the galleys and Sicily. They sent men, supplies and munitions to La Goletta. 'I have decided', wrote Philip to Chantonnay on 31st March,[321] 'to levy two regiments of Germans, one for Milan, and the other for Naples', for 'my States in Italy are at present so short of troops' that the Turk would be able to inflict severe damage upon them. The regiment sent to Naples, embarking in Gian Andrea Doria's galleys at Genoa, arrived on 3rd May[322] and was immediately directed to the coastal positions. But by the end of June, when the Turkish threat was seen to have materialized in a specific area, the men were dismissed[323] after Venice had declined to accept them on her payroll.[324]

As for the galleys, after Doria's arrival at Naples in April, about sixty were assembled in the south of Italy – sixty out of the hundred Spain had available at the time, representing the sum total of the squadrons of Sicily, Genoa and Naples, the last commanded by the Marquis of Santa Cruz. If we are to believe the complaints of the commanders, the galleys were poorly manned, with an inadequate supply of convicts[325] and few if any troops. In July, Gian Andrea Doria obtained permission to levy two thousand Italians at Naples to man his ships. Meantime the fleet had already had to make two trips to La Goletta.[326]

The second of these voyages had been aimed at Euldj 'Alī, who was known to have put in to Bizerta with 24 or 25 galleys. Gian Andrea Doria

[320] Simancas E° 1058, f° 35.

[321] *CODOIN*, CIII, p. 480–481.

[322] G. Andrea Doria to the king, Naples, 3rd May, 1570, Simancas E° 1058, f° 31.

[323] Viceroy of Naples, 19th July, 1570, Simancas E° 1058, f° 80, the Germans were dismissed on 26th June.

[324] Viceroy of Naples to the king, Naples, 2nd August, 1570, Simancas E° 1058, f° 91.

[325] Santa Cruz to the king, 26th April, 1570, Simancas E° 1058, f° 46.

[326] Simancas E° 1060, f°s 1, 39, 49, 51, 206, E° 1133, 19th March, 1570, 6th May, 1570.

was hoping to capture them with his 31 reinforced galleys, but the coast of Bizerta had been fortified by the Calabrian in the meantime and the game was too well protected. The Christian galleys turned about, called briefly at La Goletta, then made for Sardinia where they were to relieve some troops before returning to Naples.[327] Before long they received orders to cross to Sicily, then to sail eastwards. Philip II, despite the pleas of Pescara, the new viceroy of Sicily who was anxious to make an attempt on Tunis,[328] had yielded to the requests of the Pope and the Venetians. There would be an attempt to relieve Cyprus.

The relief of Cyprus. The Turks had landed on the island in July. On 9th September the capital Nicosia fell into their hands. Only Famagusta, better fortified, remained in Venetian possession, and here there were considerable forces, capable of resisting for quite some time. What with delays in relaying information, it was quite late, about midsummer, before the full implications of the problem of Cyprus became clear: would Venice go to the rescue of the island? Could she confront the Turk alone, to save her precious i land with its sugar, salt and cotton? It was in her interest, at all events, to incite the West against the Turk, to make this local war dependent upon a more general conflict, the threat of which might discourage her enemy, oblige him to loosen his hold and perhaps accept a compromise. She also wished to avoid a direct alliance with the other colossus of the Mediterranean: to avoid signing a league like that of 1538[329] which had left a bitter and lasting memory. Venice had suspected a degree of collusion between Doria and Barbarossa on that occasion; Paruta was to claim, in 1590, that there had been a betrayal. But how could she remain unattached to either side when she was directly under attack?

To persuade Spain to take up arms was no easy matter. Granvelle on first hearing the news had declared himself opposed to giving any assistance to Venice. The negotiations went through Rome and were originally three-cornered discussions, but the astonishing and overpowering personality of Pius V quickly dominated the talks; he was soon acting alone, with all the more energy since his Catholic policy had always been a militant one. Disappointed in 1566 as we have already seen, the Pope's zeal found recompense in these later developments in the Mediterranean. He brought all his dynamic energy to the problem and by his rapid assumption of command, sweeping aside all obstacles, he forced a decision upon the two parties between whom he was in theory merely a mediator.

[327] Pescara to the king, Palermo, 17th June, 18th June, 26th June, 1570, 16th July, 1570, Simancas E° 1133. G. A. Doria to the king, La Goletta, 2nd July, 1570, Simancas E° 484.

[328] Pescara, 17th April, 1570, Simancas E° 1133 and other letters dated 6th May; see also, letters from the viceroy of Naples, 14th August, 1570 (Simancas E° 1058, f° 97), Philip II to the viceroy of Sicily, 23rd September, 1570 (E° 1058, f° 217), 18th October (*ibid.*, f° 220).

[329] The point is well made in H. Kretschmayr, *op. cit.*, III, p. 53.

He had little time, as we may imagine, for the petty calculations in which Venice, preoccupied with saving her precious plantations and salt pans, was engrossed. He had already put pressure on her through his nuncio in March, 1570 on the occasion of the famous Senate session, and had immediately granted her tithes from the Venetian clergy, to aid the Republic's war effort. He had agreed without delay to creat a papal fleet, of which the Tuscan galleys were in 1571 to form the nucleus. He had agreed without delay to the transport of timber to Ancona for the construction of galleys.[330] And without delay he had dispatched to Philip II Luis de Torres, his confident and intimate friend, one of the many impassioned ecclesiastics surrounding the Holy Father.[331] Specially selected for the task since he was Spanish and had personal friends in Philip's Council, he was sent off with all haste: his instructions are dated 15th March. In April he was being received in Córdoba by Philip II: in April – that is at the height of the war of Granada;[332] and in Córdoba – that is only a few days' journey from the scene of war. So he arrived in an atmosphere of religious passion, at a time of high emotion over the destiny of Christendom,[333] which was under attack both on northern shores (with the Reformation) and on the very coasts of the southern sea, dramatically cornered by these two wars which the far-flung hostilities in the Atlantic were soon to unite. The exalted spirit of the times shines forth not only in the letters written by Pius V – where it would be surprising to find the contrary – but also in the letters written by his envoy. One has only to read them to discover the living reality of passion and crusading fervour which formed the background to the war of Granada.

The negotiations however were slow. What Pius wanted was a league: not a makeshift alliance but a formally constituted league. 'It is clear', ran Torres' instructions, in words which he was bidden to repeat to the king, 'it is clear that one of the principal reasons why the Turk has quarrelled with the Venetians is that he thought he would find them unaided, without any hope of uniting with Your Majesty, who is so occupied with the Moors of Granada. . .'. But the best calculations can sometimes produce the wrong answer. If Philip eventually agreed to join the League against Islam, it may have been precisely because in 1570 – the year of Granada – age-old passions had been revived in Spain.

Naturally neither political, nor financial interests were forgotten.

[330] Cardinal of Rambouillet to Charles IX, Rome, 8th and 12th May, 1570, E. Charrière, *op. cit.*, III, p. 112–114.

[331] L. Serrano, *op. cit.*, I, p. 51.

[332] Nobili to the prince, Córdoba, 22nd April, 1570, Florence, Mediceo 4899, f° 74. But the newcomer was not very well received by those Spanish circles 'tenendolo de razza non molto antica'; Fourquevaux, *op. cit.*, II, p. 219–220, Torres' audience on 21st April, the promise of 70 Spanish galleys; the king's final reply to be given at Seville. 50 galleys according to the cardinal of Rambouillet (to the king, Rome, 22nd May, 1570, B.N., Paris, Fr. 1789, f° 176).

[333] The Italians (the duke of Savoy and the duke of Urbino) were urging Philip II to join an alliance with Venice, Paul Herre, *op. cit.*, p. 88.

Venice was the frontier of Christendom. Would it be the stronger for her fall – or if she agreed with the Turks one of the separate treaties that the French government was offering to negotiate for her? The Pope did not neglect any of these arguments. 'The Venetian fortresses', he wrote to the king, 'are the *antemural* of the fortified positions of the Catholic King'; joining the League would mean, for Philip 'having at his service the states of Venice, her men, her arms, her fleet'. As for financial interests, they were plain to see. Since 1566, the king of Spain had been receiving the large sums of the *subsidio* (500,000 ducats a year paid by the Spanish clergy) but the bull granting the *cruzada* had not been renewed, which meant a loss of more than 400,000 ducats every year,[334] and the *excusado*, granted in 1567, had never been levied, for fear of violent opposition. Luis de Torres brought with him, in the name of the Supreme Pontiff, the grant of the *cruzada*, which had been withheld until this moment because of scruples of conscience. This substantial addition to the Spanish budget no doubt weighed heavily in the king's decision.

Philip II, generally accounted so slow to make up his mind, gave his agreement in principle one week after his interview with Torres[335] to what was certainly, both for Spain and for Venice, and indeed for Turkey and the whole Mediterranean, the greatest military initiative for many years.

The relief of Cyprus[336] was attempted even before the formal conclusion of the League, and in the most inauspicious circumstances for the entire operation had to be improvised: the Papal fleet, hastily assembled, despite the protests of Zuñiga in Rome, was certainly something of a scratch force. It was probably the Pope's intention, by having his own galleys present, to prevent the Spaniards from conducting the expedition as they pleased and as they had in 1538–1540. Another hasty decision, more serious this time, was the appointment of the commander of the allied fleet, Marcantonio Colonna,[337] a Roman gentleman, Grand Constable of Naples and therefore a vassal of Philip II,[338] much preoccupied with favours to be obtained from Spain (and which he had already tried to procure by two rather unsuccessful trips to Madrid). He was a soldier, not a sailor; for a brief period during his youth he had commanded his own galleys. But Pius V, once having chosen him, obstinately insisted on keeping him at the head of the Christian fleet. It was a grave decision: in war-time one cannot afford to improvise a commander-in-chief.

Nor should one improvise a war fleet, and yet this is what Venice had to do. She rapidly assembled a large fleet, but it was less impressive than it sounded, for the Venetian war-machine had deteriorated during the long years of peace. Her naval forces were moreover dispersed among various

[334] L. Serrano, *op. cit.*, I, p. 53, note 3. [335] *Ibid.*, p. 54.

[336] For the whole of this paragraph see B.N., Paris, Ital. 427, f°s 197 v° and 198.

[337] Cardinal of Rambouillet to Charles IX, Rome, 5th–30th June, 1570, E. Charrière, *op. cit.*, III, p. 115–116. Instructions given to M. de l'Aubespine (June, 1570), B.N., Paris, Fr. 1789, f° 181: 'The lord Marc Antonio Colonna whom the Pope has made general of all his galleys to the great wonder of many. . . .'

[338] L. Serrano, *op. cit.*, I, p. 70.

squadrons and her reserve fleet, in dry dock in the Arsenal, was unmanned. If Venice was amply provided with material – ships, galleys and galleasses, artillery, she was extremely short of manpower and totally lacking in food supplies. An army required oarsmen, crewmen, soldiers aboard the galleys, barrels and chests of supplies. They were all lacking and the Signoria was unable either to act quickly or to compensate for her lack of manpower by altering her tactics. She had only briefly glimpsed the immense future of the roundship, the galleon or the super-galley, when armed with artillery. The emergency relief sent to Cyprus in roundships in February had arrived safely, despite the season, and on the way the ships' cannon had blasted the Turkish galleys.[339] But the lesson had not yet been learnt. And it was too late now, in view of the imminence of the danger, to liberate herself from centuries of tradition and from the notion that the fighting man was the perfect instrument and engine of combat.

The 60 galleys placed under the command of Hieronimo Zane which left Venice on 30th March, had had great difficulty in reaching Zara on 13th April.[340] Here they remained for two months, until 13th June, doing virtually nothing unless one counts a few patrols against the Uskoks and pirate raids on Ragusan territory,[341] consuming to no effect the supplies which it was a problem to replenish in Zara, a port without a hinterland, and failing even to prevent the Turkish galleys from pillaging the Albanian coast. There were perhaps good reasons for this inaction: the desire to protect Venice, the fear of winter storms in the Adriatic which might permit swift raids and rapid crossings, but not the sailing in formation of large fleets; and also the fear of encountering the Turkish armada which was based in Greece, or so it was said, in order to protect its communications with Cyprus.

When summer came, Zane received orders to sail straight for Crete where he was to assemble all the Venetian galleys, which had orders to rendezvous at the island. The admiral was expressly forbidden to stop on the way for any engagement whatsoever: that would have given the Turkish fleet both the idea and the time to reach the Adriatic. On arriving at Corfu the Venetians were informed of the impending arrival of the papal and Spanish fleets. They carried on towards Crete where they expected, according to instructions, to find plenty of men and supplies. But when they arrived in the island in August, nothing was ready. Was it simply negligence and lack of foresight, or had there been real difficulties, an impression one receives from the abundant complaints in a number of Venetian reports? The recruitment of crews and galley slaves from the Archipelago for Venetian ships was becoming increasingly difficult. The Spanish and papal forces entered the port of Suda on 31st August. The

[339] J. López to the king, Venice, 1st August, 1570, Simancas E° 1327.

[340] L. Serrano, *op. cit.*, I, p. 75.

[341] Numerous letters from Ragusa to the general of the Venetian fleet, Hieronimo Zane, 7th, 13th, 17th, 18th April; 3rd, 16th May, 1570, Ragusa Archives, L.P., I, f°s 168, 168 v°, 169, 171, 174, 175 v°, 177, 198, 200 v°, 201, 202.

papal ships were hardly any better manned than the Venetians. The Spanish galleys however had been assembled at Messina where they had taken on their full complement of oarsmen and troops.[342]

What was to be done? In his talks with Luis de Torres, Philip II had at first promised only what he was asked for: the retention of his galleys in Sicily. When after further talks the decision was made to send ships to the Levant,[343] new orders were dispatched to Gian Andrea Doria. He received them on 9th August along with notice of his subordination to Marcantonio Colonna.[344] The news that he was to serve as second-in-command cannot have pleased the admiral, since it meant risking his fleet in the seas of the Levant, in particular his own galleys, which under the terms of the *asiento* were not replaced in case of loss. And leaving in August with orders to proceed all the way to Cyprus would certainly mean being exposed to autumn storms on the way home.

So it was not with a very good grace that the orders were executed. Doria's 51 galleys joined the little papal squadron at Otranto[345] on 20th August. The assembly of the entire fleet on the north coast of Crete was not completed until 14th September.[346]

The port of Suda, which had been chosen for the rendezvous, was convenient but short of supplies. At the first general review the lack of preparation of the Venetians was patent and disputes broke out. The Venetians, instead of displaying their strength on the open sea, where there was no possibility of falsifying the number of troops, chose to hold the review in the harbour, with poops turned landward, thus making it possible during the review for troops to cross from one ship to another.[347] And disagreements between the commanders were also becoming apparent. But since the Venetian admiral had received orders to attempt the relief of Cyprus, whatever the cost and alone if necessary, the allied fleet decided to sail eastwards. It was not intended to make straight for Cyprus however for fear of clashing with the Turks. The admirals were thinking in terms of an attack on Asia Minor or the Dardanelles, so as to divert the enemy fleet away from Cyprus, then to prevent its return by positioning their fleet between the Turks and the island.

So the fleet set sail for Rhodes.[348] It was a huge force: 180 galleys and 11

[342] 2000 men levied in Naples for the galleys of Gian Andrea Doria and Santa Cruz, viceroy of Naples to the king, Naples, 19th July, 1570, Simancas E° 1058, f° 82; Nobili, 16th May, 1570, A.d.S. Florence, Mediceo 4899, f° 99 v° and 100.

[343] Sauli to Genoa, Madrid, 13th July, 1570, A.d.S. Genoa, L.M. Spagna, 4/2413.

[344] The order was given after intervention of the nuncio, L. Serrano, *op. cit.*, Vol. III, p. 448, cf. also *ibid.*, III, p. 461–463; on Doria's discontent, cardinal of Rambouillet to the king, 28th August, 1570, E. Charrière, *op. cit.*, III, p. 118.

[345] Simancas E° 1058, f°s 98 and 99.

[346] Gian Andrea Doria to the king, 17th September, 1570, Simancas E° 1327. The Rectors of Ragusa to the ambassadors at Constantinople, 11th September, 1570, Ragusa Archives, L.P.. I, f° 242 v°, the papal and Spanish galleys had passed by Corfu on 18th August.

[347] Gian Andrea Doria, 17th September, 1570, see preceding note.

[348] *Memorias del Cautivo*, p. 7.

galleasses (not to mention smaller ships and transports), 1300 cannon and 16,000 troops. Despite the deficiencies in both the Venetian and the papal squadrons, a successful action could have been fought, had it not been for the deep divisions among the commanders, if Marcantonio Colonna, the last-minute admiral, had been a real leader and if the prudent advance of the fleet had not been further slowed down by Doria, who cunningly doubled his tactical precautions.

So when, on reaching the coasts of Asia Minor, the news reached them that Nicosia had been captured on 9th September,[349] that virtually the entire island was controlled by the Turks, except Famagusta which was still holding out, the leaders of the armada decided to turn for home. The autumn storms gave the fleet a terrible tossing on the way back to Crete, just as they were tossing the victorious Turkish fleet on its way back to Constantinople. The allies would not consider wintering in the island, for lack of supplies; they were obliged to retreat to Italy. There again, navigational difficulties were very considerable; Doria successfully managed to bring all his galleys home to Messina,[350] but the Venetians suffered substantial losses (13 galleys on the way back to Crete and some sources say 27)[351] and Marcantonio succeeded in bringing back in November only three of the twelve galleys he had led out.[352]

One may imagine the disagreeable impression made by this unsuccessful voyage, the recriminations and debates it inspired.[353] In Rome and Venice, the enire blame was laid on Doria's shoulders – it was a good opportunity to attack the king through his admiral. In Venice, the peace party was gaining the upper hand. The Spaniards naturally lost no opportunity of launching counter-accusations. And Granvelle wrote to his brother Chantonnay: 'Colonna knows no more about the sea than I do'.[354] The Signoria meanwhile with its habitual severity was punishing all its own unsuccessful servants, from the troop commander Pallavicino and the admiral, Hieronimo Zane, who died shortly after being imprisoned, down to subalterns – punishments and disgraces which contributed not a little to the smart turn-out of the Venetian fleet in 1571.

During the winter of 1570 then the future of the League seemed seriously compromised. Agreed in principle, without as yet a single ruler having signed it, it seemed to have dissolved of its own accord without ever having existed.

[349] L. Voinovitch, *op. cit.*, p. 33; the news reached Madrid in November; Eligio Vitale, *op. cit.*, p. 127: probably on the 16th.

[350] Pescara to the king, Palermo, 22nd October, 1570, Simancas E° 1133.

[351] Francisco Vaca to the Duke of Alcalá, Otranto, 1st November, 1570, Simancas E° 1059, f° 6.

[352] L. Serrano, *Liga de Lepanto*, Madrid, 1918–1919, I, p. 81.

[353] E. Charrière, *op. cit.*, III, p. 122–124; F. de Alava to the king, Paris, 12th November, 1570, A.N., K 1518, B 28, no. 42.

[354] *Correspondance de Granvelle*, IV, p. 51, 14th December, 1570.

CHAPTER IV

Lepanto

Lepanto was the most spectacular military event in the Mediterranean during the entire sixteenth century. Dazzling triumph of courage and naval technique though it was, it is hard to place convincingly in a conventional historical perspective.

One certainly cannot say that this sensational feat was the logical outcome of preceding events. Should one then, with Hartlaub, glorify the role of Don John of Austria, a Shakespearean hero snatching victory single-handed from the jaws of fate? Regretfully I cannot accept that as an adequate solution either.

People have always found it surprising – and Voltaire found it amusing – that this unexpected victory should have had so few consequences. The battle of Lepanto was fought on 7th October, 1571; the following year the allies were unsuccessful at Modon. In 1573, Venice abandoned the struggle, exhausted. In 1574, the Turks triumphed at La Goletta and Tunis, and all dreams of crusade vanished into the wind.

But if we look beneath the events, beneath that glittering layer on the surface of history, we shall find that the ripples from Lepanto spread silently, inconspicuously, far and wide.

The spell of Turkish supremacy had been broken.

The Christian galleys received a tremendous crop of prisoners to man the oars, enough to bolster their strength for many years to come.

Everywhere active Christian privateers re-emerged and asserted themselves.

Finally, after its victory in 1574, and above all after the 1580s, the great Turkish armada fell into decay. The naval peace which reigned until 1591 proved disastrous, as the fleet simply disintegrated in port.

To claim that Lepanto was entirely responsible for these multiple consequences is perhaps going too far. But it certainly contributed. And part of its interest to us, as a historical experience, is perhaps as a glaring example of the very limitations of 'l'histoire événémentielle'.

1. THE BATTLE OF 7TH OCTOBER, 1571

The League, that is the alliance formed to wage a united struggle against the Turk, was eventually signed on 20th May, 1571. After the clashes of the previous summer, the mutual distrust of the future allies, their widely divergent interests and disharmony not to say hostility, it is extraordinary that agreement was reached at all.

A delayed start. The Spaniards accused Venice of intending to make peace with the Turks as soon as her interests should warrant it; and, like the Spaniards, the papal party distrusted the Signoria with its reputation for duplicity. The Venetians meanwhile, as they wrestled with crushing problems, recalled without enthusiasm the parallel of 1538–1540. Even when all these obstacles had been overcome, during the summer of 1571, among the flood of reports passing through Venice, the most improbable rumours were still circulating: the Spaniards were said to be preparing to take action against Genoa, then against Tuscany and even Venice herself. The origin of these reports, who was inventing them and to what end they were being broadcast are of course matters of speculation. Perhaps they were no more than the spontaneous products of popular suspicion.

The commissioners responsible for concluding the alliance had a heavy task. Spain was represented by Cardinals Pacheco and Granville and Juan de Zuñiga, who were all three in Rome when on 7th June, 1570,[1] they received the royal order (dated 16th May) investing them with their new functions. Venice dealt through her regular ambassador at Rome, Michael Suriano, who was replaced by a new man, Giovanni Soranzo, in October. The Pope for his part nominated Cardinals Morone, Cesi, Grasis, Aldobrandino, Alessandrino and Rusticucci, the last two attending the meetings in an unofficial capacity. The negotiations did not proceed smoothly. Between the first session on 2nd July, 1570 and the last, they were suspended three times: between August and October, 1570; in January–February, 1571; and finally, just when everything seemed to be settled, between March and May, 1571. During the latter two suspensions, Venice, despite her denials, was trying to reach a separate agreement with the Porte. In January, following the disappointments of the autumn campaign, she had dispatched the secretary of the Senate, Jacopo Ragazzoni,[2] to Constantinople, where contacts between Sokolli and the Venetian *bailo* had never been broken off. This initiative delayed the signing of the western alliance: Venice would not give her final consent until she was absolutely sure there was no hope at Constantinople.

So the diplomats who met more or less regularly in Cardinal Alessandrino's salons did not dictate the course of events, experienced and skilful though they might be. Their role consisted of watching their neighbours, composing long reports and then, depending on the orders they received, either smoothing away or provoking difficulties. Bound by their instructions, their hands were tied even further by the need to consult their respective governments for all major decisions, so to the normal processes of diplomacy were added the long delays caused by distance.

Had they been left to themselves, they would have reached agreement fairly quickly, especially in the presence of the Pope who was bent on the

[1] L. Serrano, *op. cit.*, I, p. 86, note 2.

[2] Silva to the king, Venice, 21st April, 1571, Simancas E° 1329; Silva to the king, 21st June, 1571, *ibid.*; *Relatione sull'Impero Ottoman di Jacopo Ragazzoni*, 16th August, 1571, Albèri, *Relazioni . . .*, III, 2, p. 372 ff.

alliance. This much is clear from the first phase of the talks: at the outset, the papal negotiators smoothed the way by suggesting that the text of the 1537 agreement, updated, should be used as a basis for discussion. At the beginning of September, it was rumoured that the League had been formed.[3] And it was true that all the major problems had been discussed. It had been agreed that the League would be concluded for a period of at least twelve years. It would be both offensive and defensive; although aimed in the first place against the Turks, it would also be directed against the vassal states in North Africa, Tripoli, Tunis, and Algiers. This was at the express request of the Spaniards who wished by this means to preserve their future freedom[4] in what was their own particular sphere of action. Other points on which agreement was reached were that the supreme command of the allied fleet would be given to Don John of Austria; that the expenses incurred in joint action would be divided into six parts (as in 1537); three to be the responsibility of the Spanish king, two that of Venice and one that of the Pope. As far as supplies were concerned, Spain would open her Italian markets to Venice, promising reasonable purchasing prices and undertaking not to increase taxes and other export duties. Venice, who could not do without Turkish grain unless she had access to Apulian or Sicilian supplies had particularly insisted on this point.[5] Finally, all parties to the alliance were forbidden to negotiate separate treaties with the Turks without the previous consent of the other signatories.

The proposed agreement was sent to the respective governments for examination and amendment, hence the first suspension of sittings, August to October.

On 21st October, the talks re-opened. But in the interval, the fruitless naval campaign in the Levant had taken place. Philip II, nevertheless, after prolonged examination, decided to empower his negotiators to sign the treaty, with some amendments. Venice on the contrary reacted violently, going back on all that had been agreed before. She sent a new representative and insisted on reopening discussions on every point with less goodwill than ever, deliberately introducing rhetorical quarrels and pointless digressions. Every detail which might provide a pretext for not signing the treaty was questioned again: the price of food supplies, the limits on the powers of the commander-in-chief, the operational strategy and the division of financial responsibilities among the allies. The Spanish delegates, Granvelle in particular, responded with increasing irritation. They could afford not to be conciliatory, for the long winter lay ahead.

[3] Commissioners to the king, Rome, 8th September, 1570, L. Serrano, *op. cit.*, IV, p. 6.

[4] On a plan to attack Bizerta and Tunis, Pescara to the king, Palermo, 20th March, 1571, Simancas E° 487.

[5] The Venetians had already made some approaches in 1570, viceroy of Naples to the king, 4th February, 1571, Simancas E° 1059, f° 178. Orders were given to provide grain for the Venetian fleet which was wintering in Crete. On the complexities of military supplies, Don Juan de Zuñiga to the Duke of Alva, 17th July, 1571, Simancas E° 1058, f° 81.

Finally, in December, over the minor question of Don John's second-in-command (should it be the commander of the Pope's galleys or the Grand Commander of Castille?) the talks broke down altogether, to the Pope's bitter disappointment.

This second interruption lasted several weeks: the Catholic King having finally agreed that the Pope should select Don John's lieutenant from three names put forward by Spain, the commissioners began to sit again in February. A new text was produced at the beginning of March. But Venice, still waiting to see the results of Ragazzoni's mission to Turkey, postponed her signature, under various pretexts, until May. Finally, on the 20th of that month, signatures were exchanged and five days later, on 25th May, 1571, the League was officially proclaimed from the Basilica of St. Peter in Rome.[6]

The signatories had agreed in principle to a *foedus perpetuum*, a *confederación perpetua*. Having made this declaration, they were content with providing for a military agreement of three years (1571–1573) the allies undertaking to send out every year a combined fleet of 200 galleys and 100 roundships, carrying 50,000 infantry and 4500 light horse. Although directed principally against the Levant, the League did not exclude but on the contrary envisaged possible expeditions against Algiers, Tunis or Tripoli. Concerning expenses, in the event of non-payment by the Holy See, Venice and Spain would pay respectively two-fifths and three-fifths of the total costs.[7] As for supplies, the text merely stipulated that 'reasonable' prices would be asked and export taxes and other duties would not be increased. The allies were not supposed to negotiate separate peace treaties.

This, very briefly is the tale of the lengthy negotiations preceding the formation of the League of 1571, beset by problems and watched from afar with great attention and anxiety. That France should have observed it more critically than other powers is sufficient in itself to suggest that French policy was once more preparing to take a path hostile to the House of Habsburg. On 5th August, 1570 (that is between the armistice of 14th July and the peace of 8th August), Francés de Alava wrote: 'The French hope that no agreement will be reached. The Venetians, they say, are complete fools if they sign that treaty and do not retain their freedom to treat with their great enemy the Turk. Here in France, everyone is doing his utmost to prevent the league from taking shape and to encourage an entente between Venice and the sultan. If they continue in this manner,

[6] Original text in Latin, L. Serrano, *op. cit.*, IV, p. 299 ff. Spanish translation, B.N., Madrid, Ms. 10454, f° 84; the text is also given in D. Dumont, *Corps universel diplomatique*, V, p. 203 ff; Simancas Patronato Real, no. 1660, 25th May, 1571; H. Kretschmayr, *Geschichte von Venedig*, *op. cit.*, III, p. 59; L. Voinovitch, *op. cit.*, p. 3; Cardinal de Rambouillet to Charles IX, Rome, 21st May, 1571, E. Charrière, *op. cit.*, p. 149–150.

[7] *Relatione fatta alla maestà Cattolica, in Madrid, alli XV di Luglio 1571, di tutta la spesa ordinaria che correrà per la lega in 200 galere, 100 navi et 50 mila fanti ogn'anno*, Rome, undated, in-4°, B.N., Paris, Oc. 1533.

I would not be surprised if next year they offer Toulon to the Turks'.[8] On 28th August, when optimistic rumours about the speedy conclusion of a Christian alliance were circulating, the cardinal de Rambouillet maintained that he had not changed his opinion, which is, he wrote, that 'it will look very fine on paper, but . . . we shall never see any results from it'.[9] It is true that much the same was thought in Venice at the time[10] and the emperor too was extremely sceptical about the Signoria's conduct after signing the treaty, if indeed she signed it at all.[11]

Then, after the return of the fleet from the Levant, when serious difficulties were becoming apparent, most people inclined to the view that nothing would come after all of the interminable talks at Rome. At the end of December, the Pope himself, usually so optimistic, did not conceal his discouragement from the cardinal de Rambouillet, the French ambassador.[12] The nuncio in Madrid was equally downcast. The Tuscan agent in the same city declared 'frankly' that he had very little hope of the League being formed. The Spaniards, it seemed to him, were only participating in the talks with the aim of receiving the *cruzada* and the *escusado*: whenever the talks approached definite settlement, they withdrew; they would make fresh overtures when the nuncio appeared to have 'cooled off' a little too much, but in such general terms that the latter could not 'make head nor tail of them'.[13] This was at the end of January: March brought better news. But as we have already seen, now it was the turn of the Venetians to cause delays. In April and May, there was great impatience at Madrid where everyone wondered what could be holding things up. It was not until 6th June that a courier brought the great news.[14]

France: an unknown diplomatic factor. The long gestation of the alliance gave the new French policy time to take shape. I say new, for since at least 1559, throughout ten years of eclipse and discredit, France had been absent both militarily and politically from the Mediterranean. Her absence is evident in many details: the Venetian attitude after 1560; that of Joseph Micas who had persuaded the sultan to have French vessels seized at Alexandria in 1568; and, that of France herself, who had apparently lost all interest in the Mediterranean. But although ravaged by civil war, France was still far from the morass into which she looked ready to sink at the end of Henri III's reign. There were elements of a determined royal policy still alive and ready to rise again.

[8] Francés de Alava to Philip II, Poissy, 5th August, 1570, A.N., K 1516, B 27, no. 55.

[9] 28th August, 1570, B.N., Paris, Fr. 23 377, copy.

[10] Julian López to the king, Venice, 10th August, 1570, Simancas E° 1327.

[11] Count of Monteagudo to the king, Speyer, 30th October, 1574, *CODOIN*, CX, p. 98–110.

[12] On the Pope's optimism, Rambouillet to the king, Rome, 4th December, 1570, B.N., Paris, Fr. 17989; on his discouragement, the same writer, 19th December, *ibid.*, copy.

[13] Nobili to the grand duke, Madrid, 22nd January, 1571, A.d.S. Florence, Mediceo 4903.

[14] Nobili and del Caccia to the prince, Madrid, 12th April–14th June, 1571, *ibid.*, F. Hartlaub, *op. cit.*, p. 71.

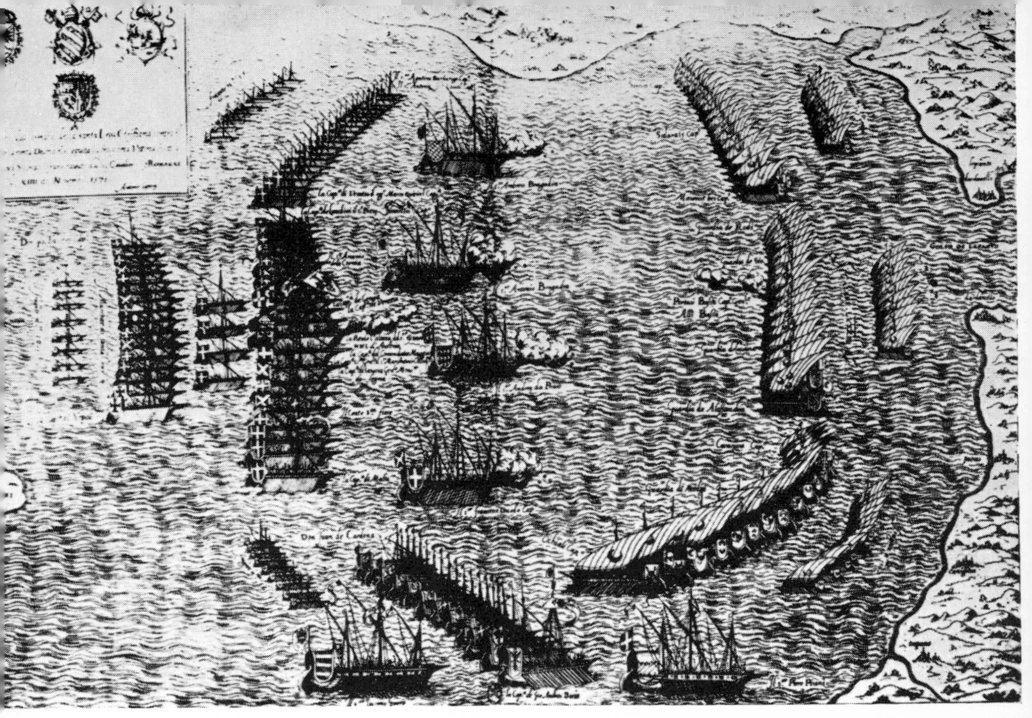

38 & 39. THE BATTLE OF LEPANTO. *Top*, plan of the battle, B.N., Paris, C 6669. *Bottom*, artist's impression, mural in Palace of the Marquises of Santa Cruz, Ciudad Real

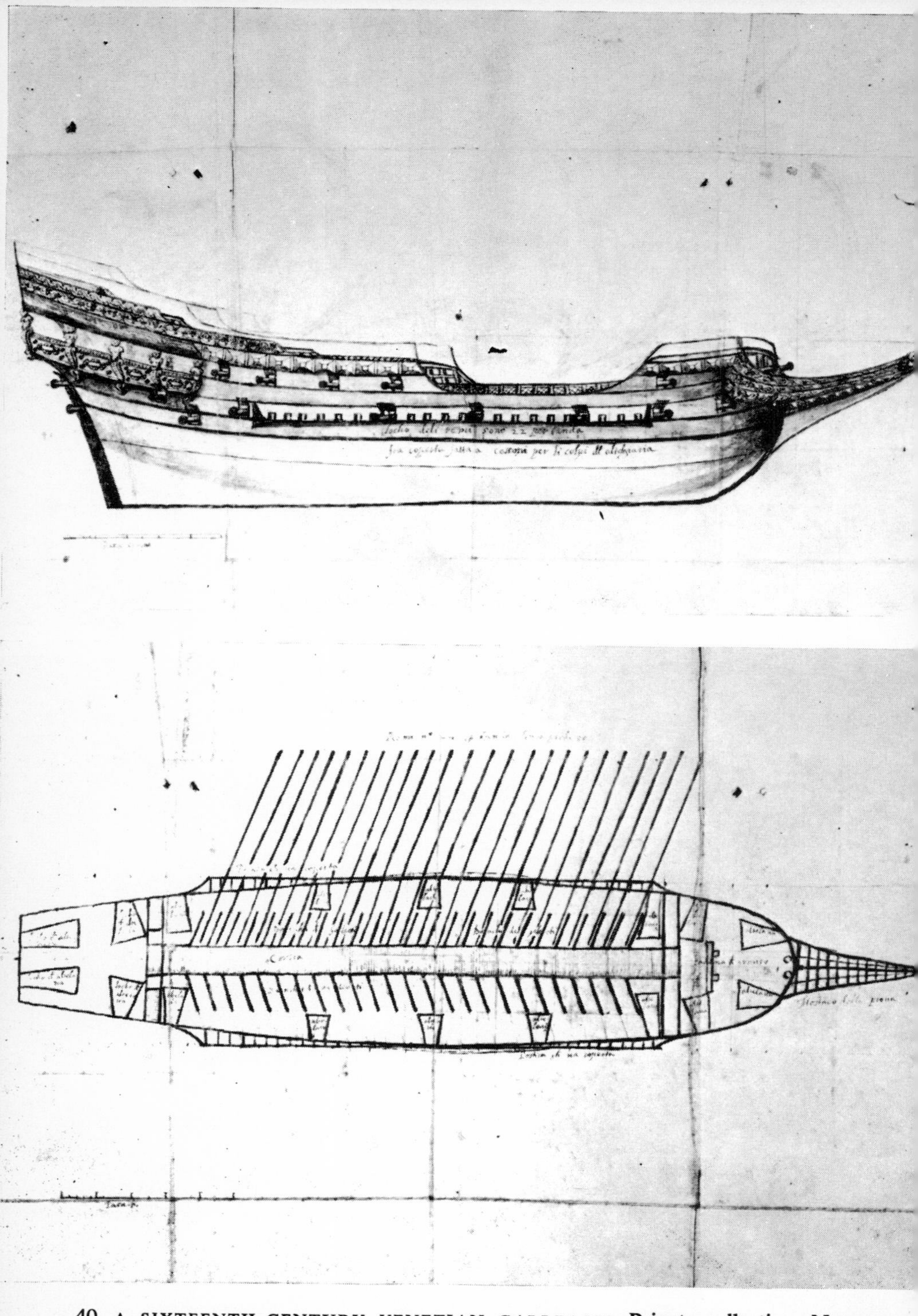

40. A SIXTEENTH-CENTURY VENETIAN GALLEASSE. Private collection. Note the disposition of the guns. The superior fire-power of the Venetian galleasses was decisive in the victory at Lepanto

As early as April 1570, France, despite her cautious handling of the Claude Du Bourg affair, had offered herself to Venice as a mediator – with surprising rapidity; after all the war of Cyprus had hardly begun and the French civil war had not yet ended. It is true that a policy of national reconciliation was already being formulated. The Protestants and the 'Politiques' were unanimously in favour of it, as were all those who represented the French king abroad. In 1570 and even at the beginning of 1571, these new tendencies had barely begun to emerge and were concealed behind general considerations relating to peace and the higher interests of Christendom. But a change is detectable very early in the royal correspondence which reveals a firmness of tone lacking for many years.[15] Charles IX had been converted to the ideas of Téligny and the Admiral. He dreamed passionately of breaking with Spain and intervening in the Netherlands. Undoubtedly he still kept such intentions secret – but not entirely successfully, for the Tuscans noticed the change. It was a change which affected the whole country where, as the commander Petrucci wrote to Francesco de' Medici, 'l'umore è contro al re di Spagna'.[16]

It is true that in this matter the Tuscans were not merely observers or confidants but accomplices if not instigators. The grand duke of Tuscany, his brand new title conferred on him by Pius V in 1569, felt himself to be isolated, kept at a distance both by the emperor and the Catholic King. Being particularly anxious about the intentions of the latter, he had for some time been seeking support throughout Europe, including Protestant countries.[17] One thread of this network leads to La Rochelle and the circle around the Admiral and Teligny. He may even have tried to work on the Turks: certain malicious tongues accused him of pulling strings behind the Jew Micas and thus being responsible for the war of Cyprus. The Spaniards in any case kept a close watch on him, feeling some anxiety for their *presidios* in Tuscany.[18]

Meanwhile, at the beginning of the year 1571, everyone continued to act as if peace was uppermost in his thoughts. Philip II and Charles IX exchanged ambassadors. At the end of January, Enrique de Guzman, count of Olivares, set out from Madrid.[19] Five months later, in June, Gondi arrived at the Spanish court to seek assurances that Philip did not seek any quarrel with France.[20] Even Alva, in words at least, was conciliatory.[21] Perhaps these words were indeed a little too honeyed, intended to contradict

[15] See for instance memorandum addressed to Fourquevaux, 15th February, 1571, Célestin Douais, *Lettres à M. de Fourquevaux . . . 1565–1572,* 1897, p. 314–343.

[16] Abel Desjardins, *Négociations diplomatiques avec la Toscane*, 1859–1886, III, p. 655 ff; 10th May, *ibid.*, p. 669.

[17] Alava to Zuñiga, 24th June, 1571, concerning Coligny and Téligny, A.N., K 1520, B 29, no. 24.

[18] Viceroy of Naples to Philip II, Naples, 3rd March, 1571, Simancas E° 1059, f° 60.

[19] His instructions, dated 29th January, 1571, Madrid, A.N., K 1523, B 23, no. 51.

[20] Philip II to Alava, Madrid, 30th June, 1571, *ibid.*; Nobili and del Caccia to the prince, Madrid, 30th June, 1571, A.d.S. Florence, Mediceo 4903.

[21] Reply by Duke of Alva to the Sieur de Montdoucet, B.N., Paris, Fr. 16127, f°s 3 and 4 (not published by L. Didier, whose collection begins in 1571).

the rather different story told by actions (like the assurances which the cardinal of Lorraine in Rome felt obliged to repeat to the Venetians, that all rumours of an approaching war between France and Spain were untrue).[22] The rumours were so widespread in Italy that the king of France in turn had to deny them. But was it merely coincidence that in March, 1571, the frontier of Navarre was being inspected by Vespasiano Gonzaga and Il Fratino, the architect of La Goletta and Melilla,[23] while in response to the cries of alarm of the Biraga at Saluzzo and the reinforcement of the French garrisons of the said Saluzzo and in Piedmont,[24] the governor of Milan, the Duke of Albuquerque on 11th April occupied the marquisate of Finale? The news reached Paris on about 9th May, where 'both Catholics and non-Catholics felt it very deeply' wrote Francés de Alava to the governor of Milan.[25] But from what I hear, they intend to complain not to His Majesty but to the Duke of Savoy and to Your Excellency'.[26] It would be excessive to claim that by this gesture the Spanish had effectively blocked any future French attack on Italy. But it was a warning. In France public opinion was outraged.

Now real talk of war began. The Duke of Savoy complained to Philip II that some enterprise against Piedmont was being planned at the other side of the Alps.[27] It was learned in Spain that the French galleys had been ordered from Bordeaux to Marseilles,[28] that troops were being moved towards the Alps[29] and lastly that some of the 'most prominent Protestants were urging the king to attempt some enterprise in the Netherlands'[30] where, moreover, the stream of French Huguenot immigrants was reaching alarming proportions.

Was it war at last? The answer was no, for outright war would have required a Protestant league and massive European support. Rumours at which Europe scoffed, such as that of the marriage of the queen of England and the Duke of Anjou, were no substitute.[31] War would only have been possible if Germany had been ready to rise[32] and Tuscany resolved on fighting, which was not the case. Tuscan agents, all smiles, schemed at Madrid as elsewhere, tunnelling right to the heart of the Spanish govern-

[22] Paul Herre, *op. cit.*, p. 163, note 1.

[23] Nobili to the prince, Madrid, 31st March, 1571, A.d.S. Florence, Mediceo 4903.

[24] Note by Antonio Pérez, Madrid, 8th May, 1571, A.N., K 1521, 830, no. 56.

[25] A.N., K 1521, B 20, no. 58.

[26] The culprit was Alonso de la Cueva, Duke of Albuquerque, who had acted without royal permission, the ambassadors were instructed to tell the French if the latter complained: Zayas to Alava, Madrid, 16th May, 1571, A.N., K 1523, B 31, no. 75. See also 31st May, 1571, Fourquevaux, *op. cit.*, II, p. 355, 'not a word is breathed about Finale in this court', F. de Alava to the king, 1st June, 1571, A.N., K 1520, B 29, no. 2.

[27] Nobili and del Caccia to the prince, Madrid, 10th May, 1571, Mediceo 4903.

[28] F. de Alava to the Duke of Albuquerque, Paris, 27th April, 1571, A.N., K 1519, B 29, no. 69.

[29] *Ibid.*

[30] Philip II to F. de Alava, 17th April, 1571, A.N., K 1523, B 31, no. 67.

[31] F. de Alava to the Duke of Albuquerque, Paris, 17th May, 1571, A.N., K 1521, B 30, no. 68.

[32] Alava to Albuquerque, 27th April, 1571, *ibid.*, no. 69.

ment through their network of 'confidants'.[33] But if war was not actually declared, there were plenty of alarms. Chiapin Vitelli, who travelled through France on his way back to Flanders in May, 1571, received an unfavourable impression from his journey and saw fit to warn Francés de Alava as soon as he reached Paris.[34] The latter had just learnt for his part that the king of France was sending to Turkey a man of no less standing and experience than the bishop of Dax. What was this semi-heretic, friend of the Protestants, likely to do in Turkey, but use his influence against Spain and Christendom, try to settle the Transylvanian problem (the occasion for every imperial intervention) and resolve any conflicts between the emperor and the Turk or between the Venetians and the Ottomans? Once more, French diplomacy was taking the distant road to Turkey.

But the new ambassador did not seem to be in any hurry to take up his post. On 26th July he was at Lyons;[35] on 9th September[36] he was presenting to the Doge of Venice the letter entrusted to him by Charles IX. The king offered to 'bring about' through the individual whom the texts insist on calling the bishop 'd'Acqs', 'a good peace, or else a truce so durable that the said peace would follow after'. Would that not be better than 'remaining at odds with such a powerful enemy, so near the lands in your obedience', continued the king; and the Signoria, only a short time after the official signing of the alliance and only some thirty days before Lepanto, did not turn a totally deaf ear.[37] So two years before she finally left the alliance, Venice was already giving hints of her defection or, as some called it, her betrayal.

Nevertheless, the League had been formed. Its importance can be measured by the depth of French antagonism. No words were strong enough in France to condemn the Pope, who was the chief architect of the alliance. And what envy there was in France at the financial concessions made to Spain by the Holy See![38] Rumours of war were so persistent that the Spanish merchants in Nantes and Rouen asked Philip II's ambassador to give them adequate warning so that they might remove their goods and their persons to safety. 'They give me no peace, and are forever plaguing me with their requests', complained Francés de Alava.[39] The French merchants in Seville showed similar anxiety.[40] And on the frontiers of

[33] Nobili and del Caccia, Madrid, 16th April–6th June, 1571, A.d.S. Florence, Mediceo 4903; Philip II to Granvelle, San Lorenzo, 13th June, 1571, Simancas E° 1059, f° 13; F. de Alava to Philip II, Paris, 26th June, 1571, A.N., K 1520, B 29, no. 31. Credit note for 150,000 crowns from the Grand Duke of Tuscany to Louis of Nassau on Frankfurt? F. de Alava to the king, 4th–9th August, 1571, A.N., K 1519, B 29, no. 69.

[34] F. de Alava to the king, Paris, 1st June, 1571, A.N., K 1520, B 29, no. 2.

[35] Bishop of Dax to Charles IX, Lyons, 26th July, 1571, E. Charrière, *op. cit.*, III, p. 161–164. B.N., Paris, Fr. 16170, f°s 9–11, copy.

[36] E. Charrière, *op. cit.*, III, p. 178.

[37] The king to the Signoria of Venice, 23rd May, 1571, B.N., Paris, Fr. 16170, f°s 4–5, copy.

[38] Francés de Alava to Zayas, 'Lubier', 19th June, 1571, A.N., K 1520, B 29, no. 12.

[39] Alava to Alva, Louviers, 25th June, 1571, *ibid.*, no. 20.

[40] 17th August, 1571, Fourquevaux, *op. cit.*, II, p. 371.

Flanders, there was such agitation 'that the people of the lowlands, both on the French side and our own', wrote the Duke of Alva,[41] 'are withdrawing with [their possessions] to the safety of the towns'.

The French were enough to exhaust anybody's patience: 'they would be happy to lose an eye as long as we lost two', exclaimed the Duke of Alva,[42] exasperated at their agreement with Louis of Nassau and the Prince of Orange, while Granvelle advised the king to let Don John strike a blow at Provence on his way past. We seem to be transported back to the time of Charles V, so readily do the old curses against the French spring to the lips of the former servants of the emperor. And indeed war had actually broken out in the Atlantic, where the privateers of La Rochelle and the Sea Beggars were already joining forces. In August, 1571, the Spaniards had good reason to fear for the safety of the Indies fleet.[43]

Will Don John and the fleet arrive in time? Meanwhile the signing of the League was bearing fruit in the Mediterranean.

The Spanish and papal parties had promised the Venetians in a separate agreement to combine their forces at Otranto before the end of May.[44] This was merely an earnest of good will, for how could the necessary orders be transmitted in ten days (the treaty was signed on 20th May)? News of the signature only reached Spain on 6th June. And the inexplicable delays which had preceded it had slowed down naval preparations on the Spanish coast even more than usual. Further, the poor harvest of 1570 caused supply problems at Barcelona and in the ports of Andalusia.[45]

This bad harvest had only one happy result for Philip: it ended the guerilla war in Granada. 'There is such great hunger among them [the rebels]', wrote Fourquevaux on 18th February, 'that they are leaving the mountains and coming down to sell themselves to the Christians as slaves, in order to eat'. In March the death of the rebel king was announced, as well as the surrender of many Moriscos who had been obliged to turn to marauding and cattle-stealing in order to survive. One indication of the shortage was that on learning, at Cartagena, that they were to be sent to Oran, some of the Spanish troops immediately deserted. It was well known what life in the *presidios* was like in times of famine. Happily the situation was better in Italy: in Naples, where the Venetians sought their supplies: in Sicily, where Pescara, still hankering after an attack on Tunis and Bizerta, gave assurances in May that 20,000 quintals of biscuit could be made available; it was being made at the rate of 7000 quintals a month.[46] Without this grain, without the barley and cheeses of Italy, and

[41] Alva to Alava, Antwerp, 11th July, 1571, original in French, A.N., K 1522, B 30, no. 16a.

[42] Alva to Alava, Brussels, 7th June, 1571, A.N., K 1520, B 29, no. 6.

[43] Nobili to the prince, Madrid, 2nd August, 1571, A.d.S. Florence, Mediceo 4903.

[44] L. Serrano, *Correspondencia*, I, p. 102.

[45] Fourquevaux to the queen, Madrid, 18th February, 1571, *op. cit.*, II, p. 331.

[46] To the king, Palermo, 20th March, 1571, Simancas E° 487; the harvest had been good in Sicily, Ragusa, 28th May, 1570, Ragusa Archives, L.P., 2, f°s 97 and 98.

the wines of Naples, who knows if Lepanto would even have been conceivable? For there had first to be assembled and fed at the southern tip of Italy, the equivalent of an entire city of soldiers and sailors, every one with a healthy appetite.

If Don John had had his way, the galleys would have left the coast of Spain with very little delay. He was anxious to take up his command. In April it was rumoured that after visiting his adoptive mother, the wife of Don Luis Quixada, at Valladolid, he would be travelling to Italy.[47] The Marquis of Santa Cruz, arriving in Barcelona on 30th April with the Naples galleys, heard a report there that Don John was on the point of embarking at Cartagena and decided at once to join him, hoping to pick up a few corsair ships on the way.[48] But at Cartagena, there was no sign of Don John.[49] For no news of the League had yet been received: on 17th May, Don John was still at Aranjuez, wondering when he would be able to leave.[50] Several other difficulties also held up the fleet. First the need to increase the numbers of soldiers to be embarked aboard the galleys; then the decision to send along with Don John, as far as Genoa, the Austrian archdukes who had been staying in Spain since 1564 and were on their way back to Vienna:[51] Philip II explained this on 7th May in a letter to Granvelle who since the death of the Duke of Alcalá, had been acting viceroy of Naples.[52] In short the anticipated delays were so great that it was judged preferable not to wait for the Spanish galleys and those of Santa Cruz to embark the thousand Italians and eight thousand Germans who were waiting on the Genoese coast to be carried south, so the Sicilian galleys were ordered to Genoa to pick them up.

However the long-awaited news of the signing of the League hastened the last preparations. The very same day, Don John of Austria left Madrid to take up his command. Gian Andrea Doria took a single galley from Barcelona to Genoa to prepare the city to receive Don John[53] who was by the 16th in Barcelona, where he was joined in turn by the galleys of the marquis of Santa Cruz, Alvaro de Bazán, those commanded by Gil de Andrade and a number of other ships; aboard this fleet were embarked the two Spanish *tercios* of Miguel Moncada and López de Figueroa[54] which had been withdrawn from Andalusia.

[47] Nobili and del Caccia, 16th April, 1571, A.d.S. Florence, Mediceo 4903.
[48] Santa Cruz to the king, 1st May–2nd May, 1571, Simancas E° 106, f°s 81, 82.
[49] Santa Cruz to the king, two letters dated 17th May, 1571, *ibid.*, f°s 83 and 84.
[50] Don John to Ruy Gómez, *CODOIN*, XXVIII, p. 157.
[51] Erwin Mayer-Löwenschwerdt, *Der Aufenthalt der Erzherzöge Rudolf und Ernst in Spanien, 1564–1571*, Vienna, 1927.
[52] Simancas E° 1059, f° 129. The Duke of Alcalá died on 2nd April, 1571, Simancas E° 1059, f° 84. On Granvelle's temporary post there are several documents, see esp. 10th May, 1571, Mediceo 4903; dispatches from Spain, 31st May, 1571, Fourquevaux, *op. cit.*, II, p. 355; *CODOIN*, XXIII, p. 288. N. Niccolini, 'La città di Napoli nell'anno della battaglia di Lepanto', in *Archivio storico per le provincie napoletane*, new series, Vol. XIV, 1928, p. 394.
[53] Nobili and del Caccia to the prince, Madrid, 6th June, 1571, Mediceo 4903.
[54] L. van Der Essen, *Alexandre Farnèse, op. cit.*, I, p. 161.

Finally on 26th June, instructions were sent from Madrid to Don John who, upon discovering the limits they imposed on his authority, fell into a rage bordering on despair. In a letter written in his own hand, in the first flush of emotion on 8th July,[55] he asked Ruy Gómez 'as I would a father' the reason for his apparent disgrace. The tone of the letter is passionate, both moving and disturbing. Hartlaub,[56] rightly in my view, proposes that these days must have marked a major watershed in Don John's life. They showed him that his subordinate position as a bastard was irremediable and that the king placed little confidence in him. Otherwise, why when he was approaching Italy where ceremony was a serious matter, was he refused the title of Highness, receiving only that of Excellency? Why else had his royal command been fenced about with so many restrictions that it was a hollow title? On 12th July,[57] he wrote to the king in person to voice his grievances.

Another problem was the lengthy assembly of the fleet.[58] He had had to wait for the galleys which were taking on troops, supplies and biscuit at Málaga and Majorca. And the archdukes Rudolph and Ernest did not arrive until 29th June.[59] The bulk of the fleet put to sea however on 18th July[60] and on the 26th, despite the bad weather, reached Genoa. Don John stayed there only until 5th August,[61] just as long as it took to embark the men, supplies and materiel waiting for him and to attend the magnificent festivities given in his honour. On the 9th he arrived in Naples.[62] Here receptions and preparations for departure kept him until the 20th.[63] On 24th August he was at last at Messina.[64]

It was much too late, in the opinion of Requesens and Gian Andrea Doria, who advised him to maintain a strictly defensive attitude. And the aged Don García de Toledo sent from Pisa a pessimistic warning about the apparent superiority of the Turkish fleet.[65] But Don John listened only to the Venetian leaders and those Spanish captains of his entourage who urged action and, having made his decision, threw himself into his task with all the single-minded passion of his temperament.

The Turks before Lepanto. The Turkish galleys, which had been marshalled more quickly, had been at sea since the beginning of the summer.

Notice of their coming had been given in advance as usual. Reports

[55] Simancas E° 1134.
[56] *Op. cit.*, p. 79.
[57] *Ibid.*, p. 78, note 2.
[58] *CODOIN*, III, p. 187.
[59] Erwin Mayer-Löwenschwerdt, *op. cit.*, p. 39.
[60] *Ibid.*, p. 40; *Res Gestae* . . ., I, p. 97.
[61] L. van der Essen, *op. cit.*, I, p. 162, puts the date of departure on 1st August.
[62] Padilla to Antonio Pérez, Naples, 15th August, 1571, received on 12th September, Simancas E° 1059, f° 91.
[63] L. van der Essen, *op. cit.*, I, p. 163, says the night of the 22nd; Juan de Soto to García de Toledo, 21st August, 1571, Don John is still in Naples, *CODOIN*, XXVII, p. 162.
[64] Don John to García de Toledo, Messina, 25th August, 1571, *CODOIN*, III, p. 15; L. van der Essen, *op. cit.*, I, p. 163, says the 23rd.
[65] To Don Luis de Requesens, Pisa, 1st August, 1571, *CODOIN*, III, p. 8.

reached Italy in February that 250 galleys and 100 other boats were being fitted at Constantinople;[66] in March that Famagusta, which was still under siege, had received reinforcements from Venice;[67] in April that the town was still holding out; that the Turks were making preparations on land to advance on Albania and Dalmatia and that the bulk of the armada had sailed under the command of the Kapudan Pasha.[68] But it was reported that only 50 galleys had made for Cyprus, since the fleet had been too short of crews to man more than 100 galleys in all.[69] But slaves escaping from Constantinople brought word of 200 galleys which they said would head for Corfu if the League was not concluded. If it was, the Turks would be content merely to patrol their own waters, meanwhile consolidating the conquest of Cyprus.[70] There was still talk of an expedition overland to Albania or Dalmatia, troops to be assembled at Sofia before departure.[71]

In fact 196 Turkish galleys had set sail from the capital and were divided between Negropont, a major supply base, and Cyprus. Biscuit was being manufactured as far away as Modon and Prevesa[72] a sure sign that something was planned in the West. Sure enough in June, leaving Cyprus where there was little more to do[73], the main force of the Turkish fleet, reinforced by Euldj 'Alī's ships (a total of 300 sail: 200 galleys and 100 foists) made for the island of Crete. On the 15th, it reached the bay of Suda and ravaged the villages and small towns along the coast. But it twice attacked without success the port of Canea (Khania),[74] where 68 Venetian galleys assigned to carry aid to Famagusta lay at anchor protected by the artillery of the forts. The rumour spread that they had been captured[75] but in fact Euldj 'Alī had only been able to take the little port of Rethimnon. After pillaging and several skirmishes, the Turkish armada sailed on westwards.

At its approach, Veniero, in order to avoid being trapped in the Adriatic with the rest of Venice's navy, left the coast of Morea and Albania, where he had made several successful assaults on Durazzo and Valona. With his 6 galleasses, three ships and 50 *galere sottile*, he sailed for the base of Messina, where he arrived on 23rd July. This withdrawal safeguarded his freedom of manoeuvre, but left the Adriatic wide open to the Turks.[76] They took immediate advantage, laying waste the coast and the Dalmatian

[66] Corfu, 3rd February, 1571, Simancas E° 1059, f° 62.

[67] Corfu, 29th March, 1571, dispatch received at Venice on 11th April, 1571, Simancas E° 1060, f° 13.

[68] Constantinople, 10th April, 1571, *ibid.*, f° 125.

[69] Messina, 23rd April, 1571, *ibid.*, f° 11.

[70] Corfu, 27th April, 1571, Simancas E° 1059, f° 56.

[71] Constantinople, 5th May, 1571, passed on from Corfu (more escaped prisoners), Simancas E° 1060, f° 133.

[72] Negropont, 3rd June, 1571, *ibid.*, f° 137.

[73] L. Voinovitch, *op. cit.*, p. 39.

[74] G. de Silva to Philip II, Venice, 6th July, 1571, Simancas E° 1329.

[75] F. de Alava to the king, Melun, 1st August, 1571, A.N., K 1520, B 29, no. 37; 60 galleys lost: Nobili to the prince, Madrid, 2nd August, 1571, Mediceo 4903.

[76] L. Voinovitch, *op. cit.*, p. 40.

islands, seizing Sopoto, Dulcigno, Antivari and Lesina and attacking Curzola (Korčula) which its inhabitants defended with spirit. Meanwhile Aḥmad Pasha's troops who had gone overland, took anything they could lay their hands on. Euldj 'Alī launched a raid on Zara; and another corsair, Kara Hodja, even sacked the gulf of Venice.[77]

The Signoria began recruiting in haste, with Granvelle's permission, in Apulia and Calabria.[78] But the Turkish fleet, thinking no doubt, as Veniero had done, that the Adriatic could be a trap, did not commit itself entirely there and sent the bulk of its force against Corfu. The island, evacuated by its inhabitants, was laid waste; only the great fortress, an island within an island, remained safe from the assailants. The Turkish fleet then fanned out from Corfu to Modon, to wait and see what the allies intended to do. For news of the formation of the League had reached Turkish territory in June by way of Ragusa.

For once the early sortie of their fleet did not help the Turks. They had exhausted their strength and their supplies during these months spent in petty skirmishes. And while they were being tempted by easy prey, burning and looting the villages of the Adriatic, they had forgotten something far more essential: the Venetian fleet in Crete. At the end of August, under the two *provveditori* Agostino Barbarigo and Marco Quirini, the sixty galleys sailed without hindrance to meet the great allied fleet.[79]

The battle of 7th October.[80] When Don John arrived at Messina, morale among the allies was not high, nor were the galleys so far assembled in good order. But the arrival of Don John's ships, in admirable condition, made a strong impression. Moreover, the first contacts between the prince and his immediate subordinates, Veniero and Colonna, were excellent. Don John could be very charming when he tried. The sorcerer cast his spell: on this initial contact the future of the whole expedition so dear to

[77] L. Voinovitch, *op. cit.*, p. 41.

[78] Granvelle to Philip II, 20th September, 1571, Simancas E° 1060, f°s 57 and 58. In Calabria and Bari.

[79] Don John of Austria to García de Toledo, Messina, 30th August, 1571, *CODOIN*, III, p. 17.

[80] There are innumerable accounts of this battle both by contemporaries and by historians. The latter give few details and are rarely unbiased. Was Lepanto a Spanish, a Venetian or an Italian victory? Some contemporary accounts: Ferrante Caracciolo, *I commentarii delle guerre fatte co' Turchi da D. G. d'Austria dopo che venne in Italia*, Florence, 1581. *Discurso sobre la vit^{a} naval sacada de la armada turquesca traduzido del toscano en sp^{ol}*, A.E., Esp. 236, f°s 51–53. *Relatione della vittoria navale christiana contro li Turchi al 1571, scritta dal Signor Marco Antonio Colonna alla Santità di N^{ro} Sign^{r} Pio V, con alcune mie* (by Francesco Gondola, Ragusan ambassador in Rome) *aggionte ad perpetuam memoriam*, ed. L. Voinovitch, p. 107–112. *Relacion de lo succedido al armada desde los 30 del mes de setiembre hasta los 10 de octubre de 1571*, f°s 168–169; *Relacion de lo succedido a la armada de la Santa Liga desde los 10 de octubre hasta los veynte cinco del mismo*, f°s 169–171, B.N., Madrid, Mss 1750. For a complete list of historical works on the subject see bibliographies in L. Serrano, Alonso Sánchez or G. Hartlaub; worth an individual mention: Guido Antonio Quarti, *Lepanto*, Milan, 1930.

his heart might depend. He was able to take a miscellany of naval forces and turn it into a united fleet. Seeing that the Venetian galleys were short of troops, he sent them four thousand soldiers, both Spanish and Italian, paid by the Catholic King. This in itself was something of a *coup*. We may imagine the apprehensions that had to be overcome before the suspicious Venetians would accept such assistance. At a stroke, the galleys of the fleet became identical and interchangeable; without being totally amalgamated, the squadrons could exchange ships if necessary and so they did, as the sailing order of the fleet proves.

The armada realized it had a leader as soon as the Council of the League met. Not every cloud disappeared miraculously from the horizon, but disagreement was reduced. Don John did not wish to cheat the Venetians, as they quickly realized, of a direct campaign against the Turk, nor did he wish to force them instead to join an expedition against Tunis as Philip II, so many Spaniards and the whole of Sicily hoped. In his own mind, Don John would have favoured a bold advance on Cyprus and beyond, through the Aegean to the Dardanelles. The final decision, a more prudent one, was that the fleet would sail in search of the Turkish armada and give battle.

The armada left Messina on 16th September, its first objective Corfu, where it hoped to receive recent information about the precise position of the enemy. In Corfu it was reported – and the news was confirmed by ships sent out on reconnaissance – that the Turks were in the long narrow gulf of Lepanto. The reports incidentally underestimated the size of the Ottoman fleet. But the other side committed the same mistake. As a result the Turkish admiral and his advisers decided to pounce on the Christian ships off the coast of Corfu; meanwhile the allied council of war, in a somewhat stormy session, was deciding to fight, despite the objections of the cautious and the pusillanimous. The insistence of the Venetians, who threatened to fight alone if they had to, the eagerness of the papal commanders and the enthusiasm of Don John, who did not hestitate to go beyond the limited powers he had been granted by Philip II, won the day.

There is no doubt that on this occasion Don John was the instrument of destiny. He took the view that he could not in all conscience disappoint Venice and the Holy See without losing both face and honour. To avoid a battle would have been to betray Christendom; to fight and perish did not necessarily, if the friendship of Venice was preserved, mean compromising the future, for with her aid the Christian fleet could be rebuilt. So Don John argued after the event[81] to explain his decision and so he no doubt thought at the time. Don García de Toledo, the following year, still shuddered at the thought that Don John had in one single action imperilled the only defence of Italy and Christendom. The smoke had hardly lifted from Lepanto, before wise men were shaking their heads over what they viewed as Don John's folly and temerity, imagining with horror the

[81] *Las causas que movieron el Sor Don Juan para dar la batalla de Lepanto* (1571). B.N., Madrid, Mss 11268/35.

dreadful possibility of a Christian defeat and the Turks pursuing the allies all the way to Naples or Civitavecchia.

The two fleets, each looking for the enemy, came upon one another unexpectedly on 7th October at sunrise, at the entrance to the gulf of Lepanto in which the Christian fleet, by a brilliant tactical move, at once succeeded in trapping its adversary. Face to face with each other, Christian and Moslem were then able, to their mutual surprise, to see their true numbers: 230 warships on the Turkish side, 208 on the Christian. Six galleasses well supplied with artillery reinforced Don John's galleys, which were on the whole better provided with cannon and arquebuses than the Turkish galleys, where many of the soldiers were still using bows and arrows.

The many extant accounts of the battle – including the study by Vice-admiral Jurien de La Gravière – are not always totally objective. It is difficult to gather from them who deserves most credit for the brilliant victory. Don John, the admiral? Yes, undoubtedly. Gian Andrea Doria, who on the eve of the battle had the idea of taking off the *espolones* (battering-rams) from the galleys, so that the bows should ride lower in the water and the trajectory of the forward guns would thus carry straight at the timber flanks of the Turkish ships? The powerful Venetian galleasses, placed in front of the Christian lines, which divided the enemy fleet, scattering its ranks with the astonishing force of their cannon? Although unwieldy and almost immobile, they acted on this occasion as ships of the line, floating fortresses. Nor should we underestimate the excellent Spanish infantry which played a major role in battle very similar to a land action; nor the admirable ordnance of the Spanish galleys, most feared by the Turks of all the *ponentinas*,[82] and the particularly fierce fire of the Venetian galleys. It must also be recognized that, as the Turks later pointed out, and as the victors themselves admitted, the Turkish fleet was tired and not on its best form.[83]

But wherever the credit lies, the Christian triumph was a resounding one. Only thirty Turkish galleys escaped: led by Euldj 'Alī, they circled lightly and with extraordinary skill around the formidable galleys of Gian Andrea Doria. Perhaps (as slanderers whispered),[84] it was because the Genoese commander was once more reluctant to take chances, a little too careful of 'his capital'. All the other Turkish galleys were captured and divided among the victors or sunk. In this encounter, the Turks lost over 30,000 dead and wounded and 3000 prisoners; 15,000 galley-slaves were freed. The Christians had lost 10 galleys, 8000 men killed and 21,000 wounded. Their success was dearly bought in human terms, more than half the combatants being put out of action. To the exhausted soldiers the sea seemed suddenly to run red with blood.

[82] *Ponentinas* does not refer (as Hartlaub assumes, *op. cit.*, p. 182) merely to Spanish vessels.

[83] On this point, see Granvelle to the king, Naples, 26th May, 1571, Simancas E° 1060, f° 30; his remarks carry the more weight for being made before the battle.

[84] Nobili and del Caccia to the prince, Madrid, 24th December, 1571, A.d.S. Florence, Mediceo 4903; L. Voinovitch, *op. cit.*, p. 42.

A victory that led nowhere? This victory seemed to open the door to the wildest hopes. But in the immediate aftermath of the battle, it had no strategic consequences. The allied fleet did not pursue the enemy, partly because of its own heavy losses and partly because of the bad weather, which may have saved the undefended Ottoman Empire. Even in September the Venetians had considered it too late to sail eastwards and try to recapture Cyprus. It was unthinkable in the autumn, when reports of the fall of Famagusta had reached the commanders of the fleet just before Lepanto. Venice herself did not hear the news until the 19th,[85] the day after word of the allied victory had arrived with the galley *Angelo Gabrielle* sent by Veniero.[86]

Don John, for his part was tempted to lead an expedition straight away to the Dardanelles, which would have enabled him to block the straits. But he was short of men and supplies, and Philip II had given orders that unless some large port in the Morea were ceded to them, the galleys were to winter in Italy. It was impossible to set off on a wild adventure without the materials needed to lay a siege. The papal ships and the Venetians, who were inclined to linger beneath the walls of several minor towns in the Adriatic, derived neither glory nor profit from their efforts. Don John returned to Messina on 1st November. A few weeks later, Marcantonio Colonna was back in Alcona and Veniero in Venice.

On the strength of this, historians have joined in an impressively unanimous chorus to say that Lepanto was a great spectacle, a glorious one even, but in the end leading nowhere.[87] The bishop of Dax skilfully developed this theme on behalf of his master, and among the Venetians with whom he heartily commiserated on their great losses, and not so much as an inch of ground gained',[88] but then the bishop of Dax had his own reasons for turning a blind eye to the victory.

However if instead of considering only what happened after Lepanto, we turn our attention to what had gone before, the victory can be seen as the end of a period of profound depression, the end of a genuine inferiority complex on the part of Christendom and a no less real Turkish supremacy. The Christian victory had halted progress towards a future which promised to be very bleak indeed. With Don John's fleet destroyed, who knows, Naples and Sicily might have been attacked, the Algerines might have tried to revive the flames of Granada or carried them into Valencia.[89] Before joining Voltaire in his ironic comments on Lepanto, we would do

[85] G. de Silva to the king, Venice, 19th October, 1571, Simancas E° 1329. Cf. B.N., Paris, Ital. 427, f°s 325–333. F. de Alava to the king, 2nd August, 1571, A.N., K 1520, B 29, no. 38; L. Voinovitch, *op. cit.*, p. 102.

[86] H. Kretschmayr, *op. cit.*, III, p. 69.

[87] See for instance H. Wätjen, *Die Niederländer im Mittelmeergebiet*, *op. cit.*, p. 9; H. Kretschmayr, *op. cit.*, III, p. 75 ff; L. Serrano, *op. cit.*, I, p. 140–141.

[88] To the Duke of Anjou, Venice, 4th November, 1571, B.N., Paris, Fr. 16170, f°s 57 to 59, copy.

[89] Henri Delmas de Grammont, *Relations entre La France et la Régence d'Alger au XVIIe siècle*, I, p. 2, note 2.

well to measure the immediate impact of the victory – which was breathtaking.

An extraordinary round of rejoicings followed – Christendom could hardly believe its luck – and a no less extraordinary crop of plans. Grandiose campaigns all over the Mediterranean were proposed, expeditions to North Africa, the traditional sphere of Spanish influence; to Egypt and Syria, faraway possessions which brought the Turks so much revenue (it was the level-headed Granvelle, oddly enough, who championed this project); to Rhodes, Cyprus, and the Morea, emigrants from which, scattered everywhere, were ready to promise anything in return for arms and sympathy. Their claims found willing listeners in Spain and Italy where it was firmly believed that the Christians of the Balkans would rise up the moment a Christian force descended. More romantic souls, the Pope and Don John of Austria, dreamed of delivering the Holy Land and capturing Constantinople itself. Indeed one timetable suggested an expedition against Tunis in the spring of 1572, the Levant in the summer and Algiers the following winter.

Philip II, to do him justice, did not take part in these extravagances. Unlike his father, his brother and his nephew, Dom Sebastian of Portugal, he was not haunted by dreams of crusades.[90] As usual he pondered, weighed alternatives, asked the advice of persons in authority, listened to their discussions, sent plans suggested by Granvelle and Requesens to Don John with instructions to comment upon each proposal point by point. Talks resumed in Rome to lay down plans for the next year's campaign. The victory had not made the negotiators any less secretive or suspicious.

We tend to smile at these grave discussions as much as at breathless schemes of conquest. With the benefit of hindsight it is too easy to explain as Serrano does in his study of Lepanto (the most recent and the best to date), that this victory could bear no fruit nor serve any purpose whatsoever. All one can say is that after all Lepanto was only a naval victory and that in this maritime world surrounded and barred by landmasses, such an encounter could not destroy Turkey's roots, which went deep into the continental interior. The destiny of the League was decided not only in Rome but in Vienna, in Warsaw, the new Polish capital, and in Moscow. If the Turkish empire had been assaulted on this landward frontier – but how could it be?[91]

[90] L. Pfandl, *Philippe II*, p. 366–367.

[91] I have not mentioned the fruitless diplomatic campaign for such an initiative, in Vienna, Poland, Moscow and Lisbon, which was already under way in 1570, see Paul Herre's book, p. 139 ff; nor even the curious negotiations by Rome in Muscovy (P. Pierling, S.J., *Rome et Moscou, 1547–1579*, Paris, 1883; *Un nonce du Pape en Moscovie, Préliminaires de la Trêve de 1582*, Paris, 1884). On Portugal's firm refusal in 1573: *Le cause per le quali il Sermo Re di Portugallo nro Sig^{re}* . . ., Vatican Archives, Spagna, 7, f°s 161–162. In April, 1571, the imperial tribute was paid to the Turks, F. Hartlaub, *op. cit.*, p. 69; Cardinal Rambouillet, Rome, 7th May, 1571, E. Charrière, *op. cit.*, III, p. 148–149. The Emperor was unable to act without the support of the king of Poland (Michiel and Soranzo to the Doge, Vienna, 18th December, 1571, P. Herre, *op. cit.*, p. 154) at least so he told the Polish ambassador: 'Ohne Euch kann

And lastly, Spain was unable to devote herself as wholeheartedly and for as long as she ought to have, to Mediterranean affairs. Here as usual we find the true answer. Lepanto would probably have led to other things if Spain had seriously taken it upon herself to pursue the struggle.

It is not always realized that the victory at Lepanto itself was only possible because Spain did for once throw her entire weight into the balance. By a happy coincidence, all her problems were temporarily but simultaneously out of the way in these years 1570–1571. The Netherlands seemed to be under Alva's stern control; England had domestic problems – first the rising of the northern earls in 1569, then the Ridolfi plot, many if not all the strings of which were being pulled by Spain. Philip II even thought of attacking Elizabeth at this time, so incapable of resistance did she appear.[92] French policy was more disturbing, but it had not yet come to a head. The bishop of Dax had gone no further than Venice in October, 1571. And Tuscany was hesitating. The anti-Spanish policy skilfully engineered by Cosimo seemed to be in eclipse: the grand duke had even advanced to the Duke of Alva money needed for the action in the Netherlands.[93] Was this a genuine political about-face or were the Tuscans playing a double game? For whatever reason, Spain was suddenly relieved of her external problems.

She took advantage of the breathing-space to act in the Mediterranean. All over Europe, regiments of soldiers and adventurers of every nation began flocking to the South where there was now employment. Dulcigno was defended in July, 1571 – but without great enthusiasm apparently – by French soldiers, probably Huguenots, paid by Venice according to one report.[94] And Frenchmen mingled with the Spaniards boarding Philip II's galleys at Alicante.[95] In the spring of 1572, two thousand French mercenaries were in Venice, serving the Signoria – all clear signs of the sudden importance of the Mediterranean battlefield.

So Spain took the opportunity of peace in the West to strike at the East. But it was no more than a brief episode. Spain was never able to deal more than a quick blow here and there, dictated by circumstances more than by her own wishes. She was never able to commit all her forces in a single

---

man nichts tun', to which the latter replied 'Und wir wollen ohne Eur. Majestät nichts tun'. Nobili to the prince, Madrid, 18th November, 1571, Mediceo 4903, had it on good authority that the Emperor was unprepared and would be unable to attack that year. Broadly speaking there were two serious or at any rate persistent attempts on the diplomatic level, one in 1570 and one at the beginning of Gregory XIII's pontificate (with the missions of Ormaneto to Spain and the archbishop of Lanciano to Portugal).

[92] For a summary of the many documents relating to this in Simancas, see O. von Törne, *op. cit.*, I, p. 111 ff. Arrival of an ambassador in Spain, Coban, *CODOIN*, XC, p. 464 ff; Nobili and del Caccia, Madrid, 10th May, 1571, Mediceo 4903.

[93] The money was borrowed but never used, see above, Vol. I, p. 485.

[94] G. de Silva to Philip II, Venice, 25th or 26th November, 1571, Simancas E° 1329.

[95] Requesens to Philip II, 8th December, 1571, A.E., Esp. 236, f° 132; Fourquevaux, *op. cit.*, II, p. 243, troops crossing the Alps; Alava to the Duke of Albuquerque, Paris, 27th April, 1571, A.N., K 1519, B 29, n° 69.

campaign and this must be the ultimate explanation of her 'victories that led nowhere'.

2. 1572: A DRAMATIC YEAR

The French crisis up to the St. Bartholomew Massacre, 24th August, 1572. French animosity towards Spain had not ceased to grow since the beginning of the Cyprus war. It simmered throughout 1571, reached boiling-point after the signing of the League and finally erupted after Lepanto.

The year 1572 began with both countries anticipating war. It threw its shadow ahead over a Mediterranean already up in arms, over the north of Europe where trouble was brewing. The embassy of the bishop of Dax who had finally left Venice (he reached Ragusa in January)[96] was of minor significance beside the threat the Spaniards could see growing in the North. They had already learned how Louis of Nassau had been received at the French court in the spring of 1571. They were not ignorant that the king had thrown in his hand with the rebels. France was surrounded and infiltrated by a network of Spanish intelligence and espionage[97] that contributed little to goodwill between the two nations, whose diplomatic conversations are a tissue of falsehoods from beginning to end – and a morass for the historian. To attribute everything, as Serrano does[98] with unconscious national prejudice, to the cunning Florentine duplicity of Catherine de Medici, whom he describes as skilfully slipping from one attitude to another, first denying any military preparations then attributing them to her rebel subjects, finally presenting them as legitimate self-defence – is to grasp only one thread of a tangled skein.

Meanwhile the English threat was looming once more. The Duke of Medina Celi, who was preparing to sail to the Netherlands via the Atlantic and English channel, was advised by Alva at the beginning of 1571 to put into French ports rather than English ones.[99] But at the beginning of 1572, Spain's two enemies drew closer together, burying the long-standing differences which Spain had always carefully fostered. An alliance was in sight – as the Spanish knew by January[100] and the placatory explanations made by the French king[101] deceived no one.

The articles of the Anglo-French 'league' were signed at Blois on 19th April.[102] The stipulations of the treaty were common knowledge in

[96] E. Charrière, *op. cit.*, III, p. 245, note; B.N., Paris, Fr. 16170, f° 70 ff.

[97] H. Forneron, *Histoire de Philippe II*, II, p. 304 ff.

[98] *Op. cit.*, I, p. 228 ff, esp. p. 228, note 2.

[99] 31st January, 1571, *CODOIN*, XXXV, p. 521 (101a), A.N., K 1535, B 35, 10a; Catherine de Medici to the queen of England, 22nd April, 1572, Count Hector de La Ferrière, *Lettres de Catherine de Médicis*, 1885, IV, p. 97, 98.

[100] G. de Spes, Canterbury, 7th January, 1572, *CODOIN*, XC, p. 551.

[101] Cavalli to the Signoria of Venice, Blois, 24th February, 1572, *C.S.P.*, *Venetian*, VII, p. 484.

[102] Articles of the defensive league between the king of France and the queen of England, 19th April, 1572, A.N., K 1531, B 35, no. 10a.

March, in particular that the English would transfer the staple for their exports to the continent from the Netherlands to Rouen and Dieppe, while the French undertook to supply the island kingdom with salt, spices and silk.[103] Such an agreement would evidently benefit the routes of the French isthmus and contribute to the decline of Antwerp.[104] It was even said at Venice that the French had only signed the treaty in order to revive the declining trade of Rouen.[105] The secretary Aguilón, who, while awaiting the arrival of Don Francés de Alava's successor, Diego ed Zuñiga,[106] was handling Spain's day-to-day affairs in France, did not hide his bitterness and in a letter to the secretary Zayas expressed himself freely, openly regretting that Philip II had not, for reasons of 'reputacion', made a pact with England. Either the negotiations should be re-opened (the agreement with France had not yet been signed), or the Netherlands were lost.[107]

The English ambassador, Walsingham, viewed things with more discrimination. He wrote to Lord Burghley just after the signing of the treaty: 'They of the Robe longue do fear that it will breed a pike between this crown and Spain. And they would be loath that the King should enter into wars for that they doubt that the managing of the affairs should come then to the hands of others'.[108] This was a more balanced view of the uncertain nature of French politics, that of a dispassionate observer. The court of Madrid, as ever well-informed, and on this occasion perhaps a little too well-informed, tended to see things in the blackest colours.

It is true that French preparations on the Atlantic coast were deliberately much publicized.[109] Fourquevaux heard talk of them at Madrid in February.[110] In Paris, Aguilón met with the Portuguese ambassador who was as worried as he was. The Duke of Alva in Brussels was concerned about them.[111] And tongues wagged everywhere.[112] Would the Atlantic galleys commanded by La Garde be brought back to the Mediterranean? And what was the destination of the tall merchant ships now being equipped as men-of-war and known to be placed under the command of Philip Strozzi?[113] Saint-Gouard said that they were intended to patrol

[103] Aguilón to Alva, Blois, 8th March, 1572, A.N., K 1526, B 32, no. 6.

[104] Saint-Gouard to the king of France, Madrid, 14th April, 1572, B.N., Paris, Fr. 1640, f°s 16 to 18.

[105] G. de Silva to the king, Venice, 22nd May, 1572, Simancas E° 1331.

[106] For his instructions, 31st March, 1572, A.N., K 1529, B 34, f° 33 and 34.

[107] Blois, 16th March, 1572, A.N., K 1526, B 32, no. 19.

[108] Walsingham to Burghley, 22nd April, 1572, quoted in Conyers Read, *Sir Francis Walsingham*, Oxford, 1925, I, p. 204–205.

[109] Where was the fleet bound for? Ch. de La Roncière, *Histoire de la marine française*, 1934, p. 68, says: 'for the West Indies, Guinea, Florida, Nombre de Dios, Algeria (sic)'. This fleet was used to blockade La Rochelle during the fourth war of religion.

[110] II, p. 241.

[111] Alva to secretary Pedro de Aguilón, Brussels, 19th March, 1572, A.N., K 1526, B 32, no. 15. Also indicates Aguilón's conversations with the Portuguese ambassador.

[112] Saint Gouard to the king, 14th April, 1572, B.N., Paris, Fr. 1610, f°s 22 and 23; Aguilón to the duke of Alva, Blois, 26th April, 1572, A.N., K 1526, B 37, no. 57.

[113] Francisco de Yvarra to the king, Marseilles, 26th April, 1572, Simancas E° 334.

against pirates in the Atlantic. But as Philip II well knew, it was not unheard of for the patrolman to turn thief.[114] It was difficult to accept the official French assurance that this fleet would respect Spanish territorial rights[115] when Diego de Zuñiga reported that Catherine de Medici had burst out laughing when she said: 'the Bordeaux fleet will not touch anything of yours! You may rest as easy as if your own king was on board!'[116]

And it was not only the Bordeaux fleet which inspired anxiety. Incidents on the southern frontier of the Netherlands were the occasion of reciprocal complaints, aggravated by the nervous tension of the population in this troubled region.[117] Had the Duke of Alva really strengthened the fortresses along the border? Had he placed artillery on the ramparts? Had he set up road blocks?[118] Had he pressed into guard duty the people of the flat lands as in wartime? The French population of the Netherlands who, according to the treaties were supposed to enjoy full possession of their property, were constantly on the alert, especially since matters of religion were not included in the treaties.[119] Meanwhile it was impossible for the Spanish not to believe that the French, whenever they could, encouraged the rebels in the Spanish provinces.

On 1st April, the Sea Beggars under William de La Marck had taken Brill on Voorne Island at the mouth of the Meuse.[120] The insurrection had reached Flushing and spread north and east throughout the Waterland. The true revolution of the Netherlands was beginning in a region of fanatical religious differences where many people were poor. The French were inevitably involved and the English too, needless to say.[121] Alva complained to the representative of the king of France at Brussels, that the rebel ships were sailing alongside newly fitted French vessels and that their arms, ammunition and supplies came from France.[122] Strozzi's fleet was practically a Huguenot outfit, wrote Aguilón, and English and Scottish vessels were sailing alongside. It was quite capable of plundering the Spanish coast on its travels.[123]

[114] Aranjuez, 10th May, 1572, A.N., K 1528, B 35, no. 48.

[115] Zayas to Diego de Zuñiga, Madrid, 20th May, 1572, A.N., K 1529, B 34, no. 54.

[116] Fourquevaux, *op. cit.*, II, p. 309.

[117] Aguilón to Alva, Blois, 3rd May, 1572, A.N., K 1526, B 32, no. 69.

[118] Alva to the king, 27th April, 1572, copy A.N., K 1528, B 33, no. 43.

[119] Memorandum and proposals by the ambassador Saint-Gouard to Philip II, April, 1572, A.N., K 1529, B 34, no. 44 (Spanish transl.). Other minor incidents, Saint-Gouard to Catherine de Medici, autograph, Madrid, 14th April, 1572, B.N., Paris, Fr. 16104, f°s 22–23; H. Forneron, *op. cit.*, II, p. 302.

[120] C. Pereyra, *Imperio* . . ., p. 170; G. de Spes to the king, Brussels, 15th April, 1572, *CODOIN*, XC, p. 563–564. Mondoucet to the king, Brussels, 27th April, 1572, orig. B.N., Paris, Fr. 16127, disarray of the Duke of Alva. See Antonio de Guaras' account to Alva, London, 18th May, 1572, *CODOIN*, XC, p. 18–19, H. Pirenne, *Histoire de la Belgique*, IV, p. 29 ff; Mondoucet to the king, Brussels, 29th April, 1572, B.N., Paris, Fr. 16127, f° 43; Aguilón to Alva, Blois, 2nd May, 1572, A.N., K 1526, B 32.

[121] See previous note, *CODOIN*, XC.

[122] Mondoucet to Charles IX, Brussels, 29th April, 1572, B.N., Paris, Fr. 16127, f° 43 ff.

[123] Aguilón to Alva, Blois, 2nd May, 1572, A.N., K 1526, B 32.

Further cause for alarm was that the attack on Brill in the north was soon to be followed by action in the south. The Huguenots entered Valenciennes with La Noue on 23rd May and on the 24th, Louis of Nassau surprised Mons at dawn.[124] This was in execution of the plan worked out by Orange and his French allies.[125] It did not surprise the Duke of Alva who was only too ready to underestimate the danger from the seaward side and the Sea Beggars, but hypnotized by the French threat and the continental southwest frontier.[126]

So war between France and Spain appeared imminent. In May, Vespasiano Gonzaga, the viceroy of Navarre, gave disturbing news of French activity in the region.[127] The whole Pyrenean frontier from the Atlantic to the Mediterranean was on the alert. Saint-Gouard's messages were as suspicious of the Spanish as those sent by the Spanish ambassadors were of the French. 'I see that they are making all sorts of preparations', he wrote on 21st May, 1571.[128] 'They have raised over five millions at the most recent fair of Medina del Campo' and have brought from Italy a quantity of arms to be distributed along the frontiers. Finally, they are maintaining 'an infinity of captains' who are not allowed to leave for the Levant. There is no reason to think, he concludes that these preparations, supposedly intended for the League, will not in fact be used in Languedoc and Provence. . . . Meanwhile the king of France for his part had raised over 80 vessels and 'a good number of Gascon infantrymen';[129] what could this fleet be waiting for? To see where the Spanish front cracks first, explained Diego de Zuñiga. And if our affairs mend in the Netherlands, it will not go there.[130]

The Portuguese were also alarmed about the possible enterprises which might tempt an unoccupied French fleet. They had equipped a large fleet, with over 20,000 troops. This is because they are afraid, says Sauli, passing on this information, that the French may think of attacking Cabo da Guer in Morocco (in what is now the bay of Agadir), which might considerably interfere with Portuguese shipping, both in the East Indies and in the 'Indies of the West'. Unless, that is, they have notions about another Moroccan expedition themselves.[131]

The Spaniards had not lagged behind: naval measures to meet the threat of these famous French ships of Bordeaux, Brittany and La Rochelle, had already been taken. In May, the Duke of Medina Celi left

[124] H. Pirenne, *op. cit.*, IV, p. 31–32; *CODOIN*, LXXV, p. 41; H. Forneron, *op. cit.*, II, p. 312.

[125] H. Pirenne, *op. cit.*, IV, p. 31.

[126] R. B. Merriman, *op. cit.*, IV, p. 294.

[127] Nobili and del Caccia, Madrid, 19th May, 1572, A.d.S. Florence, Mediceo 4903.

[128] To the king, Madrid, 21st May, 1572, B.N., Paris, Fr. 1604, f°s 58 ff.

[129] Nobili and del Caccia to the prince, Madrid, 19th May, 1572, A.d.S. Florence, Mediceo 4903.

[130] Don Diego de Zuñiga to Alva, 24th May, 1572, A.N., K 1529, B 34, no. 96a, copy.

[131] 18th May, 1572, A.d.S. Genoa, L.M. Spagna, 5–2.414.

Laredo with 50 ships;[132] orders had been given for 30 to be fitted in Flanders. And in Biscay and Galicia, permission to take up arms and to join Medina's fleet 'to observe the movements' of the French, was granted to any volunteers – an unprecedented step.[133]

As for Italy, the sensitive point was of course Piedmont and, surrounding Piedmont, the Savoy states and the great military centre of Milan, of which the Grand Commander of Castile had just been named governor. Soldiers from Germany were pouring into the city. If Saint-Gouard dared to inquire about them, he was told of course that 'they are to be sent to Don John of Austria'.[134] But at the same time,[135] Diego de Zuñiga wrote to his master with equal urgency that the French had found a way to get back some of the positions occupied by the Spaniards in Piedmont. The alarm was catching: it reached Genoa. The Signoria of Genoa wrote that the four galleys equipped and ready to leave Marseilles with eight infantry companies were, according to public report, intended for Corsica.[136]

Real intentions were carefully concealed. Every power, anxious and threatening by turns, had to snatch at straws. When on 11th May, the Algerines – rather unexpectedly – requested protection from the king of France, Charles IX hastened to agree.[137] He thought that the Duke of Anjou, paying tribute to the Turk, would make an excellent king of Algiers. How many crowns was this poor duke invited to hope for! It was perhaps a foolhardy decision, capable of upsetting the alliance with the sultan, and 'mistaking chaff for grain' as the bishop of Dax said in reply to this odd proposal. [138] But if the king was so easily attracted by the insincere promises of Euldj 'Alī – who was at the very same moment offering his city to Philip II![139] – it was because he thought they offered a chance for a good move by one of the pawns in his game against the king of Spain. One should not exaggerate the incident as Serrano and Hartlaub have done, suggesting that the key to Strozzi's instructions lies in Algiers[140] and that by alarming the Cantabrian coast, Charles IX wished to distract Philip's fleet from Algiers, so that he could eventually obtain the city for himself.

French policy was far from possessing such coherence. It was not sure

[132] He arrived at Sluys on 11th June, 1572, Medina Celi to the king, Sluys, 11th June, 1572, *CODOIN*, XXXVI, p. 25.

[133] Saint-Gouard to the king, Madrid, 31st May, 1572, B.N., Paris, Fr. 1604, f°s 75 ff. Other rumours: the French fleet would go to the Indies '. . . and only yesterday this was heard at the Duke of Sesa's', *ibid.* In France the roads were thronged with soldiers. A large fleet including 14 ships of 600 tons was anchored at Bordeaux, 20th June, 1572, A.N., K 1529, B 34, f° 9.

[134] Saint-Gouard to the king, Madrid, 21st May, 1572, B.N., Paris, Fr. 1604, f°s 58 ff.

[135] 24th May, 1572, see note 130.

[136] To Sauli, 6th June, 1572, A.d.S. Genoa, L.M. Spagna, 6.2415.

[137] The king to the bishop of Dax, 11th May, 1572, B.N., Paris, Fr. 16170, f° 122 ff. E. Charrière, *op. cit.*, III, p. 291, note.

[138] Eugène Plantet, *Les consuls de France à Alger*, 1930, p. 9.

[139] L. Serrano, *op. cit.*, IV, p. 516–517.

[140] *Ibid.*, I, p. 226; F. Hartlaub, *op. cit.*, p. 56.

of its support abroad. In August, Protestant Germany had still not decided to move. England did not particularly wish to see France installed in the Netherlands and for her it was an opportune moment to begin conversations with Spain. And the Turks had their own problems. Nor was France even sure of herself. Her anti-Spanish policy was that of Charles IX, an impulsive man, and of Coligny, a dreamer. It was opposed by the men of 'the Robe longue' mentioned by Walsingham, by the king's noble advisers and by the entire French clergy. For many reasons, notably that war was expensive, there was naturally some thought of seizing Church possessions. A project of secularization was even drawn up: this project quite as much as the flag-waving attitudes of the admiral apropos of the cross of Gastines, was to inflame Paris and the clergy.[141] On the other hand, attacking Philip II was no small decision and Catherine de Medici was beginning to have serious doubts about a war which might provoke unforeseen hazards. In the council on 26th June, Tavannes said quite clearly that the power of the Protestants might well become so great in case of conflict that 'if those who now lead them with good intentions should die or be changed . . . the king and his kingdom would still be led by the nose, and it would be better not to have Flanders and other conquests, than always to be ordered by someone else'.[142]

So French policy was wavering, ready to draw back from the brink and certainly not yet prepared to break with Spain. When in June news of the capture of Valenciennes and Mons reached Madrid, the French ambassador offered the regrets and as it were the excuses of his king. Everything had happened without his knowledge. He sincerely deplored the unfriendly actions of the Duke of Alva and would have Philip II assured once more of his desire for peace. To which Philip replied that among the rebels in Flanders, there were many subjects of the king of France and people who had lived in France. Franco-Spanish friendship as he saw it, could exist only on one condition: that words were matched by actions. In any case, the Duke of Alva had orders not to break the peace unless the French broke it first.[143] Right up until the news of St. Bartholomew's Day Massacre, Saint-Gouard continued to protest the peaceable intentions of his government. Perhaps he added his own glosses. On 28th June, in conversation with Philip II concerning the satisfaction felt by his master over the punishment inflicted on his own vassals in Flanders, the name of Coligny was mentioned. The ambassador did not hesitate to declare that the Admiral was a very bad man and that the king put no trust in him, although be it said, he was doing less harm inside the court than outside. He concluded with optimism that the king's officers on the Picardy frontier, being Catholics, would do all in their power to avoid conflict. So

[141] Pierre Champion, *Paris au temps des guerres de religion*, 1938, p. 198.

[142] E. Lavisse, *Histoire de France*, VI, 1, p. 122.

[143] *Lo que el embassador de Francia dixo a Su Mag^{d} en S. Lorenzo*, A.N., K 1529, B 29, no. 83. On the same subject, letter from Giulio del Caccia to the prince, Madrid, 19th June, 1572, A.d.S. Florence, Mediceo 4903, or Sauli to the Republic, Madrid, 4th July, 1572, A.d.S. Genoa, L.M. Spagna, 5.2414.

will the Duke of Alva, replied Philip. And Saint-Gouard left him quite satisfied according to the report in the Spanish Chancellery.[144] But when pressed hard on the question of the French fleet, Saint-Gouard had eventually replied that 'the sea is like a great forest, common to everyone, where the French feel free to seek their fortune'.[145]

Nevertheless it is quite true that French diplomacy continued to talk peace not only in Madrid, but in Vienna and Rome. After the failure of the attempt on Valenciennes and the defeat of Genlis, it was as if French policy was put into reverse. On 16th June, Charles IX, informing his ambassador in Vienna of the defeat of the rebels in Mons, 'a just judgement of God upon those who rise against the authority of their prince', claimed that he was doing all he could to see that no help reached the Sea Beggars 'so greatly do I deplore their unfortunate designs . . .'.[146] In July he sent to de Ferrals in Rome a long letter explaining how the Duke of Nassau had deceived him and testifying once more to his desire for peace.[147]

So France's policy was extraordinary for its *volte-faces*. By sixteenth-century standards it was not as immoral as historians in later years have described it, for every power broke the rules and took it for granted that the language of diplomacy was insincere. But what inconsistency it reveals! France could decide neither on a firm rupture of relations, nor on the cessation of acts of hostility along the frontiers in which Spain was interested. A message was sent to the Duke of Alva from Paris on 27th June, warning him that the French were levying three companies of light horse in Provence, that they were fortifying Marseilles 'a gran furia', that troops were being sent to the Piedmont garrisons. On 26th June, there had been a great council meeting to discuss whether to break off relations with Spain. No decision was taken, 'but is this not sufficient indication that if the opportunity arises, they will seize it? It is not safe to trust them without a sword in hand.'[148] French actions revealed such unmistakable duplicity to outside observers, including the Spanish, that almost anything became credible. At the end of June, for instance, a courier from Savoy brought more food for rumours to Madrid: the nuncio claimed that the king of France had requested free passage through the territory of the Duke of Savoy.[149] Consequently Spanish couriers travelling from Italy to Spain almost all gave up using the overland route and took to the sea.[150]

What must Philip have thought of the French protestations of friend-

[144] 28th June, 1572, A.N., K 1529, B 34, no. 100.

[145] *Relacion de lo que el S° Çayas passo con el embassador de Francia, viernes primero de agosto* (*1572*), A.N., K 1530, B 34, no. 2.

[146] Comte H. de La Ferrière, *op. cit.*, IV, p. 104, note 1.

[147] *Ibid.*, p. 106, note 2; B.N., Paris, Fr. 16039, f° 457 v°.

[148] Diego de Zuñiga to Duke of Alva, Paris, 27th June, 1572, A.N., K 1529, B 34, no. 78.

[149] G. del Caccia to the prince, Madrid, 30th June, 1572, A.d.S. Florence, Mediceo 4903.

[150] *Ibid.*, 'cosi tutti [the couriers] sono venuti per acqua'.

ship when he had before his very eyes a report on the raid led by Genlis, who had left on 12th June to relieve Mons, had been routed on the 17th and had fallen into the hands of Alva with papers on him signed by the king of France himself?[151] Or when he received information, on the same date, of the supposed movements of the fleet and the actual movements of the troops in Picardy? The admiral, he was told, was at Metz, in search of 800,000 *livres* to accomplish some mischief in Germany. He and Montmorency were in league to relieve Mons. 'There has arrived here Altucaria, this king's agent in Constantinople', wrote Zuñiga;[152] 'it is said in the streets that he had brought back two millions in gold. But that is not certain and no one knows what the motive could be. I gather that there is talk of sending him back [to Turkey] and it is announced that the Venetians have already signed a truce with the Turk' – a rumour energetically denied by the Venetian ambassador. Thus reports circulated in Europe, reports that were not always carefully sifted, both true and false, sometimes so circumstantial as to indicate treachery. Thus Don Diego de Zuñiga was informed by Hieronomo Gondi of what happened at the king's Council: neither the gentlemen of the long robe nor those of the short robe wanted war. The pacific policy of Catherine de Medici seemed to have won the day.[153] The admiral, excluded from the council, is described on the 13th as conspiring with the queen of England.[154]

The events of Saint-Bartholomew's Day did not then put an end to a determined, wholehearted and vigorous policy of intervention. Neither Rome, nor Madrid, although they welcomed the news, had had any hand in planning the massacres in Paris.[155] And while these killings, the 'novedad' from Paris, had an immediate effect on both French and international politics, while they contributed to spread the terror of the name of Spain throughout Europe and left the field open once again for the more prudent than forceful politics of the Spanish King, I do not share the view that they thereby marked the permanent diversion of French policy into new paths. I disagree with Michelet and other historians who regard 24th August, 1572 as the great turning-point of the century. The 'novedad' had only short-lived consequences. After 24th August, French policy, although temporarily diverted from its previous preoccupations and somewhat at a loss, was to remain true to form, as we shall see.

[151] *CODOIN*, CXXV, p. 56.

[152] D. de Zuñiga to the Duke of Alva, Paris, 17th July, 1572, A.N., K 1529, B 34, no. 128.

[153] Zuñiga to Philip II, Paris, 10th August, 1572, A.N., K 1530, B 34, no. 13.

[154] Zuñiga to Alva, 13th August, *ibid.*, no. 15, copy.

[155] The massacre was premeditated neither in Spain nor in Italy (where this point has long been established): Edgar Boutaric, *La Saint-Barthélemy d'après les archives du Vatican, Bibl. de l'Ecole des Chartes*, 23rd year, Vol. III, 5th series, 1862, p. 1–27; Lucien Romier, 'La Saint-Barthélemy, les évènements de Rome et la préméditation du massacre', in *Revue du XVIe siècle*, 1893; E. Vacandard, 'Les papes et la Saint-Barthélemy', in *Etudes de critique et d'histoire religieuse*, 1905.

Don John's orders and counter-orders, June–July, 1572. But nonetheless in August, 1572 the Massacre of St. Bartholomew constituted a surrender. It was as if one of the players, the king of France, had suddenly thrown in his cards. One is tempted to say that he was bound to lose anyway. But did his opponent know that and had he foreseen this resignation? The strange Mediterranean policy pursued by Philip II in June and July, 1572 would seem to suggest that on the contrary he was overestimating French strength.

Nothing could be more straightforward than Philip's Mediterranean policy towards the League up until June, 1572. To introduce some order into the many projects born of the victory of 1571, the Pope had called a new conference at Rome. The first meeting took place on 11th December.[156] This time, it took only two months to produce an agreement signed on 10th February, 1572 by the Spanish commissioners, the Grand Commander of Castille and his brother, Don Juan de Zuñiga, joined on this occasion by Cardinal Pacheco, and by the Venetian commissioners, Paolo Tiepolo and Giovanni Soranzo.[157] The result this time was favourable to Venice. The treaty stipulated that the allies would proceed to action in the Levant[158] which automatically ruled out any of the projects in North Africa so dear to Spain. Supplies for seven months were to be ordered and a massive shipment of military equipment was to be embarked for the Greeks, who were thought to be on the point of insurgency in the Morea. A base camp of 11,000 men would be set up at Otranto, to be used as a sort of military reserve, drawn on if required. In number the fleet would be if anything larger than that of the previous year: 200 galleys, 9 galleasses, 40 roundships and 40,000 men. The timetable, optimistic as usual, planned the rendezvous of the papal and Spanish fleets at Messina at the end of March; without wasting time they would sail to Corfu to meet the Venetian fleet.

So Spain was once again to defend Venice's vital interests in the Levant, to sacrifice herself for her, as she had in 1570 and 1571; or so it seemed. Granvelle wrote to Philip II that he did not consider it such a bad thing. Since the League certainly would not last long (because of the French and the impatience of the Venetians, who were chafing at the loss of the Levant trade), it would be just as well to take advantage of it while it lasted to smash Turkish power. Only an expedition to the Levant would receive effective cooperation from Venice. It would give great satisfaction to the Pope and Spain's attitude would appear to him and to Italy and the rest of Christendom, one of unselfish solidarity, which might be an advantage.[159] All this was no doubt true, but Granvelle's tone was that of one who is consoling his correspondent (or perhaps himself) for a disappointment.

[156] The commissioners to the king, Rome, 12th December, 1572, L. Serrano, *op. cit.*, IV, p. 351.

[157] *Ibid.*, IV, p. 656–659.

[158] *Ibid.*, p. 657.

[159] Granvelle to Zuñiga, Naples, 20th March, 1572, Simancas E° 1061, f° 16.

Philip II cannot have resigned himself willingly to this 'advantageous' decision, since he had already ordered Don John to attempt an expedition against Bizerta or even against Tunis, at the beginning of spring. It was supposed to take the form of a lightning raid, preceding the voyage to the Levant. In order to prepare for it, Don John left Messina for Palermo, where he arrived on 8th February.[160] He had to try to obtain money from the Duke of Terranova, which promised from the start to be difficult;[161] and then make for the southern coast, the centre for supplies and departures for the African coast. How was he to launch an expedition without supplies and money? As for galleys and troops, he wrote to Naples to ask Granvelle for them.[162] But Granvelle, who did not share Philip's enthusiasm, put it bluntly to Don John on 21st February: 'I do not see how such an important expedition can be arranged without money and without sufficient soldiers. I would not like, as I think I have written to Your Excellency before, to see you set foot in Africa without suitable forces.'[163] Don John did not at first accept these reasons. On 2nd March, he was still hoping to receive the Naples galleys and to sail for Corfu via the long detour along the Barbary coast.[164] But in mid-March, his decision had been reached. After consulting his chiefs of staff, the viceroy Terranova, Don Juan de Cardona, Gabrio Serbelloni and the *veedor general*, Pedro Velásquez,[165] he wrote to the king: 'I shall not recount in detail . . . the reasons which oblige me to return to Messina, as you will be amply informed of them in the accompanying letters written by other hands than mine. These reasons appeared so convincing, that I was absolutely unable to do otherwise, although I desire nothing more than to undertake the Tunis expedition, for the pleasure it would give Your Majesty to see the enemy dislodged from the city.'[166]

Then on 1st May, Pius V the saintly Pope died.[167] This death alone was sufficient to throw into question the entire future of the League. Coming as it did at a time of great political tension, it prompted a radical

[160] Don John to the Commander of Castile, Messina, 27th January, 1572, Simancas E° 1138; another letter, *ibid.* On Don John's stay in Palermo, 8th February, 17th April, I have followed the information given by Palmerini, B. Com. Palermo, Qq D 84.

[161] Don John to the Commander of Castile, 14th February, 1572, see preceding note.

[162] Don John to Granvelle, Palermo, 14th February, 1572, Simancas E° 1061, f° 11.

[163] Granvelle to Don John, Naples, 21st February, 1572, *ibid.*, f°s 12, 13, 14, copy.

[164] Don John to the Commander of Castile, Palermo, 2nd March, 1572, Simancas E° 1138.

[165] L. Serrano, *op. cit.*, I, p. 180, note 2.

[166] Don John to Philip II, Palermo, 17th March, 1572, Simancas E° 1138, autograph. L. Serrano, *op. cit.*, I, p. 180, says 18th March.

[167] Henry Biaudet, *Le Saint-Siège et la Suède durant la seconde moitié du XVIe siècle*, 1906, I, p. 181. Duke of Florence to Philip II, Pisa, 3rd May, 1572: condolences on the death of the Pope and offer of eleven galleys and two galleasses 'como estava obligado à S. S.', summary in Spanish, Simancas E° 1458. Sauli to Genoa, Madrid, 18th May, 1572, A.d.S. Genoa, L.M. Spagna, 5.2414: 'La morte di S. Sta dispiace a tutti universalmente et a S. Mta forse più che a niun'altro.' Nobili and del Caccia to the prince, Madrid, 19th May, 1572, Medicei 4903: 'The League will have lost much of its vigour with the death of this holy man.'

volte-face by Spain, one which subsequent events were soon to cancel out, but which was nevertheless total. On 20th May, Philip II sent his brother peremptory orders (actually written on the 17th) to postpone the departure of his galleys for the Levant. If he had already left Messina by the time the orders reached him, he was to return at once. The following despatches dated 2nd and 24th June reiterated the orders, which were only countermanded on 4th July. His action caused a crisis lasting six weeks which passed through Italy like a hurricane. The dust had still not settled when Don John joined the allies at Corfu.

20th May–4th July: these dates situate the problem within a precise temporal framework to which Serrano has not paid sufficient attention. Having been the first historian to read all the Spanish documents, he not unnaturally reproduced the defensive arguments contained in them. For Philip II had to justify himself to the Holy See, to the Venetians and to the European courts who were taken aback by his action, as well as to his own representatives in Italy. But one is under no compulsion to take this special pleading literally, the more so since Philip II himself distinguishes quite unequivocally between his true motives and the official reasons given for the action.

Officially Philip II claimed that he was keeping Don John at Messina for fear of a break with the French.[168] It cannot be denied that Philip believed there was a threat from this quarter; but there was no reason to be particularly anxious in May 1572.

This explanation simply does not fit the dates. In the history of Franco-Spanish tension, seen from Madrid, neither 17th–20th May, nor 4th July are particularly significant dates. Nothing of any importance occurred during this brief period, either in connection with the Strozzi fleet or in the secondary theatres of operations. Serious news could only concern the Netherlands. But on 20th May Madrid had heard only of the Brill landing. The attack on Valenciennes and Mons did not occur until 23rd–24th May.[169] And on the other hand why should the threat from France have appeared *less* grave to Philip II on 4th July, when on the contrary it was increasing? Indeed Philip admitted as much in the preamble to the counter-order sent on the 4th to Don John. St. Bartholomew was still to come and as yet unsuspected by anyone.[170]

The answer lies elsewhere and becomes clearer if, instead of studying Philip's correspondence with Don John, we examine the letters he sent to Rome, in particular those dated 2nd and 24th June, to Don Juan de Zuñiga.

[168] R. Konetzke, *Geschichte Spaniens . . .*, p. 181, emphasizes the reality of the threat from France. L. Pfandl, *Philippe II*, p. 377–378, who does not examine the facts very closely, considers the king's chief motive was a desire to humiliate Don John.

[169] Don Fadrique, son of Alva, was still, on 15th April, affecting derision for the events on the island of Walcheren, H. Pirenne, *op. cit.*, IV, p. 31.

[170] The interpretation put forward by Gonzalo de Illescas, *Historia pontifical y catolica*, Salamanca, 1573, Part II, p. 358 ff, that the allies waited for news of the St. Bartholomew Massacre before sailing for the Levant, cannot be accepted.

Philip II had pleaded another reason to explain his attitude: the death of the Pope. It was an excellent pretext, for everyone knew that the election of a new Pope always had its effect on European politics and *a fortiori* on the policy of the Papal state. And on this occasion the military budget of Spain, and therefore the Spanish fleet, depended on Rome. The new Pope might be a man more anxious 'to put his own house in order than that of the League', as Saint-Gouard said.[171] True; but this was still only a pretext, as Philip himself admits. The Pope's death was certainly the immediate trigger for the king's decision, in the sense that it liberated him, or permitted him to act as if he were liberated from any previous commitments. But the real reason was that Philip had no desire to send an expedition to the Levant. On the contrary, he was eager for another one: against Algiers.

In his first letter to Don John he had mentioned Algiers. And in his letter of 2nd June to Zuñiga, he wrote: 'You are aware of the orders I have sent my brother and the reasons I have told him to give for keeping the fleet at Messina, using the death of the Holy Father as a pretext and not mentioning the orders I had issued.' But 'since the new election was carried out so rapidly, we can no longer plead this excuse', particularly since the choice of the cardinals was altogether 'holy and good'. However I do not intend to change my mind and as far as Algiers is concerned, I am still convinced that 'this expedition is what is required for Christendom in general and for the good of my . . . states in particular, if I wish to see them gain any benefit from this League and all the expenses it entails, instead of frittering them away on something as doubtful as an expedition to the Levant'. The reasons to be given to the Pope are the rebellion in Flanders, the suspicion that France and England might intervene, and the information received about French rearmament. Above all, not a word about Algiers![172]

The explanation is only too evident: Philip II wished to take advantage of the forces he had massed to strike a blow at the Turks, but only where that blow could be of service to Spain: in North Africa which had always been coveted by the Spanish, and against Algiers which was a vital outpost of Islam, its western base, supply centre for men, ships and the matériel necessary for piracy, the point of departure for the corsairs who intimidated the Spanish states. It was this traditional policy which guided Philip's decision.[173]

[171] 21st April, 1572, B.N., Paris, Fr. 3604, f°s 58 ff.

[172] Philip II to Don Juan de Zuñiga, San Lorenzo, 2nd June, 1572, Simancas E° 920, f°s 95–98.

[173] And not French intrigues – dealings with Algier, the subject of an article by Berbrugger, 'Les Algériens demandent un roi français' in *Revue Africaine*, 1861, p. 1–13 – nor their increased armaments: on 12th February there is a single mention by Fourquevaux (*op. cit.*, II, p. 421) of 24 French galleys in Marseilles (but how reliable is this figure?). As for the fleet which Paulin de la Garde was to lead from the Atlantic to the Mediterranean, it consisted of two great galleys, four small ones, and two brigantines and was still at Bordeaux on 28th June, La Garde to Saint-Gouard, Bordeaux, 28th June, 1572, copy, Spanish translation A.N., K 1529, B 34, no. 103.

Why then did Philip issue the counter-order of 4th July, which meant a return to the original agreement made by the allies of the League? It appears that he yielded to the vehement and unanimous reactions of Italy, the Venetians, the new Pope, Gregory XIII, and the Spanish 'ministers' themselves, Don Luis de Requesens in Milan, his brother in Rome, Granvelle in Naples, not to mention Don John who protested vigorously, for he had taken to heart the success of the League. This general offensive finally got the better of the Prudent King. His critics told him that such a course would discredit Spain, drive Venice straight into the arms of the Turk and would thus expose the king to the loss of both prestige and strength. The suggestion of an expedition against Algiers was on every lip, before the Spanish representatives had dared breathe a word of it in Italy. Besides the new Pope was master of the *escusado* and the *subsidio*. He kept writing burning epistles in favour of the League. Supple and accommodating he might be, but all the excuses in the world would not convince him. The French peril was not taken seriously in Rome or anywhere else. Granvelle himself thought that the Most Christian King of France could be brought to reason if he was handled firmly enough. And Don John thought that the best course as far as the French king was concerned was to embark on a successful expedition to the Levant, to meet his sniping with another Lepanto.

On 12th June, Don John, who was between Messina and Palermo, sent by way of the islands a galley which took only six days to reach Barcelona.[174] It was thought on receiving news of its arrival and the unusual speed of the voyage, that some catastrophe had occurred at Messina. A few days later the galley was on its way back. On 12th July, it brought Don John the precious counter-order of the 4th.[175] He did not receive total carte blanche, for the king asked that Gian Andrea Doria's galleys should be kept back from the Spanish fleet in order that they might be sent to Bizerta – or to Toulon if the king of France should misbehave.[176]

The Morea expeditions. Major hostilities in the Mediterranean occupied the end of summer and the beginning of autumn, over the long stretch from the gulf of Corinth to Cape Matapan, along the western coast of the Morea, that inhospitable, mountainous coast, scattered with reefs, possessing infrequent watering spots and faced by the blank wastes of the Ionian sea. At the end of summer, sudden gales accompanied by tornadoes blow in these regions, flooding the coastal plains and making the sea inhospitable for such vessels as the slim, low-lying galleys. The only shelter was to the north at Corfu, or better along the Dalmatian coast, or on the other side of Cape Matapan, on the coast facing the Aegean. One of the privileges of the narrow seas, the Adriatic and the

[174] See above, Vol. I, p. 358, and note 20.
[175] L. Serrano, *op. cit.*, I, p. 363.
[176] Granvelle to Philip II, 29th August, 1572, Simancas E° 1061; L. Serrano, *op. cit.*, II, p. 70, note 2.

Aegean, is that the autumn storms do not reach them until late in the season by comparison with the open reaches of the Mediterranean, which are the first to suffer.

It was a strange place for a war, and one where the allies had few convenient bases. None of the Venetian islands in the vicinity of this unwelcoming coast (neither Cerigo, too narrow, difficult of access, desperately poor except for its vineyards, nor Cephalonia, mountainous and depopulated) offered a safe harbour or the vital supplies necessary to the fleet. All of them were too far from the coast in any case for an armada to use them as a base against the opposite shore. The only useful supply and shelter points were held by the Turks: Santa Maura, north of the Gulf of Corinth, Lepanto on the gulf, and Navarino and Modon to the south.

The allies, it is true, relying on the reports and promises of a few outlaws, had counted on the insurrection of the population of the Morea. But nothing had been organized in connection with the assumed rising. The discussions within the League had merely concerned the landing of arms for the rebels. And Don John had sent particular messages only to the Greeks on the island of Rhodes. When the allies arrived, they found no rebellion, not even the murmur of unrest. When at the end of the seventeenth century, Venice succeeded in taking the Morea, she owed her victory in part to the collaboration of the Greek population.[177] But in the sixteenth century it did not budge. We do not know if the reason was Turkish vigilance or that the Greeks knew the Latins (and particularly the Venetians, who had once been their masters) too well to let themselves be killed on their behalf. It is true that in the event of a Greek rising the Morea would have been a good choice for a war on land. Its mountains effectively eliminated the possibility of massed action by the dreaded Turkish cavalry; it was far removed from the road network controlled by the Turks, and indeed it looked an obvious candidate for independence. But it does not seem on this occasion that the allies made any conscious choice, and chance as much as calculation about native risings brought them to the coast of the Morea.

Don John had received his orders to join the allies at Corfu on 12th July. He immediately sent word to Marcantonio Colonna, who had just arrived at the rendezvous with the Papal galleys and the Spanish galleys under the command of Gil de Andrade, entrusted to Colonna by Don John. The Venetians had already arrived. On receiving Don John's letter, the allies should have waited for the commander in chief. But fearing fresh delays, hoping to win a victory single-handed and armed with a good excuse to produce later if necessary, Colonna and Foscarini set sail on 29th July for the south, where there was disturbing news. The Turkish fleet, out early as usual, was said to be laying waste the coasts of Crete, Zante and Cephalonia. How could the victors of Lepanto let the Turks plunder their islands and spoil their precious crops? The fleet left Corfu in battle order: it

[177] H. Kretschmayr, *op. cit.*, III, p. 342 ff.

was reinforced on the way by the Cretan galleys under the command of the *provveditore* Quirino. On the evening of the 31st it was at Zante,[178] where it spent three days. Then it was learnt that the Turkish fleet was at Malvasia and the fleet set off to find it. Thus the Christian armada was drawn to the extreme southern tip of Greece, where the two fleets met on 7th August off the island of Cerigo.

Euldj 'Alī's fleet was by no means negligible. The Turkish Empire had built it up again from scratch, not indeed with the effortless ease generally assumed, but at the cost of enormous labour and fatigue, all its resources strained to the utmost by danger. The result of this mighty effort, accurate reports of which had moreover reached Christendom at intervals throughout the winter, was an armada of at least 220 vessels, galleys, galliots and foists. A good half of them were newly built: they had been launched during the winter of 1571–1572 and carried few troops. But Euldj 'Alī had modernized their weapons: equipped now with artillery and arquebuses, their firing power was superior to that of 'Alī Pasha's fleet which had been manned by numbers of slingers and archers. And Euldj 'Alī had succeeded in creating on the model of the Algiers navy an extremely mobile fighting force. The galleys, lighter but solidly constructed, and burdened with less artillery and baggage than the Christian ships they attacked, could outmanœuvre the latter with disconcerting regularity.

And the Turkish fleet possessed two or three signal advantages. It had learnt to its cost that galleasses could be a formidable asset and it was a lesson never again to be forgotten. In Euldj 'Alī it possessed a remarkable admiral who was not troubled for a moment by rivalry from his fellow-commanders, a leader who was visibly more in control of his force than either Marcantonio Colonna at the beginning of the campaign, or Don John at the end. This fleet, moreover, was operating in the shadow of a friendly hinterland, near its supply stations, its troop reserves and its onshore batteries, while the Christians depended on what had been loaded in Messina, Apulia or Corfu in the way of biscuit, and was lying more or less well preserved in the hold of the roundships, those bulky floating warehouses which the galleys had to tow along with the galleasses. The slender oarships wore themselves out on this superhuman labour; and Euldj 'Alī, again to his credit, adopted a policy of maximum efficacy at least cost: his aim was to prevent the Christian fleet from reaching the Archipelago, the parade ground of Turkish power; and at the same time to preserve the fleet so miraculously rebuilt during the winter. If on 7th August he accepted the challenge of Marcantonio Colonna, he did not do so blindly. He knew very well that the allies were without their full complement, that he would not, in particular, have to deal with the stout Spanish galleys, nor the infantry of the *tercios*. He was opposed entirely or almost entirely by the Venetians, who were as short of men as he was himself and handicapped by their heavy keels. And he even possessed superiority of numbers.

[178] L. Serrano, *op. cit.*, II, p. 32.

It was late on the 7th, near the islands of Elafónisos and Dragoneras, not far from Cerigo, that the two sides met. The allied fleet was deployed with a lack of speed which did not augur well. It was four in the afternoon when it challenged the Turkish galleys. The wind was not in its favour and it was restricted in movement. All the offensives, or pseudo offensives were launched by the light Turkish galleys. The allies had with great prudence placed in front of their lines, as they had at Lepanto, the galleasses and roundships, carrying heavy artillery and crammed with soldiers. The rest of the fleet waited behind this powerful screen. Euldj 'Alī had hoped to get round these floating fortresses and fight galley to galley. Being unable to do so, he broke off the engagement. Part of his fleet returned to Malvasia while he remained in formation with 90 of his best galleys. In order to conceal his actions, he had his guns fire blanks and disappeared behind a great cloud of smoke. He rejoined the rest of his fleet without difficulty while a few of his galleys with lanterns set, sailed towards Cerigo to make the Christians suspect he was heading west to cut off Don John's galleys. A second encounter on the 10th followed the same pattern: the Christians sheltered behind their 'ships of the line' and the Turks, unable to break up the formation, stole away.

The chief strategic factor in this early part of the campaign then was the awkwardness and inertia of the allied fleet. The great ships of the line were tied to the towing galleys and thus slowed the latter down to their own speed. The allies were victims of the military routines of the Mediterranean. A more imaginative course would have been to let these large and powerful ships use the wind.

The second encounter had ended with Euldj 'Alī's retreat behind Cape Matapan. Meanwhile the allies made for Zante to meet Don John. The admiral had arrived at Corfu on 10th August to find everyone gone without leaving him the slightest indication of the whereabouts of the fleet. Beside himself with anger, he talked of returning to Sicily. But alarmed by the rumours of bad news, by dint of order and counter-order, he eventually managed to assemble the fleet at Corfu on 1st September. Much time had been wasted – but is Serrano right when he says that Colonna's expedition served no useful purpose? It seems beyond question that the allies had at least saved Crete or Zante from the plundering raids of the Turkish fleet, if not worse.

At the general review which took place once the fleet was assembled, the official figures list 211 galleys, 4 galliots, 6 galleasses, 60 transports, with between 35,000 and 40,000 troops. These figures undoubtedly conceal errors: there was no way of keeping an exact record of private ('aventureras') foists and galleys, those of the volunteer adventurers, the flower of the Italian nobility. On the other hand, this huge fleet had only about two hundred horses on board, and precious little in the way of supplies or money. It is true that its requirements were breathtaking – the armada was a tremendous drain both on the granaries of Italy and on Philip's finances.

Don John did not leave Corfu immediately, but held talks. The Venetian galleys were short of troops and Foscarini refused to accept Spaniards, so he had to be given Papal troops who were themselves replaced by soldiers in the service of Spain. At last the fleet set sail with the same aim as in July: to meet the Turkish fleet. Don John was thus drawn on 12th September to Cephalonia, then to Zante and finally to the coast of the Morea. Reports suggested that Euldj 'Alī was at Navarino. Don John tried to take his ships south of this port in order to cut the enemy off from Modon and trap him in the poorly fortified harbour of Navarino. But faulty manœuvres spoiled the attempt, although the Christian fleet had sailed by night to surprise the enemy. Spotted in time by the enemy patrols, it was unable to prevent Euldj 'Alī from leaving Navarino with his 70 galleys and withdrawing to Modon. The Christians followed him: another Morean campaign was beginning – more dramatic than the first but equally disappointing in the end.

During the day of 15th September, Don John had not allowed his fleet to challenge the retreating enemy. On 16th Euldj 'Alī offered battle, but at nightfall took refuge behind the guns of Modon. If that night instead of withdrawing to the moorings in Puertolongo and allowing the enemy to attack his rearguard (without success), Don John had turned to face him, there was every chance that he could have taken Modon and destroyed the Turkish fleet. For chaos reigned in the blockaded port. Cervantes was later to say that the Turks were all ready to leave their ships: 'tenian à punto su ropa y passamaques, que son sus zapatos, para huirse luego sin esperar ser combatidos' – they had prepared their belongings and 'passamaques' – their shoes – to flee ashore at once, without waiting to be attacked.[179]

The lost opportunity never presented itself again. Euldj 'Alī had acted quickly. While the Christians thought he was simply lying low, he had disarmed part of his fleet and had the guns from the ships mounted on the hills surrounding the town. Already a fortified position, the little stronghold now became impregnable – small though it was according to Philippe de Canaye, who passed through not long afterwards; the breakwater afforded protection for no more than twenty galleys.[180] But the ships were now safe there, even outside the harbour. The Christians by contrast could not ride indefinitely on the high seas. They had to find some anchorage near by. Euldj 'Alī was hardly more a prisoner of circumstances than Don John. There were plenty of schemes for taking the town: each more daring than the last, they were opposed not only by prudence, but often by the sheer facts of the position. Meanwhile it would not be long before Euldj 'Alī could look forward to the bad weather of autumn. And this secret ally is the key to the improvised campaign which with great good fortune, he was able to conduct to a happy conclusion. It was only gradually, and in the end too late, that it dawned on the allies exactly what was implied by his delaying tactics. They thought they might be able

[179] *Don Quixote*, I, Ch. XXXIX. [180] *Op. cit.*, p. 170.

to force the Turk out of his retreat by capturing Navarino through which all supplies for Modon from the north had to pass. The position was poorly fortified and difficult to relieve. But the attempt was dogged by ill luck. Insufficient acquaintance with the terrain, a difficult landing under the command of the young Alexander Farnese, heavy rain which flooded the plain where the Spanish infantry was advancing, the alarm raised in the town and the effective fire of its guns, the arrival of enemy cavalry, while in the distance long convoys of camels and mules could be seen approaching the town, inadequate supplies of food and ammunition, the lack of shelter in a treeless plain swept by the storm – all these things made progress difficult for the landing party, 8000 men, who, it was clear, would have to be re-embarked as a matter of urgency.

The small force returned to their ships during the night of the 5th to the 6th October. During the next two days, the allied fleet once again offered battle in vain before Modon, on the anniversary of Lepanto. It was not even able to take advantage of the sortie of about twenty galleys that set off in pursuit of a Christian transport ship, Don John's galleys being too unwieldy to prevent the hasty retreat of these ships, which Euldj 'Alī had immediately recalled. Should the siege be lifted – or should the allies try once more to storm the port? As an indication of willingness to attack, Don John had ordered the galleys left behind with Doria and the Duke of Sessa, to join him (in fact they did not reach Corfu until 16th October, by which time he was on the way back). For the Venetians wished to continue the blockade for a few more weeks, thinking that the Turkish defences were bound to collapse in the end and that at the very least Euldj 'Alī would be obliged to make for home in very poor weather conditions. Perhaps they were not entirely mistaken. It has been suggested that the Turkish garrison was sorely tried and might have given way.[181] On the other hand Foscarini was later to confess that his real reason was fear for himself, on his return to Venice, remembering the terrible justice meted out to Zane on his return from the fruitless expedition to Cyprus, in 1570.

In the end the allied fleet abandoned the siege of Modon on 8th October and withdrew to Zante where it arrived on the 9th. By the 13th it was at Cephalonia, by the 18th off Corfu. Two days later, the two fleets parted company. It had been decided not to winter either in the Levant or in Corfu or Cattaro, nor to launch any punitive expeditions within the Adriatic against Turkish strongholds. Marcantonio Colonna sided with Don John to overrule the Venetians. 'They seemed content enough from their appearance', wrote the Duke of Sessa.[182] But on 24th October, Foscarini wrote to the Republic: 'The sole cause of the little accomplished in this expedition, is the attitude of the Spaniards, who instead of helping the League, have done nothing but seek to ruin it and

[181] F. Hartlaub, *op. cit.*, p. 156.

[182] Sessa to the king, 24th October, 1572, Simancas E° 458, quoted by L. Serrano, *op. cit.*, II, p. 147.

weaken Venice. Don John's hesitation and dilatoriness throughout the campaign were all part of this plan gradually to exterminate the forces of the Republic and to assure the profit of the king in Flanders, by neglecting the interests of the League and even harming them: the ill will of the Spaniards has been evident in everything touching the interests of the Venetian states.'[183] Could anything be more unjust? But all parties did indeed have the feeling, that autumn, that the League had ceased to exist. The Spanish fleet returned rapidly to Messina, divided into three squadrons. Don John arrived on the 24th with the first; on the 26th, he made his solemn entry into the town.[184]

A contemporary observer, if asked which were the 'great' events of the year 1572, would probably not have mentioned this expedition at all. The year 1572 saw the disappearance of several of the major figures in the world and he would most likely have thought of them. We have already referred to the death of Pope Pius V on 1st May. In June, the queen of Navarre died and with her the Protestant party lost its moving spirit. Among those massacred on 24th August, was the Admiral, Coligny; Granvelle in Naples concluded that the bishop of Dax, 'that Huguenot' who depended on Coligny, would no longer play the same role as during the lifetime of his protector.[185] Another important figure, Cardinal Espinosa, president of the Council of State and grand inquisitor, grossly vain and overwhelmed with honours and responsibilities, succumbed to an attack of apoplexy on 16th September, leaving his desk still littered with unopened letters of state.[186] He was in any case in semi-disgrace and had been shattered by the blow. At the other end of Europe, the Transylvanian prince had died early in the year and on 7th July, the king of Poland departed this world,[187] opening the extraordinary succession crisis which ended the following year with the election of the Duke of Anjou.[188]

Meanwhile one Miguel de Cervantes, wounded at Lepanto, lost, partly through the incompetence of the surgeons, the use of his left hand. And in Lisbon, 'em casa de Antonio Goça Luez',[189] an unknown writer, called Camoëns, was bringing out the *Lusiads*, a book of maritime adventures which embraced that great and faraway Mediterranean, the Indian Ocean of Portuguese enterprise.

[183] L. Serrano, *ibid.*

[184] My source for the details in this paragraph is the full account given by Serrano.

[185] Granvelle to Philip II, Naples, 8th October, 1572, Simancas E° 1061, f° 65.

[186] Mondoucet to the king, 29th September, 1572, and Saint-Gouard, 7th November, 1572, L. Didier, *op. cit.*, I, p. 52 and note 2; G. del Caccia to the prince, Madrid, 20th September, 1572, A.d.S. Florence, Mediceo 4903. The cardinal had been suffering from low fevers. A robust man, a great eater and drinker, 'in due hore l'aggravó il male per un catarro che lo suffocò'.

[187] Monteagudo to Philip II, Vienna, 20th July, 1572, *CODOIN*, CX, p. 483–489, H. Biaudet, *op. cit.*, p. 178.

[188] Charles IX to the bishop of Dax, Paris, 17th September, 1572, E. Charrière, *op. cit.*, III, p. 303–309.

[189] Jean Auzanet, *La vie de Camoëns*, Paris, 1942, p. 208.

3. VENICE'S 'BETRAYAL' AND THE TWO CAPTURES OF TUNIS 1573-1574

What had been only too likely in autumn – the defection of Venice from the League – occurred on 7th March, 1573. The news became public in Italy in April and did not reach Spain till the following month. Rather than a 'defection' or 'betrayal', it was really more of a resignation. Imagine the situation of the Venetian Republic, her trade, industry and finance in confusion, exhausted by this naval war which of all wars was the most costly, her everyday life made miserable by food shortages and high prices. Those who suffered the most, the poor, were by no means the least courageous. It was not for nothing that Don John became the almost legendary hero of the songs of the gondoliers. But the poor did not run the affairs of the Republic and firms owned by rich Venetians could not long be content with indirect commercial links with Turkey.

Venice's case: In addition, Venice had watched the war reach her gates on the frontiers of Istria and Dalmatia. If she continued the struggle, would she not lose all her possessions on the shifting Dalmatian frontier? Sebenico, to take but one example, was doomed according to the experts. And this war which had already lasted three years, had brought her no substantial advantage. She had lost first Cyprus in 1571 and then a whole series of outposts in her Adriatic territories. From the expeditions of 1571 and 1572 she was left with nothing but enormous bills to pay and the bitterness we have already seen. And nothing had shaken her conviction that Spain was seeking to weaken her and make use of her, a long established conviction, for Venice, far more than surrounded and supervised Tuscany or the partly occupied state of Savoy, was the only truly independent state in Italy, free of Spanish rule and influence. Venice had always been on guard against a move from Milan. And her first gesture in spring 1573, when she had 'defected', was to fortify her positions in the western *Terraferma*: better safe than sorry.

These then were Venice's reasons. That, in contravention to the terms of the treaty, Venice did not consult her allies first is really a point of detail. And by the standards of the period, it was a venial offence. After all Philip II had not hesitated to go back on his word in May, 1572.

It was through the good offices of the bishop of Dax, who had been patiently working on it since early days, that the agreement with Turkey was finally reached. We have already referred to his departure for the East in May, 1571, his late arrival in Venice in September and his long stay in the city of lagoons. Having arrived at an inopportune moment, just before Lepanto and the flowering of hopes and dreams that followed the great victory, he continued indefatigably to explain and repeat the object of his mission before leaving for Ragusa in January, 1572. The attraction of the negotiations for the level-headed Venetians was the possibility, favoured by the victory itself perhaps, of obtaining acceptable peace terms – even of recovering Cyprus. This hope lay behind all the negotiations entered into by Venice, whether directly or by intermediary. Cyprus could only be recovered in a manner of speaking of course: Venice would have to accept that she would no longer be the colonial power, but would have to destroy the fortifications[190] and become a Turkish vassal on the island. That is, she hoped to recover not Cyprus itself, but the Cyprus trade – and to embark on the slippery slopes of a relationship with the Turk not very different from that accepted by Ragusa.

These hopes were soon to be dashed by the enormity of the Turkish demands.[191] The talks were protracted under the pretext, true or false, that the sultan was personally opposed to a peace with Venice. In the circumstances, the intervention of the bishop of Dax was decisive: he persuaded the Turks to modify their demands and on 7th March obtained the sultan's consent to a peaceful settlement. On 13th March, the Turkish terms were sent to Venice, where they arrived on 2nd April;[192] they were indeed harsh, not to say dishonourable for Venice, it was generally thought. Venice would give up Cyprus, and surrender the positions captured by the Turks in Dalmatia; she would give back the territory she had herself captured in Albania, free the Turkish prisoners without ransom and pay a war indemnity of 300,000 sequins, to be delivered before 1576, otherwise the treaty would be considered null and void. She would limit her fleet to 60 galleys and pay increased tribute (2500 sequins) for Cephalonia and Zante. The council of Pregadi, convoked by the Doge Mocenigo, was presented with a *fait accompli*. In exchange, it was true, Venice obtained peace, a peace which although uncertain and precarious for years to come,[193] brought with it all the blessings, profits and possibilities of normal life.

In any event the 1573 negotiations prove that the St. Bartholomew Massacre, despite its violence and the brutality of its aftermath, did not alter the chosen direction of Charles IX's foreign policy. The French government did not choose to bind itself to Spain, to bow to Spanish

[190] L. Serrano, *op. cit.*, II, p. 296, note 1.

[191] They claimed Cattaro, *ibid.*, II, p. 303.

[192] *Ibid.*, p. 311.

[193] The peace was not finally ratified until 1574, hence the wording of the document in Ms Ital 2117 (B.N., Paris), *Relatione del Turco doppo la pace conclusa con la Signoria di Venetia l'anno 1574*. On the delay in ratifying the treaty, G. de Silva to Philip II, Venice, 6th February, 1574 (information from Ragusa), Simancas E° 1333; de Silva to the king again, 13th February, 1574, *ibid.*; 12th March, 1574, *ibid.*, and 16th March, 1574, *ibid.*; Don Juan de Zuñiga to Philip II, Rome, 18th March, 1574, *CODOIN*, XXVIII, p. 185. The jurats of Messina to Philip II, 30th March, 1574, Simancas E° 1142; in April, 1574, Philip II at the request of the Vatican offered the support of his fleet in the event of a Turkish attack on Zante or Corfu, Philip to G. de Silva, San Lorenzo, 5th April, 1574, Simancas E° 1333, but only on condition, the king warned, that the French did not make a move. Memories were revived of the summer of 1572. The peace agreed on 7th March was only provisional, see Simancas E° 1333, 12th March; 16th March, 1574, *ibid.*; *CODOIN*, XXVIII, p. 185; 30th March, 1574, Simancas E° 1142.

control in a so-called Catholic coalition. For Catherine de Medici and her sons, the struggle against their over-powerful neighbour continued, as was soon to be seen in the election of the Duke of Anjou to the throne of Poland; or in French pressure on the German electors, particularly the Elector of the Palatinate, the first Calvinist of the *Pfaffengasse*. It was evident too in France's moves towards England, the Netherlands and after 1573, Genoa. It appears only too clearly, sometimes embroidered by imagination, in the reports of Spanish intelligence agents: the informants of the Catholic King and his ministers, not content to describe things as they stood, looked into the future, as indeed it was their job to do. When the peace of La Rochelle (1st July, 1573) put an end to the disturbances following St. Bartholomew,[194] they were most concerned. If only the king of France would now forbid his Huguenots to go to the aid of the rebels in the Netherlands! Amassed by the powerful but short-sighted bureaucracy of the Spanish empire, these rumours, true or false, grew and travelled along many channels and on their journey often took on form and substance. An undeclared war was being waged once more.

The capture of Tunis by Don John of Austria: another victory that led nowhere: Without wishing to re-open the dossier of justified complaints by Spain against Venice, a dossier which has been increased beyond measure by contemporaries and historians, let us acknowledge that Spain had never made such loyal and determined efforts on behalf of the League as she did in the winter of 1571–1572. She had raised the number of galleys by building new ships at Naples, Messina,[195] Genoa and Barcelona. A report by Juan de Soto, Don John's secretary even proposed the incredible target of 300 or 350 galleys.[196] Was this sheer madness? Not entirely. For Juan de Soto also mentioned a rather sensible solution: to have these galleys manned by militiamen and more especially to have an arsenal built at Messina, or rather to extend the arsenal at present under construction and cover it over so that the galleys could winter there – in short to imitate Venice.

So Spain was making her greatest effort yet. Ormaneto, the bishop of Padua, whom Gregory XIII had sent to Madrid as nuncio, received a warm and sympathetic welcome, although there was resistance to his suggestion that a hundred galleys or so be sent in March to the Aegean, to plunder the coasts of the islands. Was it the Turkish example which now tormented imaginations?[197] Philip II and his advisers, more aware

[194] On Alva's displeasure on hearing news of the peace, Mondoucet to the king, 17th July, 1573, L. Didier, *op. cit.*, I, p. 329.

[195] According to Philip's letter to the duke of Terranova, San Lorenzo, 20th June, 1573, Simancas E° 1140.

[196] The king gives a summary of this report in his letter referred to in note 195.

[197] An assumption suggested by the letter from the archbishop of Lanciano to the cardinal of Como, Madrid, 24th January, 1573, Vatican Archives, Spagna, no. 7, f° 10–11; and by that of the bishop of Padua to the cardinal, 25th January, 1573, *ibid.*, f° 22.

of the importance of naval matters, did not accept the plans suggested to them. Once more the king opted for the possible, not for the grandiose.

And news of the Venetian defection was greeted by all the king's ministers with steadfastness and moderation. The Pope, normally so gentle and amiable a man, had on hearing the news fallen into a great rage against the treacherous Republic. He gave the Venetian ambassador a terrible talking-to and immediately, without hesitation, withdrew all the concessions on Church revenues, great and small, that he had ever granted Venice. Then he calmed down and forgot all about it. Don John, Granvelle and Don Juan de Zuñiga on the contrary, remained cool. They had probably foreseen as much many times before. Even the most volatile of them, Don John, on this occasion displayed exemplary self-control.

Meanwhile the Turkish armada left port as predicted, but rather later than usual. An eye-witness reported that it had sailed from Constantinople on 1st June.[198] But Philippe de Canaye says that it did not pass the castles until the 15th[199] and the Dardanelles not until 15th July.[200] Probably the key to the puzzle is the dispatch from Corfu dated 15th June, which reported that the fleet had left in two squadrons, so that Karagali reached Negropont on the 3rd with one group (200 galleys according to the report), where he waited for Piāle Pasha to join him with a hundred more. This delay explains a certain degree of optimism in Italy. 'The Turk will not attack this year', wrote Don Juan de Zuñiga on 31st July;[201] 'he has left port merely to prevent Don John from attempting anything'. Besides, he added, 'his fleet is in poor shape' – a detail which recurs in all subsequent dispatches. The Turkish fleet however continued westwards and on 28th July anchored off Prevesa. A report from Corfu on 3rd August suggested that it would probably raid the Apulian coast before making for La Goletta, having the double task of preventing the Spanish from sending an expedition to Barbary and of quelling the Albanian rebels who had just risen again.[202] On 4th August, according to Canaye, it was steering for the Abruzzi, with the doubtful aim of meeting Don John's ships or making for Palermo.[203] In fact it seems to have hesitated. It headed briefly for Messina, sailing past the Capo delle Colonne on the Neapolitan coast on 8th August.[204] But by 14th August, it seems to have been back at Prevesa, tallowing the ships on the shores of the island of Sapienza. However, it put back to sea on the 19th.[205]

[198] Report by Juan Curenzi, an envoy sent by Granvelle to Constantinople, who returned on 30th June, 1573, Simancas E° 1063, f° 35.

[199] Philippe de Canaye, *op. cit.*, p. 158.

[200] According to a Genoese traveller from Chios aboard a French barque, Simancas E° 1063, f° 42.

[201] To Philip II, *CODOIN*, CII, p. 207–208.

[202] Simancas E° 1332.

[203] *Op. cit.*, p. 180; Granvelle to Don John, Naples, 6th August, 1573, Simancas E° 1063, f° 45.

[204] Granvelle to Philip II, 12th August, 1573, Simancas E° 1063, f° 49. Another report, which arrived rather late from Venice, said that the fleet might be going to attack the Tremiti islands.

[205] Philippe de Canaye, *op. cit.*, p. 181.

It was after this departure in massive formation that real worries began to be felt on the Spanish side. It was rumoured that the Turkish fleet would continue westwards and winter in a French port. Don John gave orders that the infantry which had been on guard in Sardinia during the summer should not be relieved.[206] Similar rumours that the fleet would winter in France, circulated in Rome on about 25th August,[207] but Zuñiga refused to countenance them. We shall never know whether he was right or not, for the Turkish fleet had run into a fierce storm. Some galleys were lost, others badly damaged[208] and oars and masts had to be sent from Preveza.[209] Was this what forced the fleet to turn back? A dispatch from Corfu on 29th August reports that it was at the 'Gumenizos'.[210] At the beginning of September, rather battered and in some disarray, the Turkish armada was lying not off Christian shores, but in Valona.[211] On 5th September however, it sailed to Cape Otranto, where it captured the little fortress of Castro.[212] Then on the 22nd, it turned towards Constantinople[213] with its 230 sail, having achieved little or nothing. At the end of the month, it was at Lepanto, taking on fresh supplies.[214]

This rather erratic voyage had nevertheless had its long range effect on Don John's plans for the summer. In Spanish circles there had been talk for some time of an expedition against North Africa. The forces assembled during the winter, although no match for the entire Turkish fleet, were nevertheless considerable. 'It is thought here', writes the Florentine agent, del Caccia from Madrid,[215] 'that there will be some expedition against the Turk or against Algiers or some other place, in view of the great preparations under way, the money[216] that is being collected, the fresh levies in Spain and the launching of the new galleys in Barcelona. For it cannot be supposed that all this effort is intended simply for defence against the Turkish fleet'.[217]

But as long as the Turkish fleet was within striking distance of the coasts of the kingdom of Naples, there could be no thought of such a venture. No one wanted to run the risk of another Djerba. As for the

[206] Don John to Philip II, Messina, 20th August, 1573, received 3rd September, Simancas E° 1062, f° 117.

[207] Zuñiga to Philip II, 25th August, 1573, *CODOIN*, CII, p. 229.

[208] Philippe de Canaye, *op. cit.*, p. 181, 186, eight galleys were lost and eight more damaged:

[209] Zuñiga to Philip II, 28th August, 1573, *CODOIN*, CII, p. 231.

[210] Simancas E° 1063, f° 87.

[211] Philippe de Canaye, *op. cit.*, p. 186.

[212] Granvelle to Zuñiga, Naples, 11th September, 1573, *CODOIN*, CII, p. 258–259.

[213] Philippe de Canaye, *op. cit.*, p. 195.

[214] Granvelle to Zuñiga, Naples, 8th October, 1573, *CODOIN*, CII, p. 307–311.

[215] Mediceo 4904, f° 86.

[216] The money also went to Flanders, Saint-Gouard to the king, Madrid, 14th July, 1573, B.N., Paris, Fr. 16105.

[217] But the expedition mentioned in diplomatic correspondence is usually one against Algiers; the bishop of Padua to the cardinal of Como, Madrid, 15th July, Vatican Archives, Spagna 7, f° 372; Sauli to Genoa, Madrid, 14th July, 1573, L.M. Spagna, 5.2414.

target of the expedition, there seems to have been some hesitation between Algiers – favoured by Don John, Philip II[218] and public opinion in Spain – and Bizerta-Tunis, the target preferred by Sicily. The latter was counselled by the proximity of the naval bases and also, it seems, favoured by the Council in Madrid. This choice eventually won the day. The expedition against Algiers was postponed until some future date.[219] Meanwhile an attack would be made on Tunis.

But even before the shore of Africa was in sight, there were problems to be resolved. Taking Tunis was all very well: but what was to be done with it once it was captured? Would it be wise to install a local puppet ruler, as Charles V had in 1535 (though without any illusions)? Don John wrote to Philip on 26th June: 'It is felt here that we should attempt the conquest of Tunis but without handing the town over to the king Mawlāy Hamīda'.[220] According to a letter written by Gian Andrea Doria, Philip had left it to Don John to make up his mind in the last resort.[221] The same letter indicates that the expedition against Algiers had been abandoned because of the lateness of the season. And unless haste is made, it adds, the same danger will arise for Tunis, 'for although it is but a brief voyage, it is as we know so difficult between Trapani and La Goletta, that if by chance it has been raining in Barbary, the galleys must stay in Trapani more than two months without being able to cross'. Don John has spoken of fetching cavalry from the states of the Archduke Ferdinand: that will take too long, says Doria. Let us use what we have to hand and bring the Germans from Lombardy – that should be sufficient.

To make haste was good advice on 2nd July, before the arrival of the Turkish fleet. But soon the alarm accompanying its movements had to be reckoned with. This was a fresh cause of delay, and the usual difficulties remained: grain supplies, preparation of the galleys, the shipment of troops or money – the latter being once again a matter of some urgency.[222] It was for this reason that Don John had travelled to Naples, which was a bigger supply centre than Messina. Although he was ill, 'plagued by three or four complaints', he made all possible haste, hurrying to return to Sicily, not now to Messina, where he did not stop,[223] but to Palermo and Trapani, the gateway to Africa. While he was busy with these preparations, the Duke of Sessa, his second in command, was disturbed, not without reason, by rumours that the Turkish fleet intended to winter at Valona. He asked Philip II if it would not be appropriate as he had suggested to

[218] Sauli, see preceding note; Saint-Gouard, see note 216 above.

[219] Algiers now became an indefinite project, Don John to Philip II, Naples, 25th July, 1573, Simancas E° 1062, f° 112.

[220] Simancas E° 1062, f° 96.

[221] G. Andrea Doria to Don John, Messina, 9th July, 1573, orig. in Algiers, G. A. A. Register no. 1686, f° 191.

[222] Don John to Philip II, Naples, 10th July, 1573, Simancas E° 1062, f° 105 and again on 4th August, 1573, *ibid.*, f° 113.

[223] Don John to Philip, Naples, 5th August, 1573, *ibid.*, f° 114, on his departure for Messina; Don John to Philip, Messina, 10th August, 1573, Sim. E° 1140, he had arrived in Messina on 9th August.

Don John, 'to send 12,000 foot soldiers, Spanish and German, to La Goletta, in some good roundships'.[224] Philip, for his part, must have considered the Turkish threat a serious one, for he wrote to Terranova on 12th August, that the Tunis expedition would not take place unless the Turkish fleet gave the opportunity.[225]

Don John was in need of this prudent advice, for at about this time, on 15th August, he was declaring to the king that he intended to lead the expedition, even if the Turkish fleet did not retire.[226] Was this merely his own enthusiastic nature or was he responding to the prompting and promises of the Papacy? In the first place the Pope offered his galleys.[227] And he had even spoken of the crown of Tunis for Don John. On this little matter, which has been much discussed and obscured by chroniclers and historians, I do not believe, despite the authoritative opinion of Törne,[228] that we should entirely discount the malicious gossip of Antonio Pérez. Don John was undoubtedly tempted by the desire for a princely throne, an overwhelming passion which gave him little rest. Törne claims that the Papacy waited until Tunis had been captured before mentioning crowns. This is possible. But in his letter written in June and quoted above, Don John specifically rejected the idea of restoring Mawlāy Hamīda. He would appoint a native governor in Tunis, not a king. Was this entirely without ulterior motive? The Pope for his part, was saying in Rome on about 20th October (before it was known that Don John had captured the town) that 'if Tunis were taken it would be best to keep this kingdom in Christian hands rather than bestow it on some Moorish king'.[229] And Pius V had already promised Don John the first Infidel state to be captured. Any state, no matter which. In fact what tempted Don John more than effective power was the title. In a Europe besotted with precedence and hierarchy, all young princes dreamed of crowns. The Duke of Anjou had just obtained one in Poland, after considering Algiers. Don John, bitterly resentful of his bastard status, granted only the inferior rank of Excellency, dreamed longingly of the French crown when it was briefly unclaimed on the death of Charles IX in 1574; and his last years in the Netherlands were haunted by fantasies about an English throne.

So it is not impossible that Pope Gregory XIII, in his anxiety to fight the Infidel, sought to secure the success of the enterprise by a promise of this nature. In fact it makes much more sense before than after the victory at Tunis. For Don John had to work near-miracles to assemble

[224] 4th August, 1573, Simancas E° 1063, f° 3, 167.

[225] From San Lorenzo, Simancas E° 1140 M.

[226] Simancas E° 1140.

[227] Zuñiga to Philip II, Rome, 13th August, 1573, *CODOIN*, CII, p. 209.

[228] *Op. cit.*, I, p. 243 ff.

[229] Zuñiga to the king, 23rd October, 1573, *CODOIN*, CII, p. 330. The letter is evidence that the Pope still had no news: 'Dijome el otro dia el Papa hablandome en la jornada del señor D. Juan que si ganaba a Tunes. . . .' News of the fall of Tunis (11th October) reached Naples on 22nd or 23rd October, Granvelle to Philip II, Naples, 23rd October, 1573, Simancas E° 1063, f° 110.

all the material necessary for the expedition and obtain it virtually without money.[230] He was only provided with the essential funds in miserly instalments.[231] And even these were not always much use. He received two bills of exchange for instance, one for 100,000 crowns and one for 80,000 – the first of which was payable at the end of December and the second in the forty days following the 1st January! He must use them to obtain advances, wrote Escovedo. Were his paymasters dragging their feet? Not exactly, but the Spanish treasury was in a disastrous shape. The total sums of money borrowed at Medina del Campo are frightening, Escovedo wrote to him. 'Flanders is ruining us . . .'.

At the beginning of September, Don John left Messina for Palermo where he arrived on the 7th.[232] He had left behind the Duke of Sessa and the marquis of Santa Cruz, while he sent Juan de Cardona to Trapani. The report that the Turkish fleet had been seen off Cape Santa Maura on 2nd September turning back toward the Levant, decided him.[233] But the Spanish fleet was not yet ready to leave. The president of Sicily (the island had had no viceroy since Pescara) used the interval to compile a long memorandum on the enterprises planned for Barbary.[234] On the 27th Don John arrived in Trapani from Palermo.[235] At this point the Turkish armada returned again, and the threat appeared sufficiently dangerous to set in train the usual dispositions for the defence of the kingdom of Naples.[236] The North African expedition was once more in the balance.

Nevertheless Don John, who was also faced with contrary winds, set sail for Africa on 7th October, the anniversary of Lepanto.[237] Leaving Marsala, he had sailed for Favignana whose shelter he left at four in the afternoon. On the 8th at sunset he stood before La Goletta. The landing on the 9th took all day, there being disembarked 13,000 Italians, 9000 Spaniards, 5000 Germans. The fleet numbered 107 galleys, 31 ships, plus the galleon of the Grand Duke of Tuscany and various barques laden with supplies, frigates and other small boats owned by private individuals.[238] On the 10th, he approached the town, whose inhabitants had fled without fighting and the next day occupied it without difficulty;[239] in the casbah the invaders discovered only old people left behind.

[230] Don John to Granvelle, Messina, 19th August, 1573, copy, B.N., Madrid, Ms 10.454, f°s 114–115; Zuñiga to Philip II, Rome, 21st August, 1573, *CODOIN*, CII, p. 219–220.

[231] Escovedo, to Don John, Madrid, 5th September, 1573, A.E. Esp. 236, f° 122.

[232] Duke of Terranova to the king, Palermo, 7th September, 1573, Simancas E° 1139.

[233] Granvelle to Don John, 6th September, 1573, Simancas E° 1062, f° 118.

[234] *Parere del Duca di Terranova, Presidente di Sicilia, sopra le cose di Barberia* 17th September, 1573, Simancas E° 1139.

[235] Duke of Terranova to Philip II, Palermo, 30th September, 1573, Simancas E° 1139.

[236] Granvelle to Philip II, Naples, 9th October, 1573, Simancas E° 1063, f° 94.

[237] Duke of Terranova to Philip II, Palermo, 9th October, 1573, Simancas E° 1139.

[238] According to the report in B.N., Florence, Capponi, *Codice*, V, f° 349.

[239] Formal confirmation of this date, 11th October, Jorge Manrique to Philip II, Palermo, 7th November, 1573, Simancas E° 1140.

What would Don John do with his conquest? Orders had been sent from Madrid to destroy the town, but Don John did not receive them until his return. On the spot, for want of instructions, he had called (on the 11th or possibly the 12th) a council of war in the casbah of Tunis. Besides his usual advisers, he had invited the colonels of the Spanish, Italian and German infantry, as well as some captains and other individuals whose opinion was valued. Was this entirely without significance or did Don John wish to swamp his official advisers with large numbers? At any rate the impromptu council decided by a majority vote to hold the town for the king of Spain.[240] Don John drew up his commands accordingly. As an essential measure, he left in occupied Tunis a garrison of 8000 men, 4000 Italians and 4000 Spaniards, under the orders of a 'gunner', Gabrio Serbelloni.[241] This action entailed others, notably the appointment of a civilian governor, the Hafsid heir, Mawlāy Mahamet, brother of the former king Mawlāy Hamīda (the equivalent perhaps of setting up a protectorate) and the construction of an enormous fort dominating the town.[242]

It remained to be seen – and Granvelle was anxious as day after day he watched the stormy sea – whether Don John would succeed as easily in capturing Bizerta, where some Turks were said to have taken refuge, a port which was suitable for piracy and partly fortified.[243] But Bizerta too surrendered without a struggle.[244] Don John had only spent a week in Tunis which his troops sacked. After four days of preparation at La Goletta, he had taken ship once more on the 24th and took Porto Farina the same day. On the 25th he was at Bizerta. He left it five days later and had fair weather on his voyage to the island of Favignana; on 2nd November he entered Palermo where a fortnight earlier the town had been torchlit in celebration of the fall of Tunis.[245] By the 12th he was in Naples, where Granvelle paraphrased Caesar's 'veni, vidi, vici' in his honour.[246]

Undoubtedly in military terms, the Tunis expedition had been a straightforward exercise. A spell of fine weather at the end of the summer, when 'the figs were ripe' in the orchards, had made things easy. But was it a victory?

The loss of Tunis: 13th September, 1574. There was more to the conquest of Tunis than the mere capture of the city: it had to be held. But the victorious army had only occupied a small part of the Hafsid kingdom.

[240] *Relacion que ha dado el secretario Juan de Soto sobre las cosas tocantes a la fortaeza y reyno de Tunez*, 20th June, 1574, Simancas E° 1142, copy.

[241] *Instruccion a Gabrio Cerbellon*, Simancas E° 1140.

[242] Don John to Granvelle, La Goletta, 18th October, 1573, Simancas E° 1063, f° 114.

[243] Granvelle to Philip II, Naples, 23rd October, 1573, *ibid.*, f° 110, received 11th November.

[244] Marquis de Tovalosos, B.N., Paris, Esp. 34, f° 44.

[245] Palmerini, 20th October, 1573, B. Com. Palermo, Qq D 84, Don John arrived on 2nd November.

[246] Granvelle to Don John, Naples, 24th October, 1573, A. E. Esp., 236, f°s 88–90.

There was never for a moment any suggestion of an expedition to the interior to attempt to subdue the vast hinterland.

Under such conditions, holding on to the large city presented considerable problems. The most serious was the maintenance of the 8000 soldiers posted to guard it, as well as the thousand or so men in the *presidio* at La Goletta. It was a heavy burden on the supply stations of Naples and Sicily: wine, salt meat and grain could not be had without money, nor could the hired vessels by which increasingly the transport of supplies and ammunition was handled. The financial exhaustion of both Sicily and Naples made these comparatively straightforward operations almost insoluble problems. Granvelle's complaints were no mere jeremiads, but simple statements of fact. This was the true problem of the kingdom of Tunisia, far more than the nuncio Ormaneto's efforts to obtain the title of king of Tunis from Philip II for Don John of Austria, efforts which soon came to nothing and are trivial by comparison.

There are several ways of explaining Don John's failure. The split between Philip and his half-brother was widening: gossip, espionage by his informers, probably the malevolence of Antonio Pérez (though we cannot accept it without investigation for these years) and the natural distrust of the king all contributed. But it is also true that Don John, confined to his own sphere of action, did not fully appreciate the overall situation of Spain. Much more than he suspected, from the day that Venice had left the League, Philip II, despite anything he might write or appear to do to the contrary, had abandoned any major initiatives in the Mediterranean. He was wrestling with the formidable financial crisis which was to lead to the second bankruptcy of 1575. Without the resources and the convenience of the Antwerp money-lenders, he depended more and more on Genoa and the Genoese moneymarket. But troubles had broken out there in 1573 between the 'old' and the 'new' nobility, between those who dealt in finance and those whose interests were commercial and industrial. It was a social crisis, but with political overtones. Was the king of France not behind the new nobility? And it was a nasty moment for the Spanish Empire, for Genoa was the turntable not only for money remittances, but also for troop movements.

So just when the North African problem was reaching a critical stage, Philip's calculations were all concentrated on the north, on Genoa and the Netherlands; and on France where intrigue was raising its head again. In the circumstances, remaining in Tunis was unwise. It meant opening a fresh series of expenses, weakening La Goletta since from now on the military effort would have to be divided between the new *presidio* and the city. It meant risking what was already in hand for very doubtful advantages. The king pointed this out to Ormanetto, but the latter was unwilling to understand that shortage of money in Madrid meant a change in attitude and in intentions. Apart from the king who declared 'I would rather die than consent to anything contrary to my honour and reputation', everyone wanted peace, even in the North.[247] It may be gathered from this that Spain was in no mood for adventures. However since Don John had presented his brother with a *fait accompli* by holding on to Tunis, the king judged it preferable to give his consent, but only provisionally – until the end of the year.

What Don John had in effect created was a cumbersome colonial machine. If only it had been able to function unaided, if only the conquest could feed the occupation army! Its supporters claimed that it did; Soto went to Madrid in May, 1574, on behalf of his young master to tell the king so; Granvelle had claimed as much since January of the same year. He thought that if Bizerta and Porto Farina were fortified as Don John asked, Philip II would be assured of possession of the eastern sector of North Africa, which would pose obstacles to Turkish communications with Algiers by sea and would entirely disrupt communications overland. Once the fort was built, the revenues enjoyed by the sovereigns of Tunis could be appropriated and would serve to maintain not only the said fort, but any others that might be built. And these revenues could be increased by encouraging Christian trade in the area. In order to do so it would be necessary to assure the goodwill of the local population and choose with great care the form of government they would receive, in order that they might appreciate the merits of Spanish rule.[248]

But this was just the problem. The inhabitants of Tunis had returned to the town at the end of October (though not the leading citizens, whose houses were still occupied by soldiers) but there is no evidence that economic life had returned to normal. It must have revived to some extent, since the occupation authorities and the native governor had discussions concerning the customs at La Goletta, the said governor asking to re-establish his duty of 13 per cent, chiefly on hides.[249] (The Spaniards who passed the information to Don John expressed their disappointment in their protégé on this occasion). But there is no evidence that the bulk of the country with its nomadic and settled populations had accepted the Christian conquest. In Constantinople, the Turks affected to underestimate the victory at Tunis and claimed that the 'Arabs', by which they meant the nomads, would be sufficient to reduce this conquest to size – the nomads whom the winter cold drove far south, but who would return in summer to the Mediterranean coast, at the same time as the Turkish fleet left port.

At Madrid, no one was disposed to tackle such complicated questions. There were too many other problems. Just as Don John was preparing, once the accounts of the fleet were more or less settled, to go to Spain, he received orders on 16th April to go to Genoa and Milan; at the same

[247] Saint-Gouard to Charles IX, Madrid, 3rd February, 1574, B.N., Paris, Fr. 16106, f° 304.

[248] Granvelle to Philip II, Naples, 27th January, 1574, Simancas E° 1064, f° 7, published by F. Braudel in *Revue Africaine*, 1928, p. 427–428.

[249] Simancas E° 488.

time the king appointed him his lieutenant in Italy, with authority over his ministers.[250] In Genoa, it was thought that his presence would help to settle political differences and he remained in the city from 29th April until 6th May. But his chief mission was to Lombardy, where, it was thought in Madrid, his arrival would alarm France sufficiently to prevent her from picking a quarrel with Spain. In Milan the presence of the king's brother would hasten the dispatch of reinforcements to the north – for Flanders, 'lo de Flandes', remained Spain's major concern.

To Don John, Philip's orders seemed tantamount to disgrace. The lamentable condition of the fleet which for lack of money was disintegrating with alarming speed, added to his distress. In the beautiful castle at Vigevano, he fretted as he waited for the return of the secretary, Juan de Soto, whom he had sent to the king with detailed reports. Tired, resentful, worried about his health, he refused to apply himself to any matters not even the question of Tunis, the fleet, the supplies, referring everything to the king's ministers. Not that the latter did not protest at this transfer of responsibility on to their shoulders (or as Granvelle put it, 'passing the buck'). Meanwhile Juan de Soto arrived in Madrid in May. His memoranda were carefully examined by the Council of War and the Council of State. Don John's dreams were buried under the sheaf of carefully penned documents of the Chancellery and came out in the form of questionnaires submitted for advice to the Councillors. 'Concerning "lo de Tunez" ', said the *Consulta* of the Council of War, 'it appears to everyone that the season is so far advanced that there is no need to discuss whether we should stay there or not. For this summer, the occupation force will of course be maintained. We need only recommend that Lord Don John and the ministers do all in their power to supply the position with necessities'. 'Agreed', noted Philip in the margin, 'but see that my brother is particularly reminded of La Goletta, so that it continues to be supplied just as well as if there were no fort in Tunis'.

If everyone was agreed that the Spanish soldiers, without whom everything would be lost were to remain in Tunis,[251] opinions differed as to the future role of Don John. The Duke of Medina Celi considered that if the prince had been sent to Lombardy on account of Flanders and France, his presence there was no longer necessary, since the commander in chief had now arrived in the Netherlands and in France domestic difficulties entirely occupied the horizon. So Don John ought to be once more tackling naval problems: the defence of Naples, Sicily and the Barbary positions, once more assuming command of the fleet. The Duke of Francavila shared this view. The marquis of Aguila on the contrary thought that 'many things should be taken into consideration on both sides.' If he could not be provided with the indispensable money and

[250] O. von Törne, *op. cit.*, I, p. 216, or better, L. van der Essen, *op. cit.*, I, 1, p. 181 ff.

[251] *Lo que se ha platicado en consejo sobre los puntos de los memoriales que el sec° Juan de Soto ha dado de parte del Sor D. Juan*, undated, Simancas E° 488 (May or June, 1574).

troops, what good would it do to send Don John back to the fleet? The bishop of Cuenca spoke on the same lines, stressing the poor condition of the fleet: there might well be 120 galleys, but was that enough to face the Turk? In an unequal encounter would the victor of Lepanto be reduced to attacking the enemy's rearguard, perhaps to flight, or even, more seriously, to some rash exploit to which his excess of youth and ardour may incline him? The president accepted the wisdom of all these points of view and Philip concluded on this prudent note: 'Inform my brother that the affairs of Tripoli and Bougie do not seem of such importance as to warrant risking the armada in winter'. From these few lines, which summarize a wordy document, we can see the wealth of reports and advice, the minute bureaucratic labour on which Philip II based his foreign policy.[252] It was well worth considering all possible contingencies rather than taking sudden decisions, when it took over a month for orders to reach their destination.

But as far as Tunis was concerned, in the end nothing went according to plan. Don John was forced to act on his own initiative. The Turkish fleet, whose size, departure, and slow progress hindered by the not yet fully functional new galleys, had been faithfully relayed by western agents, arrived in the gulf of Tunis on 11th July, 1574. It numbered 230 galleys, a few dozen small ships[253] and carried 40,000 men. Euldj 'Alī was in command of the fleet and the army was under the orders of the same Sinān Pasha (not to be confused with the victor of Djerba) who had in 1573 finally subdued the Yemen, which had been in revolt for years. To general astonishment La Goletta was captured on 25th August after barely a month's siege.[254] Puerto Carrero had not defended it, but handed it over. The fort at Tunis held out scarcely any longer: Serbelloni capitulated on 13th September.

How is this double defeat to be explained? The fortifications at Tunis were unfinished, and this greatly handicapped the defenders. The two positions were too far apart to be able to assist each other. And the Turks received help from local auxiliaries: the nomads assisted them with transport and the digging of trenches, providing Sinān Pasha with an army of navvies. Possibly the Christian garrisons were on the other hand manned by inferior troops. Granvelle suggested that the over-frequent levying and renewing of troops by Spain had led in the end to a loss of quality compared to the old regiments. But then Granvelle was looking

[252] See preceding note.

[253] Granvelle to Philip II, 22nd July, 1574, Simancas E° 1064, f° 46, refers to 320 ships.

[254] Puerto Carrero to Granvelle, La Goletta, 19th July, 1574, trenches had just been dug on the Carthage side, Simancas E° 1064, f° 46; *Relacion del sargento G° Rodriguez*, La Goletta, 26th July, 1574, Simancas E° 1141. On the siege and capture of La Goletta, see also E° 1064, f°s 2, 4, 5, 5, 25, 54, 57, 58; the anonymous booklet, *Warhaftige eygentliche beschreibung wie der Türck die herrliche Goleta belägert*, Nuremberg, Hans Koler, 1574; *Traduzione di una lettera di Sinan Bassà all'imperatore turco su la presa di Goleta e di Tunisi*, undated, B.N., Paris, Ital. 149, f°s 368–380.

for arguments in his own defence, since he was one of the men responsible for the disaster.

The rapidity with which his troops had surrendered did not make it any easier for Don John. He had done what he could. He had shaken himself out of his torpor on 20th July, before there were any definite reports of the Turkish landing in the gulf of Tunis. But he was far from Africa. And how could he mobilize a fleet which was short of funds and therefore in a much neglected state? On 3rd August, a sheaf of orders from the king finally reached him at Genoa; now at least he could select from among them orders which coincided with his own intentions. On 17th August, he arrived at Naples with 27 galleys.[255] On the 31st he was at Palermo, but it was already too late.[256] All that had been done so far, in response to the appeals for help of Puerto Carrero, was to send two galleys, whose oarsmen had been promised that they would be freed if they succeeded in their mission. 'I doubt whether they will get through knowing how difficult it is to enter the gulf of Tunis', wrote Don Juan de Cardona on 14th August.[257] Don John's adviser, Don García, wrote to him on the 27th that the best solution would be to have the troops in Tunis sent to La Goletta in small detachments: but La Goletta had fallen two days before he wrote.[258] By an ironic coincidence it was at this moment that Madrid decided to send back Juan de Soto whom his master had awaited for so long. He arrived at Naples on 23rd September and, again ironically, brought permission for Don John to rejoin his fleet. 'They' [the Spanish], adds Giulio del Caccia, who is our source for these details, 'have provided seven thousand crowns and given further orders, upon discovering the full extent of the danger. May God deliver us!'[259]

To crown his misfortunes, Don John was held up in September by bad weather. But he did what he could. On 20th September, he sent Gian Andrea Doria with 40 reinforced galleys[260] to Barbary, meanwhile dispatching Santa Cruz to Naples to pick up some German troops.[261] By 3rd October, he had succeeded in assembling at Trapani, besides the papal galleys, half his fleet, that is about sixty galleys. He was on the point of sailing for La Goletta, against the advice of Don García, when he received simultaneously news of the disasters in Africa and of Juan de Soto's arrival in Naples. 'What welcome dispatches can he be bringing after a five-month absence', Don John bitterly reflected: 'he will give me news of what is already past history and instructions on how to prevent a disaster which has already taken place'.[262] Don John was all the more bitter since

[255] Or 22 or 23, O. von Törne, *op. cit.*, I, p. 279, note 6.

[256] Duke of Terranova to Philip II, Palermo, 31st August, 1574, Simancas E° 1141.

[257] Simancas E° 1142.

[258] *CODOIN*, III, p. 159.

[259] Madrid, 28th August, 1574, Mediceo 4904, f° 254.

[260] Duke of Terranova to Philip II, Palermo, 20th September, 1574, Simancas E° 1141; Saint-Gouard to the king, Madrid, 23rd October, 1574, B.N., Paris, Fr. 16106.

[261] Granvelle to Philip II, Simancas E° 1064, f° 66.

[262] O. von Törne, *op. cit.*, I, p. 280, note 1.

he knew he would be held responsible for what had happened. He had no more illusions on this score than Granvelle. We can sense this from the tone of his letter, vehement and emotional, recognizing his own errors and even more readily those of other people, including the king.[263] On 4th October, he wrote once more to his brother, this time informing him not so much of his regrets but of his hesitations, the plans he had considered and his final decision not to do anything.[264] It would indeed have been imprudent at the approach of winter to attempt the raid on Djerba he had briefly contemplated. It would be but a small local victory to set against a spectacular Turkish success, even supposing that all went according to plan on this voyage of 300 miles, 200 of which were out of reach of any harbour, and with the unfavourable winds of autumn. Would it be wise on the other hand, to return to Tunis and dislodge the four or five thousand Turks who were there? And (though he did not say as much, he must have thought it) anyway what would be the point? At best he would only be repeating the 1573 expedition and would be exposed once more to the same consequences. He would act only on the king's orders, he concluded. Don John was becoming prudent.

He did not reach Palermo until 16th October because of bad weather. In the Sicilian capital, he found Juan de Soto, convened his council and asked for advice on a possible course of action; but in view of the approach of winter and the exhaustion of Sicily, he deemed it useless to wait for the king's reply. Actually he had but one desire: to return to Spain, see his brother and explain his actions to him. Continuing his journey by the longest route, along the coast,[265] he was in Naples on 29th October. On 21st November he left for Spain.[266]

Peace at last in the Mediterranean: Meanwhile the Turkish fleet was making its way unhindered back to Constantinople, where it arrived on 15th November, 247 galleys strong, not counting other vessels, according to a Genoese agent. The expedition, although apparently so successful, had been accomplished at the cost of great loss of life. '15,000 oarsmen and soldiers have died of disease and over 50,000 have been lost in action at Tunis and La Goletta' wrote the same agent, reproducing rumours which must surely have been exaggerated if not completely untrue.[267] But in any case what did the loss of even thousands of lives matter to the great Ottoman Empire? With this victory it had recovered its self-respect.

[263] Don John to Philip II, Trapani, 3rd October, 1574, *ibid.*, p. 283; Granvelle's letter of 27th September, 1574, which was dramatic in tone (Simancas E° 1064, f° 61 cf. F. Braudel, in *Revue Africaine*, 1928, p. 401, note 1) only concerned the first disaster the fall of La Goletta.

[264] Don John of Austria to Philip II, Trapani, 4th October, 1574, Simancas E° 450.

[265] Don John to Philip, Naples, 12th November, 1574, *ibid.*

[266] According to van der Hammen and Porreño, quoted by Törne, *op. cit.*, I, p. 288, notes 4 and 6. The preceding paragraph is based on Törne's account and on my article 'Les Espagnols et l'Afrique du Nord' in *Revue Africaine*, 1928.

[267] Bª Ferraro to the Republic of Genoa, 15th November, 1574, A.d.S. Genoa, Cost. 2.2170.

'They stand in awe of no Christian fortress now', says a dispatch from Constantinople.[268] What observer could have predicted then that this was the last triumphal entry a Turkish fleet would ever make to Constantinople?

At the same moment, both in Madrid and in Italy, the Spaniards were in despair before the immensity of the Turkish peril, whether those in command, like Don John, the Duke of Terranova, or Granvelle, or the advisers through whose hands passed the countless government memoranda. Elated by their victory, what might the Turks not do now? 'By the grace of God, by the blood of our Lord Jesus Christ, let not the Turks install themselves and fortify Carthage' wrote the military *veedor* at Milan, Pedro de Ibarra.[269] Let no reliance be placed, for the love of God, wrote Granvelle in his own hand, 'on the advice of those who always claim that things are possible which patently are not; let us not burden the subjects of the Empire to the point of extreme despair. I swear to Your Majesty that when I see the state to which we are everywhere reduced, I would rather lose my life than fail to spend it in seeking some remedy'. But what could be done without money? The naval arms race had undoubtedly been pure folly: 'In order to make himself superior, the Turk has increased his forces by the same amount. Whereas in the past his greatest fleet was 150 galleys with which he could not carry enough men (for his strength lies in numbers) for grand schemes, today he sends 300 ships with so many troops aboard that there is no fortress capable of resisting him!' . . .[270]

In Rome, Gregory XIII, overcome with emotion, tried to tempt Venice back into the League – in vain needless to say.[271] In Madrid there were rumours that the king himself, in order to present a better front to the Turks, would go to Barcelona and from there to Italy.[272] This course had frequently been suggested by Rome. The Council of State on 16th September, 1574,[273] debated whether or not Oran should be abandoned and the king referred the matter to the *Consejo de Guerra* for preliminary examination, the second reading to be reserved for the Council of State. On 23rd December, 1574, Vespasiano Gonzaga, who had been sent on a mission to Oran, wrote a magnificent report, proposing that the position should be evacuated and the Spaniards withdrawn to the fort of Mers-el-Kébir.[274] Melilla appears to have been the subject of a similar enquiry and

[268] Constantinople, 15th and 19th November, 1574, *ibid.*

[269] *Veedor general* of His Majesty in Piedmont and Lombardy, Milan, 6th and 23rd October, 1574, Simancas E° 1241.

[270] Granvelle to Philip II, Naples, 6th December, 1574, Simancas E° 1066, autograph.

[271] G. de Silva to Philip II, Venice, 16th October, 1574, Sim. E° 1333; de Silva to Philip, 30th October, *ibid.*

[272] Giulio del Caccia to the grand duke, Madrid, 25th October, 1574, Mediceo 4904, f° 273 and v°.

[273] *Consulta del Consejo de Estado*, 16th September, 1574, Simancas E° 78.

[274] To Philip II, Oran, 23rd December, 1574, Sim. E° 78. On Vespasiano Gonzaga's mission, see bishop of Padua to Cardinal of Como, 9th November, 1574, Vatican Archives Spagna, no. 8, f° 336.

the Genoese ambassador mentions a mission of the military engineer Il Fratino to Majorca.[275] All fortresses facing Islam were carefully re-examined: fear made the Christians watchful – and prudent. The emperor in December signed a new eight-year truce. Meanwhile in the extreme west, after a voyage of inspection of the *presidios* on the Straits, the young king Dom Sebastian of Portugal, decided not to attack the Sharif.[276] Spain was more alone than ever in facing her adversary in the East and continued to talk of plans against Bizerta, Porto Farina and Algiers. . . . But Saint-Gouard, reporting one of these rumours on 26th November, 1574, declared: 'I shall believe it when I see it'.[277]

So this was Spain, three short years after Lepanto. If that victory had been 'useless', the fault lies not in men so much, as in the uneasy equilibrium of the Spanish Empire, the complex system of forces which could not properly be centred on the Mediterranean. At the end of 1574, Spanish statesmen had no time to think of the problems of the sea, nor even to remedy the Tunis disaster. After his trip to Lombardy, it was suggested that Don John be asked to travel to the Netherlands. Saint-Gouard, always keeping an ear open for gossip in Madrid, wrote on 23rd October:[278] 'I have heard that they propose, if La Goletta is saved and the Turk does not attack anywhere else of importance, to send Don John with eighteen thousand Italians to Flanders. But if the Turk takes La Goletta, he will surely upset their plans'. La Goletta had already fallen. But Don John nonetheless went to Flanders.

Curiously enough, if Lepanto had not accomplished anything, neither did the Turkish victory at Tunis. Von Törne in his well-documented book on Don John of Austria, describes with his usual precision the disasters of 1574, then attempts briefly to move outside the limits of the history of events. 'The victory of 1574', he writes, 'which the Turks won in Tunisia, was the last major triumph of the Ottoman Empire before it fell into rapid decline. If Don John had led his African expedition a few years later, Tunis would perhaps have remained in Spanish hands and he would have convincingly won the argument against those who advised the king not to hold on to the newly won city'.

It is a fact that the *naval* (and I mean naval) decline of Turkey, was to be hastened if not after 1574 then at least after 1580. And it was a swift and brutal decline. Lepanto was certainly not its direct cause, although it had dealt a terrible blow at an empire whose resources were not inexhaustible, except in the imaginations of historians, the fears of Europe and the boasts of the Turks. What killed the Ottoman navy was inaction, that peace in the Mediterranean upon which, although hardly expecting it after following the thread of events day after day, we have now arrived.

[275] Sauli to the Republic of Genoa, Madrid, 16th November, 1574, A.d.S. Genoa, L.M. Spagna, 6.2415.

[276] Saint-Gouard to the king, Madrid, 23rd October, 1574, B.N., Paris, Fr. 16106.

[277] Saint-Gouard to the king, 26th November, 1574.

[278] B.N., Paris, Fr. 16.106.

Without warning, the two political giants of the sea, the Habsburg and Ottoman empires (as Ranke would say) abandoned the struggle. Was the Mediterranean no longer sufficiently interesting a prize? Was it too hardened to war for the latter to be as profitable as in the age of Barbarossa, the golden age of the Turkish armadas, groaning under the weight of their spoils? Whatever the reason, it is a fact that left alone in the closed vessel of the Mediterranean, the two empires no longer blindly clashed with each other. What Lepanto was unable to accomplish, peace was to bring about within a few years. It dealt the death blow to the Turkish navy. This fragile instrument, when not employed, not maintained and renewed, simply rotted away. No more seamen crowded to the docks; no more sturdy oarsmen sat on the benches. The hulls of the galleys decayed under the *volte* of the arsenals.

But to consider as Törne does that Don John had missed the chance of a lifetime is quite unjustified. If Spain had not turned away from the Mediterranean, the Turks would have kept up their naval effort. It was the mutual withdrawal of the two enemies which brought peace or the pseudo-peace of the end of the century. If Spain missed an opportunity in North Africa, I would be inclined to say that (as far as one can have an opinion at all about re-writing history), that opportunity was missed at the beginning of the sixteenth century rather than in the years following Lepanto. It was perhaps because she was at that time engaged in claiming America that she did not continue the new war of Granada on African soil, thus failing in what used to be called her 'historical' mission and would today be called her 'geographical' mission. The culprit, if there was a culprit, was not Philip II, still less Don John of Austria, but Ferdinand the Catholic. But for what they are worth, these questions have yet to receive a proper hearing. Tomorrow's historians of political change will have to reconsider them and perhaps make some sense of them.

CHAPTER V

Turco-Spanish Peace Treaties: 1577–1584

Literature has accustomed us to think of Spain as the very embodiment of inflexible Catholicism. Such was already the view taken by the French ambassador Saint-Gouard[1] when, in 1574, Spain was accused by the king of France of conspiring with the French Protestants – and may very well have been considering it. But the faith militant did not always inspire Spanish policies. Where did arguments of *raison d'état* carry more weight than in the councils of Philip the Prudent? There is plenty of evidence when one looks for it: the squabbles and disagreements with Rome; the attitude, so clearly anticlerical in many respects, of Alva in the Netherlands; even the policy which, at least until 1572, Philip adopted towards Elizabeth's England. 'Paradoxical though it may sound, the king of Spain has been called the involuntary ally of the English Reformation'.[2] And his religious policy in the Portuguese dominions in the Indian Ocean of which he became master after 1580 was remarkable for its tolerance.

But nothing more clearly illustrates the attitude of the Spanish government than its negotiations and compromises with the states and powers of Islam. The decision to seek the assistance of the Sophy (as Pius V himself was prepared to do) or to ally oneself with the Moroccan Sharif, as Philip II did only a few years after the Portuguese débâcle of Alcázarquivir, hardly betokens a crusading spirit. The emperor for his part had always maintained diplomatic relations with the Turks. In its dealings with Constantinople Spanish diplomacy inherited the methods and even the files of imperial diplomacy, which were placed at its service for both family

[1] The accusation is made in a letter by Charles IX. Saint-Gouard replied at length (to the king, Madrid, 24th February, 1574, B.N., Paris, Fr. 16106). If any Spanish agents had been present at rebel gatherings, let them be arrested immediately, or their names reported. If he has their names, he will set the Inquisition 'so hot on their trail that it must either lose all credit or else attach itself to the king himself, if he wished to employ such practices against Your Majesty, which I cannot imagine or believe'. No doubt they wished to stir up trouble in France: '. . . I think they would be well pleased to see Your Majesty's house still troubled, thinking that this circumstance will help them to remedy and set their own house in order'. No doubt too, adds Saint-Gouard, 'matters of state allow or at least suffer a little honesty'. How could Philip II be plotting with the Huguenots when he has declared of the Netherlands 'that he would rather lose them than agree to anything contrary to the Catholic religion and faith', and if he was in league 'against Your Majesty and inside his kingdom, I believe it is more likely to be with a few brigands who are not committed to either party and at heart serve neither God nor Your Majesty, since they are found in the armies or in the provinces, calling themselves Catholics, arms at the ready, gratifying their insane avarice with the utmost insolence'. What about the affair of Henry of Navarre and Claude Du Bourg? (see above, Vol. I, p. 376–377).

[2] A. O. Meyer, *England und die katholische Kirche*, I, p. 28, quoted by Platzhoff, *Geschichte des europ. Staatensystems*, p. 42.

and financial reasons: the war in Hungary, intermittent though it was, was waged partly on the voluntary contributions of Spain. So Spain was able to take advantage of the emperor's contacts in the East to send a steady stream of Spanish agents along the roads to Constantinople in the wake of the imperial ambassadors. And for every one whose name is known to us (through Hammer's investigation of the Vienna Archives) ten more were probably at work: but the threads they wove vanish into the web of events where they are well nigh invisible to the historian.

Nevertheless, our first step will be an attempt to trace these threads, using this most oblique of approaches to explore the secrets of the great reversal in Mediterranean politics which took place between 1577 and 1581.

Only when we have done this shall we examine the problems of these crucial years as a whole. If Turkey had not been diverted eastwards to Persia by her passion for conquest after 1579, and if Philip's Spain had not likewise turned westward to Portugal and the New World in 1580, what would history have made of Margliani's long and eventful visit to Constantinople (virtually unmentioned in textbooks) which is the subject of the first section of this chapter?

1. MARGLIANI'S PEACE MISSION, 1578–1581

We have already noted the peace attempt made in 1558–1559 by Nicolo Secco and Franchis, in liaison with Vienna and Genoa, and the later attempt in 1564 and 1567, both directed from Vienna. But I have not mentioned, in the narrative up to this point, the mission assigned in 1569–1570 to a Knight of the Order of Malta, Juan Bareli.

Back to the beginning: Philip II's first peace moves. Juan or Giovanni Bareli arrived at Catania in December, 1569, carrying instructions from Philip II dated 27th October. In the course of his service to the Grand Master, he had been party to a complicated scheme devised by the Greek patriarch of Rhodes, Joan Acida (the spelling given by the Spanish documents) who, together with a certain Carnota Bey, living in the Morea, had undertaken to organize an uprising against the Turks. He also promised to arrange for the Arsenal in Constantinople to burn down. The dates (the Venetian Arsenal went up in flames on 13th September, 1569) are too close together, bearing in mind the delays in communication, for the idea to have been suggested by the Venetian catastrophe. In any case offers of this kind were not infrequently made to the Spanish secret service, sometimes concerning Algiers, sometimes Constantinople. Pescara, who was given the task of examining Bareli's proposals, came to the conclusion that the knight had indeed enjoyed the confidence of the late Grand Master, and that he was acquainted with the Morea affair, but not in great detail; nor was he aware of the means envisaged for setting fire to the Turkish fleet. He had given the king what was no more than second-hand information and had overcommitted himself to say the least, relying heavily on the anticipated cooperation of the Greek patriarch.

Pescara himself now contacted the Greek of Rhodes and sent him to the Levant where he was only waiting for one of his brothers to be sent out to him, in order to accomplish the agreed plan. But the brother, a Venetian subject, who was to have come to fetch the ship *El Cuñado* carrying the explosives necessary to blow up the Turkish fleet, was reluctant to leave. Having previously been in trouble with Venice, he wanted a safe-conduct. To ask for one from the ultra-suspicious Signoria was out of the question. In the end then, it was to Juan Bareli, disguised as an honest merchant that *El Cuñado* with her cargo was entrusted, along with instructions for the patriarch and for none other than our old acquaintance Joseph Micas. As an extra precaution, the viceroy had registered the boat in his own name so that it could pass as a ransom vessel.

With the explosives and 50,000 crowns worth of merchandise aboard, the ship left Messina on 24th January. But in the Levant, the plans misfired. When it was all over, the patriarch blamed the knight for the fiasco. The attempt to fire the Arsenal had indeed failed; in the Morea, Carnota Bey, on whom the plan depended had died; the patriarch who was supposed to have sent to Zante for the gifts to offer him had not done so. The marquis of Pescara, in the somewhat confusing report on the affair which he dispatched in June 1570, declined to pronounce on those responsible for this 'grave' affair.[3]

There is no mention in all this, the reader may be thinking, of unofficial peace-talks – indeed the reverse. But there is another document. Nine years later, in Constantinople, the aged Sokolli, out of all patience, or so he claimed, with the subtleties and subterfuges of Spanish diplomacy, was interviewing the Spanish agent Giovanni Margliani, of whom more later. 'The pasha said to me: Tell me then what was (the Spaniards') motive in sending ————[4] Losata; what was their motive in sending me a knight of Malta (I have since learned that this was the knight Bareli) and later Don Martín [d'Acuña]?'[5]

So Bareli must have been entrusted with some peace mission – surely not in June 1570? We must assume that having come the first time to blow up the Arsenal and incite the Morea to rebellion, Bareli came on a second occasion bearing the olive branch; or else that he accomplished (or rather failed to accomplish), the two tasks at the same time. This can only mean that Spain used, for her diplomatic initiatives in the East, her army of spies and hired ruffians who had contacts with the twilight world of the renegades. In this connection the Bareli affair (which offers an opportunity for some more detective work) may be a significant episode.

The next emissary on Sokolli's list, whom we know better, Martín

[3] Pescara to the king, 12th June, 1570, Simancas E° 1133.

[4] First name illegible in my copy.

[5] Report by Margliani, 11th February, 1578, Simancas E° 489.

d'Acuña, was in Constantinople in 1576. He was a man of little fortune on good terms with the renegade community. When he returned to Italy, having handled the vital talks of 1576, he proceeded to boast of having set fire to the Turkish fleet. Sokolli on being informed of this ambassador's extraordinary claims, not unnaturally complained. Spanish diplomatic circles, to whom Don Martín was well-known, admitted that he might indeed have set fire to a galleon. The episode has strange parallels with the Bareli mission of 1569–1570 and indicates that only about one in ten missions of this nature may ever have come to light. Renegades hoping to be received back into grace, former captives claiming to be experts on the Levant, Greeks 'who must always be watched', (as a Spanish report puts it)[6] knights of Malta, Albanians, imperial envoys; and their interlocutors; Jews, Germans like Doctor Solomon Ashkenasi, dragomans like Horembey: in the no-mans-land between two civilizations, it was this strange army not always officially recognized by its employers which handled diplomatic matters. Later, in the seventeenth century, it would be the turn of the Jesuits to act as intermediaries.[7]

After 1573, these circles were busier than ever before. Spanish demand sent up prices on the information market; good money would be paid to anyone who could offer useful intelligence or his services in the East. Was this why the extravagant French adventurer, Claude Du Bourg, offered to handle the affairs of the king of Spain in Constantinople, where it was common knowledge how he had acquitted himself in 1569? (He asked a mere 100,000 ducats: more than would be required to bribe the Grand Vizier himself!)[8]

Negotiations in Don John's time. In 1571, Don John of Austria was himself in correspondence with the Turks. Such procedures were not unusual in the sixteenth century. Sultan Selīm sent him a letter and gifts, probably after Lepanto.[9] Don John replied that he had received the missives through the eunuch Acomato of Natolia, that he was sending back a Greek spy, 'who came here on your orders to discover the resources of the Christians, and although I could have had him put to death, not only did I spare his life, but afforded him every leisure to see all my provisions and plans – which are to wage perpetual war against you'. We cannot vouch either for the date or the strict authenticity of these documents, whose origin is uncertain. But the exchange of letters, referred to by contemporaries, undoubtedly took place: these documents are one piece of evidence. Compliments on one side, romantic challenges on the other – the 1571 contacts cannot be called negotiations. But two years later genuine talks were under way.

[6] Report on Estefano Papadopoulo, Madrid, 21st June, 1574 '. . . y es menester mirar les mucho a las manos', Simancas E° 488.

[7] Cf. for instance H. Wätjen, *op. cit.*, p. 67–69.

[8] Memorandum from Du Bourg, Spanish translation, 1576, A.N., K 1542.

[9] Letter from Selīm II, Emperor of the Turks, to Don John of Austria, 'sending him gifts when he was general of the Christian army', B.N., Paris, Fr. 16141, f° 440 to 446.

On 30th June, 1573, one of Granvelle's agents, Juan Corenzi, was returning from Constantinople; he was unquestionably an intelligence agent, but not necessarily a negotiator.[10] In June–July however, Spanish agents really were on their way to Turkey. If these had been the first to make the journey, we might think that the decision to send them followed Venice's withdrawal from the league. Philip II received news of the Venetian peace treaty on 23rd April, 1573, Don John on 7th April. In fact it seems more than likely that this trip was merely one of a series. But it was extraordinary on any count. For the Spaniards were preparing during June, 1573 to do no less than follow in Venice's footsteps.

By 16th July, the bishop of Dax had already word of their voyage from his agent in Ragusa.[11] Ten days later[12] he had full details of the mission. Don John had taken prisoner the son of 'Alī Pasha, the grandson, by a 'sultana' of Selīm himself. He had treated him with the utmost courtesy, refusing the gifts sent him by the sultana, and sending her in return magnificent presents which she immediately passed on to Selīm, or so Du Ferrier, the French ambassador in Venice informs us.[13] Formalities and courtesies provided a façade for serious negotiations. And when 'Alī Pasha's son, freed without a ransom, arrived in Constantinople on 18th July, he was accompanied by four Spaniards, including one of Don John's secretaries, Antonio de Villau (Vegliano) and a Florentine, Vergilio Pulidori, 'cortegiano del duca di Sessa'. Muḥammad Sokolli hinted in reply to the bishop's inquiries that these were the machinations of his enemies, in particular of Joseph Micas. But if the king of Spain wants peace, he added, he will have to pay tribute and sacrifice some 'forts' in Sicily. The bishop was amazed that the king of Spain should have taken such a step without seeking previous assurances: 'which makes me think that besides the great desire and necessity the king of Spain feels for peace on this front, so that he can bring affairs in Flanders to a conclusion, he foresees some more difficult pinch ahead; or else there is some other scheme afoot greater than all this'.[14]

After informing themselves of the Turkish terms, the Spaniards did not appear to reject out of hand the notion of tribute: the imperial ambassador could pay it along with the emperor's. They were waiting for Piāle Pasha and Euldj 'Alī and counted on their strong influence to win the day. The first pasha, Muhammad 'Alī was totally hostile to the idea of the talks, at least so he said. But one could not stake too much on it. The sultan was known to be avaricious and anxious to put an end to the heavy expense of naval war, the more so since after Lepanto he was terrified of its consequences. And he was waiting for the aged Sophy in the East to die. So the Spanish scheme was not doomed from the start. The French

[10] *Lo que refiere Juan Curenzi* . . . , 30th June, 1573, Simancas E° 1063, f° 35.

[11] Bishop of Dax to the king, Constantinople, 16th July, 1573, E. Charrière, *op. cit.*, III, p. 405.

[12] Bishop of Dax to the king, Constantinople, 26th July, 1573, *ibid.*, p. 413–416.

[13] To the king, Venice, 26th February, 1574, *ibid.*, p. 470, note.

[14] See note 11.

ambassadors had in the past successfully blocked several such attempts, in the time of Charles V and even of Philip II. But this time the Spaniards talked of major commercial treaties, notably the opening up of the Levant trade to the whole of Italy, excluding the Venetians and the French. Would not the increased volume of trade swell the tax revenues of the Grand Turk? This at any rate was the prospect dangled before his eyes.[15]

Other participants in this attempt were, once more, Joseph Micas, the Tuscans (we have ample evidence of their involvement), the Jewish community of Turin, encouraging the Duke of Savoy who was hoping to achieve in Nice, with the help of the Jews, what Cosimo de' Medici had accomplished at Leghorn;[16] and later the Lucchese. Was this widespread desire for trade a sign of the times? The prospect of ousting Venice (and what an inheritance she would leave!) excited hopes in many quarters, hopes which were to take definite shape and were even fulfilled to some extent when, between 1570 and 1573, Venice was effectively absent from commercial competition. The appetite for commerce accompanied and reinforced Spanish policy. Ambassadors, agents, gifts and promises flowed into Constantinople. This massed attack, as the bishop later noted, ultimately assisted 'the introduction of the Spaniards'[17] and as the official representative of France, but one with little money to offer, he could not be other than dismayed. 'The pasha laughs at us for wanting to tie his hands without putting anything into them'.[18]

But to make any sense of the attempt, we must pay careful attention to the differences between the two years, 1573 and 1574, during which the negotiations took place. In September, 1573, the bishop of Dax[19] still believed that Spain would succeed. She had considerable forces at her command; the Turks were faced with serious problems in the Yemen, details of which seem a hopeless tangle to the historian (in any case Sinān Pasha forcefully resolved them later in the same year).[20] But the capture of Tunis by Don John at the very end of the campaigning season, seems to have compromised the talks which the bishop of Dax was following with extreme anxiety. If the talks had been successful, as the parallel negotiations with the emperor promised to be, the result would have surprised the Venetians before the ratification of their own treaty, which although agreed in principle on 7th March, 1573, was not signed until February, 1574, when the Turk had finally abandoned his claim for more concessions, notably Cattaro and Zara.[21] All this helps to explain Venice's anxiety and military precautions, and the care taken by Venetian diplo-

[15] Bishop of Dax to Catherine de Medici, Constantinople, 17th February, 1574, E. Charrière, *op. cit.*, III, p. 470 ff.

[16] Pietro Egidi, *Emanuele Filiberto*, *op. cit.*, II, p. 128 ff.

[17] Bishop of Dax to the king, 18th September, 1574, E. Charrière, *op. cit.*, III, p. 572.

[18] *Ibid.*, p. 572.

[19] To the king, *ibid.*, p. 424–427, Constantinople, 4th September, 1573.

[20] To the king, *ibid.*, p. 470–475, 24th March, 1574.

[21] For the major incident of the 'fort' at Sebenico, see *ibid.*, 17th February, 1574, p. 462–470.

macy to keep in step behind the bishop of Dax. On the outcome of these dealings depended Venice's livelihood and her neighbours were quite prepared to put that livelihood in jeopardy. Crete was an obvious target, insufficiently fortified; and 'the inhabitants of this island are dissatisfied . . . and have sought for many years to escape from their bondage'.[22]

The signing of a treaty between Spain and Turkey would have been a profound shock for France. So the bishop was extremely pleased with himself for managing to avert a rupture between Venice and the Turks in February, 1574, over the frontier limits of Zara and the fort of Sebenico. France for the second time came to the rescue of Venice, the prime target of everyone else. In the end the Spanish and imperial parties were the losers. 'Since the signing of the Venetian treaty', writes the French ambassador, 'I have begun to fear less from the Spaniards'.[23] What had saved the day, as he well knew, was the occupation of Tunis, by which the Spanish had barred the Turks from access to their possessions in Barbary, possessions which 'may not be theirs for long if things remain as they are at present',[24] he wrote in February, 1574.

These are the few facts we know for certain. Somewhere in Simancas there must be a full dossier on these negotiations. The problem as it emerges from the French diplomatic correspondence is that we do not know whether Spain genuinely wanted a truce or whether she was merely tempted to outmanoeuvre Venice (at the very moment when she was making generous offers to the Venetians of help in the event of a Turkish attack). In any case neither aim was achieved.

But that contacts continued, there cannot be the slightest doubt. And they were handled by the same persons – a little embarrassed by the lack of precise instructions and by the course of events culminating in the recapture of Tunis by the Turks in September, 1574. On the 18th of that month, the bishop of Dax wrote to Catherine de Medici: 'the Spaniard and the Florentine, who have been here for fifteen months', are ready to leave; their passports are in order, but they are always held up at the last moment, being as much hostages as ambassadors'. 'Negotiations (in this country) are always dangerous', Margliani was later to write. They were above all extremely complicated. A report from the Spanish embassy in Venice refers in 1574, that is at the same time as the Constantinople talks, to a meeting in Venice between a *cavass*, 'secretario del Turco' and a certain Livio Celino, on the subject of peace between Spain and the Turks. Unfortunately it is impossible to determine the date of the meeting, though it was no doubt after the signature of the Venetian treaty, otherwise the Venetians would not have allowed the interview to take place on their territory.[25] Again in February, 1575,[26] Granvelle brings up the question of

[22] See note 20. [23] E. Charrière, *op. cit.*, III, p. 467.

[24] 17th February, 1574, *ibid.*, III, p. 462–470.

[25] *Relacion que hizo Livio Celino . . .*, 1574, Simancas E° 1333.

[26] Granvelle to the king, Naples, 6th February, 1575, Simancas E° 1066. The cardinal's letter was rather pessimistic: with the change of reign new contacts will have to be found and bribed, and it will all cost money, as it does the Emperor, who

peace with the Turks on the occasion of the accession of Murād III. But Selīm's death did not make a great deal of difference since the vizier Sokolli continued in power under the new sultan until his assassination by a fanatic in 1579. It is true that Murād, an ostentatious and puerile prince, allowed foreigners freer access to his states. But above all times were changing ultimately forcing Turkey to accept new conditions.

Martín de Acuña: the outsider who succeeded: Were the talks suspended after Spanish diplomacy's major assault on Constantinople in 1573? Possibly they were; there was at any rate reduced activity until the arrival in the Turkish capital of a new ambassador, Martín de Acuña.

I have been unable to find any documents which give precise details on the curious figure of Don Martín. But there is certainly more evidence about him than is reported by Charrière and those who have followed him (Zinkeisen and Iorga) to whom he is merely a name or rather a nickname (Cugnaletta). Don Martín makes his first appearance on the stage of history in 1577. Leaving Naples with 3000 ducats provided by the viceroy, he arrived in Constantinople on 6th March, according to a Venetian dispatch.[27] His stay must have been very brief, for he was back in Venice on 23rd April. Guzmán de Silva explains in a letter that Don 'García' de Acuña had left with a safe-conduct in order to ransom prisoners at Constantinople, but 'his sole aim was to negotiate a treaty with the Turk, and indeed he obtained one for the space of five years'. And sure enough there is in Simancas a copy of the projected agreement which Don Martín had managed to draw up together with the Turkish ministers, a document bearing the date of 18th March[28] and the first lines of which: 'The supreme and unfathomable God having inspired and illuminated the hearts of the two emperors', clearly indicate a Turkish original. He also carried a letter from the pasha for Philip II, with a promise that the Turkish armada would not be sent out in 1577.

Quick work then – but was it well done? Not all the Spaniards thought so. When, in April, Don Martín stopped in Naples on his way to Spain, the marquis of Mondéjar, Granvelle's successor as viceroy, received him – with such bad grace that he apologizes for it in his report of the meeting. The fact is, he explains, that Don Martín is 'one of the most disreputable Spaniards ever to have come to Italy'.[29] He was extremely indiscreet:

---

wants to have his truce confirmed. The new ruler, Murād is 28 years old, 'bellicose and liked by his subjects'.

[27] Constantinople, 8th March, 1577, A.d.S. Venice, Secreta Relazioni Collegio, 78; Guzmán de Silva to the king, Venice, 28th April, 1577, Simancas E° 1336, reports the sailing of Don Martín whom he calls Don García de Acuña; he had left for Constantinople with a safe-conduct for the ransoming of prisoners, actually intending to negotiate a truce 'y a salido con la resolucion dello por cincos años'. The French dispatches erroneously give the date of Don Martín's arrival in Constantinople as 15th March.

[28] Simancas E° 159, f° 283.

[29] Mondéjar to Antonio Pérez, Naples, 30th April, 1577, Simancas E° 1074, f° 31.

after having made Mondéjar swear not to reveal what he had told him about his mission, he succeeded in making it public knowledge all over Naples the very next day; through either his own indiscretion or that of his companions, says Mondéjar, whose irritation can be imagined. It seems that neither Don Martín's companions, nor the partiality of the viceroy were at fault, since Don Martín had done exactly the same at Constantinople. He has shunned the company of the respectable, writes the imperial ambassador with some surprise, and sought out the most notorious renegades in the city. 'The street urchins all know him and his secrets'.[30] Not only that but he was a spendthrift, a gambler and a drinker. Of the 3000 crowns that Mondéjar had entrusted to him on his departure for Constantinople, he had it is true, sent half back to Spain in the shape of silk and silver coins. Then he gambled away the rest at Lecce on the way back. When he reached Naples, Mondéjar was obliged to advance him more money to enable him to continue the journey to Spain. But this time he insisted on an account both of past and future allocations.[31]

By June, Don Martín was back in Spain and reporting to Antonio Pérez in person on what he had accomplished in Constantinople.[32] For it seems his negotiations had been successful. Perhaps he had arrived at an opportune moment. In any case, he had obtained from the pasha an assurance that the Turkish armada would not be sent out, despite the protests of Euldj 'Alī.[33] Above all he had undoubtedly furthered the question of the truce. In the end after all his indiscretion, disreputable acquaintances and lack of scruples may have served him well. Certainly he was no career diplomat. And he was not over-concerned to protect Spanish susceptibilities. Did the bureaucrats of Madrid find this too hard to swallow? At any rate he was not sent back to the East, 'for health reasons' according to the official documents – a phrase evidently concealing some political veto. A letter written by him to the king in 1578 in which he explains that for one thing his successor will be ill-equipped to remind the pasha of his promises and for another will not hesitate to claim the credit for everything, he, Don Martín, has achieved, is evidence both of Acuña's rancour – and of his excellent health.[34] He probably exaggerated the value of his own services, for in August another emissary had come back first to Otranto then to Naples, from Constantinople, and this man, Aurelio de Santa Cruz, also carried accommodating proposals from the Grand Turk. He explained that the latter 'greatly desired the truce, for he was a peace-loving man, and a friend of peace, devoted to letters and abhorring war . . . and anything else that might disturb his tranquillity'. As for Muḥammad Pasha, the greatest authority in the land after the sultan, he was now over seventy-five years old and hated war.

[30] Cost., 2nd May, 1577, probably passed on by G. de Silva, Simancas E° 1336.

[31] See note 29.

[32] Martín de Acuña to the king, Madrid, 6th June, 1577, Simancas E° 159, f° 35.

[33] Silva to Philip II, Venice, 19th June, 1577, Simancas E° 1336.

[34] Don Martín de Acuña to H.M., Madrid, 1578 (month unknown), Simancas E° 159 f° 283.

Among the other ministers only Sinan Pasha showed signs of bellicosity, but he was one of the least influential.[35]

One last word, rather sadly, about Martín de Acuña. According to a document in the British Museum, he was executed on the king's orders on 6th November, 1586, that is some time after the dealings just described, in a room of the castle of Pinto near Madrid. On this occasion there is a cryptic reference to one of his misdeeds in Turkey (the denunciation of a Spanish agent) and a long and moving account of the Christian circumstances of his death.[36]

Giovanni Margliani. At the end of 1577, the king sent to Constantinople, on the recommendation of the Duke of Alva, a Milanese knight related to Gabrio Serbelloni, Giovanni Margliani. He had fought in Tunisia in 1574; wounded – he had lost an eye – he was taken prisoner, then ransomed from the Turks in 1576 through the good offices of a Ragusan merchant, Niccolò Prodanelli.[37] The instructions given him by the king, of which only fragments have survived, date from some unspecified time in 1577. They are couched in very general terms. The bearer was to travel by way of Naples, but not to reveal his mission to Mondéjar. He was to be accompanied by a certain Bruti, whom the documents suggest was either an Albanian, a 'dignitary' of the imperial court, or a retainer of the Signoria of Venice[38] – perhaps he was all of these at once. Margliani was informed that he was to take the place of Martín de Acuña, now prevented from returning by ill-health, and negotiate the truce. He was to seek to have Malta and the Italian princes included in the peace treaty. This is the very little we know of the origin of these important negotiations.

Was Margliani chosen for his personal qualities, which are not in doubt (he was capable, honest, tactful, tenacious – and an indefatigable correspondent); or did the Spanish administration simply wish to take the affair out of the hands of such men as Don Martín and place it on a more respectable and dignified level? It is hard to say without seeing either the official or the secret instructions (referred to by Margliani in his letters) carried by the new Spanish envoy to Constantinople.

From the coast of Naples, Margliani and his companions arrived at Valona on 8th November.[39] They left on the 13th. By the 25th they were at Monastir (Bitola) and on 12th December at Rodosto; from here Margliani sent word of his arrival to the dragoman Horembey, who had been a party to the talks with Don Martín. The interpreter's reply was

[35] Mondéjar to Philip II, Naples, 13th August, 1577, Simancas E° 1073, f° 136.

[36] Fernand Braudel, 'La mort de Martín de Acuña' in *Mélanges en l'honneur de Marcel Bataillon*, 1962. Cf. F. Ruano Prieto, 'Don Martín de Acuña' in *Revista contemporanea*, 1899.

[37] G. Margliani to Antonio Pérez, Constantinople, 30th April, 1578, Simancas E° 489.

[38] Cf. Gerlach, *Tagebuch*, p. 539; E. Charrière, *op. cit.*, III, p. 705.

[39] The following is based on Margliani's long report, February, 1578, Simancas E° 488.

brought him by special messenger at Porto Piccolo, some distance outside Constantinople, which the party entered that evening. Lodging in the house of the *cavass* who had brought him to the city, the head of the mission had an immediate meeting with Horembey. This meeting was confined to the exchange of compliments, serious business being postponed until the next day. But in the morning Margliani's troubles began. As soon as he had explained his mission, Horembey interrupted him with these words: 'If I were a Christian I would cross myself at the falsehoods imagined by Don Martín. The pasha is expecting an ambassador, that is what was written to His Majesty, that is what Don Martín promised on this very spot and what Don Martín sent word of by a messenger who came here. The pasha will be exceedingly angry to hear of the change of mind. God grant that no grave harm to your persons results from this.'

The reproach was reiterated vehemently by the pasha, and by one Doctor Solomon Ashkenazi, said to be a German Jew, in any case an influential person in government circles, who more than once received lavish payment (not that he was necessarily bought in the vulgar sense of the word). Was this vehemence sincere? Had Don Martín been lying? From Margliani's reports and from what we know of Don Martín's character, one might be tempted to say that he had, but the complicated world of negotiations in Constantinople holds as many traps for the unwary historian as it did for the western ambassador or minor diplomatic agent. And play-acting was not unknown. What was certain was that once more the Spaniards wanted the dealings to be kept secret. If Don Martín really had committed himself too deeply in 1577, that would certainly explain his anxiety at seeing the mission transferred to another. The Turks on the contrary had been hoping for a spectacular Spanish embassy. And now they were being sent some obscure individual, only lately a prisoner – and one-eyed into the bargain (the occasion for some rather obvious sarcasm, both from Euldj 'Alī, who was opposed to the truce, and from the French). His companions were equally undistinguished. Aurelio Bruti whose origins were a matter of conjecture and the other Aurelio (de Santa Cruz) who was known to be merely a merchant, a specialist in ransom deals, and a Spanish informer, if not a spy.

As if this were not enough the little mission seemed determined to be as inconspicuous as possible. 'They do not want to be seen or recognized',[40] remarks a French correspondent. The abbé de Lisle noted on 22nd January: '. . . the said Marrian appearing instead of the expected official gift-bearing ambassadors had come before the Turks as it were privately with authority to treat of the said truce'.[41] For a whole year, Margliani even had his attendants go about 'clad as slaves'. He frequently disguised his appearance. On one occasion while waiting for an audience with the pasha, he saw the Venetian *bailo* in the distance. He immediately hid,

[40] E. Charrière, *op. cit.*, III, p. 705.

[41] To Henri III, Constantinople, 22nd January, 1578, E. Charrière, *op. cit.*, III, p. 710.

entering the room where the pasha had received him in the past, to the fury of the Turks inside. He reports the incident himself and if he claims the credit for it, the reason is not that he was personally given to conspiratorial tendencies but that he had been ordered by his masters to observe the utmost discretion. Gerlach, whose *Tagebuch* is an excellent source for the entire period of Margliani's negotiations says as much at the beginning of his narrative: the Spaniards certainly want peace, but at the same time they 'wish to preserve secrecy and not to appear to have humbled themselves before the Turk'.[42] It is not difficult to understand the irritation felt by the pasha and betrayed by his extravagant outbursts, his slighting references to the Pope, his allusions to the war in Flanders and his demands for the cession of Oran.

But the Turks needed peace just as badly as the Spaniards did, so for want of an ambassador, they agreed to talks with Margliani. Since an agreement was necessary by the spring, interview succeeded interview. After 1st February, Margliani reported a visible detente. On the 7th, a one-year truce was signed[43] – a sort of armistice or gentlemen's agreement. The text bears a certificate of attested translation by the dragoman Horembey and Doctor Solomon Ashkenazi, who both played a decisive role in the negotiations. The pasha promised that for the year 1578 and on condition that the Spaniards reciprocated, the Turkish armada would not leave port. The truce was extended to cover a list of other states, some named by the king of Spain, the others by the Grand Turk. On the Turkish side the list included the king of France, the emperor, Venice and the king of Poland, as well as the 'prince' of Fez, 'although this is not strictly necessary' says the document (thus incorporating a somewhat exaggerated Turkish claim) 'since he bears the banner of the Most Serene Grand Signior and pays his homage'. Philip II undertook the participation in the truce of the Pope, 'the island of Malta and the knights of Saint John residing on the island', the republics of Genoa and Lucca, the dukes of Savoy, Florence, Ferrara, Mantua, Parma and Urbino, and the lord of Piombino. As for the king of Portugal, it was agreed that the Turkish armada should not proceed against his states beyond Gibraltar, 'by way of the White Sea' i.e. the Aegean. The undertakings were less formal concerning the Red Sea and the Indian Ocean, where anything might happen.

All in all then it was a triumphant success. The truce had been obtained without undue delay, without financial outlay and without fanfare, just as Spanish diplomacy had hoped. Little time had been lost: the talks lasted from 12th January until 7th February and had been dispatched as quickly as those handled by Don Martín de Acuña; perhaps because to the Turks it was vital to avoid the expensive mobilization of the fleet in

[42] *Op. cit.*, p. 160; J. W. Zinkeisen, *op. cit.*, III, p. 499.

[43] *Lo que se tratto y concerto entre el Baxa y Juan Margliano*, 7th February, 1578, Simancas E° 489. Copy of the same document made possibly in 1579, *Capitoli che si sono trattati fra l'illmo S^re^ Meemet pascià (di) buona memoria . . .*, Simancas E° 490.

good time. The operation would be worthwhile only if everything was settled before the spring – hence the dates of the two truces: 18th March, 1577 and 7th February, 1578.[44]

But the Turks continued to call for an official Spanish embassy in Constantinople: they wanted a brilliant diplomatic triumph which would have repercussions all over Europe. This they persistently pressed for, and the agreement signed on 7th February carries in conclusion a formal undertaking that ambassadors will be exchanged. Combined with other circumstances, this clause was to keep Margliani for three more years among the Vines of Pera and to bring him much trouble.

Should he not have returned to Europe in the spring of 1578 now that his mission was accomplished? Probably he was not particularly anxious to, hoping as is revealed in his dispatch of 30th April to Antonio Pérez,[45] to obtain single-handed the result desired above all: namely an armistice scheduled to last two or three years. The feverish atmosphere and illusions engendered by the facility of the earlier negotiations may have encouraged him to feel hopeful. Having heard via his Ragusan friend Prodanelli of Don John's victory at Gembloux, confirmed from other sources, he tried to take immediate advantage of it at the end of April to use the 'Dottore' once more. 'I have always maintained to Horembey and to Your Lordship' he told the latter, 'that I did not believe that his Majesty the King my Master was in favour of sending an ambassador. Horembey preferred to take the word of my companion Aurelio [de Santa Cruz] rather than my own on this matter. God knows what will come of this. For my part, I am still of the same opinion, the more so since Don John is winning victories in the West, while the Grand Turk is engaged in a war against Persia which is known to be full of perils and trials: it more than outweighs the war in Flanders. So tell Mehemmed Pasha that it would be to his advantage to be certain of the non-intervention of the forces of the king of Spain my master for two or three years and to agree to an armistice formula negotiated on behalf of Spain by myself'.

His initiative was both daring and premature. The pasha's reply was prompt and, at least as conveyed by the doctor, not unfriendly. He said he did not dispute the arguments of the envoy. But the sultan was young and ambitious for military glory. Margliani had described him in February as open to suggestions, more accessible than Selīm, relying on first impressions. But for this very reason, the pasha said, the sovereign was not unreceptive to the daily urgings of Euldj 'Alī. The 'Captain of the Sea' boasted that even with a less than perfect fleet he could overcome Spain whose forces were so overstretched at the moment. And, the pasha had confided to the doctor, 'I am in a delicate situation, having employed my credit so readily on behalf of Don Martín, who let us down, that I

[44] *Lo que ha de ser resuelto sobre lo de la tregua* (1578), Simancas E° 489; on the failure to conclude economic treaties, Margliani (to Antonio Pérez?), 11th February, 1578, Simancas E° 489.

[45] Simancas E° 489. Victory of Gembloux, 31st January, 1578.

cannot begin all over again'. Thereupon (the doctor said) he heaved a sigh and muttered 'This Empire has at present neither head nor tail'. Fine phrases which were duly reported to Margliani by the doctor, who took care too to describe the pasha's reception of the outline of the Spanish proposals: a heartfelt: 'Doctor, you are quite right!'

But in the end the pasha had returned to the question of the ambassador. If Philip II would send one, he would be only too glad to help further his demands. But if an ambassador was not sent, added Muḥammad, 'I too shall take the advice of the Captain of the Sea'. He then swore, on the head of the Grand Turk, that he had had the utmost difficulty in having the terms of the treaty respected, that is in preventing the armada from leaving port. Threats and compliments mingled in the pasha's conversation. Neither Christian nor Turk was taking any chances. Their interviews, faithfully reported in Margliani's voluminous correspondence leave an impression of suspicion, of tortuous diplomatic manoeuvres, skilful if unscrupulous on both sides, with neither scorning to employ the most subtle deceits.

'Margliani's man', sent to Spain with the text of the provisional agreement, had left Constantinople on 12th February, 1578. The Council of State in Madrid met several times to discuss the news he brought.[46] Unanimity was quickly reached on the necessity of agreeing to the truce 'in view of the state of His Majesty's affairs and of his finances, and the need to settle matters and proceed to strengthen his kingdoms . . . .'. There had to be an understanding with the Turk, everyone was agreed on that, if no one wished to tackle the root of the matter. But the councillors were divided about 'questions of protocol and prestige': should an ambassador be sent or not? If so, would it be sufficient to send credentials to Margliani? This point now became the focus of debate. By September, it had been decided to send an ambassador; and a certain Don Juan de Rocafull,[47] a rather faceless individual who is mentioned in a letter in 1576 as commanding several of the Naples galleys,[48] received instructions to this effect. His formal instructions, which carry no precise date, contained an account of the foregoing negotiations. They were accompanied by a 'second' set of instructions, dated 12th September, 1578,[49] concerning

[46] *Relacion de lo que ha passado en el neg^o de la tregua y suspension de armas con el Turco y lo que para la conclusion della llevo en com^{on} don Juan de Rocafull y el estado en que al presente esta* (1578), Sim. E° 459, f° 28 (or 281). These undated documents must be situated somewhere between the beginning of June and 12th September, 1578, bearing in mind the delay in communications: for example, a letter written by Margliani on 9th December, 1578 in Constantinople addressed to Antonio Pérez, reached its destination on 31st March, 1579, after a journey lasting 3 months and 22 days.

[47] The date of the second set of instructions was 12th September, 1578, see below note 49. Don Juan de Rocafull is the same person referred to as Don Juan de Rogua, de Valenza by Gerlach, quoted by Zinkeisen, *op. cit.*, III, p. 500.

[48] Don Juan de Cardona to Philip II, Barcelona, 1st November, 1576, Simancas E° 335, f° 58 '. . . y con correo por tierra ordenando a Don Juan de Rocafull hizieze despalmar las nueve galeras'.

[49] *Instruccion secunda a Don Juan de Rocafull*, Madrid, 12th September, 1578, Simancas E° 489.

what was to be done should Rocafull be 'prevented' from going to Constantinople. He was to send Captain Echevarri, his companion, with orders that the truce be negotiated through Margliani. So the decision to send an ambassador to the sultan was not very definite: the possibility of keeping him back at the last moment was always entertained.

It was another three or four months before Margliani at his distant post received the news. According to Gerlach's *Tagebuch*, 'Margliani's man' did not arrive back in Constantinople until 13th January, 1579.[50] The delay was perhaps the result of the inclement season; or perhaps of some calculation by Spain, anxious now to repeat for the year 1579 the manœuvre which had now enabled her to obtain the non-intervention of the Turkish fleet for two years running. The French immediately jumped to this conclusion and indeed the arrival of the good tidings, the announcement that an ambassador was being sent, made Margliani's task easier. In addition, as the Turks became increasingly committed in Persia they became more accommodating. Juyé wrote to Henri III on 16th January, 1579: the Turks 'have as great need of peace as the Catholic King, because of the war in Persia, where they will have more trouble than they care to admit'.[51] Juan de Idiáquez, who was then Philip's representative in Venice, learnt from the French ambassador on 5th February, 1579, that in Constantinople Margliani was no longer living a sequestered life, had bought new clothes both for himself and his suite and was talking of renting a house in Pera. 'It is thought here in Venice therefore that the envoy from Your Majesty whom the Turks have been awaiting in order to sign the truce, cannot be far away'.[52]

But Rocafull seemed to be in no hurry. On 9th February, he was stiil in Naples. It was reported from Venice on 4th March that he was nearing Constantinople and that Margliani had sent two men out to meet him.[53] But the news was premature. Rocafull was 'unwell'. Knowing as we do the contents of the second set of instructions and Spanish reservations about the whole affair, we may well, like the Turks, have doubts about this 'illness' and even about the 'relapse' said to have been suffered by the unfortunate ambassador. Did Margliani nevertheless sign a truce similar to those of the previous years? No documents that I have read mention one. Some French correspondence refers to one indirectly but the evidence is very vague.[54] In any case, in April, the Turkish fleet or at least that part of it which could easily be mobilized, was on its way to the Black Sea under the command of Euldj 'Alī. So it was soon confidently assumed in Naples that the Turkish fleet would not 'come out'. Is it possible (as the

[50] J. W. Zinkeisen, *op. cit.*, III, p. 500.

[51] E. Charrière, *op. cit.*, III, p. 777.

[52] Juan de Idiáquez to Philip II, Venice, 5th February, 1579, A.N., K 1672, G 1, no. 22.

[53] J. de Idiáquez to Philip II, Venice, 4th March, 1579, A.N., K 1672.

[54] E. Charrière, *op. cit.*, III, p. 852, note, but the dispatch dated 9th January, 1580 could apply as much to the future as to the past. And what is the proper meaning of the 1579 text mentioned above, note 43?

merest of hypotheses) that this news had something to do with the interruption of Rocafull's journey?

Unwell or not, Rocafull did not cross the Adriatic. On 25th August, Captain Echevarri landed at Ragusa. Accompanied by one Juan Estevan, he carried gifts for the Grand Turk and his ministers and, for Margliani 'todos los poderes y recaudos necessarios' for the signing of the truce,[55] which in effect elevated Margliani from the status of a mere envoy to that of an accredited ambassador. At the same time there arrived in the Turkish capital a new French ambassador.[56] Learning from the aged Sokolli that the truce with Spain was likely to be signed,[57] he naturally took immediate steps to jeopardize it: in the first place by innuendo. While Margliani was explaining that the king of Spain's military preparations, news of which had reached Constantinople, were intended for Portugal, where the succession crisis had opened even before the death of the cardinal-king,[58] the Frenchman was encouraging a rumour current in Constantinople[59] that they were to be used against Algiers. He also hinted at the inevitability of war in Italy as a consequence of the incidents in the marquisate of Saluzzo. It was a waste of time. Germigny was later accused of failing to put obstacles in the way of Spanish diplomacy. But he had no better weapon available than this war by insinuation. France's prestige in the Levant had been severely damaged in the years following the St. Bartholomew Massacre: evidence of her impotence and exhaustion in the West reduced her credit in Constantinople. Negotiators should not come empty-handed. And French policy had just resulted in the arrest of perhaps the only man capable of drawing Turkey into European affairs: Claude Du Bourg, an agent of the Duke of Anjou, who had been apprehended in Venice in February, 1579 and transferred to Mirandola.[60] His plan had been to interest Turkey in the conquest of the Netherlands by the Duke of Anjou, in league with William the Silent, the Protestants of all Europe and the English, who were represented, as Margliani reports, in Constantinople. It might have been a tempting prospect for the Turks. But already heavily committed in the exhausting Persian war they could hardly launch a western offensive at the same time.

The year 1580 proved for Margliani to be one of hard work crowned by success. Now officially attached to Naples and taking instructions from Don Juan de Zuñiga who had become Commander of Castile on the death of his nephew, and viceroy of Naples on the death of Mondéjar,

[55] Echevarri to Margliani, Gazagua, 2nd September, 1579, A.N., K 1672, G 1, no. 117. Echevarri to Margliani again, Caravançara (sic), 2nd September, 1579, *ibid.*, no. 118, complains of Brutti, 'bellaco'.

[56] Margliani to Antonio Pérez, Pera, 2nd September, 1579, Simancas E° 490.

[57] Germigny to the king, Vines of Pera, 16th September, 1579, *Recueil*, p. 8 ff.

[58] Which was not reported in Constantinople until the beginning of April, 1580, G. Margliani to the viceroy of Naples, Vines of Pera, 9th and 14th April, 1580, A.N., K 1672, G 1, no. 166.

[59] Constantinople, 4th July, 1579, Italian copy, A.N., K 1672, G 1, no. 81a.

[60] E. Charrière, *op. cit.*, III, p. 782 ff, note. On the exploits of 'General' Du Bourg, see above, Vol. I, p. 376–377.

Margliani found his mission more manageable than in the past, freed from the lengthy communication back and forth with Spain. As we have seen, in Mondéjar's time he had been forbidden even to reveal details of his mission in Naples. That is not to say that the task seemed an easy one at the time to Margliani himself. There were anxious moments, if most of the time was occupied in idle gossip, long conversations followed by equally long information reports, and even at one point in quarrels of precedence with Germigny over the choice of seat in the principal church in Pera[61] – perhaps out of sheer futility, perhaps to prove to the Turks the impossibility of maintaining a permanent representative in Constantinople, a concession to which Philip II refused to consent.

There were other problems for Margliani too: major figures in Turkish politics were changing. Muḥammad Pasha was assassinated in October, 1579 and replaced by Aḥmad Pasha, a lightweight who might have been favourable to Spain[62] but who in turn died on 27th April, 1580, and was succeeded by Muṣṭafā Pasha. These changes were matched by a multiple reshuffle among minor figures: Doctor Solomon was still there, but Horembey had disappeared; we still find however the strange figure of Bruti, a double if not a triple agent, whom Margliani accused of treason but was unable to dismiss although his indiscretion and duplicity threatened to compromise not only Margliani but a number of agents in his pay, Sinān, Aydar, Inglès, Juan de Briones.[63] Two new faces appear: Benavides and Pedro Brea, employees in the Turkish chancellery, the first extremely well acquainted with the documents prepared there, a Jew (his religion forbade him to take ship on Saturdays); the second more difficult to assess, but both certainly double agents. And we catch glimpse of the *baili* of Nicolò Prodanelli the Ragusan merchant and his brother Marino, whose ship was supposed to be in Naples in October 1580.[64]

In fact Margliani was master of the situation, but did not know it – hence his difficulty in maintaining his advantages, in standing up to the outbursts and threats of Euldj 'Alī who made a terrifying scene in the presence of the Grand Vizier himself. Perhaps this was a scene stage-managed by the Turkish ministers, but it was disconcerting when accompanied by other forms of intimidation. At the Arsenal, Euldj 'Alī proclaimed that 'the peace talks were broken off, that he had orders to arm 200 galleys and 100 mahonnas'. But Margliani stood his ground. He spoke with authority and did not avoid risks, 'being determined', he says, 'to agree to nothing in His Majesty's name nor to give any letters or presents before the capitulation is signed.[65] And he undertook to obtain that the armada would not sail in the spring. The violent uncontrolled behaviour of Euldj 'Alī towards him,[66] is merely proof of the Turkish

[61] *Ibid.*, p. 885 ff.

[62] Commander of Castile to Philip II, 9th June, 1580, Simancas E° 491.

[63] Margliani to Don Juan de Zúñiga, 3rd February, 1580, Simancas E° 491.

[64] Margliani to the viceroy of Naples, 15th October, 1580, Simancas E° 1338.

[65] Margliani to the viceroy, 2nd February, 1580, summary in Chancellery, Simancas E° 491.

[66] E. Charrière, *op. cit.*, III, p. 872 and 876, note.

admiral's frustration and anger: obviously he was not getting his way. A dispatch from Constantinople dated 26th February, 1580, says that Margliani is finding it impossible to reach an agreement with 'these Turkish dogs',[67] which would satisfy his own honour or the king's service, but in fact the treaty was well on the way to being agreed before this date. On the morning of the 18th, Doctor Solomon came to see him with a compromise text. Since this was not to be a capitulation between sovereigns, but 'an arrangement between the Pasha and Margliani', problems of protocol were very quickly solved[68] – although Margliani had received rough treatment in the process and still sensed hovering above him 'the danger' he writes on 7th March, 'in which I have felt myself to be for the last 50 days'.[69] He was not completely confident even at this date: 'I am much afraid that all our agreements will be broken so violently that we shall wish we had never begun talks on this treaty' he wrote on the same day.

Yet agreement was now near, imposed on both sides by circumstances, by the wars of Persia and Portugal and by the terrible famine ravaging the East.[70] So every time poor Margliani thought that all was lost, contact would be made again. The doctor or some intermediary would come back. The pasha had consented to reopen the discussion. Margliani would breathe again.[71] Then some fresh stumbling block would appear concerning the kingdom of Fez which Margliani refused to recognize as belonging to the Grand Turk;[72] or some new dispute over Portugal.[73] The rumour circulated in Venice in March that Margliani was in danger of being impaled and that Euldj 'Alī had threatened to tear out the one eye he possessed.[74] But on the 21st of that month he signed a new truce with the pasha, in the usual form, valid for 10 months until January, 1581. To avoid any dispute about the wording, the Italian text remained in the hands of the pasha, while the Turkish text, in gilt lettering, was handed to Margliani, who sent it to Zuñiga.[75]

After this, with the immediate future assured, the conversations lapsed for a while and both sides paused for rest. Juan Estefano went to Spain to carry the news and bring back instructions. This time, all Christendom was informed. In early May the news reached Rome, where it was noted that the truce hardly corresponded to previous Spanish declarations,

[67] Constantinople, 26th February, 1580, Simancas E° 1337.

[68] Margliani to Commander of Castile, Vines of Pera, 27th February, 1580, Simancas E° 491, copy.

[69] Margliani to the Commander, 7th March, 1580, Simancas E° 491.

[70] Margliani to the Commander, 29th October, 1580, Simancas E° 1338; Germigny to the king, 24th March, 1580, E. Charrière, *op. cit.*, III, p. 885.

[71] Margliani to the Commander, 12th March, 1580, copy, Simancas E° 491.

[72] See preceding note.

[73] Margliani to the Commander, 18th March, 1580, Simancas E° 491.

[74] Ch. de Salazar to Philip II, Venice, 18th March, 1580, Simancas E° 1337.

[75] Margliani's letters to the Commander, 23rd and 25th March, 1580, (Simancas E° 491), do not give the precise date of signing. But Germigny is categorical, 24th March, 1580. E. Charrière, *op. cit.*, III, p. 884–889.

notably that the ambassador was actually being sent to break off negotiations.[76] But Rome did not press the point: in the year 1580, she too had turned away from the Mediterranean and the crusade against Islam, being much taken up with the problem of Ireland and the war against the Protestants.

Germigny, who had closely followed Margliani's movements, claimed that the agreement had been paid for in gold. In fact Margliani had had to trade in mere promises.[77] He owed his success above all to circumstances. The latest event to put pressure on the Turks when the treaty was on the point of being signed, was the alarming news of a rising in Algiers. Much stood at risk if Philip (whose fleet had been newly prepared for the Portuguese campaign) were to have a free hand in the Mediterranean.[78] Venice realized this too: until then reticent not to say hostile, she changed her attitude and made strenuous efforts to be included in the forthcoming peace treaty.

We might note in passing that the truce of 1580, no doubt because it has received its due both in Charrière's collection of documents and in Zinkeisen's study (first published in Gotha in 1855 and still useful) is mentioned by most conscientious historians.[79] But oddly enough it is always presented as an isolated and exceptional agreement, when in fact it is merely one link in a long chain – and without the rest of the chain virtually incomprehensible.

The 1581 agreement. In Constantinople, it was generally considered inevitable that peace would follow soon. But it took almost a year to be achieved. The summer went by in intermittent conversations. The quarrels now concerned not the usual endless controversies over titles and precedence but the events reported by couriers. On 5th April, news arrived via Ragusa of the death of Cardinal Henry. 'This news', wrote Margliani,[80] 'has brought some alteration to the spirits of the people here. It seems to them that with the annexation of these kingdoms, accomplished without great bloodshed or a long war, his Majesty's forces have become so great that there is reason to fear them. They are also persuaded that [from now on] His Majesty will place more obstacles in the way of a peace or ceasefire, or whatever they desire'. Margliani for his part was disturbed by the movements of Euldj 'Alī. It was said that he was to sail to Algiers with 60 galleys to restore order. 'But others claim that though he will indeed go with those instructions, he is also to attack the king of Fez. I am ready to try to prevent this . . . when we have confirmation of his voyage'.[81] Otherwise, all the provisions of the truce would be threatened.[82] So too

[76] 2nd May, 1580, Vatican Archives, Spagna no. 27, f° 88.

[77] To the king, 17th May, 1580, E. Charrière, *op. cit.*, III, p. 910–911.

[78] M. Philippson, *Ein Ministerium unter Philipp II*, p. 404; L. von Pastor, *Geschichte der Päpste*, Vol. IX, 1923, p. 273; H. Kretschmayr, *op. cit.*, III, p. 74.

[79] J. W. Zinkeisen, *op. cit.*, III, p. 107.

[80] 9th and 14th April, 1580, A.N., K 1672, G 1, no. 166.

[81] *Ibid.* [82] *Ibid.*

thought the commander of Castile and wrote to Margliani, having information no doubt from other sources.[83] In the end the expedition to Algiers did not take place, but it had been the subject of much talk, as was the news, which arrived in October, of Alva's victory over Dom Antonio. The pasha, on learning that on this occasion the duke had distributed 200,000 doubloons to his soldiers, immediately sent to Margliani to ask why this distribution had been made and how much a doubloon was worth. A doubloon is worth two crowns, Margliani hastened to reply and by way of illustration, sent a dozen to his interlocutor. And what is Juan de Estefano up to, the latter wanted to know? Why so much delay?[84] Such was the conversation, at once trivial and suspicious, exchanged by the two negotiators. This was summer; only in winter would talks begin in earnest.

In December, the situation suddenly became more tense.[85] Don Juan de Estefano still had not arrived and Margliani was gravely embarrassed as the Pasha pressed him to say whether the king of Spain had sent orders to make peace, yes or no. Unfortunately some of Margliani's letters concerning the last months of his embassy are missing. Apparently between 10th and 20th December, the Turks made more precise demands – placing Margliani in a very embarrassing situation for at this very point he received the long-awaited orders (with or without Juan de Estefano, he does not say). He read the instructions in some perplexity. The king sent word that he had abandoned the notion of a truce in proper legal form, in view of the difficulty of proceeding with the 'desired equality'. What he meant was that he refused an agreement of the type accepted by the emperor, accompanied by permanent diplomatic representation.[86]

Margliani, who now at least knew what he was supposed to do, went ahead as quickly as possible. A letter from him dated 28th December describes his three-hour long conversation with the Agha of the Janissaries. Before daybreak on 27th December, the latter had sent a *caïque* to fetch him over from Pera to Istanbul. Niccolò Prodanelli acted as interpreter during the interview. Margliani was grateful: 'he is more intelligent and capable than any of the others', he wrote. And no doubt he had chosen him precisely because he was rather embarrassed about his mission and about what he had to convey to the Agha. The latter could make neither head nor tail of the interview. When he asked whether Margliani would go to kiss the sultan's hands, the Italian replied that he would go if there was a capitulation, but not if there was merely a temporary truce. It is important to note the difference between the two terms: temporary truce and formal capitaluation: it was the latter which Philip refused to countenance. The sultan will grant the capitulation, said the Agha; but

[83] (April, 1580), Simancas E° 491.
[84] Margliani to the viceroy of Naples, Pera, 29th October, 1580, Simancas E° 1338.
[85] Margliani to the viceroy, Pera, 10th December, 1580, Simancas E° 1338.
[86] Margliani to the viceroy, Pera, 20th, 21st, 26th (29th or 30th) December, 1580. Chancellery Summary, Simancas E° 491.

in that case what about 'equality', asked Margliani. And his interlocutor was once more in the dark. Then he in turn began to ask more simple and direct questions. Would Margliani stay in Turkey? 'I told him no. He asked why. I said that since there would be no trade, according to the decisions taken, my presence would no longer be necessary. I pronounced these words with a slight smile and added that I wanted to tell him the whole truth: that two reasons had determined me in my resolve: on one hand the lack of courtesy I had found here and on the other the rumour spread by the secretary of the French embassy throughout all the parts of Christendom he passed through [on his return] concerning the declaration of precedence he was taking back with him', to France.

And since he was not very sure of his case, the Spanish ambassador took the precaution of paying the Dowager Sultaness 5000 crowns (whereupon she took the opportunity of asking for more). At the same time he took care not to reveal the powers the king had conferred upon him, on the pretext that he had sent the document back to Naples.[87] He must have manoeuvred with some skill, for having been alerted on 10th December, he had more or less won over his adversaries by the end of the year. On 4th February, several letters and dispatches left Constantinople for various destinations carrying the news that the truce had been agreed for three years.[88] On the same day, Margliani wrote to Don Juan de Zuñiga:[89] 'On the feast of St. John, 27th December, I went to see Chā'ūsh Pasha, to whom I explained my task in language I thought fitting, having before my eyes the honour of his Majesty. I thereafter had several interviews with the said Pasha and finally, on 25th January, he called me to him to tell me of his sovereign's decision to grant me leave to depart and go to tell His Majesty. He hoped I would do my best to see that good relations were established and meanwhile there would be suspension of the war for a further three years'. From what Germigny says, it appears that the treaty was virtually identical with the previous agreements except that this time it was to last three years.[90] The viceroy of Naples who received the news on 3rd March, hastened to pass it on to Philip II, adding that in his opinion Margliani had done very well, but also wondering if the Pope would not use this as a pretext to pull in the purse strings.[91]

How indeed would the Papacy react? Don Juan de Zuñiga, who had

[87] Margliani gives these details in his letter to the Commander of Castile (end of December, 1580), A.N., K 1672, G 1, no. 169.

[88] Bartolomé Pusterla to Don Juan de Zúñiga, dispatch from the Levant, 4th February, 1581, in *Cartas y avisos . . .*, p. 53–54. Germigny to the king, 4th February, 1581, *Recueil . . .*, p. 31; E. Charrière, *op. cit.*, IV, p. 26–28, note, mentions the 'new-minted crowns stamped with the mark of Aragon' with which Margliani paid the Pashas. Dispatches from Levant, 4th February, 1581, Simancas E° 1339.

[89] Margliani to Don Juan de Zúñiga, 4th and 5th February, 1581, *Cartas y avisos . . .*, *op. cit.*, p. 55; 5th February, 1581, Simancas E° 1339. I read 'Sciaous Pasha' not 'Scianus' as the anonymous editor of the *Cartas* has it.

[90] See note 88.

[91] Don Juan de Zúñiga to Philip II, Naples, 3rd March, 1581, received at Tobarra on 23rd March, Simancas E° 1084.

once been ambassador at Rome, was particularly concerned about this. He judged it wisest to make the first move and on 4th March,[92] he penned the following singular version of the news for Rome's benefit. He said that he had in the past told Margliani that Philip II did not want a truce, invoking the best possible reasons to excuse the king. But when the Turks were informed, there was immediate talk of impaling Margliani, who was accused of having lied in all his dealings with the sultan until the conquest of Portugal should be achieved. Since it was well known that the Turks were perfectly capable of such atrocities, the poor knight in order to save his skin, had offered a truce of one year. The Turks insisted that the peace last for three years and only on receiving satisfaction had they allowed him to return to Christendom. But of course if any of the Christian powers wished to attack the Turks it would be easy to break the truce, in the first place because it had been imposed 'under duress', and in the second because the corsairs would always provide a pretext for renewing hostilities. Unfortunately, says the letter in passing, we have little leisure to launch an attack on the Turks at the moment, with all these other matters on our hands. But the truce itself is of no consequence at all. . . .

The viceroy's version of events was repeated in the same year 1581, by a Venetian ambassador.[93] Did he believe it? And was it believed in Rome? Probably no great efforts were made to verify it. The Papacy was concerned above all with action in Ireland against England. And who was better placed to strike a blow against England than Spain?[94]

So there was no reference by contemporaries to Spain's 'betrayal' of the West, as there had been to the betrayal by Venice. The one exception was the Spanish clergy which protested loud and long. Not that it was particularly attached to the crusade against the Infidel, but since the war was supposed to be over, it demanded to be released from the taxes created or maintained for this purpose. The demands went unheard.

Only later historians have put Spain on trial for betraying her allies. Perhaps trial is rather a strong word to describe the few lines devoted to the subject by Wätjen and Konetzke. 'The war against the Turks',[95] writes the latter 'was now abandoned once and for all. With this break a centuries-old Spanish tradition was interrupted. The religious war against Islam, which had stimulated and assembled the spiritual forces of the Peninsula, ceased to exist. True the *Reconquista* and the conquering raids which extended it into North Africa had not been entirely religious conflicts. But it was the religious spirit which had constantly animated and impelled these enterprises and had caused them to be felt in Spain as a

[92] Don Juan de Zúñiga to the marquis de Alcaniças, 4th March, 1581, Simancas E° 1084.

[93] E. Albèri, *op. cit.*, I, V, p. 328.

[94] To the nuncio in Spain, Rome, 11th July, 1580, Vatican Archives, Spagna 27, f° 123 '. . . il passar con silentio nel fatto de la tregua è stata buona risolutione poiché il farne querella in questo tempo non potria sinon aggiungere travaglio a S. Mtà senza speranza di frutto'.

[95] *Op. cit.*, p. 181.

mighty communal endeavour. The most powerful driving force of Spanish progress was paralysed'.

This judgement has some force if one considers the evolution of Spain in general terms, but is less convincing on several particular points. The powerful religious urges in Spain found a different outlet after the 1580s. The war against heresy quite as much as the crusade against Islam, was a religious war, with all the basic components of war. Besides there were to be several more attempts in the direction of North Africa and against Turkey – the so-called war of 1593 (of minor significance it is true).

It is a fact nonetheless that the 1580s mark a turning-point in the history of Spain's relations with Islam, even if the past course of this history had been more untidy, disconnected and vacillating than it is usually presented. After Margliani's embassy, a *de facto* peace reigned. The 1581 truce seems to have been renewed in 1584 and even in 1587.[96] And later hostilities, when they occurred, bore no comparison with the mighty confrontations of the past. The truce was more than a clever expedient of Spanish diplomacy.[97]

But was it enough to justify the charge that Spain walked out on her allies in 1581? If Spain betrayed anyone, she betrayed herself, her own traditions and spiritual identity. But such betrayals, when they concern a country, are often little more than an intellectual concept. In any case, she did not betray Mediterranean Christendom: she did not deliver Venice up to possible vengeance, nor did she abandon Italy, the burden of whose protection she continued to support. If she negotiated with the Porte, who can blame her? It was not Spain who had introduced Turkey to the European free-for-all. Major war in the Mediterranean exceeded the resources even of the greatest states, the political monsters who had found it so difficult to retain their respective halves of the sea. There is a difference between betrayal and giving up the struggle. The swing of the pendulum which carried war outside Mediterranean confines during these years of delicate and obscure peace talks, was twofold: on one hand it carried Spain towards Portugal and the Atlantic, into a maritime adventure incomparably greater than those she had known in the Mediterranean, while on the other it flung Turkey towards Persia and the depths of Asia, the Caucasus, the Caspian, Armenia and, later, the Indian Ocean itself.

2. WAR LEAVES THE CENTRE OF THE MEDITERRANEAN

While we cannot always explain them, we are able to recognize the major rhythmic phases of the Turkish wars. In even the briefest summary of the

[96] In 1584 possibly negotiated by Margliani himself, if we may assume as much from a remark by Hammer, *op. cit.*, VI, p. 194–195. The truce was extended for two years in 1587, but no source is named.

[97] As M. de Brèves thought in 1624, E. Charrière, *op. cit.*, IV, p. 28, note.

reign of Sulaimān the Magnificent, they are impossible to ignore.[98] These rhythms, rather than the personal will of the sultan, governed the course of his glorious reign. Over the years, Turkish power was drawn in turn towards Asia, Africa, the Mediterranean and Europe north of the Balkans. To each of these movements corresponds an irresistible impulse. If ever there was a pattern in history, this is one. But it is unknown territory chiefly because historians have persistently overestimated the power of individuals. They have paid little attention to deep-seated, underlying movements (for example those the Ottoman empire inherited when it simultaneously destroyed and enlarged the empire of Byzantium), to that physics of international relations which in the sixteenth century was busy establishing the necessary compensations between the major war fronts along which Turkish power impinged upon the outside world.

Turkey and Persia. Between 1578 and 1590, the internal workings of Turkish history are a mystery to us and the chronicles upon which Hammer's account, for instance, depends present major historical problems only in terms of events.

We are woefully ignorant not only about Turkey – which by comparison is almost coherent and comprehensible – but about the great expanses of Persia, another form of Islam and another civilization. We know little too about the intervening areas between Persia, Turkey and Orthodox Russia. And what was the role played by the no-man's land of Turkestan? Beyond these territories, to the south, stretched the vast Indian Ocean, this too an unknown quantity, its trades imperfectly controlled by the Portuguese (after 1580 assisted by the Spanish, but more in theory than in practice).

But it is in these areas that we must seek for evidence of the major reorientation of the Ottoman Empire in the years after 1577–1580, a movement as strong as that which propelled Spain towards the Atlantic. The Ocean represented a vast new source of wealth for Europe. Was Turkey drawn to Asia by a similar prospect of gain? There are no documents to support the assumption and our information is so incomplete that we can do no more here than suggest a few impressions.

What the chronicles do tell us is that Persia was in the grip of a great political upheaval. The assassination in May, 1576 of Shah Tahmāsp[99] who had reigned in Persia since 1522; the almost immediate murder of the new sovereign Haidar; the accession of Prince Ismā'īl, who was brought out of his terrible dungeon and placed on the throne to reign only for the space of sixteen months, until 24th September, 1577; and finally the coming to power of the near-blind prince Muhammad Khudābanda, the father of the future 'Abbās the Great; all these events and several

[98] I am thinking in particular of Franz Babinger's 'Suleiman der Mächtige', in *Meister der Politik*, Stuttgart and Berlin, 1923.

[99] J. von Hammer, *op. cit.*, VII, p. 70. On all these points see W. E. D. Allen's short but authoritative book already mentioned, p. 105, note 2.

others (in particular the obscure role played by the Georgian, Circassian, Turcoman and Kurdish tribes) explain the vulnerability of Persia at this time; they explain the temptation for the leaders on the Turkish frontier, notably Khusraw Pasha, and the policy of the Turkish 'Military', the land commanders whose careers had for years been sacrificed to the navy: Sinān Pasha, Muṣṭafā Pasha. Persia seemed to be disintegrating at the centre: someone ought to take advantage of it.

In 1578, letters left Constantinople, addressed to all princes of the northern zone of Persia, whether they were in power, obeyed by their followers and strong, or not, in Shirvān, Daghestan, Georgia, and Circassia. About a dozen of these missives are preserved in the historian 'Alī's *Book of Victory*, the account of the first campaign in the new war against Persia.[100] They are addressed to 'Shabruk Mirza, son of the former sovereign of Shirvān; to Shemkhal, prince of the Kumuks and the Kaitaks; to the governor of Tabazeran, in Daghestan, on the banks of the Caspian Sea; to Alexander son of Levend, ruler of the country between Eriwan and Shirvān; to George, son of Lonarssab, lord of the district of Baksh Atshuk [Imereti] to the ruler of Guria and to the Dadian, the prince of Mingrelia [Colchis]'. This list of names maps out a recognizable area, between the Black Sea and the Caspian, the same area which in 1533–1536 and in 1548–1552 could already be seen in the background of Sulaimān's campaigns against Persia.

Although we know very little about these intermediary zones, little too about the Turkish frontier in the region of Lake Van or indeed about Persia itself, where the blood of princes flowed freely in the years 1576–1578, it seems likely that Turkish imperialism had its eye on the Caspian. The aim was not necessarily control of the sea itself: access to it would be sufficient to threaten directly the Persian shores of Mazandaran; in this sea where they were virtually unknown, galleys would be all the more effective. This strategic aim had already been mentioned in western diplomatic correspondence on the occasion of the 1568 war and the projected Don-Volga canal. But is it not also likely that the Turks desired access to Turkestan, to the internal routes of Asia which the Russians had interrupted by occupying Astrakhan in 1566? Turkestan after all commanded the silk route. Persia was to owe part of the economic revival she experienced at the end of the century under the great reign of Shah 'Abbās, to these routes to the Asian interior. They were also the origin of that first phase of Persian expansion, visible in the growth of the towns, which was capable of attracting English traders from far away, and was illustrated by the amazing dispersion of Armenian merchants throughout all the countries bordering the Indian Ocean, throughout all the Turkish states in Asia and Europe; some of them even reached Danzig in 1572.[101] Tabriz, an important centre of this world-wide trade, must have been a tempting prey.

[100] J. von Hammer, *op. cit.*, p. 77.
[101] B.N., Paris, Ital., 1220.

The opportunity afforded by the sudden weakness of Persia was even more likely to attract the Turks at a time when they possessed clear technical superiority over their adversary. The Persians had no artillery and few arquebuses; the Turks were not richly equipped but the little they had gave them the upper hand. And facing them stood no fort worthy of the name. The only defence on the long Turco-Persia frontiers was provided by the deserts, some natural, others man-made, deliberately depopulated by the prudent Iranian rulers.[102]

Religion inevitably played some part in any war between Turks and Persians: the *fetwas* (legal pronouncements by the *mufti*) consecrated the pious and near-sacred character of the struggle against the 'Shiite dogs',[103] renegades and heretics with their 'red caps',[104] the more vehemently since the Shiites, worshippers 'of the Persian religion' were to be found throughout the Asian part of the Turkish empire, even in the heart of Anatolia. There had been a Shiite uprising in 1569.[105] But in the East as in the West, no war was fought for purely religious reasons. The Turks when they took the road to Persia were swayed by many other passions: to those we have already mentioned, might be added the attractions of Georgia, a land rich in potential slaves both men and women, trade routes and fiscal revenues.

The notion of a concerted long-term Turkish policy has not always been accepted. But surely it was evident in this war. The much-proclaimed decadence, whose beginnings are often detected at the death of the great Sulaimān is a miscalculation. Turkey remained an immense force, by no means untamed but on the contrary organized, disciplined and deliberate. If she suddenly abandoned the familiar battle ground of the Mediterranean to turn eastwards, that is no reason to assume the onset of 'decline'. The Ottoman empire was merely following its destiny.

The war against Persia. The war was nonetheless an exhausting ordeal for the Turks.

The very first campaign, led in 1578 by the *serasker* Muṣṭafā, the victor of Cyprus, was a foretaste of the hardships ahead. The Turks won some major victories, all dearly bought (the victory of the Devil's Castle on the river Kura on the Georgian frontier, for instance).[106] If the Turkish army had no difficulty in entering Tiflis (Tbilisi), the long march from Tiflis to the Kanak and beyond this river through forests and swamps, proved a different matter. Famine combined with exhaustion to decimate

[102] B.N., Paris, Ital., 1220. f° 317 v° (about 1572).

[103] J. von Hammer, *op. cit.*, VII, p. 75.

[104] *Ibid.*, p. 80; *Voyage dans le Levant de M. d'Aramon, op. cit.*, I, p. 108.

[105] De Grantrie de Grandchamp to M. de Foix, Constantinople, 30th August, 1569, E. Charrière, *op. cit.*, III, p. 62–66.

[106] J. von Hammer, *op. cit.*, VII, p. 81. On the Persian war Hammer uses the precious sources of Minadoi and Vicenzo degli Alessandri, as well as eastern sources, the historians 'Alī and Pechevi. Once more I am drawn to acknowledge the superiority of this early work over those of Hammer's successors, Zinkeisen and Iorga.

the army, which was constantly harassed by the Persian khans. However in September, on the banks of the Kanak, the Turks were once more victorious. Most of Georgia was now in their hands. In September, the *serasker* divided it into four provinces, leaving a *beglerbeg* in each, with troops and artillery, and orders to collect the dues which the Persians in these rich provinces, notably Shirvān, had been levying on silk. At the same time, the *serasker* managed to obtain the submission of the local princes who had received the Turkish conquest more or less grudgingly. At the approach of autumn, he retired with his troops, decimated 'by five battles and by sickness',[107] to Erzurum where they spent the winter.

What problems had been revealed by this opening campaign? In the first place the tenacity of the enemy; the inconstancy of the local tribes who were capable of launching cruel attacks without warning in the mountain passes; and above all distance, the number of forced marches, their extreme severity, the near-impossibility of supporting life in these lands of varying fertility, with their mountains, forests and marshes, and the severe continental winters. Distance, as in the 'Russian' campaign of 1569, again operated against the Turks. Between Constantinople, the beginning of the army's journey, and Erzurum, there were sixty-five stages; between Erzurum and Arash (the furthest point reached) sixty-nine; and the same distance back of course. For long-distance wars like this, a cavalry force, with very little baggage, was the best weapon – not an army in the western style with its cumbersome supply services, infantry and cannon.[108] The ideal instrument was the Tartar cavalry which had served in the 1568 campaign. But it was not one on which the Turks could automatically rely, and was of most service not in the mountainous zones where it was quite powerless, but in the wide plains to the north and south of the Caucasus, especially the north (where it was effectively used by Osman Pasha in 1580). But how could the army even survive its victories in these devastated regions, let alone occupy them?

The Persians meanwhile made the weather their ally and during the winter of 1578–1579 moved into the offensive. They were better equipped than their adversaries – now far from home, camping out in improvised billets and besides accustomed to a Mediterranean climate – to endure the terrible Asian cold. The Turkish strongholds resisted the first attack. But under the second, some gave way: Shirvān was evacuated and the garrison withdrew to Derbent. It was a terrible winter. It is hardly surprising that the dispatches from Syria carried alarming reports.[109] The Spanish agents in Constantinople were jubilant. 'It has been learned', wrote one,[110] 'that an ambassador from H.M. [the Catholic King] is to be sent. I am not pleased to hear it, for this is not the time to be sending ambassadors.

[107] Pera, 9th December, 1578 (Margliani to Pérez, received 31st March, 1579), Simancas E° 489.

[108] What would Émile-Félix Gautier have thought of it?

[109] Venice, 7th January, 1579, A.N., K 1672, G 1.

[110] Constantinople, 4th February, 1579, A.N., K 1672, G 1.

If he came it should be at the head of a mighty armada'. The war, observers concluded, would go on.[111] Persian demands were considered exorbitant.[112] On 8th July, 1579, the Spanish ambassador in Venice wrote that 'not content with demanding Mesopotamia, the Persians want the Turks to forsake the rites of their religion'.[113] For by now the Turkish reverses had begun to look like a rout. The combatants whom the fortunes of this terrible winter campaign brought back to Constantinople horrified all who set eyes on them,[114] so nakedly did they present the image of human suffering. The sultan however did not intend to give up yet. The entire year, or at any rate the campaigning season of 1569, was used by the *serasker* to construct the great fortress of Kars. Once more then, troops had to be massed, supplies had to be assembled at Erzurum,[115] while 40 galleys, munitions, artillery and wood were sent to Trebizond,[116] and envoys were dispatched to negotiate with the Tartar ruler and several Indian princes. The threat from Persian cavalry troops was still serious near Kasvin and in Shirvān, particularly since the Georgians were said to be in league with them and to have given them hostages.[117]

Meanwhile in the south the fortress of Kars was rising from the ground at the cost of heroic labours.[118] Eye witnesses reported in Constantinople that it was already impregnable against enemy attack, 'which news', wrote Margliani, 'is worth much consideration and will be prized with reason by the Grand Turk. For he will have achieved something never accomplished by his grandfather, the sultan Sulaimān, who, as no one can deny, was a very great soldier. These last two nights, there have been fireworks and much rejoicing in the Seray of the Grand Signior. I am much afraid that this news will make him even more haughty and intractable'. But, he wrote a few days later, 'I console myself with the hope that Kars may suffer the same fate as Servan [Shirvān?] which was captured and fortified by the Turks with the same enthusiasm and later recovered by the Persians to the great loss of the said Turks'.[119] In Venice – but then Venetian reports were notorious – the fortress was said to be half the size of the city of Aleppo and to measure three miles round![120]

[111] Constantinople, 24th March, 1579, *ibid.*
[112] Juan de Idiáquez to Philip II, Venice, 21st March, 1579, *ibid.*, no. 35.
[113] X. de Salazar to Philip II, Venice, 8th July, 1579, *ibid.*, no. 84.
[114] Margliani (exact reference mislaid).
[115] J. de Idiáquez to Philip II, Venice, 29th April, 1579, A.N., K 1672, no. 56, copy.
[116] Germigny to the king, Pera, 16th September, 1579, *Recueil*, p. 10; *Relacion de lo que ha succedido al capitan de la mar Aluchali desde los 17 de Mayo que partio de aqui de Constantinopla asta los 6 de agosto sacada de las cartas que se han recibido de Juan de Briones y Aydar Ingles*, A.N., K 1672, G 1, no. 115. Same report, Simancas E° 490. *Relacion de lo que ha sucedido de los 9 de agosto hasta los 28*, A.N., K 1672, G 1, no. 116. Euldj 'Alī returned to Constantinople on 10th September (cf. Germigny reference, above, this note) with 13 galleys.
[117] Constantinople, 29th April, 1579, A.N., K 1672, no. 56, copy.
[118] Margliani to Antonio Pérez, Pera, 2nd March, 1579, Simancas E° 490.
[119] Margliani to Antonio Pérez, 5th September, 1579, *ibid.*
[120] J. de Cornoça to H.M., Venice, 17th October, 1579, A.N., K 1672, G 1, no. 142a.

In any case the Persians, during the summer of 1579 seem to have deliberately remained on the defensive: for fear of the plague, raging in the Turkish ranks, according to reports in Venice;[121] out of respect for the Turkish forces and artillery, we may prefer to think, as they waited for the return of their ally, winter. But the threat was always there. In Venice it was suggested that 250,000 Persians were mobilized on the frontier[122] – though this is perhaps another Venetian exaggeration. But even in Constantinople it was reported that while the Turks had succeeded in establishing a solid position at Kars, Tiflis, the focus of the 1578 conquest was now under enemy siege.[123] There were reports in Venice in September – one must allow for delays in transmission – of the difficulties the *serasker* had encountered in forcing his troops to march from Erzurum to Kars and even of mutinies among the janissaries and the *Sipahis*. It was suggested that they might have been deliberately provoked by the *serasker*, to provide a pretext for going no further.[124] In Constantinople in October there was official optimism: peace with the Persians could be obtained at any time. But Muṣṭafā nevertheless received orders to go into winter quarters and to withdraw his troops not to Erzurum, it was reported, but much further west to Amasya.[125] Tiflis, although closely surrounded, was relieved by Ḥasan Pasha, son of Muḥammad Sokolli, and with his assistance received plentiful supplies.[126] But winter was approaching. And soon the bulk of the Tartar cavalry would leave Daghestan which they had ravaged throughout the summer months.[127] We may note incidentally that this was but a small force (about 2000 horseman according to Hammer) and that it had managed to cross the desert wastes of the north Caucasus separating the Crimea from Derbent on the Caspian Sea in barely a month. This was perhaps an indication of an easier invasion route than the cruel mountains of Armenia.

The death of Sokolli, the short vizierate of Aḥmad,[128] the appointment of Sinān Pasha to the command of the army in Erzurum,[129] then his promotion, while he was leading the army into Georgia, to the rank of Vizier, did not materially alter the conditions of the war. During the summer, Sinān marched his army along the road from Erzurum to Tiflis. He reorganized the Ottoman occupation of Georgia. Then, to avenge an attack on a foraging party, he decided to launch a major attack on the powerful city of Tabriz. He had to accept soon afterwards that the raid could not be followed up and withdrew to Erzurum. By now in any case,

[121] Salazar to Philip II, Venice, 7th September, 1579, *ibid.*

[122] *Ibid.*

[123] Germigny to the king, Pera, 16th September, 1579, *Recueil*, p. 10.

[124] See note 121.

[125] Germigny to the Grand Master of Malta, Pera, 8th October, 1579, *Recueil* . . ., p. 17–18. Only to Erzurum, Hammer, *op. cit.*, VII, p. 96.

[126] J. von Hammer, *op. cit.*, VII, p. 97.

[127] *Ibid.*, p. 98.

[128] He died on 27th April, 1580, E. Charrière, *op. cit.*, III, p. 901.

[129] Three letters from Margliani to Zúñiga, 27th and 30th April, 1580, Simancas E° 491 (Chancellery Summary).

peace talks were under way. Sinān obtained permission to come to discuss them in Constantinople. They presently resulted in a sort of armistice, valid for the year 1582. Ibrāhīm, the Persian ambassador, entered Constantinople on 29th March, 1582, 'with a suite composed of as many people as there are days in the year'.[130]

But difficulties in Georgia imposed certain obligations upon the Turkish army. Tiflis had to be supplied from Erzurum during the summer of 1582 for the forthcoming winter.[131] But the supply convoy was ambushed by the Georgians and by Persian guerillas. The situation in Tiflis was becoming desperate. Meanwhile the Persian mission to Turkey came to nothing. This series of setbacks led to the disgrace and banishment of Sinān Pasha who was said to be opposed to the Persian war, and the appointment on 5th December, 1582 of the new vizier, the same Chā'ūsh Pasha whom we have already seen in conversation with Margliani in the later stages of his negotiation, in January, 1581.

This domestic crisis meant that the war went on. Chief command was now in the hands of the *beglerbeg* of Rumelia, Ferhād Pasha (promoted to the rank of vizier) who was responsible for the campaigns of 1583 and 1584. His principal task on the sultan's own orders, was to fortify the contested frontiers – hence the erection of the great fort at Eriwan in 1583 and the construction or repair in 1584 of a certain number of castles as well as the fortification of Lori and Tomanis. So oddly enough, the eastern side of the Ottoman empire was gradually turning into a fortified frontier on the western pattern with its strongholds, garrisons and supply convoys. Counselled by prudence, such a policy required much patience and brought little glory and many hardships to the soldier.

Meanwhile, north of the Caucasus, there had already begun, at the urging of Osmān Pasha, governor of Daghestan, another war, more hotly contested, on the great routes over the Tartar steppes. Beginning in 1582 (for the armistice had not been very strictly observed here) and renewed in 1583–1584, it was a war which spread without difficulty from the Black Sea to the Caspian. On the sultan's orders,[132] considerable forces were assembled at Kaffa; besides troops, materiel and supplies, eighty-six tons of gold were transported to the base: the Persian wars were not only exhausting and burdensome from the human point of view, but needless to say swallowed up huge sums of money as well. There was soon talk of forced loans from the property of the mosques. Meanwhile an English traveller in 1583 describes the Persians preparing loads of silver ingots and coins for the army's pay.[133] This distinction – gold in Turkey, silver in Persia – is one we have already met.

The expeditionary force, led by Djafar Pasha, took two weeks to cross the Don. In order to advance it had to pay indemnities to the tribes it met north of the Caucasus and travel for long distances over deserts

[130] J. von Hammer, *op. cit.*, VII, p. 104.
[131] *Ibid.* [132] *Ibid.*, p. 112.
[133] R. Hakluyt, *op. cit.*, II, p. 171.

inhabited only by herds of wild deer.[134] After a march of eighty days, it arrived at Derbent on 14th November, 1582, exhausted, and prepared to go into winter quarters. In the spring the small army left again under the command of Osmān Pasha, defeated a Persian force and pushed on to Baku. The Osmān, having installed Djafar Pasha in Daghestan, withdrew the rest of his troops to the Black Sea. On the way back he encountered unheard-of difficulties; after repeated engagements with the Russians near Terek and the Kuban, he found the way blocked by the Tartars when he reached Kaffa. The latter, unreliable and at best reluctant allies, refused to depose their khan as Osmān demanded. To bring them under control, no less than the intervention of a squadron of galleys under Euldj 'Alī was required. Let us note in passing that if the figures given by the sources are correct, Osmān had no more than 4000 men with him: this indicates the measure of this extraordinary campaign. On his arrival in Constantinople he received an unusually warm welcome from the sultan, who sat for four whole hours listening to his long tale. Three weeks after this audience, he was appointed Grand Vizier, and the sultan gave him command of the army of Erzurum, with orders to take Tabriz.

Preoccupied during the winter with further pacification of the Crimea, which finally settled down of itself, the new commander of the Turkish army left Erzurum with the first fine weather at the head of an army whose size had deliberately been reduced; at the end of the summer (September, 1585) he descended upon Tabriz and took it. Tabriz, a trade and craft centre in the middle of a fertile plain of crops and fruit trees, was a godsend to the exhausted and starving Turkish army. But after a spell of atrocious pillage, the city had to be fortified in haste, for the Persians surrounding the position had not abandoned the struggle. After his spectacular victory, Osmān Pasha died on the evening of one of these encounters (29th October, 1585). The army was led back to its winter quarters by Cighāla-zāde. The Persians still did not give up. During the winter of 1585–1586, all the positions along the Turkish *limes* from Tiflis to Tabriz came under attack from the subjects of the Sophy and their local accomplices. Once more, the line held and the road to Tabriz was opened in time by the *serasker* Ferhād Pasha, who came for the second time to take command in Asia. Slowly but surely the Turks were gaining the upper hand. During the next two years, the war was to change in character. For the Persians were suddenly confronted with a new enemy, the Uzbegs of Khurāsān. Their defences were therefore surprised from the rear, while at the same time recruitment of cavalry forces was becoming more difficult. The Turks moved beyond Tabriz and advanced southwards. The centre of war was for a brief spell to move from Erzurum to Baghdad. It was near this city that in 1587 the army of Ferhād Pasha, its ranks swelled with hastily recruited Kurdish soldiers, crushed the Persians on the Plain of Cranes. The following year the Ottomans concentrated their efforts once more on the north, around Tabriz in the Kara-Bagh.

[134] J. von Hammer, *op. cit.*, VII, p. 113, note 1.

They seized Ganja which they immediately fortified in preparation for further campaigns.

But in the meantime, the young 'Abbās, whether by persuasion or by force, had taken over the reins of government from his father, during the latter's lifetime (June, 1587). He had the wit to see that with his kingdom threatened by the Uzbegs on one side and the Ottomans on the other, his best course was to make come concessions in the West. Another magnificent Persian embassy arrived in Constantinople in 1589, led by Prince Haidar Mīrzā. In the Turkish capital where the accession of the new sultan Murād had inaugurated an age of pomp and splendour, sumptuous official receptions were held. The negotiations themselves took time, but eventually bore fruit. The peace treaty was signed on 21st March, 1590, bringing to an end a war which had lasted for twelve years. The relentless pressure of the Turks at last received its reward: the sultan kept all the conquered provinces, that is Georgia, Shirvān, Luristan, Sharazūr, Tabriz and 'the portion of Azerbaydzhan' dependent on it.[135] In effect the Ottoman ruler now controlled the whole of Transcaucasia, the entire populated part of the Caucasus and possessed a large window on to the Caspian Sea.

This was no small triumph. Indeed it was a singular sign of Ottoman vitality and by no means the only one. But to the historian of the Mediterranean, its most significant aspect is the new focus of Turkish ambition: now the Caspian, far from the Mediterranean. This centrifugal urge explains Turkey's absence from the Mediterranean arena, at least until 1590.

The Turks in the Indian Ocean. The Turks had not only been fighting in Persia: they had been deeply involved in a war for control of the Indian Ocean, where once more our information is far from complete.

The Indian Ocean, at least the western part, had for centuries been the preserve of Islam. The Portuguese had never totally succeeded in driving out the Moslem powers, indeed since 1538 they had suffered repeated attacks from them, attacks in which the Turks had played a major role. But in the end perhaps it was because its advance southward was not entirely successful that the Ottoman Empire failed to prove a match for Europe. The missing element was seapower. True, the Turks possessed a formidable fleet. But they were obliged to approach the Indian Ocean by way of the narrow Red Sea and their naval techniques were Mediter-

[135] J. Von Hammer, *op. cit.*, VII, p. 223. So the Turks could count it a victory, G. Botero, *op. cit.*, p. 188 v° sees it as follows: 'for although the Turk had been undone and routed more than once, he had nevertheless, by gradually fortifying his own positions occupied a very large area: and having captured finally the great city of Tauris, he made certain of it with a great and mighty citadel. So the Persians for lack of citadels and fortresses lost the campaign and their cities too'.

[136] Karl Brockelmann, *History of the Islamic Peoples*, 1949, p. 313; on Piri Reīs as an individual and his oddities, see Erich Bräunlich, *Zwei türkische Weltkarten*, Leipzig, 1937.

ranean in origin. So it was with Mediterranean ships – galleys which had to be taken to pieces, transported overland to Suez, then reassembled and launched – that Turkey sailed to meet her competitors in the Indian Ocean. It was with a fleet of galleys that the aged Sulaimān Pasha, governor of Egypt, took Aden in 1538 and in September of that year, advanced on Diu, which he failed to capture. And it was with galleys too that Piri Rē'īs in 1554 tried his luck against the Portuguese sailing-ships, ocean-going vessels which got the better of the Turkish oarships. Based at Basra, at the entrance to another narrow sea, the Persian Gulf, the galley fleet commanded by the admiral-poet 'Alī, was driven in 1556 on to the shore of the Gujerat peninsula, there to be abandoned by commander and crews. The Indian Ocean thus witnessed a strange combat between galley and sailing-ship.[137]

Turkish thrusts in this direction were as a rule linked to the Persian wars, often following close on their heels. War with Persia dominated the years 1533 to 1536, and was followed by Sulaimān Pasha's expedition: the capture of Aden and first siege of Diu took place in 1538. A second war with Persia lasted from 1548 to 1552 (but was hard fought only during the first year) and in 1549 came the second siege of Diu; then in 1554 the expedition of Piri Rē'īs and in 1556 that led by 'Alī. Similarly in about 1585, as the war with Persia became less important, the battle for the Ocean broke out with renewed vigour along the east coast of Africa, the seaboard which the Portuguese called the 'Contra Costa'.[138]

The Turco-Spanish truce in fact only applied in the Mediterranean. Philip II in vain urged the Portuguese officials – rather late in the day it is true – to exercise their rule with restraint and tolerance lest the local princes should carry their grievances to the Turks.[139] The latter did not wait for the call. After 1580 they continued their profitable raids on the Portuguese traders. In 1585, a fleet commanded by Mir 'Alī Bey[140] even reached the African coast. It took Mogadiscio, Brava, Giumbo and Ampaza with little difficulty. The prince of Mombasa declared himself a vassal of the Porte. The following year Mir 'Alī Bey carried off all the coastal positions except Malindi, Patta and Kilifi, which remained loyal to Portugal. Was this, as Philip II thought, a result of Portuguese ill-treatment of the natives?[141]

Portuguese reaction was slow. One fleet went astray on the shores of

[137] This is something of a simplification; but to go into detail would take too much space here. Vitorino Magalhães Godinho who has been working on a full-length account of the Indian Ocean in the sixteenth century tells me that the Portuguese fleets were made up of sailing-ships, of Atlantic type, locally constructed ships and some galleys . . . a composite and multi-purpose fleet.

[138] M. A. Hedwig Fitzler, 'Der Anteil der Deutschen an der Kolonialpolitik Phillips II in Asien', in *Vierteljahrschrift für Sozial- und Wirtschaftsgeschichte*, 1936, p. 254–256.

[139] Lisbon, 22nd February, 1588, *Arch. port. or.*, III, no. 11, quoted by M. A. H. Fitzler, *art. cit.*, p. 254.

[140] Cf. W. E. D. Allen, *op. cit.*, p. 32–33 and notes, who corrects the error I made in the first edition of this book.

[141] 14th March, 1588, *ibid.*, no. 43, quoted by M. A. H. Fitzler, *art. cit.*, p. 256.

southern Arabia in 1588.[142] In this, the year of the Invincible Armada, the weighty Iberian war machine had more urgent affairs to attend to than the distant struggle in the Indian Ocean. But the stakes were enormous: behind Mombasa, which the Turks hoped to fortify, lay the gold mines of Sofala; more important still, access to Persia and India, which the Portuguese fleet had tried in vain to defend at Bab-el-Mandeb in 1588. Luckily for the Portuguese, the Turks too had reached the limit of their endurance, exhausted by distance. In 1589, Mir 'Alī Bey attacked with only five ships. The Portuguese fleet under Thomé de Souza succeeded in trapping him in the river at Mombasa; meanwhile a furious rebellion of the black inhabitants of the entire coast broke out and finally swept all before it, local rulers and Turkish invaders alike. Only those Turks who took refuge on board Portuguese ships, including Mir 'Alī Bey himself, escaped the massacre. Thus ended in 1589 one of the most bizarre and most obscure of Ottoman ventures.

The invasion of Portugal: turning point of the century. For Michelet, the year of the St. Bartholomew Massacre was the turning point of the century. A greater divide, to my mind, is marked by the years 1578–1583, when the war of Portugal opened the great battle for control of the Atlantic and world domination. Spanish thoughts turned to the ocean and western Europe; while following the bankruptcy of 1575, which ended the first part of Philip's reign, the influx of precious metals suddenly improved the state of the Spanish treasury. These critical years saw the beginning of what one historian has called 'the royal silver cycle'[143] which was to last from 1579 to 1592. In the Netherlands and elsewhere too. Philip's policy, emboldened by this sudden swelling of his resources, was to become increasingly aggressive and less prudent.

This dramatic change has certainly not gone unremarked by historians, those of Portugal in particular, who are more familiar with this material than the others but perhaps for that very reason inclined to take a narrow view. Their national destiny was undoubtedly central to this phase of the ocean's history, but not exclusively so. The many incidents in the course of the Atlantic war, taken together, reveal the extent of the mighty struggle in which the combatants were locked. Nor can I entirely agree with those who see these struggles as preparing the way for 'modern times'. Such a claim is premature: for many years to come they actually held up progress in the Atlantic.

The Spanish change of course was very marked. Granvelle arrived in Madrid in 1579. He remained there until his death in 1586, for seven years during which he occupied first unofficially and later officially the position of chief minister. Many historians (even Martin Philippson) have been tempted to see a connection between these governmental changes and the switch from a defensive, prudent phase of Philip's reign to a new

[142] Fitzler, *art. cit.*, p. 256.

[143] Pierre Chaunu, *art. cit.*, in *Revue du Nord*, 1960, p. 288, and *Conjoncture*, p. 629 ff.

aggressive, imperialist phase. Until 1580, Spanish policy had been closer to that advocated by the 'peace party' led by Ruy Gómez than to that of Alva and his followers, the 'war party'. There had been some exceptions of course: Alva's expedition in 1567 for instance – and Lepanto. We should not regard these as parties in any modern sense: they were more akin to cliques from which the king remained aloof while at the same time making use of them. He was not entirely displeased at the quarrels among his subordinates since they provided him with fuller information and allowed him to exercise more adequate supervision and in the last resort more complete authority. By playing them one against the other, by fostering their mutual suspicions, Philip wore out many of the men who served him. The heavy responsibilities of the reign took their toll too. By 1579, only a few survivors remained from the parties of the first phase of the reign. Ruy Gómez had died in 1573 and his clique, regrouped around Antonio Pérez, lacked its former coherence. The Duke of Alva who left the Netherlands in December, 1573 had never regained his former eminent position in Spain, and his abrupt disgrace in 1575 had removed him from political life altogether.

It was in March, 1579 that Philip recalled Granvelle. 'I need your presence here', he wrote, 'and your help in the tasks and cares of government. . . . The sooner you arrive the better I shall be satisfied.'[144] The cardinal was by then living in Rome. Despite his age – he was sixty-two – he accepted the challenge, but was several times delayed; he had to wait first at Rome then at Genoa. Only on 2nd July, with Don Juan de Idiáquez at his side, did he set eyes on the coast of Spain. On the 8th they landed at Barcelona. The cardinal left immediately with wagons and pack animals, travelling by night to avoid the heat. On the express orders of the king who was already at the Escorial, he did not go to Madrid but arrived in the first week of August at San Lorenzo, where the king welcomed him as a saviour.[145]

For this was indeed his role. Philip had waited until the cardinal was already on his way to the Escorial before dropping the mask of trust and brutally pouncing on Antonio Pérez and his accomplice the princess of Eboli. They were both arrested during the night of 28th to 29th July. It is worth noting the date. For although the king had long been suspicious of Antonio Pérez, it was not until the new government was almost in position that he decided to confront a group still powerful at court. Granvelle's arrival set the seal on the downfall of the peace party. Personal motives, sensational, complicated and to us extremely obscure, played a part: but events had also conspired to bring change.[146] In the Netherlands, the

[144] M. Philippson, *op. cit.*, p. 62; *Correspondance de Granvelle*, VII, p. 353.

[145] Granvelle to Margaret of Parma, 12th August, 1579, Philippson, *op. cit.*, p. 71.

[146] In no way illuminated by Louis Bertrand's hasty and partial monograph, *Philippe II, Une ténébreuse affaire*, Paris, 1929. The main problem is the question of the authenticity of the Hague manuscript. G. Marañon's admirable *Antonio Pérez*, 2 vols., Madrid, 2nd ed. 1948, raises these problems but does not I think entirely resolve them.

attempt at reconciliation by Requesens had ended in a failure even more abject than that of Alva. Over the matter of the Portuguese succession first tackled in the summer of 1578, the peaceful approach appeared unproductive. Some have said that Antonio Pérez betrayed his master over the affair: there are many possible interpretations and the evidence produced is unconvincing. There is always the hypothesis that the sovereign was unhappy with the policy adopted.

So there was a major change. Granvelle's arrival brought to the heart of Philip's empire an energetic and intelligent statesman with breadth of vision. Determined, honest and devoted to the king and the grandeur of Spain, the cardinal was also an old man, of another generation, much given to harking back to the age of Charles V for examples and comparisons, and inclined to shake his head over the sad times through which he was living. A man of decision and imagination, at first his influence was very great: he was the architect of the victories of 1580. But after these triumphs, once Philip had returned from Lisbon, Granvelle's influence was more apparent than real. He too wore himself out in the king's service.

There was then no real correlation between the major reshaping of events, that is the great swing of the pendulum that carried Spain from the Mediterranean to the Atlantic, and Granvelle's arrival in power. The biographical approach is likely to prove as misleading, for us as it did even for that scrupulous scholar, Martin Philippson, who failed to recognize that a major redistribution of energy had taken place. He chose to disregard Spain's withdrawal from the Mediterranean theatre of war merely because he came across a declaration by the cardinal of his opposition to the truce with Turkey. But there is absolutely no reason to suppose that Granvelle meant what he said on this occasion.[147] The fact is that there actually was a truce, repeatedly extended throughout Granvelle's 'ministry': and if Spain withdrew from the Mediterranean, it was neither because of, nor in spite of, the cardinal's personal views.

Alcázarquivir. The last Mediterranean crusade was not Lepanto, but the Portuguese expedition seven years later which was to end in the disaster of Alcázarquivir (4th August, 1578) not far from Tangier, on the banks of the river Loukkos which reaches the sea at Larache.[148] King Sebastian, still a child at twenty-five, a visionary with the child's irresponsibility, was obsessed with the idea of a crusade. Philip whom he had met before the *Jornada de Africa* had tried in vain to dissuade him from carrying war into Morocco. The expedition was prepared too slowly to possess even the advantage of surprise. The Sharif, 'Abd al-Malik, receiving information first of Portuguese rearmament and the departure of the fleet, then of its

[147] M. Philippson, *op. cit.*, p. 104 and 224.

[148] See General Dastagne, 'La bataille d'Al Kasar-El-Kebir', in *Revue Africaine*, Vol. 62, p. 130 ff, and especially the narrative by Queiroz Velloso, *D. Sebastião*, 2nd ed. Lisbon, 1935, Chap. IX, p. 337 ff, another version of which by the same author appears in Vol. V of Damião Peres' *Historia de Portugal.*

stay in Cadiz, had time to prepare countermeasures and to declare a holy war. So the small Portuguese army, landing at Tangier, and transferred to Arçila on 12th July, was invading a land determined to defend itself and moreover equipped with excellent cavalry forces, artillery and arquebusiers (often of Andalusian origin). The long column of Portuguese waggons having advanced inland, the two sides met at Alcázarquivir on 4th August, 1578. The king's incapacity for command contributed to the weakness of the Christian army, already ill-fed and exhausted by the marches under the hot sun. Against it, Morocco had 'risen up in the mass'.[149] The Christians were overwhelmed by numbers. Mountain tribesmen of the neighbouring regions looted their baggage train. The king was among the slain, the dethroned Sharif who had accompanied the Christians drowned himself, while the reigning Sharif succumbed to illness on the evening of the 'Battle of the Three Kings' as it is sometimes called. Between ten and twenty thousand Portuguese remained in the hands of the infidel.

It may not have been the greatest disaster in Portuguese history, but the importance of the battle of Alcázarquivir should not be underestimated, for it was heavy with consequences. It reaffirmed the power of Morocco, now so enriched with the ransoms of the Christians that its ruler Aḥmad, the brother of 'Abd al-Malik, was named not only 'the victorious', al-Mansūr, but also 'the Golden', al-Dahabi. Further, the battle opened the question of the Portuguese succession. Sebastian had left no direct heir. His uncle, Cardinal Henry, succeeded him on the throne, but the reign of this infirm and consumptive old man was unlikely to last long.

Portugal was far from fit to meet such a harsh blow. Her empire was essentially based on a series of commercial transactions, shipments of gold and silver crossing the Atlantic to be exchanged for pepper and spices. But the African trade had been a link in the chain. With Alcázarquivir, the circuit was interrupted. In addition a large section of the country's nobility had remained in enemy hands. In order to provide ransoms so huge that they could not be paid in cash, the country had to send all its available coinage as well as jewels and precious stones to Morocco and Algiers. And without those imprisoned, the Kingdom was bereft of its administrative and military elite. The combination of blows left Portugal more defenceless than at any other time in her history. It is no easy matter for the historian to sift the many versions of the decline of Portugal and measure the degree of real distress in this small kingdom. But if it had been subject for some time to a wasting disease, it was as if it suffered a stroke in the summer of 1578. Events were to aggravate the sickness even further.

As the height of misfortune, the invalid fell into the hands of an incompetent physician. The ageing cardinal – he was sixty-three – last surviving son of Emmanuel the Fortunate, suffering from gout and tuberculosis, was by his vacillations to add to the growing troubles of the kingdom. He had old scores to settle too: he had suffered too much under the haphazard

[149] Ch. A. Julien, *Histoire de l'Afrique du Nord*, p. 146.

and capricious government of Dom Sebastian. No sooner was he on the throne than he took his revenge. One of his first victims was the all-powerful secretary of the Fazenda, Pedro de Alcoçaba, whom he stripped of office and banished, though he lacked the force to eliminate this man and his large clientele from political life altogether.

Such inept conduct smoothed the way for Spanish intrigue. Through his mother, Philip had an indisputable claim to the Portuguese throne; between him and the coveted object, stood the rival – and equally indisputable – claims of the duchess of Braganza. But this 'feudal' dynasty was a poor match for the Catholic King. There was also the bastard son of Dom Luis, another son of Emmanuel the Fortunate. But the Prior of Crato's claim was damaged by his illegitimate birth. In fact only the aged sovereign in Lisbon now stood between Philip and the crown of Portugal. And even in autumn 1578, the cardinal's age and precarious health were giving rise to speculation about the succession. Philip immediately dispatched to Portugal a smooth-tongued diplomat, Christoval de Moura, who brought gold and promises. But more than these the cardinal's own mistakes gave the Spanish party its first recruits. For Christoval de Moura made contact with Pedro de Alcoçaba.

Moreover, the Cardinal's sincere religious devotion delivered him, along with the whole of Portugal up to the spiritual guidance of the Company of Jesus. How far were the Jesuits in league with Spain? Documents concerning this matter published by Philippson are sometimes overlooked. It is a fact that the Jesuits, who had previously been kept at a distance by Philip II, agreed to collaborate with him in Portugal. Cardinal Henry, who was at the start opposed to his Spanish nephew and favoured his niece, Catherine of Braganza, gradually allowed himself to be drawn into semi-official pronouncements in favour of Philip. Many reasons could be advanced for his change of heart, but the possibility of intervention by the Jesuits cannot be ruled out. The letter from the General of the Order, Mercuriano, agreeing to Philip's request, is dated January 1579.[150] It took a little time before his followers (who were originally more inclined to favour the duchess of Braganza according to Spanish reports) came round and began to work on behalf of Philip, from whom they no doubt hoped to receive favours which he was better equipped to give them than any other ruler inside or outside Europe.

In the circumstances, Portuguese national independence was scarcely likely to be preserved. In order to defend it, the country would have to be armed and its ruler decide upon a domestic solution to the succession crisis: in short recognize the house of Braganza or if need be the Prior of Crato. But the cardinal refused to take steps towards national defence. It would have meant great expense; and the only payments the old king would agree to were the ransoms of the *fidalgos* still captive in Morocco.

[150] Mercuriano to Philip II, 11th January, 1579, Simancas E° 934; Mercuriano to Philip, Rome, 28th April, 1579, *ibid.*; M. Philippson, *op. cit.*, p. 92, note 2 and p. 93, note 1.

If he never counted the cost for the prisoners of Alcázarquivir, he never granted a penny for the country's defence. Nor perhaps were his subjects, or at any rate the richer among them, particularly the merchants, ready to make the sacrifices necessary if the country was to embark on such a course.

A further necessity was a quick decision by the cardinal about the succession. But he wasted precious time negotiating his own marriage with the former queen of France, the widow of Charles IX, a marriage which would only be possible with a papal dispensation, and this Gregory XIII hesitated to grant him. The Spanish ambassador in Rome had very little difficulty in thwarting these negotiations. The old king reluctantly bowed to necessity – not that he was in any sense a *vert galant*; the idea of marriage was dictated to him solely by *raison d'état*. It proves, as not uncommonly happens in such cases, that the cardinal was the only person who did not believe in his own approaching death.

After this setback, he concerned himself no more with the succession. He did, it is true, recall the Cortes and attempted to set up a commission of arbitration to which all pretenders to the throne could submit their claims. But the little energy he had left was mostly expended on the Prior of Crato whom he hounded with implacable hatred, seeking to discredit him with his illegitimate birth, even expelling him from the kingdom and obliging Dom Antonio to take refuge briefly in Spain, and to go into hiding when he returned to his native land.

Philip II at any rate was determined to defend his claim. He had been rearming since 1579, perhaps a little too obviously: he firmly intended that Europe and Portugal in particular, should be aware of his strength. Not that he had assembled a mighty army, but it would suffice for the invasion. He had however to amass a great deal of money, notably borrowing 400,000 crowns[151] from the Grand Duke of Tuscany, as well as levying men from the garrisons in Italy. The concentration of troops, supplies and war matériel threw alarm into all hearts from Constantinople (where it was thought that an attack on Algiers was imminent) to England, where Elizabeth feared that Philip was planning an invasion of the island. But nowhere did the war of nerves have more effect than in Portugal.

The mass of the country was violently opposed to Spanish domination. The urban population and the lower clergy hated the Spaniard with a vehemence which made the rich and powerful shudder. Popular indignation prevented them from open defection, hence the particular form taken by the 'betrayal', the hypocritical stances, deceptive speeches, patriotic rhetoric and prudent progress. The ordinary people were delivered up to the enemy by the rich and educated classes. How indeed could the rich have been other than capitulationist, since for the most part they were foreigners, Flemings, Germans and Italians; and moreover reluctant to pay the extra taxes required for war, which would fall heaviest on them?

[151] Grand Duke of Tuscany to Philip II, Florence, 17th June, 1579, Simancas E° 1451. See also R. Galluzzi, *Istoria del Gran Ducato di Toscana*, III, p. 345 and 356.

The high clergy was very similarly disposed, as was the nobility (which virtually put the army out of action). Portugal, to the east, was to some extent defended by its natural frontiers. The roads from the plateau of Castile to the plains of Portugal were difficult and guarded by solid fortresses. But the frontier could only be held if there was sufficient determination within the country to resist the invader. The money distributed by Cristoval de Moura had not quite dispersed all spirit of combat, any more than the attempts by Spanish 'feudal' lords in the frontier regions to suborn their Portuguese neighbours, the masters of the castles, villages and military posts which provided the nation's security. But more potent than the puny efforts of individuals, whether de Moura, himself Portuguese by birth, or the Duke of Osuna, one-time ambassador of the Catholic King in Lisbon, more potent even than treachery, that weapon beloved of the Habsburgs, was the crushing force of circumstances. Portugal needed American silver, and a large section of her navy was already serving Spain on the high seas.[152] The rallying to the Spanish cause of the rich and powerful men of Lisbon was firmly linked with the desperate need for the Portuguese Empire, already embarrassed by Protestant privateers on the Ocean and shortage of specie, above all not to quarrel with her powerful neighbour, but on the contrary to obtain her support. This is borne out perhaps not so much by the events of 1578–1580, as by their sequel: Portugal's long subordination, her symbiosis, so to speak, with Spain, which only the disasters of the 1640's were to interrupt, or rather allow to be interrupted. We should not forget either that Spain was now united, no longer broken up into hostile provinces (Aljubarrota, often mentioned in this context, lay in the past), so that Portugal would only have been able to assert her independence of the larger power by some alliance with the Protestant powers – La Rochelle, the Dutch or the English. The Spaniards were careful to point out this inescapable fact, but the Portuguese were already well aware of it. If the Prior of Crato failed in his attempt to return to his native land, it was because he came in an English ship, because he had been conspiring with the enemies of Rome (and even, in 1590, negotiating with the Turks).

1580: the Coup. Cardinal Henry died in February, 1580. Of the regents provided for in his will, two or three were already won over to Philip.[153] Would the Spanish king allow them to settle the succession, or would he defer to the judgement of the Pope who wished to act as arbitrator? In fact Philip considered himself to possess an inalienable right to the throne, a divine right which there could be no question of submitting to the regents or the Cortes. Nor did he desire papal arbitration, being reluctant

[152] Portugal had been obliged to cooperate with Spain since the crisis of 1550 and the triumph of American silver. There was heavy Portuguese immigration into Spanish cities, particularly Seville.

[153] R. B. Merriman, *op. cit.*, IV, p. 348, some illuminating remarks, based on the correspondence of the Fuggers, *The Fugger News-Letters*, ed. Victor von Klarwill, 1920, Vol. II, p. 38.

to recognize the temporal authority of the Supreme Pontiff. Moreover, assured now of peace in the Mediterranean, confident regarding the Netherlands, England and France, he knew he could count on a moment's respite amid the turbulent stirrings of Europe. Portugal was within his grasp – but on one condition: he had to act swiftly, a course in which he was encouraged by Granvelle from the moment of his arrival at the Escorial. It was the cardinal, even more than the king, who precipitated the events in Portugal, he too who engineered the appointment to the head of the army of the old Duke of Alva, now in disgrace, but whose military reputation seemed a sure token of success. One of Granvelle's virtues was his ability to forget past grudges when the occasion called for it. Perhaps too he felt the need, as a foreigner, to tread carefully among the Spaniards with their formidable cliques and easily injured sensibilities. It was he after all who pleaded for the reduced sentences passed on Antonio Pérez and the princess of Eboli.

Little more than a military exercise, the invasion of Portugal went according to plan. The frontier barriers fell quickly and on 12th June the Spanish army set foot on Portuguese soil near Badajoz. First the mighty fortress of Elvas, then that of Olivenza yielded without a fight: the road to Lisbon along the Zatas valley lay open. Meanwhile on 8th July, a Spanish fleet of roundships and galleys left Puerto de Santa Maria, took Lagos on the coast of the Portuguese Algarve, and was soon sighted at the mouth of the Tagus. Dom Antonio, Prior of Crato, who had been proclaimed king at Santarem on 19th June, had entered Lisbon in triumph thanks to the support of the common people. But to hold out in this huge city, where supplies were short and plague had been decimating the population for some months, now entirely cut off from the outside world by the Spanish fleet, required emergency measures and above all time. Emergency measures were taken, particularly of a fiscal nature: the confiscation of money from churches and convents, the devaluation of the currency, forced loans from the merchants. But time was too short. It was the rapidity of the Spanish advance more than any supposed weakness on the part of Dom Antonio (notably in his indirect negotiating with Alva, a deliberate manœuvre to gain time) that led to the pretender's downfall. Around him the movement towards defection and surrender was gathering speed. Setubal, attacked on land and sea, capitulated without a fight on 18th July (but did not escape being sacked). Thus the invasion army arrived on the south side of the Tagus estuary, which is at this point as broad as a miniature sea; it was a considerable obstacle, but not to the fleet which transported to Cascais on the north bank, a number of Spanish units. The operation culminated in an attack on the capital from the west and the right bank of the Tagus. Dom Antonio sought to defend access to Lisbon with a few troops on the Alcantara bridge. But that very evening, the capital surrendered. The victor spared the city from a sack; or at least limited looting to the suburbs.

The prior, wounded in the fighting, fled through the city, stopping to have his wounds dressed at the small village of Sacavem. He gathered more partisans, moved on to Coimbra, then entered Oporto by force, remaining there over a month. Here again, he tried to rouse resistance but encountered the same gradual erosion of his support as in Lisbon. A cavalry raid by Sancho Davila obliged him to leave his final retreat, on 23rd October and to seek refuge in northern Portugal until an English ship arrived to take him off.

So it had taken only four months to crush all resistance in Portugal. Granvelle, in his advice to Philip, reminded the king that Caesar, in order not to lose time never occupied captured towns, but took hostages. It seems that the invader in 1580 was content merely to move on, wherever treason opened the gates to him: the Portuguese themselves provided reliable guardians of his conquests. It proved unnecessary to send massive reinforcements, or to call up the reserves from the estates of the frontier lords, as had been decided previously. There is no doubt about it: Portugal gave herself up, or was given up to the enemy.

Philip II had been wise enough, before 1580, to uphold the ancient privileges of Portugal; he now granted further privileges both political and economic. Portugal was not to be incorporated to the Crown of Castile; but to keep its own administration, machinery and councils. Like Aragon, perhaps even more than Aragon, it retained its identity, in spite of the union of the crowns in the person of Philip II. It was only a 'Spanish dominion'.[154] That certainly does not justify the invasion of 1580 – such is not our concern – but helps to explain why the conquest lasted and partnership proved a long-lived solution.

On receiving the news, the Indies rallied to Spain without a struggle, as did Brazil, where, in view of its western frontier, the union of the two crowns was far from unwelcome. The only serious problems concerned the Azores. For the sudden extension of Philip's realms (the Portuguese *Ultramar* combined with the Spanish gave him the two greatest colonial empires of the century) raised the question of control of the Atlantic. Whether consciously or unconsciously, Philip's composite empire by force of circumstance became centred on the Atlantic, that vital sea connecting his many dominions, the base of the claims to what was known even in Philip II's lifetime, as his 'Universal Monarchy'.[155]

Spain leaves the Mediterranean. We have come a long way from the Mediterranean now.

The day Philip took up residence in Lisbon, he placed the centre of his composite empire on the shores of the Atlantic. Lisbon, where he remained from 1580 to 1583, was moreover an excellent headquarters from which to rule the Hispanic world, certainly better placed and better equipped than

[154] On this subject see remarks by Juan Beneyto Pérez, *Los medios de cultura y la cantralización bajo Felipe II*, Madrid, 1927, p. 121 ff.

[155] A major question, intelligently perceived by Jacques Pirenne, *Les grands courants de l'histoire universelle*, II, 1944–1945, p. 449 ff.

Madrid in the wilds of Castile, particularly when the king's chief concern was the new battle for control of the ocean. The constant procession of ships which he could see from his palace windows and which he described in charming letters to his daughters, the Infantas, must have been a daily object lesson in the economic realities on which his empire was based. If Madrid was better situated to coordinate the intelligence network covering the Mediterranean, Italy and Europe, Lisbon made a magnificent observatory from which to scan the ocean. How great the number of delays and even of disasters which might have been avoided if Philip had completely understood this at the time of preparing the invincible Armada, if he had not remained tied to Madrid, far from the realities of war!

The re-orientation of Spanish policy towards the West, where it was caught up in the powerful currents of the Atlantic; the affair of the Azores in 1582–1583, when the Archipelago was saved while at the same time, with Strozzi's failure, the dream of a French Brazil crumbled in ruins; the war in Ireland, patiently rekindled from 1579 until the end of the century; the preparations for war with England, leading up to the sailing of the Invincible Armada in 1588; Philip II's expeditions against the English in 1591–1597; Spanish interference in French affairs, including the partial occupation of Brittany; English and Dutch counter-measures; the Protestant privateers now roaming throughout the whole Atlantic – all of these though external to the Mediterranean, are not entirely foreign to it. If peace returned to the inland sea, it was because war had moved to its outer confines: to the Atlantic in the West, to Persia and the Indian Ocean in the East. Turkey's swing to the East was balanced by Spain's swing to the West: gigantic shifts which the history of events cannot by its very nature explain. No doubt explanations other than those advanced in this book, are possible. But the phenomenon itself is visible for all to see: the Hispanic bloc and the Ottoman bloc, so long locked together in a struggle for the Mediterranean, at last disengaged their forces and at a stroke, the inland sea was freed from that international war which had from 1550–1580 been its major feature.

CHAPTER VI

Out of the Limelight: The Mediterranean after 1580

Roger Bigelow Merriman's *The Rise of the Spanish Empire*,[1] an excellent traditional history, concludes with the end of Philip II's reign, in 1598. It contains no mention of any event in Mediterranean history after 1580. This silence, typical of almost all histories of Spain, is significant. For Merriman as for other narrative historians, the Mediterranean which was the scene of no major wars or diplomatic initiatives after Margliani's mission to Turkey, is suddenly plunged into darkness as other locations steal the limelight.

Mediterranean life did not of course come to a standstill. But what kind of world was it now? The answer does not lie in our usual sources, the Spanish or even the Italian archives. Like the press in the twentieth century, the information collected in the various chancelleries – even in Italy – records only events of world-wide importance. There is little mention from now on of the Mediterranean. In Venice, Florence, Rome or Barcelona, people talked and wrote about happenings far from its shores. Would peace be signed between Turkey and Persia? Would the king of France and his subjects be reconciled? Had Portugal been completely subdued? Were Philip II's military preparations in the years leading up to the invincible Armada aimed at the Atlantic or Africa? From time to time some report or discussion concerning the Mediterranean finds its way into a diplomatic letter. But for some reason these references always turn out to be unaccomplished dreams: like the projects which crop up time and again of a league between Venice, Rome, Tuscany and Spain, against the Turks. They never seem to amount to any more than La Noue's anti-Turkish scheme of 1587. (Why then did Paruta attach so much importance to them in 1592?[2])

In Rome too, thoughts and actions remained, and would long remain drawn to the Atlantic. The papacy backed Spain in her struggle against the northern heresy. First Gregory XIII then Sixtus V granted Philip II considerable church revenues to make war on Elizabeth and her allies, just as Pius V on the eve of Lepanto had contributed to the struggle against Islam. All Italy associated itself with the fight to preserve the Roman church. In short, all the attention and the better part of the political efforts of Mediterranean countries were focussed on regions marginal to the Mediterranean world; everyone seemed to be looking elsewhere: the Turks as they pressed on towards the Caspian, the Moroccan mercenaries

[1] *The Rise of the Spanish Empire*, New York, 1918–1934, 4 vols.

[2] Paruta to the Doge, 7th November, 1592, *La legazione di Roma* . . ., ed. Giuseppe de Leva, 1887, I, p. 6–9.

capturing Timbuktu in 1591[3] or Philip II himself as he sought to become or rather remain master of the Atlantic.

So it was until about 1590. But the death of Henri III of France (1st August, 1589)[4] unleashed a serious crisis the effects of which were felt throughout the Mediterranean, particularly in Venice where much anxiety was caused by the possible withdrawal of France, an indispensable element in the European balance of power and as such a safeguard of the Venetian Republic's freedom in the face of her many enemies. 'We do not know what to believe now', writes a Venetian merchant,[5] 'nor what to do next. The rumours from France are very damaging to trade.' The Signoria regarded the threat as so great that she did not hesitate to ally herself with the Protestants in the Grisons and to accept in principle in August and receive in person in January, 1590, M. de Maisse, the ambassador of Henri IV.[6] Why such haste, asked Sixtus V. 'Does the Republic fear something from Navarre? She should have no fear. If necessary we are ready to defend her with all our might.'[7] But he was mistaken. The Republic was taking sides in advance against the mighty Catholic bloc which could only reinforce the intolerable supremacy of Spain.

The crisis caused by the succession of Henri III was to take full effect gradually from the year 1590. In the same year the sultan was finally released from the distant war with Persia. Would he now turn to the Mediterranean or to the Balkans, in other words Hungary? Or would he instead (and this was what actually happened after 1593) try to wage war on two fronts, attacking Christianity from two separate directions? This policy and its results are responsible for the renewal of activities in the Mediterranean but there was to be no repetition of the dramatic events between 1550 and 1580. The plans and conflicts after 1593 were pale shadows of the earlier struggles. True, they were accompanied by much fanfare, but the gap between words and actions was large. The war was no longer completely in earnest. Deals between the two sides were frequent. The Turkish Kapudan Pasha, Cigala, in 1598 visited the Venetian captain of the Adriatic[8] and suggested amongst other things that Cyprus might be restored to Venice. Another time,[9] with the permission of the viceroy of Sicily (Cigala was a Sicilian renegade who had been captured as a child from the ship of his father, a well-known Christian privateer), he received his mother and his large family on board his own

[3] Emilio Garcia Gómez, 'Españoles en el Sudan', in *Revista de Occidente*, October–December, 1935, p. 111.

[4] Muerte del Rey de Francia por un frayle dominico, Simancas E° 596; E. Lavisse, *op. cit.*, VI, 1, p. 298 ff.

[5] A. Cucino to A° Paruta, Venice (September–October), 1589, A.d.S. Venice, Let. Com. XII c.

[6] H. Kretschmayr, *op. cit.*, III, p. 42–43, says August and November. But the ambassador seems to have arrived in January according to F^co de Vera, to Philip II, Venice, 20th January, 1590, A.N., K 1674.

[7] L. von Pastor, *op. cit.*, X (German ed.), p. 248.

[8] I. de Mendoza to Philip III, Venice, 19th December, 1598, A.N., K 1675.

[9] G. Mecatti, *op. cit.*, II, p. 814.

ship. Such official indulgence would have been unheard of twenty years earlier.

1. PROBLEMS AND DIFFICULTIES FOR THE TURKS

Between 1580 and 1589, while a bitter war raged in the Atlantic, Mediterranean chroniclers have little to say. The punitive expeditions sent by the Turks against Cairo, Tripoli or Algiers were little more than police operations and it is hard to discover anything accurate even about these. In the Christian basin of the Mediterranean there is little to remark apart from the constant voyages of the Spanish galleys (or of galleys in Spain's service), endlessly ferrying soldiers from Italy to Spain: Italians recruited locally, lansquenets who were levied beyond the Alps and marched down to Milan or Genoa,[10] Spanish veterans on their way home from Sicily and Naples where they were succeeded by young recruits from Spain, themselves to be relieved in a few years' time when they had completed their training. During these years, Milan was Spain's chief military base; from Milan, Philip II's soldiers were sent in every direction, including the long overland route to Flanders. Simply by studying troop movements through the great Lombard city one could discover what Spain's military preoccupations were at any given moment and plot the rhythms of the Spanish Empire.

On the return trip from Spain, along with the recruits, the galleys carried huge quantities of silver. All Italy was enriched by the flow of bullion and through Italy the entire Mediterranean. This influx was one of the outstanding features of a period in Mediterranean life which could almost be described as happy were it not for the unofficial warfare of the privateers, unremarked in conventional histories, but no less cruel for that. Privateering too was undergoing certain transformations be it said, as two inconspicuous incidents remind us. The first is symbolic: in July, 1587, Euldj 'Alī died at the age of 67.[11] He had no successors: last in the line of great corsairs (Dragut, Barbarossa), his passing marked the end of an era. The other incident pointed the way to the future: in 1586, five English merchant ships sent the Sicilian galley squadron packing![12] Unrecognized at the time, this was a foretaste of the career ahead for the ship-of-the-line.[13]

[10] Doing the usual damage on the way '. . . come è il loro solito'. Republic of Genoa to H. Picamiglio, Genoa, 17th July, 1590, A.d.S. Genoa, L.M. Spagna, 10.249.

[11] Simancas E° 487.

[12] R. Hakluyt, *op. cit.*, II, p. 285–289, in an encounter off Pantelleria.

[13] Usually the incidents reported were even less significant. See Hammer, *op. cit.*, VII, p. 192–194 and 194 note 1, and L. C. Feraud, *op. cit.*, p. 86: according to one the wife, according to the other the sister of Ramadan left Tripoli in 1584 following the assassination of her husband, or brother, the pasha of Tripoli. She took with her in her galley 800,000 ducats, 400 Christian slaves and 40 young girls. She was well received in Zante, on the way to Constantinople, but was attacked shortly afterwards near Cephalonia by Emo, the commander of the Venetian fleet. The galley was captured and the Moslems aboard massacred. The incident was settled without further

After 1589: rebellion in North Africa and in Islam. In 1589, the silence in the Mediterranean was broken: fresh alarms were heard, in Europe with the French crisis, but in Islam, as well.

Behind the minor disturbances in North Africa, imperfect though the evidence is, one senses a general crisis throughout the eastern and central regions of North Africa, regions which the Turks had controlled since the recapture of Tunis in 1574, but extending even further, perhaps to cover all the Islamic regions of the Mediterranean. The rebellions and disturbances were not altogether new. Several times during the previous years there had been problems with the lieutenants whom Euldj 'Alī, combining the functions of admiral and *beglerbeg*, had left in Algiers to replace him. Perhaps his death in 1587 aggravated the situation. The Turkish government in any case decided to abolish the system of *beglerbegs* – who were virtually kings of their little domains – replacing them with pashas appointed for three years at a time.[14]

For this was essentially a challenge to Turkish authority. The corsairs were asserting their freedom, or trying to. Besides, Turk and 'Moor' as Haëdo says, had remained virtually strangers to each other, even inside the city of Algiers, the Moor being kept in a position of inferiority by his conqueror. Some texts hint at Marabout and native movements, a religious reaction which might take various forms in different places but was always hostile to the Turkish invader. 'Wherever the Turks set foot', said a rebel of Tripoli, 'the grass withers and decay comes after.'[15] These uncoordinated and to us obscure movements were certainly linked to the beginning of the gradual erosion of bonds between the Maghreb and Turkey as the latter's seapower declined. The turning-point was not Euldj 'Alī's death in 1587, so much as the failure in 1582 of his attempt on Algiers[16] (directed in fact through Algiers at Fez). Historians have put it down to Ottoman decline. But was the problem not rather that throughout all the Islamic countries linked to the Turkish system, to Turkish currency, finance and authority, there was a general malaise and unrest, though as yet undeveloped?

If this was not the end of Turkish power, it at least betrayed the shelving indefinitely of the costly Mediterranean policy of the past. At the beginning of 1589, Venetian correspondents were still suggesting that a major expedition was planned under Ḥasan Veneziano, the former *beglerbeg* of Algiers who had succeeded in sailing, in winter, with five galliots, from Algiers to Constantinople, 'to the shame of the Christian vessels' which failed to intercept him on the way.[17] Ḥasan had arrived in the Turkish

clashes, thanks to the intervention of the sultana; Emo was however beheaded and his prize restored, or compensation granted. 150 prisoners were released during the incident, according to Hakluyt, *op. cit.*, II, p. 190, who dates it in October, 1585. Aboard the galley were two Englishmen whom Ramadan's son had forcibly circumcised on Djerba.

[14] Charles André Julien, *Histoire de l'Afrique du Nord*, *op. cit.*, p. 538.

[15] L. C. Féraud, *op. cit.*, p. 83.

[16] Ch. André Julien, *op. cit.*, p. 537.

[17] Juan de Cornoça to Philip II, Venice, 4th February, 1589, A.N., K 1674.

capital on 10th January and presently it was rumoured that the new Kapudan Pasha intended to arm 50 or 60 galleys and sail for Fez, hoping to succeed where Euldj 'Alī had failed. Dispatches arriving in Naples spoke of provisions of grain and biscuit in the Morea and of 100 galleys bound for Tripoli.[18] Such a large fleet had not been heard of for years and reports from Venice seemed to confirm the figure.[19] The news was sufficiently disturbing for the Spaniards to send reconnaissance vessels, strengthened galleys, out to the Levant in spring, to discover something of the enemy's movements.[20] But in April it was learnt that little work was being done in the Constantinople Arsenal and that there would be fifty galleys at most for the expedition to Barbary, if indeed it took place at all.[21] A month later, it was reported from Venice that there would be no Turkish fleet of any consequence.[22] But at the end of June, there was a fresh report of 30 to 60 westbound Turkish galleys, so it was decided to assemble the royal galleys at Messina, as in the old days.[23]

In fact, Hasan had left port on 18th June; by the 22nd he was at Negropont where the next day he was joined by the 'guards' of Rhodes, Alexandria and Cyprus. His ships were short of oarsmen, reported one dispatch which also gave a total figure of 80 galleys and galliots[24] (an overestimate according to later messages). On 28th July, he was still waiting at Modon for ten galleys from Coron before sailing on.[25] The fleet probably left on 1st August, with 30 to 44 galleys according to figures reported by Venice,[26] 46 galleys and four galliots, 'in poor shape apart from the flagship',[27] according to sources in Palermo.[28] The fleet made straight for Tripoli, carrying 8000 troops. It was sighted off the coasts of Sicily but made no move against them.[29]

The scale of this Turkish effort can only be appreciated when one bears in mind that the fleet had left port in the face of numerous obstacles, in particular the unrest which had wrought havoc in Constantinople since late spring.[30] This unrest, born of poverty and indiscipline in the army, had reached such a pitch by May that the pashas no longer felt safe

[18] Miranda to Philip II, Naples, 18th February, 1589, Simancas E° 1090, f° 21.

[19] Juan de Cornoça to Philip II, Venice, 9th May, 1589, A.N., K 1674.

[20] Miranda to Philip II, Naples, 12th April, 1589, Simancas E° 1090, f° 35.

[21] Miranda to Philip II, *ibid.*, f° 53. The Adelantado of Castile to Philip II, Gibraltar, 13th May, 1589, Simancas E° 166, f° 72.

[22] J. de Cornoça to Philip II, 9th May, 1589, A.N., K 1674. 30 galleys only. Constantinople, 22nd June, 1589, A.N., K 1674; Miranda to Philip II, 8th July, 1589, Simancas E° 1090, f° 83; F^co^ de Vera to Philip II, Venice, 8th July, 1589, A.N., K 1674.

[23] Miranda, see preceding note.

[24] Miranda to Philip II, Naples, 14th July, 1589, Simancas E° 1090, f° 89.

[25] Dispatches from the Levant, 27th July and 1st August, 1589, A.N., K 1674.

[26] F^co^ de Vera to Philip II, 5th August, 1589, A.N., K 1674, same information; Miranda to Philip II, Naples, 12th August, 1589, Simancas E° 1090, f° 105.

[27] See note 25.

[28] Viceroy of Sicily (to Philip II?), Palermo, 17th August, 1589, Simancas E° 1156.

[29] So the defence measures taken by the viceroy of Sicily the count of Alva, were unnecessary. Alva to Philip II, Palermo, 22nd May, 1589, Simancas E° 1156.

[30] J. de Cornoça to Philip II, Venice, 13th May, 1589, A.N., K 1674.

in their own houses. 'They were surrounded by bodyguards as if in an enemy camp'.[31] Such a crisis at the very heart of the Turkish empire suggests that the disturbances throughout Islam were not as independent of each other as they might seem. 'Concerning the Turkish armada which has gone to Tripoli', wrote the count of Miranda on 8th September, 'we know little at present except that it arrived in the city on the 12th of last month and that Hasan Agha was making all speed to subdue the Moors there who have risen up. Meanwhile the Marabout, who is the rebel leader, is doing all he can to resist. He has many soldiers but wants a Christian armada to come to his assistance. This much has been learned from a galley of the Religion which went to take assistance to the Moors'.[32] His account is confirmed by a report from the Knights of Malta.[33]

So in spite of all the fears it had aroused, Ḥasan's expedition was not directed against the Christians. The Turks had no intention of picking a quarrel with Spain. Neither on the outward nor on the homeward voyage did they attack the coasts of Naples or Sicily. There was in any case a semi-official representative of Philip II still in Constantinople and the series of successive truces had not been interrupted. In August–September, Philip ordered Gian Andrea Doria to take about forty galleys to Spain to pick up troops, evidence that while it was still considered necessary to replenish the *tercios* in Naples and Sicily, as a precaution against trouble in the future, no undue anxiety was felt about the present.[34]

A few months later, a dispatch from Constantinople announced that Ḥasan Pasha's 35 galleys had come limping home 'in such disorder that all the Christians residing here were grieved to see that they had been allowed to get through and a great opportunity missed. Indeed the fleet had been given up for lost after the death of the greater part of the oarsmen and a good number of the soldiers'.[35] The tone of this bellicose dispatch is significant: if work was still going on in the Arsenal, it was to 'espantar el mundo'. The writer of this letter for one refused to be taken in; after all he had before his eyes evidence of the many difficulties besetting the Turkish empire. And my reason for singling out Ḥasan's expedition at all is simply that (besides marking the immense contrast between Turkish military efforts in the 1580s and those of twenty years earlier) it is one of the few elements to emerge from the clouded Turkish horizon of these critical years.

Tripoli, Constantinople: the two rebellious cities were at least separated by distance. But now trouble was taking hold in Tunisia too, where an

[31] Cornoça to Philip II, Venice, 10th June, 1589, *ibid.*

[32] Miranda to Philip II, Naples, 8th September, 1589, Simancas E° 1090, f° 124.

[33] *Relacion del viaje que hizieron las galeras de la religion de Sant Juan que estan al cargo del comendador Segreville en ausencia del General de la Religion*, 1589, Simancas E° 1156.

[34] Miranda to Philip II, Naples, 18th September, 1589, Simancas E° 1090.

[35] Constantinople, 8th December, 1589, A.N., K 1674, F^co^ de Vera to Philip II, Venice, 2nd December, 1589, A.N., K 1674; Vera to Philip II, 22nd December, 1589, *ibid.*

informant, Mahamet Capsi,[36] had reported as early as November, 1589 that there was growing irritation against the Turks among the native inhabitants. Was his warning to be dismissed as the invention of an isolated adventurer or visionary, bent on settling a personal grudge and hoping for a few kegs of gunpowder or simply a few ducats? Apparently not, for the unrest of the Tunisian regions erupted in 1590 with revealing violence. All the *Annals* record for the month of Hadja in the year 999 of the Hegira (1590) the uprising in Tunis and the massacre of almost all the Boulouk Bashis, the officials, hated both by the army and the populace, to whom all administration had been entrusted.[37] Meanwhile in Tripoli, the rebellion broke out again with the new year. A dispatch from Constantinople in March reported the death of the pasha of Tripoli and the desperate situation of the Turks who had taken refuge in the port. It was thought that they could be saved only by a cavalry expedition from Cairo or by the arrival of 50 or 60 galleys as in the previous year.[38] But it was no longer child's play for the sultan to fit a fleet of 50 galleys. According to a dispatch from Constantinople dated 16th March, 1590, the Grand Turk in order to cut costs had offered the post of governor of Tripoli to any individual who would agree to arm five galleys at his own expense and use them for the relief expedition. No one had come forward. So now there was talk of fitting 30 galleys and adding those of the guard in the Aegean, and a few foists to be requisitioned in Greece.[39]

That is not to say there was not much anxiety in Constantinople over the persistent troubles in Africa. But it was difficult to take immediate action. The Arsenal could not build a fleet overnight. The soldiers were underpaid and mutinous. Not that this prevented the Turks (on the contrary) from boasting of a fleet of 300 galleys or the chief pasha from hurling challenges at Spain, the emperor, Poland, Venice and Malta. And since it looked as if peace might be signed with Persia, his threats caused a certain stir: Venice put Crete into a state of alert and stepped up work in her Arsenal.[40]

Meanwhile the crisis in North Africa continued. According to two Christian captives who had managed to escape from Tripoli and reported to the count of Alva at the beginning of April, 'the Marabout is still battling on. And although in Italy it is thought that the Turkish fleet will not sail this year, [Tripoli] thinks the opposite, since it will soon be a matter of necessity. For if the enemy takes Tripoli from them they [the Turks] stand to lose all their possessions in Barbary as far as Algiers. Their African empire would collapse if they lost Tripoli'.[41] From Venice, developments in the rebellion were watched with interest, while at the very same time the Turks were embarking on stormy negotiations with

[36] Palermo, 25th November, 1589, E° 1156.
[37] Alphonse Rousseau, *Annales Tunisiennes, op. cit.*, p. 33.
[38] Constantinople, 2nd March, 1590, Simancas E° 1092, f° 18.
[39] Constantinople, 16th March, 1590, A.N., K 1674.
[40] F^co de Vera to Philip II, Venice, 31st March, 1590, A.N., K 1674.
[41] Count of Alva to Philip II, Palermo, 7th April, 1590, Simancas E° 1157.

Poland. The news has just arrived, writes the Spanish ambassador in Venice, 'that the rebel Moors of Barbary have taken Tripoli and beheaded the soldiers from Tunis and the city garrison, including the pasha. The fortress is still in the hands of the Turks who are now besieged there.' It was thought that they would be relieved by an overland expedition from Cairo.[42] But as the count of Miranda said, another Turkish fleet might well be sent to the rescue.[43]

In any case there had been no cause for alarm as far as Spain was concerned, particularly since news had now reached Madrid that the Turco-Spanish truce had been renewed for a further period of three years[44] 'which was much desired' in Madrid, notes the French envoy. The count of Alva insisted on alerting the Sicilian coasts, but the North African affair alone was quite sufficient to occupy the entire strength of the barely convalescent Turkish fleet as well as any Egyptian cavalry which might be sent.[45] The rebellion appeared to be spreading. The count of Alva had sent an agent, Juan Sarmiento, to Tabarka, the Genoese coral-fishers' island which made an excellent observation post. 'All of Barbary', he was told by the governor Spirolo and his factor, de Magis (secretly of course, for if the Turks got wind that the honest Genoese were handing information over to the Christians they might pay with their lives) 'all of Barbary has risen against the Turk with much bitterness, especially at Tunis, where the pasha is in great financial difficulty: he owes his troops six months' back pay and cannot find the money. So his lieutenant and council are prisoners. All over Barbary the Turks are fleeing and taking ship for Algiers.' 'I was also told' added Sarmiento, 'to suggest that Your Excellency and the Infant [the Hafsid prince whom the Spaniards were holding in reserve] send 70 galleys: if they so much as appeared in the Gulf of Goletta, the Moors would cut the Turks to pieces, provided that the Infant had agreed with His Majesty that the property of the Moors would be respected'.[46] The Genoese had also expressed surprise that 'when Ḥasan Agha came to Tripoli and landed a force of 3000 Turks, leaving his galleys unprotected', the Sicilians had not sent 'twenty galleys which could have seized them all and burned them'.[47]

It seems likely that the Spaniards abstained from action quite deliberately. They had no intention of rekindling war with the Turks over Tripoli or indeed any other port of Barbary. A letter written by the count of Alva in April, 1590, makes this quite plain.[48] Some galleys from Florence had just arrived in Palermo, 'well-supplied with oarsmen and infantry troops. . . . It was said that they were on their way to join the ships of the Grand Master of Malta and intervene in Tripoli, thus giving

[42] Fco de Vera to Philip II, Venice, 14th April, 1590, A.N., K 1674.

[43] Miranda to Philip II, Naples, 14th August, 1590, Simancas Eo 1090, fo 15.

[44] Longlée to the king, Madrid, 15th August, 1590, ed. A. Mousset, *op. cit.*, p. 401.

[45] Constantinople, 27th April, 1590, A.N., K 1674.

[46] 25th April, 1590, *Relacion q. yo Juan Sarmiento hago para informacion de V. Exa del viaje que hize para la isla de Tabarca en Berveria de orden de V. Exa*, Simancas Eo 1157.

[47] *Ibid.*

[48] To the king, Simancas Eo 1157.

the Turk an excuse for sending his fleet into western waters and inconveniencing His Majesty'. 'Inconveniencing' the king of Spain: one could hardly be more explicit. Alva later learned to his relief that the Florentine galleys merely intended to join the Knights in a pirate expedition against the 'caravan' of galleons between Rhodes and Alexandria.[49] The Spaniards were if anything more anxious than the Turks to avoid a confrontation.

This attitude helped to limit rebellion in North Africa, which was often a matter of ill-equipped native fighters, with at most a few arquebuses, pitted against fortified towns guarded by entire companies of arquebusiers and often possessing heavy artillery. While the position of the Turkish 'presidios' was certainly not easy, it was by no means untenable, provided the rebels received no outside help. Thus it was that during the summer of 1590, the Marabout of Tripoli, who had been such a thorn in the flesh of the Turks, was betrayed to the enemy. It had been reported at the beginning of May that he had withdrawn to 'Cahours' leaving the entire coastal strip to the Turks, a normal precaution once the arrival of the new season's fleet was possible.[50] Shortly afterwards word reached Naples, in a dispatch dated 21st May, that he had been assassinated.[51] From Constantinople on 8th June came further details: 'the skin of the Marabout who had incited Tripoli and several other places to revolt had been exposed on a cross in one of the most frequented public squares both as a sign of triumph and to humiliate the Christians. Afterwards it was nailed on an ordinary gallows'.[52] True, it was announced on 2nd June that the Marabout had been succeeded by an even more aggressive leader.[53] But the Turks appear to have had less trouble from Tripoli thereafter. Apart from the small squadrons of galleys which made regular trips between Constantinople and North Africa – the ten galleys which escorted Djafar Pasha, the new governor of Tunis for example[54] – the Turks launched no major naval expedition that year. Nor did they in 1591 or 1592 if our sources are reliable. Was it no longer absolutely necessary – or was it now actually beyond the Turk's strength? It is at any rate a fact that the North African crisis ultimately benefited the native peoples less than the small Turkish colonies and garrisons in Africa, Ottoman outposts now left in virtual autonomy. They were obliged to become increasingly self-supporting and independent. In Algiers where this development can certainly be detected, the final victor was the republic of the *re'īs*, the *taïfa*, and the immediate result was an increase in piracy. So too in Tunis, where the pirates were already becoming more active before the end of the century. The future face of the Barbary Regencies was already emerging as they gradually shook off outside control.[55] A

[49] Constantinople, 25th May, 1590, A.N., K 1674; Alva to Philip II, Palermo, 2nd June, 1590, Simancas E° 1157.

[50] Alva to Philip II, Palermo, 5th May, 1590, Simancas E° 1157.

[51] Simancas E° 1092, f° 32. [52] A.N., K 1674.

[53] Alva to Philip II, Palermo, 2nd June, 1590, Simancas E° 1157.

[54] Constantinople, 8th June, 1590, A.N., K 1674.

[55] F. Braudel in *Revue Africaine*, 1928.

further consequence of the loosening of ties with Constantinople was that towards the end of the century North Africa became much more accessible than in the past to Christian trade and intrigue, a world attracting the business and covetous designs of its neighbours over the water. A French merchant offered to arrange the transfer of a whole city, Bougie.[56] The king of Cuco was prepared to trade several ports in exchange for help against Algiers.[57] Bône fell an easy prey to the Tuscan galleys in 1607. For North Africa, it was the end of an era: no longer was it an outpost of the East.

The Turkish financial crisis. We have yet to investigate the connection between the crisis of the years 1590–1593 and Turkish history as a whole. Apart from local grievances (in North Africa for instance) there must have been some more fundamental general causes, for in spite of repression, we find signs of crisis appearing, disappearing and reappearing throughout almost all the provinces of the Turkish empire. Neither Asia Minor, a hotbed of dissension and revolt, nor Constantinople, where after the events of 1589, the *sipāhis* mutinied again on January, 1593[58] was spared.

There may be some relation between these troubles and the Turkish financial crisis dating from 1584. In that year,[59] the Turkish government first attempted currency manipulation on a grand scale, following the example of Persia which had devalued the currency by 50 per cent at a stroke. At any rate the gold coins (still made of African gold) which the Turks obtained from Cairo at a rate of one *sultanin* to 43 *maidin* were paid out to the soldiers at a rate of 85 *maidin*, indicating a devaluation equivalent to the Persian 50 per cent. Meanwhile the Venetian sequin, which had the same value as the *sultanin*, now became worth 120 aspers instead of 60. Furthermore the aspers, the small silver coins which were the standard currency of Turkish countries and normal medium of soldiers' pay, were melted down and reissued with a gradually increasing proportion of copper – and as thinner coins: they were 'as light as the leaves of the almond tree and as worthless as drops of dew', said a contemporary Turkish historian.[60] Did the disturbances among the *sipāhis* in 1590 not have something to do with these issues of debased coinage? A report from Venice in the same year[61] reveals that there was some interference too with the Turkish thaler (the 'piastre' or *grush*) which had been worth 40 aspers at the beginning of the century. Under the reign of Muḥammad III (1595–1603) the

[56] G. A. Doria to Philip II, 6th June, 1594, Simancas E° 492.

[57] On this king of Cuco, see preceding note and analysis in F. Braudel, 'Les Espagnols en Algérie' in *Histoire et Historiens de l'Algérie*, 1930, p. 246. On a similar incident concerning the fortifications of the little town of Africa which were to be pulled down, see the duke of Maqueda to Philip II, Messina, 12th August, 1598, Simancas E° 1158.

[58] J. von Hammer, *op. cit.*, VII, p. 264.

[59] J. W. Zinkeisen, *op. cit.*, III, p. 802.

[60] *Ibid.*, p. 803.

[61] F^co de Vera to Philip II, Venice, 14th April, 1590, A.N., K 1674, 'con que havian baxado los talleres diez asperos cada uno'.

collapse of the currency continued, the sequin rose from 120 to 130 and later 220 aspers, while the tax office continued imperturbably to regard it at the old rate of 110. 'The empire is so poor and so exhausted', wrote the Spanish ambassador in Venice,[62] 'that the only coins now circulating are aspers made entirely of iron'. He was of course exaggerating but all the same it is an indication of the domestic collapse about which general history books have little to say.

Between 1584 and 1603, there were at least two currency crises and apart from currency fluctuations, several grave financial crises. With a certain time-lag, the eastern half of the Mediterranean was encountering difficulties already experienced in the west. But it was not rescued by a fresh source of income such as the Iberian peninsula now found in silver from South America. We might draw the tentative conclusion then that the bankruptcy and economic weakness of Turkey engendered in about 1590 a crisis which spread rapidly both because of the failure to pay the army and because of the diminished authority of the central power. Through the shaky and in places broken barriers emerge demonstrations of many kinds of dissatisfaction: political, religious, ethnic and indeed social. A series of revolts and disturbances followed the collapse of the currency in the Turkish empire.[63]

But this can be no more than a provisional explanation. There are many gaps to be filled in, nuances to be marked and, I have no doubt, corrections to be made. Only a searching investigation of the Turkish archives will remedy these shortcomings.

1593–1606: the resumption of major offensives on the Hungarian front. Along Turkey's long European frontier, from the Adriatic to the Black Sea, the war had never really come to a standstill even after the 1568 truce. It was not so much war perhaps as piracy on land. Certain peoples had come to specialize in this kind of warfare and derived their livelihood from it: the Uskoks and Martolosi on the Dalmatian and Venetian borders; the Akinjis (forerunners of the Bashi-Bazouks) and Haiduks on the great plains of Hungary; Tartars and Cossacks throughout the vast disputed area between Poland and Muscovy on the one hand and the Danube and Black Sea on the other. This unceasing guerrilla war had prospered during the long interval following the 1568 truce, originally agreed for eight years and renewed in 1579 and 1583.

It was further encouraged after 1578 by the withdrawal of Turkish forces from these frontiers to Asia: the Turkish frontier was now left virtually unmanned and disturbances had broken out not strictly comparable to those in North Africa but encouraged by the same lack of firm

[62] D. Iñigo de Mendoza to Philip II, Venice, 9th September, 1590, autograph, A.N., K 1677.

[63] For the present state of the question, see Vuk Vinaver, 'Der venezianische Goldzechin in der Republik Ragusa', in *Bollettino dell'Istituto di Storia della Società e dello Stato veneziano*, 1962.

control. In June, 1590 Sinān Pasha put it in the following terms in a letter to the queen of England about the Polish borders: Taking advantage of the Persian wars, 'in regard of which [the sultan] would not goe in battell against any other place . . . certaine theeves in the partes of Polonia called Cosacks, and other notorious persons living in the same partes ceased not to trouble and molest the subjects of our most mightie Emperour'. Once the war in Persia was over, the sultan intended to punish these raiders. In the case of Poland however, he was prepared to accept a peaceful solution (and an agreement was eventually signed in 1591).[64] but only on the queen's intervention: Elizabeth had expressed an interest in Poland because it provided British subjects with both grain and gunpowder. Sinān does not mention the gifts brought by the Polish ambassadors, along with their king's promise that he would himself punish the Cossacks.

The situation on the Turco-Polish border, devastated by continuous raids on flocks and villages, was typical of the whole long frontier. Even worse was the border between Turkey and the Habsburg Empire which in the centre and west was a far more dangerous neighbour than either the Poles or the Russians, who had access in the south to deserted lands for the most part or regions of Rumania imperfectly controlled by the Turks. We should not of course imagine that the Turks were always the victims; guerilla wars operate in both directions and according to the Imperials, the Turks were always the aggressors. There was not a single bey on the Turkish frontier, lord of the smallest castle, who did not take part in this endemic war, in an area where he was his own overlord, with his own army, allies and personal headquarters. The Christian landowners on the frontiers, the 'bans' of Slavonia and Croatia, were not content, whatever may have been said, merely to counter and contain enemy raids or to 'forestall' them as count Joseph von Thurn and the ban Thomas Eröddy did in October, 1584 and again two years later in December, 1586, when they wiped out several Turkish bands operating in Carinthia.[65] Both sides participated energetically and a local incident often led to a pitched battle during which hundreds and even thousands of prisoners were taken. In the course of such combats, the whole of Hungary, whether Christian or Moselm, was laid waste, as were Carinthia, the borders of Styria, the marches of Slavonia, Croatia and Carniola, where the line of castles and fortified towns, the mountain barriers and marshy river beds did not create an impassable barrier.[66] There was no peace as long as the campaign season lasted and even the enforced ceasefire during winter could not be taken for granted. The result we may guess: the degeneration of these marcher regions into horrifying deserts. Any large-scale operations here posed impossible problems. To supply the forward position of Gran for instance, long ox-drawn caravans had to be sent from Buda. But

[64] 12th June, 1590, R. Hakluyt, *op. cit.*, II, p. 294–295. The agreement finally made in 1591, J. W. Zinkeisen, *op. cit.*, III, p. 657.

[65] Zinkeisen, *op. cit.*, III, p. 582.

[66] *Ibid.*

if the Christians intercepted them, the peasants of the Hungarian plains from whom the oxen had been borrowed had no beasts to pull their ploughs – a problem they solved by putting their wives in harness. This was a cruel, inhuman war. When it flared up again during the last decade of the sixteenth century, it was like the Persian war all over again for the Turks, equally harsh, equally costly, and in the end equally long (1593–1606).

Until then, although the Christians had often had the best of the guerilla fighting, it had been the policy of the emperor to abide by the terms of the 1568 truce. In 1590, he had negotiated to have it renewed for a further eight years, against payment of the usual tribute of 30,000 ducats, accompanied on this occasion by a supplementary gift of silver ware. This policy of apparent weakness was an inherited reflex, the result of an inferiority complex for which explanations are unnecessary.

More difficult to understand is the attitude of the Turks. After the Persian peace treaty had been signed, it was assumed that they would reappear on the western front and there was no lack of threats or boasts to that effect. But was it not possible that the full force of the Turkish attack might be directed at for example Venice? The Signoria put the fleet on the alert, hastened to fortify Crete in the spring of 1590, and in 1591 sent over 2000 foot soldiers.[67] The French and English ambassadors urged the Grand Turk to send his fleet into Mediterranean waters. As early as 1589[68] and again at the beginning of 1591, dispatches from Constantinople talked of 300 galleys under preparation and due to sail to the assistance of the Moriscos in Spain who were thought to have risen gain.[69] But it was in the North that the blow fell.

Perhaps it was because of the defeat inflicted in 1593 upon Ḥasan the governor of Bosnia, before the walls of Sisak in Croatia. In previous years, Hasan had launched massive attacks on the Uskoks;[70] in 1591 he had devastated the countryside between Kreuz and Suanich and repeated the performance in the spring of 1592.[71] It may have been a case of deliberate provocation. In any event in June, 1593 it was learned in Constantinople that the usual spring sortie had ended in total defeat on the banks of the Kupa. Hasan himself was among the many thousand Turkish dead and great booty had fallen into the hands of the victors.

This news tipped the hitherto uncertain balance between those who wanted peace and the war-mongers, chief among whom was Sinān Pasha, the implacable enemy of Christians and the Empire, the veteran of the Hungarian war and the man whom the army had imposed as Grand Vizier. It would be a mistake to underestimate the role played by this relentless Albanian, now an old man, wily, obstinate and insatiable for

[67] F^co^ de Vera to Philip II, Venice, 3rd March, 1590, A.N., K 1674; J. W. Zinkeisen, *op. cit.*, III, p. 623.

[68] De Vera to the king, Venice, 3rd September, 1589, A.N., K 1674.

[69] Constantinople, 5th January, 1591, A.N., K 1674.

[70] J. W. Zinkeisen, *op. cit.*, III, p. 581.

[71] *Ibid.*, p. 585.

treasure. Perhaps the Imperials had failed to take him sufficiently seriously during the negotiations which began in 1591. However Sinān's return to power had not brought the immediate severance of relations. Talks had continued with the imperial ambassador, von Kreckwitz. And Sinān's own son, the *beglerbeg* of Rumelia, had even acted as intermediary.

Did the news from Sisak do any more than hasten the outbreak of a storm which had been impending for some time and which it was perhaps necessary to unleash now in Constantinople? The end of the Persian war faced the sultan's government with the familiar problem of what to do with the demobilized troops. It had riots from unpaid soldiers on its hands and in 1590 they seemed to be increasing to revolutionary proportions. This situation, as much as the changing humours of Murād III weighed upon the destiny of the empire and led to a succession of vizierates or, as we should say, ministries. The need to rid the capital of unemployed troops drove Turkey towards another European war. In Hammer's study of the Ottoman Empire which draws heavily on primary sources and is a mine of anecdote, the Grand Vizier Ferhād (who was in office briefly in 1594) is said to have been besieged in the streets of the capital by a throng of angry *sipāhis* demanding their pay. The Vizier replied: 'Go to the frontier, you will be paid there'.[72] In 1598 the janissaries mutinied again because of the debased money with which they were paid, and a dispatch dated 18th April declares it is becoming impossible to live in the city among such persons.[73] Three years later it was again the turn of the *sipāhis*. There were similar incidents between 20th and 25th March[74] and over a month after these riots, dispatches from Constantinople reported that 'the licence and insolence of the men of war' had obliged most merchants to shut up shop.[75]

So the Hungarian war of 1593 had at least the merit for the Turks of giving work to the unemployed soldiers of Constantinople.

This war, which was to last fourteen years (1593–1606) is known to us only through scattered incidents, military and diplomatic.[76] Hammer,

[72] J. von Hammer, *op. cit.*, VII, p. 297.

[73] A.N., K 1677.

[74] Constantinople, 18th April, 1601, A.N., K 1677.

[75] Constantinople, 4th May, 1601, *ibid.* On the 1601 troubles, their causes and antecedents, see also Constantinople, 27th March, 1601, A.N., K 1630; Iñigo de Mendoza to Philip III, Venice, 13th May, 1600, autogr. K 1677; Lemos to Philip III, Naples, 8th May, 1601, K 1630; F^co^ de Vera to Philip III, Venice, 5th May, 1601, K 1677 and Constantinople, 29th November, 1598, K 1676.

[76] J. von Hammer, *op. cit.*, Vol. VII and J. W. Zinkeisen, *op. cit.*, Vol. III. A few dates: 1594, capture of Novigrad by the Imperials; 1595, Giavarino captured by the Turks, great stir in Christendom, G. Mecatti, *op. cit.*, II, p. 799; 1598, Giavarino recaptured, Simancas E° 615; 16th May, 1598, Iñigo de Mendoza to H.M., Venice, A.N., K 1676, the sultan's wrath at hearing of the recapture of Giavarino; 11th April, 1598, I. de Mendoza to H.M., news of the recapture of Giavarino had reached Venice on 6th April, A.N., K 1676; 5th December, 1598, I. de Mendoza to the king, Venice: the Venetians' satisfaction on hearing that the Imperials had raised the siege of Buda; 28th November, I. de Mendoza, to H.M., the Turks have lifted the siege of Vadarino, the Imperials that of Buda, A.N., K 1676; 20th October, a false report (not sent as such) of the fall of Buda, I. de Mendoza, A.N., K 1676; 4th November, 1600, F^co^ de

who sticks very closely to the accounts in the sources does not give a very clear picture and the later works by Iorga and Zinkeisen are equally disappointing. I do not propose here to attempt to improve on their narrative, for only the main outline need concern us.

Even to discover the outline is not easy however, for this was a monotonous war, dictated by the nature of the terrain, fought in various places in the vast zone scattered with castles and strongholds between the Adriatic and the Carpathians. Each of the adversaries would field a more or less large army every year. The side that emerged first would without difficulty capture a string of forts and castles: the troops stationed within, being of varying reliability, very often evacuated positions as soon as they came under pressure or even surrendered a string of forts without a struggle. Whether the victor then kept these newly acquired positions was largely a matter of available troops and money. But never once – and this is of great importance – did a breach opened in the line of fortresses lead to a major invasion of enemy territory. There were several reasons why this should be so. In the first place, it was easy to starve to death in these devastated and inhospitable regions. Transporting supplies was out of the question. And there was also the risk that fortresses left intact either side of the breach would join forces to cut off the invader's retreat. Above all although the imperial army had made great strides in this direction with the help of the Hungarian horsemen, they were not yet equipped with a cavalry force conceived as a weapon in itself. Even the Turks had less cavalry than one might suppose. They had to beg from their allies: in 1601[77] Turkish galleys went in search of Tartar horseman to ship to Hungary. It was of course by their powerful cavalry raids that the Turks had originally won the Balkans; and it was by cavalry raids that in later times first Charles of Lorraine then Prince Eugene were to push the Christian frontier forward in the south. The armies of 1593 lacked the wherewithal to mount large-scale operations.

The period between 1593 and 1606 seems to have been largely taken up with a series of minor actions: positions were besieged, surprised, surrendered, saved, blockaded or relieved. The net result was rarely significant. There were only two or three really important events: the capture of Gran and Pest by the Christians, the capture of Erlau and recapture of Gran by the Turks (1605). The armies seldom met face to face. So there was only one major battle, which lasted three days: the issue was at first uncertain, but the sultan finally emerged the victor, after leading his troops in action from 23rd to 26th October on the plain of Mezö-Keresztes.

Vera to H.M., Venice, A.N., K 1677, Canisia captured by the Turks, 22nd October; 11th August, 1601, di Viena, A.N., K 1677, the Transylvanians defeated by the Imperials near Goroslo; 21st October, 1601, defeat of the 'Scribe' celebrated with much feasting, Constantinople, A.N., K 1677; 10th November, 1601, the defeat of the 'Scribe' not certain. Fco de Vera to H.M., Venice, A.N., K 1677; 1st December, 1601: failure of imperial assault on Canisia, F. de Vera to H.M., Venice, A.N., K 1677.

[77] Constantinople, 4th May, 1601, A.N., K 1677. Only four galleys went, it is true.

Even this battle was not decisive: the enforced ceasefire of the winter months obliged the sultan to withdraw his troops to Buda and Belgrade.

But in the course of these repetitive engagements, a fairly clear battle zone emerged. The imperial frontier, supported in the west by the natural barrier of the Alps and in the east by the Carpathians, ran from one range of rugged and forested mountains to another. It was in the intervening region that the war was fought, in the great open plain of Hungary where the major routes were the rivers Danube and Tissa, along which small boats carried troops and supplies. Now and again the war would cross the rivers, over hastily built bridges. Of the two waterways running north, the Danube was the more deserted and exposed. The Tissa valley was no better as a route, but it did afford better lodging and was more convenient for supplies. It also had the advantage of running through pacified territory.

The outstanding feature of the war was the undeniable progress made by the imperial troops, whose first successes were enthusiastically (perhaps over-enthusistaically) reported[78] in Europe and celebrated in 1595. True the Imperials had not been taken by surprise. They had been aware of the impending danger and had sought help in good time from the Reich and the Erbland. They had also received timely contributions from Italy, the Papacy and Tuscany – quite substantial ones for Italy at the end of the century was rich and aware of the covetous eyes of the Turks. The Pope granted the emperor financial aid and permission to levy tithes. The Grand Duke of Tuscany offered to send an army[79] and much pressure was put on Venice to ally herself with the emperor – in vain. The Signoria refused to abandon her policy of armed neutrality and continued to allow the Turks to seek supplies on her very doorstep, to the profound annoyance of Spain.[80] Similarly, fruitless attempts were made to draw Poland and Muscovy into the war on the imperial side. One point that deserves emphasis is that Germany, more or less free from internal strife after 1555, officially at peace with the Turks since 1568, and since 1558 relieved of possible trouble in the North, had been enjoying a long period of peace and economic growth. Her strength was felt along the frontier which the additional presence of Italian and French troops made the bulwark of Christendom.

But there was activity on other frontiers too. Besides the principal arena, there were secondary theatres of war: Croatia-Slavonia on one side and on the other – of rather greater significance for the war in the long run – the eastern lands of Moldavia and Wallachia, the rich granaries and stock-raising lands upon which Constantinople drew for her own use; then there was Transylvania, that complex world part Hungarian, part

[78] G. Mecatti, *op. cit.*, II, p. 789, 809.

[79] *Ibid.*, p. 790. On Cardinal Borghese's mission to Spain, see Clement VIII's instructions, 6th October, 1593, ed. A. Morel Fatio, *L'Espagne au XVIe et au XVIIe siecle*, p. 194 ff.

[80] *Consejo sobre cartas de Fco de Vera*, May, 1594, Simancas E° 1345. Spain also reproached Venice for her policy in favour of Henri IV.

Rumanian, part German (the German element being the series of fortified and industrial towns, strange grafts from the west whose role in history was to be so important). And indeed it was in these secondary arenas – the area roughly covered by present-day Rumania – that the result of the Hungarian war was ultimately decided. Their sharp intervention at the beginning of the war on the imperial side led to the extremely serious crisis of 1594–1596, from which the Turkish empire was rescued only by the providential victory of Keresztes. On the other hand the unilateral intervention of Transylvania in 1605, this time against the imperial army, allowed the Turks to recapture their lost terrain in a single campaign and thus to obtain without difficulty the stalemate peace of Zsitva-Torok (11th November, 1606).

It was in 1594, when the situation in Hungary was still unclear, that the three tributary countries, Transylvania, Moldavia and Wallachia rebelled against the sultan and made a pact with the emperor Rudolph.[81] Michael the Brave massacred the former rulers of the country in Wallachia. This triple revolt caused a serious diversion in the Turco-Imperial war. But on the role played by this Balkan bloc between Poland, Russia and Danube, traditional historical accounts can offer us commentaries only on the principal actors in the drama, rather than on the action itself. The chief actors in this case were Sigismund Bathory, the stern ruler of Transylvania whom the Pope assisted with money and who dreamed of leading the Crusade now being formulated on the banks of the Danube;[82] Aaron, voivode of Moldavia; and finally, the most enigmatic and most difficult of all to assess, Michael the Brave, ruler of Wallachia and the vast regions bordering it.

The coincidence of this uprising with the accession of Muhammad III, aggravated its impact on events. So it was against Michael the Brave that Sinān Pasha led a fierce campaign in summer, 1595. He crossed the Danube in August, took Bucharest then Targoviste, the old capital of Wallachia. But confronted with the aggressive boyars and their cavalry, he was unable to maintain his conquests. He had to burn down the hastily constructed wooden ramparts at the approach of winter, and his retreat became a débâcle. Only a few tattered remnants of his army followed him back across the Danube. Meanwhile the victorious enemy pressed southwards along the snow-covered roads, taking Brăila and Izmail, the latter town, recently founded by the Turks, being the strongest position on the lower Danube.[83] The Turks fared little better in Transylvania: one of their expeditionary forces was lost there, men, supplies, artillery and all.[84] Meanwhile the Imperials crushed the little army which tried to relieve Gran on 4th August. The town surrendered on 2nd September, 1595.

The sultan came in person to retrieve the situation and halted the collapse by his victory on the plain of Keresztes, 23rd–26th October, 1596.

[81] J. W. Zinkeisen, *op. cit.*, III, p. 587.
[82] G. Mecatti, *op. cit.*, II, p. 800 (1595), N. Iorga, *op. cit.*, III, p. 211.
[83] N. Iorga, *Storia dei Romeni*, p. 213.
[84] G. Mecatti, *op. cit.*, p. 801.

So one should not exaggerate the 'Germanic' rally of the early days of the campaign, undeniable though it was in other respects. Nor above all should we talk of the irreversible decline of the Ottomans, although this was a phrase already to be heard in the West. The Turkish Empire is beginning 'to fall apart, piece by piece', wrote a Spanish ambassador,[85] but his evidence has not quite the meaning we may be tempted to lay upon it. Besides the Turks were prudent. Faced with the hostility of Transylvania and the 'Danube provinces' they temporized and even negotiated. Schooled by experience, they were wary of interfering in the hornets' nest of Wallachia. They urged the Poles towards these rich plains temporarily out of their control[86] with the intention of neutralizing them as far as possible. They were unable to avoid receiving some rude blows from time to time from the troops of Michael the Brave,[87] but they were able to turn to face the emperor with a much improved position.

The last years of the war saw two much more equally matched forces wearing themselves out in this costly and monotonous conflict. Resources both financial[88] and human were exhausted. On the Turkish side soldiers were failing in their duty[89] but then so were the imperial troops.[90] According to the experts, both armies were inadequate.[91] And on both sides the exalted spirit with which the war had begun was considerably abated.[92] In 1593, the emperor had ordered that every day at noon the *Türkenglocke*, the Turkish bell, be sounded as a daily reminder that war was being fought against Christendom's great enemy. In 1595 Sultan Murād had had the green banner of the Prophet transported in great ceremony from Damascus, where it was normally kept, to Hungary. But by 1599 no one had the stomach for such gestures and the Grand Vizier Ibrāhīm embarked upon serious peace talks.[93] They continued as the war dragged on. On both sides the rear stood up less well than the front. In about 1600 an obscure rebellion led by a certain Yaziji[94] (known in the West as the 'Black Scribe') broke out in Asia Minor, eventually

[85] F^co de Vera to Philip III, Venice, 5th May, 1601, A.N., K 1677.

[86] Constantinople, 17th March, 1601, A.N., K 1677.

[87] For instance at the beginning of 1600, near Temesvar, I. de Mendoza to Philip III, Venice, 26th February, 1600, A.N., K 1677, and as in 1598, this was with the help of the Transylvanians and in winter (3rd January, 1598), A.N., K 1676.

[88] Vienna, 28th March, 1598, A.N., K 1676.

[89] P. Paruta, *op. cit.*, p. 15 and 16.

[90] I. de Mendoza to Philip III, Venice, 19th December, 1598, A.N., K 1676.

[91] Mendoza to Philip III, 11th July, 1598, *ibid.* (not 18th July as classified).

[92] Juan de Segni de Menorca to Philip II, Constantinople, 3rd November, 1597, A.N., K 1676. Turkish soldiers were deserting and taking refuge in Christian villages.

[93] J. W. Zinkeisen, *op. cit.*, III, p. 609. Rumours of peace: the duke of Sessa to Philip III, Rome, 14th July, 1601, A.N., K 1630; Iñigo de Mendoza to Philip III, Venice, 1st August, 1600, K 1677; Mendoza to Philip III again, Venice, 27th May, 1600: if the Emperor does not receive financial aid soon, he will make peace, *ibid.*

[94] Constantinople, 17th July, 1601, A.N., K 1677; Golali according to a letter from Ankara, 10th December, 1600, *ibid.* And earlier, Iñigo de Mendoza to the king, Venice, 8th August, 1598, K 1676, but was it the Scribe who on this occasion called himself, or passed as, the sultan Muṣṭafā?

leading to the disruption of trade and a virtual blockade of Ankara.[95] So successful was it that Bursa itself was threatened.[96] The defeat of the 'Scribe' by Ḥasan Pasha in 1601 was celebrated with much feasting in Constantinople.[97] More seriously, war had broken out again with Persia in 1603. It brought unbelievable expense and obliged the Ottomans to take extremely seriously the threat to the empire from the endemic revolts of Asia Minor.

It was however at this moment of all-round weakness that the Turks succeeded in restoring the situation in the north. They had merely to obtain the defection of Transylvania[98] by promising in 1605 to its ruler of the moment, Stephen Bocskai, the crown of Hungary, that is of Turkish Hungary, except for the frontier positions facing the emperor. This poor prince of the mountains was tempted by the prospect of the rich plains. It was only a trick, but it created the diversion the Turks needed. Aided by the Tartars whom they sent to the borders of Styria and Croatia in the west, they were able to advance victoriously along the valley of the Danube. Gran was recaptured on 29th September, 1605, shortly after Visegrad. Next fell Veszprem and Palota to mention only the major triumphs of the Grand Vizier, Lala Muṣṭafā.

The peace talks now began to make progress, as the Turks hastened, because of the costly war with Persia, to make the most of their victories in 1605. Peace was finally signed on 11th November, 1606. The status quo was restored and positions and prisoners of war handed back. The prince of Transylvania, who had now made a separate treaty with the emperor, gave up the crown of Hungary, the sultan received a parting gift from the emperor of 200,000 ducats, but in exchange agreed to forgo the tribute in future. The 1606 peace treaty was the first settlement between the Turks and the empire in which both sides treated each other as equals.

2. FROM THE FRENCH CIVIL WARS TO OPEN WAR WITH SPAIN: 1589–1598

In the West, equally marginal to the Mediterranean, but affecting it from time to time, another war was in the making: the French war, closely linked to the general crisis of the Atlantic and the western world. Here again, our concern will not be to describe this war in detail, but simply to mark the connections between these events and the now calmer history of the Mediterranean. Even within these limits our task is not easy. The French wars of religion are part of a general European drama, both religious and political, even if we leave aside for a moment the social and

[95] Ankara, 10th December, 1600, copy, A.N., K 1677.

[96] Constantinople, 8th and 9th September, 1601, A.N., K 1677.

[97] Constantinople, 21st October, 1601, A.N., K 1677, his defeat by Ḥasan Pasha. The Duke of Sessa to Philip III, Rome, 9th December, 1601, A.N., K 1630, Ḥasan Pasha was one of the sons of Muḥammad Ṣoḳolli.

[98] J. W. Zinkeisen, *op. cit.*, III, p. 613–614.

economic background. Is it possible at all, in such a complex area, to map out a precise and limited zone of inquiry?

Between 1589 and 1598, France suffered two crises: until 1595 a predominantly domestic crisis, though the worst the country had known since the troubles first began; then from 1595 to 1598 a foreign crisis, open war with Spain. Both profoundly disturbed France itself, but concern us here only as events with a marginal bearing on the Mediterranean.

The wars of religion in Mediterranean France. The Mediterranean regions of France in the end played only a minor role in the wars of religion. The Protestant heresy, cause and pretext of the troubles, was more concerned to move into the Dauphiné and thence to Italy, or into Languedoc and thence to Spain, than to convert Provence which bordered only the sea. So between the troubled provinces on either side, Provence was relatively calm. But frequent echoes of the disturbances reached it and there were several major alarms in 1562, 1568 and 1579. During the last of these years especially, the region was ravaged by repeated *jacqueries.*[99] The war was by this time becoming endemic, as massacres and looting grew commonplace. As in the rest of France, from the 1580s on, trouble was simmering and order disintegrating in Provence, a region still uncertainly linked with the rest of the kingdom,[100] poor, attached to its liberty, with its bitter local rivalries, its towns jealous of their privileges and its turbulent nobility. But is there any way of classifying these minor disturbances, or of establishing responsibility for these sordid local wars between *carcistes* and *razas*, or later between *ligueurs* and *bigarrats*, the widespread troubles which were to increase with the last years of Henri III's reign and his assassination on 1st August, 1689?[101]

After 1589 of course and until at least 1593, the bulk of the fighting was in the North, between the Netherlands and Paris and between Paris and Normandy–Brittany. But violence was increasing in the South. Here as elsewhere, the beginning of Henri IV's reign saw the disintegration of the kingdom into partisan towns, noble factions and autonomous armies. There followed a fairly rapid period of reconstruction: the grains of sand came together again to form once more the solid stones of the old building. This basically simple pattern is impossibly complicated in its detail. Every grain of sand has its chronicler, every person of note his biographer.

In the Mediterranean South as a whole, six or seven enterprises were launched: the opposed strategies of Montmorency and the Duke of Joyeuse in Languedoc; that of the Duke of Épernon in Provence, of the constable Lesdiguières[102] in and around the Dauphiné; of the Duke of Nemours in the Lyonnais,[103] of the Duke of Savoy from Provence to

[99] Paul Moret, *Histoire de Toulon*, 1943, p. 81–82.
[100] Maurice Wilkinson, *The last phase of the League in Provence*, London, 1909, p. 1.
[101] *Muerte del rey de Francia*, Simancas E° 597.
[102] Charles Dufayard, *Le Connétable de Lesdiguières*, Paris, 1892.
[103] He died on 15th August, 1595, E. Lavisse, *op. cit.*, VI, I, p. 399.

the shores of Lake Geneva; and to cap everything, the complicated manœuvres of Philip of Spain. Only three of those mentioned were serving the cause of Henri IV: Montmorency, Lesdiguières and Épernon. Service is a rather too simple and in one case at least (Épernon's) quite inaccurate word to describe their activities. The latter, like so many of his contemporaries, was serving himself if anyone and in November, 1594, he rallied to the cause of the foreigner.[104]

To disentangle these separate enterprises is easier said than done since they overlap and collide. As a very rough guide, they fall geographically into two fairly distinct wars, with very dissimilar courses: the first, in Languedoc, was practically over by the end of 1592; the second, in Provence, did not end until 1596. This seems to run in the face of any *a priori* assumptions one might make: in Languedoc, that is on her very doorstep, Spain was unable to prolong the troubles beyond 1592, while she could keep a war going in faraway Provence until 1596. This apparent paradox is explained by local circumstances.

In Languedoc, Henry IV's adversaries were well placed. They were supported by Spain, now installed in Cerdaña and Roussillon, well advanced north of the Pyrenees and also possessing what was generally considered to be naval supremacy in the Mediterranean. And they had other sympathizers too: in Guyenne in the west where the *ligueurs* had a good deal of support, and in the east in Provence, which had declared itself massively against the heretic on the throne.

However the Duke of Montmorency, who was devoted to the new king had a strong force at Montpellier. In addition he would be able if necessary to join forces via Pont-Saint-Esprit with Lesdiguières in the Dauphiné, who was always prepared to intervene. Controlling the route along the Rhône, or at any rate in a position to cut it at will, the 'royalists' had a means of pressure on all the countries of the Mediterranean. And the part of the sea bordered by Languedoc was most inhospitable: the Gulf at Lions, dangerous for galleys and tossed by storms every winter, as seamen were always telling Philip II, who sent them on difficult and sometimes impossible missions[105] – the transport of troops or supplies, the pursuit of French pirates, or the long requested but totally impracticable destruction of the fort of Briscon. Moreover since 1588[106]

[104] On d'Épernon, see Léo Mouton, *Le Duc et le Roi*, Paris, 1924.

[105] D. Pedro de Acuña to Philip II, Rosas, 19th September, 1590, Simancas E° 167, f° 218. Bad weather made it impossible to dismantle the fort of Briscon. Dispatch from D. Martín de Guzmán, advised by the coastal pilots: the galleys should not return on this mission because of the bad weather which will last two or three months 'y entrar en el golfo de Narbona y costearle es mucho peor'. The marquis of Torilla (Andrea Doria) to H.M., Palamos, 28th September, 1590, Simancas E° 167, f° 223, explains the difficulty of blockading the Languedoc coast in bad weather. Torrilla to the king again, *ibid.*, f° 221, on the difficulty of attacking the fort of Briscon.

[106] The councillors of Barcelona to Philip II, 17th July, 1588, Simancas E° 336, f° 157. *Lista del dinero y mercadurias que han tomado los de Mos. de Envila a cathalanes cuyo valor passa de 30 U escudos* (1588), Simancas E° 336 (no folio number), Manrique? to Montmorency, 26th April, 1588, Simancas E° 336, f° 152.

Montmorency had had at his disposal a fleet of brigantines and frigates, small fast vessels which plundered Catalan ships and were usefully employed after 1589 for the blockade of the port of Narbonne. Against these little ships, the heavy Spanish galleys were no more effective than those of Venice against the Uskoks. So Montmorency who did not possess 'naval supremacy' was able to receive, by sea, reinforcements from Corsica[107] and oars from Leghorn.[108]

In fact the *ligueurs* had a more difficult task than at first appeared and unfortunately for them it was entrusted to the less than capable hands of the Duke of Joyeuse, son of the marshal. At first all went well. He had turned immediately towards nearby Spain, straight away taking the important town of Carcassonne, although the 'royalists' had kept the 'burgo' according to a Spanish dispatch dated 8th May, 1590.[109] Meanwhile Montmorency was massing his troops at Pont-Saint-Esprit, with a view to attacking the surrounds of Narbonne. Joyeuse anxiously wrote to Philip II: 'the affairs of the Catholics in Languedoc are in such a state that if there is no remedy soon, by mid-June at latest, it is to be feared that the heretic king will be master of all; since Monsieur de Montmorency who is his chief adviser and commands for him the said region, is preparing a large army . . .'.[110]

Joyeuse painted an over-gloomy picture perhaps, since his object was to obtain the assistance and subsidies which always took so long to come from the wealthy coffers of Spain. But the situation was certainly serious. By 12th June, he had still received nothing and was worried lest the enemy should 'harass the said Catholics during harvest time and deprive them of the means of maintaining themselves in their holy resolution'.[111] He complained again and appealed once more on 22nd June, sending an agent, the archdeacon Villemartin to the Catholic King;[112] and we find him complaining once again to Philip II on 10th July concerning the promised aid which had still not arrived:[113] 'I humbly beg Your Majesty to pardon me for daring to importune him so frequently with this statement of our needs, which I should not do were I not persuaded of his zeal for the preservation of the Catholic faith and the honour he has graciously done me of assuring me that he would support it in this province'.[114] Did help arrive in August? No, or at least not as much as had been hoped. True a letter does refer to the expected arrival of Spanish galleys carrying supplies for the German troops near Narbonne[115] and another letter

[107] Spanish dispatch, 8th May, 1590, A.N., K 1708.

[108] The Spaniards captured a small boat carrying arms loaded at the castle of Leghorn, Andrea Doria to Philip II, Rosas, 13th August, 1590, Simancas E° 167, f° 219. [109] A.N., K 1708. [110] May, 1590, A.N., K 1708.

[111] Joyeuse to Martín de Guzmán, Narbonne, 12th June, 1590, A.N., K 1708.

[112] Joyeuse to H.M., 22nd June, 1590, Simancas E° 167, f° 154.

[113] Joyeuse to D. Martín de Idiáquez, Narbonne, 10th July, 1590, A.N., K 1449, identical note to D. J. de Idiaquez.

[114] Joyeuse to Philip II, Narbonne, 10th July, 1590, A.N., K 1449.

[115] D. Pedro de Acuña to Philip II, Rosas, 13th August, 1590. Simancas E° 167, f° 220.

tells us that some of the promised ammunition had arrived.[116] But at the same time we know that the German lansquenets, who had not been paid, refused to enter 'enemy territory', in other words to fight.[117] So it dragged on, with repeated requests on one side and slow responses on the other, extravagant promises followed by the usual fiascos, and occasionally the odd success.

The war continued after a fashion for two years. But in 1592 came the Aragon affair south of the Pyrenees. Part of the region had risen up in defence of Antonio Pérez. The fugitive and his friends had taken refuge in Béarn where Henri IV's sister Catherine took advantage of the incident to send armed bands from her side of the Pyrenees on raids throughout Aragon.[118] Philip II was therefore obliged to keep his troops south of the mountains and abandoned Joyeuse and the Catholics of Languedoc. The latter, now staking everything on one last desperate throw, tried in September to take Villemur on the Tarn, in the hope of pushing on to Quercy and Guyenne, that is of moving out[119] and continuing the struggle elsewhere. The attempt ended in disaster and the Catholics according to one report, dated 4th November, 1592, lost all their infantry and artillery.[120] The conquered side had no one to appeal to except his Catholic Majesty, 'most Christian of Christians and most Catholic of Catholics, on whom after God all hope must be founded'.[121] But the appeal was unsuccessful.

In the new year a truce was agreed, reports of which reached Paris in mid-February, 1593.[122] The civil war in Languedoc had ended more rapidly and successfully than the royalists had dared to hope.[123] In the inland part of Languedoc, around Toulouse, fighting went on until 1596. But the victory in the east, in 1593, made a big difference: it cut in two the area of revolt which at the beginning of Henri IV's reign had extended from the borders of Italy to the Atlantic Ocean. And where the division

[116] Pedro de Ysunça to the king, Perpignan, 13th August, 1590, A.N., K 1708.

[117] D. J. de Cardona to Philip II, Madrid, 30th August, 1590, Simancas E° 167, f° 189.

[118] E. Lavisse, *op. cit.*, VI, 1, p. 353. Cf. Samazeuilh, *Catherine de Bourbon, régente de Béarn*, 1868. Antonio Pérez and his friends levied troops in Béarn . . . 'Antonio Pérez y otros caballeros que benieron a bearne hazen hazer esta gente en favor de los Aragoneses . . .', Dispatch, 1592, Simancas E° 169.

[119] E. Lavisse, *op. cit.*, VI, 1, p. 352.

[120] Dendaldeguy, envoy of Villars, to the Catholic King, Brionnez, 4th November, 1592, copy, A.N., K 1588.

[121] *Ibid.*

[122] Diego de Ibarra to Philip II, Paris, 15th February, 1593, A.N., K 1588. The entire League was tottering at this time, see the letter from the marquis of Villars to Philip II, Auch (?), 5th February, 1593, A.N., K 1588.

[123] Joyeuse died in January, 1592. The new duke (his son or brother?) Ange, who left his Capuchin monastery to take up the succession, had an interview with Montmorency at the Maz d'Azille et d'Olonzac. The truce signed for a year was not broken in the south. Joyeuse retired to Toulouse where he took up the struggle against Henri IV once more (Joyeuse to Philip II, Toulouse, 10th March, 1593, A.N., K 1588). He remained in the pay of Spain.

came, the 'royalists' found themselves, as in Béarn, up against the Spanish frontier.

In Provence however, the conflict which had begun even before the death of Henri III, lasted considerably longer than it did in nearby Languedoc and in its later complicated stages was prolonged until the end of the war between France and Spain in 1598 (the date of the evacuation of Berre by its small Savoyard garrison).

In April, 1589, that is before Henri III's death, Provence had seceded from the kingdom: or to put it more accurately, the Parlement of Aix had sworn allegiance to the Catholic Union and recognized Mayenne as 'Lieutenant General of the Kingdom'.[124] The small 'royalist' minority in the Parlement had withdrawn to Pertuis in July of the same year.[125] As for the larger towns, Aix, Arles and Marseilles (the last outside Provence strictly speaking, but in the Provençal region) they were all for the League. So the whole Provençal region, dominated by its towns with their protective privileges, had declared itself even before the accession of the new king. As for the governor, the Duke of Épernon, appointed in 1587, he had given up his post to his brother, Bernard de Nogaret de Lavalette. The new governor, who was a man of action and energy, did not shrink from the fresh threat. Loyal to the crown and supported by Lesdiguières' troops and the popular and peasant masses, he stood firm and even succeeded in reoccupying southern and central Provence. A Spanish dispatch of 1590 reports that he was fortifying Toulon[126] (as much against the Duke of Savoy as against any threat from the sea). But Lavalette had no more success than any of the other men on either side, during the ten last terrible years of fighting, in imposing his undisputed rule in Provence. Until at least 1596, there were always two separate and hostile Provences, one centred on Aix, the other on the provisional royalist capital of Pertuis, divided by a fluctuating and often imprecise frontier.

Marseilles, the greatest city in the entire region, had rallied to the cause of the League after the assassination of the second royalist consul Lenche (April, 1588)[127] with a fervour which continued unabated. And to declare for the League meant sooner or later linking arms with Spain.

But the summer of 1590 saw only one foreign conspiracy in Provence and that launched by the Duke of Savoy, a small enemy but one better placed to interfere in Provence than the powerful but distant Catholic King. Charles Emmanuel invaded in July following an appeal from the *ligueuse* plotter Christine Daguerre, countess of Sault. On 17th November, 1590, he arrived in Aix, where the Parlement received him and handed over to him military command of Provence, without however offering him the crown of the county, the object of his ambition.[128]

[124] Victor L. Bourrilly and Raoul Busquet, *Histoire de Provence*, Paris, 1944, p. 92

[125] *Ibid.*

[126] 8th May, 1590, A.N., K 1708.

[127] V. L. Bourrilly and R. Busquet, *op. cit.*, p. 91, R. Busquet, *Histoire de Marseille*, Paris, 1945, p. 224 ff.

[128] Bourrilly and Busquet, *op. cit.*, p. 92–93.

In that winter of 1590, all the elements of the Provençal drama were present. If things did not begin to move quickly at once it was because instead of bringing her weight to bear in this zone of influence (where the Duke of Savoy in turn found he lacked the strength to impose his rule alone) Spain had concentrated on Languedoc until the Aragonese crisis of 1592 and the débâcle of Villemur on 10th September that year. But in 1592, as the secondary theatre of Provence emerged as the only possible zone for intervention in Mediterranean France, the Spaniards put more effort into this area than they ever had before – though without much haste or speed, and without displacing the local protagonists, Savoy, Lesdiguières and Lavalette.

The weakness of the Duke of Savoy was proved when, during the winter of 1592, Lesdiguières first with the assistance of Lavalette then alone (Lavalette having been mortally wounded on 11th January, 1592 at the siege of Roquebrune)[129] was able to drive the Savoyard troops back over the Var and in the spring to surprise the duke on his own territory in the Niçois. The Savoyard garrisons scattered throughout Provence, cut off though not besieged, had an anxious time of it[130] but in the summer, Lesdiguières returned to the Alps and the Savoyards were able to launch another summer campaign through Provence, taking Cannes and Antibes on the way in August, 1592.[131] But these successes were no more decisive than the preceding ones. War in this now impoverished land degenerated into a series of surprise attacks and the victor gained only hollow triumphs. The Duke of Épernon, who took over the government on the death of his brother, arrived as if in occupied territory with his troop of Gascon adventurers. As autumn approached, a series of direct attacks and hard-fought engagements, remarkable for cruel atrocities, allowed him to recapture Cannes and Antibes from the Duke of Savoy. Had the situation been reversed? The deputies of 'anti-royalist' Provence in September turned for assistance to the Catholic King.[132] The count of Carcès who had governed Provence in the name of the League since his father-in-law Mayenne had appointed him in 1592, repeated the appeal early in 1593. His alarm was unnecessary: total success which had twice escaped the Duke of Savoy now eluded Épernon, who in June–July, 1593 tried unsuccessfully to take the city of Aix.[133]

Then in France, also in July 1593, the king announced his decision to return to the Catholic church; the situation immediately changed. There was a heartfelt mass move back towards the king and domestic peace. The Parlement of Aix swore allegiance to the king on 5th January, 1594. The first of the pro-League Parlements to recognize Henri IV,[134] it had

[129] Bourrilly and Busquet, *op. cit.*, p. 93.

[130] Don Cesar d'Avalos to Philip II, Aix, 4th March, 1592, Simancas E° 169, f° 103.

[131] Don Cesar d'Avalos to Don J. de Idiáquez, Antibes, 7th August, 1592, Simancas E° 169, f° 45.

[132] Don Jusepe de Acuña to D. D° de Ibarra, 13th September, 1592, copy, A.N., K 1588.

[133] V. L. Bourrilly and R. Busquet, *op. cit.*, p. 93.

[134] E. Lavisse, *op. cit.*, VI, I, p. 384.

begun the new year with an action which looked – but was not – decisive. In Provence the year certainly saw the return to the fold of many former anti-royalists, but it also saw the last flickers of conspiracy and revolt, some miscalculations and violent incidents and a good deal of bargaining.

One event stands out, the political incident of the season as it were: the former Leaguers, now rallying to Henri IV, joined forces to vent all their fury on Épernon. In the light of this, the latter's actions become clear. He knew himself to be disliked by Henri IV (indeed it was by forcing the king's hand that he had taken over the government of Provence in 1592) and mortally detested by the local nobility. He realized far ahead of time that the peace now on the horizon would spell the end of his authority and the dreams he had probably entertained of an independent principality. So the duke had his reasons for not wishing to be reconciled with the people of Aix and the Provençal nobility, and for being disturbed by the activities of the strange agent whom Henri IV sent to faction-torn Provence, Jacques de Beauvais La Fin. But he was forced, under the double impulsion of Lesdiguières and Montmorency to settle, on the king's orders, with the people of Aix. However Lesdiguières' bad faith and the news of the appointment to the government of Provence of the Duke of Guise (the son of 'le Balafré'), determined Épernon to rebel, to save as he put it, his honour and his life,[135] which meant joining forces with Savoy and Spain. He had already made up his mind by November, 1594, according to his letters and Spanish sources.[136] But the treason was not down in black and white in a formal agreement until November, 1595, a year later.[137] The duke went over to the enemy with his Gascons, and the few towns he still held in Provence and even apparently outside Provence. In the Spanish papers concerning the incident at any rate, there is a list of the property and towns which Épernon claimed to possess in every part of France.[138]

But he had changed sides very late in the day. By the time his agreement with Spain was down in writing in November, 1595, the fate of the South of France had been settled. However in 1594, Spain had resolved to make a special effort. The constable of Castile, Velasco, governor of Milan,

[135] Léo Mouton, *op. cit.*, p. 40.

[136] A.N., K 1596, nos. 21 and 22, quoted by Léo Mouton, *op. cit.*, p. 42 and note, p. 43. For the demands of the 'Catholics' of Provence, 8th December, 1594, A.N., K 1596. The matter discussed in the Council of State, 1st February, 1595, A.N., K 1596.

[137] *Accordi di Mon^e de Pernone con S. Mtà*, copy in French, Saint-Maximin, 10th November, 1595, A.N., K 1597, see Léo Mouton, *op. cit.*, p. 44, note 2. He had made an agreement with Savoy by August at the latest (*Disciffrati del Duca de Pernone*, Saint-Tropez, 11th December, 1595, A.N., K 1597). Cf. the undated document in A.E., Esp. 237, f° 152.

[138] *Estat des villes qui recongnoissent l'authorité de Monseigneur le duc d'Épernon*, A.N., K 1596 (also a list of the 'towns' he claimed to possess in Dauphiné, Touraine, Angoumois, Saintonge). The same document in Spanish, *Lista de las villas de Provenza*, in A.E., Esp. 237, f° 152. *Mémoire sur ce qui est sous le commandement de M^r d'Épernon*, undated, A.N., K 1598.

had assembled a large army and was preparing to launch an expedition through Savoy and over the Jura towards Dijon. Marshal de Rosne[139] even advised him in the interests of his cavalry to set up headquarters at Moulins in the Bourbonnais on the Allier.[140] So the attack planned in summer 1595 was directed at the very heart of France. The victory of Fontaine-Française on 5th June brought the invasion to a halt. Thus although insignificant from the military point of view it had profound consequences. If Henri IV by moving south had left the North unprovided[141] he had at least consolidated as far as the sea the positions which might have been in serious danger from the flanking manœuvres of his enemy.

In 1596, both Épernon and the city of Marseilles were brought to heel. The Duke of Guise had little difficulty in crushing both obstacles. In February the 'royalists' routed Épernon's small force at Vidauban.[142] Fighting went on even in the waters of the Argens and many soldiers were drowned. Within a month, on 26th March, the duke signed a peace with the king[143] and two months later left Provence.[144] As for Marseilles, traitors opened the gates of the city to the Duke of Guise during the night of 16th to 17th February.[145]

Thus ended an eventful period in this city's history which deserves a few words to itself. Like many other French towns during these troubled years, Marseilles had found itself virtually autonomous. Self-confident, Catholic, pro-League, in 1588 the city allowed itself to be governed by its passionate sentiments. But how easy was it to survive on the borders of the kingdom or rather outside the kingdom altogether, for France was entirely dislocated and divided by the troubles? Appeals to Spain for grain tell part of the story.[146] And while the war which encircled the city on all sides was not a modern war (it was more wasteful of men than of matériel) it was nevertheless expensive. In Marseilles, guards and military precautions had to be paid for. A determined policy was necessary if the city was to consent to such sacrifices. Such a policy was embodied for five years in the person of Charles de Casaulx, a man whom his recent biographer, Raoul Busquet, while not whitewashing him, has shown in a new light.[147] Although it was by a revolutionary act of violence that this energetic leader of men seized the Hôtel de Ville in February, 1591, in the event he was to prove an attentive and efficient administrator for the city. Totally devoted to the interests of the municipality, his policy was quite

[139] Charles de Savigny, sieur de Rosne: on the part he played at Fontaine-Française, T. A. d'Aubigné, *op. cit.*, IX, p. 55 ff.

[140] 12th September, 1594, copy, A.N., K 1596. Henri IV had to go to Lyons. *Nuevas generales que han venido de Paris en 26 de noviembre* (1594, A.N., K 1599).

[141] E. Lavisse, *op. cit.*, VI, 1, p. 401.

[142] *Ibid.*, p. 405; Léo Mouton, *op. cit.*, p. 47.

[143] Léo Mouton, *op. cit.*, p. 47.

[144] *Ibid.*, p. 47–48.

[145] Étienne Bernard, *Discours véritable de la prise et reduction de Marseille*, Paris and Marseilles, 1596

[146] An undated document (A.N., K 1708) refers to requests from Marseilles for grain from either Oran or Sicily. The city also asked to be relieved from Épernon's galleys which were cruising offshore.

[147] R. Busquet, *op. cit.*, p. 226 ff.

independent of the menacing intrigues of the Duke of Savoy, who was anxious to use Marseilles as a direct link with Spain. The duke stopped in the city in March, 1591, but in vain; in vain too he tried to capture Saint-Victor by treachery (16th–17th November, 1591).[148] Casaulx also remained firmly aloof from the intrigues and quarrels of the Provençal nobility, although Marseilles did grant refuge briefly to the countess of Sault; but the dictator skilfully rid himself of her later.

If one remembers Casaulx's policies within Marseille itself, the welfare arrangements, as we should call them, which he initiated, the printing-presses he introduced, the public buildings he had erected, and above all the popularity he enjoyed, his 'tyranny' begins to emerge in a rather different light. Undoubtedly it was, like all dictatorships, wary, highly policed and vindictive, at least towards the *bigarrats* who were unhesitatingly thrown into prison, exiled and stripped of their wealth. But it was a popular dictatorship, unusually favourable towards the urban masses, the poor. In 1594, a Spanish dispatch reports the war being waged in Marseilles against the rich merchants and nobles. 'It is not clear why this is so', says the report, 'probably to get some money out of them'.[149] The town might be mistress of her own destiny, but finance was still an enormous burden. The Pope and the Grand Duke of Tuscany, who were approached in 1594, refused to advance a single *blanca*.[150] Necessity as much as religious allegiance obliged Casaulx in the end to turn to powerful Spain to obtain grace and favour and the means of survival.[151]

Events pushed Marseilles towards the Spanish cause and the city was soon fully committed to it. On 16th November, 1595, the Provost and Consuls of Marseilles wrote Philip II an extraordinary letter, still prudent but at the same time quite categorical. It is well worth quoting from:[152] 'God having inspired in our souls the sacred fire of zeal for his cause and seeing the great and perilous shipwreck of the Catholic faith in France, having stood firm against all the assaults we have suffered from the enemy of the faith and of this state, by the particular favour of heaven, the religion and state of this city are intact and preserved to this day, and we have the unshakeable desire to continue in the same at the cost of our lives and those of all our citizens who are permanently united in this holy resolution. But foreseeing the stormclouds approaching as the affairs of Henry of Bourbon prosper and since the public funds are already exhausted and the purses of private individuals are no longer sufficient for the execution of this great and beneficial enterprise, we have dared to raise our eyes towards Your Catholic Majesty and appeal to him . . . as the protector of all Catholics to beg him most humbly to cast the rays of

[148] *Ibid.*, p. 231.

[149] *Nuevas de Provenza*, 1594, Simancas E° 341. [150] *Ibid.*

[151] 10,000 *salme* of grain were obtained from Sicily in 1593, A.N., K 1589.

[152] Louis d'Aix, Charles de Casaulx, Jehan Tassy to Philip II, Marseilles, 16th November, 1595, A.N., K 1597. This letter had been preceded by a letter of recommendation from Andrea Doria to Philip II, Genoa, 13th November, 1595, A.N., K 1597, B 83.

his natural goodness upon a city so full of merit, for the sake of its ancient faith and loyalty'.

According to this document at least, Marseilles was not making a gift of herself to the king of Spain. There are degrees of 'treason'. The city (or rather Casaulx) merely declared that it was not willing to abandon the good fight. A propaganda pamphlet, a fairly long anonymous document printed between 1595 and 1596, gives the same impression. The 'Reply of the French Catholics of the city of Marseilles to the advice of certain neighbouring heretics, *politiques*, anti-Christians and atheists'[153] is a rather rambling document, adding nothing new to the well-known propaganda campaigns of the French wars of religion, making little pretence at objectivity and lumping royalists with atheists and Huguenots with debauchees; as polemic it is of little interest and its violent, virulent tone rings hollow to us. The one point worth remarking is that no mention is anywhere made of the relations between the city and Spain.

An entente was however inevitable: the city had either to take refuge behind the might of Spain or reach agreement with the king's representative, the president Étienne Bernard who had arrived in Marseilles and was prodigal in promises to Casaulx and his companion in arms Louis d'Aix. But fearing some trap behind the glowing prospects he painted,[154] the rulers of Marseilles preferred to make an agreement with Philip II. Three 'deputies' from the city, including Casaulx's son, made the journey to Spain, where they recounted at length the events in Marseilles between 1591 and 1595,[155] stressing the part played by the so-called dictators Casaulx and Louis d'Aix, both sons of established families in the city, who had been able with the aid of their relatives, their friends and popular support, to restore Catholic peace and order within its walls. But only at great cost: they had had to arm themselves, levy mercenaries, occupy the fortresses of Notre Dame and Saint Victor, the Tour Saint Jehan, guard the 'porte Reale', the great platform and the Aix gate, the most 'defendable' sites; at the entrance to the harbour, they had been obliged to build the 'fort Chrestien' – still uncompleted – and maintain a number of horses for the security of the region and to allow the people of Marseilles to 'gather in their fruits without being troubled by their enemies'.[156] Now that Henry of Bourbon has been absolved by the Pope, has asserted his rule and is master of Arles (and therefore of Marseilles' grain supply) and with the city meanwhile packed with refugees, including 'that great and learned person Monseigneur de Gembrard, archbishop of Aix, dispossessed by Henry of Bourbon' – in such an extremity, despite the offers made by Henry IV, the city can find true succour 'only under the wing' of the Catholic King. The latter was implored to send help and

[153] A.N., K 1597, B 83.

[154] R. Busquet, *op. cit.*, p. 240. In any case and above all, the entente had been in the making for a long time, see note 151 and Cardinal Albert to Philip II, Marseilles, 7th September, 1595, A.N., K 1597.

[155] Undated, A.N., K 1597, B 83.

[156] *Ibid.*

quickly: money, ammunition, men and galleys. The situation was becoming increasingly tense as royal troops approached the gates of the city and conspiracies were being hatched within its walls.

Help reached Marseilles in December, 1595,[157] in the shape of the galleys of Prince Doria's son and two Spanish companies, just in time to prevent the entry of royal troops. But inside Marseilles itself, there were complications as the townspeople became suspicious even of their friends. On 21st January, 1596,[158] the deputies from Marseilles left the court of Spain having obtained satisfaction. The city was to open its arms to the Catholic King though stopping short of total surrender: his galleys would be allowed free access to the port, he had permission to send in troops and he had the word of the citizens of Marseilles that they would not settle with Henry of Béarn and would recognize as king of France only a candidate acceptable to Spain. The people of Marseilles, said their declaration, 'will not recognize Henry of Bourbon, and will not pay allegiance to him or to any other enemies of Your Catholic Majesty, but will keep and remain in the Catholic faith and in their present state until it shall please God to give France a truly Christian and Catholic king who is bound by friendship, fraternity and common aims to Your Catholic Majesty.' On 12th February, the Marseilles deputies were still in Barcelona and writing to Don Juan de Idiáquez, asking for Catalan grain.[159] But five days earlier, on the 7th, a successful plot had been carried out in the city: Casaulx had been assassinated and the town delivered up to Henri IV.[160] 'I was not king of France till now', he is reported to have said on hearing the good news.[161]

It is quite possible to take this fragment of French history as a starting point for more general speculation, and along with many reputable historians see Provence as a microcosm of France during the last years of the Wars of Religion: with the inflation, the terrible poverty in town and countryside, the spread of brigandage, the political ruthlessness of the nobility. In Épernon we might see an example of the provincial 'kings', like Lesdiguières in the Dauphiné (although the difference in character between the two could not be more marked), Mercoeur in Brittany, or Mayenne in Burgundy. It would be even more tempting to seek to understand, through the example of Marseilles, the vital part played by the towns first in the disintegration then in the reconstruction of France as a unit.

The League was more than an alliance of passionate Catholics. And it was more than a tool in the hands of the Guises. It marked a deliberate return to a past which the monarchy had fought and in part overcome: above all it was a return to an age of urban independence, the age of the city-state. The lawyer Le Breton who was strangled and hanged in

[157] Antonio de Quinones to Philip II, Marseilles, 1st January, 1596, A.N., K 1597; Carlos Doria to Philip II, Marseilles, 1st January, 1596, *ibid.*

[158] *Puntos de lo de Marsella*, A.N., K 1597.

[159] *Los diputados de Marsella a Don Juan de Idiáquez*, Barcelona, 12th February, 1595, Simancas E° 343, f° 92 (chancellery summary).

[160] R. Busquet, *op. cit.*, p. 245. [161] *Ibid.*

November, 1586[162] was no doubt an eccentric individual. But it is significant that his plans envisaged a return to urban privilege, his dream being to divide the country into a number of small Catholic republics which would be responsible for their own actions. Quite as serious as the defection of the Guises then was the defection on the part of the towns, which were invariably passionately committed to the struggle, from their bourgeois leaders to the humblest artisan. And Paris was the enlarged model of all these towns. In 1595, the Duke of Feria proposed to the Archduke Albert that they should try once more to set up a League in France on the same principles as that which had existed during the reign of Henri III, 'which was not founded by the princes of the House of Lorraine, but by a few citizens of Paris and other cities, only three or four at first, but in such respectable Christian conditions that most of the rest of France, and that the best, followed suit'.[163] Some of these men were still in Brussels: the mistakes and betrayals of its leaders had certainly not destroyed the cause.

This is eloquent – perhaps exaggerated – testimony to the role played by the towns. But how long would they really have been able to survive in a state of insurrection? By cutting the roads and hence the flow of goods, they were cutting their own throats. If they rallied to Henri IV after 1593, was this not, in addition to the reasons usually quoted, because they desperately needed access to the rest of France? The case of Marseilles, finding it impossible to live off the sea alone without aid from the interior, should alert us to the inescapable symbiosis of land and sea routes in the Mediterranean.

And we shall certainly never fully understand the Casaulx episode if we do not place it in the narrow context of municipal life. His task, as Casaulx saw it from beginning to end, was to stand firm, not to fail *his* city. His attitude can only be judged, if one seeks to pass judgement, in this light. For proof, one has only to read the memorandum his envoys took to Spain: 'The gentlemen of Marseilles' it says, 'have borne in mind that from its foundation, their city was almost always governed by its own laws and in the form of a Republic, until the year one thousand two hundred and fifty-seven, when it came to an agreement with Charles of Anjou, count of Provence, and recognized him as its sovereign, with many reservations, pacts and conventions, chief among which was the stipulation that no Vaudois heretic (a sect prevalent in those days) nor anyone whose faith was suspect, would be sheltered in Marseilles . . .'.[164]

The Franco-Spanish war: 1595–1598. A general outline of the war between France and Spain can be given in a few words: officially lasting from 1595 to 1598, it had actually begun in 1589 if not before, for there were few periods during the fifty years we have chronicled year by year when France and Spain were not at odds.

[162] E. Lavisse, *op. cit.*, VI, 1, p. 264; *ibid.*, p. 342 ff, on the revival of trade.

[163] Undated, about 1595, Simancas E° 343.

[164] Undated, A.N., K 1597, B 83.

This time war was formally declared by Henri IV, on 17th January, 1595. The text of the declaration, printed in Paris by Frédéric Morel, even reached the Spanish authorities. In it the French king outlined his grievances against Philip, who had 'dared, under the pretext of piety, to interfere openly with the loyalty of the French towards their natural Princes and sovereign Lords, ever the object of admiration of all other nations of the world, publicly and wrongfully seeking to claim this noble Crown either for himself or for one of his kinsmen'.[165] To place the conflict on the level of the rights of princes was astute and to some extent justified. But it made little or no difference to the actual course of events. The Duke of Feria had long before prophesied that the Franco-Spanish war would once again be 'peripheral' as in the time of François I and Henri II, and that it would end with a truce when both sides were exhausted.[166]

And indeed the war was fought chiefly on the frontiers of the kingdom: on the Somme, in Burgundy, Provence, the Toulouse–Bordeaux region, and Brittany. France was apparently surrounded; but the circle was not unbroken. For Spain was unable, despite her proximity to the fighting zones and despite her fleet, to maintain her strongholds on the outer edges of France. Toulouse had been lost in 1595 and Marseilles the same year; the Duke of Mercoeur, the last to resist,[167] capitulated in 1598. He had in any case been virtually out of action for two years. It is important then to note that war spared the heart of the kingdom: France was protected by her size. If the Catholic King found city gates whose locks he could pick, consciences he could buy or even Protestants willing to sell themselves (one named Montverant is recorded in the region of Foix)[168] it was invariably in some marginal area.

True there was the expedition by the Spanish army in Italy over the Alps: but this turned out to be merely a trip to the Franche-Comté and back. As we have already pointed out, Fontaine-Française led to the withdrawal of the Spaniards and the final submission of Mayenne: only the determined opposition of the Swiss cantons on this occasion prevented the occupation of the Comté.

A rather more serious struggle was taking place in the north, on the Netherlands border. The Spaniards won some major victories, capturing several fortified towns, Cambrai, Doullens, Calais and Amiens (taken by surprise on 11th March, 1597). Their chief problem was to maintain the positions they had seized and the coexistence within them of both garrisons and civilian populations without trouble and fighting, as is explained in a

[165] A.N., K 1596. [166] Reference in note 163.

[167] In the text of the prorogation of the truce, 3rd July, 1596, A.N., K 1599, it was agreed that each side should levy funds from the regions it controlled. Money must be sent or Mercoeur will sue for peace, M° de Ledesma to Philip II, 20th January, 1598, A.N., K 1601. Mercoeur to Philip II, Nantes, 24th March, 1598, (A.N., K 1602) informs him that he has made peace with Henri IV and asks for employment in Hungary.

[168] Philip II to the Duke of Albuquerque, Madrid, 10th July, 1595, Simancas E° 175, f° 290.

dispatch written to the count of Fuentes as early as 1595.[169]

The capture of Amiens however alarmed the French: the road to Paris beyond the valley of the Somme now lay open. A riposte was called for and Henri IV resolved to recapture the town. His decision meant a frantic search for funds and renewed appeals to his allies – England (which had officially declared war on Spain in 1596)[170] and the United Provinces. So in the French army which marched to recapture Amiens there were, according to Spanish reports, 2000 English and 2000 Dutch soldiers.[171] Amiens was not taken until 24th September, 1597, after a six months' siege, and the failure of a Spanish attempt to relieve the town nine days before it fell.[172] Its fall was regarded as a great victory and was much commented on throughout Christendom. But the victor had hardly entered the town before his army melted away. Fortunately for France, her poverty and exhaustion were matched only by the irremediable fatigue of the Spanish Empire and its financial distress following the bankruptcy of 1596.

Insolvency paralysed Spanish actions. At the centre of the military effort, in the essential relay post of Milan, troop movements were being accomplished with great difficulty if at all, from spring 1597.[173] Italian troops were supposed to be brought from Naples to Genoa then sent on to Flanders. Troops quartered in the Milanese were also supposed to move north. But would there be enough money? A further worry was the question of taking assistance to the reckless Duke of Savoy, Charles Emmanuel. Should the Spanish troops at present aboard galleys leaving Spain for Genoa be used for this mission? It would clearly be useful to tie the French down in this quarter so that the archduke should have some relief on the Netherlands front.[174] When at last everything was ready and some of the troops were on the point of leaving for Flanders, a fresh problem cropped up: would the route through Savoy be safe? Chambéry and Montmélian might be lost as Lesdiguières went into the offensive. Worse, all of Savoy, Piedmont and even Milan were at risk if money did not arrive in time, as the Constable of Castile wrote in his own hand and informed the king.

So the Spanish military authorities were overwhelmed with problems,

[169] *Advis a Monsieur le Comte de Fuentes*, 12th March, 1595, A.N., K 1599.

[170] *Déclaration des causes qui ont meu la royne d'Angleterre a déclarer la guerre au roy d'Espagne*, Claude de Monstr'oeil, 1596, A.N., K 1599.

[171] Mendo de Ledesma to Philip II, Nantes, 25th June, 1597, A.N., K 1600.

[172] E. Lavisse, *op. cit.*, VI, 1, p. 410.

[173] The lack of money was felt by February (*Relattione summaria del danaro che si presuppone manca nello stato di Milano*, 12th February, 1597, Simancas E° 1283). On troop movements, Philip II to the Constable of Castile, Madrid, 7th April, 1597, Simancas E° 1284, f° 126. Philip II to the Constable, 2nd May, 1597, *ibid.*, f° 125. The Constable to Philip II, Milan, 12th May, 1597, *ibid.*, f° 86.

[174] On assistance to be taken to the Duke of Savoy, Philip II to the Constable of Castile, San Lorenzo, 28th April, 1597, Simancas E° 1284, f° 116; the Constable to the king, Milan, 12th May, 1597, f° 83; the Constable to the king, 23rd July, f° 55; the king to the Constable, 8th August, 1597, f° 122.

whether in the North or in the Swiss cantons where negotiations with Appenzell were proving costly. Moreover after the capture of Amiens, the French party in Italy regained confidence and Savoy was threatened. At this point the death was announced of the Duke of Ferrara: Pope Clement VIII at once claimed the succession for the Holy See. 'It pains me to the bottom of my heart', wrote the Constable of Castile to Philip II on 16th November, 1597, 'to see so much movement of arms in Italy. I shall remain here without taking action apart from manning our frontier until I receive orders from Your Majesty. Indeed, even if I had orders to the contrary, I should be forced by necessity to adopt the same course. I therefore beseech Your Majesty to consider the poverty and misery of this state, while the Pope is assembling a mighty army. He is naturally drawn to France: he loves the man of Béarn as a son and as a protégé. He has made clear in many conversations and circumstances the lack of goodwill he bears your Majesty's affairs and his dissatisfaction at the greatness of your states. He is a Florentine . . . As the Venetians and other princes who do not love us are also taking up arms, all these people may turn upon the state of Milan . . . from which there is a general desire in Italy to expel the Spaniards. Our salvation can only lie in more troops, money and above all dispatch, which I leave to the great prudence of Your Majesty'.[175]

The peace of Vervins. To whose advantage was this long-drawn-out war? Beyond question the only ones to benefit were the Protestant powers, whose ships were ravaging the oceans. The United Provinces gained in stature through the very poverty of the southern provinces which had remained Catholic: they grew rich on the collapse of Antwerp which was accomplished by the occupation of the mouths of the Scheldt by the States General and in spite of the recapture of the city by Alexander Farnese. All these blows were essential for the future growth of Amsterdam. London and Bristol were now expanding too. For everything had conspired to favour the rising northern powers. Spain was still open to them in spite of her attempts at a continental blockade; the gates of the Mediterranean had been forced; the Atlantic was already in their hands; the Indian Ocean was to be theirs before the century ended, in 1595. These

[175] Velasco to Philip II, Milan, 16th November, 1597, Simancas E° 1283, f° 2. On the Ferrara affair, see his letter of 4th November (f° 5) and (E° 1283, no folio number) 5th November, 1597. *Relacion de las prevenciones que S. S[d]. . . .* On the Swiss, letters from Philip II, 31st July (E° 1284, f° 123) and from the Constable, 23rd July, (E° 1283, f° 55), 7th October (*ibid.*, f° 4). On Amiens, his letter of 25th October, 1597 (E° 1283). I note that in 1597, troops were transferred from Italy to Spain, notably a *tercio* of Neapolitans under D. Cesar de Eboli, who arrived on 7th August in Alicante, on board Ragusan ships, D. Jorge Piscina? to Philip II, Alicante, 8th August, 1597, *ibid.* (6 Ragusan ships). Subsequently these ships 'q. llevan el tercio de Cesar de Eboli', were sent to El Ferrol (prince Doria to Philip II, Cadiz, 21st August, 1597, Simancas E° 179). On the arrival of a convoy (2 ships which had sailed from Spain to Calais, 40 ships expected with 4000 Spaniards under D. Sancho de Leyva and possibly money): Frangipani to Aldobrandino, Brussels, 27th February, 1598, *Corresp.*, II, p. 298–299.

were the truly important events of the last years of the century; alongside them the many incidents in the Franco-Spanish war appear trivial. While France and Spain were fighting over townships, forts and hillocks of ground, the Dutch and the English were conquering the world.

This fact seems to have been grasped by the Papacy by 1595. Clement VIII offered to act as mediator and was vigorous in his efforts to obtain peace between France and Spain for by now Rome, the Church and the whole Catholic world was distressed by this conflict between two Christian powers both faithful to Rome. Since the reign of Sixtus V, the Papacy had sought to salvage a Catholic France: it had finally played the card of French independence and by so doing embraced the cause of Henri IV who received absolution from Rome two years after abjuring the Protestant faith in 1593.

The rest of free (or semi-free) Italy[176] supported Rome, exerting pressure in the same direction, except for the Duke of Savoy who was distracted by his greed for French territory. And this Italy, only too glad to loosen the grip of Spain, was both rich and active. Venice had been the first to challenge Spanish policy by receiving an ambassador from Henri IV as early as 1590; the Grand Duke of Tuscany was financing the actions of the French king, whose debts very quickly ran into many 'écus au soleil'.[177] His creditor shrewdly took pledges, occupying the Château d'If and the Iles Pomègues near Marseilles. The marriage to Marie de Medici a few years later was arranged for several reasons, but in part it had to do with payment of arrears.

By September, 1597, at any rate, after the recapture of Amiens, Henri IV was beginning to look like a conqueror. Was he seriously threatening the Netherlands? Did he even secretly hope to push forward in that direction? That is another question. But even if he had wanted to pursue this audacious policy or to send an expedition against Bresse or Savoy, success would have been impossible without the support of his allies. And the allies had no desire to see him winning decisive victories either in Netherlands or on the approaches to Italy. What they wanted was for the war to continue, immobilizing Spain in huge military operations on the continent, while they continued to reap the benefits of long distance maritime trade. It is quite possible that as Émile Bourgeois once suggested,[178] Henri IV felt that at this vital moment he was abandoned or at any rate inadequately supported by his allies and thereby propelled towards the peace, which, in view of the state of his kingdom, was so necessary.

But was it not equally necessary for Spain? The further state bankruptcy

[176] See the curious remarks made by William Roger in 1593, to 'Burley' and Essex, (near Louviers, 1st May, 1593, A.N., K 1589) or even more significant, re the financial assistance that came from Italy for Holland and Henri IV, J. B. de Tassis to Philip II, Landrecies, 26th January, 1593, A.N., K 1587, annotated by Philip.

[177] R. Galluzzi, *op. cit.*, V, p. 302, Berthold Zeller, *Henri IV et Marie de Medicis*, Paris, 1877, p. 17.

[178] In a remark on a manuscript work of mine on the Peace of Vervins, 1922, written under his supervision.

of 1596 had once more brought the mighty Spanish war machine to a halt. Realizing that his end was near, Philip was anxious to establish in the Netherlands his favourite daughter, Clara Isabella Eugenia, his preferred companion in the last sad years of his life, his reader and secretary, confidant and private joy. His secret desire was to marry her to the Archduke Albert, who had recently (in 1595) been appointed to the precarious government of the Netherlands. An urgent decision about his daughter's future appeared the more necessary since new influences were already at work on his son, the future Philip III, influences opposed to the solutions the old king so anxiously desired. The archduke for his part, writes the historian Mathieu Paris, 'was burning with the desire to be married'. Was this one of the reasons great or small which led to the peace of Vervins? Exhaustion, the need to draw breath and perhaps too one of those designs by which Spanish diplomacy was so frequently governed, from necessity rather than from choice. Is it not likely that Philip II's diplomacy sought to obtain a rapid settlement with France, even at a high price, in order to be free to turn to those other two enemies, England and the United Provinces? We know that Cecil visited Henri IV to try to prevent him at the last moment from making peace with Spain. We should not forget that at that very moment a Spanish armada was sailing towards the British Isles. Nor should we forget that the immediate action of the archduke, once peace was signed with France, was to turn his forces to the north. It is by no means impossible that calculations of this kind influenced Spain's decision.

But in order fully to understand the sequence of events, it is rather to Rome that we must turn: the blossoming of the Counter-Reformation, the monetary prosperity of Italy and the exhaustion of her western adversaries had brought new eminence at the end of the century to the city of the Popes. Some inkling of this new-found force can be seen from the rapid fashion in which Clement VIII solved to his own advantage the small but thorny problem of Ferrara, This bustling town, one of the great ports of Italy, in the centre of a large region, in a key position on the Italian chessboard, the Pope annexed for himself before France, Spain or even Venice had had either the time or the temerity to intervene.[179]

Rome's growing importance can also be accounted for by the fact that she had her own solution, her own clear policy, dictated not so much by the calculations of a few shrewd tacticians as by the circumstances of these years and the unanimous desires of all Catholicism. Making itself heard through Rome was the general will, a profound upsurge of Catholicism directed now against the Protestant enemy in the North, now the infidel in

[179] On the Ferrara affair, I. de Mendoza to Philip II, Venice, 3rd January, 1598, A.N., K 1676 (excommunication of D. Cesare). *Lo platicado y resuelto en materia de Ferrara en consejo de Estado* . . ., 7th January, 1598, *que Cesare d'Este se soumette*, Simancas E° 1283, I. de Mendoza to Philip II, Venice, 10th January, 1598, A.N., K 1676. *Accordi fatti tra la Santa Seda Apostolica et D. Cesare d'Este*, 13th January, 1598, *ibid.*, I. de Mendoza to Philip II, Venice, 24th January, 1598, A.N., K 1676; Mendoza to the king again, 31st January, 1598, *ibid.*

the East. In 1580, Rome had been carried along by the general movement and had enthusiastically agreed to turn from the crusade against Islam to war on the Protestant heresy. Towards the end of the century, the anti-Protestant movement had spent itself and the most convincing proof of the fact is that Rome was now trying to revive a holy war against the Turks in the East.

So for Catholics, the sixteenth century ended and the seventeenth began in an atmosphere of crusade. After 1593, war against the Turks once more became a reality, in eastern Europe, in Hungary and on the waters of the Mediterranean. Without ever becoming a general conflict, it was for thirteen years, until the peace of 1606, to loom as a constant threat. In 1598, the Duke of Mercoeur left Brittany, the western tip of France, to go to war in Hungary. His voyage is symbolic: untold thousands of devout Catholics were dreaming of carving up the Turkish empire, which many believed to be on the point of collapse. The nuncio in the Netherlands, Frangipani, wrote to Aldobrandini in September 1597:[180] 'if only one quarter of the combatants in Flanders were to march against the Turk . . .'. It was a wish now beginning to be entertained by more than one Catholic or indeed non-Catholic: witness La Noue's anti-Turkish project which goes back to 1587.

The peace of Vervins, signed on 2nd May, was ratified by the king of France on 5th June, 1598:[181] it restored to Henri IV the kingdom as defined by the peace of Cateau-Cambrésis in 1559. Spain had therefore to give up certain conquered territory, notably to withdraw from the positions captured in Brittany and on the northern frontier, including Calais, the return of which was of some significance. On the whole the peace terms favoured France. Bellièvre, Henri IV's future Chancellor, said with some exaggeration that 'it was the most advantageous peace France had secured for five hundred years' – an official but not entirely inaccurate verdict. While the peace of Vervins entailed no acquisition of territory abroad, it had unquestionably saved the integrity of the kingdom. Its chief advantage was to bring France the peace she so desperately needed, the peace which could heal the wounds of a country delivered for so many years to foreign powers, a kingdom which had so passionately and blindly torn itself apart. There can be no doubt that the downward turn in the overall economic situation after 1595 made its contribution to the ending of hostilities[182].

3. THE END OF NAVAL WAR

The local wars described in this chapter, some in the western, some in the eastern regions of the Mediterranean, were not connected. No doubt they

[180] 25th September, 1597, *Correspondence*, II, p. 229.

[181] Ratification of the peace of Vervins by the king of France, Paris, 5th June, 1597, A.N., K 1602.

[182] For a general view of this question see Pierre Chaunu, 'Sur le front de l'histoire des prix au XVIe siècle: de la mercuriale de Paris au port d'Anvers', in *Annales, E. S. C.*, 1961.

had faraway repercussions, but there was never any direct link. Why not? Because the sea which separated them remained adamantly neutral, refusing its assistance to general conflict which could only have been possible if the Mediterranean itself had been involved.

There was some fighting at sea between 1589 and the end of the century and even later. But this was no more than the usual 'peace-time' Mediterranean fighting, that is to say the privateers' war: minor and anarchic, fought by individuals usually within a very small compass and rarely occupying more than a few ships at a time. That this should be so is doubly significant, in that after 1591, and in particular in 1593, 1595 and 1601, attempts were made to launch major naval offensives on which we shall try to shed some light. Their short-range and failure show how times had changed.

False alarm in 1591. As early as 1589, when negotiations with Persia were under way, but above all in 1590 when the Turco-Persian peace treaty had been signed, Turkey turned her attention once more to the West. We have already noted Ḥasan Veneziano's modest expedition to Tripoli in the summer and autumn of 1590. This expedition despite its limited objective marks the revival of Turkish naval activity in the Mediterranean.

But the long interval, the years of inactivity had consumed away the basic structures of the Turkish navy, structures and foundations which could be rebuilt only slowly and then not completely. Experienced seamen were in short supply; there were few expert craftsmen in the Arsenal and even the necessary naval infantry were not to be found.[183] The ten years since Euldj 'Alī's last expedition to Algiers in 1581 had been sufficient for the whole machinery to rust away. The effort of reconstruction was made even more difficult by lack of money and by the Christian privateers who had laid waste the shores of the Aegean from which the Turkish navy had for so long drawn sustenance.

It is not surprising that the Tripoli expedition of 1590, strictly confined as it was to a punitive action, did not renew the Turco-Spanish conflict. Neither Spaniards nor Turks were seeking a pretext to declare war. When, on leaving Modon, Ḥasan's galleys headed for the African coast, contrary to all past tradition they made no attempt on the coasts either of Sicily or Naples. During these years moreover, the presence in Constantinople of a Spanish representative, Juan de Segni, many of whose letters are preserved in the archives (though they are alas more remarkable for their personal complaints than for any information about the object of his mission) seems to suggest that the truce was extended on a more or less formal basis until 1593. At any rate, despite the repeated appeals and efforts of French and English agents,[184] the attempts by all the anti-

[183] J. W. Zinkeisen, *op. cit.*, III, p. 124.

[184] Spanish dispatches, 5th January, 1591, A.N., K 1675. Many details of these interventions which can I think be passed over here (J. W. Zinkeisen, *op. cit.*, III, p. 629 ff).

Spanish powers in Europe to incite the Moslems to attack all-powerful Spain on sea were unsuccessful. In 1591, the English representative for instance, backed up by Hasan Pasha, the 'general of the sea' was explaining to the sultan that Philip II had withdrawn the better part of the usual garrisons from the Italian coastal positions, in order to reinforce his troops in France. It would be easy in the circumstances, argued the Englishman, for Turkey to seize large expanses of territory.

They may have been fruitless attempts, but they awoke some response, particularly since a whole series of threats, loudly voiced in speeches evidently intended for external consumption, seemed to confirm the success, rather than the failure of the French or English agents. Once again, contradictory rumours flew about the Mediterranean, partly inspired no doubt by the traumatic memory of the awe-inspiring Turkish fleets of years gone by. It was whispered that three hundred galleys would be sailing in the spring for Apulia and the Roman coast; they would winter at Toulon and then move to Granada, where the Moriscos (this was equally untrue) had already risen up. Alternatively, they might be content merely to attack Venice or Malta (whose knights had just seized another galleon crowded with pilgrims for Mecca).[185] Some Venetians feared for Crete, where, they said, Christian pirates making for the Levant found shelter and assistance[186] (despite strenuous efforts by the Venetian authorities, needless to say . . .).

Juan de Segni, who was no doubt better informed, wrote from Constantinople that plans were indeed afoot, but not for that year.[187] The Turks had given written undertakings to the king of France and the queen of England, but these bound them to act only in the spring of 1592.[188] All the preparations appeared to be aimed at the long term. The sultan was for instance preparing – but as yet only preparing – to introduce a series of fiscal measures: 'voluntary' contributions by the *sandjakbegs* and pashas, special levies from Jews and other taxes, details of which are difficult to interpret from the western version.[189]

And indeed after mid-January, there was less talk of the great spring armada.[190] Would the predicted small fleet come out instead?[191] Some thought so. Yes, said one report, it would be sent out, but only to Egypt and Barbary and possibly to Provence.[192] And it was unlikely to number

[185] See preceding note.

[186] J. de Segni de Menorca to Philip II, Constantinople, 7th January, 1591, A.N., K 1675.

[187] 7th January, 19th January, 1591, *ibid.* From Constantinople, 19th January, 1591, *ibid.* The sultan to the king of France, letter intercepted, Italian copy, January, 1591, *ibid.*; similarly the sultan to the queen of England, *ibid.*

[188] See preceding note.

[189] Dispatches from the Levant, passed on by the ambassador in Venice, A.N., K 1675.

[190] Dispatches from Constantinople, 16th February, 1591, A.N., K 1675.

[191] Fco de Vera to H.M., Venice, 2nd March, 1591, A.N., K 1675.

[192] Fco de Vera to the king, Venice, 16th March, 1591, *ibid.*; dispatches from Constantinople, 16th March, 1591, passed on by the imperial ambassador, *ibid.*

more than 40 to 60 galleys and those in poor trim – or so someone from Pera told Francisco de Vera in Venice.[193] In early May there was even more optimism. It was learned that the Turks, although the season was now advanced, were in no hurry to fit their galleys and did not even seem anxious to have them ready for the next year.[194] Only about twenty sail were expected to leave to guard the Aegean, that is to protect the Turkish trade routes.[195]

However the same dispatch also reported that rumours were still current concerning the three hundred Turkish galleys which might possibly be joined by two hundred English ships.[196] And this report coincided with others: the Constantinople Arsenal is back at work and has had *maestranza* brought in from the Archipelago (this in early March, 1591).[197] In April, the Turks are said to have ordered enormous quantities of linen and hemp from Transylvania. What else could this be for but the sails and rigging of the future armada?[198] In June fresh stir is reported in the Arsenal. Galleys are under construction in the Black Sea. Old ones are being overhauled. Cargoes of sails are arriving in Constantinople.[199] There might be no cause for alarm for the present. But what did the future hold? Although the Turks had sent no expedition of any size in 1591 (only a few galleys had been sent out on guard duty and reports dated 15th June had been received of the return to Constantinople of six weary galleys 'consumidas de hambre' from Barbary)[200] anxiety was growing in Christendom. Mecatti[201] writes that Venice lived in fear of the Turk throughout the summer of 1591. It is certainly true that Venice fitted and prepared a number of galleys and sent troops to Crete.[202]

Such fears are understandable. With Turkey at last relieved of war with Persia, the possibility of her intervention was very real, and indeed the Turks seem to have employed a policy of bluff and blackmail, keeping the Christian powers on tenterhooks and obliging them to take the usual precautions. Perhaps their intention was to hasten the dispatch of another Spanish representative to negotiate a renewal of the truce, bringing with him, for the pashas at any rate, the considerable sums of money which normally accompanied the negotiations. 1591 was a year in which the Spanish truce was due for renewal. We have already noted the presence in Constantinople of Juan de Segni of Minorca, who combined the functions of informer, spy and official representative.[203] A letter from Francisco de Vera[204] refers to another Spanish agent, Galeazzo Bernon (as he is described in the Spanish document). This man, 'que avisa y sirve en

[193] F. de Vera to H.M., Venice, 30th March, 1591, A.N., K 1675.
[194] F. de Vera to H.M., Venice, 4th May, 1591, A.N., K 1675.
[195] 11th May, *ibid.* [196] *Ibid.*
[197] Constantinople, 2nd March, 1591, A.N., K 1675.
[198] F. de Vera to Philip II, Venice, 17th April, 1591, A.N., K 1675.
[199] Constantinople,. 12th June, 1591, *ibid.*
[200] Constantinople, 15th June, 1591, *ibid.* [201] *Op. cit.*, II, p. 785.
[202] J. W. Zinkeisen, *op. cit.*, III, p. 623. [203] See notes 186, 187.
[204] Venice, 8th June, 1591, A.N., K 1675.

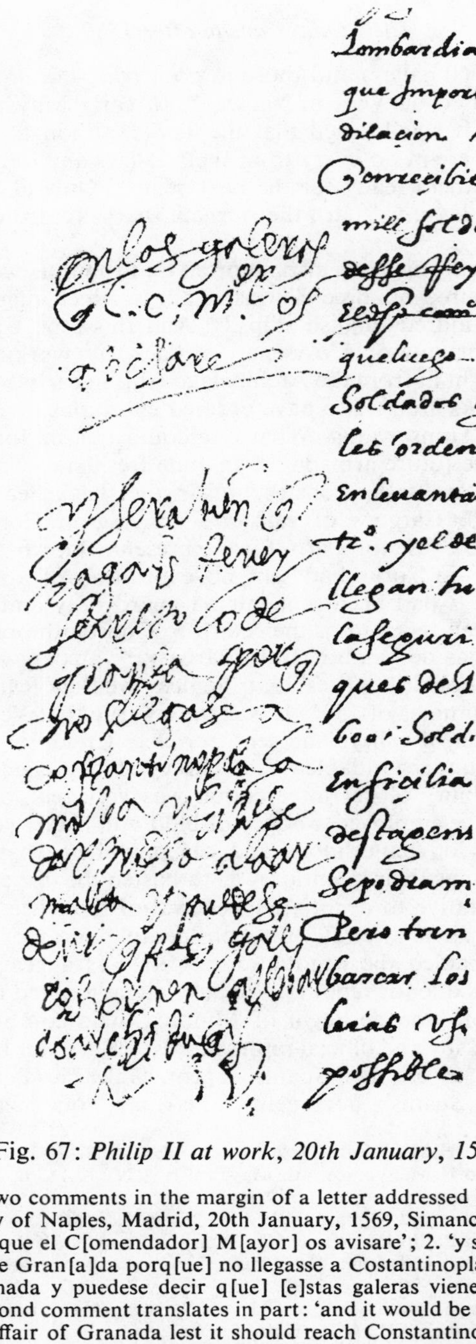

Fig. 67: *Philip II at work, 20th January, 1569*

The king wrote two comments in the margin of a letter addressed by him to the Duke of Alcalá, viceroy of Naples, Madrid, 20th January, 1569, Simancas E° 1057, f° 105: 1. 'en las galeras que el C[omendador] M[ayor] os avisare'; 2. 'y sera bien que hagais tener secreto lo de Gran[a]da porq[ue] no llegasse a Costantinopla la nueva y hiziese dar priesa al armada y puedese decir q[ue] [e]stas galeras vienen a llevar al archi duq[ue]'. The second comment translates in part: 'and it would be as well if you could keep secret the affair of Granada lest it should reach Constantinople and hasten the sortie of the Turkish fleet'. Granada had risen up on Christmas night, 1568. But the news was already common knowledge in the streets of Naples by the time the king's letter arrived.

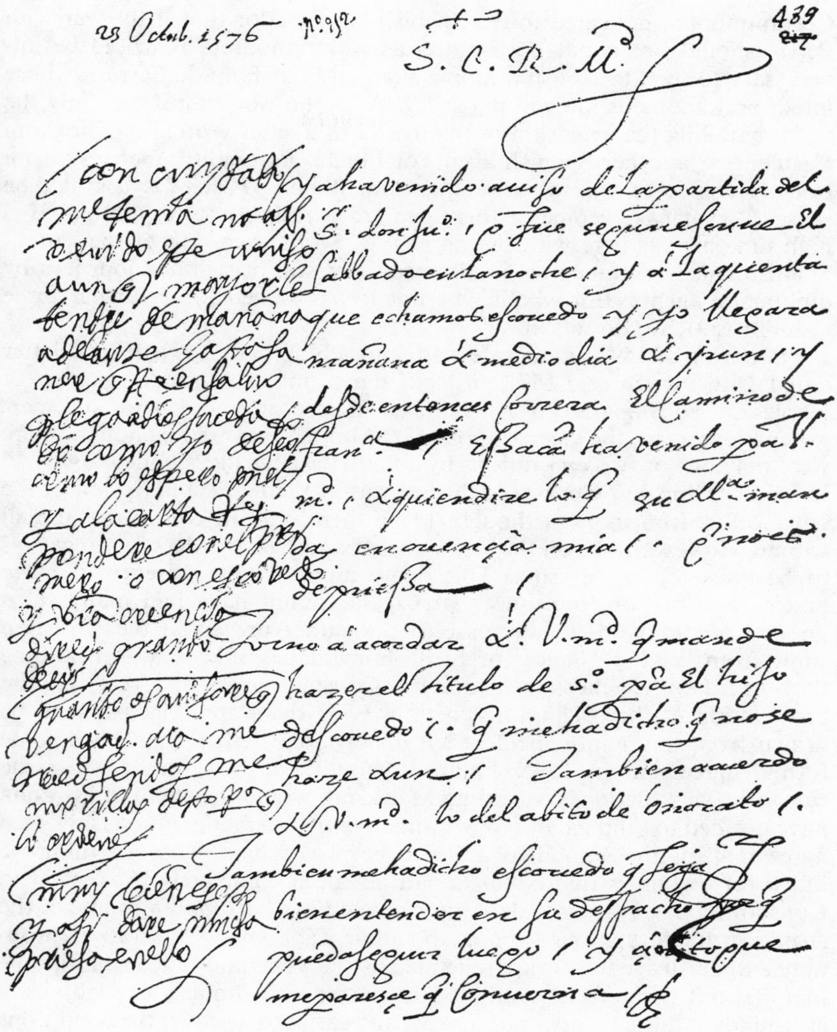

23 Octub. 1576 Nº 952

489

S. C. R. M.

Fig. 68: *Philip II at work, 23rd October, 1576*

This letter written by Antonio Pérez in his own hand to the king on 29th October, 1576, informs the king of Don John of Austria's departure for the Netherlands and contains two requests on behalf of Escovedo. Philip II replies to each paragraph in the margin. To the last (Escovedo seeks permission to go to the Netherlands himself), Philip replies: 'muy bien es esto y asi dare mucha priesa in ello'. Simancas Eº 487. The page has been considerably reduced in size in this photocopy. Escovedo was assassinated on 31st March, 1577. The text is published in facsimile by J. M. Guardia, in his edition of Antonio Pérez, *L'art de gouverner*, Paris, 1869, following p. LIV.

Costantinopla', indicated to the Spanish ambassador that if the war with Persia should break out once more, as was rumoured, it would be unnecessary to send to Constantinople Juan and Estefano de Ferrari 'whose intention to come is known in Turkey'. Was this not almost certainly the agent handling the extension of the truce? In a letter written by Giovanni Casteline,[205] another Spanish agent, of Italian nationality, there is also a brief reference to this obscure question. 'Sinān [Pasha] is asking', he says,[206] 'what has become of this Spaniard who has not yet arrived. It is high time he was here with the money'.[207] So the question of the Turco-Spanish truce continued to be discussed and still occasioned journeys by diplomatic agents. But whether or not it was signed for a further three years in 1591, we do not know.

The spring of 1592 at any rate was calm. Philip had, it is true, in a letter dated 28th November, 1591, ordered the count of Mirānda, viceroy of Naples, to be prepared to assist Malta in case of emergency. No emergency is expected, the viceroy replied.[208] And that year, sure enough, there were only one or two expeditions by a few Turkish galleys under Cigala.[209] In October, the new General of the Sea was in Valona,[210] but in view of the advanced season, he was thought to be there only to collect the annual tribute. However to be on the safe side, Miranda gave orders for 16 galleys to be mustered at Messina. This figure alone, in the absence of more precise information, indicates that Cigala cannot have had many ships under his command. In any case, bad weather prevented the Christian ships from leaving Naples for Sicily immediately and in the meantime they heard of Cigala's departure for Constantinople and received the cancellation of the original instructions.[211] Is there any conclusion to be drawn from this minor incident? Probably that if there had been some formal agreement in 1591, Philip II would not have needed to issue categorical instructions concerning Malta, nor would the count of Miranda have decided to send galleys to Messina. Neither in fact mentions any such agreement, or the amount of faith to be placed in it. The one thing we know for certain is that contacts had not been completely broken off in Constantinople. Francisco de Vera seems to have been pulling some of the strings in the mysterious Lippomano affair. Lippomano[212] who was probably a Spanish agent, was apprehended by the Venetians in Constantinople in 1591, but chose to commit suicide on the way home. The obscurity surrounding this case does not make it any easier to discover the conditions under which negotiations in the Turkish capital were conducted.

[205] Quoted above, see note 200. The agent's name I decipher as 'Castelie' with a ligature over the i, but have not been able to verify this anywhere else.

[206] I quote the chancellery summary on the back of the letter, not the text of the dispatch. [207] '. . . Que venga con dineros'.

[208] Naples, 15th February, 1592, Simancas E° 1093, f° 8.

[209] Miranda to Philip II, Naples, 8th September, 1592, *ibid.*, f° 79.

[210] Miranda to Philip II, Naples, 25th October, 1592, *ibid.*, f° 91.

[211] Miranda to Philip II, Naples, 16th November, 1592, *ibid.*, f° 93.

[212] F^co de Vera to Philip II, Venice, 29th June, 1591, A.N., K 1675.

It is possible that the failure of some such talks was responsible for the arrival out of the blue in 1593 of a Turkish fleet which set about looting the Calabrian coast. Sicily and Naples received warning in time to take the usual security measures. But after a false start for Sicily, about a hundred Turkish ships suddenly appeared off Messina in the Fossa San Giovanni, sacked Reggio and fourteen neighbouring villages,[213] then without inflicting further damage on the coast which was bristling with defences, returned to Valona. Thus began an unofficial war which was to continue virtually uninterrupted, a debased version of the official war between Turkish and Spanish armadas. At some time, possibly in 1595 (the exact date is unknown)[214] the galleys of Sicily and Naples took spectacular revenge, sacking Patras and to even greater effect participating in pirate raids in the East. Alonso de Contreras, who served aboard the galley of the Duke of Maqueda in these profitable privateering expeditions as a soldier whose official pay was three crowns, on one occasion received on his return home 'a hat filled to the brim with reals'.[215] This was petty war, a multitude of individual actions unrecorded by history, sometimes (on the Turkish side) involving Morisco emigrés and frequently concerning restless Calabria, where dispatches report suspicious ships standing off the coast and sailing past at night with lanterns set.[216] But this war was very different from the real thing.

In 1594, Cigala took out a fleet;[217] it cannot have left Constantinople later than July and arrived at Puerto Figueredo on 22nd August.[218] At the news of the Turkish approach, panic seized Sicily,[219] which was at the time unprotected. Indeed, Prince Doria's galleys were still anxiously awaited in Naples on 9th September.[220] If the attack had been pursued it would have found the Spanish defences in considerable disarray. But by mid-September, sufficient confidence had been restored for the militia to be withdrawn from the Sicilian coasts, leaving them protected only by the 'gente ordinaria'.[221] The Turkish fleet, which was said to consist of 90, 100 or even 120 galleys, made an early return to Constantinople.[222] On 8th October, the 2500 Neapolitans originally levied to join the army in

[213] Pietro Giannone, *Istoria civile del Regno di Napoli*, The Hague, 1753, Vol. IV, p. 283, 1593, not 1595 as given in A. Ballesteros y Baretta, *op. cit.*, cf. note 226 below.

[214] See note 226 below.

[215] *Les aventures du capitaine Alonso de Contreras*, trans. and ed. Jacques Boulenger, *op. cit.*, p. 14.

[216] 21st September, 1599, *Archivio storico italiano*, Vol. IX, p. 406.

[217] F^{co} de Vera to Philip II, Venice, 6th August, 1594, Simancas E° 1345 and 20th August, *ibid.*

[218] Carlo d'Avalos to Philip II, Otranto, 25th August, 1594, Simancas E° 1094, f° 89.

[219] Olivares to Philip II, Palermo, 8th September, 1594, Simancas E° 1138. This is when we must date the burning of Reggio and the robbing of several ships off Messina, Carlo Cigala to the count of Olivares, Chios, 3rd November, 1594, Simancas E° 1158 and retrospective statements, 15th January, 1597, Simancas E° 1223, G. Mecatti, *op. cit.*, II, p. 789–790.

[220] Miranda to Philip II, Naples, 9th September, 1594, Simancas E° 1094, f° 99.

[221] Olivares to Philip II, Palermo, 15th September, 1594, Simancas E° 1158.

[222] Miranda to Philip II, Naples, 11th October, 1594, Simancas E° 1094, f° 110.

Lombardy but kept back until now, set off for the North.

In 1595, there were further false alarms. On 31st July, the Turkish armada was at Modon, its usual campaign base and lookout post.[223] Italy immediately took fright. The Spaniards considered massing the galleys at Messina.[224] But secret reports, later confirmed, suggested that the Turk would not depart 'de sus mares'.[225] The Spanish fleet therefore continued as planned with its usual transport operations in the West, while the Turkish threat turned out to be all bark and no bite.[226]

Gian Andrea Doria refuses to fight the Turkish fleet: August–September, 1596. 1596 as we already know was a crucial year for the Turkish army on the battlefields of Hungary, the year of the gruelling battle of Keresztes. The Turks nevertheless took up their usual positions on the Greek coast, with the more reason this year since they had to reckon not only with the Christian West but with the Albanian regions which were in a state of upheaval, as so often before in a century which had rarely found them at peace.[227] This time it looked as if Albania was on the point of revolt. In Rome and Florence, there was talk of attempting a landing. But Venice had too many interests at stake in this possible conflict on her borders; she was too anxious to remain neutral to have any part in such a venture or even to allow it to take shape. Meanwhile the Spaniards were under pressure from the Papacy which wanted to see them challenge the Turkish fleet.[228] In the past, in 1572, Rome had been unable to arrange for a continental war to accompany the naval attack on Turkey. Now, in 1596, it was proving impossible to organize a naval action to coincide with the continental war in Hungary. When requested to intervene during the summer of 1596, Gian Andrea Doria took refuge behind his official instructions. When pressed further, he consulted Philip II, but in terms which showed that he had already made up his sovereign's mind. 'I wrote to Your Majesty on the last day of the past month. I add today that on the 2nd of this month, the galleys of the Grand Duke and those of His Holiness arrived. The first came with intentions which will be seen from the copies of the letters from the Grand Duke which they brought me. I do not know what the president of the Kingdom of Sicily, whom I immediately informed, will decide. His Holiness wants me to go in search of the enemy and give battle, but since the enemy fleet is greatly superior to us in the number of galleys, and since in addition to the troops it is

[223] F^co^ de Vera to Philip II, Venice, 19th August, 1595, Simancas E° 1346.

[224] Miranda to Philip II, Naples, 19th August, 1595, Simancas E° 1094, f° 181.

[225] Miranda to Philip II, Naples, 24th August, 1595, Simancas E° 1094, f° 170.

[226] I take note of A. Ballesteros y Baretta's reference to the sack and looting of Reggio by Cigala in 1595 and Spanish reprisals on Patras, but I have found no mention of these events in the documents I have consulted. Since no source is quoted in support of these claims, I cannot pronounce upon them. See note 219 above.

[227] V. Lamansky, *op. cit.*, p. 493–500.

[228] Which had just arrived at Navarino. Olivares to Philip II, Naples, 24th September, 1596, Simancas E° 1094, f° 258.

carrying it can embark all the men of arms it needs from points along the coast, I did not judge it wise to obey him on this point. Moreover, His Holiness says I should land troops in Albania, in view of intelligence reports he has received from that region. I replied that I had no orders from Your Majesty on this subject, except that I was to remain with his galleys on the coast of Christendom . . .'.[229]

And indeed, Gian Andrea Doria was content merely to send a few contingents of galleys to the Levant as a diversion. Then he calmly settled down in Messina to await developments. He reported to Philip II on 13th August[230] that 'if it was not necessary to face the enemy', he would set out with the entire fleet for Spain, which he did a little later in September.[231] In the same month the Turkish galleys arrived at Navarino, in poor shape, as the Pope had reported. They went no further abroad[232] and at the first bad weather, sailed for home.[233]

1597–1600. The Christians had a few more alarms at the beginning of 1597 when it was reported that 35 or 40 Turkish galleys[234] were about to leave port. For once the report proved true, but perhaps this was merely coincidence, since other, more optimistic reports had suggested that there would be no fleet at all.[235] It was learned in Venice in early August that a fleet had already left Constantinople, one of modest size intended to punish the galleys of Malta whose raids in the Levant 'have woken the sleeping dogs' as the Spanish ambassador in Venice put it.[236] The Spaniards would evidently have preferred to let them lie. The fleet might well be smaller than usual, but that need not prevent it being both as mobile and as effective as the corsair flotillas, causing as much or more damage than a regular armada. Word soon came that it consisted of 30 galleys and four galleons, under the command of Mami Pasha and that it had left Constantinople on 2nd July. Its aim was to attack western privateers but also, if the opportunity arose, to do a little privateering on its own account. War between mighty states was now taking the form of pirate raids. Iñigo de Mendoza began to wonder if it was not the inertia of the Spanish fleet and its evident reluctance to fight which was encouraging the Turks to arms.[237]

However in the event Mami Pasha's fleet was not perhaps as strong a fighting force as the true pirate fleets of Algiers and had not been suitably armed for such an expedition. For in spite of its original plans, it quickly returned to port, incidentally receiving a severe buffeting on the way home.[238]

[229] Gian Andrea Doria to Philip II, Messina, 8th August, 1596, Simancas E° 1346.
[230] Doria to Philip II, Messina, 13th August, 1596, *ibid.*
[231] Olivares to H.M., Naples, 24th September, 1596, Simancas E° 1094, f° 258.
[232] *Ibid.*
[233] Inigo de Mendoza to H.M., Venice, 7th December, 1596, A.N., K 1676.
[234] Mendoza to the king, Venice, 5th April, 1597, A.N., K 1676.
[235] Mendoza to the king, Venice, 14th June and 5th July, 1597, *ibid.*
[236] Mendoza to the king, Venice, 2nd August, 1597, *ibid.*
[237] 9th August, 1597, *ibid.* [238] 18th October, 1597, *ibid.*

1598: and once again there is nothing to report, surprisingly since a Turkish fleet had once again been sent out.[239] It had left Constantinople on 26th July under Cigala.[240] After passing the Seven Castles it had been delayed by shortage of supplies and money, but pressed on despite the plague which was said to have broken out aboard the ships.[241] This fleet numbered 45 galleys, better fitted than those of the previous year, not that that meant very much. Cigala arrived at Zante[242] in September, but made no attack on Christendom presumably because his galleys were in no fit state to be risked on long expeditions.[243] So once more in 1598 the Turkish fleet had been sent out to no purpose and still war was not declared. And still in Constantinople, efforts were being made to negotiate a truce on behalf of Spain, with this time Jewish residents in the Turkish capital as intermediaries.[244]

In 1599, once more all was quiet. In 1600, Cigala sailed with 19 galleys which by the time they reached the Seven Castles had dwindled to ten – the other nine having been stripped to reinforce the rigging and troops of the remainder.[245] In the West the peace was unbroken. It was even suggested in Spain that the galleys be sent to Flanders, in reply to the appeals of the Archduke Albert.[246]

False alarm or missed opportunity in 1601? So it is surprising to find Spain making fresh naval preparations the following year. Had Henri IV's war against Savoy over Saluzzo[247] or the threat to Tuscany drawn Spanish attention back to the Mediterranean? Or was it the need to safeguard the route between Barcelona and Genoa? Or merely the fact that now the war with France was over, the Peninsula had more forces available for the Mediterranean? For whatever reason, in 1601, the Mediterranean saw the deployment of Spanish forces on a scale unknown for years. The whole of Spanish Italy was mobilized.[248] Venice was the more disturbed since German soldiers were marching through her territory, without permission to get to Milan, which was already crammed with troops. Venice naturally armed in turn.[249] The count of Fuentes equally naturally and to no effect attempted to reassure her.[250] This rearmament, the naval and troops movements, quickly brought Italy, which was perhaps over-sensitive to any threat to peace, to the point of a general crisis. What master plan was

[239] Letters dated 14th February, 1598, 14th April, 4th July, 18th July, 8th August, on Cigala's movements, *ibid.* [240] Venice, 29th August, 1598, *ibid.*

[241] Letters, 12th September, 1598, *ibid.* [242] 30th September, 1598, *ibid.*

[243] Maqueda to Philip II, Messina, 28th September, 1598, Simancas E° 1158.

[244] J. von Hammer, *op. cit.*, VII, p. 362 and note 2.

[245] Mendoza to Philip III, Venice, 19th August, 1600, A.N., K 1677.

[246] Archduke Albert to Juan Carillo, Brussels, 14th September, 1600, *Affaires des Pays-Bas*, Vol. VI, p. 33. Henri IV. to Rochepot, Grenoble, 26th September, 1600, *Lettres inédites du roi Henri IV à M. de Villiers*, ed. Eugène Halphen, Paris, 1857, p. 46. [247] Henri IV to Villiers, 27th February, 1601, *ibid.*, p. 12–13.

[248] J. B. de Tassis to Philip III, Paris, 5th March, 1601, A.N., K 1677.

[249] F^co de Vera to H.M., Venice, 31st March, 1601, *ibid.*

[250] F. de Vera to the count of Fuentes, Venice, 14th April, 1601, *ibid.*

the envoy from Ibrāhīm Pasha, Bartolomé Coreysi[251] a renegade from Marseilles, carrying to France and England when he was seen passing through Venice in April on his way to Florence and Leghorn?[252]

Henri IV did not think there would be war: if the count of Fuentes were to disturb Italy, he wrote to M. de Villiers[253] he would be opposed by the Pope, 'without whom the said king would find it very difficult to accomplish anything'. Besides Philip III did not need any more trouble. 'In truth', wrote the French king, again to Villiers, on 16th May, 1601,[254] 'I never thought the Spaniards would want to make war in Italy or anywhere else, since they already have one on their hands in the Netherlands which is a great burden for them and they have hardly any more money than anyone else'. By now in any case, the Italian panic had calmed down. Venice was demobilizing her troops.[255] And the crisis was virtually over when Philip III finally 'swore' the peace of Vervins on 27th May.[256]

But the alert that had first centred on Italy was suddenly to move to the waters of the sea during the summer. Dispatches from Constantinople (on this occasion providing information about the West) refer in mid-June to the preparation of a strong Spanish fleet, to the great disarray of Cigala who was unable to muster more than 30 to 50 galleys to meet the threat, even with the help of the corsairs.[257] France was aware of this concentration of the Spanish fleet. 'The naval force now under preparation in Genoa', wrote Henri IV on 25th June,[258] 'is a threat to the Turkish Empire and intimidates Spain's neighbours, but I hope its bark will be worse than its bite, like all the other actions of the count of Fuentes'. The Turks took a few precautions, sending thirty galleys to the island of Tenedos,[259] at the mouth of the Dardanelles. The king of France did not take the sortie of the Turkish fleet very seriously. 'It is my opinion', he wrote on 15th July, 'that report will outrun reality'.[260]

As for the Christian armada, it seemed to be a more serious threat, but no one knew where it would strike, although Venice claimed to know that it was bound for Albania and intended to capture Castelnuovo.[261] On 5th August, Gian Andrea Doria left Trapani with his ships.[262] In Constantinople, where the peril was exaggerated by report, the tables were turned for once. There were rumours of a Christian fleet of 90 galleys and

[251] Or Coresi, F. de Vera to Philip III, Venice, 21st April, 1601, *ibid.*; he was to leave Venice on 2nd May (F. de Vera to Philip III, 5th May, *ibid.*).

[252] F. de Vera to Philip III, 14th April, 1601, *ibid.*

[253] 24th April, 1601, *Lettres inédites du roi Henri IV à M. de Villiers*, *op. cit.*, p. 19.

[254] *Ibid.*, p. 29.

[255] F. de Vera to Philip III, Venice, 5th May, 1601, A.N., K 1677.

[256] Henri IV to Villiers, 16th May, 1601, *Lettres . . .*, *op. cit.*, p. 26.

[257] Constantinople, 17th and 18th June, 1601, A.N., K 1677.

[258] To Villiers, *Lettres . . .*, *op. cit.*, p. 36.

[259] Constantinople, 2nd and 3rd July, 1601, A.N., K 1677.

[260] To Rochepot, *Lettres*, *op. cit.*, p. 98.

[261] F. de Vera to Philip III, Venice, 28th May, 1601, A.N., K 1677, Henri IV to Villiers, 3rd September, 1601, *Lettres . . .*, *op. cit.*, p. 44–45.

[262] Sessa to Philip III, Rome, 17th August, 1601, A.N., K 1630.

40 galleons.[263] And Cigala, on arriving at Navarino, very prudently remained in port with his 40 galleys.[264]

But there was to be no second Lepanto. Doria's port of departure, Trapani, was sufficient indication that Spain was concerned not with the Levant but with North Africa. The great Spanish fleet had in fact been assembled for an attack on Algiers. It had been hoped to surprise the Barbary port,[265] but once more, weather thwarted all such hopes. This, combined with the commander's distaste for risks obliged the armada to turn about. The French ambassador in Spain[266] filed a report as early as 14th September on the failure of the mission which they 'say was caused by a storm which assailed the fleet only four leagues from the place where it hoped to land, and so scattered and shook the galleys that they were obliged to abandon their plans'. Was this one more to be added to the long list of opportunities to take Algiers missed by the Christians before 1830? So it was thought at Rome, where the Duke of Sessa reported that His Holiness had told him he was 'greatly pained by the disgrace which had befallen the said armada'.[267] The Holy Father thought in particular that the African diversion had prevented a fruitful expedition to the Levant. So at the beginning of the seventeenth century it is curious to find once more the eternal dispute between the Spaniards whose eyes were fixed on North Africa and the Italians who looked to the East.

We might note that even if this expedition had succeeded – and in this respect it reveals the changed climate of the Mediterranean – it would have led merely to a local war. The Spanish fleet would not have encountered any Turkish fleet. Major confrontations between armadas of galleys and galleons would not again be seen in the sea. Overriding events, men's plans and calculations, the powerful tide of history was opposed to the revival of these costly engagements. In its own way, the degeneration of official war was a warning sign of the general decline of the Mediterranean, which, there can be no doubt, was becoming clearer and more apparent with the last years of the sixteenth century.

The death of Philip II, 13th September, 1598.[268] In this account of events on the Mediterranean stage, we have not mentioned in its proper place one piece of news which quickly travelled to every corner of the world: the death of Philip II of Spain, which occurred on 13th September, 1598 at the

[263] Constantinople, 26th and 27th August, 1601, A.N., K 1677.
[264] Constantinople, 8th and 9th September, 1601, *ibid.*
[265] A. d'Aubigné, *op. cit.*, IX, p. 401 ff.
[266] Henri IV to Villiers, Fontainebleau, 27th September, 1601, *Lettres . . ., op. cit.*, p. 48.
[267] Sessa to Philip III, Rome, 6th October, 1601, A.N., K 1630.
[268] The most detailed narrative source is P. de Sepuldeva, *Sucesos del Reinado de Felipe II*, in ed. J. Zarco, *Ciudad de Dios*, CXI to CXIX. *Historia de varios sucesos y de las cosas*, Madrid ed., 1924. Among accounts by contemporary modern historians, see Jean Cassou, *La vie de Philippe II*, Paris, 1929, p. 219 ff, and Louis Bertrand, *Philippe II à l'Escorial*, Paris, 1929, Chap. VII, 'Comment meurt un roi', p. 228 ff.

Escorial, at the end of a long reign that to his adversaries had seemed interminable.

Is this a serious omission? Or did the personal disappearance of the Prudent King from the international scene not really mark any great change in Spanish policy? Vis-à-vis the East, this policy (notwithstanding Doria's attempt on Algiers in 1601) remained prudent to a fault, clearly unwilling to reopen the conflict with the Turks.[269] Spanish agents continued their intrigues in Constantinople, trying to negotiate the impossible, ever-elusive peace treaty and meanwhile adroitly avoiding clashes. Any talk of war referred only to the Barbary corsairs, and this as we have seen could be but a limited war. There was not even any significant change within Spain itself. Forces which had been long at work continued to act. We have already said as much of the so-called reaction of the nobility under the new reign. The whole mood was one of continuity, including, despite the long delays in accomplishing it, the return to peace, which was bound to come after the uncoordinated but strenuous efforts of the last years of the reign of Philip II. The peace of Vervins in 1598 was the work of the late king; peace with England followed six years later (1604) and peace with the United Provinces had to wait another eleven years (1609). But both were the result of a trend which had long been under way.

There is perhaps nothing so revealing of the enigmatic figure of Philip II as his manner of dying: his exemplary death has been described so often and so touchingly that one hesitates to repeat the moving details. It was undoubtedly the death of a king and of a Christian, singularly assured of the virtue of the intercessionary powers of the Church.

After the first painful attacks of his illness in June, the king ordered, against the advice of his doctors, that he be carried to the Escorial to die. He continued to struggle nevertheless against the septicaemia which finally killed him after fifty-three days of sickness and suffering. His death was not in any way marked by pride, that symbol of the century of reformation.[270] The king did not come to the Escorial in order to die alone; he was returning to his own people, his buried kinsmen waiting for him to join them, and he came accompanied by his son, the future Philip III, his daughter the Infanta who would later leave for Flanders, by the dignitaries of the Church and the dignitaries of this world who were all to attend him throughout his long agony. It was as public a death, as social and ceremonious (in the best sense of the word) as it is possible to imagine. This

[269] The presence of Juan de Segni de Menorca in Constantinople we know from one of his letters to Philip II, written 3rd November, 1597, A.N., K 1676. Just before the Thirty Years War there was an attempt by the Imperials to free Spain for good from this burden or rather from the constant threat, action over Baron Mollart. In 1623, negotiations were being handled by Giovanni Battista Montalbano, there was a suggestion of a lasting peace with the Turks and for the diversion of the spice trade through the Middle East, with the help of Poland. Cf. H. Watjen, *Die Nederlander . . .*, *op. cit.*, p. 67–69.

[270] Jean Cassou, *op. cit.*, p. 228.

deathbed was attended not as has been said, by 'Pride, Solitude and Imagination', but by the royal family, the army of the Saints and by the universal prayers which accompanied his last moments, in an orderly procession which in its own way was a work of art. This man, of whom it has often been said that his whole life consisted of distinguishing the temporal from the spiritual, whom his enemies shamelessly slandered with the most absurd calumnies, and on whom his admirers were perhaps a little too ready to confer a halo, can only be understood in relation to a life of the purest religion, perhaps only in the atmosphere of the Carmelite revolution.

But what of the sovereign, the force of history symbolized by his name? This is a very different matter from the solitary and secretive man himself. To historians he is an enigma: he receives us as he did his ambassadors, with the utmost courtesy, listening to us, replying in a low and often unintelligible voice, never speaking of himself at all. For three whole days just before he died, he confessed the errors of his lifetime. But who could truly imagine these errors, numbered before the tribunal of a conscience whose judgements may or may not have been just, as it searched the recesses of a long life? Here lies one of the great mysteries of his life, the shadow which if we are truthful we must leave across his portrait. Or rather his many portraits. What man does not change in the course of his life? And Philip's life was a long and disturbed one, from the painting by Titian of the prince in his twentieth year to the terrible and moving portrait by Pantoja de la Cruz which shows us the king at the end of his reign, the shadow of what he once had been.

The self who is most accessible to us is the sovereign, applying himself to the work of a monarch, sitting at the centre, the intersection of the endless reports which combined to weave the great web of the Spanish Empire and the world. This is the man who sits silently reading at his desk, annotating reports in his hasty handwriting, far from other men, distant and pensive, linked by the threads of information to that living history which makes its way towards him from every corner of the world. He is the sum of all the weaknesses and all the strengths of his empire, he alone sees the balance sheet. His deputies, first Alva then Farnese in the Netherlands, Don John in the Mediterranean, could see only one section, their personal sphere of action in the vast enterprise. The king was to them as the conductor is to the players in the orchestra.

He was not a man of vision: he saw his task as an unending succession of small details. Every one of his annotations is a small precise point, whether an order, a remark or the correction of a spelling mistake or geographical error. Never do we find general notions or grand strategies under his pen. I do not believe that the word Mediterranean itself ever floated in his consciousness with the meaning we now give it, nor that it conjured up for him the images of light and blue water it has for us; nor even that it signified a precise area of major problems or the setting for a clearly conceived policy. Geography in the true sense was not part of a prince's education. These are all sufficient reasons why the long agony which ended in September, 1598 was not a great event in Mediterranean history; good reasons for us to reflect once more on the distance separating biographical history from the history of structures, and even more from the history of geographical areas.

Conclusion

This book has been in circulation now for almost twenty years: it has been quoted, challenged, criticized (too seldom) and praised (too often). In that time I have frequently had occasion to elaborate its explanations, to defend its arguments, to reflect upon the attitudes it embodies, and to correct its mistakes. I have now re-read it thoroughly in order to prepare the second edition and have made very extensive alterations. But inevitably a book has its own career, its own existence, independent of its author. One can improve it here and there, embellish it with a whole apparatus of footnotes and details, maps and illustrations, but never radically change its outline. Not uncommonly in the sixteenth century, a ship purchased elsewhere would be brought to Venice, where it was overhauled from stem to stern and refitted by skilful carpenters; underneath it was still the same ship, built in the yards of Dalmatia or Holland, and recognizable at a glance.

So in spite of the lengthy process of revision, readers of the first edition of this book will have little difficulty in recognizing it here. Conclusion, message and meaning have remained the same. It is the result of research into a very considerable number of previously unpublished documents from every corner of the broad Mediterranean stage during these ambiguous years at the threshold of modern times. But it also represents an attempt to write a new kind of history, *total history*, written in three different registers, on three different levels, perhaps best described as three different conceptions of time, the writer's aim being to bring together in all their multiplicity the different measures of time past, to acquaint the reader with their coexistence, their conflicts and contradictions, and the richness of experience they hold. My favourite vision of history is as a song for many voices – but it has the obvious disadvantage that some will drown others: reality will not always adapt conveniently into a harmonized setting for solo and chorus. How then can we consider even one single moment in time and perceive simultaneously, as though in transparent layers, all the different histories that coexist in reality? I have tried to meet this problem by using certain words and explanations as key themes, motifs which reappear in all three sections of the book. But the worst of it is that there are not merely two or three measures of time, there are dozens, each of them attached to a particular history. Only the sum of these measures, brought together by the human sciences (turned retrospectively to account on the historian's behalf) can give us that total history whose image it is so difficult to reconstitute in its rich entirety.

I

None of my critics has reproached me for including in this historical work the very extended geographical section which opens it, my homage to those timeless realities whose images recur throughout the whole book, from the first page to the last. The Mediterranean as a unit, with its creative space, the amazing freedom of its sea-routes (its automatic free trade as Ernest Labrousse called it) with its many regions, so different yet so alike, its cities born of movement, its complementary populations, its congenital enmities, is the unceasing work of human hands; but those hands have had to build with unpromising material, a natural environment far from fertile and often cruel, one that has imposed its own longlasting limitations and obstacles. All civilization can be defined as a struggle, a creative battle against the odds: the civilizations of the Mediterranean basin have wrestled with many often visible obstacles, using sometimes inadequate human resources, they have fought endlessly and blindly against the continental masses which hold the inland sea in their grip, and have even had to contend with the vast expanses of the Indian and Atlantic Oceans.

I have therefore sought out, within the framework of a geographical study, those local, permanent, unchanging and much repeated features which are the 'constants' of Mediterranean history; the reader will not find here all the unspectacular structures and recurrent patterns of life in the past, but the most important of them and those which most affect everyday existence. These provide the reference grid as it were, the most easily recognizable part of the book as well as its most haunting images – and one has not far to search for more of the same. They are to be found unchanged in Mediterranean life today: one may stumble across them in a journey, or in the books of Gabriel Audisio, Jean Giono, Carlo Levi, Lawrence Durrell or André Chamson. All western writers who have at some time in their lives encountered the Mediterranean, have been struck with its historical or rather timeless character. Like Audisio and Durrell, I believe that antiquity lives on round today's Mediterranean shores. In Rhodes or Cyprus, 'Ulysses can only be ratified as an historical figure with the help of the fishermen who to-day sit in the smoky tavern of *The Dragon* playing cards and waiting for the wind to change.' I also believe, with Carlo Levi, that the wild countryside which is the true subject of his evocative book *Christ stopped at Eboli*, takes us back into the mists of time. Eboli (from which Ruy Gómez took his princely title) is on the coast near Salerno, at the point where the road turns inland towards the mountains. Christ (in other words civilization, the rule of law, the gentle arts of living) never reached the highlands of Lucania, and the village of Gagliano, crouching among the barren treeless slopes 'above the cliffs of white clay'. Here the poor *cafoni* are ruled now as in the past by the privileged men of the time: today these are the pharmacist, the doctor, the schoolteacher, people whom the peasant avoids, fears and shies away from. Here vendettas, brigandage, a primitive economy and tools are still the rule. An emigrant may return to his almost deserted village from America laden with the strange new gadgets of the outside world: but he will never change the way of life in this isolated, archaic little universe. This is the deep bone-structure of the Mediterranean and only with the eye of the geographer (or the traveller or novelist) can one truly discern its rugged contours and oppressive reality.

II

The second undertaking of this book – to discover the collective destiny of the Mediterranean in the sixteenth century, its 'social' history in the fullest sense of the word, immediately and continuously brings us face to face with the insidious and still unsolved problem of the decline of material existence, of the decadence one after another of Turkey, the whole of Islam, Italy and the Iberian supremacy, as older historians would say or, as today's economists would put it, of the malfunctioning and collapse of its vital sectors (public finance, investment, industry, shipping). There have always been historians, whether of German inspiration or not – the last to date being perhaps Erich Weber[1] a disciple of Othmar Spann and his universalist school – who have argued that in any unit there is an inbuilt process of decline, of which the destiny of the Roman empire provides the perfect example. Amongst other laws, all decline (*Verfall*) according to Weber, is compensated for by some simultaneous advance elsewhere (*Aufstieg*); in the life of mankind, it would seem, nothing is ever created or destroyed. Equally categorical theories have been advanced by Toynbee and Spengler. I have made it my business to challenge these to my mind over-simple theories and the sweeping explanations they imply. How could the course of Mediterranean destiny easily fit into any of these classifications? There is surely no such thing as a model of decadence. A new model has to be built from the basic structures of every particular case.

Whatever meaning one attributes to the very imprecise term decadence, I cannot accept the view that the Mediterranean fell an easy and resigned victim to a vast, irreversible and above all early process of decline. In 1949 I said that I could not detect any visible decline before 1620. Today I would be inclined to say, though without any guarantee of certainty, that 1650 is a likelier date. At any rate the three most significant books on the fortunes of the Mediterranean regions published in the last ten years – René Baehrel on Provence, E. Le Roy Ladurie on Languedoc and Pierre Vilar on Catalonia – do not contradict my original thesis. To my mind, if one were to reconstruct the new panorama of the Mediterranean after the great divide marking the end of its prosperous youth, one would have to choose a date as late as 1650 or even 1680.

[1] *Beiträge zum Problem des Wirtschaftverfalles*, 1934.

One would also, as local research permits more precision, have to proceed further with the calculations, estimates and pursuit of orders of magnitude to be found in this book, which have brought me nearer than my very imperfect essays in this direction might appear to suggest to the thought of economists whose chief concerns are the problems of growth and national accounting (in France for example François Perroux, Jean Fourastié and Jean Marczewski). As we advance along this path, one thing will become strikingly evident: the Mediterranean in the sixteenth century was overwhelmingly a world of peasants, of tenant farmers and landowners; crops and harvest were the vital matters of this world and anything else was superstructure, the result of accumulation and of unnatural diversion towards the towns. Peasants and crops, in other words food supplies and the size of the population, silently determined the destiny of the age. In both the long and the short term, agricultural life was all-important. Could it support the burden of increasing population and the luxury of an urban civilization so dazzling that it has blinded us to other things? For each succeeding generation this was the pressing problem of every day. Beside it, the rest seems to dwindle into insignificance.

In Italy for instance, as the sixteenth century came to an end, a great deal of investment went into land. I hesitate to label this a sign of premature decline: rather it was a healthy reaction. In Italy a precious equilibrium was being preserved – material equilibrium that is, for in social terms, the development of large privately-owned estates everywhere brought its particular forms of oppression and long-lasting constraints. The same was true of Castile.[2] Nowadays, historians tell us that material equilibrium was maintained in Castile until the mid-seventeenth century. This must affect our former assumptions. I had previously supposed that the short but violent crisis of the 1580s could be explained entirely by Spain's new reorientation towards Portugal and the Atlantic. This, it seems was too 'noble' an explanation! Felipe Ruiz Martín[3] has shown that it was simply the result of a process arising basically from the great grain crisis in the Iberian Peninsula in the 1580s. This was what Labrousse would describe as an 'ancien régime' type of crisis.

In short, even in the investigation of short-term crises, we are often obliged to look to structural history for an answer. This is the sea-level by which we must measure everything else, the achievement of progressive cities for example (in 1949 I was perhaps over-impressed by them – civilization can be a blinding spectacle) but also the short-term historical phenomena which we sometimes tend to explain away too promptly, as if their shallow eddies were responsible for the deeper currents of history instead of the other way about. The fact is that economic history has to be rethought step by step, starting from these slow, sometimes motionless

[2] Work under preparation.

[3] Felipe Ruiz Martín in *Anales de Economia, segunda época*, July–September, 1964, p. 685–686.

currents. Still waters run deep and we should not be misled by surface flurries.

At all events, neither the reversal of the secular trend in the 1590s, nor the sharp shock of the brief crisis of 1619–1621, marked the end of Mediterranean splendour. Nor can I accept, without further evidence, that there was a catastrophic discrepancy between the 'classic' conjunctures' obtaining in northern and southern Europe, a discrepancy which if it really existed would have sounded both the death knell of Mediterranean prosperity and the summons to supremacy of the Northerners. This solution conveniently kills two birds with one stone, but I see no corpses.

The division of history into the slow- and fast-moving levels, structure and conjuncture, remains at the centre of a still unresolved debate. We have to classify these movements in relation to each other, without knowing before we start whether one type has governed the other or vice versa. Identification, classification and comparison must be our first concern then. Unfortunately, it is not yet possible to trace the general fluctuations of 'national incomes' in the sixteenth and seventeenth centuries and this is a grave drawback. But we are at least in a position now to study the changing fortunes of the cities, thanks to the work of Gilles Caster[4] on Toulouse, and of Carlo Cipolla and Giuseppe Aleati on Pavia.[5] The complex economic life of the cities provides us with an indicator of economic change at least as reliable as the familiar wage and price curves.

Finally we have the problem of reconciling contradictory chronologies. How for instance did short-term changes in the economic climate affect states or whole civilizations, prominent actors on the historical stage with a will of their own? It appears from our study of the states that difficult times tended to favour their relative expansion. Was the same true of civilizations? Their brightest flowerings often seem to thrive under lowering skies. It was during the autumn of the city state or even (in Venice and Bologna) its winter, that there flourished the last blossoms of an Italian Renaissance. And it was in the waning years of the great maritime empires, whether of Istanbul, Rome or Madrid, that the mighty imperial civilizations set out on their conquering path. As the sixteenth century ended and the seventeenth began, these brilliant shadows floated where great bodies politic had stood fifty years before.

III

Alongside such problems, the role of the individual and the event necessarily dwindles; it is a mere matter of perspective – but are we right to take so Olympian a view? 'Under the formal pageant of events which we have dignified by our interest, the land changes very little, and the structure

[4] *Op. cit.*, p. 382 ff.

[5] 'Il trend economico nello stato di Milano durante i Secoli XVI e XVII. Il caso di Pavia', in *Bollettino della Società Pavese di Storia Patria*, 1950.

of the basic self of man hardly at all,' to quote a twentieth-century novelist, Durrell, who is, like me, passionately attached to the Mediterranean. Well yes, but the question is frequently put to me, both by historians and philosophers, if we view history from such a distance, what becomes of man, his role in history, his freedom of action? And indeed, as the philosopher François Bastide once objected to me, since history is always a matter of development, of progression, could we not say that a secular trend is an 'event' too? I grant the point, but my definition of the 'event' is closer to that of Paul Lacombe and François Simiand: the pieces of flotsam I have combed from the historical ocean and chosen to call 'events' are those essentially *ephemeral* yet moving occurrences, the 'headlines' of the past.

That is not to say that this brilliant surface is of no value to the historian, nor that historical reconstruction cannot perfectly well take this micro-history as its starting-point. Micro-sociology, which it calls to mind (wrongly, I believe), is after all not generally frowned upon in the world of scholarship. Micro-sociology, it is true, consists of constant repetition, while the micro-history of events consists of the unusual, the outstanding, and atypical; a series in fact of 'socio-dramas'. But Benedetto Croce has argued, not without reason, that any single event – let us say the assassination of Henri IV in 1610, or to take an example outside our period the arrival in power in 1883 of the Jules Ferry government – contains in embryo the entire history of mankind. To put it another way, history is the keyboard on which these individual notes are sounded.

That having been said, I confess that, not being a philosopher, I am reluctant to dwell for long on questions concerning the importance of events and of individual freedom, which have been put to me so many times in the past and no doubt will be in the future. How are we to interpret the very word freedom, which has meant so many different things, never signifying the same from one century to another? We should at least distinguish between the freedom of groups, that is of economic and social units, and that of individuals. What exactly is the freedom today of the unit we call France? What was Spain's 'freedom' in 1571, in the sense of the courses open to her? What degree of freedom was possessed by Philip II, or by Don John of Austria as he rode at anchor among his ships, allies and troops? Each of these so-called freedoms seems to me to resemble a tiny island, almost a prison.

By stating the narrowness of the limits of action, is one denying the role of the individual in history? I think not. One may only have the choice between striking two or three blows: the question still arises: will one be able to strike them at all? To strike them effectively? To do so in the knowledge that only this range of choices is open to one? I would conclude with the paradox that the true man of action is he who can measure most nearly the constraints upon him, who chooses to remain within them and even to take advantage of the weight of the inevitable, exerting his own pressure in the same direction. All efforts against the prevailing tide of history – which is not always obvious – are doomed to failure.

So when I think of the individual, I am always inclined to see him imprisoned within a destiny in which he himself has little hand, fixed in a landscape in which the infinite perspectives of the long term stretch into the distance both behind him and before. In historical analysis as I see it, rightly or wrongly, the long run always wins in the end. Annihilating innumerable events – all those which cannot be accommodated in the main ongoing current and which are therefore ruthlessly swept to one side – it indubitably limits both the freedom of the individual and even the role of chance. I am by temperament a 'structuralist', little tempted by the event, or even by the short-term conjuncture which is after all merely a grouping of events in the same area. But the historian's 'structuralism' has nothing to do with the approach which under the same name is at present causing some confusion in the other human sciences.[6] It does not tend towards the mathematical abstraction of relations expressed as functions, but instead towards the very sources of life in its most concrete, everyday, indestructible and anonymously human expression.

26th June, 1965

[6] Cf. Jean Viet, *Les méthodes structuralistes dans les sciences sociales*, 1965.

Sources

I. UNPUBLISHED SOURCES

This book was written primarily from manuscript sources. The plentiful published literature on the subject was studied only after research in the archives and then for the purposes of confirmation, correlation of results and complementary evidence.

But my investigations, in past years and more recently, have by no means exhausted the endless stocks of unpublished material lying in the archives. I have uncovered only a fraction of the documents relating to the enormous subject I had chosen.

While I have explored almost all the Parisian collections (in particular the precious K series in the Archives Nationales which has now been returned to Spain) and studied virtually all the accessible documents in the political series at Simancas (a mine of information), delved deeply into the major Italian collections and gained access to the precious information contained in the Archives of Ragusa, my efforts have not always been rewarded. They were unequal to the impossibly dispersed Archives of Venice and to the enormous bulk of notarial archives in Marseilles, Spain and Italy; nor could I fully explore the superabundance of documents in Genoa, Florence and Turin, not to mention Modena, Naples and Palermo.

The largest gap in my research occurs in the southern and eastern part of the Mediterranean, the great unknown of all studies of the sixteenth century. There are in Turkey magnificent archives of great richness whose documents refer to no less than half the Mediterranean region. But as yet these precious archives are unclassified and access to them is not easy. For the area they cover, we have been obliged to turn to historical literature, travellers' tales, to the archives and published studies of the Balkan countries – and above all to western sources. The history of Turkey and the countries under Turkish influence has therefore been approached from outside, deduced rather than ascertained from the random evidence of the dispatches written by westerners from the Levant, now forming long series in the Italian and Spanish archives. But as any historian knows, there can be a wide gap between the history of a country seen from the outside (France, for instance, as seen through the *Relazioni* of the Venetian ambassadors, which have been so often quoted as to be commonplace in history books) and the history of the same country illumined from inside. There is, then, a tremendous lacuna in historical knowledge of the Turkish countries. For this book such a gap is a weakness of some consequence and I can only appeal to all Turkish, Balkan,

Syrian, Egyptian and North-African historians to help us fill in the spaces and cooperate in a task which requires lengthy collective research. Substantial progress has already been made by Ömer Lutfi Barkan and his pupils, by Robert Mantran, Glisa Elezovic, Bistra A. Cvetkova, Vera P. Mutafčieva, L. Feketé and Gy Kàldy-Nagy, the last two the Hungarian co-authors of the excellent *Rechnungsbücher türkischer Finanzstellen in Buda, 1550–1580*, 1962.

On the Turkish archives (in the broad sense) there is some information now somewhat dated, in *Histoire et Historiens*, Paris, 1930, contributed by J. Deny; see also by the same specialist, 'A propos du fonds arabo-turc des Archives du Gouvernement-Général de l'Algérie', *Revue Africaine*, 1921, p. 375–378. On Turkish archives in general, see P. Vittek, *Les archives de Turquie*, *Byz.*, vol. XII, 1938, p. 691–699 and a guide to the archives in Turkish: *Topkapi, Sarayi Müzesi Archivi Kilavuzu*, Istanbul. On the Egyptian archives, see J. Deny, *Sommaire des archives turques du Caire*, 1930.

A useful summary of the problems of eastern research is to be found in J. Sauvaget, *Introduction à l'histoire de l'Orient musulman*, 1943.

ABBREVIATIONS OF ARCHIVE COLLECTIONS

A.C.	Archives Communales
A. Dép.	Archives Départementales
A.d.S.	Archivio di Stato
A.E.	Affaires Étrangères, Paris
A.H.N.	Archivo Histórico Nacional, Madrid
A.N. K.	Archives Nationales, Paris, series K
B.M.	British Museum, London
B.N.	Bibliothèque Nationale, Paris
B.N. F.	Biblioteca Nazionale, Florence
B.N. M.	Biblioteca Nacional, Madrid
G.G.A.	Former Government-General Building, Algiers
P.R.O.	Public Record Office, London
Sim.	Simancas
Sim. E°	Simancas, series Estado

THE SPANISH ARCHIVES

ARCHIVO GENERAL DE SIMANCAS: The old guide by Mariano Alcocer Martinez, *Archivo General de Simancas, Guía del investigador*, Valladolid, 1923, has now been superceded by Angel Plaza's recent *Guía del investigador*, 1961. There are also a number of specialized catalogues for which the researcher is indebted to the magnificent work of the former archivist Julián Paz. These catalogues contain summaries of documents or at the very least detailed titles, and I have often used them as sources in themselves. Among Paz's catalogues I would especially mention *Diversos de Castilla* published by *Revista de Archivos, Bibliotecas y Museos*, 1909; *Negociaciones con Alemania*, published by the *Kaiserliche Akademie der*

Wissenschaften, Vienna; *Negociaciones con Francia*, published by the *Junta para ampliación de estudios* (includes what was at one time the K series in the French *Archives Nationales*, since 1943 back in Simancas); *Negociaciones de Flandes, Holanda y Bruselas 1506–1795*, Paris, Champion, 1912; *Patronato Real*, 1912, *Revista de Archivos, Bibliotecas y Museos*, 1912. With the exception of *Diversos de Castilla*, there are recent editions of these valuable catalogues (1942–1946). From the archivist Angel Plaza we have *Consejo y Juntas de hacienda* (1404–1707); from the archivist G. Ortiz de Montalban, *Negociaciones de Roma* (1381–1700) 1936; from the present director of the Simancas archives, Ricardo Magdaleno, *Nápoles*, 1942; *Inglaterra*, 1947; *Sicilia*, 1951; and from C. Alvarez Teran, *Guerra y Marina* (in the time of Charles V), 1949.

My investigations at Simancas were further extended by prolonged scrutiny of microfilms. But there still exists in these vast archives a wealth of unexplored material, particularly financial, economic and administrative documents. In the early stages of my research I had access only to a few files, concerning the export of wool to Italy, and was particularly sorry that documents relating to the pepper trade could not be traced for me. Even today many of these documents are still unclassified, so numerous are they, and are bound to remain so for some time.

During my later visits to Simancas (in 1951 and 1954) I made frequent use of the following important series (see no. 12 below).

Files consulted at Simancas

1. *Corona de Castilla* Correspondencia (1558–1597) legajos 137–179
2. *Corona de Aragón*, Correspondencia (1559–1597), legajos 326–343
3. *Costas de Africa y Levante*, a series not entirely in order, 1559–1594, legajos 485–492
4. *Negociaciones de Nápoles* (1558–1595), legajos 1049–1094
5. *Negociaciones de Sicilia* (1559–1598), legajos 1124–1158
6. *Negociaciones de Milán* (1559 and 1597–1598), 3 legajos, 1210 (1559), 1283–1285 (1597–1598)
7. *Negociaciones de Venecia* (1559–1596), legajos 1323 to 1346 (much of the Venetian correspondence was formerly in Paris)
8. *Negociaciones con Génova* (1559–1565), legajos 1388–1394
9. *Negociaciones de Toscana*, very incomplete, legajos 1446–1450
10. *Nápoles Secretarias Provinciales* (1560–1599), legajos 1–8 (Some documents published by Mariano Alcocer), legajo 80 (1588–1599), composed of original decrees
11. *Contaduria de rentas*. Saca de Lanas. Lanas, 1573–1613
12. *Consejo y Juntas de Hacienda; Contaduria Mayor de Cuentas* (Primera epoca – Segunda epoca); *Expedientes de Hacienda*

ARCHIVO HISTÓRICO NACIONAL DE MADRID: I was unable to make use of *Confederación entre Felipe II y los Turcos* (*Guía*, p. 40) since the only classified documents are those of the seventeenth century.

I therefore confined my research to the papers of the Inquisition, that is to what is left of these immense archives. Inquisition of Barcelona (libro I); of Valencia (libro I); of Calahorra and Logroño (legajo 2220); of the Canary Islands (legajos 2363–2365); of Córdoba (legajo 2392); of Granada (legajos 2602–2604); of Llerena (legajo 2700) of Murcia (legajo 2796); of Seville (legajo 2942); and of Valladolid (legajo 3189).

BIBLIOTECA NACIONAL DE MADRID (Manuscript Section): This is a very heterogenous collection. Its total contribution is modest. I would just say here that Sánchez Alonso, in his classic *Fuentes de la Historia española e hispano-americana*, Madrid, 1927, gives a numbered record of the scattered documents stored here. I chose not to spend time on the very numerous copies of documents (notably those for which Antonio Pérez was responsible) which have been published or commented on elsewhere. The following documents were consulted:

Memorial que un soldado dio al Rey Felipe II porque en el Consejo no querian hacer mercedes a los que se havian perdido en los Gelves y fuerte de la Goleta (undated), G. 52, 1750.

Instrucción de Felipe II para el secretario de estado Gonzalo Pérez, undated, G. 159–988, f° 12.

Memorial que se dio a los teologos de parte de S.M. sobre differencias con Paulo IV, 1556, KK. 66 V 10819 22.

Carta de Pio V a Felipe II sobre los males de la Cristiandad y daños que el Turco hacia en Alemania, Rome, 8th July, 1566, G. 52, 1750.

Relación del suceso de la jornada del Rio de Tituan que D. Alvaro de Bazan, Capitan General de las galeras de S.M. hizo por su mandado la qual se hizo en la manera siguiente (1565), G. 52, 1750, published (rather disconcertingly) by Cat in *Mission bibliographique en Espagne*, Paris, 1891.

Carta de Felipe II al principe de Melito sobre las prevenciones que deben hacerae para la defensa de Cataluña, 20th March, 1570, 476.

Capitulos de la liga entre S. Santidad Pio V, el Rey católico y la Señoria de Venecia contra el Turco, 1571, *ibid.*

Las causas que movieron el Sr D. Juan de Austria para dar la batalla de Lepanto, 1571, *ibid.*

Advertencias que Felipe II hizo al Sr Covarrubias cuando le eligió Presidente del Consejo, 1572 (eighteenth century copy), 140–11261 b.

Carta de D. Juan de Austria a D. Juan de Zuñiga, embajador de Felipe II en Roma sobre la paz entre Turcos y Venecianos y sobre los aprestos militares hechos en Italia para ir a Corfu contra los Turchos, Messina, April (1573), KK. 39, 10454, f° 1080.

D. J. de Zuñiga to D. John of Austria, Rome, 7th April, 1573, *ibid.*, f° 1070.

Instrucción al Cardenal Granvela sobre los particulares que el legado ha tratado de la juridición del Reyno de Nápoles (in Italian), seventeenth century copy, 8870.

Representación hecha al Señor Felipe II por el licienciado Bustos de

Nelegas en el año 1570 (on the transfer of church property), eighteenth century copy, 3705.

Correspondencia de D. Juan de Gurrea, gobernador de Aragón con S.M. el Rey Felipe II (December, 1561–September, 1566), sixteenth century copy, V 12.

Instrucción de Felipe II al Consejo Supremo de Italia 1579, E 17, 988, f° 150.

ACADEMIA DE LA HISTORIA (Manuscript Section): The following list refers to documents concerning Algiers, the *presidios* and North Africa, read for me by D. Miguel Bordonau. References are to the catalogue by Rodriguez Villa.

Relación del estado de la ciudad de Argel en 1600 por Fr. Antonio de Castañeda $\frac{(12\text{–}11\text{–}4)}{111}$ (f° 21 v°).

Berberia siglo XVI (11–4–4–8).

Instrucción original dada por Felipe II a D. Juan de Austria sobre los fuertes de Berberia y socorro de Venecianos, año 1575, *ibid.*

Carta del conde de Alcaudete a S.M., año 1559, *ibid.*

Relación de Fr. Jeronimo Gracia de la Madre de Dios sobre cosas de Berberia, año 1602, *ibid.*

Carta de Joanetin Mostara al Duque de Medina Sidonia con noticias de Fez, año 1605, *ibid.*

Carta original del Duque de Medina Sidonia a Felipe III con noticia de la muerte del Jarife, año 1603, *ibid.* (f° 42 v°).

Relación del estado de la ciudad de Barberia (*sic*) escrita en italiano, año 1607 con 2 planos, *ibid.*, f° 45.

Sobre asuntos de Africa en tiempos del Emperador, años 1529–1535. A 44 (f° 55).

Cartas y documentos sobre la conquista de Oran par Cisneros. Letra del siglo XVII 11–2–1 – 11 (f° 65 v°).

Documentos y cartas dirigidas o emanadas de Felipe II, 12–25–5 = C – 96.

Papeles originales sobre gobierno de Oran. A. 1632–1651, 12–5–1 – K. 63–66–65, 3 volumenes (f° 102 v°).

Sobre Oran, Berberia, Larache . . . comienços del siglo XVII. 11–4–4 = 8. Instrucción a D. Juan de Austria sobre cosas de Argel, año 1573 11–4–4 (f° 172 v°).

Relación de las dos entradas hechas en 1613 por el Duque de Osuna en Berberia y Levante, *ibid.* (f° 176 v°). Aviso sobre Argel, año 1560, Colección Velásquez, tomo 75 (f° 242 v°).

OTHER SPANISH COLLECTIONS CONSULTED OR MENTIONED: Communal Archives of Valladolid, Málaga, Tarragona (Archivo diocesano), Barcelona, Valencia (on recommendation of Earl J. Hamilton), Cartagena, Burgos, Medina del Campo (Simon Ruíz Hospital). The collection last mentioned, containing the correspondence of Simon Ruíz, is now in the Archivo Provincial of Valladolid and perfectly in order.

THE FRENCH ARCHIVES

Paris

ARCHIVES NATIONALES: For the K series, I have followed the order of Paz's catalogue (roughly chronological). All the documents are now filed in Simancas, under the same numbers. The Archives nationales in Paris has a complete microfilm of the series which is easy to consult.
K 1643.
K 1447–1448–1449–1450–1451–1426.
K 1487–1488–1489–1490–1491–1492.
K 1493 to 1603; K 1689 to 1707–1629–1708.
K. 1692. Correspondence of the marquis of Villafranca, viceroy of Naples, 1534–1536; K 1633; K 1672 to 1679; letters from Venice, K 1630–1631; letters from Rome, 1592–1601.

The papers of the Mission Tiran, AB XIX 596 are of limited interest. I have merely glanced through the meagre Marine Archives for the Mediterranean in the sixteenth century. A2, A5 IV, V, VI, B4 1, B6 1, 77, B7, 204, 205, 473, 520, B8 2–7, D 2, 39, 50, 51, 52, 53, 55.

BIBLIOTHÈQUE NATIONALE: The documents are so widely dispersed that the list could become impossibly long. Basically there are three collections, the French collection (Fonds français, F. Fr.), the Spanish (Fonds espagnol, F. Esp.) and the Italian (Fonds italien, F. Ital.)

1. *Fonds français:* I have read the autograph correspondence of the French ambassadors in Spain, Sébastien de l'Aubespine, 1559–1562, and Saint-Gouard, 1572–1580. Copies of the important letters of Sébastien de l'Aubespine, made by M. Hovyn de Tranchère from originals in the Imperial Library of Saint Petersburg, are filed in Affaires Etrangères under the numbers Esp. 347 and 348.

I endeavoured (with a view to possible publication) to trace in the B.N. all the letters and original papers of Sébastien de l'Aubespine, abbé de Basse-Fontaine and later bishop of Limoges; it adds up to a long list. Miscellaneous papers 4398 (f° 133); 4400, f° 330; 15877, 15901, 16013, 20787. His embassy to Brandenburg 3121. Missions to Spain: 3880, f° 294; 3899, f° 82; 3951, f° 26; 4737, f° 91; 10753, 15587, 16013, 16121, 16958, 17830, 20991, 23406, 23517; embassy to Switzerland, 20991, 23227, 23609, diplomatic code, *n.a.* 8431; 2937 (37), 3114 (102), 3121, 3130 (52 and 88), 3136 (10), 3158 (51, 54, 59, 76), 3159, 3174 (90), 3185 (102), 3189 (19), 3192, 3196 (26), 3216 (27), 3219 (2, 117), 3224 (82), 3226 (27), 3249 (73, 92), 3320 (96), 3323 (76, 119), 3337 (144), 3345 (55, 70), 3390 (15), 3899 (11), 3902 (88), 4639, 4641, 6611, 6614, 6616, 6617, 6618, 6619, 6620, 6621, 6626, 15542, 15553, 15556, 15557, 15559, 15784, 15785, 15876, 15902, 15903, 15904, 16016, 16017, 16019, 16021, 16023.

For Saint-Gouard's correspondence, see 16104 to 16108.

Further research:

a) French diplomatic correspondence with Rome, 17987; extracts from various negotiations by ambassadors in Rome (1557–1626). 3492 to 3498: dispatches from M. de Béthune, ambassador in Rome.

b) I tried to fill the gaps in Charrière's published diplomatic correspondence with Turkey and Venice, only to find very often that I was reading letters already in his collection. I therefore read 16142, f°s 7–8, 32 and 32 v°, 34, 43 to 44 v°, 48 to 49, 58, 60–61, Germigny's letters, 16143, letters from Paul de Foix, 16080 and du Ferrier, 16081.

I also read the letters sent by the bishop of Dax during his eventful mission of 1571–1573 (16170).

c) I consulted a series of moderately useful documents, 6121, f° 2 to 15, on Constantinople, Dupuy, II, 376, on Claude du Bourg, 16141, f° 226 to 272 v°. Advis donné au Roi Très Chrétien par Raymond Mérigon de Marseille pour la conqueste du royaume d'Alger; *na* 12240. The official documents and correspondence concerning the Bastion de France touch very little on the sixteenth century and are less important to this book than Giraud's work from the Lenche family papers. I drew on two contracts between Philip II and the Genoese merchants, May, 1558, 15875, f° 476 and 476 v°, 478 to 479.

2. *Fonds espagnol:* the documents have been inventoried, for the most part accurately, by A. Morel Fatio, in the duplicated catalogue of the Section. This is a very mixed bag. It contains a great many copies of dispatches, reports and letters from the Duke of Alva. The pride of the collection is the long discourse on Neapolitan affairs (Esp. 127) which is difficult to date and obscure in origin.

3. *Fonds italien:* contains even more material than the Spanish collection, but a great deal of it consists of copies, of documents of uncertain date, many of which have been published or used by historians. Of this collection I read: 221, 340 (Cyprus, 1570), 427, 428 (Corfu in 1578), 687, 772, 790 (on D. Pietro di Toledo's administration in Naples), 1220, 1431, 2108.

MINISTRY OF FOREIGN AFFAIRS (AFFAIRES ETRANGERÈS): The copies and authentic documents in the Spanish collection were catalogued impeccably in 1932 by Julián Paz, *Colección de documentos españoles existentes en el Archivo del Ministerio de Negocios Extranjeros de Paris*, Madrid. These documents compose a series of thick registers, referred to in this book as A. E. Esp. (Affaires Étrangères, Espagne). The following registers were used: 138, 216, 217, 218, 219, 222, 223, 224, 227, 228, 229, 231, 232, 233, 234, 235, 236, 237, 238, 261, 264 (f°s 51, 60, 70, 120), 307.

I have already mentioned the files Esp. 347 and 348 containing copies of the letters of the bishop of Limoges.

I also read the volumes *Venice* 46, 47, 48 of the correspondence of Hérault de Maisse, French ambassador in Venice from 1589 to 1594 (copies; originals in B.N., Paris).

Marseilles

ARCHIVES DE LA CHAMBRE DE COMMERCE: Contains a few miscellaneous letters dating from the end of the sixteenth century, unnumbered; examples, Coquerel, consul in Alexandria to the consuls of Marseilles, Alexandria, 29th November, 1599, original; Louis Beau, consul at Aleppo to the consuls of Marseilles, Aleppo, 1st September, 1600, original, etc.

ARCHIVES COMMUNALES: Deliberations of the Town Council, 1559 to 1591, BB 40 to BB 52 (registers).

A series of documents at present being classified in the valuable HH series, notably files 243, Letters Patent from Charles IX permitting nobles to trade in merchandise without derogating; 272, the *droit du tiers* (royal levy on timber); anchorage rights at Antibes, 1577–1732; 273, Arles, 1590–1786; 284, letters from Henri III authorizing the export of wool; 307, trade with England, 1592–1778; 346 *bis*, letter from the 'king' of Morocco requesting the appointment of a consul; 350, Tunis, miscellaneous correspondence; 351, 2 per cent duty at Constantinople, 1576–1610; 367, cargoes or ships seized; 465, harbour traffic, entry of goods 1577. In addition there is a series of letters from Aleppo, Tripoli, Syria and Alexandretta which I have quoted giving their dates (notably in Part II, chapter 3), but which had not been given numbers when I was working in Marseilles; I have temporarily called them the Ferrenc collection.

ARCHIVES DÈPARTMENTALES DES BOUCHES-DU-RHÔNE: Three precious documents:

1. The first archives of the Consulate at Algiers, to which R. Busquet has drawn the attention of researchers.

2. The register containing the declaration of goods leaving Marseilles to be distributed in the interior of the kingdom in the year 1543. Amirauté de Marseille, IX B 198 *ter*. The document contains three separate lists a) from 15th January, 1543 to 21st May, 1543, list 'des marchandises de Marseille pour pourter dans le royaulme', in other words goods exported from Marseilles to the rest of France overland, f° I to f° XLV; b) from 15th January, 1543 to 28th May, 1543, a record of goods brought into Marseilles by land and water from the rest of the kingdom (f° XLVI to LXXVI); c) record of exports from Marseilles to foreign destinations by sea between 16th January, 1543 and 18th May, 1543 (f° LXXVIII to LXXXIX).

3. Register of unloading certificates (for goods landed from ships between 1609 and 1645), pages unnumbered, Amirauté de Marseille, B IX 14.

Algiers

The archives of the former Government-General building of Algieria contained a rather odd collection of Spanish documents, both copies and originals, obtained from the Tiran mission between 1841 and 1844. The catalogue (which contains a few errors) is by Jacqueton.

I have read the entire collection, including material outside the period of this book; in the course of my research I discovered several items wrongly attributed in the catalogue as well as omissions in the free translations by E. de la Primaudaie, *Documents inédits sur l'histoire de l'occupation espagnole en Afrique 1506–1564*, Algiers, 1866, 324 p., which in any case contains less than a fifth of the documents in Algiers.

Other archives used: Besançon, Archives départementales du Doubs; Toulon, Archives communales; Cassis, Archives communales; Orange, Archives communales; Perpignan, Archives départementales des Pyrénées-Orientales; Rouen, Bibliothèque Municipale, Archives départementales, Archives communales; Aix, Bibliothèque de la Méjanes.

THE ITALIAN ARCHIVES

Genoa

ARCHIVIO DI STATO: There are four important series: 1) correspondence with Spain: 1559–1590, file no. 2.2411 (which contains only a few documents by the agent Angelo Lercaro for the year 1559) to file 8.2417; I was later able to read further in this series, up to file 38.2247 (1647); 2) secret correspondence with Constantinople, extremely interesting, 1558–1565, contained in two files, 1.2169 and 2.2170; both series belong to the *Lettere Ministri* and are listed as *Lettere Ministri Spagna* and *Lettere Ministri Costantinopoli* respectively; 3) the copious but disappointing series of the *Lettere Consoli*: Messina 1529–1609 (no. 2634), Naples 1510–1610 (no. 2635), Palermo 1506–1601 (no. 2647), Trapani 1575–1632 (no. 2651), Civitavecchia 1563 (no. 2665), Alghero 1510–1606 (no. 2668), Cagliari 1519–1601 (no. 2668), Alicante 1559–1652 (no. 2670), Barcelona 1522–1620 (no. 2670), Ibiza 1512–1576 (no. 2674), Majorca 1573–1600 (no. 2674), Seville 1512–1609 (no. 2674), Pisa 1540–1619 (no. 2699), Venice 1547–1601 (no. 2704), London, from 1651 (no. 2628), Amsterdam 1563–1620 (no. 2567) – in fact fifteenth-century documents referring to Antwerp; 4) the important series of papers of the *Magistrato del Riscatto degli Schiavi* (atti 659) for the last years of the century.

Apart from these I also consulted *Lettere Ministri*, *Inghilterra*, filza 1.2273 (1556–1558); *Diversorum Corsicae*, filza 125, and above all *made an inventory* of the huge registers *Venuta terrae* (1526–1797), *Caratorum occidentis* (1536–1793, *Caratorum Orientis* (1571–1797), *Caratorum veterum* (1423–1584) which give an extremely clear picture of movements in the great port of Genoa during the last thirty years of the sixteenth century and later. In October, 1964 I had the good fortune to find a register of marine insurance (see volume I, page 618–620).

ARCHIVIO CIVICO: In the Palazzo Rosso are to be found documents more nearly concerning the city of Genoa itself, the activity of the *Arti* (on the

Arte della Lana from 1620), and the seventeenth century traffic of boats from Cap Corse bringing the island's timber to Genoa. There is also an amazing file of papers concerning the activity of the small barques of Marseilles: *Consolato francese presso la Repubblica – Atti relativi 1594–1597*, no. 332. The French consulate which curiously fell by escheat under the administration of the municipality of Genoa between 1594 and 1597, is by far the richest source to my knowledge for information about the trade of Marseilles in the western Mediterranean. Many of the documents refer to Marseilles' trade with Spain: ships' masters inquire whether or not it is permitted to go to Spain, whether they are supposed to be at war with Spain or not. Hence inquiries are made of captains returning from Sardinia and Spain which show that the waters of the western Mediterranean were absolutely crawling with Marseilles boats.

Venice

In the Archivio di Stato, the following files were consulted:

Senato Secreta, Dispacci Costantinopoli 1/A, 2/B, 3/C, 4/D, 1546–1564 (archives of the *bailo*). Senato Secreta Dispacci Napoli I. Cinque Savii alla Mercanzia. *Buste* 1, 2, 3, 4, 6, 8, 9, 26, 27. Relazioni Collegio Secreta 31, 38, 78. Capi del Conso dei Dieci, Lettere di ambasciatori, Napoli, 58. Lettere ai Capi del cons° dei X. Spalato 281. Lettere commerciali XII *ter*, a large miscellaneous collection of commercial letters dating from the end of the century written in Venice, Pera, Aleppo and Tripoli in Syria. The *Archivio generale di Venezia*, Venice, 1873, in–8° gives an outline of the principles on which the collection has been organized. By a fortunate chance I discovered in the *Archivio Notarile* (here as elsewhere adjoining the *Archivio di Stato*), among the papers of the notary Andrea de Catti, 3361 (July, 1590) a document establishing the existence of a company formed by 12 marine insurance handlers.

Since the first edition of this book was published, I have spent many months in Venice reading the greater part of the documents in the series Senato Mar, Senato Terra, Senato Zecca, Cinque Savii alla Mercanzia for the years 1450 to 1650, as well as the entire correspondence of the Venetian ambassadors in Spain until 1620 and, lastly, the precious *Annali di Venezia* which is the chronicle, based on official documents, of events in the city and in the world, continuing the record we have in Sanudo's *Diaries*. I have also made forays into the collections housed in the *Archivio Notarile* and the *Quarantia Criminale*.

I have also explored the collections in the Marciana and the Museo Correr (the Donà delle Rose and Cicogna collections).

Florence

ARCHIVIO DI STATO: Here I worked exclusively on the Mediceo collection. I read the series of *filze* containing the correspondence with Spain from 1559 to 1581, nos. 4896 to 4913 inclusive (as well as the file no. 4897 a) and prospected as far as 1590 using the long manuscript resumé of the series. Besides this very long series I consulted *filze* 1829, 2077, 2079, 2080, 2840, 2862, 2972, 4185, 4221, 4279, 4322. Files 2077, 2079, 2080 correspond to the *portate* of Leghorn from which I have frequently quoted. Among the Misc. Medici, file 123 contained nothing of great significance, 124 (f° 44) contains a prohibition from the emperor, dated 1589, on his subjects' trading with England. I have spent more time since 1949 on the Mediceo collection, in particular reading family papers deposited in the Archives.

BIBLIOTECA NAZIONALE: The extremely rich Capponi collection was acquired by the Library in about 1935. It contains the commercial ledgers of this powerful family. Our period is represented by registers 12 to 90; 107 to 109; 112 to 129, the last series containing the books of various companies. This collection is a world in itself, but the great size and weight of the registers are an obstacle to the researcher; the difficulties of handling make it very desirable that their contents be put on microfilm. I have consulted only no. 41, *Libro grande debitori e creditori di Luigi e Alessandro Capponi, Mariotto Meretti e compagni di Pisa 1571–1587*. This magnificent volume giving prices, exchange rates, marine insurance and freight charges, opens our eyes to a trading business which, like others of the time, had far-flung connections.

The Capponi collection also contains a number of varied political, economic and historical documents of which I could no more than scratch the surface. The titles of these documents on the Medici family, the Republic of Florence, Spain, Poland, China, Turkey and the great events of the century would fill an entire volume. A few examples: a chronicle of Florence from the year 1001 to 1723 (Codice C CXXX, vol. 2); on Flanders and Philip II in 1578 (XXXIX, p. 360–375); on Sicily in 1546 (LXXXII, no. 18), in 1572 (XV, p. 63–112), from 1600 to 1630 (CLXXXIX, p. 148 to 196); on galley-slaves in Venice (XI, p. 153–157); on the abuses of the exchanges, 1596 (XLIII, p. 274–287); on Lucca in 1583 (II, p. 357–366); on Transylvania in 1595 (XLV, p. 423–428); on Genoa, a description in the form of a dialogue between Philip II and the Duke of Alva (XXXVI, p. 205–269); on Genoa in about 1575 (LXXXI, XVIII, II); a Venetian report on Cyprus in 1558 (XIII, p. 266–293); a report by D. Filippo Peristen, the imperial ambassador to the Grand Prince of Muscovy (XIV, p. 232–253); a description of Portuguese boats arriving at Villefranche (XXXIX, p. 61–67); speech by the ambassador of the Grand Turk to the Diet of Frankfurt, 27th November, 1562 (XV, p. 274–277); on the alum contract between the Pope and Cosimo de' Medici, 16th June, 1552 (cassetta 7 a, no. XVIII); a letter from Tommaso Scierly to Ruggiero Goodluke, an English merchant in Leghorn, Naples, 14th July, 1606 (LXXXI, no. 23 a); on Spain under Philip II (XI); on the court of Philip II in 1564 (LXXXII), 1576 (LXXXI, no. 9 a). 1577 (LXXXII, no. 3); on Antonio Pérez and Escovedo (XV, p. 262–269); on the contract for the iron mines on Elba, 1577 (casetta 8 a, no. 11), and a report on the island (CLXI); the death of Vicenzo Serzelli (1578), one of the most celebrated bandits of the time (CLXI, CCLVI, CCXXXVII); on Tuscany in the time of Ferdinand I (CCL, CXXIV); on the New Christians of Portugal in about 1535 (XXXVI); on Portugal in 1571 (XXV, p. 109–127); on naval operations between 1570 and 1573 (CCXV, CLXXII); the gift sent by Selīm to Don John after Lepanto (XL, p. 41–44); on Famagusta (LXXXII); on the peace between Venice and the Turks, the Signoria's justification of her action – and on Don John's projects; the quarrel between Marcantonio Colonna and Gian Andrea Doria in 1570; the planned conquest of Portugal, 25th May, 1579 (XV, p. 1–61); on the income of Naples, 1618 (CCLVIII); on the income of Charles V (XI, p. 216–220); on the trade of Ancona (XXI, p. 257–298); on exchange and currencies (cassetta 10 a, no. xvi) on Milan (XV); on the opposition of the Holy See to the ceding of the Empire to Ferdinand (LIX); the causes of the war of Hungary (LIX, p. 436–469); the Turko-Persian war of 1577–1579 (LXXXII, no. 7); Cipriano Saracinelli implores Philip II to abandon the war against France and turn against the Turk (LVIII, p. 106–151).

LA LAURENZIANA: This too is a very mixed bag, particularly rich in collections of historical documents amassed in the seventeenth and eighteenth centuries concerning Tuscany, Naples, Crete (Candia), the Valtellina, autograph letters by Alexander Farnese (1518–1585) (Ashb. 1691) an unpublished minor work on exchange by (Giovanni) Sili, *Pratica di Cambi*, 1611, Ashb. 647, a report on Egypt and the Red Sea, *La retentione delle galee grosse della Illustrissima Signoria de Venecia in Allessandria con le navigazioni dell'armata del Turco dal Mar Rosso nell'India nel anno* MD XXXV I, 37 f° Ashb. 1484–1508 (Relatione d'Allessandria con la navigatione del Turco dal Mar Rosso nell'India, 1536, Ashb. 1408), a *Libro de Mercanzia* dating from the sixteenth century (Ashb. 1894); a seventeenth-century shipping treaty, *ibid.*, 1660; a sixteenth-century voyage to the Holy Land, 1654; maps and documents of seventeenth-century Portugal, 1291.

GUICCIARDINI-CORSI-SALVIATI FAMILY ARCHIVES: Thanks to the kindness and generosity of Marquis Guicciardini, I was able to spend some time studying the invaluable archives of this family. Libri mercantili 1, 7, 8, 9 to 15, 21 to 25 (1550–1563), 26 (libro di magazino de Messina 1551–1552), 27 to 32 (1552–1571); II, 33 to 48 (1542 to 1559): III, 49 to 59 (1554 to 1559); IV, 60 to 64 (1565–1572); V, 65 to 67 (1582–1585); VI, 68 to 71 (1579–1590); VII, 72 to 102 (1587–1641); VIII, 103–130 (1582 to 1587); IX, 131–135 (1588–1591); X, 136 to 155 (1590–1602); XI, 156 to 166 (15..? to 1617); XII, 167 to 172 (1589–1608); XIII, 173–202 (1592–1597). Like the company books of the Capponi family, these archives are a world in themselves, over 200 thick registers for our period. There is a wealth of material on prices, transport, purchases and resale on credit, on the silk trade, the Sicilian grain trade, and the pepper and spice trade. These papers have now been transferred to the Archivio di Stato of Florence.

Rome

ARCHIVIO DI STATO: I had hoped to study two collections here, but was unable to do so. The first is rather an unknown quantity, *Annona e Grascia* 1595–1847 (Busta 2557), the second the very precise records of the *portate* of Civitavecchia, which are however of limited interest for our period since most of the surviving documents date from the first half of the century.

Naples

ARCHIVIO DI STATO: In the *Carte Farnesiane*, I studied the correspondence of the agents of Margaret of Parma and the Duke of Parma, from 1559, in the series Spagna, fascio I to fascio VII, and made an integral copy of the record of exports from the port of Bari in 1572, Dipendenze della Sommaria, fascio 417, fascico I°. My greatest advance since my original research in these archives has been among the documents of the Sommaria, of which A. Silvestri compiled a detailed catalogue for me, thus enabling me to have many papers photographed. These documents are of capital importance for the history both of Naples and of the Mediterranean.

ARCHIVIO MUNICIPALE: There is a detailed catalogue by Bartolomeo Capasso, *Catalogo ragionato dei libri, registri, scritture esistensi nella sezione antica o prima serie dell'archivio municipale di Napoli* 1387–1808, Naples, vol. I, 1877, vol. II, 1899 – remarkable for the entries concerning institutions but containing only occasional glimpses of the city's commercial and industrial life. Important documents on supplies of grain and oil in the great city, *Acquisti de' grani*, 1540–1587, N 514; 1558, N 515; 1590–1803, N 516; *Acquisto e transporto de' grani*, 1600, N 518; 1591–1617, N 532; 1594, N 533.

PALERMO: When I visited Palermo in 1932, the Archivio di Stato and the Archivio Comunale had closed their doors, and I was only able to spend a few days in the former and view the outside of the huge registers of the latter. My research was concentrated on the rich Biblioteca Comunale where I consulted: the descriptions of Palermo, Qq E 56 and Qq E 31 (the latter by Auria Vicenzo in the seventeenth century); Successi di Palermo, Palmerino, Qq D 84; lettere reali al vicere di Sicilia dal 1560 al 1590, 3 Qq E 34; accounts of Sicily, Qq F 221, Qq C 52, Qq F 80, Qq D 186, 3 Qq C 19, f° 212 (in Spanish, Qq D 190) (1592), Relazione del Conte di Olivares (avertimenti lasciati dal Conte Olivares, 1595), Qq C 16, memorandum on the government of the island, Qq F 29; letters of the kings and viceroys of Sicily, 1556–1563, 3 Qq C 35; miscellaneous letters 1560–1596, 3 Qq E 34; letter from Duke of Albuquerque, July, 1570, 3 Qq C 45, n° 25; Qq H 113 (nos. 15, 17) and Qq F 231; 3 Qq C 36, no. 22; 3 Qq E 11 Camilliani, *Descrizione delle marine*, Qq F 101; *Itinerario* . . ., Qq C 47;

Pugnatore, *Istoria di Trapani* Qq F 61. On Sicilian trade, 2 Qq E 66, no. 1; on the grain trade, sixteenth and seventeenth centuries, Qq D 74; letter from Duke of Feria on grain (1603), 2 Qq C 96, no. 18; memorandum on Sicilian Jews by Antonino Mongitore, Qq F 222, f° 213; deputati del Regno 1564 to 1603, 3 Qq B 69, f° 339. On trade with Malta, Qq F 110, f° 295; on variations of Sicilian silver currency between 1531 and 1671, Qq F 113, f° 22; ordinances by Duke of Medina Celi, 1565, Copy, Qq F 113, f° 32 to 40; on Greeks who had come from Albania to Sicily, memorandum by Antonino Mongitore, Qq E 32, f° 81; biographical note on Covarrubias, Qq G 24, no. 43; eighteenth century memorandum on the value of Castilian currency, Qq F 26, f° 87; on the Sicilian population between 1501 and 1715, memorandum by Antonino Mongitore, Qq H 120; on Sicilian bandits (in the seventeenth century), Qq E 89, no. 1; on food shortages and especially the famine in Palermo in 1591, Qq H 14a, f° 144. Carta al Rey nuestro Señor de Filiberto virrey de Sicilia sobre traer carne de Berveria, 10th April, 1624, Qq D 56, no. 21, f° 259.

OTHER COLLECTIONS NOTED OR CONSULTED: Turin, Pisa, Ancona, Milan, Leghorn, Cagliari, Messina; in particular Mantua (Archivio di Stato) and Modena (Archivio di Stato).

THE VATICAN ARCHIVES

In the Vatican Archives, with the aid of the publications by Hinojosa and above all Luciano Serrano's *Correspondencia diplomática entre España y la Santa Sede durante el Pontificado de S. Pio V*, 1566–1572, I concentrated on the correspondence of the nuncios in Spain from 1573 to 1580 (which has some gaps), Spagna 7 to 27. My research benefited from the advice of Mgr. Tisserand and the generous assistance of Mgr. Mecati, in particular a series of valuable photocopies.

THE RAGUSA ARCHIVES

For reasons which I have frequently mentioned, the Ragusa Archives are far and away the most valuable for our knowledge of the Mediterranean. Here as elsewhere there are shelves full of political documents, consisting for the most part of letters from the Rectors and their advisers to Ragusan agents and ambassadors and the letters they received in return; this vast mass of documentation is divided into two series, the letters from the West (Lettere di Ponente) and from the Levant (LP and LL, the latter known under the general title of *Lettere et commissioni di Levante*). I have myself read LP 1 (1566) to LP 7 (1593) and among the letters from the Levant only the register LL 38, corresponding to the year 1593. Those whose acquaintance with these Ragusan documents is confined to the papers of Francesco Gondola, Ragusan ambassador at Rome at the time of Lepanto, published by Count L. Voinovitch as *Depeschen des Francesco*

Gondola, Gesandten der Republik Ragusa bei Pius V. und Gregor XIII., Vienna, 1909, can have little idea of the Republic's normal handling of diplomatic negotiations; this was an association of traders as well as a political unit. Ragusan agents were at the same time merchants who took orders for grain, cloth, velvet, copper or kerseys according to needs and circumstances. Their correspondence therefore contains none of the general observations on human behaviour and great men to be found in Venetian documents, but is full of the banal but useful details of everyday life.

But the true interest of the Ragusa Archives lies elsewhere again. To anyone with the time and patience to study the voluminous *Acta Consiliorum*, they afford an opportunity to observe the extraordinarily well-preserved spectacle of a medieval town in action. For reasons of registration or through legal disputes an amazing collection of documents has survived: bills of exchange, notes, marine insurance contracts, regulations for participation, founding constitutions of companies, wills, domestic servants' contracts. These documents are divided into three series: the *Diversa de Foris*, the *Diversa di Cancellaria*, and the *Diversa Notariae*. I have no more than glanced through the last two, *Diversa di Cancellaria* registers 132 to 145 (1545–1557); *Diversa Notariae*, register 110 (1548–1551) but have read quite widely among the *Diversa de Foris* for the period 1580–1600 (the documents are chronologically divided, rather inconsistently, among the three large registers which make up the series); I studied nos. I to XVI. Other documents studied: *Libro dogana* no. 10, 1575–1576, XXI, 1, 12; XXI, 7, 3 and especially XXII, 7, 4, on imports of Spanish wool (rather difficult to decipher, or rather understand, I have copied it out in full). Finally a recent acquisition, 1935, no. 44, *Quadernuccio dove s'ha da notare le robe che vanno o venghono alla giornata, cossi d'amici comme le nostre*, 20th December, 1590 – 2nd April, 1591, throws an odd light on the activities of Ragusan travellers and merchants actively concerned with the Balkan routes.

I have already expressed my debt to M. Truhelka, the archivist at Ragusa while I was working there in the winter of 1935 – and my friendly appreciation of his kindness. But his ban on photographing documents made my research a hundred times more difficult. By a stroke of good fortune I have obtained a complete film of the series *Lettere di Ponente*, *Diversa di Cancellaria*, *Noli e Sicurtà*, now housed in the Centre Historique (VIe Section), 54 rue de Varenne, Paris VIIe. Together with Alberto Tenenti, I have viewed this long series of invaluable photographs.

EUROPEAN ARCHIVES OUTSIDE THE MEDITERRANEAN AND FRANCE

I could do no more than take cognizance, from a distance, of the German, Austrian and Polish archives. My intention was to supplement the many published works available by taking samples from documents in the archives, in order to confirm the trade currents leading to the Mediterranean, particularly overland – in other words to do for other towns what Mlle von Ranke has admirably established in the case of Cologne. In the Rhineland there are few surviving documents: those of Aachen disappeared in the city fire of 1656, those of Worms with the disturbances of 1689, the documents of Speyer have been made known through Hans Siegel's work and they are in any case few in number. I have no information on the resources of Coblenz, Mainz, and above all of Frankfurt, the rise of which during the last decades of the century poses such enormous problems. In southern Germany, there is nothing in Stuttgart or Munich, but Nuremberg and Augsburg have much to offer. It would have been interesting, after Strieder and Ver Hees, to formulate a clearer picture of the role of German merchants in their dealings with Lyons and Marseilles. I know nothing about Ulm. But further east, Leipzig and Dresden have plenty to offer the historian. The fairs of Frankfurt-an-der-Oder were frequented until 1600 by French and Italian merchants. The trade with the south included wines, silks and *Boysalz*, the Spanish or Biscay salt (Boy = Biscaya) which came in all probability through Hamburg. *Boysalz* was the subject of imperial privilege, and the emperor had a *Boyfactor* in Frankfurt-an-der-Oder. Two documents concerning this trade are preserved in the Stadtarchiv of Frankfurt-an-der-Oder, dated 1574 and 1597. A direct trade link between Breslau and Italy does not appear to have survived beyond 1450; after this approximate date it seems to have been diverted westwards, to the advantage of the towns of Bohemia and Austria, if H. Wendt's theory is correct.

The best collection of archives in this central European region which looked to the Mediterranean is in Vienna, at that time a political centre rather than an economic crossroads, but it made, for political or dynastic reasons, an excellent listening-post. The extremely rich collections of the Haus- Hof-und Staats-Archiv are by no means exhausted in the many books and publications which have drawn upon them, the Spanish correspondence (Hof Korrespondenz; Korrespondenz Varia), the correspondence with Venice, Turkey, Rome, Malta, Faszikel 1 (1518 *sic*–1620), Ragusa, f° 1 (1538–1708), Genoa, f° 1 (1527–1710), Italy (Kleine Staaten), Fasz. 7 (Neapel 1498–1599), Sicily, I (1530–1612); Hetrusca, I (1482–1620), Lusitana I (1513–1702); not to mention the *Kriegsakten* series (Fasz. 21–33 for the period 1559–1581). Circumstances prevented me from making this trip from which I had hoped to obtain information about the Mediterranean as seen from the northern interior of Europe. The preceding lines merely indicate the plan I would have followed. No serious study could ignore the cities of Danzig, Lübeck, Hamburg and Bremen with their maritime and *overland* links with the Mediterranean. In this connection we should not fail to consult either the Antwerp archives or those of the English ports (which are even more instructive than the Spanish collection in the British Museum, catalogue by P. Gayangos); or, indeed, those of the Netherlands and the Scandinavian countries, even the Polish archives, for Europe in its entirety was associated with the life and influence of the Mediterranean. Since 1949, I have visited Vienna, very briefly; I have spent a few days in Antwerp, Danzig, Warsaw and Cracow. I made a long and fruitful visit to London (British Museum and Public Record Office). In Geneva, thanks to the large set of microfilms, I was able to study the whole of the Édouard Fabre collection (Archives of the House of Altamira) which is housed in the Public and University Library (cf. *Inventaire* . . ., ed. Léopold Michel, in *Bulletin Hispanique*, 1914).

2. CARTOGRAPHICAL SOURCES

Under this heading I have grouped all maps, sketch-maps, plans and descriptions of shorelines and routes: these may be divided into modern works of reference and original sources.

A. MODERN REFERENCE WORKS:* For present-day maps, my usual work of reference is the *Géographie Universelle*, vol. VII, 1 and 2, VIII, XI, 1. For an account of maps of Spain, I used the information in *Revue de géographie du Sud-Ouest et des Pyrénées*, 1932. On transhumance see the map constructed by Elli Müller, *Die Herdenwanderungen im Mittelmeergebiet*, *Petermanns Mitteilungen*, 1938, reproduced in vol. I, p. 98. On the Berber Atlas, see J. Dresch's essays with their most original presentation of human geography.

I found two convenient, though not highly scientific, maps helpful for verification purposes: Map of Asia Minor (Turkey-Syria-Transjordan, Palestine, Iraq, Lower Egypt) scale 1 : 1.500.000 2nd ed., Girard and Barrère; and Mittelmeer, 1 : 1.500.000 Munich, Iro-Verlag, 1940.

I made great use of the magnificent series *Instructions Nautiques* (published by the Service hydrographique de la marine française), nos. 405 (Spain, N. and W.); 356 (Africa, W,); 345 (Spain S. and E.); 360 (France S., Algeria, Tunisia); 368 (Italy W.); 408 (Adriatic); 348 and 349 (Eastern Mediterranean); 357 (Black Sea and Sea of Azov).

B. ORIGINAL SOURCES:

a) *Bibliothèque Nationale*, Paris, Map Section (*Département des Cartes et Plans*)

Ge B 1425 *Italo-Catalan Portolan* (sixteenth century)
The Mediterranean, excluding coast of Syria; Atlantic coast from Scotland to the Canaries.
(In very battered condition, decorated with Spanish and Portuguese flags and animals illustrating Africa.)

Ge AA 640 *Portuguese map attributed to the Reinels*
Recto: Mediterranean.
Verso: Atlantic.

* For place names in the English edition, the reader is referred to the *Times World Atlas, Comprehensive Edition*, and the *Times World Index Gazeteer*.

(Very richly ornamented: boats decorated with the Portuguese cross, coats of arms, sketches of towns, Tower of Babel, towers of Jerusalem.)

Ge B 1132 *Map by Gaspar Viegas*, 1534
Western Mediterranean and Atlantic.
(Ornamented with compass-roses.)

Ge B 1134 *Map by Gaspar Viegas*, 1534
Eastern Mediterranean.

Ge C 5097 *Mediterranean, Red Sea, Black Sea*, 1534?
Atlantic from Scotland to Bojador.
(Decorated with green dome-shaped mountains.)

Ge AA 567 *Archipelago* (Aegean) from the Bosphorus to southern Crete.
The outline of the coast very stylized in scallops (attributed to Viegas).

Ge C 5096 *Attributed to Viegas by a note on the back*, 1534.
Western Mediterranean, Atlantic (from Taranto to the Azores).

Ge D 7898 *Attributed to Viegas by note on the back*, 1534.
Greece, part of the Aegean.

Ge C 5086 *Collection of 8 anonymous portolans*
Portuguese, considered according to a pencilled note to be copies of Diego Homem's map (1558) in the British Museum.
Folio 4, Mediterranean, Western Europe, Azores.

Ge DD 2007 *3 folios, Italian workmanship of sixteenth century, in leather binding.*
Folio 1, Aegean.
Folio 2, Mediterranean.
(Very beautifully drawn with figures of sovereigns, palm-trees, conventional drawings of cities, more detailed views of Venice and Marseilles.)

GG FF 14 410 *Atlas by the Genoese Battista Agnese* (1543), 12 folios.
Folio 6: Mediterranean, three sections.

Ge FF 14 411 *Idem*, larger format.

Ge C 5084 *Map by Vesconte di Maggiolo*, Genoa, 1547.
Mediterranean, Alexandria to Gibraltar.

Ge AA 626 *Andreas Homem: Universa ac Navigabilis tolius terrarum orbis descriptio*, Antwerp, 1559.
10 folios.
Folio 4: Mediterranean, Arabia, Caspian Sea.

Ge DD 2003 *Atlas by Diego Homem*, 1559.
Folio 2: Western Mediterranean.
Folio 3: Mediterranean.
Folio 4: Eastern Mediterranean.
Folio 6: Adriatic.
Folio 7: Greece, Archipelago.
(Beautifully ornamented: drawings of mountains, flags, view of Genoa, etc.)

Ge D 4497 *Portolan by the Cretan Georgio Sideri dicto Calapodo*, 1566.
Mediterranean.

Ge DD 2006 *Atlas by Diego Homem* (Venice, 1574), 7 folios.
Folio 2: Western Mediterranean.
Folio 3: Italy, Adriatic S., Central Mediterranean.
Folio 4: Adriatic.
Folio 5: Eastern Mediterranean.
Folio 6: Archipelago.
Folio 7: Black Sea.
(Very different from DD 2003: only ornaments are wind roses.)

Ge DD 682 *Atlas by Joan Martines* (Messina, 1589), 7 folios.
Folio 6: Sicily, Western Calabria.
Folio 7: Mediterranean.
(Views of towns, African bestiary, imaginary river linking the Black Sea to the Rhine and Rhône.)

Ge B 1133 *Portolan by Bartolomeo Olives* (Messina, 1584).
Europe, Mediterranean.
(Fine lettering, mentions Regina Saba, Prete Jani de las Indias.)

Ge AA 570 *Portolan by Mateus Prunes*, Majorca, 1586.
Mediterranean coasts of Western Europe, Africa as far as Gambia.
(Ornamented with drawing of sovereigns under royal canopies, from the king of Fez to the great Khan of Tartary who sits to the N.E. of the Black Sea, various African bestiaries, views of Marseilles, Genoa, Venice.)

Ge C 5094 *Mateus Prunes*, 1588.
Mediterranean from Alexandria to Morocco, Portuguese coast.
(Very simple, decorated only with wind roses.)

Ge C 234 *Carta Navigatoria by Joan Oliva*, Messina 1598?
Mediterranean, views of Barcelona (with Marseilles, Venice, Genoa and other ports conventionally represented, palm-trees, lions, elephants).

Ge C 5095 *Portolan by Vintius Demetrei Volcius Rachuseus* (in terra Libuani, 1598).
Mediterranean, very detailed of Dalmatia.

Fe FF 14 409 *Anonymous Portuguese Atlas, known as 'the Duchess of Berry's'* (end of sixteenth or early seventeenth century).
20 maps, unbound.
Folio 3: Spain and the western coasts of Algeria, Morocco and West Africa.

Ge DD 2012 *Anonymous Portolan, sixteenth century.*
2 folios: Mediterranean, Aegean.
(Ornamented with wind roses, drawing of the Crucifixion, imaginary rivers).

Ge DD 2008 *Portuguese portolan, late sixteenth century.*
Folio 1: Mediterranean, in binding.
Folio 2: Aegean.
(Ornamented with plumes and banners.)

Ge DD 2009 *Anonymous portolan, Franco-Italian, sixteenth century* (?)
4 folios pasted back to back.

Mediterranean, Archipelago, Western Mediterranean.
(Pious images pasted on to map.)
Ge C 5085 *Anonymous Portolan, sixteenth century.*
Mediterranean.
(Figure of monk cut out and pasted on Spain.)
Ge C 5100 *Italian portolan, sixteenth century.*
Mediterranean.
Ge C 5083 *Anonymous portolan, sixteenth century.*
(Very finely drawn.)
Ge C 2341 *Genoese portolan, sixteenth century.*
Shows Tripoli in Spanish hands.
(Handsomely decorated with flags, coats of arms; on the map of Spain a king bears the Spanish arms.)
Ge DD 2010 *Anonymous portolan.*
Mediterranean.
2 folios pasted back to back.
Ge D 7887 *Portolan of the Archipelago.*
Small views of towns, drawing of Troia.
Ge C 5093 *Portolan by Franciscus Oliva*, Messina, 1603.
Mediterranean.
(Highly ornamented, views of Marseilles, Barcelona, Genoa, Venice; warriors, flags, etc.)
N.B. There is a portolan by Salvador Oliva, Ge D 7889, the correct date of which (1635) has been scratched out and 1535 substituted.

b) *Archivo general*, Simancas, a series of plans and maps.
1. Costas tocantes a Argel y Bujía, 1602, E° 1951, a, 769 m
2. Plan of a fort sent by a soldier-slave in Barbary (possibly Sūs/Sousse), 22nd March, 1576, 0 m 490 × 0·461, Planos, Carpeta, II, f° 48
3. Diseño del Golfo de Arzeo, ink and wash, 28th December, 1574, 0·490 × 0·424 m, *ibid.*, f° 102
4. Plan of Algiers about 1603, 0·426 × 0·301 m, Carpeta, I, f° 53
5. Plan of the imperial castle at Bougie, 1548, 0·418 × 0·309 m, Carpeta, II, f° 61–62
6. Plans of the fortifications of Bougie, *ibid.*, f° 166 (0·392 × 0·294 m) f° 167 (0·326 × 0·284 m), f° 168 (0·514 × 0·362 m), 9th January, 1543
7. Plans of Mers-el-Kebir (Oran, 20th December, 1574), *ibid.*, f°s 98 and 99, 1·174 × 0·432 m and 0·580 × 0·423 m
8. Nuovo disegno dell'arsenale di Messina, *ibid.*, III, f° 58
9. Plan of Melilla, ink and wash, 0·445 × 0·320 m, E° 331
10. Plan of the fortifications of Malta, E° 1145 (Planos Carpeta, III, f° 61)
11. Disegno de la città di Siragusa (Syracuse), E° 1146, *ibid.*, III, f° 63
12. Traza del Reino de Murcia (about 1562), E° 141 a, f° 183, 0·908 × 0·214 m
13. The new fort of Tunis, ink and wash, Planos, III, f° 59, 1574 (0·694 × 0·585 m)

14. Dizeño del fuerte de Túnez y la Goleta, Rome, 7th August, 1574 (0·457×0·310 m), *ibid.*, f° 21
15. Traza de la Goleta de Túnez, ink and wash, about 1554 (0·488×0·348 m), *ibid.*, II, f° 126
16. Plano de Biserta, ink and wash, 1574 (0·627×0·577 m), *ibid.*, f° 60
17. Pianta de la città di Palermo, E° 1146.
18. Dizeños (4) de la laguna de Melilla, hechos por el Fratin, Madrid, 4th October, 1576 (II, f°s 134 to 137) (0·533×0·431 m, 0·483×0·315 m, 0·439×0·318 m, 0·313×0·216 m)
19. Plano de la Fortaleza de Argel, 1563, 0·606×0·448 m, I, f° 72.
20. Plano de los fuertes de la Goleta, 29th November, 1557 (0·320×0·217 m). Plans, E° 483, f° 174
21. Traza de los torreones de Melilla, 24th February, 1552 (0·442×0·315 m), III, f° 56
22. Plan of the Peñon de Velez, 1564, ink and wash, 0·30×0·209 m, *ibid.*, f° 19
23. Map of the Adriatic Sea, E° 540

3. PRINTED SOURCES

I cannot pretend to draw up an exhaustive catalogue of the literature devoted to the Mediterranean: it would fill volumes and the list would still not be complete. Since 1949 a great many more works have been published and for Spain alone, the *Indice Histórico Español* established in 1953 by J. Vicens Vives, gives the measure of the increase in our knowledge. Nor shall I list every work read in the course of my research, for even this would not easily be contained in the space allotted for a bibliography.

I have confined myself therefore to listing in section A the *major* collections of published documents, in section B the books and studies which essentially guided my research and as it were provided the architecture of the book, and in section C an alphabetical list of works mentioned in the text itself and the footnotes.

A. MAJOR COLLECTIONS OF PUBLISHED DOCUMENTS

STANDARD COLLECTIONS: Every country linked directly or indirectly with the Mediterranean has its own series of sixteenth-century documents.

Most remarkable of all for its size and convenient arrangement is the *Calendar of State Papers*.

The monumental *Colección de documentos ineditos para la historia de España* (abbreviated as *CODOIN*), 112 vols., in-8°, was my richest single published source. Since 1930–1931, thanks to Julián Paz, we have a magnificent catalogue of the collection in two volumes, *Catalogo de la coleccion . . .*, vol. I, Madrid, 1930, 728 p., in-8°; vol. II, Madrid, 1931, 870 p. in-8°, with an index of persons, places and subjects. The subject matter is systematically analysed in the invaluable bibliographical manual by R. Sanchez Alonso, *Fuentes de la historia española e hispano-americana*, which should be consulted in either the second edition (1927) or the third (1946). Italy has the no less monumental collection by Albèri, *Relazioni degli ambasciatori veneti al senato*. This has been so frequently plagiarized or paraphrased by historians that it finds its way into all studies of the sixteenth century and one tends unjustly to overlook its very considerable contribution. Is it not true though that blind confidence in the Venetians – said to be among the best, if not the very best, judges of sixteenth-century man – may be somewhat misplaced? One can trust the *dispacci* perhaps, but the *relazioni* were prepared reports, with all the faults and weaknesses to which such documents are prone, frequently plagiarizing previous orators. These are perhaps superficial criticisms, but they should not be forgotten. And I would unreservedly criticize the collection for its imperfect arrangement and the lack of a good index. A good one, at least for Italian economic history, is provided, it is true, by: A. Pino Branca, *La vita economica degli stati italiani nei secoli: XVI, XVII, XVIII, secondo le relazioni degli ambasciatori Veneti*, Catania, 1938, 515 p., in-16°.

Among the collected documents of France concerning the Mediterranean between 1550 and 1600 are several weighty volumes, foremost among which are E. Charrière's *Négociations de la France dans le Levant*, Paris, 1840–1860, 4 vols. in-4°; the equally classic *Négociations diplomatiques de la France avec la Toscane*, Paris, 1859–1886, 6 vols. in-4°; edited by A. Desjardins; the *Papiers d'État du cardinal Granvelle*, Paris, 1842–1852, 9 vols. in-4°, a useful collection for many purposes, and the *Lettres de Catherine de Médicis*, Paris, 1880–1895, 10 vols., in-4°, ed. H. de la Ferrière.

From Belgium – that part of the Netherlands which in the end remained faithful as much to the Catholic Church as to Philip II – we have L. P. Gachard's publications, *Correspondance de Philippe II sur les affaires des Pays-Bas* (until 1577), 1848–1879, 5 vols. in-4°; *Correspondance de Marguerite d'Autriche avec Philippe II* (*1559–1565*), 3 vols. in-4°, 1867–1881 (recently continued by J. S. Theissen), and the publication by Edmond Poullet and Charles Piot of the *Correspondance du Cardinal Granvelle, 1566–1586*, 12 vols. in-4°, 1877–1896, all of which have an indirect bearing on our subject.

There are four essential works in German: *Die römische Kurie und das Konzil von Trient unter Pius IV*, by the Czech historian J. Susta, Vienna, 1904–1914, 4 vols. in-8°; G. Turba's *Venetianische Depeschen vom Kaiserhofe*, in the rich collection *Nuntiaturberichte aus Deutschland*; Lanz's classic *Correspondenz des Kaisers Karl V*, III, 1550–1556, xx+712 p. in-8°.

Among Portuguese publications, the most useful is the *Archivo diplomatico Portuguez*, in the 10 volumes of which can be found in particular the edited correspondence of the Portuguese ambassadors in Rome with their government, 1550–1580.

OTHER PUBLISHED COLLECTIONS OF DOCUMENTS: Alongside these Herculean labours, a great deal of work has been devoted to the publication of diplomatic documents. Better classified than the other papers, often written in a legible and ornamental hand, they seem to have been waiting to tempt editors. In Belgium, one particularly voracious editor in the nineteenth century is said to have simply sent the documents from the archives straight to the printers. For the last fifty to a hundred years or even more, historians have devoted themselves to this task and the results of their labours will come to mark an era of historiography, from 1850 to the present day.

In Spain, scholars have published many documents concerning the relations between the Peninsula and the Roman curia, an area not fully covered by the *Colección de documentos inéditos.* Ricardo de Hinojosa however only produced one volume of his *Los despachos de la diplomacia pontifical en España*, in 1896. Luciano Serrano on the other hand gave us four magnificent volumes: *Correspondancia diplomática entre España y la Santa Sede durante el Pontificado de Pio V*, Madrid, 1914, containing both the correspondence of the nuncios in Spain and that of the Spanish ambassadors in Rome. In another area, the relations of Spain with the Vienna Habsburgs, there is, in addition to the papers published in *CODOIN*, the *Correspondancia inédita de Guillén de San Clemente, embajador en Alemania, sobre la intervención de España en los successos de Polonia y de Hungria* (1581–1608), ed. Marquis de Ayerbe, Saragossa, 1892. Unquestionably the most important Spanish contribution is the publication by the Real Academia de la Historia of the diplomatic correspondence between Spain and France (from the documents of the K series, returned by the French government in 1943) under the title *Negociaciones con Francia*, 9 vols. published (1950–1955) from 1559 until 21st October, 1567.

For Italy, we have only scattered collections: G. Berchet, *La Repubblica di Venezia e la Persia*, Turin, 1865; and by the same editor, *Relazioni dei consoli veneti nella Soria*, 1886. An excellent contribution is *La Lega di Lepanto nel carteggio inedito di Luys de Torres*, ed. A. Dragonetti de Torres, Turin, 1931; Mario Brunetti and Eligio Vitale, *Corrispondenza da Madrid di Leondardo Donà*, 1570–1573, Venice, Rome, 1963, 2 vols., in-4°.

On Ragusan diplomacy there is *Die Depeschen des Francesco Gondola Gesandten der Republik Ragusa bei Pius V. und Gregor XIII.*, 1570–1573, 1909.

In German there is Matthias Koch's *Quellen zur Geschichte des Kaisers Maximilian II*, Leipzig, 2 vols., 1857, Viktor Bibl's *Familienkorrespondenz Maximilians II.*, and Döllinger's *Dokumente zur Geschichte Karls V., Philips II, und ihrer Zeit*, Ratisbon, 1862, in-8°.

The famous letters of Busbecq, the imperial ambassador to Sulaimān's court, I read in the French edition, *Lettres du Baron de Busbec*, ed. Abbé de Foy, Paris, 1748, 3 vols., in-12°, since this happened to be the only one easily available to me [For English readers the most accessible version is

Edward Seymour Forster's translation, *The Turkish Letters of Ogier Ghiselin de Busbecq*, Oxford, 1927 (reprinted 1968) and wherever possible references and quotations are from this text)].

Among Belgian publications to note are vol. I of *Correspondance de la Cour d'Espagne sur les affaires des Pays-Bas, 1598–1621*, Brussels, 1923, ed. Lonchay and Cuvellier, in-4° and the *Correspondance d'Ottavio Mirto Frangipani, premier nonce de Flandre (1596–1606)*, 2 vols. published (1596–1598), I, 1924, by L. Van der Essen, II, 1932, by Armand Louant.

Head and shoulders above the rest however, in quality, quantity and relevance to the Mediterranean because of France's geographical position, are the French collections. It can be said without chauvinism that the ambassadors, churchmen and nobles whom France dispatched to foreign parts were alert and judicious, quick to understand and to gather information. A Fourquevaux in Spain for instance (and he is by no means the most outstanding) is an informant easily equal to his Italian colleagues and rivals. Among the most important publications are: Alexandre Teulet, *Relations politiques de la France et de l'Espagne avec l'Écosse au XVIe siècle*, 5 vols., in-8°, Paris, 1862 (covering the period 1515–1588), a work only marginally relevant to this study since it concerns the far north; similarly the *Correspondance politique de MM. de Castillon et de Marillac, ambassadeurs de France en Angleterre (1537–1542)*, Paris, 1885, ed. Jean Kaulek; the *Correspondance politique d'Odet de Selve*, French ambassador to England (1546–1549), ed. Germain Lefèvre-Pontalis, Paris, 1888; *Ambassades de MM. de Noailles en Angleterre*, ed. Abbé Vertot, 1763, 5 vols., the *Correspondance de la Mothe Fénelon, ambassadeur de France en Angleterre de 1568 à 1575*, 7 vols., Paris and London, 1838–1840 (under the direction of Charles Puzton Cooper); the *Mission de Jean de Thumery, sieur de Boissise*, 1598–1602 (to England), ed. P. Laffleur de Kermaignant, Paris, 1886 (gives several precise details on general policy). More directly relevant to our subject is *Jean Nicot, ambassadeur de France en Portugal au XVIe siècle. Sa correspondance diplomatique inédite*, ed. Edmond Falgairolle, Paris, 1897, and central to it is *Correspondance de Babou de la Bourdaisière, évêque d'Angoulême, depuis cardinal, ambassadeur de France à Rome*, ed. Henry et Loriquet, Paris, 1859. *Dominique du Gabre, trésorier des armées à Ferrare (1552–1554), ambassadeur de France à Venise (1555–1557); correspondance politique*, Paris, 1905, ed. A. Vitalis; the *Ambassade en Espagne de Jean Ébrard, seigneur de Saint-Sulpice*, ed. E. Cabié, Albi, 1903; *Dépêches de M. de Fourquevaux, ambassadeur du roi Charles IX en Espagne, 1565–1572*, 3 vols., Paris, 1896–1904, ed. C. Douais and the *Lettres de Charles IX à M. de Fourquevaux, 1565–1572*, Paris, 1897, by the same editor, two excellent publications; see also the admirable *Dépêches diplomatiques de M. de Longlée*. French resident in Spain, 1582–1590, ed. Albert Mousset, Paris, 1912. I would add to this list two shorter collections, the *Lettres inédites du roi Henri IV à Monsieur de Villiers, ambassadeur à Venise*, 1601, ed. Eugène Halphen, Paris, 1887; *Lettres à M. de Sillery, ambassadeur à Rome*, from 1st April to 27th June,

1601, ed. Eugène Halphen, 1866, and the *Lettres de Henri IV au comte de la Rochepot, ambassadeur en Espagne*, 1600–1601, ed. P. Laffleur de Kermaingant, Paris, 1889. These introduce us to a new climate in France – that of the reign of Henri IV – on which our study barely touches. Finally I would end this survey of French publications with the *Correspondance du Cardinal François de Tournon*, Paris, 1946, ed. Michel François, occupying a place midway between political and diplomatic history; parts 4, 5, and 6 directly concern our period and our areas of interest. I could not incorporate all the useful references of this handsome work into the body of my text (see Part I, chapter 1 and Part III, chapter 1): p. 318, 15th May, 1556, the deplorable morals and ignorance of the clergy in Corsica; p. 277–281, the peace treaty with Julius III, 29th April, 1552, to which Charles V gave his assent on the evening of 15th May, freed the French force engaged at Parma and explains the eruption of the Siena affair on 26th July, 1552; both the French and the Imperials were at this time in search of solid positions on the Italian chequer board: 'The success of the enterprise of Siena has singularly increased the prestige of the king in Italy and the Pope and the Signoria of Venice are particularly glad of it', says a letter from the cardinal dated August, 1552, p. 281. Was it not prestige too which dictated the Corsican expedition in 1553? Michel François repeats apropos of this 'untimely' operation (p. 289, note 1) the explanation he gives in his monograph, *Albisse del Bene, surintendant des finances françaises en Italie, 1551–1556* (Bibliothèque de l'École des Chartes, 1933), emphasizing the role played by the cardinal of Ferrara, who advanced the necessary money for the expedition out of his personal fortune. Was it 'untimely'? Many details and references to the Siena question make it possible to discern the policy of Paul IV's nephews and the atmosphere of Rome, during the 'unreasonable' reign of Paul IV. I would particularly mention the remarkable change of heart after 1559 of cardinal de Tournon or rather the evidence he gives of a French change of heart. Another age was beginning. See the vehement letter from the cardinal to the king, 14th June, 1559 (p. 397) against the 'punaisye' of the heretics in France, a word he repeats the same day (p. 398) in a letter to the constable, or the 'collaborationist' letter sent to Philip II on 31st January, 1561, p. 426–427. Lastly see (p. 373) an excellent comment on Venice, 17th September, 1558, and her policy of the *bilancia*: the Venetians want peace, 'for fear that the Victor might be too strong for them'.

The only important publication since 1949, and this is still in its infancy, is the *Acta Nuntiaturae Gallicae*, edited by the Gregorian University and the French School in Rome. So far there have appeared: *Correspondance des nonces en France Carpi et Ferrerio (1535–1540)*, ed. J. Lestocquoy, 1961; *Girolamo Ragazzoni, évêque de Bergame, nonce en France. Correspondance de sa nonciature, 1583–1586*, ed. Pierre Blet, 1962.

DOCUMENTS OUTSIDE DIPLOMATIC HISTORY: the foregoing paragraphs indicate the tremendous effort put into the publication of diplomatic documents. Unfortunately a great deal less has been accomplished in other domains. While there is something to show for the investigation of political history or the biography of major figures, there is little or nothing available relating to economic, social or cultural life or the history of technical progress.

1. However from Portugal we do have the *Historia tragico-maritima*, by Bernardo Gomes de Brito, 1st ed. Lisbon, 2 vols., 1735–1736, 2nd ed. 1904–1909, 12 vols. (B.N., 8° Z 18.199 (40)); it relates very largely of course to the Atlantic and Indian Oceans, but these oceans determined the life of the Mediterranean.

2. For Spain, there is the valuable old work, *Nueva Recopilación de las Leyes*; the collected *Actas de las Cortes de Castilla, 1563–1623*, 39 vols., Madrid, 1861–1915 (for the years before 1563 see vol. V of the *Cortes de los antiguos reinos de León y Castilla*); I made little use of the collection *Documentos inéditos para la historia de Aragón*; but would certainly mention the valuable collection of *Libros raros ó curiosos*, vol. XIX, *Tres relaciones históricas* (Gibraltar, los Gelves, Alcazarquivir, 1540, 1560, 1578), Madrid, 1889; *Cartas y avisos dirigidos à D. J. de Zúñiga, virrey de Nápoles en 1581*, Madrid, 1887. The two volumes of previously unpublished letters, *Retraite et mort de Charles-Quint au monastère de Yuste*, 2 vols., in-8°, Brussels, 1854–1855 (ed. Gachard), should perhaps be included in the previous category, and also to Gachard we owe the *Lettres de Philippe II à ses filles, les infantes Isabelle et Catherine, ècrites pendant son voyage au Portugal, 1581–1583*, Paris, 1884, excellent material for any apologia of the Prudent King.

I drew quite extensively on two publications of regional history, Carlos Riba y Garcia's *El consejo supremo de Aragón en el reinado de Felipe II*, Madrid, 1914, based on the rich Spanish collection of the British Museum; and volumes II and III of the re-edition (by Dario de Areitio) of Fidel de Sagarminaga's classic *El gobierno y regimen foral del señorio de Viscaya*, vol. II, 1577–1589, Bilbao, 1932, vol. III, 1590–1596, Bilbao, 1934.

I also consulted the large disparate collection of documents which while not strictly political and economic have some bearing both on political and economic matters, Lerruga, *Memorias políticas y económicas sobre los frutos, comercio, fabricas y minas de España*, 45 vols., in-4° Madrid, 1745.

However the most useful publications in this area are still:

a) The project started to mark the fourth centenary of Philip II's birth (1927) of a new collection of unpublished documents devoted to his reign (the programme, like all programmes very ambitious and still not completed, can be found in *Boletín de la Comisión de Monumentos históricos y artisticos de la provincia de Valladolid*, I, 2nd July–September, 1925). The work, under the imprint *Archivo Histórico Español* has the same title as the *CODOIN*. Published so far are: vol. I on the Council of Trent (1530–1552), vol. II on the Invincible Armada (1587–1589), xv–488 p., 1929, ed. G. P. Enrique Herrera Oria; *Consultas del consejo de*

Estado, 1930, ed. Mariano Alcocer, but I have no information about the volumes said to be forthcoming on *Portugal*, *Expediciones a Levante*, *Lepanto*, *Moriscos* etc. Jaime Salvá has edited *La orden de Malta y las acciones españolas contra Turcos y Berberiscos en los siglos XVI y XVII*, 1944, 448 p. in-4°.

b) The *Relaciones topográficas* (the surveys conducted on Philip II's orders in the *pueblos* of Spain in 1575 and 1578) are preserved in the Escorial. There is a catalogue, by the Augustinian Father Miguélez, *Las relaciones histórico-geográficas de los pueblos de España hechas par orden de Felipe II*, Madrid, 1915; a summary by Ortega Ribio, *Relaciones topográficas de España. Lo mas interesante de ellas escogido . . .*, Madrid, 1918. And the material concerning the diocese of Cuenca and the province of Guadalajara has been published separately, *Relaciones topográficas de España. Relaciones de pueblos que pertenecen hoy a la provincia de Guadalajara*, ed. Juan Catalina Garcia and Manuel Perez Villamil, Madrid, 1905–1915, 7 vols. (vol. XLI to XLVII of the *Memorial Histórico español*). The Bibliothèque de la Sorbonne has the *Relaciones de pueblos de la diocesis de Cuenca, hechas par orden de Felipe II*, ed. Father J. Zarco Cuevas, Cuenca, 1925, 2 vols. N.B. the remarkable publications by Carmelo Viñas and Ramon Paz, *Relaciones de los pueblos de España ordenadas por Felipe II: Provincia de Madrid*, 2 vols., 1949; *Provincia de Toledo*, 3 vols., 1951–1963; Noël Salomon, *La campagne de Nouvelle Castille à la fin du XVI siècle d'après les Relaciones topográficas*, 1964, summarizes the findings so far.

c) For North Africa, the enormous *Collection des sources inédites de l'histoire du Maroc*, is oriented largely towards the Atlantic and in any case barely extends to 1550; concerning the Spanish *presidios* on the Mediterranean coast, La Primaudaie's publication, previously mentioned, stops in 1564; the publication of the archives of the French consulate in Tunis, by Pierre Grandchamp, *La France en Tunisie à la fin du XVIe siècle, 1582–1600*, Tunis, 1920, is the most important North African contribution to Mediterranean history; vol. VI (1551–1575) of the *Histoire d'Oran* by General Didier, can only be read by those familiar with the sources previously published by the author.

d) Of the vast numbers of published documents in Italy, I would single out in particular: Marco Formentini, *Rivista storica della dominazione spagnuola sul ducato di Milano colla pubblicazione di 500 e più documenti ufficiali inediti*, Milan, 1872; Vladimir Lamansky, *Secrets d'État de Venise, documents, extraits, notices et études*, St. Petersburg, 1884, is a collection of exceptional value; as is the volume of miscellaneous documents on Naples, vol. IX of the *Archivio storico italiano*, Florence, 1846. Documents of social history are sufficiently rare for one to be grateful to Nini Cortese for his publication of *Feudi e Feudatari napoletani della prima metà del Cinquecento* (from documents in Simancas), Naples, 1931. There are several useful publications concerning Sicily: *Corrispondenza particolare di Carlo d'Aragona, Presidente del Regno, con Filippo II* (*Doc.*

per servire alla storia di Sicilía, 1st series II), ed. S. V. Bozzo and G. Salvo Cozzo, 1879 and *ibid.*, 4th series, IV, *Le fortificazioni di Palermo nel secolo XVI. Relazione delle cose di Sicilia fatta da D. Ferdinando Gonzaga all'imperatore Carlo V* (1546), 1896.

e) From France, M. Pardessus' *Collection des lois maritimes antérieures au XVIII*e *siècle*, Paris, 1837, 6 vols., is still an excellent work of reference.

f) Of the documents from the Balkans, Egypt, Syria and Turkey (and there are Turkish texts, bibliographies and periodicals) I have been able to discover only the titles. N. Iorga, in his *Ospiti romeni in Venezia, 1570–1610*, Bucharest, 1932, would have rendered historians a greater service if he had published the documents made available to him instead of presenting them in the form of a narrative which truncates the actual text used.

g) I consulted Hakluyt's marvellous collections, *The principal navigations, voiages, traffiques and discoveries of the English nation*, 3 vols., London, 1598, 1599, 1600 (References in the text are usually to this edition) and John Harris, *Navigantium atque itinerantium bibliotheca*, 2 vols., London, 1745.

h) The major collections of sixteenth-century economic documents come from the northern countries, from Antwerp and Augsburg. For Antwerp we have the controversial but valuable publications by Denucé. In Augsburg we have the various Fugger documents: most interesting for our purposes are Aloys Schulte, *Die Fugger in Rom*, 2 vols., Leipzig, 1904; Weitnauer, *Venezianischer Handel der Fugger*, 1931; Johannes Kleinpaul, *Die Fuggerzeitungen 1568–1605*, Leipzig, 1921. On the origin and interpretation of the documents in the National Library of Vienna. M. A. H. Fitzler, *Die Entstehung der sogenannten Fuggerzeitungen in der Wiener National-bibliothek*, Vienna, Rohrer, 1937; the economic value of the Fugger newsletters is assessed in Kempter, *Die wirtschafliche Berichterstattung in den sogenannten Fuggerzeitungen*, Munich, 1936. An English translation of them is available, *The Fugger News-Letters, 1568–1605*, 2 vols., London, 1924 and 1926, vol. I by Viktor von Klarwill and Pauline de Chary, and vol. II by L. S. R. Byrne.

i) Finally, let me recommend the great contribution made to our knowledge of economic life in the sixteenth century by an international team of researchers, on the initiative of the French Centre de Recherches historiques, VIe Section of the École Pratique des Hautes Études, 54, rue de Varenne, Paris, VII. Among these, in no particular order of merit, I would mention: Fernand Braudel and Ruggiero Romano, *Navires et marchandises à l'entrée du port de Livourne, 1547–1611* (1951), (to be completed and extended by Maurice Carmona who is at present working on it); Huguette and Pierre Chaunu, *Séville et l'Atlantique, de 1504 à 1650*, 12 vols., 1955–1960; Alberto Tenenti, *Naufrages, corsaires et assurances maritimes à Venise d'après les notaires Catti et Spinelli, 1502–1609*, 1959; Renée Doehaerd, *Études anversoises*, 3 vols., 1962; Micheline Baulant, *Lettres de négociants marseillais, les frères Hermite* (*1570–1616*)

1953; José Gentil da Silva, *Lettres marchandes des Rodrigues d'Evora et Veiga (1595–1607)*, 1956; Ugo Tucci, *Lettres d'un marchand vénitien Andrea Berengo, 1553–1556*, 1957; José Gentil da Silva, *Lettres de Lisbonne, 1563–1578*, 1959; Valentín Vázquez de Prada, *Lettres marchandes d'Anvers*, 4 vols., 1960; Domenico Gioffré, *Gênes et les foires de change*, 1960; Corrado Marciani, *Lettres de change aux foires de Lanciano*, 1962; Felipe Ruiz Martín, *Lettres marchandes échangées entre Florence et Medina del Campo*, 1965; Édouard Baratier, *La démographie provençale du XIIIe au XVIe siècle*, 1961; Léopold Chatenay, *Vie de Charles Esprinchard, Rochelais, et journal de ses voyages au XVIe siècle*, 1957; Xavier A. Flores, *Le 'peso politico de todo el mundo' d'Anthony Sherley*, 1963.

To these publications I would also add the monumental work by the Cuban historian Modesto Ulloa, *La hacienda real de Castilla en el reinado de Felipe II*, Rome, 1963.

B. ESSENTIAL WORKS

1. FOR THE GENERAL ORIENTATION OF THE BOOK: Top of the list are the works of Henri Pirenne, *Les Villes du Moyen Age* (Eng. trans. *Medieval Cities*), *Mahomet et Charlemagne* (Eng. trans. *Mahomet and Charlemagne*). And I have read and re-read the pages which Vidal de la Blache devotes to the Mediterranean in *Principes de géographie humaine*, ed. E. de Martonne 1922. (English trans., *Principles of Human Geography*, see section C.)

On the Mediterranean region as a whole, I based my reading from the start on Alfred Philippson's classic work, *Das Mittelmeergebiet*, which I have studied over and over again (in the 1904 edition through force of circumstances; there is a 4th edition, 1922). I consider this book a masterpiece of documentary precision. I also owe a great deal to Charles Parrain's *La Méditerranée: les hommes et leurs travaux*, Paris, 1936; to the monumental work by Maximilien Sorre, *Les fondements de la géographie humaine*, Paris, 4 vols., 1943–1952, and to André Siegfried's *Vue générale de la Méditerranée*, 1943.

2. HISTORY AND THE HUMAN ENVIRONMENT: Under the heading of 'history with roots in the soil' or rather the environment, come the complete works of Victor Bérard, Hellenist, diplomat and traveller, the complete works of Alfred Jardé, which open so many windows on to geographical reality, and the works of Jules Sion, which consist for the most part of scattered articles.

I am even more indebted to the complete works of Emile-Félix Gautier, which are much criticized today on points of detail whereas I feel it is perhaps more important to press forward in the same spirit. Of his books I would particularly mention: *Siècles obscurs du Maghreb*, 1927 (re-issued as *Le passé de l'Afrique du Nord*, 1952), *Mœurs et Coutumes des Musulmans*, 2nd ed. 1959, and the brief and simple confession of faith, *Le cadre géographique de l'histoire en Algérie, Histoire et Historiens de l'Algérie*, 1931, p. 17–35. I would also recommend a long and interesting article by Alfred Hettner, 'Der Islam und die orientalische Kultur', in *Geogr. Zeitung*, 1932; to this article as indeed to the whole German school of geographers, I am much indebted. (Bibliography in the useful manual by Hugo Hassinger, *Geographische Grundlagen der Geschichte*, Fribourg-im-Brisgau, 1931, 2nd ed., 1953).

3. HISTORY OF STRUCTURES: Since little or nothing existed before 1949 in the immense domain of 'structural history', I had to risk an individual initiative, and was therefore grateful for the pioneering work of some hardy prospectors: Pietro Sardella's book (then unpublished) *Nouvelles et Spéculations à Venise*, 1948 (based on Sanudo's *Diarii*) which I read in manuscript; Julius Beloch's huge posthumous work *Bevolkerungsgeschichte Italiens*, 3 vols.; Earl J. Hamilton's monumental study of prices in Spain, *American Treasure and the Price Revolution in Spain, 1501–1550*, 1934; Frederic C. Lane's fresh study *Venetian Ships and Shipbuilders of the Renaissance*, Baltimore, 1934; Richard Ehrenberg's still useful *Das Zeitalter der Fugger*, Jena, 1922, 2 vols. (A.-E. Sayous' criticism of this book is so excessive that it cannot be taken seriously); Ernst Schäfer's *Der könige. span. oberste Indienrat*, vol. I, 1936; the various studies of the nobility published in *Annales*; Marcel Bataillon's huge thesis, *Érasme et l'Espagne*, Paris, 1937; and the classic studies by Ernst Schäfer, *Beiträge zur Gesch. des span. Protestantismus*, 3 vols., 1902; Benedetto Croce, *La Spagna nella vita italiana durante la Rinascenza*, Bari, 1922; Ludwig Pfandl, *Geschichte der span. Literatur in ihrer Blütezeit*, 1929; and Emile Mâle, *L'Art religieux après le Concile de Trente*, 1932.

On the relationship between structure and conjuncture, the essential book as an introduction to the debate is still Ernest Labrousse, *La crise de l'économie française à la fin de l'ancien régime et au début de la Révolution*, 1944.

4. THE HISTORY OF EVENTS: here a large number of books jostle for the historian's attention, some as they become out of date being replaced by others. The best biographical studies published since the First War are O. von Törne, *Don Juan d'Autriche*, 2 vols., Helsinki, 1915 and 1928, a beautifully written work of remarkable erudition; and Van der Essen, *Alexandre Farnèse, 1545–1592*, 1933; the best studies of actual events are Charles Monchicourt, *L'expédition espagnole de 1560 contre l'île de Djerba*, Paris, 1913, and Felix Hartlaub, *Don Juan d'Austria und die Schlacht bei Lepanto*, Berlin, 1940; for the best sustained narrative see Lucien Romier's classic *Les origines politiques des guerres de religion*, 1913, 2 vols.; *La conjuration d'Amboise*, 1923; *Catholiques et huguenots à la cour de Charles IX*, 1924; *Le royaume de Catherine de Médicis*, 2 vols., 1925; *La Liga de Lepanto*, by P. L. Serrano, 2 vols., 1918–1919, and Martin Philippson's still useful study, *Ein Ministerium unter Philipp II., Kardinal Granvella am spanischen Hofe*, Berlin, 1895.

I must also mention that inexhaustible source of information, the weighty volumes of L. von Pastor's *Geschichte der Päpste*. (References in the text are to the French translation *Histoire des Papes*, by Furcy-Raynaud, 1888–1934, 16 vols., except for vol. X which is quoted from the German edition).

5. REFERENCE WORKS AND GENERAL STUDIES OF THE MEDITERRANEAN: I have consulted all the current well known manuals by Fueter, Platzhoff, C. Lozzi, Barbagallo, Kulischer, Doren, Georg Mentz, Stählin, Luzzatto, Segre, Zinkeisen, Hammer, Lavisse, Ballesteros, Agnado Bleye, Altamira, R. Konetzke (*Gesch. des spanischen und portugiesischen Volkes*, vol. VIII of the *Grosse Weltgeschichte*, Leipzig, 1941), Damião Peres, Mercier, Charles-André Julien, Henri Pirenne (*Bibliographie* and *Histoire de Belgique*), Henri Hausser (the *Sources* . . . and *Prépondérance espagnole*), Trevelyan, Hans Delbrück (*Geschichte der Kreigskunst* and *Weltgeschichte*, vols. III, 1926), vol. III of the *Neue Propyläen Weltgeschichte*, ed. Willy Andreas, Berlin, 1942, Karl Brandi (*Deutsche Geschichte im Zeitalter der Reformation und Gegenreformation*, 2nd ed.), W. Sombart (*Der Moderne Kapitalismus*, and *Vom Menschen*, 1940 ed.), Farinelli (*Viajes por España*. . .). The only general history which brings a totally fresh approach is J. Vicens Vives, *Historia social y económica de España y América*, which I read with much pleasure, particularly vol. III, 1957.

6. THE GENERAL HISTORY OF THE MEDITERRANEAN

a) Carl Rathlef, *Die Weltistorische Bedeutung der Meere, insbesond, des Mittelmeers*, Dorpat, 1858.

b) Count Edouard Wilczek, *Das Mittelmeer, seine Stellung in der Weltgeschichte und seine historische Rolle im Seewesen*, Vienna, 1895. Much influenced by the thought of Admiral Mahan.

c) Helmolt, *Weltgeschichte*, IV. *Die Randländer des Mittelmeeres*, Leipzig, 1900.

d) Giuseppe de Luigi, *Il Mediterraneo nella politica europea*, Naples, 1926.

e) Pietro Silva, *Il Mediterraneo dall'unità di Roma all'unità d'Italia*, 2 vols., Milan, 1927, re-issued in Milan in 1942 with a significant change of title, '*all'impero italiano*,' replacing '*all'unità d'Italia*'.

f) Paul Herre, *Weltgeschichte am Mittelmeer*, Leipzig, 1930, remarkable illustrations and excellent text by a political historian who is an expert on the sixteenth century.

g) Ulrich von Hassel, *Das Drama des Mittelmeers*, Berlin, 1940, develops a brilliant and original theme, the adventures of Pyrrhus, and attempts to explain the history of the Mediterranean in terms of the great central division, but unfortunately does not live up to its promises: disappointing, sketchy and inaccurate.

h) Philipp Hiltebrandt, *Der Kampf ums Mittelmeer*, Stuttgart, 1940 (very poor maps), by the correspondent of the *Kölnische Zeitung* in Rome, lively, sometimes inaccurate, often brilliant.

i) Emil Ludwig, *La Méditerranée, destinées d'une mer*, trans. from the German, Editions de la Maison Française, New York, 1943, 2 vols., is a disappointing book, inflated, with a few excellent pages and many incredible blunders.

j) Felice Vinci, *L'unità mediterranea*. 2nd ed., Milan, 1946, a slight book with insufficient historical perspective.

C. BIBLIOGRAPHY: PUBLISHED WORKS IN ALPHABETICAL ORDER

Accarias de Sérionne, Jacques, *La richesse de Hollande*, London, 1778, 2 vols

Achard, Paul, *La vie extraordinaire des frères Barberousse, corsaires et rois d'Alger*, Paris, 1939

Acta Tomiciana epistolarum Sigismundi regis Poloniae, Poznan (vol. 15, Wrocław), 1852–1957, 15 vols

Actas de las Cortes de Castilla, 1563–1623, Madrid, 1861–1915, 39 vols

Albani, Dina, *Indagine preventiva sulle recenti variazioni della linea di spiaggia delle coste italiane*, Rome, 1933

Albèri, Eugenio, *Relazioni degli ambasciatori veneti durante il secolo XVI*, Florence, 1839–1863, 15 vols. (A selection in English, ed. and trans. James C. Davis, *Pursuit of Power*, Harper Torchbook, New York, 1970)

Alberti, T., *Viaggio a Costantinopoli*, ed. A. Bacchi della Lega, Bologna, 1889

Albitreccia, L., *La Corse dans l'histoire*, Lyons-Paris, 1939

Alcocer y Martínez, Mariano, *Consultas del Consejo de Estado*, in the collection 'Archivo Histórico Español', Valladolid, 1930

Castillos y Fortalezas del antiguo Reino de Granada, Tangiers, 1941

Aleati, Giuseppe, *La popolazione di Pavia durante il dominio spagnolo*, Milan, 1957

Alemán, Mateo, *De la vida del pícaro Guzmán de Alfarache*, Milan, 1615, 2 vols

Allen, W. E. D., *Problems of Turkish Power in the Sixteenth Century*, London, 1963

Almanacco di economia di Toscana dell'anno 1791, Florence, 1791

Almeida, Fortunato de, *Historia de Portugal*, Coimbra, 1926–1929, 3 vols

Almeida d'Eça, Vincente, *Normas economicas na colonizacão portuguesa*, Lisbon, 1921

Amadei, Federigo, *Il Fioretto delle croniche di Mantova*, Mantua, 1741

Amari, Michele, *Storia dei Musulmani di Sicilia*, Florence, 1864–1868, 3 vols

Ammann, Hektor, *Schaffhauser Wirtschaft im Mittelalter*, Thayngen, 1949

Ancel, J., *Peuples et Nations des Balkans*, Paris, 1926

Andrada, F. de, *O primeiro cerco que os Turcos puzerão na fortaleza de Dio, nas partes de India*, Coimbra, 1589

Angelescu, I. N., *Histoire économique des Roumains*, Geneva, I, 1919

Anquez, Léonce, *Henri IV et l'Allemagne*, Paris, 1887

Aramon, G. d', see Chesneau, Jean

Arantegui y Sanz, José, *Apuntes históricos sobre la artilleria española en los siglos XIV y XV*, Madrid, 1887

Arbos, Philippe, *L'Auvergne*, Paris, 1932

Arco y Fortuño, Ricardo del, *La idea del imperio en la politica y la literatura españolas*, Madrid, 1944

Argenti, Philip P., *Chius vincta; or The occupation of Chios by the Turks (1566)*

and their administration of the island (1566–1912) described in contemporary reports and official despatches, Cambridge, 1941

Armstrong, H. C., *Grey Wolf, Mustafa Kemal: an intimate study of a dictator*, London, 1933

Arqué, Paul, *Géographie des Pyrénées françaises*, Paris, 1943

Arrigo, A., *Ricerche sul regime dei littorali nel Mediterraneo*, Rome, 1936

Arsandaux, H., see Rivet, P. and Arsandaux, H.

Arvieux, Chevalier d', *Mémoires du Chevalier d'Arvieux*, Paris, 1735, 6 vols

Ashauer, H. and Hollister, J. S., *Ostpyrenäen und Balearen*, Berlin, 1934, no. 11 in the collection 'Beiträge zur Geologie der westlichen Mediterrangebiete'

Aspetti e cause della decadenza economica veneziana nel secolo XVII, Venice-Rome, 1961

Assézat, J., see Du Fail, Noël

Asso, Ignacio de, *Historia de la economía política de Aragón*, Saragossa, 1798 (re-edited in 1947)

Atkinson, G., *Les nouveaux horizons de la Renaissance française*, Paris, 1935

Atti del convegno per la conservazione e difesa delle laguna e della città di Venezia (Istituto Veneto), Venice, 1960

Aubenas, Roger, *Chartes de franchise et actes d'habitation*, Cannes, 1943

Aubespine, Sébastien de l', *Négociations . . . relatives au règne de François II*, ed. L. Paris, Paris, 1941

Aubigné, Théodore Agrippa d', *Histoire Universelle*, ed. Baron Alphonse de Ruble for the Société de l'Histoire de France, 1886–1897, 9 vols

Aubin, G. and Kunze, A., *Leinenerzeugung und Leinenabsatz im östlichen Mitteldeutschland zur Zeit der Zunftkaufe. Ein Beitrag zur industriellen Kolonisation des deutschen Ostens*, Stuttgart, 1940

Audisio, Gabriel, I: *Jeunesse de la Méditerranée:* II, *Le sel de la mer*, Paris, 1935–1936, 2 vols

Aurigemma, S., see Bosio, Giacomo

Auton, Jean d', *Chroniques*, Paris, 1834–1835, 2 vols

Auzanet, Jean, *La vie de Camöens*, Paris, 1942

Avenel, Georges d', *Histoire économique de la propriété, des salaires, des denrées et de tous les prix en général depuis l'an 1200 jusqu'à l'a 1800*, Paris, 1894–1898, 4 vols

Avity, Pierre d', see Davity, Pierre

Axevedo, Lucio de, see Lucio de Azevedo, J.

Babeau, Albert, *Les voyageurs en France depuis la Renaissance jusqu'à la Révolution*, Paris, 1885

Babelon, Ernest, *Les origines de la monnaie considérées au point de vue économique et historique*, Paris, 1897

Bacchi della Lega, A., see Alberti, T.

Badaloni, Nicola, *La filosofia di Giordano Bruno*, Florence, 1955

Baehrel, René, *Une croissance: la Basse-Provence rurale (fin du XVIe siècle-1789)*, Paris, 1961, 2 vols

Balandier, Georges, *Afrique ambiguë*, Paris, 1957

Balducci Pegolotti, Francesco, *Pratica della mercatura*, Lisbon, 1765–1766, 4 vols

Ballesteros y Beretta, A., *Historia de España, y su influencia en la Historia Universal*, Barcelona, 1918–1940, 10 vols

Bandello, Matteo, *Novelle*, London-Leghorn, 1791–1793, 9 vols. (Page references in the text are to this edition. Various versions in English: *The novels of Matteo Bandello*, trans. Payne, 1890, 6 vols., and more recently a re-edition of

Fenton's translation *Bandello: Tragical Tales* ed. H. Harris, New York and London, 1924, 1 vol., 524 p.)
Baratier, Édouard, *La démographie provençale du XIIIe au XVIe siècle*, Paris, 1961
Barbagallo, Corrado, *Storia universale*, Turin, 1930, 5 vols
Bardon, Achille, *L'exploitation du bassin houiller d'Alais sous l'ancien régime*, Nîmes, 1898
Barrau-Dihigo, L., see Joly, Barthélemy
Barros, J. de, *Da Asia*, Venice, 1551
Bartoli, Daniele, *Degli uomini e de' fatti della Compagnia di Gesù*, Turin, 1847
Baruzi, Jean, *Problèmes d'histoire des religions*, Paris, 1935 (25th February, 1936)
Baschiera, Luigi, see Paruta, Andrea
Bataillon, Marcel, *Érasme et l'Espagne*, Paris, 1937
Batiffol, Louis, *La vie intime d'une reine de France au XVIIe siècle*, Paris, 1931, 2 vols
Baudrillart, Mgr. Alfred, *Philippe V et la cour de France*, Paris, 1890–1901, 4 vols
Bauer, Clemens, *Unternehmung und Unternehmnugsformen im Spätmittelalter und in der beginnenden Neuzeit*, Jena, 1936
Baulant, Micheline, *Lettres de négociants marseillais: les frères Hermite (1570–1612)*, Paris, 1953
Baumann, Émile, *L'anneau d'or des grands Mystiques*, Paris, 1924
Bayard, see Terrail, Pierre du, seigneur de Bayard
Beatis, A. de, *Die Reise des Kardinals Luigi d'Aragona durch Deutschland, die Niederlände, Frankreich und Oberitalien, 1517–1518*, ed. L. Pastor, Fribourg-im-Brisgau, 1905. (French translation by Madeleine Havard de la Montagne, *Voyage du Cardinal d'Aragon, 1517–1518*, Paris, 1913)
Beaujour, Baron Louis, Auguste Frédéric de, *Tableau du commerce de la Grèce, formé d'après une année moyenne depuis 1787 jusqu'en 1797*, Paris, 1800, 2 vols
Bechtel, Heinrich, *Wirtschaftsgeschichte Deutschlands. I: Von der Vorzeit bis zum Ende des Mittelalters; II, Vom Beginn des 16. bis zum Ende des 18. Jahrhunderts*, Munich, 1951–1952, 2 vols
Beiträge zur Geologie der westlichen Mediterrangebiete, published in *Auftrag der Gesellschaft der Wissenschaften zu Göttingen von Hans Stille*, Berlin, 1927–1939, 19 vols
Belda y Perez de Nueros, Fr., *Felipe secundo*, Madrid, undated (1927)
Bellay, Martin et Guillaume du, see Bourrilly, V. L., and Vindry, F.
Bellettini, Athos, *La popolazione di Bologna dal secolo XV all' unificazione italiana*, Bologna, 1961
Beloch, Karl Julius, *Bevölkerungsgeschichte Italiens*, Berlin, 1937–1961, 3 vols
Belon, Pierre (Belon du Mans), *Les observations de plusieurs singularitez et choses memorables trouvées en Grèce, Asie, Judée, Egypte, Arabie et autres pays estranges*, Paris, 1553
Below, G. von, *Über historische Periodisierungen mit besonderem Blick auf die Grenze zwischen Mittelalter und Neuzeit*, Berlin, 1925
Beltrami, Daniele, *Storia della popolazione di Venezia dalla fine del secolo XVI alla caduta della Repubblica*, Padua, 1954
Forze di lavoro e proprietà fondiaria nelle campagne venete dei secoli XVII e XVIII, Venice, Rome, 1961

Benedetti, B., *Intorno alle relazioni commerciali della Repubblica di Venezia e di Norimberga*, Venice, 1864

Beneyto Pérez, Juan, *Los medios de cultura y la centralización bajo Felipe II*, Madrid, 1927

Benichou, Paul, *Romances judéo-españoles de Marruecos*, Buenos Aires, 1946

Benjamin of Tudela, *Voyage du célèbre Benjamin autour du monde commencé l'an MCLXXIII*, trans. Pierre Bergeron, The Hague, 1735. (English versions: *Itinerary of Rabbi Benjamin of Tudela*, trans. A. Asher, London, Berlin, 1840, 2 vols.; *Itinerary of Benjamin of Tudela*, Hebrew text with English translation, ed. Adler, London, 1907)

Bennassar, B., *Valladolid au siècle d'or*, Paris, The Hague, 1967

Benndorf, Werner, *Das Mittelmeerbuch*, Leipzig, 1940

Benoit, Fernand, *La Provence et le Comtat-Venaissin*, Paris, 1949

Bérard, Victor, *La Turquie et l'hellénisme contemporain*, Paris, 1893
Les navigations d'Ulysse; II: Pénélope et les Barons des îles, Paris, 1928

Béraud-Villars, Jean, *L'Empire du Gaô. Un Etat soudanais aux XVe et XVIe siècles*, Paris, 1942

Berchet, G., *La Repubblica di Venezia e la Persia*, Turin, 1865

Bercken, Erich von der, *Die Gemälde des Jacopo Tintoretto*, Munich, 1942

Bergier, Jean-François, *Les foires de Genève et l'économie internationale de la Renaissance*, Paris, 1963

Bermúdez de Pedraza, Francisco, *Historia eclesiástica de Granada*, Granada, 1637

Bernaldo de Quirós, C., *Los reyes y la colonización interior de España desde el siglo XVI al XIX*, Madrid, 1929

Bernard, Étienne, *Discours véritable de la réduction de la ville de Marseille*, Paris and Marseilles, 1596

Bernardo, L., *Viaggio a Costantinopoli*, Venice, 1887

Bertoquy, P., see Deffontaine, P., *et al.*

Bertrand, Louis, *Sainte Thérèse*, Paris, 1927
Philippe II à l'Escorial, Paris, 1929

Beutin, Ludwig, *Der deutsche Seehandel im Mittelmeergebiet bis zu den Napoleonischen Kriegen*, Neumünster, 1933

Bianchini, Lodovico, *Della storia economico-civile di Sicilia*, Naples, 1841
Della storia delle finanze del Regno di Napoli, Naples, 1839

Biaudet, Henry, *Le Saint-Siège et la Suède durant la seconde moitié du XVIe siècle*, Paris, 1906

Bibl, Viktor, *Der Tod des Don Carlos*, Vienna and Leipzig, 1918
Die Korrespondenz Maximilians II. Familienkorrespondenz, Vienna, 1916–1921, 2 vols
Maximilian II., der rätselhafte Kaiser, Hellerau by Dresden, 1929

Bihlmeyer, Karl (vol. III with Tüchle, Hermann), *Kirchengeschichte*, 11th to 13th edition, Paderborn, 1951–1956, 3 vols. (French translation, *Histoire de l'Église*, Mulhouse, 1962–1964, 3 vols)

Bilanci generali, seria seconda, Venice, 1912

Billioud, Joseph and Collier, Jacques-Raymond, *Histoire du Commerce de Marseille*, vol. III, Paris, 1951

Binet, Rev. P. Étienne, *Essay des merveilles de nature et des plus nobles artifices*, 13th ed., Paris, 1657

Birot, Pierre and Dresch, Jean, *La Méditerranée et le Moyen-Orient*, Paris, 1953–1956, 2 vols

Bisschop, Eric de, *Au delà des horizons lointains . . .*, Paris, 1939. (English translation, by M. Ceppi, *The voyage of the Kaimiloa*, London, 1940)
Blache, Jules, *L'Homme et la Montagne*, Paris, 1934
Blanchard, Raoul, *Géographie de l'Europe*, Paris, 1936. (English translation by Crist, *A geography of Europe*, 1936)
Blanchet, Léon, *Campanella*, Paris, 1920
Bloch, Marc, *Les caractères originaux de l'histoire rurale française*, Paris, 1931
La société féodale, Paris, 1940. (English translation by L. Manyon, *Feudal Society*, London, 1961)
Blok, P. J., *Relazioni veneziane*, The Hague, 1909
Bodin, Jean, *Les Six livres de la République*, Paris, 1583. (English translation by Knolles, *The Six Books of the Commonwealth*, 1606, Harvard facsimile edition, 1962)
La Response de Jean Bodin à M. de Malestroict, 1568, ed. Henri Hauser, Paris, 1932. (English translation, ed. Moore, *The response of Jean Bodin to the Paradoxes of Malestroict*, Washington, 1946)
Bog, I., *Die bäuerliche Wirtschaft im Zeitalter des Dreissigjährigen Krieges. Die Bewegungsvorgänge in der Kriegswirtschaft nach den Quellen des Klosterverwalteramtes Heilsbronn*, Coburg, 1952
Bonnaffé, Edmond, *Les Arts et les moeurs d'autrefois. Voyages et voyageurs de la Renaissance*, Paris, 1895
Bono, Salvatore, *I corsari barbareschi*, Turin, 1964
Boppe, Léon, *Journal et Correspondance de Gédoyn 'le Turc', consul de France à Alep*, Paris, 1909
Borderie, Bertrand de la, see La Borderie, B. de
Borel, Jean, *Gênes sous Napoléon Ier (1805–1814)*, 2nd ed., Paris-Neuchâtel, 1929
Borlandi, Franco, *Per la storia della popolazione della Corsica*, Milan, 1942
Bory de Saint-Vincent, J. B., *Guide du voyageur en Espagne*, Paris, 1823, 2 vols
Bosio, Giacomo, *I cavalieri gerosolimitani a Tripoli negli anni 1530–1551*, ed. S. Aurigemma, Intra, 1937
Botero, Giovanni, *Relationi universali*, Brescia, 1599. (An English translation, *The Travellers Breuiat or, An historicall description of the most famous kingdomes in the World*, by R. Johnson, 1601)
Bouché, Honoré, *La Chorographie ou description de Provence*, Aix-en-Provence, 1664, 2 vols
Boué, Ami, *La Turquie d'Europe*, Paris, 1840, 4 vols
Boulenger, Jacques, see Contreras, Alonso de
Bourcart, Jacques, *Nouvelles observations sur la structure des Dinarides adriatiques*, Madrid, 1929
Bourgeois, Émile, *Manuel historique de la politique étrangère*, Paris, 1892–1926, 4 vols
Bourget, Paul, *Sensations d'Italie*, Paris, 1891. (English translation by L. Maitland, *The Glamour of Italy*, London, 1923)
Bourgoing, Baron Jean-François, *Nouveau voyage en Espagne*, Paris, 1788, 3 vols. (English translation London, 1789, *Travels in Spain*)
Bourrilly, V.-L. and Busquet, R., *Histoire de la Provence*. Paris, 1944
Bourrilly, V.-L. and Vindry, E., *Mémoires de Martin et Guillaume du Bellay*, ed. for the 'Societé de l'Histoire de France', Paris, 1908–1909, 4 vols
Bowles, William, *Introduction à l'histoire naturelle et à la géographie physique de l'Espagne*, translated from the Spanish by Flavigny, Paris, 1776

Boxer, C. R., *The Great Ship from Amacon, Annals of Macao and the old Japan Trade, 1555–1640*, Lisbon, 1959
Bozzo, S. V., *Corrispondenza particolare di Carlo di Aragona . . . con S. M. il re Filippo II* (Documenti per servire alla storia di Sicilia, 1st series, vol. II), Palermo, 1879
Bradi, Count Joseph M. de, *Mémoire sur la Corse*, Orleans, 1819
Bradi, Lorenzo de, *La Corse inconnue*, Paris, 1927
Bragadino, A., see Stefani, Fr.
Braganza Pereira, A. B. de, *Os Portugueses em Diu*, Basra, 1938
Brantôme, Pierre de Bourdeilles, abbé and seigneur de, *Œuvres complètes*, ed. Mérimée, Paris, 1858–1895, 13 vols
Bratianu, G., *Études byzantines d'histoire économique et sociale*, Paris, 1938
Bratli, Charles, *Philippe II, roi d'Espagne; Étude sur sa vie et son caractère*, Paris, 1912
Braudel, Fernand, *La Méditerranée et le monde méditerranéen à l'époque de Philippe II*, 1st ed., Paris, 1949
Capitalisme et civilisation matérielle, Paris, 1967. English translation in preparation
Braudel, F. and Romano, R., *Navires et marchandises à l'entrée du port de Livourne, 1547–1611*, Paris, 1951
Bräunlich, Erich, *Zwei türkische Weltkarten aus dem Zeitalter der grossen Entdeckungen*, Leipzig, 1937
Brémond, Gen. E., *Yémen et Saoudia*, Paris, 1937
Berbères et Arabes, Paris, 1942
Brémond d'Ars, Guy de, *Le père de Mme. de Rambouillet, Jean de Vivonne, sa vie et ses ambassades près de Philippe II et à la cour de Rome*, Paris, 1884
Brésard, M., *Les foires de Lyon aux XVe et XVIe siècles*, Paris, 1914
Bretholz, Berthold, *Lateinische Paläographie*, 2nd ed., Munich, 1912
Brèves, François Savary, seigneur de, *Relation des voyages de . . . tant en Grèce, Terre Saincte et Aegypte, qu'aux royaumes de Tunis et Arger*, Paris, 1628
Brion, Marcel, *Laurent le Magnifique*, Paris, 1937
Michel-Ange, Paris, 1939
Brockelmann, C., *Geschichte der islamischen Völker und Staaten*, Munich, 1939. (English translation by Carmichael and Perlmann, *History of the Islamic Peoples*, London, 1949)
Brosses, Président Charles de, *Lettres familières écrites d'Italie en 1739 et 1740*, Paris, 1858, 2 vols
Brückner, A., *Geschichte der russischen Literatur*, Leipzig, 1905, 2nd ed., 1909
Brulez, W., *De Firma della Faille en de internationale handel van Vlaamse firma's in de 16e eeuw.*, Brussels, 1959
Brun, A., *Recherches historiques sur l'introduction du français dans les provinces du Midi*, Paris, 1923
Brunetti, Mario, see Vitale, Eligio and Brunetti, M.
Brunhes, J.-B., *Etude de géographie humaine. L'irrigation, ses conditions, géographiques, ses modes et son organisation dans la péninsule ibérique et l'Afrique du Nord*; Paris, 1902. See also Deffontaines, Pierre, *et al*
Brunner, O., *Neue Wege der Sozialgeschichte. Vorträge und Aufsätze*, Göttingen, 1955
Brunschvig, Robert, *La Berbérie orientale sous les Hafsides, des origines à la fin du XVe siècle*, Paris, 1940, 2 vols
Bubnoff, Serge von, *Geologie von Europa*, Berlin, 1926–1930, 2 vols

Buchan, John, *Oliver Cromwell*, London, 1934
Bugnon, Didier, *Relation exacte concernant les caravanes ou cortèges des marchands d'Asie*, Nancy, 1707
Bullón, Eloy, *Un colaborador de los Reyes Católicos: el doctor Palacios Rubios y sus obras*, Madrid, 1927
Burckhardt, Jacob, *Geschichte der Renaissance in Italien*, Stuttgart, 1867; (English trans. by Samuel G. C. Middelmore, *The Civilization of the Renaissance*, London, 1944)
Busbec(q), Baron Ogier Ghiselin de, *The Turkish Letters*, translated by Edward Seymour Forster (This is the English version most easily available; some references in the text are to the French edition, *Lettres du Baron de Busbec*, ed. Abbé de Foy, Paris, 1748, 3 vols)
Busch-Zantner, R., *Agrarverfassung, Gesellschaft und Siedlung in Südost-Europa. Unter bes. Berücksichtigung der Türkenzeit*, Leipzig, *1938*
Albanien. Neues Land im Imperium, Leipzig, 1939
Busquet, Raoul, *Histoire de Marseille*, Paris, 1945
See also Bourrilly and Busquet
Cabié, Edmond, *Ambassade en Espagne de Jean Ébrard, seigneur de Saint-Sulpice, de 1562 à 1565*, Albi, 1903
Cabrera, de Córdova, L., *Relaciones de la cosas sucedidas en la Corte de España desde 1599 hasta 1614*, Madrid, 1857
Felipe segundo, Rey de España, Madrid, 1876, 4 vols
Cagnetta, Franco, *Bandits d'Orgosolo* (translated from the Italian *Inchiesta su Orgosolo*, by Michel Turlotte), Paris, 1963
Calendar of State Papers, Colonial Series, East Indies, China and Japan, London, 1862–1892, 5 vols
Calendar of State Papers and Manuscripts relating to English Affairs existing in the archives and collections of Venice and in other libraries of Northern Italy, London, 1864–1947, 38 vols
Calvete de Estrella, Juan Christóval, *El felicissimo viaje del . . . Principe don Felipe*, Antwerp, 1552
Campana, C., *La vita del catholico . . . Filippo II*, Vicenza, 1605–1609, 3 vols
Canaye, Philippe, sieur de Fresne, *Le voyage du Levant*, 1573, ed. Henri Hauser, Paris, 1897
Cano, Thomé, *Arte para fabricar . . . naos de guerra y merchante . . .*, Seville, 1611
Capasso, B., *Catalogo ragionato dell'Archivo municipale di Napoli*, Naples, 1876
Capasso, C., *Paolo III*, Messina, 1924, 2 vols
Capmany y de Montpalau, A. de, *Memorias históricas sobre la Marina, Comercio y Artes de la antigua ciudad de Barcelona*, Madrid. 1779–1792, 4 vols
Cappelletti, Giuseppe, *Storia della Repubblica di Venezia dal suo principio al suo fine*, Venice, 1850–1855, 13 vols
Caracciolo, Ferrante, *I commentarii delle guerre fatte co'Turchi da D. Giovanni d'Austria dopo che venne in Italia*, Florence, 1581
Carande, Ramón, *Carlos V y sus banqueros*, Madrid, 1949
Carcopino, Jérôme, *Le Maroc antique*, Paris, 1943
Cardauns, Ludwig, *Von Nizza bis Crépy. Europäische Politik in den Jahren 1534–1544*, Rome, 1923
Carmoly, Éliacin, *La France israélite, galerie des hommes et des faits dignes de mémoire*, Paris-Leipzig, 1855
Caro Baroja, Julio, *Los Moriscos del Reino de Granada*, Madrid, 1957
Los Judios en la España moderna y contemporánea, Madrid, 1961

La sociedad criptojudía en la Corte de Felipe IV (address on being received into Academia de Historia), Madrid, 1963

Carré, J. M., see Fromentin, E.

Carrera Pujal, Jaime, *Historia política, y económica de Cataluña*, Barcelona, 1946, 4 vols

Carreras y Candi, Franceschi, *Geografia general de Catalunya*, Barcelona, undated (1913–1918), 6 vols

Cartas y avisos dirijidos a D. J. de Zúñiga, virrey de Napoles en 1581, Madrid, 1887, vol. XVIII of the Colección de libros españoles raros o curiosos

Carus-Wilson, Eleanora, *Medieval Merchant Venturers*, London, 1954

Casa, Giovanni della, see Della Casa, Giovanni

Casanova, Abbé S. B., *Histoire de l'église corse*, Ajaccio, 1931, 2 vols

Casola, Petrus, *Viaggio a Gerusalemme*, 1494 (1855). (English translation, *Pietro Casola's Pilgrimage to Jerusalem in 1494*, Manchester, 1907)

Cassou, Jean, *La vie de Philippe II*, Paris, 1929

Les conquistadors, Paris, 1941

Castaneda-Alcover, Vicente, see Porcar, Moisé Juan

Caster, Gilles, *Le commerce du pastel et de l'épicerie à Toulouse de 1450 environ à 1561*, Toulouse, 1962

Castro, M. de, *Vida del soldado español Miguel de Castro*, Madrid-Buenos Aires, 1949

Casti, Enrico, *L'Aquila degli Abruzzi ed il pontificato di Celestino V*, Aquila, 1894

Cat, Édouard, *Mission bibliographique en Espagne*, Paris, 1891

Catalina García, Juan and Perez Villamil, Manuel, *Relaciones topográficas de España. Relaciones de pueblos que pertenecen hoy a la provincia de Guadalajara*, Madrid, 1905–1915, 7 vols. (vol. XLI to XLVII of the 'Memorial Historico Español')

Cavaillès, Henri, *La vie pastorale et agricole dans les Pyrénées des Gaves, de l'Adour et des Nestes*, Bordeaux, 1931

Caxa de Leruela, Miguel, *Restauración de la antigua abundancia de España*, Naples-Madrid, 1713

Cecchetti, B., *Informazione di Giovanni dall'Olmo console veneto in Lisbona sul commercio dei Veneziani in Portogallo e sui mezzi per ristorarlo, 1584, 18 maggio*, per nozze Thienesa, Schio, Venice, 1869

Celli, Angelo, *The history of malaria in the Roman Campagna from ancient times*, London, 1933

Cellini, Benvenuto, *Vita di Benvenuto Cellini, scritta da lui medesimo*, (English translations various)

Cervantes, Miguel, *Novelas Ejemplares*, ed. Marín, Madrid, 1944

Don Quixote (various editions)

Chabod, Federico, *Per la storia religiosa dello stato di Milano*, Bologna, 1938

Champion, Maurice, *Les inondations en France depuis le VIe siècle jusqu'à nos jours*, Paris, 1858–1864, 6 vols

Champion, Pierre, *Paris sous les derniers Valois, au temps des guerres de religion. Fin du règne Henri II. Régence de Catherine de Médicis. Charles IX*, Paris, 1938

Chardin, Jean, *Journal du Voyage en Perse et autres lieux de l'Orient*, Amsterdam, 1735, 4 vols

Charles-Quint et son temps, Paris, 1959 (International Colloquium of C.N.R.S., Sciences humaines, Paris, 30th September–3rd October, 1958)

Charliat, P. J., *Trois siècles d'économie maritime française*, Paris, 1931

Charrière, Ernest, *Négociations de la France dans le Levant*, 'Collection de documents inédits sur l'histoire de France', 1st series, Paris, 1840–1860, 4 vols
Chastenet, Jacques, *Godoï, prince de la paix*, Paris, 1943
Chateaubriand, François-René de, *Itinéraire de Paris à Jérusalem*, Paris, 1831
Chatenay, Léopold, *Vie de Jacques Esprinchard, rochelais, et Journal de ses voyages au XVIe siècle*, Paris, 1957
Chaunu, Pierre and Huguette, *Séville et l'Atlantique de 1601 à 1650*, Paris, 1955–1960, 12 vols
Chaunu, Pierre, *Les Philippines et le Pacifique des Ibériques (XVIe, XVIIe, XVIIIe siècles). Introduction méthodologique et indice d'activite*, Paris, 1960
L'Amerique et les Amériques, Paris, 1964
Chavier, Antonio, *Fueros del reyno de Navarra*, Pamplona, 1686
Chesneau, Jean, *Le voyage de Monsieur d'Aramon, ambassadeur pour le roy en Levant*, Paris, 1887
Chevalier, François, *La formation des grands domaines au Mexique. Terre et Société aux XVIe–XVIIe siècles*, Paris, 1952
Cicogna, see Tiepolo, Lorenzo
Cirillo, Bernardino, *Annali della città dell'Aquila*, Rome, 1570
Cochenhausen, Friedrich von, *Die Verteidigung Mitteleuropas*, Jena, 1940
Cock, Henrique, *Relación del viaje hecho por Felipe II en 1585 a Zaragoza*, Madrid, 1876
Codogno, O., *Nuovo itinerario delle Poste per tutto il mondo*, Milan, 1608
Coindreau, Roger, *Les corsaires de Salé*, Paris, 1948
Colección de documentos inéditos para la historia de España (CODOIN), Madrid, 1842–1896, 112 vols
Colette, *La naissance du jour*, Paris, 1942 (English translation by Rosemary Bénet, *A Lesson in Love*, New York, 1932, also published as *Morning Glory*, London, 1932)
Collier, Jacques-Raymond, see Billioud, J. and Collier, J.-M.
Colmenares, Diego de, *Historia de la insigne ciudad de Segovia*, 2nd ed., Madrid, 1640
Colonna, Marco Antonio, see Voinovitch, L.
Comines, Philippe de, *Mémoires de Messire Philippe de Comines . . . où l'on trouve l'histoire des rois de France Louis XI et Charles VIII*, ed. Messieurs Godefroy, augmented by M. l'abbé Langlet du Fresney, London and Paris, 1747. (English translation by T. Danett, re-ed. London, 1897)
Conestaggio, Jeronimo, *Dell'unione del regno di Portogallo alla corona di Castiglia*, Genoa, 1585
Congrès international des Sciences Historiques (XIe), *Rapports*, Stockholm, 1960, 5 vols
Coniglio, G., *Il Viceregno di Napoli nel secolo XVII*, Rome, 1955
Contarini, Giampetro, *Historia delle cose successe dal principio della guerra mossa da Selim ottomano a' Venetiani*, Venice, 1572
Contreras, Alonso de, *The Life of Captain Alonso de Contreras, written by himself, 1582–1633*, trans. C. A. Phillips, London, 1926. (References in text to French ed., *Aventures du capitaine Alonso de Contreras (1582–1633)*, trans. J. Boulenger, Paris, 1933)
Coornaert, Émile, *Un centre industriel d'autrefois. La draperie-sayetterie d'Hondschoote (XIVe–XVIIIe siècles)*, Paris, 1930
Les Français et le commerce international à Anvers, fin du XVe–XVIe siècle, Paris, 1961, 2 vols

Corazzini, G. O., see Lapini, Agostino
Córdova, L., see Cabrera de Córdova, L.
Cornaro, L., *Trattato di acque*, Padua, 1560
Corridore, F., *Storia documentata della popolazione di Sardegna*, Turin, 1902
Corsano, A., *Il pensiero di Giordano Bruno nel suo svolgimento storico*, Florence, 1954
Corsini, O., *Ragionamento istorico sopra la Val di Chiana*, Florence, 1742
Corte Real, J., *Successo do segudo cerco de Diu*, Lisbon, 1574
Cortes de los antiguos reinos de León y de Castilla, Madrid, 1861–1903, 8 vols
Cortese, Nino, *Feudi e feudatari napoletani della prima metà del Cinquecento*, Naples, 1931
Cossé-Brissac, Philippe de, see unpublished sources
Costa, Joaquín, *Colectivismo agrario en España*, Madrid, 1898
Crescentio, Bartolomeo, *Nautica mediterranea*, Rome, 1607
Croce, Benedetto, *Storia del Regno di Napoli*, 3rd ed., Bari, 1944
Cunnac, J., *Histoire de Pépieux des origines à la Révolution*, Toulouse, 1946
Cunningham, W., *The Growth of English Industry and Commerce*, 5th ed., Cambridge, 1910–1912, 3 vols
Cupis, C. de, *Le vicende dell'agricoltura e della pastorizia nell'agro romano e l'Annona di Roma*, Rome, 1911
Cuvelier, J. and Jadin, J., *L'ancien Congo d'après les archives romaines, 1518–1640*, Brussels, 1954
Cvijić, Jovan, *La Péninsule balkanique*, Paris, 1918
Dall'Olmo, see Cecchetti, B.
Da Mosto, Andrea, *L'archivio di Stato di Venezia*, Rome, 1937–1940, 2 vols
Dan, P., *Histoire de Barbarie et de ses corsaires*, Paris, 2nd ed. 1649
Danvila, Manuel, *El poder civil en España*, Madrid, 1885
Dauzat, Albert, *Le village et le paysan de France*, Paris, 1941
Davis, James C., *The Decline of the Venetian Nobility as a Ruling Class*, Baltimore, 1962. See also Albéri, Eugenio
Davity, Pierre, *Les Estats, empires et principautés du monde*, Paris, 1617
Debien, Gabriel, *En Haut-Poitou, défricheurs au travail, XVᵉ–XVIIᵉ siècles*, 'Cahiers des Annales', Paris, 1952
Decker, H., *Barockplastik in den Alpenländern*, Vienna, 1943
Decrue de Stoutz, Francis, *Anne, duc de Montmorency, connétable et pair de France sous les rois Henri II, François II et Charles IX*, Paris, 1889
Deffontaines, Pierre, Jean-Brunhes-Delamarre, Mariel, Bertoquy, P., *Problèmes de Géographie humaine*, Paris, 1939
Delbrück, Hans, *Geschichte der Kreigskunst im Rahmen der politischen Geschichte*, Berlin, 1900–1920, 4 vols
Weltgeschichte. Vorlesungen gehalten an der Universität Berlin 1896/1920, Berlin, 1923–1928, 5 vols
Deledda, Grazia, *La via del male*, Rome, 1896
Il Dio dei viventi, Rome, 1922
De Leva, Giuseppe, *Storia documentata di Carlo V*, Venice, 1863–1894, 5 vols. See also Paruta, P.
Della Casa, Giovanni, *Galateo*, Florence, 1561
Della Rovere, Antonio, *La crisi monetaria siciliana (1531–1802)*, ed. Carmelo Trasselli, Palermo, 1964
Della Torre, Raffaele, *Tractatus de cambiis*, Genoa, 1641
Delmas de Grammont, Henri, see Grammont, Henri Delmas de

De Luigi, Giuseppe, *Il Mediterraneo nella politica europea*, Naples, 1925
Delumeau, Jean, *Vie économique et sociale de Rome dans la seconde moitié du XVIe siècle*, Paris, 1957–1959, 2 vols
L'alun de Rome, XVe–XIXe siècle, Paris, 1963
Denucé, J., *L'Afrique au XVIe siècle et le commerce anversois*, Antwerp, 1937
Dermigny, L., *La Chine et l'Occident, le commerce à Canton au XVIIIe siècle, 1719–1833*, Paris, 1964, 4 vols
Descamps, Paul, *Le Portugal. La vie sociale actuelle*, Paris, 1935
Desdevises du Dezert, Georges, *Don Carlos d'Aragon, prince de Viane, étude sur l'Espagne du Nord au XVe siècle*, Paris, 1889
Desjardins, Abel, *Négociations diplomatiques de la France avec la Toscane*, 'Collection de documents inédits sur l'histoire de France', Paris, 1859–1886, 6 vols
Despaux, Albert, *Les dévaluations monétaires dans l'histoire*, Paris, 1936
Despois, Jean, *La Tunisie orientale, Sahel et Basse Steppe*, Paris, 1940
Dider, General L., *Histoire d'Oran. Période de 1501 à 1550*, Oran, 1927
Didier, L., see Mondoucet, C. de
Diehl, Charles and Marçais, Georges, *Histoire du Moyen Age*, III: *Le Monde oriental*, Paris, 1936, in *Histoire Generale*, under the general editorship of Gustave Glotz
Dietz, A., *Frankfurter Handelsgeschichte*, Frankfurt, 1910–1925, 4 vols
Di Giovanni, Giovanni, *L'ebraismo della Sicilia*, Palermo, 1748
Dieulafoy, Jane, *Isabelle la grande reine de Castille (1451–1504)*, Paris, 1920
Dion, Roger, *Histoire de la vigne et du vin en France des origines au XIXe siècle*, Paris, 1959
Di Tocco, Vittorio, *Ideali d'indipendenza in Italia durante la preponderanza spagnuola*, Messina, 1926
Doehaerd, Renée, *Etudes anversoises. Documents sur le commerce international à Anvers (1488–1514)*, Paris, 1962–1963, 3 vols
Doehaerd, Renée and Kerremans, Charles, *Les relations commerciales entre Gênes, la Belgique et l'Outremont d'après les archives notariales génoises des XIIIe et XIVe siècles*, Brussels–Rome, 1941–1953, 3 vols
Döllinger, I. J. J., von, *Dokumente zur Geschichte Karls V., Philipps II. und ihrer Zeit. Aus spanischen Archiven*, Ratisbon, 1862
Domínguez Ortíz, Antonio, *La sociedad española en el siglo XVII*, vol. I, Madrid, 1963
Donà, Leonardo, see Vitale, Eligio and Brunetti, Mario
Doren, Alfred, *Italienische Wirtschaftsgeschichte*, vol. I only published, Jena, 1934. (Italian version by Gino Luzzatto, *Storia economica dell' Italia nel medioevo*, Padua, 1936)
Dorez, Léon, see Maurand, Jérôme
Dorini, Umberto, *L'isola di Scio offerta a Cosimo de Medici*, Florence, 1912.
Dornic, François, *L'industrie textile dans le Maine et ses débouchés internationaux, 1650–1815*, Paris, 1955
Douais, Célestin, *Dépêches de M. Fourquevaux, ambassadeur du roi Charles IX en Espagne, 1565–1572*, coll. 'Société d'Histoire diplomatique,' Paris, 1896–1904, 3 vols
Lettres de Charles IX à M. de Fourquevaux, 1566–1572, same collection, Paris, 1897
Dozy, Reinhart Pieter Anne, *Histoire des Musulmans d'Espagne jusqu'à la conquête de l'Andalousie par les Almoravides (711–1110)*, Leyden, 1861, 4 vols

Dragonetti de Torres, A., *La lega di Lepanto nel carteggio diplomatico inedito di Don Luis de Torres*, Turin, 1931
Dresch, J., see Birot, P. and Dresch, J.
Drouot, Henri, *Mayenne et la Bourgogne (1587–1596) contribution à l'histoire des provinces françaises pendant la Ligue*, Paris, 1937, 2 vols
Du Fail, Noël, *Oeuvres facétieuses*, ed. J. Assézat, Paris, 1875, 2 vols
Dufayard, Charles, *Le connétable de Lesdiguières*, Paris, 1892
Du Gabre, Dominique, see Vitalis, A.
Dumont, Jean, *Corps universel diplomatique du droit des gens, contenant un recueil des traitez d'alliance, de paix, de trêves . . . depuis le règne de Charlemagne jusqu'à présent*, Amsterdam, 1726–1731, 8 vols
Duro, C., see Fernandez Duro, Cesáreo
Du Vair, Guillaume, *Recueil des harangues et traictez*, Paris, 1606
Eberhardt, Isabelle, *Notes de route: Maroc, Algérie, Tunisie*, Paris, 1908
Ébrard, Jean, see Cabié, Edmond
Eck, Otto, *Seeräuberei im Mittelmeer. Dunkle Blätter europäischer Geschichte* Munich, 1940, 2nd ed. 1943
Egidi, Pietro, *Emmanuele Filiberto, 1559–1580*, Turin, 1928
Ehrenberg, Richard, *Das Zeitalter der Fugger. Geldkapital und Creditverkehr im 16. Jahrhundert*, Frankfurt, 1896, 3rd ed. 1922, 2 vols. (Translated as *Capital and Finance in the Age of the Renaissance. A study of the Fuggers and their connections*, by H. M. Lucas, London, 1928, Bedford Series of Economic Handbooks)
Einaudi, Luigi, see Malestroict, sieur de
Eisenmann, Louis, see Milioukov, P., Seignobos, Charles and Eisenmann, Louis
Élie de la Primaudaie, F., *Documents inédits sur l'histoire de l'occupation espagnole en Afrique (1506–1574)*, Algiers, 1875
Emmanuel, I. S., *Histoire de l'industrie des tissus des Israélites de Salonique*, Lausanne, 1935
Encyclopedia of Islam, Ed. M. Th. Houtsma *et al.*, Leyden/London, 1913–1938. New edition, ed. J. H. Kramers, H. A. R. Gibb and E. Lévi-Provençal, Leyden/London, 1954–
Epstein, F., see Staden, H. von
Eškenasi, Eli, see Habanel, Aser and Eškenasi, Eli
Espejo de Hinojosa, Cristóbal and Paz y Espeso, Julián, *Las antiguas ferias de Medina del Campo*, Valladolid, 1912
Essad Bey, Mohammed, *Allah est grand!*, Paris, 1937
Essen, Léon van der, *Alexandre Farnèse, prince de Parme, Gouverneur général des Pays-Bas, 1545–1592*, Brussels, 1933–1934, 5 vols
Essen, Léon van der, and Louant, Armand, see Frangipani, Ottavio Mirto
Estienne, C., *La guide des chemins de France, revue et augmentée*, Paris, 1552
Estrangin, Jean-Julien, *Etudes archéologiques, historiques et statistiques sur Arles*, Aix-en-Provence, 1838
Eydoux, Henri-Paul, *L'homme et le Sahara*, Paris, 1943
Fagniez, Gustave, *L'économie sociale de la France sous Henri IV, 1589–1610*, Paris, 1897
Fail, Noël du, see Du Fail, Noël
Falgairolle, Edmond, *Une expédition française à l'île de Madère en 1566*, Paris, 1895
Falke, Johannes, *Die Geschichte des deutschen Handels*, Leipzig, 1859–1860, 2 vols

Fanfani, Amintore, *Storia economica, Dalla crisi dell'Impero romano al principio del secolo XVIII*, 3rd ed. Milan, 1948

Fanfani, Pietro, *Saggi di un commento alla Cronica del Compagni; I: La descrizione di Firenze*; II: *I Priori*, Florence, 1877

Febvre, Lucien, *Philippe II et la Franche-Comté, la crise de 1567, ses origines et ses conséquences*, Paris, 1911

Le problème de l'incroyance au XVIe siècle. La religion de Rabelais, Paris, 2nd ed., 1947

Pour une Histoire à part entière, Paris, 1963

Febvre, Lucien and Martin, Henri, *L'apparition du livre*, Paris, 1957

Féraud, Laurent-Charles, *Annales tripolitaines*, Paris, 1927

Fernández, Jesús, see García Fernández, Jesús

Fernández, Duro, Cesáreo, *Armada española desde la unión de Castilla y de Aragón*, Madrid, 1895–1903, 9 vols

Filippini, A. P., *Istoria di Corsica*, 2nd ed., Pisa, 1827–1831, 5 vols

Fisher, Godfrey, *Barbary Legend, War, Trade and Piracy in North Africa, 1415–1830*, Oxford, 1957

Flachat, Jean-Claude, *Observations sur le commerce et sur les arts d'une partie de l'Europe, de l'Asie, de l'Afrique et des Indes Orientales*, Lyons, 1766, 2 vols

Flores, Xavier A., *Le 'Peso politico de todo el mundo' d'Anthony Sherley ou un aventurier anglais au service de l'Espagne*, Paris, 1963

Floristán Samanes, Alfredo, *La Ribera tudelana de Navarra*, Saragossa, 1951.

Foglietta, Uberto, *De sacro foedere in Selimum*, Genoa, 1587

Fordham, Herbert, *Les guides routiers. Itinéraires et cartes routières de l'Europe*, Lille, 1926

Les routes de France, Paris, 1929

Formentini, Marco, *La Dominazione spagnuola in Lombardia*, Milan, 1881

Forneron, Henri, *Histoire de Philippe II*, Paris, 1881–1882, 4 vols

Forster, William, see Sanderson, John

Forti, U., *Storia della tecnica italiana*, Florence, 1940

Foscarini, J., see Stefani, Fr.

Fossombroni, V., *Memorie idraulico-storiche sopra la Val di Chiana*, Florence, 1789

Foucault, Michel, *L'histoire de la folie à l'âge classique*, Paris, 1961

Fouqueray, P. H., *Histoire de la Compagnie de Jésus en France des origines à la suppression (1528–1752)*, Paris, 1910–1925, 5 vols

Fourastié, Jean, *Prix de vente et prix de revient*, 13th series, Paris, undated (1964)

Foy, Abbé de, see Busbec(q), Baron O. G. de

Franc, Julien, *La colonisation de la Mitijda*, Paris, 1928

François, Michel, *Albisse del Bene, surintendant des finances françaises en Italie, 1551 à 1556*, 'Bibliothèque de l'École des Chartes', Paris, 1933

Correspondance du cardinal François de Tournon, 1521–1562, Paris, 1946

Le cardinal François de Tournon, homme d'État, diplomate, mécène et humaniste, 1489–1562, 'Bibliothèque des Écoles françaises d'Athènes et de Rome', Paris, 1951, fasc. 173

Frangipani, Ottavio Mirto, *Correspondance d'Ottavio Mirto Frangipani, premier nonce de Flandre (1596–1606)*, I, Rome, 1924, ed. L. van der Essen; II–III (1–2), ed. Armand Louant, 1932 and 1942

Franklin, Alfred, *Dictionnaire historique des arts, métiers et professions exercés dans Paris depuis le XIIIe siècle*, Paris, 1906

La vie privée d'autrefois: arts et métiers, modes, moeurs, usages des Parisiens, du XIIe au XVIIIe siècle, Paris, 1887–1902, 27 vols.; XIII: *Le café, le thé, le chocolat*, 1893; XV and XVI, *Les magasins de nouveautés*, 1894–1895

Franz, G., *Der Dreissigjährige Krieg und das deutsche Volk. Untersuchungen zur Bevölkerungs- und Agrargeschichte*, Jena, 1940

Fremerey, Gustav, *Guicciardinis finanzpolitische Anschauungen*, Stuttgart, 1931

Freyre, Gilberto, *Introdução à histôria da sociedade patriarcal no Brasil;* I: *Casa grande y senzala*, Rio de Janeiro, 5th ed., 1946, 2 vols. II: *Sobrados e mucambos*, 2nd ed., Rio de Janeiro, 1951, 3 vols

Frianoro, Rafaele, see Nobili, Giacinto

Fried, Ferdinand, *Wandlungen der Weltwirtschaft*, Munich, 1950; (French translation of first ed., Leipzig, 1939: *Le tournant de l'économie mondiale*, Paris, 1942)

Friederici, Georg, *Der Charakter, der Entdeckung und Eroberung Amerikas durch die Europaër*, Stuttgart, 1925–1936, 3 vols

Frobenius, Leo, *Histoire de la civilisation africaine*, trans. from the German by Dr. H. Back and D. Ermont, Paris, 1936

Frödin, J., *Zentraleuropas Alpwirtschaft*, Oslo, 1940–1941, 2 vols

Fromentin, Eugène, *Voyage en Egypte (1869)*, ed. J. M. Carré, Paris, 1935

Fuchs, R., *Der Bancho Publico zu Nürnberg*, Berlin, 1955

Fuentes Martiáñez, M., *Despoblación y repoblación de España (1482–1920)*, Madrid, 1929

Fueter, Eduard, *Geschichte des europäischen Staatensystems von 1492–1559*, Munich, 1919

Fugger, see Klarwill, V. von

Gabre, Dominique du, see Vitalis, A.

Gachard, L. P., *Correspondance de Philippe II sur affaires des Pays-Bas (jusqu'en 1577)*, Brussels, 1848–1879, 5 vols

Retraite et mort de Charles Quint au monastère de Yuste, Brussels, 1854–1855, 3 vols

Don Carlos et Philippe II, Brussels, 1863, 2 vols

Correspondance de Marguerite d'Autriche avec Philippe II (1559–1565), Brussels, 1867–1881, 3 vols

Lettres de Philippe II à ses filles, les infantes Isabelle et Catherine, écrites pendant son voyage au Portugal, 1581–1583, Paris, 1884

Gachard, L. P. and Piot, Ch., *Collection des voyages des souverains des Pays-Bas*, Brussels, 1876–1882, 4 vols

Gaffarel, P.. *Histoire du Brésil français au XVIe siècle*, Paris, 1878

Galanti, G. M., *Descrizione geografica e politica delle Due Sicilie*, Naples, 1788, 4 vols in 2

Gallardo y Victor, Manuel, *Memoria escrita sobre el rescate de Cervantes*, Cadiz, 1896

Galluzzi, R., *Istoria del granducato di Toscana sotto il governo della casa Medici*, Florence, 1781, 5 vols

Gamir Sandoval, A., *Organización de la defensa de la costa del Reino de Granada desde su reconquesta hasta finales del siglo XVI*, Granada, 1947

Gandilhon, René, *Politique économique de Louis XI*, Paris, 1941

Ganier, Germaine, *La politique du connétable Anne de Montmorency*, Le Havre, undated (1957)

Ganivet, García, Angel, *Obras completas;* I: *Granada la Bella, Idearium español*, Madrid, 1943

García, Juan, see Cataline García, Juan and Pérez Villamil, Manuel
García de Quevedo y Concellón, Eloy, *Ordenanzas del Consulado de Burgos de 1538*, Burgos, 1905
García Fernández, Jesús, *Aspectos del paisaje agrario de Castilla la vieja*, Valladolid, 1963
García Mercadal, G., *Viajes de extrangeros por España y Portugal;* I. *Viaje del noble bohemio León de Rosmithal de Blatina por España y Portugal hecho del año 1465 a 1467*, Madrid, 1952
García y García, Luis, *Una embajada de los reyes católicos a Egipto*. Valladolid, 1947
Gassot, Jacques, *Le discours du voyage de Venise à Constantinople*, Paris, 1606
Gautier, Dr. Armand, *L'alimentation et les régimes chez l'homme sain et chez la malade*, Paris, 1904
Gautier, Émile-Félix, *L'islamisation de l'Afrique du Nord. Les siècles obscurs du Maghreb*, Paris, 1927
Un siècle de colonisation, Paris, 1930
Moeurs et coutumes des Musulmans, Paris, 1931
Genséric, roi des Vandales, Paris, 1932
Le passé de l'Afrique du Nord, Paris, 1937
Gautier, Théophile, *Voyage en Espagne*, Paris, ed. 1854 and 1879. (English translation by C. A. Phillips, *A Romantic in Spain*, 1926)
Constantinople, Paris, 1853
Gavy de Mendonca, Agostinho, *Historia do famoso cerco que o xarife pos a fortaleza de Mazagão no año de 1562*, Lisbon, 1607
Gédoyn, 'le Turc', see Boppe, Léon
Gelzer, H., *Geistliches und Weltlisches aus dem türkisch-griechischen Orient*, Leipzig, 1900
Gentil da Silva, J., *Stratégie des affaires à Lisbonne entre 1595 et 1607*, Paris, 1956
Géographie Universelle, general editors P. Vidal de La Blache and L. Gallois, Paris, 1927, 23 vols
George, P., *La région du Bas-Rhône, étude de géographie régionale*, Paris, 1935
Gerlach, R., *Dalmatinisches Tagebuch*, Darmstadt, 1940
Gerometta, B., *I forestieri a Venezia*, Venice, 1858
Gévay, Anton von, *Urkunden und Aktenstücke zur Geschichte der Verhältnisse zwischen Österreich, Ungarn und der Pforte im 16. und 17. Jahrhundert*, Vienna, 1840–1842, 9 vols
Giannone, Pietro, *Istoria civile del Regno di Napoli*, The Hague, 1753, 4 vols
Gillet, Louis, *Dante*, Paris, 1941
Gioffrè, Domenico, *Genova e Madera nel I° decennio del secolo XVI*, in 'Studi Colombiani', vol. III, Genoa, 1951
Gênes et les foires de change: de Lyon à Besançon, Paris, 1960
Girard, Albert, *La rivalité commerciale et maritime entre Séville et Cadix jusqu'à la fin du XVIIIe siècle*, Paris, 1932
Girard, Paul, *Les origines de l'Empire français nord-africain*, Marseilles, 1937
Giustiniani, Girolamo, *La description et l'histoire de l'ile de Scios*, 1506 (? for 1606)
Glamann, K., *Dutch-Asiatic Trade, 1620–1740*, The Hague, 1958
Goldschmidt, L., *Universalgeschichte des Handelrechts*, Stuttgart, 1891
Goleta. Warhafftige eygentliche Beschreibung wie der Türck die . . . Vestung Goleta . . . belägert . . ., Nuremberg, 1574
Gollut, L., *Les Mémoires historiques de la république séquanoise*, Dole, 1592

Gomes de Brito, Bernardo, *Historia tragico-maritima*, Lisbon, 2nd ed., 1904–1909, 3 vols
Gomez Moreno, Manuel, see Hurtado de Mendoza, Diego
Gondola, Francesco, see Voinovitch
González, Tomás, *Censo de la población de las provincias y partidos de la Corona de Castilla en el siglo XVI*, Madrid, 1829
González Palencia, Angel, *Gonzalo Pérez secretario de Felipe II*, Madrid, 1946, 2 vols
Gooss, Roderich, *Die Siebenbürger Sachsen in der Planung deutscher Sudostpolitik. Von der Einwanderung bis zum Ende des Thronstreites zwischen König Ferdinand I. und König Johann Zapolya (1538)*, Vienna, 1940
Gosselin, E.-H., *Documents authentiques et inédits pour servir à l'histoire de la marine normande et du commerce rouennais pendant les XVIe et XVIIe siècles*, Rouen, 1876
Götz, Wilhelm, *Historische Geographie. Beispiele und Grundlinien*, Leipzig, 1904
Goubert, Pierre, *Beauvais et le Beauvaisis de 1600 à 1730*, Paris, 1960, 2 vols
Gounon-Loubens, J., *Essais sur l'administration de la Castille au XVIe siècle*, Paris, 1860
Gourou, Pierre, *La Terre et l'homme en Extrême-Orient*, Paris, 1940
Gothein, A., *Geniza.*
Grammont, Henri Delmas de, *Relations entre la France et la Régence d'Alger au XVIIIe siècle*, Algiers, 1879–1885, 4 vols
Grandchamp, Pierre, *La France en Tunisie à la fin du XVIe siècle (1582–1600)*, Tunis, 1920
Granvelle, Cardinal, *Papiers d'État du cardinal Granvelle*, edited by Ch. Weiss, 'Collection de documents inédits sur l'histoire de France', Paris, 1841–1852, 9 vols
Correspondance du cardinal Granvelle, 1566–1586, ed. Edmond Poullet and Charles Piot, Brussels, 1877–1896, 12 vols
Grataroli, G., *De regimine iter agentium, vel equitum, vel peditum, vel mari vel curru seu rheda*, Basel, 1561
Graziani, A., *Economisti del Cinque e Seicento*, Bari, 1913
Grekov, B. and Iakoubovski, A., *La Horde d'Or*, trans. from the Russian by Françoise Thuret, Paris, 1939
Grenard, Fernand, *Grandeur et décadence de l'Asie*, Paris, 1939
Grevin, Emmanuel, *Djerba, l'île heureuse, et le Sud tunisien*, Paris, 1937
Griziotti Kretschmann, Jenny, *Il problema del trend secolare nelle fluttuazioni dei prezzi*, Turin, 1935
Gröber, G., *Grundriss der romanischen Philiologie*, Strasbourg, 1888–1902, 3 vols.; 2nd ed., vol. I: 1904–1906
Grottanelli, Lorenzo, *La Maremma toscana: studi storici ed economici*, Siena, 1873–1876, 2 vols
Grousset, René, *L'empire des steppes*, Paris, 1939
Gaell, S., Marçais, G., Yver, G., *Histoire d'Algérie*, Paris, 1927
Guardia, G. M., see Pérez, Antonio
Guarnieri, Giuseppe Gino, *Un'audace impresa marittima di Ferdinando dei Medici, con documenti e glossario indo-caraibico*, Pisa, 1928
Cavalieri di Santo Stefano. Contributo alla storia della marina militare italiana, 1562–1859, Pisa, 1928
Guéneau, Louis, *L'organisation du travail à Nevers aux XVIIe et XVIIIe siècles (1600–1790)*, Paris, 1919

Guevara, A. de, *Epistolas familiares* . . ., French translation (*Epistres dorées, moralles et familières*), by Seigneur de Guterry, Lyons, 1558–1560; English translation: *Golden Epistles* . . ., by Geoffrey Fenton, London, 1577

Guglielmotto, Alberto, *La guerra dei pirati e la marina pontificia dal 1500 al 1560*, Florence, 1876, 2 vols

Guicciardini, Francesco, *La historia d'Italia*, Venice, 1568 (English translation by Fenton, *The Historie of Guicciardin*, London, 1570)
Diario del viaggio in Spagna, Florence, 1932

Guijo, G. M. de, *Diario de Gregorio Martín de Guijo*, 1648–1664, ed. M. R. de Terreros, Mexico, 1953

Guillaume de Vaudoncort, Frédéric François, *Memoirs of the Ionian Islands*, London, 1816

Guillon, Pierre, *Les trépieds du Ptoion*, Paris, 1943

Gunther, A., *Die Alpenländische Gesellschaft als sozialer und politischer, wirtschaftlicher und kultureller Lebanskreis*, Jena, 1930

Ha Cohen, Joseph, *Emek Habakha, la Vallée des Pleurs, Chronique des souffrances d'Israël dans sa dispersion jusqu'à 1575*, and *Continuation de la Vallée des Pleurs*, ed. Julien Sée, Paris, 1881

Haebler, Konrad, *Die wirtschaftliche Blüte Spaniens im 16. Jahrhundert und ihr Verfall*. Berlin, 1888
Geschichte Spaniens unter den Habsburgern, vol. I only: *Geschichte Spaniens unter der Regierung Karls I (V.)*, Gotha, 1907

Haëdo, P. Diego de, *Topographia e historia general de Argel* and *Epitome de los Reyes de Argel*, Valladolid, 1612, in one volume

Hagedorn, B., *Die Entwicklung der wichtigsten Schiffstypen bis ins 19. Jahrhundert*, Berlin, 1914

Hahn, W., *Die Verpflegung Konstantinopels durch staatliche Zwangswirtschaft, nach türkischen Urkunden aus dem 16. Jahrhundert*, Stuttgart, 1926

Hakluyt, Richard, *The principal navigations, voyages, traffiques and discoveries of the English nation*, London, 1599–1600, 3 vols

Halperin Donghi, Tulio, *Un conflicto nacional: Moriscos y Cristianos viejos en Valencia*, Buenos Aires, 1955

Halphen, E., see Henri IV

Hamilton, Earl J., *El florecimiento del capitalismo y otras ensayos de historia económica*, Madrid, 1948
American Treasure and the Price Revolution in Spain, 1501–1650, Cambridge, Mass., 1934

Hammen y León, Lorenzo Van der, see Vander Hammen y León, Lorenzo

Hammer-Purgstall, J. von, *Histoire de l'empire ottoman depuis son origine jusqu'à nos jours*, trans. from the German by J. J. Hellert, Paris, 1835–1848, 18 vols

Häpke, Rudolf, *Niederländische Akten und Urkunden zur Geschichte der Hanse und zur deutschen See geschichte*, vol. I only: *1531–1557*, Munich and Leipzig, 1913

Harris, John, *Navigantium atque itinerantium bibliotheca, or a complete collection of voyages and travels*, London, 1705, 2 vols

Hartlaub, F., *Don Juan d'Austria und die Schlacht bei Lepanto*, Berlin, 1940

Hassel, U. von, *Das Drama des Mittelmeers*, Berlin, 1940

Hauser, Henri, *La prépondérance espagnole (1559–1560)*, 2nd ed., Paris, 1940. See also Bodin, Jean

Hauser, Henri and Renaudet, Augustin, *Les débuts de l'âge moderne, la Renais-*

sance et la Réforme, 3rd ed., 1946, vol. VIII of *Histoire générale*, ed. Louis Halphen and Philippe Sagnac

Hayward, F., *Histoire de la Maison de Savoie*, Paris, 1941–1943, 2 vols

Heckscher, Eli F., *Der Merkantilismus*, Jena, 1932, 2 vols. English translation *Mercantilism*, by M. Shapiro, London, 1935, 2 vols

Heeringa, K., *Bronnen tot de geschiendnis van den Levantschen handel*, The Hague, 1910–1917, 2 vols

Heers, Jacques, *Gênes au XVe siècle. Activité économique et problèmes sociaux*, Paris, 1961

Hefele, Charles-Joseph, *Le cardinal Ximénès et l'église d'Espagne à la fin du XVIe et au début du XVIIe siècle*, trans. from the German by M. l'abbé A. Sisson and M. l'abbé A. Crampon, Paris–Lyons, 1856

Hefele, Charles-Joseph and Hergen Roether, Cardinal J., *Histoire des Conciles d'après des documents originaux*, trans. into French by Dom H. Leclerq and continued in twentieth century; vol. IX, part I, *Concile de Trente*, by P. Richard, Paris, 1930

Hefele, Hermann, *Geschichte und Gestalt. Sechs Essays*, Leipzig, 1940

Helwig, Werner, *Braconniers de la mer en Grèce* (trans. from the German *Raubfischer in Hellas*, by Maurice Rémon, Leipzig, 1942)

Hennig, Richard, *Terrae incognitae. Eine Zusammenstellung und kritische Bewertung der wichtigsten vorcolumbischen Entdeckungsreisen an Hand der darüber vorleigenden Originalberichte*, Leyden, 1936–1939, 4 vols., 2nd ed., 1944–1956, 4 vols

Henri IV, king of France, *Lettres inédites à M. de Sillery, ambassadeur à Rome, du 1er avril au 27 juin 1601*, ed. Eugène Halphen, Paris, 1866

Lettres inédites à M. de Villiers, ambassadeur à Venise, ed. Eugène Halphen, Paris, 1885–1887, 3 vols

Lettres au comte de Rochepot, ambassadeur en Espagne (1600–1601), ed. P. Laffleur de Kermaignant, Paris, 1889

Hentzner, Paul, *Itinerarium Germaniae, Galliae, Italiae*, Nuremberg, 1612

Herder, Johann Gottfried von, *Ideen zur Geschichte der Menschheit*, Riga and Leipzig, 1784–1791, 4 vols. (French trans., *Philosophie de l'histoire de l'humanité*, by Emile Tandel, Paris, 1874, 3 vols)

Hering, Ernst, *Die Fugger*, Leipzig, 1940

Héritier, *Catherine de Médicis*, Paris, 1940

Herre, Paul, *Europäische Politik im cyprischen Krieg (1570–1573), mit Vorgeschichte und Vorverhandlungen*, Leipzig, 1902

Papsttum und Papstwahl im Zeitalter Philipps II., Leipzig, 1907

Weltgeschichte am Mittelmeer, Leipzig, 1930

Herrera, Gabriel Alonso de, *Libro de Agricultura*, Alcalá, 1539 and 1598

Herrera Oria, Enrique, *La Armada Invencible*, in the collection 'Archivo Historico Español', Madrid, 1929

Herrera y Tordesillas, Antonio de, *Primera (tercera) parte de la Historia general del mundo*, Madrid, 1601–1612, 3 vols

Heyd, W., *Histoire du Commerce du Levant au Moyen Age*, trans. and ed. Furcy-Raynaud, Leipzig, 1885–1886, 2 vols

Hiltebrandt, Philipp, *Der Kampf ums Mittelmeer*, Stuttgart, 1940

Hinojosa, Ricardo de, *Los despachos de la diplomacia pontificia en España*, Madrid, 1896

Hirth, Friedrich C. A. J., *Chinesische Studien*, vol. I only, Munich, 1890

Hispanic Studies in honour of J. Gonzáles Llubera, ed. Frank Pierce, Oxford, 1959

Histoire et Historiens de l'Algérie, Paris, 1931
Historiadores de Indias, general editor Manuel Serrano y Sanz, Madrid, 1909, 2 vols. II, 'Guerra de Quito', by Pedro de Cieza de León, 'Jornada de Managua y Dorado', by Toribio de Ortiguera, 'Descripción del Perú Tucuman, Rio de la Plata y Chile', by Fr. Reginaldo de Lizárraga, in the *Nueva Biblioteca de autores españoles*, dirigida por Marcelino Menéndez y Pelayo, vols. XIV and XV
Höffner, Joseph, *Wirtschaftsethik und Monopole im 15. und 16. Jahrhundert*, Jena, 1941
Holland, Henry, *Travels in the Ionian Isles, Albania, Thessaly, Macedonia, etc., during the years 1812 and 1813*, London, 1815
Holleaux, Maurice, *Rome, la Grèce et les monarchies hellénistiques au IIIe siècle avant Jésus-Christ (273–205)*, Paris, 1921
Hollister, J. S., see Ashauer, H. and Hollister, J.
Hommage à Lucien Febvre. Éventail de l'histoire vivante, Paris, 1953, 2 vols
Hopf, Carl, see Musachi, Giovanni
Hoszowski, St., *Les prix à Lwow (XVIe–XVIIe siècles)*, trans. from the Polish, Paris, 1954
Howe, Sonia, E., *In Quest of Spices*, London, 1946 (2nd ed.)
Howe, W., *The Mining Guild of New Spain and its Tribunal General, 1770–1821*, Cambridge, 1949
Hugo, Victor, *William Shakespeare*, Paris, 1882
Hürlimann, Martin, *Griechenland mit Rhodos und Zypern*, Zurich, 1938
Hurtado de Mendoza, Diego, *De la guerra de Granada, comentarios*, ed. Manuel Gómez Morena, Madrid, 1948
Huvelin, P., *Essai historique sur le droit des marchés et des foires*, Paris, 1897
Huxley, Aldous, *Jesting Pilate, The diary of a journey*, London, 1926
Iakoubovski, A., see Grekov, B. and Iakoubovski, A.
Ibn Iyās, *Journal d'un bourgeois du Caire, Histoire des Mamlouks*, trans. and annotated by Gaston Wiet, Paris, 1955–1960, 2 vols
Ibn Verga, Salomon, *Liber Schevet Jehuda*, ed. M. Wiener, Hanover, 1855–1856, 2 vols
Illescas, Gonzalo de, *Historia pontifical y católica*, Salamanca, 1573
Imbart de la Tour, Pierre, *Les origines de la Réforme*, 2nd ed., Melun, 1944–1945, 2 vols
Imbert, Gaston-Paul, *Des mouvements de longue durée Kondratieff*, Aix-en-Provence, 1959
Indice de la colección de documentos de Fernández de Navarrete que posee el Museo Naval, Madrid, 1946
Instructions Nautiques du service hydrographique de la Marine française, in particular nos. 357, 360, 368, Paris, 1932 and 1934
Iorga, N., *Geschichte des osmanischen Reiches*, Gotha, 1908–1913, 5 vols. *Points de vue sur l'histoire du commerce de l'Orient au moyen âge*, Paris, 1924 *Ospiti romeni in Venezia*, Bucharest, 1932
Jacobeit, Wolfgang, *Schafhaltung und Schäfer in Zentraleuropa bis zum Beginn des 20. Jahrhunderts*, Berlin, 1961
Jadin, J., see Cuvelier, J. and Jadin, J.
Jäger, Fritz, *Afrika*, 3rd ed., Leipzig, 1928
Jal, A., *Glossaire nautique*, Paris, 1948
Janaček, J., *A History of the trade of Prague before the battle of the White Mountain* (in Czech), Prague, 1955

Janssen, Johannes, *Geschichte des deutschen Volkes, seit dem Ausgang des Mittelalters*, Fribourg-in-Brisgau, 1878–1894, 8 vols
Jardé, Auguste, *Les céréales dans l'Antiquité;* I, *La production*, 'Bibliothèque des Écoles françaises d'Athènes et de Rome', Paris, 1925
Jean-Brunhes-Delamarre, Mariel, see Deffontaines, P. *et al*
Jelavich, C. and B., *The Balkans in transition: essays on the development of Balkan life and politics since the eighteenth century*, ed. C. and B. Jelavich, Berkeley, 1963
Jireček, Constantin, *Die Romanen in den Städten Dalmatiens während des Mittelalters*, Vienna, 1901–1904, 3 vols
Joly, Barthélémy, *Voyage en Espagne, 1603–1604*, ed. L. Barrau-Dihigo, Paris, 1909
Joly, Henry, *La Corse française au XVIe siècle*, Lyons, 1942
Jones, W. H. S., *Malaria, a neglected factor in the history of Greece and Rome*. Cambridge, 1907
Jonge, Johannes Cornelis de, *Nederland en Venetie*, The Hague, 1852
Juchereau de Saint-Denys, Antoine, *Histoire de l'empire ottoman depuis 1792 jusqu'en 1844*, Paris, 4 vols
Julien, Charles-André, *Histoire de l'Afrique du Nord*, Paris, 1931
Jurien de la Gravière, Vice-admiral J. B. E., *Les chevaliers de Malte et la marine de Philippe II*, Paris, 1887, 2 vols
Justinian, Jérosme, see Giustinañi, Girolamo
Kellenbenz, Hermann, *Sephardim an der unteren Elbe. Ihre wirtschaftliche und politische Bedeutung vom Ende des 16. bis zum Beginn des 18. Jahrhunderts*, Wiesbaden, 1958
Kerhuel, Marie, *Les mouvements de longue duréé des prix*, Rennes, 1935
Kermaingant, P., Laffleur de, see Henri IV and Laffleur de Kermaingant, P.
Kernkamp, J. H., *De handel op den vijand, 1572–1609*, Utrecht, 1931–1934, 2 vols
Kerremans, Charles, see Doehaerd, Renée and Kerremans, Charles
Kirchner, Walther, *The Rise of the Baltic Question*, Newark, 1954
Klarwill, Victor von, *The Fugger News-Letters*, London, 1924–1926, 2 vols
Klaveren, Jacob van, *Europäische Wirtschaftsgeschichte Spaniens im 16. und 17. Jahrhundert*, Stuttgart, 1960
Klein, Julius, *The Mesta: A Study in Spanish Economic History, 1273–1836*, Cambridge, 1920
Koch, Matthias, *Quellen zur Geschichte des Kaisers Maximilian II*, Leipzig, 1857–1861, 2 vols
Konetzke, R., *Geschichte des spanischen und portugiesischen Volkes*, Leipzig, 1939
Kretschmann, Jenny, see Griziotti, Kretschmann, Jenny
Kretschmayr, H., *Geschichte von Venedig*, Gotha and Stuttgart, 1905–1934, 3 vols
Kroker, E., *Handelsgeschichte der Stadt Leipzig*, Leipzig, 1925
Kronn und Aussbunde aller Wegweiser, Cologne, 1597 (anonymous)
Kulischer, Josef, *Allgemeine Wirtschaftsgeschichte des Mittelalters und der Neuzeit*, Munich, 1928–1929, 2 vols; reprinted 1958
Kunze, A., see Aubin, G. and Kunze, A.
Laborde, Cte Alexandre-Louis de, *Itinéraire descriptif de l'Espagne*, Paris, 1827–1830, 6 vols
La Borderie, Bertrand de, *Le Discours du voyage de Constantinople*, Lyons, 1542
La Boullaye, Le Gouz, François, *Les voyages et observations du sieur de La Boullaye le Gouz où sont décrites les religions, gouvernements et situations des Estats et royaumes d'Italie, Grèce, Natolie, Syrie, Palestine, Karaménie,*

Kaldée, Assyrie, Grand Mogol, Bijapour, Indes orientales des Portugais, Arabie, Egypte, Hollande, Grande-Bretagne, Irlande, Danemark, Pologne, isles et autres lieux d'Europe, Asie et Afrique . . ., Paris, 1653

La Bruyère, René, *Le drame du Pacifique*, Paris, 1943

La Civiltà veneziana del Rinascimento, Fondazione Giorgio Cini, Venice, 1958

Lacoste, L., *Mise en valeur de l'Algérie. La colonisation maritime en Algérie*, Paris, 1931

La Ferrière-Percy, Comte Hector de, *Lettres de Catherine de Médicis (1533–1587)*, in the 'Collection de Documents inédits sur l'Histoire de France', Paris, 1880–1909, 10 vols., Index, Paris, 1943, 1 vol

Laffleur, de Kermaingant, Pierre-Paul, *Mission de Jean de Thumery, sieur de Boissise (1598–1602)*, Paris, 1886
See also Henry IV

La Jonquière, Vicomte A. de, *Histoire de l'empire ottoman depuis les origines jusqu'à nos jours*, Paris, 1914, 2 vols., in *Histoire Universelle* ed. Victor Duruy

La Lauzière, J. F. Noble de, see Noble de la Lauzière

Lamansky, Vladimir, *Secrets d'Etat de Venise, documents, extraits, notices et études*, St. Petersburg, 1884

La Marmora, Alberto Ferrero de, *Voyage en Sardaigne ou description statistique, physique et politique de cette île*, 2nd ed., Paris and Turin, 1839–1857, 4 vols

Landry, Adolphe, *Traité de démographie*, Paris, 1945

Lane, Frederic C., *Venetian Ships and Shipbuilders of the Renaissance*, Baltimore, 1934
Andrea Barbarigo, Merchant of Venice, 1418–1449, Baltimore, 1944

Lanz, Karl, *Correspondenz des Kaisers Karl V,· Aus dem kgl. Archiv und der 'Bibliothèque de Bourgogne' zu Brüssel*, Leipzig, 1844–1846, 3 vols

Lanza del Vasto, *La baronne de Carins*, Paris, 1946

Lapeyre, Henri, *Une famille de marchands, les Ruiz; contribution à l'étude du commerce entre la France et l'Espagne au temps de Philippe II*, Paris, 1955
Géographie de l'Espagne morisque, Paris, 1960

Lapini, Agostino, *Diario fiorentino di Agostino Lapini dal 252 al 1596*, ed. G. O. Corazzini, Florence, 1900

La Primaudais, see Élie de la Primaudaie, F.

La Roncière, Charles de, *Histoire de la marine française*, Paris, 1899–1932, 6 vols

Larruga, Eugenio, *Memorias políticas y económicas sobre los frutos, comercio, fábricas y minas de España*, Madrid, 1745–1792, 45 vols

La Torre y Badillo, M., *Representación de los autos sacramentales en el periodo de su mayor florecimiento*, Madrid, 1912

Lattes, E., *La libertà delle banche a Venezia*, Milan, 1869

Laval, François, Pyrard de, see Pyrard de Laval, François

Lavedan, Pierre, *Histoire de l'Art*, Paris, 1949–1950, 2 vols

Lavisse, Ernest, *Histoire de France depuis les origines jusqu'à la révolution*, Paris, 1903–1911, 18 vols

Lebel, Roland, *Le Maroc et les écrivains anglais aux XVIe, XVIIe et XVIIIe siècles*, Paris, 1927

Leca, Philippe, *Guide bleu de la Corse*, Paris, 1935

Leclerq, Dom H., see Hefele, Charles-Joseph and Hergen Roether, Cardinal

Leclercq, Jules, *De Mogador à Biskra; Maroc et Algérie*, Paris, 1881

Le Danois, Edouard, *L'Atlantique, histoire et vie d'un océan*, Paris, 1938

Lefaivre, Albert, *Les Magyars pendant la domination ottomane en Hongrie, 1526–1722*, Paris, 1902

Lefebvre, Georges, *La grande Peur de 1789*, undated (1957)
Lefebvre, Th., *Les modes de vie dans les Pyrénées atlantiques*, Paris, 1933
Lefebvre des Noëttes, Cdt., *L'attelage. Le Cheval de selle à travers les âges. Contribution à l'histoire de l'esclavage*, Paris, 1931
Lefevre-Pontalis, Germain, see Selve, Odet de
Le Glay, Dr. André, *Négociations diplomatiques entre la France et l'Autriche durant les trente premières années du XVIe siècle*, 'Collection de documents inédits sur l'histoire de France', Paris, 1845, 2 vols
Le Lannou, Maurice, *Pâtres et paysans de la Sardaigne*, Paris, 1941
Lenglet du Fresnay, M. l'abbé, see Comines, Philippe de
Leo Africanus (Leo Johannes Africanus), *Description de l'Afrique, tierce partie du Monde*, Lyons, 1556; modern edition, *De l'Afrique, contenant la description de ce pays et la navigation des anciens capitaines portugais aux Indes Orientales et Occidentales*, trans. Jean Temporal, Paris, 1830, 4 vols. (Page references in the text refer to these editions). An English version: *A Geographical Historie of Africa* . . . trans. and collected by J. Pory, 420 p., London, 1600; another ed., 1896, ed. Dr. R. Brown
Le Roy, Loys, *De l'excellence du gouvernement royal avec exhortation aux François de persévérer en iceluy*, Paris, 1575
Le Roy Ladurie, Emmanuel, *Les paysans de Languedoc*, 2 vols., Paris, 1966
Leti, Gregorio, *Vita del Catolico re Filippo II monarca delle Spagne*, Coligny, 1679, 2 vols. (French trans. by J.-G. de Chèvrières, *La vie de Philippe II, roi d'Espagne*, Amsterdam, 1734)
Lescarbot, Marc, *Histoire de la Nouvelle France*, Paris, 1611
Levi, Carlo, *Christ stopped at Eboli* (*Cristo si è fermato a Eboli*), trans. F. Frenaye, London, 1959
L'Herba, G. da, *Itinerario delle poste per diverse parti del mondo*, Venice, 1564
L'Hermite de Soliers, Jean-Baptiste known as Tristan, *La Toscane françoise*, Paris, 1661
Lilley, S., *Men, Machines and History; a short history of tools and machines in relation to social progress*, London, 1948
Lisičar, V., *Lopud. Historički i savremeni prikaz*, Dubrovnik, 1931
Livet, Roger, *Habitat rural et structures agraires en Basse Provence*, Gap, 1962
Livi, Giovanni, *La Corsica e Cosimo de' Medici*, Florence-Rome, 1885
Livi, R., *La schiavitù domestica nei tempi di mezzo e nei moderni*, Padua, 1928
Lizárraga, Fr. Reginaldo de, see *Historiadores de Indias*
Lonchay, Henri and Cuvelier, Joseph, *Correspondance de la cour d'Espagne sur les affaires des Pays-Bas, 1598–1621*, Brussels, 1923.
Longlée, Pierre de Ségusson, de, *Dépêches diplomatique de M. de Longlée, résident de France en Espagne, 1581–1590*, ed. A. Mousset, Paris, 1912
Lopez, Roberto S., *Studi sull'economia genovese nel medio evo*, Turin, 1936
Lortz, Joseph, *Die Reformation in Deutschland*, 2nd ed., Fribourg-in-Brisgau, 1941, 2 vols
Los Españoles pintados por si mismos (collective work), Madrid, 1843
Lot, Ferdinand, *Les invasions barbares et le peuplement de l'Europe, introduction à l'intelligence des derniers traités de paix*, Paris, 1937
Louant, Armand, see Frangipani, Ottavio Mirto
'Loyal Serviteur', see Bayard
Lozach, J., *Le delta du Nil, étude de géographie humaine*, Cairo, 1935
Lubimenko, Inna, *Les relations commerciales et politiques de l'Angleterre avec la Russie avant Pierre Le Grand*, Paris, 1933

Luccari, G., *Annali di Rausa*, Venice, 1605
Lucchesi, E., *I monaci benedettini vallombrosani in Lombardia*, Florence, 1938
Lucio de Azevedo, J., *Historia dos Christãos novos portugueses*, Lisbon, 1921
Luetić, J., *O pomorstvu Dubrovačke Republike u XVIII. stoljeću*, Dubrovnik, 1959
Lusignano, Stefano, *Chorografia et breve historia universale dell'isola de Cipro*. Bologna, 1573; French trans., Paris, 1580
Luzac, Élie de, see Accarias de Sérionne, Jacques
Luzzatto, Gino, *Storia economica dell'età moderna e contemporanea*, Padua, 1932
Storia economica di Venezia dall' XI al XVI secolo, Venice, 1961
Madariaga, S. de, *Spain and the Jews*, London, 1946
Maffei, Giovanni Pietro, *Historiarum Indicarum . . .*, Florence, 1588. French trans., *Histoire des Indes*, Lyon, 1603
Magalhães, Godinho, Vitorino, *Historia economica e social da expansão portuguesa*, vol. I, Lisbon, 1947
Os descobrimientos e a economia mondial, Lisbon, 1963
Les finances de l'État portugaise des Indes orientales au XVIe et au début du XVIIe siècle, typed thesis, Paris, 1958
L'économie de l'Empire portugais aux XVe et XVIe siècles. L'or et le poivre. Route de Guinée et route du poivre, Paris, SEVPEN, 1969
Maisons et villages de France, collective publication, Paris, 1945
Mal, J., *Uskočke seobe i slovenske pokrajine*, Ljubljana, 1924
Mâle, Émile, *L'art religieux après le Concile de Trente. Étude sur l'iconographie de la fin du XVIe siècle, du XVIIe siècle, du XVIIIe siècle. Italie, France, Espagne, Flandres*, Paris, 1932
Malestroict, sieur de, *Paradoxes inédits du Sieur de Malestroict touchant les monnoyes*, ed. Luigi Einaudi, Turin, 1937
Malraux, André, *La lutte avec l'Ange*, Geneva, 1945
Malynes, Gerard, *A Treatise of the Canker of England's Commonwealth*, London, 1601
Mandich, Giulio, *Le pacte de ricorsa et le marché italien des changes au XVIIe siècle*, Paris, 1953
Manfroni, C., *Storia della marina italiana*, Rome, 1897
Mankov, A. G., *Le mouvement des prix dans l'État russe du XVIe siècle*, French trans., Paris, 1957
Mans, Raphaël du, see Raphaël du Mans
Mantran, Robert, *Istanbul dans la seconde moitié du XVIIe siècle*, Paris, 1962
Marañon, Gregorio, *Antonio Pérez*, 2nd ed., Madrid, 1948, 2 vols
Marca, P. de, *Histoire de Béarn*, Paris, 1640
Marçais, Georges, see Diehl, C. and Marçais; Gsell, Marçais and Yver
Marciani, Corrado, *Lettres de change aux foires de Lanciano au XVIe siècle*, Paris, 1962
Marcucci, Ettore, see Sassetti, F.
Margaret of Austria, see Gachard, L.-P.
Mariana, Juan, *Storiae de rebus Hispaniae*, libri 25; vol. I of the continuation ed. Manuel José de Medrano, Madrid, 1741
Marliani, Giovanni Bartolomeo, *Topographia antiquae Romae*, Lyons, 1534
Martiáñez, M., see Fuentes Martiáñez, M.
Martin, Alfred von, *Sociologia del Renacimiento*, Mexico, 1946
Martin, Felipe Ruíz, see Ruiz Martín, F.
Martin, Henri-Jean, see Febvre, Lucien and Martin, H.-J.

Martínez, Mariano, see Alcocer Martínez, M.
Martínez de Azcoitia, Herrero, *La Población Palentina en los siglos XVI y XVII*, Palencia, 1961
Martínez, Ferrando, J. E., *Privilegios otorgados por el emperador Carlos V . . .*, Barcelona, 1943
Marx, Karl, *A contribution to the Critique of Political Economy* (*Zur Kritik der politischen Oekonomie*), trans. N. I. Stone, London, 1904
Mas-Latrie, Jacques-M. J. L., *Traités de paix et de commerce . . .*, Paris, 1866, 2 vols
Massieu, abbé Guillaume, *Histoire de la Poësie française avec une défense de la Poësie*, Paris, 1739
Massignon, Louis, *Annuaire du monde musulman*, Paris, 1955
Masson, Paul, *Histoire du commerce français dans le Levant au XVIIe siècle*, Paris, 1896
Histoire du commerce français dans le Levant au XVIIIe siècle, Paris, 1911
Les Compagnies du Corail, Paris, 1928
Maull, Otto, *Geographie der Kulturlandschaft*, Berlin and Leipzig, 1932
Maunier, René and Giffard, A., *Faculté de droit de Paris, salle de travail d'ethnologie juridique: Conférences 1929–30. Sociologie et Droit romain*, Paris, 1930
Maurand, Jérôme, *Itinéraire de Jérôme Maurand d'Antibes à Constantinople* (*1544*), ed. Léon Dorez, Paris, 1901
Maurel, Paul, *Histoire de Toulon*, Toulon, 1943
Mauro, F., *Le Portugal et l'Atlantique au XVIIe siècle, 1570–1670*, Paris, 1960
Mayer-Löwenschwerdt, Erwin, *Der Aufenthalt der Erzherzöge Rudolf und Ernst in Spanien, 1564–1571*, Vienna, 1927
Mayerne, Théodore Turquet de, *Sommaire description de la France, Allemagne Italie, Espagne, avec la guide des chemins et postes*, Rouen, 1615
Mazzei, J., *Politica doganale differenziale e clausola della nazione più favorita*, Florence, 1930
Mecatti, G. M., *Storia cronologica della città di Firenze*, Naples, 1755, 2 vols
Medici, Catherine de, see La Ferrière, count H. de
Medina, Pedro de, *Libro de grandezas y cosas memorables de España*, Alcalá de Henares, 1595
Medrano, José de, see Mariana
Meester, B. de, *Le Saint-Siège et les troubles des Pays-Bas, 1566–1579*, Louvain, 1934
Meilink-Roelofsz, M. A. P., *Asian Trade and European Influence in the Indonesian Archipelago between 1500 and about 1630*, The Hague, 1962
Meinecke, F., *Die Idee der Staatsräson in der neueren Geschichte*, Munich, 1924
Meister der Politik, ed. Erich Marks and Karl Alexander v. Müller, 2nd ed. Stuttgart, 1923–1924, 3 vols
Mélanges en l'honneur de Marcel Bataillon, Paris, 1962
Mélanges Luzzatto, Studi in onore di Gino Luzzatto, Milan, 1950, 4 vols
Melis, Federigo, *Aspetti della vita economica medievale*, Siena-Florence, 1962
Mellerio, Joseph, *Les Mellerio, leur origine et leur histoire*, Paris, 1895
Mendez de Vasconcelos, Luis, 'Diálogos do sítio de Lísboa', 1608, in Antonio Sérgio, *Antologia dos Economistas Portugueses*, Lisbon, 1924
Mendonça, see Gavy de Mendonça
Mendoza, Diego de, see Hurtado de Mendoza
Mendoza y Bobadilla, Cardinal Francisco, *Tizon de la nobleza española*, Barcelona, 1880

Menéndez, Pidal, Gonzalo, *Los caminos en la historia de España*, Madrid, 1951
Menéndez Pidal, Ramón, *Idea imperial de Carlos V*, Madrid, 1940
Mentz, Georg, *Deutsche Geschlichte im Zeitalter der Reformation, der Gegenreformation und des Dreissigjährigen Krieges*, 1493–1648, Tübingen, 1913
Mercadal, G. García, see García Mercadal, G.
Mercier, Ernest, *Histoire de l'Afrique septentrionale (Berbérie) depuis les temps les plus reculés jusqu'à la conquête française (1830)*, Paris, 1888–1891, 3 vols
Mérimée, Henri, *L'art dramatique à Valencia depuis les origines jusqu'au commencement du XVIIe siècle*, Toulouse, 1913
Merle, L., *La métairie et l'évolution agraire de la Gâtine poitevine de la fin du Moyen Age à la Révolution*, Paris, 1958
Merner, Paul-Gerhardt, *Das Nomadentum im nordwestlichen Afrika*, Stuttgart, 1937
Meroni, Ubaldo, I *'Libri delle uscite delle monete' della Zecca di Genova dal 1589 al 1640*, Mantua, 1957
Merriman, R. B., *The Rise of the Spanish Empire in the Old World and in the New*, New York, 1918–1934, 4 vols
Mesnard, Pierre, *L'essor de la philosophie politique au XVIe siècle*, Paris, 1936
Meyer, Arnold O., *England und die katholische Kirche unter Elisabeth und den Stuarts*, vol. I only, *England . . . unter Elisabeth*, Rome, 1911
Michel, Francisque, *Histoire des races maudites de la France et de l'Espagne*, Paris, 1847, 2 vols
Michel, Paul-Henri, *Giordano Bruno, philosophe et poète*, Paris, 1952 (extract from 'Collège philosophique': *Ordre désordre, lumière*)
La cosmologie de Giordano Bruno, Paris, 1960
Michelet, Jules, *Histoire de France*, vol. VII: *La Renaissance*, Paris, 1855
Mignet, F.-Auguste-A., *Charles Quint, son abdication, son séjour et sa mort au monastère de Yuste*, Paris, 1868
Milano, Attilio, *Storia degli ebrei in Italia*, Turin, 1963
Milioukov, P., Seignobos, Charles and Eisenmann, Louis, *Histoire de Russie*, Paris, 1932–1939, 2 vols
Milojević, Borivoje, *Littoral et îles dinariques dans le royaume de Yougoslavie* (Mémoires de la Société de Géographie, vol. 2), Belgrade, 1933
Minguijón, S., *Historia del derecho español*, Barcelona, 1933
Mira, Giuseppe, *Aspetti dell'economia comasca all'inizio dell'età moderna*, Como, 1939
Moheau, M., *Recherches et considérations sur la population de la France*, Paris, 1778
Monchicourt, Charles, *L'expédition espagnole de 1560 contre l'île de Djerba*, Paris, 1913
Mondoucet, C. de, *Lettres et négociations de Claude de Mondoucet, résident de France aux Pays-Bas (1571–1574)*, ed. L. Didier, Paris, 1891–1892, 2 vols
Monod, Th., *L'hippopotame et le philosophe*, Paris, 1943
Montagne, R., *Les Berbères et le Makhzen dans le Sud du Maroc*, Paris, 1930
Montaigne, Michel Eyquem de, *Journal de voyage en Italie*, ed. Ed. Pilon, Paris, 1932
Montanari, Geminiano, *La zecca in Consulta di Stato*, ed. A. Graziani, Bari, 1913
Montchrestien, Antoine de, *L'économie politique patronale, traicté d'œconomie politique*, ed. Th. Funck-Brentano, Paris, 1889
Monteil, Amans-Alexis, *Histoire des Français*, Paris, 1828–1844, 10 vols

Morales, A. de, *Las antigüedades de las ciudades de España*, Madrid, 1792
Morand, Paul, *Lewis et Irène*, Paris, 1931
Morazé, Charles, *Introduction à l'histoire économique*, Paris, 1943
Morel-Fatio, Alfred, *L'Espagne au XVIe et au XVIIe siècle*, Heilbronn, 1878
Études sur l'Espagne, 1st series, 2nd ed.: *L'Espagne en France*, Paris, 1895
Études sur l'Espagne, 4th series, Paris, 1925
Ambrosio de Salazar et l'étude de l'espagnol en France sous Louis XIII, Paris, 1900
Moscardo, L., *Historia di Verona*, Verona, 1668
Mousset, A., see Longlée, P. de Ségusson de
Mouton, Léo, *Le Duc et le roi: d'Épernon, Henri IV, Louis XIII*, Paris, 1924
Müller, Georg, *Die Türkenherrschaft in Siebenbürgen. Verfassungsrechtliches Verhältnis Siebenbürgens zur Pforte, 1541–1688*, Hermannstadt-Sibiu, 1923
Müller, Johannes, *Zacharias Geizkofler, 1560–1617, des Heiligen Romischen Reiches Pfennigmeister und oberster Proviantmeister im Königreich Ungarn*, Baden, 1938
Müller, K. O., *Welthandelsbräuche (1480–1540)*, Stuttgart, 1934; 2nd ed. Weisbaden, 1962
Musachi, Giovanni, *Historia genealogica della Casa Musachi*, ed. Carl Hopf in *Chroniques gréco-romaines inédites ou peu connues*, Berlin, 1873
Nadal, G. and Giralt, E., *La population catalane de 1553 à 1717*, Paris, 1960
Nalivkin, K., *Histoire du Khanat de Khokand*, Paris, 1889
Naudé, W., *Die Getreidehandelspolitik der europäischen Staaten vom 13. bis zum 18. Jahrhundert*, Berlin, 1896
Navagero, Andrea, *Il viaggio fatto in Spagna*, Venice, 1563
Nef, John U., *The Rise of the British Coal Industry*, London, 1932, 2 vols
Nelson, John Charles, *Renaissance Theory of Love, the Context of Giordano Bruno's 'Eroici furori'*, New York, 1958
Niccolin, Fausto, *Aspetti della vita italo-spagnuola nel Cinque e Seicento*, Naples, 1934
Nicot, Jean, *Jean Nicot, ambassadeur de France au Portugal au XVIe siècle. Sa correspondance inédite*, ed. E. Falgairolle, Paris, 1897
Nicolay, Nicolas de, *Navigations, pérégrinations et voyages faicts en la Turquie*, Antwerp edition, 1576
Description générale de la ville de Lyon et des anciennes provinces du Lyonnais et du Beaujolais, Lyons, 1889 ed., 2 vols
Nielsen, A., *Dänische Wirtschaftsgeschichte*, Jena, 1933
Niemeier, G., *Siedlungsgeographische Untersuchungen in Niederandalusien*, Hamburg, 1935
Nistor, J., *Handel und Wandel in der Moldau bis zum Ende des 16. Jahrhunderts*, Czernowitz, 1912
Noailles, MM. de, see Vertot, Abbé
Noberasco, F., see Scovazzi, I. and Noberasco, F.
Nobili, Giacinto (also known as Rafaele Frianoro), *Il vagabondo*, Venice, 1627
Noble de La Lauzière, J.-F., *Abrégé chronologique de l'histoire d'Arles*, Arles, 1808
Nueva Recopilación de las leyes de España, Madrid, 1772–1775, 3 vols
Nuntiaturberichte aus Deutschland nebst erganzenden Aktenstücken, 1. Abt, 1533–1559, Gotha, 1892–1912, 12 vols; *2. Abt., 1560–1572*, Vienna, 1897–1939, 6 vols.; *3. Abt. 1572–1585*, Berlin, 1892–1909, 5 vols., *4. Abt. 1585 (1584)–1590 (1592)*, Paderborn, 1895–1919, 5 vols

Obermann, Karl, see *Probleme der Ökonomie und Politik in den Beziehungen zwischen Ost und West*

Oexmelin, Alexandre O., *Histoire des aventuriers flibustiers*, Trévoux, 1775, 2 vols

Olagüe, L., *La decadencia española*, Madrid, 1950–1951, 4 vols

Oncken, Wilhelm, *Allgemeine Geschichte in Einzelerstellungen*, Berlin, 1878–1892 (1893), 43 vols

Ortega y Gasset, José, *España invertebrada*, Madrid, 1934

Papeles sobre Velázquez y Goya, Madrid, 1950

Ortega y Rubio, Juan, *Historia de Valladolid*, Valladolid, 1881, 2 vols

Relaciones topográficas de los pueblos de España, Madrid, 1918

Palatini, Leop., *L'Abruzzo nelle Storia documentata di Carlo V di Giuseppe de Leva*, Aquila, 1896

Palencia, see González Palencia, Angel

Paléologue, M., *Un grand réaliste, Cavour*, Paris, 1926

Parain, Charles, *La Méditerranée: les hommes et leurs travaux*, Paris, 1936

Pardessus, J.-M., *Collection des lois maritimes antérieures au XVIIIe siècle*, Paris, 1828–1845, 6 vols

Paré, Ambroise, *Œuvres*, 5th ed., Paris, 1598

Parenti, G., *Prime ricerche sulla rivoluzione dei prezzi in Firenze*, Florence, 1939

Paris, L., see Aubespine, Sébastien de l'

Pariset, G., *L'État et les églises en Prusse sous Frédéric-Guillaume Ier*, Paris, 1897

Parpal y Marqués, C., *La isla de Menorca en tiempo de Felipe II*, Barcelona, 1913

Paruta, Andrea, *Relazione di A. P. console per la Repubblica Veneta in Alessandria presentata nell'ecc. mo Collegio ai 16. dicembre 1609* . . ., a cura di Luigi Baschiera, per nozze Arbib-Levi, Venice, 1883

Paruta, Paolo, *Historia vinetiana*, Venice, 1605

La legazione di Roma (*1592–1595*), ed. Giuseppe de Leva, Venice, 1887, 3 vols

Pastor, Ludwig von, *Geschichte der Päpste seit dem Ausgang des Mittelalters*, 3rd and 4th editions, Fribourg-in-Brisgau, 1901–1933, 16 vols.; French translation *Histoire des Papes*, by Furcy-Raynaud, 1888–1934, 16 vols.; references in text to these editions

Paz Espeso, Julián, *Catalogo de la Colección de documentos ineditos* (*CODOIN*), Madrid, 1930, 2 vols. See also Espejo de Hinojosa, Cristóbal, y Paz Espeso, Julián

Pédelaborde, P., *Le climat du bassin parisien, essai d'une méthode rationnelle de climatologie physique*, Paris, 1957

Pegolotti, see Balducci Pegolotti

Pellegrini, Amedeo, *Relazioni inedite di ambasciatori lucchesi alla corte di Roma, sec. XVI–XVII*, Rome, 1901

Pellissier de Raynaud, E., *Mémoires historiques et géographiques sur l'Algérie*, Paris, 1844

Pereyra, Carlos, *Historia de la América española*, Madrid, 1924–1926, 8 vols

Pérez, Antonio, *L'art de gouverner*, ed. J. M. Guardia, Paris, 1867

Pérez, Damião *Historia de Portugal*, Barcelona, 1926–1933, 8 vols

Pérez, Juan Beneyto, see Beneyto Pérez, J.

Pérez de Messa, D., see Medina, Pedro de

Pérez Villamil, Manuel, see Catalina García, J. and Pérez Villamil, M.

Peri, Domenico, *Il negociante*, Genoa, 1638

Perret, Jacques, *Siris*, Paris, 1941

Petit, Édouard, *André Doria, un amiral condottiere au XVIe siècle*, 1466–1560 Paris, 1887

Petrocchi, Massimo, *La rivoluzione cittadina messinese del 1674*, Florence, 1954

Peyeff, Christo, *Agrarverfassung und Agrarpolitik in Bulgarien*, Charlottenburg, 1926

Pfandl, L., *Introducción al siglo de oro*, Barcelona, 1927

Geschichte der spanischen Nationalliteratur in ihrer Blütezeit, Fribourg-in-Brisgau, 1928

Johanna die Wahnsinnige. Ihr Leben, ihre Zeit, ihre Schuld, Fribourg-in-Brisgau, 1930 (French trans. by R. de Liedekerke, *Jeanne la Folle*, Brussels 1938)

Philipp II. *Gemälde eines Lebens und einer Zeit*, Munich, 1938; French trans., *Philippe II*, Paris, 1942

Philipp, Werner, *Ivan Peresvetov und seine Schriften zur Erneuerung des Moskauer Reiches*, Königsberg, 1935

Philippson, Alfred, *Das Mittelmeergebiet, seine geographische und kulturelle Eigenart*, Leipzig, 1904; 4th ed., 1922

Philippson, Martin, *Ein Ministerium unter Philipp II. Kardinal Granvella am spanischen Hofe (1579–1586)*, Berlin, 1895

Pieri, Piero, *La crisi militare italiana nel Rinascimento*, Naples, 1934

Pierling, Paul, *Rome et Moscou, 1547–1579*, Paris, 1883

Un nonce du Pape en Moscovie, préliminaires de la trêve de 1582, Paris, 1884

Piffer, Canabrava, Alice, *O commercio portugues no Rio da Plata, 1580–1640*, São Paulo, 1944

Piganiol, A., *Histoire de Rome*, Paris, 1939

Pino-Branca, A., *La vita economica degli Stati italiani nei secoli XVI, XVII, XVIII secondo le relazioni degli ambasciatori veneti*, Catania, 1928

Piot, Charles, see Granvelle, also Gachard, L.-P. and Piot, Ch

Pirenne, *Les villes du Moyen Age*, Brussels, 1927 (English translation *Medieval Cities*) *Histoire de Belgique*, Brussels, 1900–1932, 7 vols

Pirenne, Jacques, *Les grands courants de l'histoire universelle*, Neuchâtel, 1948–1953, 3 vols

Planhol, Xavier de, *De la plaine pamphylienne aux lacs pisidans. Nomadisme et vie paysanne*, Paris, 1958

Plantet, Eugène, *Les Consuls de France à Alger avant la conquête*, Paris, 1930

Platter, Felix, *Beloved Son Felix, the Journal of Felix Platter*, ed. and trans. by Seàn Jennett, London, 1961

Platter, Thomas, *Journal of a Younger Brother*, ed. and trans. by Seàn Jennett, London, 1963

Platzhoff, W., *Geschichte des europäischen Staatensystems, 1559–1660* Munich, 1928

Plesner, J., *L'émigration de la campagne à la ville libre de Florence au XIIIe siècle*, Copenhagen, 1934

Pohlhausen, H., *Das Wanderhirtentum und seine Vorstufen*, Brunswick, 1954

Poirson, A.-S.-J.-C., *Histoire du règne de Henri IV*, Paris, 1865–1866, 4 vols

Poliakov, Léon, *Histoire de l'antisémitisme*, I: *Du Christ aux Juifs de Cour*, Paris, 1955; II, *De Mahomet aux Marranes*, Paris, 1961

Les 'banchieri' juifs et le Saint-Siège, du XIIIe au XVIIe siècle, Paris, 1965

Poni, Carlo, *Gli aratri e l'economia agraria nel Bolognese dal XVII al XIX secolo*, Bologna, 1963

Porcar, Moisés Juan, *Cosas evengudes en la ciutat y regne de Valencia. Dietario*

de Moisés Juan Porcar, 1589–1629, ed. Vicente Castaneda Alcover, Madrid 1934
Porchnev, Boris, *Les soulèvements populaires en France de 1623 à 1648*, Paris, 1963 (French translation)
Porreño Baltasar, *Dichos y hechos del señor rey don Philipe segund el prudente . . .*, Cuenca, 1621
Pose, Alfred, *La monnaie et ses institutions*, Paris, 1942, 2 vols
Poullet, Edmond and Piot, Charles, see Granvelle, Cardinal
Pouqueville, F.-C.-H.-L., *Voyage de la Grèce*, Paris, 1820–1821, 5 vols
Presotto, Danilo, '*Venuta Terra*' e '*Venuta Mare*' *nel biennio 1605–1606*, typed thesis, Faculty of Economics, Genoa, 1963
Prestage, E., *The Portuguese Pioneers*, London, 1933
Prévost, Abbé A.-F., *Histoire générale des voyages*, Paris, 1746, 20 vols
Primeira Visitação do Santo Officio as partes do Brasil pelo Licenciado Heitor Furtado de Mendoça . . ., deputado de Santo Officio; I: *Confissões de Bahia, 1591–1592*, São Paulo, 1922; II: *Denunciacões de Bahia, 1592–1593*, São Paulo, 1925; III: *Denunciacões de Pernambuco, 1593–1595*, São Paulo, 1929
Probleme der Ökonomie und Politik in den Beziehungen zwischen Ost- und Westeuropa vom 17. Jahrhundert bis zur Gegenwart, ed. Karl Obermann, Berlin, 1960
Ptasnik, S., *Gli Italiani a Cracovia dal XVIo secolo a XVIIIo*, Rome, 1909
Puig y Cadafalc, J., *L'architectura romanica a Catalunya* (in collaboration), Barcelona, 1909–1918, 3 vols
Le premier art roman, Paris, 1928
Pugliese, S., *Condizioni economiche e finanziarie della Lombardia nella prima metà del secolo XVIII*, Turin, 1924
Putzger, F. W., *Historischer Schulatlas*, 73rd ed. by A. Hansel, Bielefeld, Berlin and Hanover, 1958
Pyrard de Laval, François, *Voyage . . . contenant sa navigation aux Indes orientales . . .*, 3rd ed., Paris, 1619
Quadt, M., *Deliciæ Galliæ sive itinerarium per universam Galliam*, Frankfurt, 1603
Quarti, Guido Antonio, *La battaglia di Lepanto nei canti popolari dell'epoca*, Milan, 1930
Queiros, Vegoso, José Maria de, *Dom Sebastião, 1554–1578*, 2nd ed., Lisbon, 1935
Quétin, *Guide en Orient, itinéraire scientifique, artistique et pittoresque*, Paris, 1846
Quevedo y Vellegas, Francisco Gómez, 'Isla de los Monopantos', in *Obras satíricas y festivas*, Madrid, 1958, vol. II, Madrid, 1639
Quinet, Edgar, *Mes vacances en Espagne*, Paris, 4th ed., 1881
Les Révolutions d'Italie, Paris, 1848–1851, 2 vols
Quiqueran de Beaujeu, P., *La Provence louée*, Lyons, 1614
Rabelais, François, *Gargantua*, ed. 'Les Belles Lettres', Paris, 1955
Le Quart Livre du noble Pantagruel, in *Œuvres de Rabelais*, ed. Garnier, Paris 1962, 2 vols
Rachfahl, F., *Le registre de Franciscus Liscaldius, trésorier général de l'armée espagnole aux Pays-Bas de 1567 à 1576*, Brussels, 1902
Raffy, Adam, *Wenn Giordano Bruno ein Tagebuch geführt hätte*, Budapest, 1956
Rahola, Federico, *Economistas españolas de los siglos XVI y XVII*, Barcelona, 1885
Ramel, François de, *Les Vallées des Papes d'Avignon*, Dijon, 1954

Ranke, Leopold von, *Die Osmanen und die spanische Monarchie im 16. und 17. Jahrhundert, 4 Aufl. des Werkes 'Fürsten und Völker von Südeuropa'*, Leipzig, 1877, 2 vols., French trans.: *Histoire des Osmanlis et de la monarchie espagnole pendant les XVIe et XVIIe siècles*, Paris, 1839

Raphaël du Mans, *Estat de la Perse en 1660 . . .*, ed. Ch. Schefer, Paris, 1890

Rau, Viriginia, *Subsidios para o estudo das feiras medievais portuguesas*, Lisbon, 1943

Raveau, Paul, *L'agriculture et les classes paysannes. La transformation de la propriété dans le Haut-Poitou au XVIe siècle*, Paris, 1926

Raynaud, E. Pellissier de, see Pellissier de Raynaud, E.

Razzi, Serafino, *La storia di Raugia*, Lucca, 1595

Rebora, Giovanni, *Prime ricerche sulla 'Gabella Caratorum sexaginta Maris'*, typed thesis, Faculty of Economics, Genoa, 1963

Recherches et Matériaux pour servir à une Histoire de la domination française aux XIIIe, XIVe et XVe siècles dans les provinces démembrées de l'empire grec à la suite de la quatrième croisade, ed. J. A. C. Buchon, 'Panthéon littéraire', Paris, 1840, vol. III,

Recopilación de las leyes destos reynos hecha por mandado del rey, Alcalá de Henares, 1581, 3 vols

Recouly, Raymond, *Ombre et soleil d'Espagne*, Paris, 1934

Recueils de la Société Jean Bodin, V: *La foire*, Brussels, 1953; VII, *La ville*, Brussels, 1955, 3 vols

Renaudet, Augustin, *Machiavel*, Paris, 1942

L'Italie et la Renaissance italienne (Lectures at the Sorbonne), Paris, 1937. See also Hauser, H. and Renaudet, A.

Reparaz, Gonzalo de, *Geografia* y *política*, Barcelona, 1929

Reparaz, Gonzalo de, jnr. (son of the above), *La época de los grandes descubrimientos españoles y portugueses*, Buenos Aires, 1931

Retaña, Luis de Fernández, *Cisneros y su siglo*, Madrid, 1929–1930

Reumont, Alfred von, *Geschichte Toscana's seit dem Ende des florentinischen Freistaates*, Gotha, 1876–1877, 2 vols

Reznik, J., *Le duc Joseph de Naxos, contribution à l'histoire juive du XVIe siècle*, Paris, 1936

Riba y García, Carlos, *El consejo supremo de Aragón en el reinado de Felipe II*, Valencia, 1914

Ribbe, Charles de, *La Provence au point de vue des bois, des torrents et des inondations avant et après 1789*, Paris, 1857

Ribier, Guillaume, *Lettres et mémoires d'estat*, Paris, 1666, 2 vols

Ricard, Samuel, *Traité générale du commerce*, 2nd ed., Amsterdam, 1706

Richard, P., see Hefele, C.-J. and Hergen Roether, Cardinal

Rilke, R. M., *Letters to a young poet*, trans. Snell, London, 1945

Rivet, P. and Arsandaux, H., *La métallurgie en Amérique précolombienne*, Paris, 1946

Riza Seifi, Ali, *Dorghut Re'is*, 2nd ed., Constantinople, 1910 (an edition in Turko-Latin alphabet, 1932)

Rochechouart, L.-V.-L. de, *Souvenirs sur la Révolution, l'Empire, et la Restauration*, Paris, 1889

Rodocanachi, Emmanuel-P., *La réforme en Italie*, Paris, 1920

Rodriguez, Domingos, *Arte de Cozinha*, Lisbon, 1652

Rodríguez Marín, Francisco, *El ingenioso hidalgo Don Quijote de la Mancha*, Madrid, 1916

Roger, Noëlle, *En Asie Mineure: la Turquie du Ghazi*, Paris, 1930
Röhricht, R., *Deutsche Pilgerreisen nach dem Heiligen Lande*, re-ed. Innsbruck, 1900
Romanin, Samuele, *Storia documentata di Venezia*, Venice, 1853–1861, 10 vols
Romano, Bartolomeo, see Crescentio, B.
Romano, Ruggiero, *Commerce et prix du blé à Marseille au XVIIIe siècle*, Paris, 1956
Romano, Ruggiero, Spooner, Frank and Tucci, Ugo, *Les prix à Udine*, see also Braudel, F. and Romano R.
Romier, Lucien, *Les origines politiques des guerres de religion*, Paris, 1913–1914, 2 vols
La conjuration d'Amboise, Paris, 1923
Catholiques et huguenots à la cour de Charles IX, Paris, 1924
Le royaume de Catherine de Médicis, 3rd ed., Paris, 1925, 2 vols
Rossi, E., *Il dominio degli Spagnuoli e dei Cavalieri di Malta a Tripoli* (*1530–1551*), Intra, 1937
Roth, Cecil, *The House of Nasi:* I. *Doña Gracia*, Philadelphia, 1948; II, *The Duke of Naxos*, Philadelphia, 1948
Roth, Johann Ferdinand, *Geschichte des nürnbergischen Handels*, Leipzig, 1800–1802, 4 vols
Roupnel, Gaston, *Le vieux Garain*, 7th ed., Paris, 1939
Histoire et destin, Paris, 1943
La ville et la campagne au XVIIe siècle. Etude sur les populations du pays dijonnais, 2nd ed., Paris, 1955
Rousseau, Baron, Alphonse, *Annales Tunisiennes*, Algiers, 1864
Rovelli, Giuseppe, *Storia di Como*, Milan, 1789–1803, 3 vols
Rowlands, R., *The Post of the World*, London, 1576
Rubio Ortega, see Ortega y Rubio, J.
Ruble, Alphonse de, *Le traité de Cateau-Cambrésis* (*2 et 3 avril 1559*), Paris, 1889; see also Aubigné, T. A. de
Rubys, Claude, *Histoire véritable de la ville de Lyon*, Lyons, 1604
Ruiz Martín, F., *Lettres marchandes échangées entre Florence et Medina del Campo*, Paris, 1965
Les aluns espagnols, indice de la conjoncture économique de l'Europe au XVIe siècle, forthcoming
El siglo de los genoveses en Castllla (*1528–1627*); *capitalismo cosmopolita y capitalismos nacionales*, forthcoming
Rumeu de Armas, Antonio, *Piraterías y ataques navales contra las islas Canarias*, Madrid, 1947, 6 vols
Rybarski, R., *Handel i polityka handlowa Polski w XVI stoleciu*, Poznán, 1928–1929, 2 vols. in 1
Sachau, Eduard, *Am Euphrat und Tigris, Reisenotizen aus dem Winter 1897–1898*, Leipzig, 1900
Saco de Gibraltar in *Tres relaciones históricas*, 'Colección de libros raros ô curiosos', Madrid 1889
Sagarminaga, Fidel de, *El gobierno y régimen foral del señorío de Viscaya*, re-ed. by Dario de Areitio, Bilbao, 1934, 3 vols
Saint-Denys Antoine Juchereau de, see Juchereau de Saint-Denys, A.
Saint-Sulpice, see Cabié, E.
Sakâzov, Ivan, *Bulgarische Wirtschaftsgeschichte*, Berlin and Leipzig, 1928
Salazar, J. de, *Política Española*, Logroño, 1617

Salazar, Pedro de, *Hispania victrix*, Medina del Campo, 1570
Salomon, Noël, *La campagne en Nouvelle-Castille à la fin du XVIe siècle d'après les 'Relaciones Topográficas'*, Paris, 1964
Salva, Jaime, *La Orden de Malta y las acciones españolas contra Turcos y Berberiscos en los siglos XVI y XVII*, Madrid, 1944
Salvestrini, Virgilio, *Bibliografia di Giordano Bruno, 1581–1950*, 2nd ed. (posthumous), ed. Luigi Firpo, Florence, 1958
Salzman, L. F., *English Trade in the Middle Ages*, Oxford, 1931
Samanes, Floristan, see Floristan Samanes, A.
Samazeuilh, Jean-François, *Catherine de Bourbon, régente du Béarn . . .*, Paris, 1863
Sánchez Alonso, Benito, *Fuentas de la historia española e hispano-americana*, 3rd ed., Madrid, 1946
Sanderson, John, *The Travels of John Sanderson in the Levant (1584–1602)*, ed. William Forster, London, 1931
Sandoval, A., see Gamir Sandoval, A.
Sansovino, Francesco, *Dell'historia universale dell'origine et imperio de' Turchi*, Venice, 1564
Sanundo, Marin, *Diarii*, Venice, 1879–1903, 58 vols
Sapori, Armando, *Studi di Storia economica medievale*, Florence, 1946, 2 vols
Saraiva, Antonio José, *Inquisição e Cristãos-novos*, Oporto, 1969 (2nd ed.)
Sardella, P., *Nouvelles et spéculations à Venice*, Paris, 1948
Sassetti, F., *Lettere edite e inedite di Filippo Sassetti*, ed. Ettore Marcucci, Florence, 1855
Sauermann, Georg, *Hispaniæ Consolatio*, Louvain, 1520
Sauvaget, J., *Introduction à l'histoire de l'Orient musulman*, Paris, 1943.
Alep. Essai sur le développement d'une grande ville syrienne des origines au milieu du XIXe siècle, Paris, 1941
Savary, François, see Brèves, François Savary, sieur de
Savary des Bruslons, Jacques, *Dictionnaire universal de commerce, d'histoire naturelle et des arts et métiers*, Copenhagen, 1759–1765, 5 vols
Sayous, A.-E., *Le commerce des Européens à Tunis depuis le XIIe siècle jusqu'à la fin du XVIe*, Paris, 1929
Scarron, P., *Le Roman comique*, Paris, 1651, ed. Garnier, Paris, 1939
Schäfer, Ernst, *Beiträge zur Geschichte des spanischen Protestantismus und der Inquisition im 16. Jahrhundert*, Gütersloh, 1902, 2 vols
Schalk, Carlo, *Rapporti commerciali tra Venezia e Vienna*, Venice, 1912
Scharten, Théodora, *Les voyages et séjours de Michelet en Italie, amitiés italiennes*, Paris, 1934
Schefer, Ch. see Raphaël du Mans
Schiedlausky, G., *Tee, Kaffee, Schokolade, ihr Eintritt in die europäïsche Gesellschaft*, Munich, 1961
Schmidhauser, Julius, *Der Kampf um das geistige Reich. Bau und Schichksal der Universität*, Hamburg, 1933
Schnapper, Bernard, *Les rentes au XVIe siècle. Histoire d'un instrument de crédit*, Paris, 1957
Schnürer, Gustav, *Katholische Kirche und Kultur in der Barockzeit*, Paderborn, 1937
Schöffler, Herbert, *Abendland und Altes Testament. Untersuchung zur Kulturmorphologie Europas, insbesondere Englands*, 2nd ed., Frankfurt-am-Main, 1941

Schulte, Aloys, *Geschichte des mittelalterlichen Handels und Verkehrs zwischen Westdeutschland und Italien mit Ausschluss von Venedig*, Leipzig, 1900, 2 vols

Die Fugger in Rom (1495–1523), mit Studien zur Geschichte des Kirchlichen Finanzwesens jener Zeit, Leipzig, 1904, 2 vols. in 1

Geschichte der grossen Ravensburger Handelsgesellschaft, 1380–1530, Stuttgart and Berlin, 1923, 3 vols

Schumacher, Rupert von, *Des Reiches Hofzaun. Geschichte der deutschen Militärgrenze im Südosten*, Darmstadt, 1940

Schumpeter, Joseph, *History of Economic Analysis*, London, 1954

Schweigger, Salomon, *Eine neue Reissbeschreibung auss Teutschland nach Konstantinopel und Jerusalem*, 4th ed., Nuremberg, 1639

Schweinfurth, G., *Im Herzen von Afrika, Reisen und Entdeckungen im centralen Äquatorial Afrika während der Jahre 1868 bis 1871*, Leipzig, 1874, 2 vols

Sclafert, Th., *Cultures en Haute-Provence: déboisements et pâturages au Moyen Age*, Paris, 1959

Scovazzi, Italo and Noberasco, F., *Storia di Savona*, Savona, 1926–1928, 3 vols

Sée, Henri, *Esquisse d'une histoire du régime agraire en Europe aux XVIIIe et XIXe siècles*, Paris, 1921

Sée, Julien, see Ha Cohen, Joseph

Segarizzi, A., *Relazioni degli Ambasciatori Veneti al Senato*, vol. III (1–2): *Firenze*, Bari, 1916

Segni, B., *Storie fiorentine . . . dall'anno 1527 al 1555*, Augsburg, 1723

Ségusson de Longlée, P. de, see Longlée, P. de Ségusson de

Seidlitz, W. von, *Diskordanz und Orogenese der Gebirge am Mittelmeer*, Berlin, 1931

Seignobos, Charles, see Milioukov, P., Seignobos, Charles and Eisenmann, L.

Sella, Domenico, *Commerci e industrie a Venezia nel secolo XVII*, Venice-Rome, 1961

Selve, Odet de, *Correspondance politique . . .*, ed. Germain Lefèvre-Pontalis, Paris, 1888

Sens et usage du terme structure dans les sciences humaines et sociales, collective publication, Paris-The Hague, 1962

Sepúlveda, P. de, *Sucesos del reinado de Felipe II*, ed. J. Zarco Cueva in *Historia de varios sucesos*, Madrid, 1922

Sercey, Count Félix-E. de, *Une ambassade extraordinaire. La Perse en 1839–1840*, Paris, 1928

Sereni, Emilio, *Storia del paesaggio agrario italiano*, Bari, 1961

Serra, Antonio, *Breve trattato delle cause che possono far abondare li regni d'oro argento . . ., con applicatione al Regno di Napoli*, Naples, 1613

Serrano, Luciano, *Correspondencia diplomática entre España y la Santa Sede durante el Pontificado de Pio V*, Madrid, 1914, 4 vols

La Liga de Lepanto, Madrid, 1918–1919, 2 vols

Serres, Olivier de, *Le Théâtre d'agriculture*, Lyons, 1675

Le Théâtre d'agriculture et mesnage des champs (selections), Paris, 1941

Servier, Jean, *Les portes de l'année, rites et symboles: l'Algérie dans la tradition méditerranéenne*, Paris, 1962

Sestini, Dom, *Confronto della ricchezza dei paesi . . .*, Florence, 1793

Sicroff, Albert-A., *Les controverses des statuts de 'pureté de sang' en Espagne du XVe au XVIIe siècle*, Paris, 1960

Siegfried, André, *Vue générale de la Méditerranée*, Paris, 1943

Signot, Jacques, *La division du monde*, Paris, 1539
Simiand, François-J.-Ch., *Ciurs d'économie politique*, Paris, 1930 and 1932, 3 vols
Le salaire, l'évolution sociale et la monnaie, Paris, 1932, 3 vols
Recherches anciennes et nouvelles sur le mouvement générale des prix du XVIe au XIXe siècle, Paris, 1932
Les fluctuations économiques à longue période et la crise mondiale, Paris, 1932
Simon, Wilhelm, *Die Sierra Morena der Provinz Sevilla*, Frankfurt, 1942, translated into Spanish as *La Sierra Morena de la provencia de Sevilla en los tiempos postvariscios*, Madrid, 1944
Simonsen, Roberto, *Historia economica do Brasil, 1500–1820*, São Paulo, 1937
Simonsfeld, H., *Der Fondaco dei Tedeschi und die deutsch-venetianischen Handelsbeziehungen*, Stuttgart, 1887, 2 vols
Singer, Charles *et al.*, *A History of Technology*, Oxford, 1954–1958, 5 vols
Sion, Jules, *La France méditerranéenne*, Paris, 1934
Siri, Mario, *La svalutazione della moneta e il bilancio del Regno di Sicilia nella seconda metà del XVIo secolo*, Melfi, 1921
Soetbeer, Adolf, *Litteraturnachweis über Geld- und Münzwesen*, Berlin, 1892
Sombart, Werner, *Krieg und Kapitalismus*, Munich, 1913
Die moderne Kapitalismus, Munich, 1921–1928, 6 vols
Die Juden und das Wirstschaftsleben, Munich, 1922. (English trans. by M. Epstein, *The Jews and Modern Capitalism*, London, 1913)
Vom Menschen. Versuch einer geistwissenschaftlichen Anthropologie, Berlin, 1938
Sorre, Maximilien, *Les Pyrénées méditerranéennes*, Paris, 1913
Méditerranée. Péninsules méditerranéennes, Paris, 1934, 2 vols. (vol. VII of *Géographie Universelle*)
Les fondements biologiques de la géographie humaine, Paris, 1943
Sottas, J., *Les messageries maritimes à Venise aux XIVe et XVe siècles*, Paris, 1938
Sources inédites de l'histoire du Maroc, ed. Philippe de Cossé-Brissac, 2nd series: *Dynastie filalienne*, Archives et Bibliothèques de France, vol. V, Paris, 1953
Souza, A. S. de, *Historia de Portugal*, Barcelona, 1929
Soveral, Visconde de, *Apontamentos sobre relacões politicas e commerciaes do Portugal com a Republica di Veneza*, Lisbon, 1893
Spenlé, Jean-Édouard, *La pensée allemande de Luther à Nietzche*, Paris, 1934
Speziale, G. C., *Storia militare di Taranto*, Bari, 1930
Spooner, Frank C., *L'économie mondiale et les frappes monétaires en France, 1493–1680*, Paris, 1956
See also Romano, R., Spooner, F. and Tucci, U.
Sprenger, Aloys, *Die Post- und Reiserouten des Orients*, Leipzig, 1864
Staden, H. von, *Aufzeichnungen über den Moskauer Staat*, ed. F. Epstein, Hamburg, 1930
Stählin, Karl, *Geschichte Russlands von den Anfängen bis zur Gegenwart*, Stuttgart, Berlin and Leipzig, 1923–1939, 5 vols
Stasiak, Stefan, *Les Indes portugaises à la fin du XVIe siècle d'après la Relation du voyage fait à Goa en 1546, par Christophe Pawlowski, gentilhomme polonais*, Lwow, 1926–1928, 3 fasc
Stefani, Fr., *Parera intorno al trattato fra Venezia e Spagna sul traffico del pepe e delle spezierie dell'Indie Orientali, di A. de Bragadino e J. Foscarini*, ed. for nozze Correr-Fornasari, Venice, 1870

Stella, C. de, *Poste per diverse parti del mondo*, Lyons, 1572
Stendhal, *Promenades dans Rome*, Paris, 1858, 2 vols
L'abbesse de Castro, Paris, 1931
Sternbeck, Alfred, *Histoire des flibustiers et des boucaniers*, Paris, 1931
Stochove, Chevalier Vincent, *Voyage du Levant*, Brussels, 1650
Stone, Lawrence, *An Elizabethan: Sir Horatio Palavicino*, Oxford, 1956
Storia di Milano, ed. La Fondazione Treccani degli Alfieri: *L'età della Riforma cattolica, 1554–1630*, Milan, 1957
Strachey, Lytton, *Elizabeth and Essex*, 2nd ed., London, 1940
Stubenrauch, Wolfgang, *Kulturgeographie des Deli-Orman*, Stuttgart, 1933
Suárez, Diego, *Historia del maestre último que fue de Montesa*, Madrid, 1889
Sully, Maximilien de Béthune, duc de, *Mémoires*, ed. Paris, 1822, 6 vols
Šusta, Josef, *Die römische Curie und das Konzil von Trient unter Pius IV*, Vienna, 1904–1914, 4 vols
Szekfü, J., *Etat et Nation*, Paris, 1945
Taine, Hippolyte-A., *Voyage aux Pyrénées*, 2nd ed., Paris, 1858
La philosophie de l'art, 20th ed., Paris, 1926
Tamaro, Attilio, *L'Adriatico, golfo d'Italia*, Milan, 1915
Tassini, Giuseppe, *Curiosità veneziane*, Venice, 1887
Tavernier, Jean-Baptiste, *Les six voyages qu'il a faits en Turquie, en Perse et aux Indes*, Paris, 1681. English trans. from first French ed. of 1676, by J. Phillips, *The Six Voyages . . .*, London, 1678
Tawney, R. H. and Power, E., *Tudor Economic Documents*, London, 1924, 3 vols
Telbis, Hans, *Zur Geographie des Getreidebaues in Nordtirol*, Innsbruck, 1948
Tenenti, A., *Naufrages, corsaires et assurances maritimes à Venise, 1592–1609*, Paris, 1959
Cristoforo da Canal, La Marine vénitienne avant Lépante, Paris, 1962
Termier, P., *A la gloire de la Terre*, Paris, 1922
Terrail, Pierre du, seigneur de Bayard, *La très joyeuse et très plaisante Histoire composé par le Loyal Serviteur des faits, gestes, triomphes du bon chevalier Bayart*, ed. J. C. Buchon, in the collection 'Le Panthéon litteraire', Paris, 1836. An English translation *The right joyous . . . History of the feasts, gests and prowesses of the Chevalier Bayard . . . by the Loyal Servant* (*J de Mailles*) by Sara Coleridge, London, 1825 and 1906
Terreros, M. R. de, see Guiso, G. M. de
Teulet, J.-B.-T.-Alexandre (ed.), *Relations politiques de la France et de l'Espagne avec l'Écosse au XVIe siècle* (*1551–1588*), re.-ed. Paris, 1862, 5 vols
Tevins, J., *Commentarius de rebus in India apud Dium gestis anno MDXLVI*, Coimbra, 1548
Tharaud, Jérôme and Jean, *La bataille à Scutari*, 24th ed., Paris, 1927
Marrakech ou les seigneurs de l'Atlas, Paris, 1929
Theissen, J. S., see Gachard, L.-P.
Thénaud, J. *Le voyage d'Outremer*, Paris, 1884
Thomazi, Cdt. A.-A., *Histoire de la navigation*, Paris, 1941
Thumery, Jean de, see Laffleur de Kermaingant, P.
Tiepolo, Lorenzo, *Relazione del console Lorenzo Tiepolo* (*1560*), ed. Cicogna, Venice, 1857
Tocco, Vittorio di, see Di Tocco, Vittorio
Tollenare, L.-F., *Essai sur les entraves que le commerce éprouve en Europe*, Paris, 1820

Tomić, S. N., *Naselje u Mletackoj Dolmaciji*, Niš, 1915
Tommaseo, Nicolò, *Relations des ambassadeurs vénitiens sur les affaires de France au XVIe siècle*, Paris, 1838, 2 vols
Tongas, G., *Les relations de la France avec l'Empire ottoman durant la première moitié du XVIIe siècle et l'ambassade à Constantinople de Philippe de Harlay, comte de Césy, 1619–1640*, Toulouse, 1942
Törne, P. O. von, *Don Juan d'Autriche et les projets de conquête de l'Angleterre, étude historique sur dix années du XVIe siècle (1568–1578)*, Helsinki, 1915–1928, 2 vols
Torres, A., see Dragonetti de Torres
Tott, Baron François de, *Mémoires sur les Turcs et les Tartares*, Amsterdam, 1784, 4 vols. English translation, *Memoirs of the Baron de Tott on the Turks and Tartars*, London, 1785
Tournon, Cardinal François de, *Correspondance* . . ., ed. Michel François, Paris, 1946
Toynbee, Arnold, *A study of History*, abridgement by D. C. Somervell of vols. I to VI, London, 1946
Trasselli, Carmelo, see Della Rovere, Antonio
Trevelyan, George Macaulay, *History of England*, re-ed. London, 1946
Tridon, M., *Simon Renard, ses ambassades, ses négociations, sa lutte avec le cardinal Granvelle*, Besançon, 1882
Truc, Gonzague, *Léon X et son siècle*, Paris, 1941
Tucci, Ugo, see Romano, R., Spooner, F. and Tucci, U.
Tudela, Benjamin of, see Benjamin of Tudela
Turba, Gustav, *Venetianische Depeschen vom Kaiserhof*, Vienna, 1889–1896, 3 vols
Geschichte des Thronfolgerechtes in allen habsburgischen Länden, Vienna, 1903
Turquet de Mayerne, Théodore, see Mayerne, Théodore Turquet de
Tyler, Royall, *Spain, a Study of her Life and Arts*, London, 1909
Uccelli, Arturo, *Storia della tecnica del Mediaevo ai nostri giorni*, Milan, 1944
Ugolini, L. M., *Malta, origini della civiltà mediterranea*, Rome, 1934
Uhágon, Francisco K. de, *Relaciones históricas de los siglos XVI y XVII*, Madrid, 1896
Ukers, William H., *All about Coffee*, New York, 1922
Ulloa, Modesto, *La hacienda real de Castilla en el reinado de Felipe II*, Rome, 1963
Usher, A. P., *The Early History of Deposit Banking in Mediterranean Europe*, vol. I only, Cambridge, Mass., 1943
Ustariz, Jerónimo de, *Theorica y pratica de comercio y de marina* . . ., 2nd ed. Madrid, 1742
Vair, Guillaume du, see Du Vair, Guillaume
Valle de la Cerda, Luis, *Desempeño del patrimonio de su Magestad y de los reynos, sin daño del Rey y vassallos y con descanso y alivio de todos*, Madrid, 1618
Van der Essen, Léon, see Essen, Léon van der; see also Frangipani, Ottavio Mirto
Vander Hammen y Léon, Lorenzo, *Don Felipe el Prudente, segundo deste nombre rey de las Españas*, Madrid, 1625
Varenius, Bernardus, *Geographia generalis*, Amsterdam, 1664
Varrennes, Claude de, see *Voyage en France . . .*
Vasconcellos, L. Mendes de, see Mendez de Vasconcelos, L.
Vaudoncourt, Guillaume, see Guillaume de Vaudoncourt, Frédéric.
Vaudoyer, J. L., *Beautés de la Provence*, 15th ed., Paris, 1926

Vaumas, G. de, *L'éveil missionaire de la France d'Henri IV à la fondation du Séminaire des Missions étrangères*, Lyons, 1941
Vayrac, Jean de, *État présent de l'Espagne*, Amsterdam, 1719
Vázquez de Prada, V., *Lettres marchandes d'Anvers*, Paris, 1960, 4 vols
Verlinden, Charles, *L'esclavage dans l'Europe mediévale. I: Péninsule ibérique, France*, Bruges, 1955
Vertot, René Aubert de, *Ambassades de MM. de Noailles en Angleterre*, ed. C. Villaret, Leyden and Paris, 1763
Vicens Vives, J., *Historia Social y Económica de España*, Barcelona, 1957, 3 vols
Manual de Historia Económica de España, Barcelona, undated (1959)
Vidal de la Blache, Paul, *Etats et nations de l'Europe*, Paris, 1889
Tableau de la géographie de la France, 3rd ed., Paris, 1908
Principes de géographie humaine, Paris, 1922. (English translation by M. T. Bingham, *Principles of Human Geography*, London, 1926)
Viet, Jean, *Les méthodes structuralistes dans les sciences sociales*, Paris, 1965
Vilar, Pierre, *La Catalogne dans l'Espagne moderne*, Paris, 1962, 3 vols
Villalón, Christóval de, *Viaje de Turquía . . .* (1555), Madrid-Barcelona, 1919, 2 vols
Villamil, M., see Catalina García, Juan and Pérez Villamil, Manuel
Villaret, C., see Vertot, René Aubert de
Vital, L., *Premier voyage de Charles-Quint en Espagne de 1517 à 1518*, Brussels, 1881
Vitale, Eligio and Brunetti, Mario, *Corrispondenza da Madrid di Leonardo Donà, 1570–1573*, Venice-Rome, 1963, 2 vols
Vitale, Vito, *Breviario della storia di Genova*, Genoa, 1955, 2 vols
Vitalis, A., *Correspondance politique de Dominique du Gabre (évêque de Lodève), trésorier des armées à Ferrare (1551–1554), ambassadeur de France à Venise (1555–1557)*, Paris, 1903
Vivoli, G., *Annali di Livorno*, Leghorn, 1842–1846, 4 vols
Voinovitch, L., *Depeschen des Francesco Gondola, Gesandten der Republik Ragusa bei Pius V. und Gregor XIII. 1570–1573*, Vienna, 1909
Histoire de Dalmatie, Paris, 1935, 2 vols
Voyage de France, dressé pour l'instruction et commodité tant des François que des étrangers, 4th ed., trans. by Cl. de Varennes, Rouen, 1647
Wahrmund, L., *Das Ausschliessungsrecht (jus exclusiva) der katholischen Staaten Osterreich, Frankreich und Spanien bei den Papstwahlen*, Vienna, 1883
Walcher, Joseph, *Nachrichten von den Eisbergen in Tyrol*, Vienna, 1773
Walsingham, Francis, *Mémoires et instructions pour les ambassadeurs* (a translation of Sir Dudley Digges's *The Compleat Ambassador*, 1655), Amsterdam, 1717
Waltz, Pierre, *La Question d'Orient dans l'antiquité*, Paris, 1943
Wätjen, Hermann, *Die Niederländer im Mittelmeergebiet zur Zeit ihrer höchsten Machtstellung*, Berlin, 1909
Weber, Erich, *Beiträge zum Problem des Wirtschaftsverfalls*, Vienna, 1934
Wee, Hermann van der, *The Growth of the Antwerp Market and the European Economy, fourteenth-sixteenth centuries*, Louvain, 1963, 3 vols
Weiller, Jean, *Problèmes d'économie internationale*, Paris, 1946–1950, 2 vols.
L'économie internationale depuis 1950, du plan Marshall aux grandes négociationes commerciales entre pays inégalement développés, Paris, 1965
Weiss, Charles, *L'Espagne depuis le règne de Philippe jusqu'à l'avènement des Bourbons*, Paris, 1844, 2 vols

See also Granvelle, Cardinal
Werth, Emil, *Grabstock, Hacke und Pflug*, Ludwigsburg, 1954
Weulersse, Jacques, *Paysans de Syrie et du Proche-Orient*, Paris, 1946
Wiet, G., see Ibn Iyās
Wilczek, Eduard Graf, *Das Mittelmeer, seine Stellung in der Weltgeschichte und seine historiche Rolle im Seewesen*, Vienna, 1895
Wilhelmy, Herbert, *Hochbulgarien*, Kiel, 1935–1936, 2 vols
Wilkinson, Maurice, *The Last Phase of the League in Provence, 1588–1598*, London, 1909
Williamson, James A., *Maritime Enterprises, 1485–1588*, Oxford, 1913
Wood, Alfred C., *A History of the Levant Company*, London, 1935
Wright, I. A., *Documents concerning English Voyages to the Spanish Main, 1569–1580*, London, 1932
Wyrobisz, Andrzej, *Budownictwo Murowane w Malopolsce w XVIe et XVe wieku* (summary in French), Cracow, 1963
Yver, G., *Le commerce et les marchands dans l'Italie méridionale au XIIIe et au XIVe siècle*, Paris, 1903
See also Gsell, S., Marçais, G. and Yver, G.
Zanelli, A., *Delle condizioni interne di Brescia dal 1642 al 1644 e del moto della borghesia contro la nobiltà nel 1644*, Brescia, 1898
Zanetti, Armando, *L'ennemi*, Geneva, 1939
Zanetti, Dante. *Problemi alimentari di una economia preindustriale*, Pavia, 1964
Zarco, Cuevas, Father J., *Historia de varios sucesos y de las cosas*, ed. Madrid, 1922
Relaciones de pueblos de la diócesis de Cuenca, hechas por orden de Felipe II, Cuenca, 1925, 2 vols
Zeller, Berthold, *Henri IV et Marie de Medicis*, Paris, 2nd ed., 1877
Zeller, Gaston, *La réunion de Metz à la France, 1552–1648*, Paris-Strasbourg, 1927, 2 vols
Le siège de Metz par Charles-Quint, oct–dec., 1552, Nancy, 1943
Les Institutions de la France au XVIe siècle, Paris, 1948
La vie économique de l'Europe au XVIe siècle (lectures at Sorbonne), Paris, 1953
Zierer, Otto, *Bilder aus der Geschichte des Bauerntums und der Landwirtschaft*, Munich, 1954–1956, 4 vols
Zinkeisen, J. W., *Geschichte des osmanischen Reiches in Europa*, Gotha, 1840–1863, 7 vols
Zweig, Stefan, *Sternstunden der Menschheit*, Leipzig, 1927

Books published too late to be fully consulted for second edition:
*Aymard, Maurice, *Venise, Raguse et le commerce du blé pendant la seconde moitié du XIVe siècle*, Paris, 1966
Gestrin, Ferdo, *Trgovina slovenskega Zaledja s Drimorskimi Mesti od 13. do Konga 16. stoletja*, Ljubljana, 1965
Manolescu, Radu, *Comertul Tãrii Romînești și Moldovei cu Brașovul (secolele XIV–XVI)*, Bucharest, 1965
Randa, Alexander, *Pro Republica Christiana*, Munich, 1964
Rougé, Jean, *Recherches sur l'organisation du commerce en Mediterranée sous l'empire romain*, Paris, 1966

*Maurice Aymard's work on Sicily between the fifteenth and eighteenth centuries, which I have followed with interest, rightly criticizes Bianchini's figures, particularly those incorporated in Fig. 49, vol. I of this book. Future research will have to take account of these corrections of detail.

Index of Proper Names

General Index

Index of Proper Names

Note: References to the Mediterranean, Philip II, Venice, Turkey (and the Ottoman Empire) are so numerous that it would be pointless to index them. Further, any entry with more than twenty-five or thirty references (Genoa and Constantinople for instance were found to have well over a hundred apiece) has with a few exceptions been abbreviated to the single word *passim*. The table of contents itself provides by its subheadings a much handier means of reference than a vast string of numerals.

A

C

D

E

F

G

H

I

J

K

L

M

N

O

P

S

T

U

V

General Index